INDUSTRIAL ELECTRONICS AND ROBOTICS

CHARLES A. SCHULER
California University of Pennsylvania

WILLIAM L. McNAMEE
University of Pittsburgh

McGRAW-HILL BOOK COMPANY

New York   Atlanta   Dallas   St. Louis   San Francisco
Auckland   Bogotá   Guatemala   Hamburg   Johannesburg
Lisbon   London   Madrid   Mexico   Montreal   New Delhi
Panama   Paris   San Juan   São Paulo   Singapore
Sydney   Tokyo   Toronto

Sponsoring Editor: Gordon Rockmaker
Editing Supervisor: Evelyn Belov
Design and Art Supervisor: Patricia F. Lowy
Production Supervisor: Kathleen Morrissey

Text Designer: Gayle Jaeger
Cover Photographer: Ken Karp

Library of Congress Cataloging-in-Publication Data
Schuler, Charles A.
    Industrial electronics and robotics.

    Includes index.
    1. Industrial electronics.   2. Robots,
Industrial.   I. McNamee, William L.
II. Title.
TK7881.S38   1986   621.3815   85–18082
ISBN 0-07-055625-3

1 2 3 4 5 6 7 8 9 0 KGPKGP 8 9 3 2 1 0 9 8 7 6

ISBN 0-07-055625-3

CONTENTS

PREFACE

Modern industry, confronted with intense competition and the need to improve productivity, costs, and product quality, has turned to automation for its survival. The heart and soul of automation technology is electronics—and that is what *Industrial Electronics and Robotics* is all about.

This book was written for students of electronics technology programs that emphasize industrial applications. These students will have completed a traditional dc and ac circuits course; basic algebra and trigonometry are the only mathematics prerequisites.

Industrial technicians must understand theory, devices, circuits, and systems. All these are covered thoroughly within the fifteen chapters of this text. It is equally important that technicians exhibit personal skills when interacting with other workers, supervisors, and customers. Interviews with industry leaders have invariably amplified this point. Because human-to-human interface must be a top priority for technical personnel, the topic is discussed in Chapter 1.

Industrial Electronics and Robotics is comprehensive in its treatment of devices and applications. It covers basic control devices, semiconductor devices, motor control circuits, amplifiers, and operational amplifiers. It covers digital electronics and microprocessors, and it applies the microprocessor to a basic control system. The treatment of these subjects is up to date and relevant to the needs of a modern electronics technology program. Students will find the material interesting, well illustrated, and practical.

Robotics is an important issue, although the industrial robot is largely misunderstood. A robot is a piece of equipment; some robots are quite simple, some rather sophisticated. Even the most sophisticated do not replace human workers on a one-to-one basis. Attempts at substituting robots for human beings have resulted in failure and lost revenues.

Robots work best in an environment designed for automation technology. Such an environment is very different from that found in previous industrial eras. In the modern plant, automation is based heavily on both computer control and an almost total integration of many pieces of equipment. Materials are moved, positioned, machined, inspected, assembled, and tested under computer control. The robot is an important component of automation technology, but not the focal point. It is true that some specialists in robotics are needed. However, the intent of this book is to prepare the electronics technician to deal with the broad concepts of automation technology and to show how electronics makes it work.

A book of this type would not be possible without the cooperation of educators and industrial organizations. We wish to acknowledge the invaluable input provided by both groups, although it is not possible to list them all here. We also wish to express our appreciation to our families for their understanding, assistance, and support, throughout the preparation of this text.

Charles A Schuler
William L. McNamee

1

INTRODUCTION

This chapter introduces industrial electronics. It covers some of the key developments and includes definitions for a few of the major terms. It also discusses the skills and knowledge that the industrial technician must possess to work efficiently. Finally, it introduces safety, which is of paramount importance for those preparing to work in the industrial environment.

1-1
PERSPECTIVE

The industrial world is where raw materials are processed, parts are manufactured, and final products are assembled. It provides us with a high standard of living in terms of the goods it offers and the employment it provides. It is central to our economy and to our general well-being. It is a changing world. Many believe that we are now experiencing the second phase of the industrial revolution or a totally new technological revolution. Industry is changing from a labor-intensive sector of our economy to a highly technical and automated environment. This change is impacting the industrial work force, and future industrial workers will work with their brains far more than with their brawn.

The merger of industry and electricity began early in this century, when motors began replacing other energy forms such as steam, water, animals, and humans. The motor was rapidly and widely applied to a variety of industrial tasks. As the applications increased, developing control systems that were more flexible and accurate than the mechanical systems based on clutches, gears, belts, and pulleys became necessary. The Ward-Leonard control system was the first to gain widespread application. In this system, the speed and direction of a direct current (dc) motor were controlled by energizing its armature circuit from a dc generator. By varying the polarity and intensity of the field current in the generator, it was possible to control the direction and speed of the motor efficiently by varying the polarity and magnitude of its armature voltage. The system offered smooth and stepless speed adjustment. When connected to a mechanical system with inertia, the motor was even able to return power to the alternat-

ing current (ac) supply upon deceleration. The efficiency and good control characteristics made the Ward-Leonard system very popular even though it is large, expensive, difficult to maintain, and not so efficient by modern standards.

Electronics began to enter the industrial arena during the 1940s. Various control devices were developed to supplement and, in some cases, replace the Ward-Leonard system. The devices of that era included thyratrons, mercury pool rectifiers, and ignitrons. Up until that time, dc motors were chosen for control applications because their torque and speed characteristics could be widely and smoothly controlled with good efficiency. Then, frequency control of ac motors opened up new possibilities. This was an attractive development since an ac motor costs only one-fourth as much as a comparable dc motor. During a later period, solid-state replacements for thyratrons, called *thyristors,* further extended the applications of ac motors. In the 1950s, magnetic and electronic amplifiers (using vacuum tubes) closed the control loop for even more sophisticated applications. These systems used feedback to control positioning automatically and accurately and were called *servomechanisms.* During the late 1950s and early 1960s, solid-state devices made their debut in industry with devices such as the transistor and the thyristor. Robots first appeared in the 1960s and were used in applications such as welding and spray painting and in environments too dangerous for human workers. The operational amplifier made its inroads in the late 1960s, and systems based on digital logic also gained in popularity. The *programmable controller,* also introduced in this period, began to eliminate electromagnetic relay logic in timing and sequencing control. During the 1970s the miracle device, the microprocessor, opened up a new world of sophisticated control for both dc and ac rotating equipment. A *microprocessor* is most of the circuitry of a digital computer reduced to a tiny chip of silicon. Microprocessors quickly boosted the sophistication of automated systems as managers and designers reacted to pressure for increased productivity and quality.

In the modern industrial plant there are a variety of dc motor control systems ranging from fractional horsepower ratings to ratings over 10,000 horsepower (hp). Applications include portable power tools, conveyors, robots, palletizers, hoists, cranes, elevators, rolling mills, tube mills, extrusion mills, and many others. Thanks to high-power thyristors, ac drive systems are available with ratings ranging from fractional horsepower to 20,000 hp and are replacing dc motors in many industrial applications. In addition to motors and motor drives, modern industry uses electronics in many processes and machines. Examples include electronic welding, electroplating, process control, and computer-controlled machines such as lathes, milling machines, drill presses, and robots.

Key Terms in Industrial Electronics

The following list of definitions is brief, but it will help in developing a perspective of modern industrial electronics:

Automation: the implementation of processes by automatic means without the need for human intervention; self-moving, self-adjusting, and self-controlling.

Process control: the control imposed on physical or chemical changes in a material.

Automatic controller: a device that operates automatically to regulate a variable in response to a command and to a feedback signal.

Programmable controller: a device that is configured by software commands (instructions) to sequence and time events.

Robot: a programmable, multifunction manipulator designed to move material, parts, tools, or specialized devices through variable programmed motions for the performance of a variety of tasks (source: Robot Institute of America).

Microprocessor: the entire central processing unit of a computer reduced to a tiny chip of silicon.

Numerical control: a programming system of letters and numbers that regulates an entire machining process and the sequence of operations to machine a work piece.

Flexible manufacturing system: a computerized production line with a high degree of programmability to allow quick and easy changes from one small production batch to another.

CAD/CAM: computer-assisted design/computer-assisted manufacturing.

Computer-integrated manufacturing: an integration of design, engineering, manufacturing, and business functions on a central computer or on a network of computers.

REVIEW QUESTIONS

1. What is the name of the early motor control system that used a dc generator to energize the armature circuit of a dc motor?

2. Thyristors are solid-state replacements for _____.

3. How much does an ac motor cost as compared to a dc motor with similar ratings?

4. A servomechanism uses feedback to automatically control mechanical _____.

1-2
KNOWLEDGE AND SKILLS REQUIRED

The industrial workplace is often a combination of old, newer, and state-of-the-art technology. The worker may be confronted with brand-new, late-technology equipment sitting beside equipment that is 50 years old. This environment demands a thorough understanding of basics. It requires a diligent technician who takes the time and makes the effort to understand all of the equipment in his or her care. Learning cannot be confined to classrooms. The best technicians are always actively involved in learning more about their technology and about their workplace.

Importance of Theory

Basic circuit theory is one of the most valuable tools for the industrial technician. Ohm's law, Kirchhoff's laws, and other basic principles make the logical process of fault isolation possible. This knowledge must be coupled with the ability to use instruments. Meters, oscilloscopes, signal generators, logic probes, logic pulsers, and other test equipment must be used safely, accurately, and effectively. The technician must also understand the theory of operation for the equipment being diagnosed. Some equipment is very complicated, and the overall picture must be clear. If the equipment interacts with other equipment, then the entire system must be understood. A familiarity with electronic circuits and devices is also necessary. An *electronic circuit* is one that contains active components such as transistors and integrated circuits. When troubleshooting at the component level, one must understand the way the active devices function. Then, when they malfunction, it is possible to diagnose their behavior as abnormal and to make the proper replacements. Troubleshooting at the board level is usually less demanding. In *board-level troubleshooting* the fault is traced to one plug-in circuit board or module, and then the entire circuit is replaced. This is fine in those cases in which a spare board is in stock. If a board is not in stock, then loss of production time can cost the company a considerable sum. It is also often wasteful to discard a defective board. Defective boards may be sent back to the manufacturer for repair, or the technician may be expected to repair them.

Importance of Mechanical Skills

In addition to electrical and electronic knowledge, the industrial technician must have considerable mechanical skill. Much of the work will involve mechanisms that interact with circuits. Sometimes it is difficult to determine whether a problem is mechanical, electrical, or both. Mechanical skills are also necessary for the proper disassembly and reassembly of equipment. All parts and fasteners must be reinstalled just as they were. Attitude has a lot to do with this aspect of the job. Sorting parts and fasteners during the disassembly process takes a little more care, but this care usually pays off handsomely during reassembly. The careful technician will also make sketches when tearing down a complex unit, especially the first time.

Interpersonal Relationships

Human skills are sometimes overlooked but are very important. A breakdown in industry can cause a lot of tension. The down time is costly, and there may be considerable pressure on the technician from several sources to get things fixed quickly. Tempers can flare and control is absolutely necessary. This is not a pleasant atmosphere to work in, but it sometimes occurs. Experienced and skillful technicians have the ability to remain calm and communicative under these conditions and are often highly valued for this reason. Another aspect of interpersonal relations is that the technician must communicate with other people who may have more operating experience with the equipment. They may have been there when it broke down or may have noticed some peculiar behavior before it broke down. Obviously, this information is invaluable. The skilled technician who calmly asks a series of logical questions can stabilize a tense situation and get on with the important task at hand.

Another important human skill for a troubleshooter is to listen attentively to everything others have to say but to acknowledge privately the possibility that they may be wrong about some or all of the information they give. First of all, a coworker who did something wrong and damaged the equipment may not be likely to tell the whole truth about it. The person reporting might be confused or biased or might fabricate some of the events for self-protection. This is not to say that people cannot be trusted, but that they do make mistakes and very few like admitting them. The best technicians learn to verify facts. They never assume anything. Once something is verified and noted (mentally or on paper), it is time to move on to the next step. This kind of logical process saves time in the long run. A disorganized technician will wind up going in circles because of the failure to verify and note conditions.

Software problems are another area in which a debate can ensue and tempers can flare. Much modern industrial equipment is programmable. Programs are called *software* and are written by programmers who are sometimes a little too quick to blame the hardware when things don't work as planned. To be fair, there are also hardware technicians who are too

quick to blame the software. Software problems can be difficult to solve. A new program that exercises the equipment in ways never tried before often creates a challenging situation, which requires a cooperative effort by both hardware and software people to solve the problem. This is why it is so important to deal skillfully with people; maintaining good working relations pays off handsomely. On the other hand, a rude technician may alienate everyone in the plant and be rendered ineffective because of a lack of communication. Think about it.

Preventive Maintenance

In addition to corrective maintenance, preventive maintenance is an important aspect of industrial electronics. This is often totally ignored, but it shouldn't be. Operation, instruction, and service manuals for the equipment in the plant are invaluable. Periodic lubrication, inspection, cleaning of air filters, running of diagnostic programs, checking for ground faults, and all the other procedures recommended in the equipment manufacturer's literature are absolutely essential for a successful maintenance program.

Organizational skills are invaluable in instituting an effective program. Maintenance logs for every piece of equipment will serve to monitor the program. Logs should contain the date and time and a brief but complete description of every procedure performed and every part replaced. Some technicians start a maintenance log by "fingerprinting" a piece of equipment. This procedure involves a complete operational profile in which all deviations from the original factory specifications are noted. Every effort is then made to restore the equipment to original parameters. Minor problems, such as burned-out indicators and faulty panel switches, are normally corrected immediately. More involved problems should be scheduled for attention as soon as possible, and any parts that are not on hand should be ordered. A maintenance log should also contain an inventory of all spare parts in stock for the equipment. A general rule of thumb is to stock an inventory worth 10 percent of the purchase price of the equipment. The technician's experience and the manufacturer's recommendations will dictate what should be stocked. Some common items are circuit boards, modules, fuses, switches, circuit breakers, panel lamps, resolvers, tachometers, motors, and solid-state parts such as diodes, thyristors, transistors, and integrated circuits.

An operational profile is really a form of preventive maintenance. It is often combined with several procedures that ensure proper operation. The equipment should be given full function and range testing. It may also involve tune-ups of servomechanisms and adjustments of feed rates and spindle speeds. Feedback devices and limit switches are also checked for proper operation. Cleaning may be done at the same time, along with replacement of hydraulic and air filters.

REVIEW QUESTIONS

5. Electronic troubleshooting may take place at the board level or at the _____ level.

6. Which level of troubleshooting demands more knowledge?

7. A good technician has electronic knowledge, mechanical skill, and _____ skills.

8. Programmable equipment may malfunction because of _____ problems in addition to electronic and mechanical faults.

1-3
SAFETY

Safety in any environment involves two major components: knowledge and attitude. Knowledge is the body of information that defines dangerous procedures, conditions, materials, and all other potential hazards. It also includes proper procedures: how and when to use protective equipment, and how to react in an emergency. Technicians should know cardiopulmonary resuscitation (CPR) and first aid treatment for burns, acid contact, and other emergency situations. They must also know the location of all safety devices, emergency showers, fire extinguishers, alarm systems, planned evacuation routes, and shelter areas. Knowledge is absolutely essential but does not guarantee safety. Most industrial workers who are injured or killed have been trained. Some people have a tendency to bypass safe procedures for one reason or another. It is very important to gain knowledge and then make it a way of life. Technicians who have the knowledge and the proper attitudes are seldom injured.

The industrial environment may be replete with many kinds of hazards. It is not possible to deal with all safety areas adequately here. Radiation, loud sounds, rotating machines, pinch points, dangerous chemicals and gases, explosive atmospheres, and lasers are some examples of potential hazards for which specific safety knowledge is acquired on the job. This section will be limited to general rules of electrical and mechanical safety.

Electrical shock can cause falls and other physical reaction injuries, permanent damage to the human body, and even death. Improper procedures may also damage expensive equipment. Moving equipment, such as a robotic arm, can cause serious injury. The following rules will help establish safe procedures and alert the technician to dangerous situations:

1. Circuits that are being serviced must be locked open and tagged. A "buddy" system may prevent any chance of anyone's inadvertently energizing a system before the appropriate time.

2. Dirt and moisture conditions must be noted. They can drastically alter conductivity and may provide unusual current paths.

3. All connections must be checked carefully. Cables that move and cables that are subject to external abuse are especially susceptible to insulation damage and connector damage.

4. Contacts that normally carry heavy current must be inspected periodically. They must be tight and clean. High current can lead to heat, sparks, and flames when it flows through a high-resistance joint or connection.

5. Care must be taken around batteries because some types produce an explosive gas. Open flames or sparks in their vicinity can result in an explosion. Battery spills can be dangerous since the electrolyte is extremely caustic.

6. When working with live line voltage both hands should not touch the equipment simultaneously. Leaning on a cabinet with one hand while working with the other is dangerous. Hand-to-hand shocks can be lethal. High-voltage mats and any other protective equipment that is recommended should be used when working on live circuits. Rings, bracelets, and similar metal items should not be worn around live circuits. Wet clothing and shoes are good conductors of electricity and must not be worn around live circuits.

7. Test equipment must be kept in good working order. Worn leads should be replaced, connectors should be checked regularly, and only exact replacements for broken knobs and probes should be used. Verify that the power cords are in good condition and that grounds are intact. Technicians must always keep their hand tools and soldering equipment clean and in good working order.

8. Manufacturer's service literature should be kept readily available. Follow the recommended procedures for each piece of equipment that is worked on.

9. All work must conform to the current National Electrical Code and any applicable local regulations.

10. Care must be taken when equipment is installed and energized for the first time. A phase error may cause motors to run backward with unpredictable results. Most equipment contains a phase-sequence lockout circuit which will prevent the contactor from closing in the event of a wiring error. Never manually override such devices in such a situation.

11. Never defeat interlocks, limit switches, overcurrent devices, and other protective features. Verify that they all operate properly and have not been tampered with. Do not overrate replacement protective devices.

12. Power devices, such as thyristors, that are subject to large surge currents may physically fail by package rupture and an expulsion of materi-

als. Eye protection should be worn when working around such devices. Electrolytic and tantalum capacitors can also explode when subject to an overvoltage or a reverse bias. A tantalum capacitor may release a liquid electrolyte. Skin, eye, or mouth exposure to this material must be treated promptly. Flush with large amounts of running water and seek medical attention immediately in the case of eye contact or ingestion.

13. The side effects of certain medication may impair judgment or equilibrium or induce drowsiness. Extreme caution must be practiced under such conditions.

14. Equipment grounds must not be removed or circumvented by the use of adaptors. The power cord should be connected before using test probes.

15. Protective clothing and eye protection should be worn when working around high-vacuum devices such as cathode ray tubes.

16. Learn all appropriate hazard symbols and warning signs.

17. Only approved fire extinguishers should be used on electrical equipment. Water is a conductor and should not be used on electrical fires. Carbon dioxide, foam, and halogen types are preferred.

18. Capacitors and hydraulic accumulators store energy after the power has been turned off. A high-voltage capacitor may present a lethal shock hazard even when disconnected from its source.

19. Be wary of moving equipment and pinch points. Loose, floppy clothing invites getting caught in machinery. Do not enter "hard hat areas" without the prescribed protective gear.

20. Use extreme care when working around robots since they can make unpredictable movements. A robot that appears to be "dead" may be awaiting input signals.

21. There are two operating zones, referred to as envelopes, in robotics: the program envelope and the absolute maximum reach envelope, including the reach of any tooling. Learn the maximum envelope and stay clear of it.

22. If servicing is required within a robot's envelope it may be necessary to use blocking posts that are set into the floor to prevent the arm from striking anyone within the envelope.

23. Verify that the robot works properly in the "teach mode." Most control systems allow only slow movements when an operator or technician is in the envelope. Make certain that the controls are working properly.

24. Many robots use an interlocking control system that prevents an automatic cycle from beginning before homing to the start position. This is a safety feature because the home position is usually the fully retracted position. This system must be checked for proper operation.

25. Robot controls may also be interlocked to other pieces of equipment such as conveyors. The entire system must be checked for proper operation to ensure that the entire cycle begins and ends properly, for example, when a part clears a die, a mold, or a fixture.

26. Eye protection is absolutely essential around robots. A welding robot may initiate an arc without any warning. Care must be exercised around lasers. In general, approved safety goggles should be worn in all potentially dangerous situations.

27. The use of vacuum is preferred to the use of compressed air for cleaning electronic equipment. Many of the air filters used in air cooling systems contain fiberglass, which may cause serious eye injury if the glass fibers are blown into the eye.

Be extremely careful with solvents. Many of them are highly volatile. Breathing the fumes can be extremely damaging to the human body. Use of carbon tetrachloride is strictly regulated by law because it is easily absorbed into the skin and causes organ damage. Some solvents should be used only in an area fitted with a flameproof exhaust hood. A respirator and other protective clothing may also be required. Solvents are also known to damage electrolytic capacitors. Electrolytics are especially susceptible to halogenated hydrocarbons, which enter the end seals and cause internal corrosion. Other solvents and degreasers that should be avoided around electrolytics include Freon, chlorine, and fluorine compounds. Spent solvents must be disposed of properly. Because it is usually unacceptable to dump them into a sewer system, they are often stored in holding tanks for eventual removal and processing.

Beware of polyvinyl chloride (PVC) insulation. It is commonly found on wires and is used in heat-shrink tubing. Gaseous hydrogen chloride is released from PVC at a temperature of 230°C. The hydrogen chloride combines with moisture in the air, and a hydrochloric acid mist results. It acts as a primary irritant and corrosive agent to the respiratory system. It is also absorbed on contact with skin. Do not use an open flame to activate heat-shrink tubing. Use a heat gun or other approved technique.

Safety in Tests and Measurements

Safe procedures may not always be obvious. One of the most dangerous and least understood procedures is in the area of floating measurements. A *floating measurement* is taken with the test equipment chassis, for example, of an oscilloscope, at some potential higher than ground. Technicians must sometimes make measurements across two points in a circuit where neither is at ground potential. The common connection for the signal may be hundreds of volts with respect to the line ground. Another related problem is that the green grounding wire on two separate line circuits may be several volts different. Although this condition, known as a *ground loop,* is symptomatic of a poor grounding system, it does cause fault current and interferes with proper test measurements.

Oscilloscopes and other test equipment have a signal common terminal that is connected to the protective ground, which is usually the chassis of the instrument. This protective ground is connected via a green wire to the power line ground when the instrument is plugged into an outlet. This is a safety feature and prevents the chassis, cabinet, connectors, probe ground, and the controls from assuming a dangerous potential with respect to ground. Without this feature, it would be possible to receive a dangerous shock just by touching some part of the instrument. Unfortunately, it also requires that all signals be measured with respect to ground, making it difficult to measure a signal across two points in a circuit when neither point is grounded. Connecting the test equipment in a case like this will ground one of the circuit points and may cause fault currents which could damage the equipment. Even if it does not cause damage, the connection will often inject noise into the circuit, causing abnormal operation and making any meaningful measurement impossible. Too many technicians improperly "float" the test equipment in these cases, creating a potentially dangerous situation. Improper floating defeats the protective ground system by using a three-to-two wire adaptor or by cutting off the ground prong on the power connector. Using an isolation transformer on the oscilloscope is also unacceptable since it also defeats the protective grounding system.

Floating an oscilloscope is dangerous because it allows the cabinet, connectors, and test leads to assume the same potential as that of the probe ground connection. One practice is to rope off the area when floating measurements are being made, to protect personnel from touching the oscilloscope and receiving a dangerous shock. This is a questionable practice, and it does not take into account that the power transformer in the oscilloscope may be unduly stressed, possibly leading to insulation failure. Another problem with this practice is that the capacitive loading effect of the floating instrument exerts a filter action on the circuit point where the probe ground is connected, and this filter action may interfere with proper operation and measurement. For these reasons, floating measurements that defeat the protec-

tive ground system are unacceptable even though commonly practiced.

One acceptable method of performing floating measurements is to use an isolation amplifier between the oscilloscope and the circuit under test. The amplifier connects between the oscilloscope and the equipment under test and passes measurement signals across an insulating barrier. Another method that may work in some applications is indirect grounding by using equipment such as a ground isolation monitor. The monitor is connected between the power lines and the test equipment. If the potential of the isolated ground exceeds 40 volts (V) peak, it disconnects power to the test equipment, sounds an alarm, and restores the protective ground circuit.

Another safe technique is to use an oscilloscope with a differential input. Matched probes are used with the negative preamp input connected to the circuit common and the positive preamp input connected to the signal test point. The common mode dynamic range and the common mode rejection ratio are important specifications of such an amplifier since the measured signal may be very small in comparison to the common mode signal. Some technicians forget this point and attempt to use a quasi-differential technique with a dual-trace oscilloscope in the ADD mode. By inverting channel 2, the two signals are electrically subtracted, effectively removing the common mode signal. Unfortunately, the dynamic range of dual-trace amplifiers is too small, and the common mode rejection ratio is only about 100 to 1. The results obtained are often poor.

Another acceptable measurement technique is to use an all-insulated oscilloscope. Battery-operated equipment avoids ground loops and is preferred by many technicians. Finally, some advanced oscilloscopes are capable of storing waveforms. If one of these is available, two separate measurements can be taken and stored; then the common mode signal is mathematically subtracted out. Such oscilloscopes are very expensive.

REVIEW QUESTIONS

9. Safe workers have the required knowledge and the proper _____ regarding their environment and practices.

10. Circuits that are being serviced should be locked open and _____.

11. Dirt and moisture can decrease electrical _____.

12. You should not work on circuits or machines while wearing _____.

CHAPTER REVIEW QUESTIONS

1-1. What is the name given to a chip of silicon containing most of the circuitry of a digital computer?

1-2. Name a device that is configured by software to sequence and time events.

1-3. A programmable, multifunction manipulator designed to move material, parts, or tools through variable programmed motions for the performance of a variety of tasks is commonly called a _____.

1-4. Name the two broad categories of maintenance.

1-5. Which category of maintenance is likely to be ignored?

1-6. An industrial technician should maintain a diary, called a _____, for each piece of equipment.

1-7. According to the rule of thumb, if a piece of equipment cost $50,000 then _____ worth of spare parts should be stocked for that piece of equipment.

1-8. A phase error may cause a piece of equipment to run _____.

1-9. A robot that appears to be "dead" may be awaiting a control _____.

1-10. Heat-shrink tubing should never be activated with a _____.

1-11. What prevents the chassis and connectors of test equipment from assuming a dangerous potential?

1-12. Isolation amplifiers, ground isolation monitors, differential amplifiers, and battery-operated equipment are examples of equipment that can be used to make _____ measurements safely.

ANSWERS TO REVIEW QUESTIONS

1. Ward-Leonard 2. thyratrons 3. one-fourth 4. position 5. component 6. component 7. human 8. software 9. attitude 10. tagged 11. resistance 12. metal items and wet clothing

2

SOLID-STATE DEVICES

This chapter investigates solid-state materials and devices. A majority of industrial systems are based on semiconductor technology rather than the older vacuum tube technology. Semiconductors are solid-state crystals that can control and amplify. Every industrial employee with responsibilities for electrical and electronic systems must have a good working knowledge of solid-state devices, including their theory of operation and their physical, electrical, and thermal characteristics.

2-1
SEMICONDUCTOR THEORY

Semiconductors are materials with electrical properties that fit in between the properties of conductors and insulators. Common semiconductor materials include the elements carbon, silicon, germanium, and the compound gallium arsenide. These materials oppose the flow of electricity and can be used to make resistors and control devices.

It will be helpful to review conduction as a prelude to examining semiconduction. Copper is an excellent conductor. It exhibits high conductivity because many, many current carriers are available to support the flow of current. One cubic centimeter (cm^3) of copper has 10^{23} current carriers. Figure 2-1 shows

the structure of atomic copper. Note that the outermost electron is called a *valence electron*. The valence electrons are the only ones that can move through the material and support the flow of current. A length of copper wire is held together by metallic bonding. In this situation positive copper ions float in a cloud of valence electrons. Any given electron in the cloud is not associated with any particular copper atom. Electrostatic forces between the positive ions and the negative electrons hold the wire together. Placing a potential difference across the ends of the wire causes a significant current since so many electrons are available to serve as carriers. This type of flow is referred to as a *drift current*. Therefore, conductors such as copper behave as they do because they have large numbers of current carriers.

Semiconductor materials, such as silicon, are held together by *covalent bonding*. In this process atoms share valence electrons with neighboring atoms to gain a total valence count of eight. Figure 2-2 shows the structure. Eight electrons fill the valence orbit and leads to stability. The resulting silicon structure is very stable and is called a *crystal*. Since the valence electrons of the crystal are locked into the covalent bonds, pure silicon conducts very poorly at room temperature. It is not possible to make absolutely pure silicon, but it can be approached with only 1 impurity atom for every 10^{13} silicon atoms. Because of the lack of current carriers, it does not even semiconduct. It is considered an intrinsic semiconductor. The resistivity of intrinsic (pure) silicon at room temperature is 200,000 ohms per cubic cen-

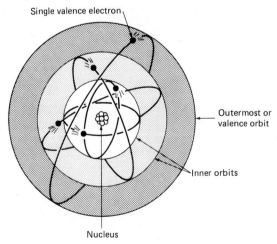

Fig. 2-1 Simplified copper atom.

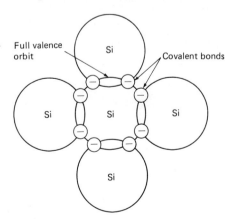

Fig. 2-2 Silicon atoms sharing valence electrons.

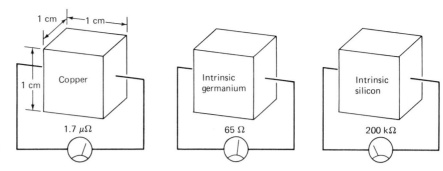

Fig. 2-3 Comparing resistivities of copper, germanium, and silicon.

timeter (Ω/cm^3). Compare this to the resistivity of copper, which is only 1.7 $\mu\Omega$/cm^3. It should be obvious how few current carriers are present in an intrinsic semiconductor as compared to a conductor. Intrinsic germanium, another semiconductor material, exhibits a resistivity of 65 Ω/cm^3 at room temperature. Figure 2-3 compares the resistivities of copper, intrinsic germanium, and silicon. It turns out that the higher resistivity of intrinsic silicon is an advantage when building most semiconductor devices and is one of the reasons that silicon is far more popular than germanium.

Temperature can affect the resistivity of semiconductors. When heat enters the crystal, some of the valence electrons will gain enough heat energy to break their covalent bonds. It is interesting to note that electromagnetic radiation (such as light) entering the crystal will also release carriers. This effect is utilized in certain devices and will be covered in the chapter on optoelectronics. The heat-liberated electrons will support the flow of current. Carriers produced in this manner are known as *thermal carriers*. As Fig. 2-4 shows, heat can release an electron from its covalent bond and leave a positive hole behind. The hole also acts as a carrier to support the flow of current. Therefore, we may expect the resistivity of semiconductors to decrease as temperature increases. This relationship is known as a *negative temperature coefficient* and is an important characteristic of solid-state devices. Compare this with con-

ductors which have a positive temperature coefficient. Their resistivity increases with an increase in temperature. This occurs because heat entering a conductor increases the activity of the electron cloud. The increased activity enhances the probability of electron collisions for the drift current. The collisions impede the current, and resistivity increases. This is why conductors become superconductors (no resistance) at absolute zero ($-273°$C). Conversely, an intrinsic semiconductor insulates at absolute zero since no thermal carriers are present in the crystal.

Doping is the process used to make crystals semiconduct at room temperature. It is possible to diffuse into the crystal impurity atoms that have five valence electrons. These materials, such as arsenic, are called *donor impurities* because they donate extra electrons to the crystal. The donated electrons are not covalently bound and serve as current carriers. Figure 2-5 shows a silicon crystal that is doped with a donor impurity. The doping level is typically very small with something on the order of 1 impurity atom for every 10 million silicon atoms. The doped material is known as *extrinsic silicon* and shows a lower resistivity than intrinsic silicon. Materials doped with donor impurities are known as *N-type semiconductors* since the current carriers are electrons and have a negative charge.

It is also possible to dope the crystal with acceptor impurities. These materials, such as boron, have only three valence electrons. When they enter the crystal, one covalent bond is not satisfied and a hole results. Think of the hole as a position for an electron. These

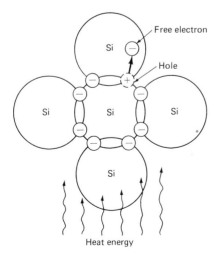

Fig. 2-4 Thermal carriers.

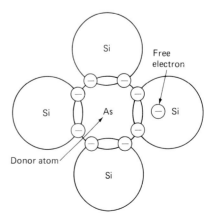

Fig. 2-5 N-type silicon.

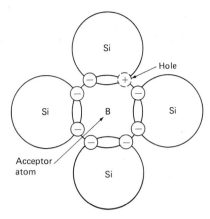

Fig. 2-6 P-type silicon.

positions will accept electrons; for this reason impurities that create holes in the crystal are called *acceptor impurities*. A semiconductor that is doped with acceptor atoms is also extrinsic but is called *P-type*. The structure of P-type silicon is shown in Fig. 2-6. The holes behave as positive charges and support the flow of current in a similar fashion to electrons. However, there are two important differences between electron current and hole current; one of the differences is indicated in Fig. 2-7. Note that at the top of the illustration a hole appears as the electron leaves the crystal at the right. Then that hole is filled, and another appears to its left as the electrons move to the right. It is apparent from Fig. 2-7 that the holes seem to be moving opposite to the direction of the electrons. The second difference is that hole current is based on position swapping, and the apparent motion of the holes is slower than simple electron motion in an N-type crystal. This provides N-type semiconductor material with a comparative speed advantage which is important in some applications.

We have discussed two ways to provide crystalline silicon with current carriers. Heat can produce thermal carriers, and impurities can provide electrons or holes to serve as carriers. At room temperature, the extrinsic carriers outnumber the thermal carriers by about 1 million to 1. Thus, the extrinsic conduction is all that concerns us and the thermal carriers are insignificant. However, somewhere in the range of 200 to 400°C, the conduction becomes intrinsic. In *intrinsic conduction* device characteristics are controlled by temperature, and the device becomes useless for most electronic applications. Intrinsic conduction must be avoided, and therefore most semiconductor devices must not exceed a temperature of 200°C. Extreme cold must also be avoided since the crystals can crack. The range of safe operating temperatures varies according to the type of device (diode, transistor, integrated circuit, etc.) and according to its classification. For example, commercial integrated circuits are rated from 0 to 75°C, industrial integrated circuits from −40 to 85°C, and military integrated circuits from −55 to 125°C. Germanium devices exhibit intrinsic conduction at a much lower temperature than silicon devices and must operate over an even more restricted temperature range. This is another reason why silicon is far more useful for most applications. Storage temperatures and soldering temperatures are also specified by device manufacturers and must be observed.

As Fig. 2-8 shows, extrinsic crystals contain majority and minority carriers. The N-type crystal at the top shows a majority of electrons. A few holes can be found and are considered minority carriers. At the bottom, the P-type crystal has a majority of holes. A few electrons act as minority carriers. Thermal carriers come in pairs. Remember, every electron that gains its freedom leaves a hole behind. This means that the number of minority carriers in any extrinsic crystal is directly related to the temperature of the crystal. For example, heating an N-type crystal will cause a few electrons to be freed. These join the other majority carriers. The holes that are created

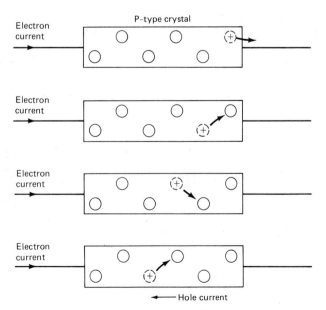

Fig. 2-7 Direction of hole current.

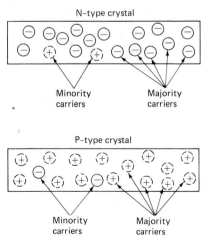

Fig. 2-8 Majority and minority carriers.

at the same time become minority carriers. Heating a P-type crystal also releases a few electrons, which become minority carriers, and the holes join the other majority carriers. It will be seen in subsequent sections of this chapter that minority carriers act to degrade the operation of devices such as diodes and transistors.

REVIEW QUESTIONS

1. *Intrinsic silicon* is another name for _____ silicon.

2. How does the resistivity of intrinsic germanium compare to the resistivity of intrinsic silicon?

3. Semiconductors have a _____ temperature coefficient.

4. Conductors have a _____ temperature coefficient.

5. Extrinsic crystals can be either N-type or _____.

6. Holes, in an N-type crystal, are considered _____ carriers.

2-2
JUNCTION DIODES

Junction diodes are two-terminal devices that utilize both P-type and N-type conduction in a single crystal. They are usually manufactured by a planar diffusion process. By using heat, pressure, and a dopant in gaseous form, it is possible to force atoms into a crystal to create carrier sites. The structure of the typical diffused junction diode is illustrated in Fig. 2-9. The *junction* is the boundary region between the N-type material on the left and the P-type material on the right. It is not a mechanical joint or a seam since a continuous and single crystalline structure is necessary for diode action. The figure also shows that a depletion region forms along both sides of the junction. This region is the result of free electrons from the N material on the left crossing the junction and filling holes in the P material on the right. Notice that there are no carriers in this depletion region. The depletion region behavior is therefore intrinsic. We may expect it to exhibit very high resistance at room temperature. Therefore, we may also expect

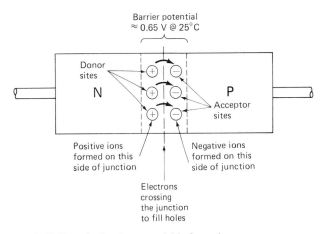

Fig. 2-10 How the barrier potential is formed.

high lead-to-lead resistance in the diode when the depletion region is present.

Figure 2-10 shows why all of the electrons from the N material do not cross the junction and fill all of the holes in the P material. As each electron leaves the N side it also leaves its parent donor atom. This donor atom becomes a positive ion since it has lost an electron. Likewise, a negative ion is formed on the P side, where each electron joins the valence orbit of an acceptor atom. These ionized atoms produce a force called the *ionization potential* that prevents any further carrier movement. Note that the negative ions shown in Fig. 2-10 will tend to repel additional electrons that might try to cross the junction to fill holes. The current that flows until the forces go into equilibrium is called a *diffusion current*. This is different from the drift current discussed in the previous section. The ionization potential that develops as a result of the diffusion current is also known as the *barrier potential*. At room temperature, this potential is approximately 0.65 V for silicon junctions and 0.25 V for germanium junctions.

It is possible, by using an external voltage source, to overcome the barrier potential and collapse the depletion region. This is known as *forward bias* and is shown in Fig. 2-11. Note in Fig. 2-11(b) that the negative terminal of an external battery has been connected to the N material. Like charges repel, and the electrons are driven toward the junction, thus collapsing the depletion region. Assuming a silicon diode, the external battery must provide more than 0.65 V to produce significant current through the diode since the barrier potential must be overcome. With sufficient forward bias applied, the diode current can be large and some load- or current-limiting device will be required to protect the circuit. The schematic symbol for the diode is given in Fig. 2-11(a). The N side is called the *cathode* since it delivers electrons to the P side. The P side is called the *anode*. The direction of conventional current flow is opposite to the electron current; the diode symbol is based on this flow, since the arrow indicates conventional direction when the diode is forward-biased. As a memory aid, note that the anode side of the symbol is shaped like the letter *A*.

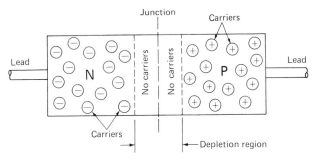

Fig. 2-9 Junction diode.

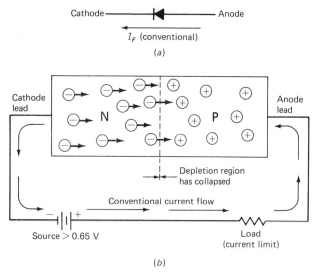

Fig. 2-11 Junction diode with forward bias. (*a*) Schematic symbol of diode. (*b*) Direction of current flow.

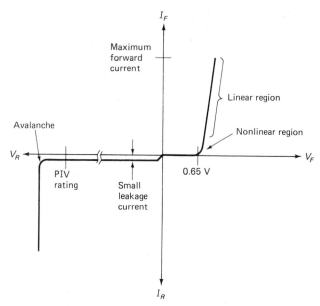

Fig. 2-13 Diode volt-ampere characteristic curve.

Figure 2-12 shows what happens when the diode is reverse-biased. The positive terminal of the external source is now connected to the N side of the diode. Unlike charges attract, and the electrons are moved away from the junction. This makes the depletion region wider. The center of the diode now exhibits intrinsic behavior, and very little current flows at room temperature. There will be some reverse current due to minority carrier action. Minority electrons do exist in the P material. These are driven by the reverse bias source to the junction, where they combine with minority holes in the N material. Since the number of minority carriers is directly related to temperature, reverse (leakage) current increases with temperature. At room temperature, the leakage current in many silicon diodes is so small that it cannot be measured with ordinary meters. It is usually in the nanoampere (nA) or microampere (μA) range. At elevated temperatures, the leakage

current can increase to the point where it adversely affects circuit performance. Germanium diodes exhibit much more leakage current at room temperature. It is typically in the microampere range for small devices and in the milliampere (mA) range in large devices. This is another reason why most designs employ silicon devices.

Some important concepts of silicon diode behavior are summarized in Fig. 2-13. It is a volt-ampere graph. It shows diode current for various levels of forward and reverse bias. Forward bias is designated as V_F and ranges from the origin to the right on the horizontal axis. Forward current is designated as I_F and ranges from the origin to the top on the vertical axis. You can see the effect of the barrier potential by examining the forward bias portion of the graph. There is no current flow until a forward voltage of 0.65 V is reached. At this point the volt-ampere graph is curved, because diode behavior is nonlinear near the turn-on point. This nonlinear region is nearly logarithmic, and some applications utilize this part of the curve when a logarithmic response is needed. As the forward bias is further increased, the graph becomes linear, as it would be for a resistor. Figure 2-13 also depicts reverse bias (V_R) performance. Notice that as V_R increases from the origin toward the left, a small leakage current flows. This current is small and remains rather constant up to the peak inverse voltage (PIV) rating of the diode. Other designations, such as V_{RRM} (peak repetitive reverse voltage) may be used to specify this point. This maximum reverse bias voltage is typically specified from 50 to 1000 V in general purpose silicon diodes. If the maximum reverse voltage is exceeded, the diode will break down. Breakdown occurs at high potential where the field is strong enough to break covalent bonds. The electrons accelerate and collide with other electrons and knock them loose. An avalanche of carriers results, and the graph (Fig. 2-13) shows a

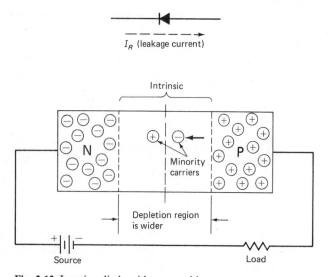

Fig. 2-12 Junction diode with reverse bias.

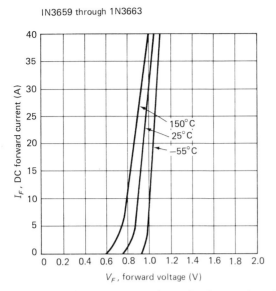

1N3659 through 1N3663

150°C
25°C
−55°C

Fig. 2-14 Effect of temperature on forward voltage and conduction.

drastic increase in reverse current flow. This reverse flow, if not controlled, will destroy the diode.

As Fig. 2-14 indicates, volt-ampere characteristic curves can also be used to depict the effect of temperature on diode performance. Note that the curves show only forward bias performance. At 150°C, the diode turns on at 0.6 V and reaches a current of 10 A at 0.8 V. Compare this to the −55° curve, which shows turn on at 0.9 V and a current of 10 A at just over 1 V. This occurs because the barrier potential is temperature-dependent and decreases approximately 2 millivolts per degree Celsius (mV/°C). The part numbers 1N3659 through 1N3663 represent a series of 30 A silicon diffused junction diodes. The 1N3659 is the least expensive part and is rated for V_{RRM} of 50 V. The 1N3663 is the most expensive in the series and is rated for 400 V.

Diodes as Rectifiers

One of the major industrial applications for diodes is in the area of rectification. *Rectification* is a process of changing alternating current to direct current. Since diodes conduct well in one direction and oppose current in the other direction, they make excellent rectifiers. The rectification frequency is typically 60 hertz (Hz) and general purpose diodes are characterized for this frequency. Chapter 6 treats rectification in detail.

Some industrial circuits require rectification at higher frequencies. As the frequency goes above 1 kilohertz (kHz), the diode will have to be derated because of the recovery time; this phenomenon is due to stored charges which prevent a diode from turning off instantaneously at the moment of reverse bias. The current that continues to flow when reverse bias is applied causes extra dissipation in the diode; for this reason it must be derated for high-frequency

operation. It is known as the *recovery current,* and it also detracts from circuit efficiency. Special fast-recovery diodes have been developed to increase efficiency and reduce diode heating when high frequencies must be rectified. The recovery current flows because a conducting diode injects electrons into the P-type anode. These injected electrons are minority carriers in the P material. At the moment of reverse bias, these minority electrons will support a reverse current flow until they move back to the N-type cathode or recombine with holes. The recovery time is longest when the forward current is high because the number of minority electrons is a function of the forward current. Fast-recovery rectifiers speed up turn off by using gold doping in addition to the ordinary donor and acceptor doping. The gold provides extra sites in the crystal where recombination can occur. This makes it possible for the diode to recover (turn off) faster. These rectifiers recover in dozens or hundreds of nanoseconds and extend rectifier performance to the hundreds of kilohertz.

Schottky Rectifier

The Schottky rectifier is another special diode that extends performance into the megahertz (MHz) region. Schottkys are not PN junction devices but are included in this section because they are used in many of the same application areas (Fig. 2-15). An N-type silicon chip is bonded to a barrier metal (typically platinum). At this semiconductor to metal barrier the diode action is achieved. When the barrier is forward-biased, the electrons from the N-type cathode gain energy to cross the metal barrier. Since the electrons gain energy in order to cross, these devices are also commonly called *hot-carrier diodes*. Once in the metal, the electrons quickly give up this

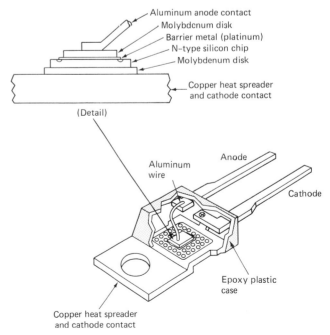

Aluminum anode contact
Molybdenum disk
Barrier metal (platinum)
N–type silicon chip
Molybdenum disk
Copper heat spreader and cathode contact
(Detail)

Aluminum wire
Anode
Cathode
Epoxy plastic case
Copper heat spreader and cathode contact

Fig. 2-15 Schottky rectifier T0-220.

extra energy and join the electron cloud associated with metallic conductors. Minority carrier storage is not a problem since the electrons are not minority carriers in the metal anode. Therefore, there is no recombination and no turn-off delay. Unfortunately, Schottky diodes are restricted to low-voltage applications since they cannot block reverse voltages much over 50 V. They are available with current ratings up to around 100 A. The volt-ampere curves in Fig. 2-16 show another interesting characteristic of Schottky diodes: They turn on at a much lower voltage. At room temperature, the turn-on point is 0.35 V, and at 10 A the forward voltage is only 0.63 V. Compare this with Fig. 2-14. You will also notice that the Schottky curves never become linear.

Zener Diodes

It was mentioned that diodes may be destroyed if allowed to avalanche. This is true for rectifiers. However, zener diodes (Fig. 2-17) are normally operated at the avalanche voltage. Fig. 2-17(*b*) shows the characteristic curve for a zener. Note that the *zener knee* is that part of the graph where reverse conduction takes place. The diode is safe if the reverse current is limited to the maximum value. Zeners are rated according to safe power dissipation. The maximum reverse current may be found by dividing the power rating by the zener voltage. For example, a 1-watt (W), 10-volt (V) zener will have a maximum reverse current rating of

$$I_{MAX} = \frac{1}{10}$$
$$= 0.1 \text{ A}$$

Thus, the diode will be safe at reverse currents up to 0.1 A. However, solid-state devices must be derated for operation above 25°C. This is why diodes are often mounted on thermally conducting assemblies called *heat sinks*.

Zener breakdown is different from avalanche breakdown. Zeners are very heavily doped PN devices. The heavy doping produces a very narrow depletion region, and even a moderate reverse bias produces an intense field across the narrow region. The intense field yields zener emission and the device begins to conduct. Actually, zener emission occurs up to around 4 V of reverse bias. Avalanche conduction is the major mechanism above about 6 V. The region from 4 to 6 V is a combination of avalanche and zener emission. However, zener diodes are available with breakdown ratings up to 200 V. What has happened is that the name *zener* has come to be applied to a broad range of diodes designed to operate in reverse breakover.

Zeners are not used as rectifiers. The characteristic curve (Fig. 2-17[*b*]) shows that ΔV_z is reasonably small from the zener knee to I_{MAX}. The impedance (Z) of a zener diode is given by

$$Z = \frac{\Delta V_z}{\Delta I_z}$$

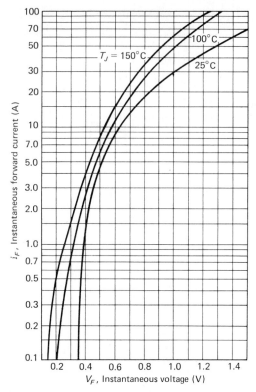

Fig. 2-16 Forward voltage for a Schottky diode.

The ideal zener diode will show no change in V_z and has an impedance of 0 Ω. The best practical zener diodes have a low impedance, which indicates that they approach the ideal. They show only a small voltage change over a range of reverse current. This makes them useful for voltage regulation and reference applications. Zeners are available in voltages from 1.8 to 200 V and in power ratings from 0.25 to 50 W. Figure 2-17(*a*) shows the schematic symbol for

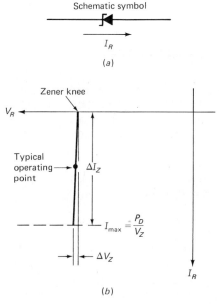

Fig. 2-17 Zener diode. (*a*) Schematic symbol. (*b*) Volt-ampere characteristic curve.

the zener diode. Note that the cathode has extra "wings" that imply the breakdown characteristic curve. Also notice that the conventional current direction is against the arrow when the device is operating in reverse breakover. If the anode of a zener is made 0.65 V positive with respect to the cathode, it will turn on. Zeners have the same forward characteristic curve as is shown in Fig. 2-13 for the rectifier diode.

The zener voltage tends to increase with temperature. Special temperature-compensated zener diodes are available for those applications in which a highly stable voltage reference is required. The temperature compensation is achieved by connecting a second diode in series with the zener. The second diode is forward-biased when the zener is operated in breakover, and it drops less voltage as temperature increases. This tends to cancel the change in zener voltage. These devices are called *reference diodes* or *stabistors*. They operate at around 6 V, and high-precision types are available with a maximum voltage change of only several millivolts (mV) and a temperature coefficient of 0.001 percent. Reference diodes do not have the same forward characteristic curve as ordinary zeners do, because of the second compensating junction.

Other Diode Types

A summary of diode schematic symbols is shown in Fig. 2-18. This illustration will give you an overview of the variety and application areas for diodes. Later chapters will provide more details. The bidirectional breakdown diode is a back-to-back zener device that can be used in transient suppression service or in ac power control. The constant current diode, which is actually a field effect transistor, is covered later in this chapter. The varicap or varactor diode utilizes

the depletion region as the dielectric portion of a capacitor. If you refer to Fig. 2-12, you will see that the reverse-biased junction diode has much in common with a capacitor. The width of the depletion region is directly related to the reverse bias voltage. Thus, diodes can serve as voltage-variable capacitors. Light-emitting diodes are also common and make excellent indicators. Photodiodes are useful in sensing and control applications.

Figure 2-19 provides some examples of the way diodes are packaged. The size of the package is related to the device ratings. For example, the DO-15 case is usually restricted to diodes having maximum current ratings of 1 A or less. When the DO-15 case is used to house zener diodes, the maximum rating will be 1 W or less. The TO-3 metal case may be

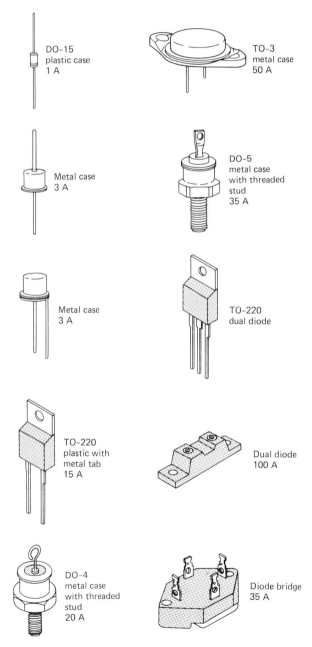

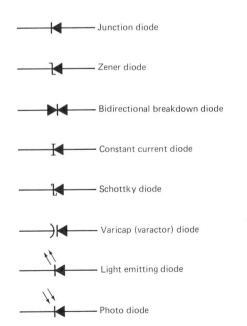

Fig. **2-18** Schematic symbols for various diode types.

Fig. **2-19** Diode packages.

used to house diodes with current ratings up to 50 A. Current ratings are directly related to the size of the PN structure and its ability to transfer heat to its environment. Therefore, the high-current devices are physically larger and must be mounted in a package that allows efficient heat transfer. For example, the TO-220 case uses a metal tab for conducting heat away from the PN junction. Some packages are designed to house more than one diode. The dual-diode and diode bridge packages shown in Fig. 2-19 are examples. Figure 2-20 shows an elaborate package designed for three-phase service and forced-air cooling. Water cooling is also used in some industrial applications.

REVIEW QUESTIONS

7. A PN junction diode will exhibit a depletion region under conditions of zero bias and _____ bias.

8. Diode leakage current is due to _____ carriers.

9. Diode avalanche occurs at high values of _____ bias.

10. Diodes turn on at lower values of forward bias as temperature _____.

11. When a diode is used to convert alternating current to direct current it is called a _____.

12. Schottky rectifiers turn off very fast because they exhibit no _____ carrier storage.

2-3
JUNCTION TRANSISTORS

Figure 2-21 shows the way the planar diffusion process can be used to build a device with two junctions. Once again, the crystal is continuous, and the junctions represent boundaries between the P- and N-type regions. The structure is called a *bipolar NPN transistor. Bipolar* means that the two polarities of current carriers are present: electrons in the N material and holes in the P material. Unipolar transistors will be covered in the next section.

The schematic symbol for the NPN transistor is also shown in Fig. 2-21(a). The leads are identified as emitter, base, and collector. These names tell us something about the functions. The *emitter* emits the carriers, the *base* controls the flow of carriers, and the *collector* collects the carriers coming from the emitter. Note that the physical structure of the transistor is arranged so that the carriers coming from the emitter must pass through the base region to reach the collector. The schematic symbol shows an arrow on the emitter lead. This arrow points in the direction of conventional current flow as it does in the diode symbol.

Another way to build a bipolar junction transistor is illustrated by Fig. 2-22. Here, the base is N ma-

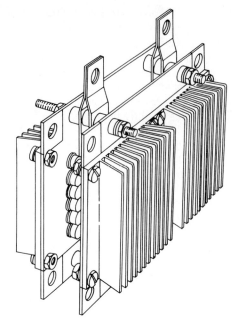

Fig. 2-20 Three-phase rectifier assembly with forced air cooling rated at 650 A.

terial, and the collector and emitter are made up of P material. This structure is called a *PNP transistor,* and the schematic symbol shows the emitter arrow pointing in. Compare Fig. 2-21 with Fig. 2-22. As a memory aid, think of NPN as denoting "Not Pointing iN." Type NPN and PNP transistors are electrical complements. They both can achieve the same basic functions but with opposite emitter-collector flows. They are operated with opposite terminal polarities as well. This feature makes them electrical complements and offers circuit designers some interesting choices.

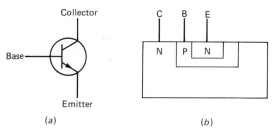

Fig. 2-21 NPN transistor. (*a*) Schematic symbol. (*b*) Physical structure.

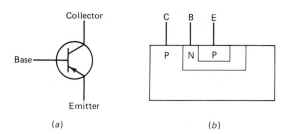

Fig. 2-22 PNP transistor. (*a*) Schematic symbol. (*b*) Physical structure.

Figure 2-23 shows an NPN transistor with both junctions biased. The transistor structure is shown in "sandwich" form to illustrate the circuit action more clearly. Voltage V_{CB} reverse biases the collector-base junction and a depletion region is shown. Voltage V_{BE} forward biases the emitter-base junction, and that depletion region is collapsed. On the basis of what we know about junction behavior, we can expect conventional current to flow from the base to the emitter and zero collector current or a small collector current due to leakage. However, Fig. 2-23 shows that the collector current I_c is substantial. This is so because of the way the transistor is fabricated. The base is a very thin region in the crystal: on the order of 0.025 millimeters (mm) (about a thousandth of an inch). It is moderately doped. The emitter is heavily doped. When the emitter-base junction is forward-biased, the emitter sends large numbers of electrons into the base region. Only a few of the emitter carriers will combine with a hole in the base material because the base does not contain a great number of holes. Most of the emitter carriers will come under the influence of the collector field and be swept through the depletion region and into the collector. Notice that the collector is positive with respect to the base and will attract any electrons that it can get. It is a statistical process whereby the emitter electrons are far less likely to combine with a hole and become base current than they are likely to be swept into the collector. Typically, only 1 or 2 percent of the emitter carriers become base current; the balance, 99 or 98 percent, become collector current.

Even though the base current is only a small percentage of the total current in Fig. 2-23, it is very important because it is a control current. Suppose that V_{BE} is adjusted to 0.3 V. If the transistor is silicon, this voltage is not adequate to collapse the emitter-base depletion region. Now, the emitter will not emit its electrons into the base region. They will never get close enough to the collector field to be swept across the collector-base junction. Therefore, all three currents are now zero. This is a very important concept. A small current (I_B) is controlling much larger currents (I_E and I_C). Another way to demonstrate the control capabilities of the base would be to place a large resistance in series with the base lead in Fig. 2-23. This resistance would reduce the base current, say, to half of what it was. Now we will find the emitter and collector currents about half of what they were before the resistor was added. This shows that a transistor is capable of gradual control: A change in the base circuit will be accompanied by a proportional change in the emitter and collector circuits.

How is the transistor controlled? We have discussed changing the base-emitter bias voltage. We have also discussed adding resistance to the base circuit to reduce the base current. Both seemed to control emitter and collector behavior. This leads us to the conclusion that either point of view can be

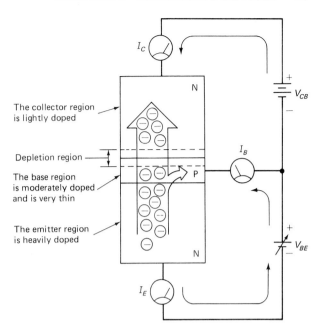

Fig. 2-23 NPN transistor currents.

taken. In other words, the transistor can be viewed as a voltage-controlled device or as a current-controlled device. Which will be used is a matter of convenience. In Chapter 6, voltage gain in transistor amplifiers is discussed. *Voltage gain* implies that a small voltage change is accompanied by a larger voltage change in another part of the circuit. In this chapter, we are more concerned with the current gain of the bipolar transistor. Even though voltage gain and current gain are both viable points of view, the bipolar transistor is definitely a current-amplifying device. When it is operating as a linear amplifier, any change in base current will cause a corresponding change in emitter and collector currents. The most important idea is this: Base current controls emitter and collector current. There are voltage-controlled transistors, and they will be discussed in the next section.

Bipolar junction transistors are the workhorse of industrial electronics. It is essential that you understand their characteristics. Refer to Fig. 2-24, which shows PNP action, and compare it to Fig. 2-23. The PNP transistor is also reverse-biased from collector to base and forward-biased from base to emitter. Note that both external sources must be reversed to accomplish proper bias for the PNP device. The holes in the P-type emitter are plentiful, and most do not recombine with electrons in the base. The collector is negative, and a majority of the holes are attracted to it. Once again, the base current is very small but critical in that it controls both emitter and collector currents. Both NPN and PNP transistors are capable of current gain, and both are useful as amplifiers and controllers. However, it should be apparent that they are not interchangeable since the applied polarities are reversed.

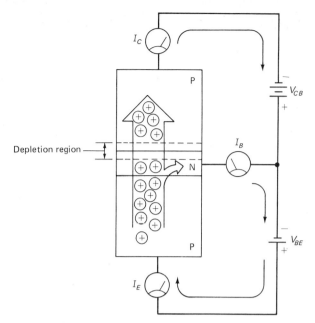

Fig. 2-24 PNP transistor currents.

Figure 2-24 shows holes moving from emitter to collector. Actually, the positions move as discussed in the first section of this chapter. The illustration makes the PNP transistor currents easier to understand. The electrons move from the collector to the emitter by moving from hole to hole. This type of flow is a little slower, and for this reason NPN transistors are better choices for high-frequency applications. The NPN transistor is also more convenient to use in many negative ground circuits. Finally, device manufacturers make a broader line of NPN transistors than of PNP transistors. For these reasons, most circuits use NPN transistors. However, the complementary characteristics of PNP devices make them very useful for certain applications, and many industrial devices will use both types.

Let's summarize some very important ideas. The collector-base junction must be reverse-biased for proper transistor action. The emitter-base junction must be forward-biased for the transistor to be on (to conduct carriers from its emitter to collector). The base current is a small percentage of the total current. The emitter current is the total current ($I_E = I_C + I_B$), and the collector current is almost as large. The base current controls the emitter and collector currents. The NPN and PNP transistors are electrical complements.

Characteristic curves are an important aid to understanding transistor behavior and are also a useful way to present data. Figure 2-25 shows transfer curves for a typical NPN transistor. The curves transfer (relate) the effect of base-emitter voltage to collector current. At room temperature (25°C), the transistor begins conducting at about 0.55 V. At 150°C, the base-emitter turn-on voltage is only about 0.25 V. For a collector current of 40 mA, V_{BE} varies from 0.4 to 0.7 V from the high temperature to the

low temperature. The transfer curves show that the transistor is a temperature-sensitive device.

A collector family of characteristic curves is shown in Fig. 2-26. These are not transfer curves since the collector voltage is plotted against the collector current. The graph shows that the collector-to-emitter voltage has little effect on the collector current over most of the graph. For example, inspect the lowest curve in the family. It is labeled for a base current of 0 mA. This curve shows that the collector current is constant at 0 over a range of collector voltage from 0 to 300 V. This is reasonable because with no base current there is no emitter or collector current. What happens as the collector voltage goes over 300 V? The curve shows that the collector current starts to increase. This phenomenon is called *reach-through*. The collector field is now intense enough to reach through the base and into the emitter. Transistors are not normally operated at collector potentials high enough to cause reach-through.

What is controlling collector current in Fig. 2-26 when the collector voltage is normal? The curves provide the answer: base current. Look at the second curve up. It represents collector behavior for a base current of 0.2 mA. The collector current is nearly

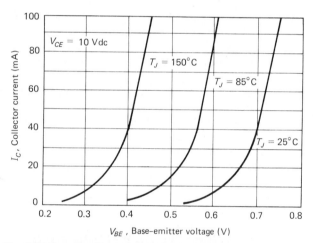

Fig. 2-25 Transfer characteristic curves.

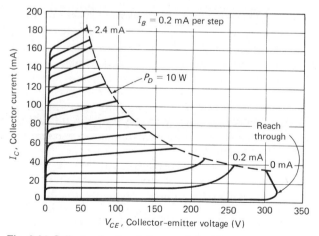

Fig. 2-26 Collector family of characteristic curves.

constant at about 12 mA for a range of collector voltages near 0 to about 200 V. The collector resistance is given by

$$R_c = \frac{\Delta V_{CE}}{\Delta I_c}$$

This equation shows that the collector resistance is high because the change in collector current is small. Once again it must be stressed that base current, not collector voltage, is the major controlling factor. The only two places on the graph where collector voltage does have significant control are near 0 V and at the high end, where reach-through occurs. However, it can be seen that the curves are not horizontal for the higher values of collector current. The slope of these curves indicates that collector voltage is playing a partial role in setting collector current and that the collector resistance is lower for the higher values of collector current.

The dc gain of the transistor can now be investigated with some numbers. *Beta*, or h_{FE}, is defined as the ratio of collector current to base current.

EXAMPLE

Calculate the gain for an operating point of 50 V and a base current of 0.2 mA by using the data from Fig. 2-26:

SOLUTION

$$h_{FE} = \frac{I_C}{I_B}$$

From Fig. 2-26, at V_{CE} = 50 V and I_B = 0.2 mA, I_C = 12 mA.

$$= \frac{12 \times 10^{-3}}{0.2 \times 10^{-3}}$$
$$= 60$$

The ac current gain, h_{fe} (ac beta), is found by

$$h_{fe} = \frac{\Delta I_C}{\Delta I_B} \Bigg| V_{CE}$$

The vertical bar in this equation signifies that V_{CE} is to be held constant. Although not plotted, the emitter current can also be found. Remember that the emitter current is the total transistor current. It can be found by

$$I_E = I_B + I_C$$

Using the values from the previous example,

$$I_E = 0.2 \text{ mA} + 12 \text{ mA}$$
$$= 12.2 \text{ mA}$$

The value of h_{FE} varies somewhat as the operating point is changed. The collector family of curves in Fig. 2-26 will now be used to calculate gain for an operating point of 50 V and a base current of 1 mA. These two conditions intersect at a corresponding collector current of 80 mA.

$$h_{FE} = \frac{80 \text{ mA}}{1 \text{ mA}}$$
$$= 80$$

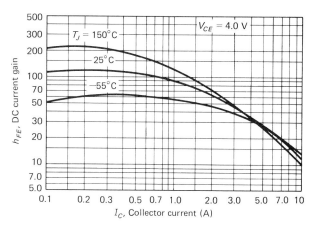

Fig. 2-27 The dc current gain.

Figure 2-27 is another type of transistor graph and shows that h_{FE} varies with collector current and with temperature. Most transistors show best gain at some moderate or intermediate value of collector current. All bipolar transistors show increasing gain as temperature increases.

Refer to Fig. 2-26 to see how a constant power curve can be used to define the safe operating area for a transistor. Transistors are rated according to maximum collector dissipation. Small transistors used to switch and amplify small currents may have a maximum collector dissipation of 100 mW. Large transistors are available with collector dissipation ratings of 300 W. The collector dissipation of a transistor is found by

$$P_C = I_C \times V_{CE}$$

The constant power curve in Fig. 2-26 represents a collector dissipation of 10 W. Any point on the power curve represents a current-voltage product of 10. For example, the curve passes through the 100-mA, 100-V intersection (100 mA × 100 V = 10 W). Since there are an infinite number of operating points, the power curve helps us decide which are safe and which are not. Any point falling on the curve or to the left will not exceed the maximum collector dissipation. Any point falling to the right of the curve will exceed the maximum collector dissipation.

If the maximum collector dissipation is exceeded for any length of time, the transistor will overheat and be damaged or destroyed. This is known as the *thermal limit of the transistor*. There are other failure modes. A high current, even if momentary, can cause a bonding wire within the transistor case to melt (open). Finally, second breakdown may occur. *Second breakdown* is a failure mode peculiar to bipolar power transistors and occurs within the safe operating area defined by a constant power curve. It is associated with high collector voltages which set up fields in the crystal. These fields can cause the current to be focused into a tiny area about the diameter of a human hair. A localized heating, which may melt a hole from the emitter to the collector, shorting out the transistor, results. Figure 2-28 combines the fail-

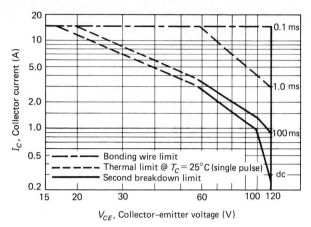

Fig. 2-28 Safe operating area graph for a transistor.

ure modes into one safe operating area graph. Any current greater than 15 A will exceed the bonding wire limit. The dc curve shows the thermal limit as the other controlling factor up to 60 collector-emitter volts. Second breakdown is the limiting factor for collector-emitter voltages from 60 to 120 V, where the maximum current is less than would be allowed by the transistor's dissipation ratings. Figure 2-28 also shows that greater voltages and currents are safe for pulse-mode service. A 0.1 millisecond single pulse will not damage the transistor unless it exceeds the bonding wire limit or involves a collector-emitter voltage greater than 120 V.

Figure 2-29 shows that transistors must be derated above 25°C. For example, at 100°C the thermal derating factor is 0.4. This means that the transistor is capable of dissipating only 40 percent of its rated power at that temperature. It can be seen that second breakdown must also be derated at elevated temperatures, but not as much as the thermal ratings. At 100°C the second breakdown ratings are 80 percent of maximum.

Transistors have three operating modes: cutoff, linear, and saturation. Figure 2-30 depicts the modes and shows an electrical equivalent for each. The transistor circuit is shown at the left and the equivalent emitter to collector circuit at the right. It is important

to notice that the base circuit conditions determine which mode is selected. Cutoff (Fig. 2-30[*a*]) is caused by zero base current and results when the base switch is open. The base current is zero, and so are the other currents. All the supply voltage drops across the transistor, which is acting as an open switch from collector to emitter. The load is off for the cutoff mode. The linear mode in Fig. 2-30(*b*) is selected by closing the base switch. The base current is now some moderate value. The collector current is now beta times the base current, and approximately half of the supply voltage drops across the collector-emitter circuit. Notice that the collector-emitter circuit is now acting as a resistance. The load is partially energized in the linear mode. Saturation (Fig. 2-30[*c*]) is caused by the base current's reaching some high value. When one of the resistors has been removed from the base lead this condition results. The collector circuit is now acting as a closed switch. The drop from collector to emitter approaches 0 V, and the current through the load is limited by Ohm's law: the resistance of the load and the value of V_{CC}.

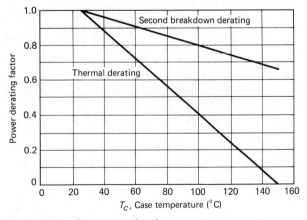

Fig. 2-29 Transistor power derating.

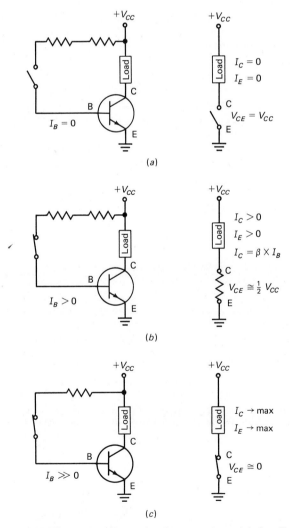

Fig. 2-30 Three operating modes for a transistor. (*a*) Cutoff. (*b*) Linear. (*c*) Saturation.

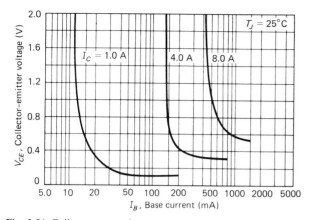

Fig. 2-31 Collector saturation.

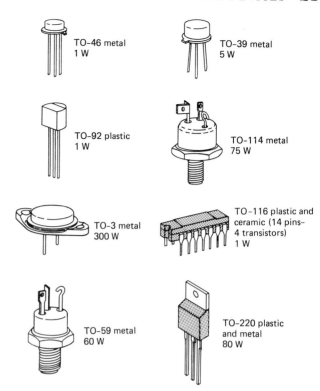

Fig. 2-32 Transistor packages.

The load is fully energized for the saturation mode.

Figure 2-30 also shows that a single source voltage is all that is required to bias both junctions of the transistor properly. When the emitter lead is grounded, a single positive supply will reverse bias the collector and forward bias the base of an NPN transistor. In actual use, the load could represent a motor, a lamp, a heating element, or a relay coil. The linear mode may not be desired in some applications. For example, in digital circuits the load is switched on or off (saturation or cutoff). In other applications, the linear mode may be desired. For example, if the load is a motor, smooth speed control can result from gradually adjusting the base current.

Suppose the circuit of Fig. 2-30 is used to provide on-off control for a dc motor. In this case, the circuit will operate in cutoff or in saturation. When in saturation, the transistor should closely approximate a closed switch from the emitter to the collector. Now look at Fig. 2-31. This graph shows the saturation characteristics of a typical transistor. It shows that perfect switching action is not possible. For a load (collector) current of 1 A, the base current must be at least 100 mA to saturate the transistor fully. Even with this high base current, the graph shows that the collector-to-emitter voltage is a little over 0.1 V. The saturated transistor is approximating a closed switch but not quite attaining it. At high load currents, say 8 A, the graph shows that even more base current is required for saturation and that even more voltage drops from collector to emitter. This is an important factor when transistors are used to switch large loads, as they often must in the industrial environment. Any drop across the transistor when it is saturated detracts from circuit efficiency since some of the circuit energy will heat the transistor, rather than all of it being available for the load. The base losses are also significant in these cases. The graph shows that the base current must be in excess of 1 A to saturate the transistor for an 8-A load current. This represents another loss which detracts from the overall circuit efficiency. The next section of this chapter investigates a newer type of transistor that improves efficiency in these cases.

The larger the transistor structure, the larger its current and thermal ratings and the better its saturation characteristics. Figure 2-32 shows some popular packages for transistors. The smaller packages are limited in wattage and current ratings. Heat flow is important in large dissipation devices. The large metal packages provide good heat transfer from the crystal to the external heat sink. Some packages house more than one transistor. The TO-116 ceramic package can house four transistors. In that case each transistor will have a maximum dissipation of 250 mW.

REVIEW QUESTIONS

13. On the schematic symbol for an NPN transistor, which lead has an arrow, and how does it point?

14. The base current in a transistor is very important because it _____ the other currents.

15. How must the collector-base junction of a transistor be biased?

16. If a transistor is to be on, how must its base-emitter junction be biased?

17. Refer to Fig. 2-25. If V_{BE} is fixed at 0.6 V and the transistor's temperature is increased from 25 to 85°C, what will happen to the collector current?

18. Use Fig. 2-26 to find the collector current when the collector voltage is 100 V and the base current is 0.6 mA.

2-4
FIELD EFFECT TRANSISTORS

The bipolar junction transistors covered in the last section are current-controlled devices. This section investigates transistors that normally have no input current; they are called *field effect transistors* and they are *voltage-controlled*. Field effect transistors (FETs) are unipolar devices. They involve a single conducting channel which can be of either N or P type material. The single conducting channel supports device current with majority carriers only. Bipolar transistors support flow with both majority and minority carriers. For example, in an NPN transistor the emitter injects electrons into a P-type base region. These injected electrons are minority carriers and delay the turn-off of the transistor since they must be swept out of the base (or recombine) before the collector current can cease. The FETs offer advantages in switching service since they do not suffer the delays associated with minority carrier storage. Because they also demand no input current, they are easier to drive. You should recall that the required base current to saturate a power bipolar transistor fully is substantial. The FETs are also less temperature-sensitive and less susceptible to second breakdown in high-power applications. These reasons explain why FETs and power FETs are gaining in popularity.

Figure 2-33 shows the structure and the schematic symbol for an N-channel junction field effect transistor (JFET). The source lead sources the electron current, the drain lead drains it, and the gate lead controls the flow from source to drain. Conventional current flow is from drain to source. The structure shows that the source lead is a contact at one end of a semiconducting N channel, and the drain lead contacts the other end of the same channel. No junctions are crossed, and only majority carriers (electrons, in this case) support the flow. The gate lead is reverse-biased with respect to the N channel. This sets up a depletion region and prevents any gate current from flowing. The negative gate repels the majority electrons in the channel and manages to push some of them down into the substrate, thus increasing the channel resistance and decreasing the source and drain currents. This is known as the *depletion mode of operation,* since the channel carriers are depleted by gate bias. If the gate is made sufficiently negative, all of the electrons are removed from the channel and the drain current ceases. This is called *cutoff.*

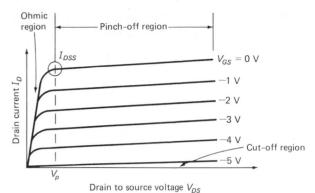

Fig. 2-34 N-channel JFET characteristic curves.

Figure 2-34 shows the drain family of characteristic curves for the N-channel JFET. It can be seen that as the voltage from gate to source (V_{GS}) is made increasingly negative, the drain current decreases. When the gate voltage is more negative than 5 V, the device cuts off and there is no drain current. This is shown as the cut-off region in Fig. 2-34. The graph also shows an ohmic region and a pinch-off region. In the *ohmic region* the drain current is a function of the drain-to-source voltage. It is due to the resistance of the channel, and the graph shows the expected linear volt-ampere relationship. The pinch-off region shows a constant current behavior, where the drain current does not vary over a wide range of drain voltage. The pinch-off voltage is shown as V_P on the graph. Most switching and amplifying operations occur in the pinch-off region of Fig. 2-34. Therefore, these applications require a drain supply greater than V_P.

Pinch-off is explained in Fig. 2-35(a). The electron current is shown inside the transistor, and conventional flow is shown outside the transistor. The gate lead is tied to the source lead, and therefore V_{GS} is equal to zero. The drain supply is adjustable. As the drain voltage is gradually increased from 0 V, the transistor initially operates in the ohmic mode. As the drain voltage is increased further, a significant voltage gradient begins to form across the N channel. A gradient is a gradual voltage change along the length of a resistive path that is supporting current flow. The channel current creates a gradually increasing positive voltage from left to right. Therefore, the right end of the channel is more positive with respect to the gate and the substrate than the left end. Or, to say it another way, the gate is more negative with respect to the channel at the right end. This causes a gradually increasing depletion region to form. At pinch-off, the channel is constricted and allows only so much flow regardless of drain voltage. In transistor applications, the pinch-off current is varied by the gate-to-source bias voltage.

The value of drain current that flows during pinch-off when $V_{GS} = 0$ is called I_{DSS} and is shown in Fig. 2-34. It is the maximum current that will flow during normal operation. Every JFET has a specific value of I_{DSS}, and it varies quite a bit from device to device, just as h_{FE} does among bipolar transistors.

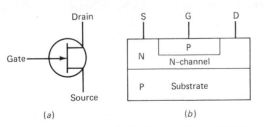

Fig. 2-33 N-channel junction field effect transistor. (*a*) Schematic symbol. (*b*) Structure.

The FETs can be used as voltage-controlled resistors. In the ohmic region, shown in Fig. 2-34, the resistance of the drain-source channel is a function of the gate-source voltage alone, and the JFET will behave as an almost pure ohmic resistor. The supply voltage is set at V_P (or less) to operate the device in its ohmic region. With zero gate bias, the current will be equal to I_{DSS}, and the drain-to-source resistance will be at its minimum. As the gate-to-source voltage is made increasingly negative, the channel resistance will increase.

Constant-current or current-limiting diodes are based on pinch-off and are actually FETs with a common gate-source (cathode) connection. Fig. 2-35(b) shows a Motorola MC1303 current-limiting diode. Its characteristic curve is shown in Fig. 2-35(c). Current I_P is the specified current (3 mA $\pm$ 0.6 mA) at the specified test voltage V_T (25 V). Voltage V_{PO} is the maximum voltage that can be applied to the device without exceeding its power or breakdown ratings. Impedance Z_K is the impedance at the limiting voltage ($V_L = 6$ V) and is equal to 50,000 Ω minimum for the diode shown. An ideal current source has an infinite internal impedance, and Z_K is a measure of how closely the actual device approaches the ideal. These diodes are designed for

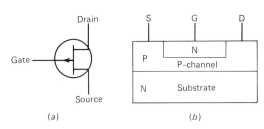

Fig. 2-36 P-channel junction field effect transistor. (a) Schematic symbol. (b) Structure.

applications requiring a current reference or a constant current over a specified voltage range.

The structure and schematic symbol for the P-channel JFET are shown in Fig. 2-36. These devices are the electrical complements of N-channel JFETs. Proper operation for the P-channel transistor includes a positive gate-to-source voltage and a negative drain-to-source voltage. The positive gate voltage will repel the majority holes from the P channel and thereby control drain current. The drain current will be opposite in direction from the N-channel JFET. Note the difference in the schematic symbols. Compare Fig. 2-36 with Fig. 2-33 and look at the arrow on the gate lead. As a memory aid, associate N channel with pointing iN.

All JFETs are operated with reverse bias on their gate leads to prevent gate current. However, there will be some temperature-dependent gate leakage. Also, a large input signal may momentarily overcome the reverse bias and turn on the gate diode. An appreciable amount of gate current will flow, and this will reduce the amplitude of the input signal. These disadvantages are overcome by insulating the gate terminal from the channel with a thin layer of silicon dioxide (metallic oxide). Those FETs that use this technique are known as *metallic oxide semiconductor field effect transistors*, or MOSFETs. They may also be called *insulated gate field effect transistors* (abbreviated *IGFETs*). Figure 2-37 shows the structure and schematic symbol for an N-channel MOSFET. The structure and schematic symbol show that there is no electrical contact from the gate to the rest of the transistor structure. These devices operate in the depletion mode as do JFETs. A negative voltage applied to the gate terminal will repel electrons from the N channel and reduce drain current. The N-

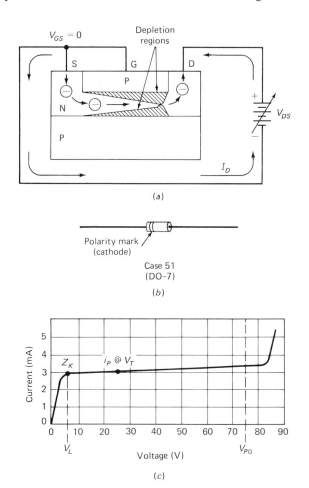

Fig. 2-35 Pinch-off and a typical field effect diode. (a) Pinch-off in an N-channel JFET. (b) MC1303 current limiting diode. (c) Characteristic curve of MC1303.

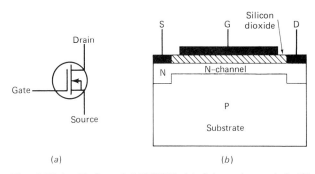

Fig. 2-37 An N-channel MOSFET. (a) Schematic symbol. (b) Structure.

channel MOSFET characteristic curves will be similar to those shown in Fig. 2-34 for the N-channel JFET.

The MOSFETs may also be operated in the enhancement mode. Notice that the P-type substrate in Fig. 2-38 extends all the way up to the oxide layer. There is no channel connecting the source and the drain. This is a normally off device. Note that the schematic symbol uses a broken line to indicate that there is normally no channel connecting the source to the drain. With zero gate-source bias, the device is cut off and the drain current will be zero. Applying a positive gate voltage causes electrons to be attracted near the gate terminal, and an enhanced N channel is the result. Now drain current will flow. The N-channel characteristic curves are shown in Fig. 2-39. It should be clear that drain current increases with positive gate-source bias. The enhancement mode is only feasible when the gate is insulated from the channel. Any attempt to operate a JFET in the enhancement mode would cause gate current, which is undesirable. In addition, P-channel MOSFETs are available, and they are electrical complements to the N-channel devices. The arrow on the schematic symbols in Figs. 2-37 and 2-38 will be reversed for P-channel transistors.

All of the FETs discussed to this point use a lateral flow of current between source and drain. A newer structure has been developed with the drain connection at the bottom of the device. This allows a vertical flow between source and drain. The vertical format allows a wide, short channel with very low resistance. Thus, vertical FETs (often called *VFETs* or *DMOS* or *VMOS devices*) are very attractive for high-power applications.

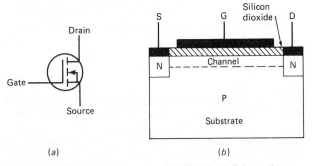

Fig. 2-38 N-channel enhancement MOSFET. (*a*) Schematic symbol. (*b*) Structure.

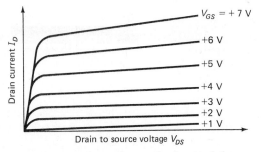

Fig. 2-39 N-channel enhancement mode characteristics.

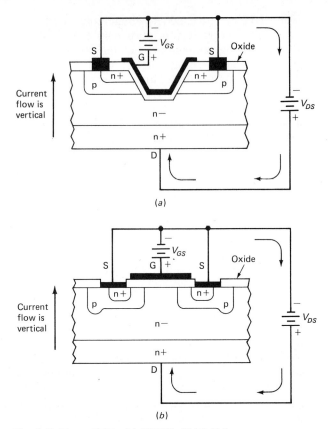

Fig. 2-40 Power FETs. (*a*) VMOS. (*b*) DMOS.

Power FETs are available in two basic structures, as shown in Fig. 2-40. The schematic symbol for either type is the same as shown in Fig. 2-38. They are also available as P-channel devices, in which case the arrow on the symbol will be reversed. The VMOS structure in Fig. 2-40(*a*) uses an etched V groove that extends through the n+ and p and slightly into the n− region. The n+ denotes heavy N-type doping (more than 10^{19} impurity atoms per cubic centimeter), and n− denotes light N-type doping (fewer than 10^{15} impurity atoms per cubic centimeter). There is an oxide layer that insulates the V-shaped metal gate structure from the crystal. The P-type region near each of the drain contacts prevents a continuous N channel between source and drain. The VFETs are enhancement mode devices and are normally off. Applying a positive voltage to the gate causes electrons to be attracted into the P regions, and an enhanced N channel results. Figure 2-40 shows that conventional current flow is vertical from the drain contact on the bottom to the source contacts on the top. The overall N channel is very short and wide, and an extremely low ON resistance can be achieved. These devices are very attractive for switching and controlling large currents in industrial circuitry.

The DMOS shown in Fig. 2-40(*b*) uses the double-diffused structure. The gate metal is in a flat plane and is insulated from the crystal by metallic oxide. It is also an enhancement mode device. A positive gate-to-source bias voltage will attract electrons into

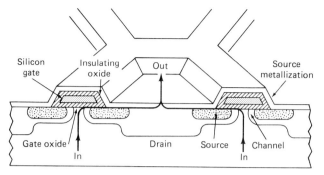

Fig. 2-41 HEXFET® structure (courtesy International Rectifier Company).

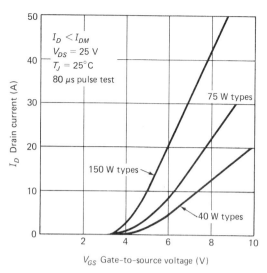

Fig. 2-42 Typical transfer characteristics for N-channel HEX-FETs.

the P-type wells and complete the N channel between the source and drain. Once again, current flow is vertical through a wide, short channel for low ON resistance. The International Rectifier Company has registered the name *HEXFET* for their DMOS power FETs. Figure 2-41 shows the HEXFET arrangement. A single HEXFET is made up of hundreds or thousands of hexagonal-shaped source cells. The cells are very dense (over 500,000 per square inch) and act in parallel to provide very low ON resistance. Figure 2-42 shows some typical transfer characteristic curves for 40-, 75-, and 150-W HEXFETs. These curves indicate the high transconductance that power FETs are noted for. Transconductance (g_{fs}) is a transfer characteristic and relates the change in drain current that occurs for a change in gate voltage. The curves for the 150-W transistor in Fig. 2-42 show a change in drain current of 10 A (from 10 to 20 A) when the gate changes 1 V (from 5 to 6 V). The transconductance is

$$g_{fs} = \frac{\Delta I_D}{\Delta V_{GS}}\bigg|\, V_{DS}$$
$$= \frac{10}{1}$$
$$= 10 \text{ siemens (S)}$$

You should recall that conductance is the reciprocal of resistance and is measured in siemens (S). Transconductance is the figure of merit for power FETs (and small signal FETs) just as h_{FE} is for bipolars.

Bipolar transistors are still very popular; FETs and especially power FETs are newer devices and have been increasing in popularity. Bipolar power transistors have several limitations which detract from their usefulness. They exhibit minority carrier storage, which makes them slower in switching applications. They suffer from second breakdown. Finally, they are very temperature-sensitive. It was shown that beta goes up with temperature. This can lead to a condition known as *thermal runaway*. When a transistor is conducting significant current and dropping significant voltage, it gets hot. This makes its gain increase and it tends to conduct even more current. Now it gets even hotter and the gain increases still

more. The transistor may run away and be damaged or destroyed by heat. Power FETs eliminate minority carrier storage problems, do not exhibit thermal runaway tendencies, and have greatly reduced second breakdown problems. Last but not least, they are easy to drive since there is no gate current. In low-power applications, bipolar devices show better gain and usually offer more performance for the money. Thus, the industrial worker will encounter both bipolar and unipolar transistors. It is important to know their schematic symbols and polarities, the way they are controlled, and whether they are normally on or normally off devices.

Field effect transistors are packaged in many of the same case styles as bipolar transistors. The TO-92 package is popular for low-power devices and the TO-220 and TO-3 packages are often used for high-power devices.

REVIEW QUESTIONS

19. Refer to Fig. 2-33. Does this device normally operate in the depletion mode or in the enhancement mode?

20. The pinch-off region in a FET is also known as the constant _____ region.

21. The P-channel FETs are electrical _____ to N-channel devices.

22. Does the arrow on the schematic symbol point in or out for P-channel FETs?

23. Metallic oxide semiconductor field effect transistors use metallic oxide to _____ the gate from the channel.

24. When the schematic symbol for a MOSFET uses a broken line to represent the drain-to-source channel, it operates in the _____ mode.

25. Enhancement mode transistors are normally _____ devices.

2-5
THYRISTORS

Both the bipolar and unipolar devices studied in the previous sections have a linear mode of operation. This section deals with a family of devices that are not capable of linear operation. Thyristors are strictly switching devices based on a PNPN structure. They are bistable (on or off) and use internal regenerative feedback. They include two, three, and four terminal devices and devices capable of unidirectional (dc) and bidirectional (ac) operation.

The *silicon controlled rectifier* (SCR) is the oldest and most popular thyristor. It is an extremely reliable device and can be expected to deliver billions of operations before failure. It has current ratings that range from 0.25 to several thousand amperes rms. The SCR voltage ratings range up to 5000 V. It is possible to operate SCRs in parallel for even higher current capacity and in series for greater voltage capacity. As a control device, it is one of the most impressive available with microwatt pulses having the ability to switch hundreds of watts. This translates to power gains on the order of 10 million times. Silicon controlled rectifiers can turn on in about 1 microsecond (μs) and turn off in about 10 to 20 μs. They represent an economical solution to many industrial control problems and are especially suited to switching applications up to several kilohertz.

The SCR is sometimes referred to as a *reverse blocking triode thyristor*. Figure 2-43(a) shows its structure. Since it is a triode, it has three external connections: an anode (A), a cathode (K), and a gate (G). Figure 2-43(b) shows that its PNPN structure can be viewed as a two-transistor structure with two connections between the transistors. Figure 2-43(c) shows the equivalent PNP and NPN transistor circuit. This view of the SCR will help you understand its operation. Bipolar transistors are normally off devices. They must be supplied some base current to turn on. Therefore, an SCR is also a normally off device. Notice that the collector lead of the NPN transistor in Fig. 2-43(c) is the base current path for the PNP transistor. Also notice that the collector of the PNP transistor is the base path for the NPN transistor. This is the way the regenerative action is achieved. Once the SCR is on, each transistor will hold the other one on. This regenerative latching

action means that an SCR must be turned off by some external circuit action. Figure 2-43(d) shows the schematic symbol.

Silicon controlled rectifiers can be turned on in five different ways:

1. By avalanche: When the anode is made much more positive than the cathode, forward breakover occurs and latches the device on.
2. By rate of change: If the forward bias voltage across the device increases very quickly, a current will flow to charge the collector-base capacitance of the PNP transistor. This charging current represents base current for the NPN transistor and turns it on.
3. By high temperature: Reverse-biased silicon junctions show a leakage current that approximately doubles for every 8°C temperature rise. At some temperature, the leakage current will reach a level that latches the SCR on.
4. By transistor action: This is the normal mode of operation for all but light-sensitive thyristors. An external gate pulse or signal is used to switch on the NPN transistor and latch the SCR.
5. By light energy: Light entering the junction area will release electron-hole pairs and latch the SCR on. Light-sensitive devices are covered in more detail in the chapter on optoelectronics.

Figure 2-44 shows the volt-ampere characteristic curves for the silicon controlled rectifier. The reverse blocking region is shown in quadrant three of the graph and is similar to that of a silicon rectifier. Normally, the maximum reverse voltage is never exceeded since reverse avalanche could destroy the SCR. It should be emphasized at this point that SCRs are unilateral devices: they normally conduct in one direction only. Forward avalanche is shown in quadrant one. It occurs at zero gate current and at some high value of forward bias. Notice that turn-on occurs at lower values of forward bias as the gate current increases above zero. Current I_{G2} in Fig. 2-44 represents the greatest gate current and ensures turn-on at even a low value of forward bias. This is the normal turn-on mode for the SCR. Once the SCR is on, it drives each of its transistor components into saturation, and its internal resistance drops to a low value. The graph shows that the high conduction

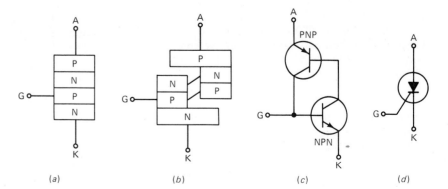

Fig. 2-43 Silicon controlled rectifier. (*a*) Basic structure. (*b*) Viewed as a two-transistor structure. (*c*) Equivalent two-transistor circuit. (*d*) Schematic symbol.

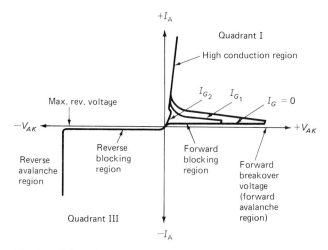

Fig. 2-44 SCR volt-ampere characteristic curves.

most pronounced at high current levels. To turn an SCR off in a minimum amount of time, a reverse bias must be applied across its anode and cathode terminals. This bias will diffuse the holes and electrons to the end junctions. A reverse current flows while this is happening. After the carriers have been removed, the reverse current will cease, and the two outer junctions will assume a blocking state. Recovery is not complete, however, until the center junction is cleared of its carriers by recombination. If forward voltage is reapplied before the center junction is cleared, the SCR will gate itself back on. The time that elapses between the cessation of reverse current, and the point at which forward voltage can safely be reapplied is the turn-off time. It ranges from several microseconds to several hundred microseconds, depending on device design.

In ac power control, commutation is automatic. Figure 2-46 shows the way that an SCR can be used

region is associated with a low forward voltage drop. This makes the SCR efficient in switching operations.

Figure 2-45 shows a test circuit that demonstrates the latching ability of the SCR. The external battery is arranged to forward bias the SCR. It is not large enough to avalanche the SCR, so there is no current flow. When S_2 is pressed, the gate circuit is completed, and the base of the NPN transistor is now forward-biased. The NPN transistor turns on and supplies current to the PNP base circuit so it also turns on. When S_2 is released, the current continues to flow through the load and the SCR. The SCR has latched on since each transistor is now supplying the base current for the other, and S_1 must now be opened to remove forward bias and turn off the SCR. This is an important concept. Silicon controlled rectifiers can be gated on, but they cannot be gated off. Gate turn-off devices (GTOs) have been developed but are expensive and not very popular. When SCRs are used to control dc power, additional circuitry is required to achieve turn-off. Turn-off is often called *commutation* and is achieved by momentarily zero biasing or reverse biasing the device. For example, one popular commutation circuit uses a second SCR to switch in a charged capacitor across the first SCR to reverse bias it.

When an SCR is conducting, each junction is forward-biased, and both base regions are heavily saturated with holes and electrons. The saturation is

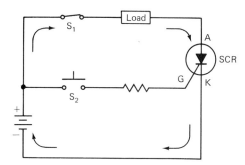

Fig. 2-45 SCR test circuit.

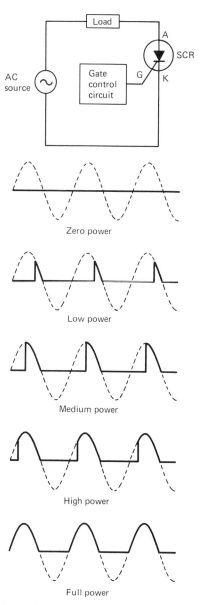

Fig. 2-46 Using an SCR to control ac power.

to control a load efficiently. The waveforms represent load current. The SCR is in series with the load. A gate control circuit is used to pulse the gate lead at the desired moment of turn-on. Turn-off is automatic when the ac line passes through zero. If the SCR is never gated on, the load power is zero. If the SCR is gated on late in the positive alternation, the load power is low. Load power is controlled by conduction angle. With a large conduction angle the circuit is on much of the time and the power will be high. However, since the SCR is a unilateral device only half of the waveform can be utilized. It is possible to achieve full waveform control by using two SCRs connected in inverse parallel or by using rectifier circuits in conjunction with the SCR control. Rectifier circuits are covered in the power sources chapter.

Circuits such as the one shown in Fig. 2-46 are very widely applied in industry. They are commonly used to control motor speed, output from light and heat sources, and battery charging. They are popular because they are efficient. Silicon controlled rectifiers either block current flow or support it with a very low resistance. When they are blocking, there is no current flow and therefore no dissipation in the SCR. When they are on, their low resistance ensures a low voltage drop, and the dissipation in the SCR is small. This means that almost all of the energy is expended in the load and very little in the control device.

Some of the packages used to house silicon controlled rectifiers are shown in Fig. 2-47. As with transistors, the larger devices are mounted in larger packages and are capable of greater dissipation. Notice that the DO-200, or so-called hockey puck package, can be used for devices rated as high as 3000 A. The molded packages at the bottom of the illustration show that more than one SCR can be mounted in a single package. These packages are convenient when bilateral control is needed.

Triac

Bilateral control may also be achieved with another type of thyristor that is equivalent to two SCRs in a single crystalline structure. It is known as a *triode ac semiconductor switch,* or *triac* for short and its structure and schematic symbol are illustrated in Fig. 2-48. It can support the flow of current in both directions. It is also regenerative and latches on once gated. Its connections are labeled *main terminal 1* (MT_1), *main terminal 2* (MT_2), and *gate* (G). Figure 2-49 shows the triac's characteristic curves. It is a symmetrical device and capable of the same performance in quadrant three of the graph as it is in quadrant one. Current I_H, called the *holding current,* is the minimum value of current required to hold the triac on (keep it latched). SCRs also have some minimum holding current value but only in one quadrant of operation. The graph of Fig. 2-49 does not show it, but the forward and reverse turn-on voltages can

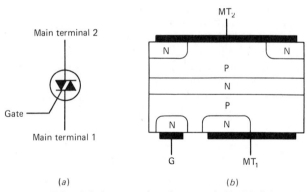

(a)   (b)

Fig. 2-48 Triac (triode ac semiconductor switch). (a) Schematic symbol. (b) Structure.

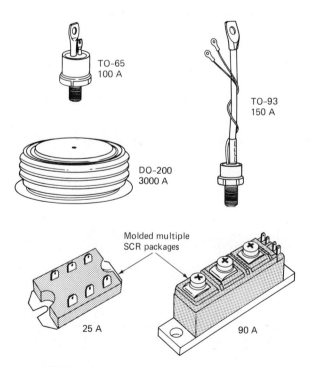

Fig. 2-47 SCR packages.

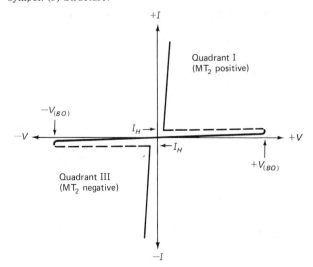

Fig. 2-49 Triac volt-ampere characteristic curve.

be greatly reduced with gate current. As with the SCR, a triac is normally gated on and not operated at breakover.

Figure 2-50 shows the advantage of a triac in ac power control. The circuit uses the main terminals connected in series with the source and the load. A gate control circuit supplies pulses to gate the triac on. The waveforms represent load current and show that conduction angle controls the load power. Compare this illustration with Fig. 2-46. It should be clear that a single triac provides full wave control.

Triacs are very appropriate for some applications but suffer several limitations for others. They are not available with current ratings beyond about 50 A and voltage ratings above 600 V. Also, they have less time to turn off than a pair of SCRs would have in a full waveform control circuit. Each SCR would have an entire half cycle to achieve turn-off if necessary.

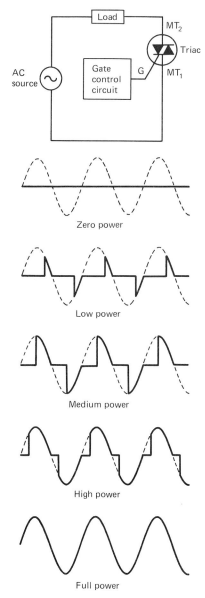

Fig. 2-50 Using a triac to control ac power.

On the other hand, a triac must turn off during the brief moment when the line passes through zero. If the load is inductive (as in a motor), turn-off can be difficult in a triac control circuit. The current lags the line voltage in an inductive circuit. Commutation is attempted when the load current drops below the holding current value. Unfortunately, because of the phase shift, there is a bias voltage across the triac at this time, and the triac recovery current acts as a gate current and tries to turn it back on. The recovery current is due to the recombination of holes and electrons as the device attempts to reestablish its depletion regions and enter the blocking mode. This complicates circuit design. For these reasons, the SCR is still the most popular thyristor for industrial control systems.

Radio Frequency Interference (RFI)

All thyristors cause radio frequency interference (RFI). They can cause a sudden rise in line current at the moment they are gated on. Waveforms with sudden changes in current are rich in harmonic energy. Harmonics are integer multiples of some fundamental frequency. They can extend into the radio frequency spectrum and cause interference with radio receivers, television receivers, logic circuits, and other sensitive equipment. For this reason, thyristor control circuits must include filters to prevent RFI.

REVIEW QUESTIONS

26. Silicon controlled rectifiers are normally turned on by applying a brief pulse to their _____ terminal.

27. Refer to Fig. 2-44. Which value of gate current would be used to ensure device turn-on?

28. Refer to Fig. 2-45. The SCR is gated on by closing switch _____.

29. Refer to Fig. 2-45. The SCR is turned off by opening switch _____.

30. Commutation in an SCR circuit refers to some method of turning the device _____.

31. Commutation is automatic when the power source is _____.

32. Refer to Fig. 2-46. As the SCR is gated on earlier, the power dissipated in the load _____.

2-6
TRIGGER DEVICES

Thyristors can be triggered (gated on) by simple divider circuits consisting of resistors or capacitors across the ac line. One of the divider components can be made adjustable to produce earlier or later firing to accomplish the conduction angle control discussed in the last section. Such simple divider circuits are seldom used, however. They have the disadvantage of temperature instability. Thyristors fire at lower gate currents as their temperature increases. Another problem is that the gate characteristics of the thyristors vary from device to device, even though the part numbers are identical and all have been made by the same manufacturer. Negative resistance devices are generally used to establish a more predictable and stable trigger behavior.

Unijunction Transistor

The *unijunction transistor* (UJT) exhibits negative resistance and is a popular trigger device (Fig. 2-51[a]). It is made from a bar of lightly doped N-type silicon with a heavily doped P zone alloyed into the bar. The P zone forms the emitter section of the transistor and its only junction, hence the name *unijunction*. The B_1 and B_2 (base 1 and base 2) contacts at the ends of the bar are ohmic (no diode action). In the equivalent circuit (Fig. 2-51[b]) the emitter diode is connected at the junction of two resistors. The bottom resistor is variable. This is where the negative resistance effect occurs. It is called a *negative resistance* because it decreases abruptly when the UJT emitter diode becomes forward-biased with respect to B_1. Complementary UJTs are also available but are less popular. They use a P-type bar and an N-type emitter. The emitter arrow is reversed on the schematic symbol for the complementary UJT.

The volt-ampere characteristic curve of Fig. 2-52 shows the negative resistance behavior of the UJT. As the forward bias is gradually increased across the emitter to base 1 section, a point, called V_P, is reached; at it the diode becomes forward-biased, and the transistor fires. It then enters its negative resistance region, and the voltage quickly drops to a much lower value. One normally expects the voltage drop to increase as current increases. However, the UJT curve shows a region where voltage decreases as the current increases. This is due to a sudden resistance drop inside the transistor. The curve shows that the emitter current at the firing point is called I_P. It also shows a higher current, called the *valley current* or I_V. This valley current is similar to the holding current in a thyristor. The UJT cannot assume any stable operating point between I_P and I_V. It switches rapidly between the two. This characteristic makes the UJT useful as a trigger device.

The negative resistance behavior of the UJT can be understood by referring to Fig. 2-51. When the emitter diode becomes forward-biased with respect to base 1, it injects minority carriers into the region between the emitter and the base 1 contact. These minority carriers decrease the resistance of the lower part of the bar. This is why the equivalent circuit shows the base 1 resistor as a variable component. When this resistance decreases, the diode forward bias increases, increasing the diode current, and still more minority carriers are injected into the lower part of the bar. This regenerative action rapidly decreases the resistance from the emitter to base 1.

The firing voltage of a UJT is predicted by the voltage across the base leads and the intrinsic stand-

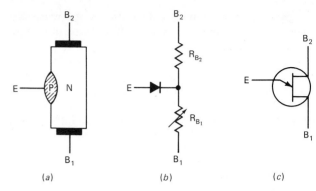

Fig. 2-51 Unijunction transistor (UJT). (*a*) Structure. (*b*) Equivalent circuit. (*c*) Schematic symbol.

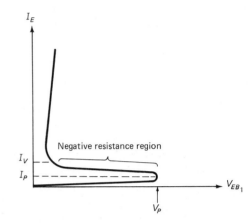

Fig. 2-52 UJT volt-ampere characteristic curve.

off ratio of the transistor. This ratio is set by the off resistance from base 1 to the emitter (R_{B1}) and the resistance from the emitter to base 2 (R_{B2}). The following equation is the familiar voltage divider relationship:

$$\text{Intrinsic standoff ratio} = \frac{R_{B1}}{R_{B1} + R_{B2}} = \eta$$

When a supply voltage (V_{BB}) is impressed across the base leads of a UJT, the intrinsic standoff ratio determines the voltage at the cathode of the emitter diode. Since the diode is silicon, an additional 0.6 V is required to turn the diode on. Therefore, the firing voltage, or V_P, is given by

$$V_P = \eta \times V_{BB} + 0.6$$

The intrinsic standoff ratio of UJTs ranges from 0.5 to 0.8.

EXAMPLE

The supply is 12 V and the intrinsic standoff ratio is 0.6. Find the firing voltage.

SOLUTION

Use the formula

$$V_P = \eta \times V_{BB} + 0.6$$
$$V_P = 0.6 \times 12 + 0.6$$
$$= 7.8 \text{ V}$$

Figure 2-53(*a*) shows a UJT relaxation oscillator circuit that is useful in many industrial timing and control applications. An *oscillator* is a circuit that changes dc to ac. A relaxation oscillator is one type that uses RC time constants to control the frequency of oscillations. When the supply voltage is applied, the capacitor begins charging through R_1. Eventually, the capacitor voltage reaches the firing point of the UJT. The emitter diode turns on, the internal resistance of the transistor from base 1 to the emitter drops, and the capacitor is rapidly discharged through R_3 (it is usually less than 100 Ω) and the transistor. When the capacitor discharge current reaches I_v, the UJT switches off and the next cycle begins. The waveforms in Fig. 2-53(*b*) and Fig. 2-53(*c*) show an exponential sawtooth at the emitter terminal and a pulse waveform at the B_1 terminal. The period of the waveforms is approximately equal to the R_1C time constant. Since a capacitor reaches 63 percent of its final charge during the first time constant, the approximation is good when the intrinsic standoff ratio is near 0.63.

EXAMPLE

Suppose R_1 is 100 kΩ and C is 0.1 μF. Find the frequency of oscillation.

SOLUTION

$$\text{Period} = R \times C$$
$$= T$$
$$= 100 \times 10^3 \times 0.1 \times 10^{-6}$$
$$T = 0.01 \text{ s}$$

The frequency (*f*) of oscillation is found by the reciprocal of the period:

$$f = \frac{1}{T} = \frac{1}{0.01}$$
$$= 100 \text{ Hz}$$

Figure 2-54 indicates one way that a UJT can be used to control an SCR, which, in turn, controls the power delivered to its load. This circuit controls load power by conduction angle. The sooner the SCR is gated on, the larger the conduction angle and the greater the load power. Circuits of this type are called *phase control circuits* since the phase angle of the gating pulse in relation to the source phase determines conduction angle. The zener diode clips the positive peaks of the source at its breakdown voltage. The negative alternations are clipped near 0 V since they forward bias the zener. You should recall from the previous section that the negative alternations are not used by an SCR control circuit of this type, and therefore there is no need to energize the UJT circuit during negative alternations. The breakdown voltage of the zener is reached early during the positive alternation, and the voltage across the UJT circuit is constant for nearly the entire alternation. The capacitor charges through R_2 until the firing voltage of the UJT is reached. When it fires, it develops a pulse across R_4, which gates the SCR on. The phase of the gating pulse can be advanced or retarded by decreasing or increasing R_2. If the period of the

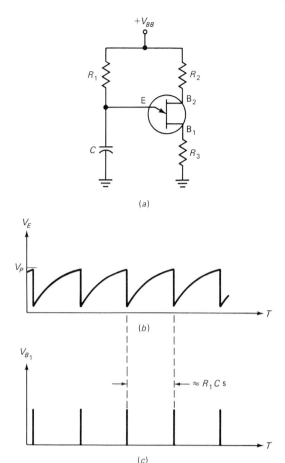

Fig. 2-53 UJT relaxation oscillator and waveforms. (*a*) Circuit. (*b*) Sawtooth waveform at the emitter *E*. (*c*) Pulse waveform at base B_1.

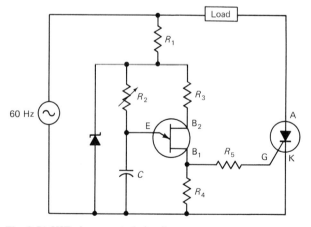

Fig. 2-54 UJT phase control circuit.

UJT oscillator is small, extra gating pulses may be delivered during the positive alternation. These will not cause any effect since the first pulse will gate the SCR on and any subsequent pulses will be ignored.

Programmable UJT

The *programmable unijunction transistor* (PUT) is an improved trigger device. It is a small thyristor

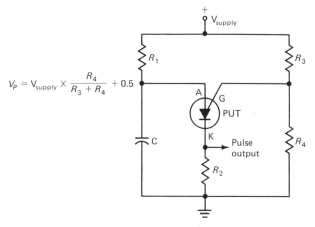

$$V_P = V_{supply} \times \frac{R_4}{R_3 + R_4} + 0.5$$

Fig. 2-55 Programmable UJT (PUT) relaxation oscillator.

structure with an anode gate. Its schematic symbol and application in a relaxation oscillator circuit are shown in Fig. 2-55. Resistors R_3 and R_4 program the firing voltage of the PUT and form a voltage divider for the supply voltage. If, for example, they are equal in value, the gate voltage of the PUT will be equal to half the supply voltage. The firing voltage will be equal to half the supply plus 0.5 V in this case. This programming feature gives the PUT a greater range than the UJT and allows the designer to select the desired V_P. It also eliminates the problem with "batch spread" since UJTs of the same manufacturing batch will show a considerable spread in their intrinsic standoff ratios. The PUT can be programmed with precision resistors to eliminate this problem. The peak current I_P is also a function of the programming resistors. By using large resistors, it is possible to obtain a very low peak current in a PUT timing circuit. This makes it possible to increase the size of the timing resistor (R_1 in Fig. 2-55) to a much greater value than it can have in a UJT circuit. The UJT current's being larger is a disadvantage with long time constants since the current will cause a drop across the timing resistor, and the capacitor will never reach the firing voltage if the timing resistor is made too large. The only way to achieve long time constants in a UJT circuit is to use a large timing capacitor since timing resistors above 1 MΩ are not practical.

Shockley Diode

Figure 2-56 shows the schematic symbol for a four-layer diode (also called the *Shockley diode*). It is essentially a miniature SCR with no gate lead. It exhibits negative resistance once its firing voltage is reached. Its volt-ampere characteristic curve is the same as for an SCR, except that there is no possibility of varying the breakover point with gate current. Four-layer diodes are available with breakover ratings from about 10 V to 400 V. They usually exhibit peak currents of about 100 μA and holding currents of 1 mA or so. They are used in some thyristor gating circuits.

Fig. 2-56 Four-layer diode schematic symbol.

Fig. 2-57 Silicon unilateral switch schematic symbol.

Silicon Unilateral Switch

A *silicon unilateral switch* (SUS) is shown in Fig. 2-57. It is equivalent to a miniature anode gate SCR with a built-in zener diode from its gate to its cathode. This trigger device is more flexible than the four-layer diode because its gate terminal can be used to alter its forward breakover voltage. As all other trigger devices do, it exhibits a negative resistance characteristic. The SUS can be fired at low anode-to-cathode potentials. The major difference between SUS and UJT operations is that the SUS switches at a voltage determined by its internal zener, and the UJT fires at some fraction of its supply voltage. The SUS can also be synchronized or locked out by applying pulse signals or a bias to its gate terminal.

Diac

All the trigger devices investigated to this point are unilateral. They fire in one direction only. As such, they are more appropriately applied in SCR control circuits than they are in triac control circuits. They can be used to gate triacs by adding rectifiers or pulse transformers, but we will investigate bidirectional trigger devices which are well suited to triac control.

The diac (Fig. 2-58) is a transistor-type structure with a bidirectional negative resistance characteristic. Diac breakover current is typically around 100 μA and occurs at approximately 30 V. A diac phase

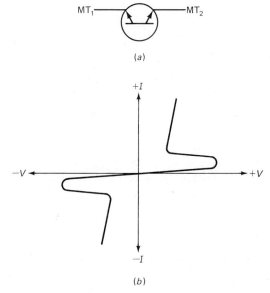

Fig. 2-58 Diac. (*a*) Schematic symbol. (*b*) Characteristic curve.

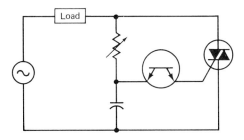

Fig. 2-59 Diac phase control circuit.

control circuit is shown in Fig. 2-59. On either alternation, the capacitor will begin charging through the variable resistor. When the capacitor voltage reaches the breakover potential, the diac fires and gates the triac on. The phase angle can be advanced by decreasing the variable resistor. This shortens the time constant, and the triac gates sooner for a larger conduction angle and greater load power. Increasing the variable resistor will delay firing for a smaller conduction angle and lower load power.

Silicon Bilateral Switch

Another full-wave trigger device, the *silicon bilateral switch* (SBS) is shown in Fig. 2-60. The SBS is equivalent to two SUS devices connected in inverse parallel. It typically exhibits a breakover of around 8 V and has a more pronounced negative resistance region than the diac. It is also more temperature-stable than the diac. The gate lead provides additional capabilities and can be used to alter the breakover characteristics. The gate lead is also useful for elim-

inating the hysteresis effect, or "snap-on" effect, found in many triac control circuits. This effect is noticed when the phase angle control is slowly advanced from the zero power position. The load "snaps" to some intermediate power level. Then the phase control can be backed off for low-power operation. Any circuit that exhibits a different threshold when a control is moved in one direction than it does when the control is moved in another direction is said to have *hysteresis*. Hysteresis can be very desirable in some applications but is often undesirable in phase control circuits. It is just not possible to smoothly adjust the load for low power when starting from zero power if the circuit has hysteresis.

The hysteresis effect can be understood by referring to Fig. 2-59. At zero power, the capacitor has been somewhat charged by the prior alternation at the beginning of any positive or negative alternation. This charge is a reverse charge as far as the current alternation is concerned. It takes time to reverse the charge on the capacitor. This time delays the firing of the diac. However, once the diac does fire, it gates on the triac, which in turn tends to drain any residual charge from the capacitor. Therefore, on all subsequent alternations, the reverse charge is absent, and the capacitor charges more quickly to the firing voltage of the diac. The phase angle is now advanced, and the load has snapped to some intermediate power level. It is now necessary to increase the variable resistor to back the power level down to the desired condition.

Asymmetrical ac Trigger

Some phase control circuits take advantage of the gate terminal of the silicon bilateral switch to overcome the hysteresis, but an asymmetrical ac trigger device can also be used (Fig. 2-61). This device presents a forward breakover voltage different from its reverse breakover voltage. It acts as a zener diode in series with an SBS. Note that the schematic symbol includes a zener curve to show this effect. The typical asymmetrical trigger device switches at 8 V in one direction and at 16 V in the other. It will fire first at its lower breakover voltage when being adjusted from zero power. Then it will fire at its higher breakover voltage on the next alternation. This delays firing and tends to offset the tendency for the circuit to snap to an intermediate power level. The asymmetrical trigger is a simple solution to snap-on, but its inherent asymmetry develops a dc component in the load circuit which may not be acceptable in some applications.

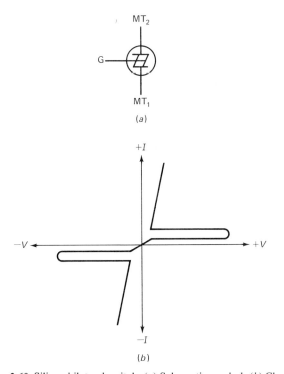

Fig. 2-60 Silicon bilateral switch. (*a*) Schematic symbol. (*b*) Characteristic curve.

Fig. 2-61 Asymetrical ac trigger schematic symbol.

REVIEW QUESTIONS

33. A unijunction transistor has an intrinsic stand-off ratio of 0.75. What will its firing voltage be with a 15-V supply?

34. Refer to Fig. 2-53. Assume an intrinsic stand-off ratio of 0.63 and calculate the period of oscillation if the timing resistor is 470 kΩ, and the timing capacitor is 0.01 μF.

35. Calculate the frequency of oscillation for question 34.

36. Refer to Fig. 2-54. What happens to the load power when R_2 is adjusted for lower resistance?

37. Refer to Fig. 2-54. What would happen to the load if the period of the UJT oscillator were made longer than 8.33 ms? (Hint: Calculate the period of the positive alternation.)

38. Refer to Fig. 2-55. If the supply is 10 V, R_3 is 220 kΩ, and R_4 is 47 kΩ, what voltage will the capacitor charge to?

2-7 INTEGRATED CIRCUITS

The same planar process that is used to manufacture most of the discrete devices covered in this chapter is also used to manufacture most integrated circuits. The typical integrated circuit uses components that are all formed at the same time and are not individually accessible. This arrangement contrasts to that of a *discrete circuit,* in which individual resistors, diodes, capacitors, transistors, and other components are interconnected to form a working circuit. Discrete circuits require assembly and some system of interconnection such as a printed circuit board. This means that discrete circuits are larger and more costly than an equivalent integrated circuit (IC). The ICs are also usually more reliable and more power-efficient than equivalent discrete circuits. Therefore, designers choose ICs wherever their use is feasible. Many industrial systems are heavily dependent on ICs. They do suffer power-handling limitations, so it is common to find equipment based on both integrated and discrete circuits. The ICs handle most or all of the low-level signal processing, and the discrete circuits use large power devices to control the high-level signals.

The ICs may be classified according to the technology used to manufacture them. Figure 2-62 shows a typical monolithic (*monolithic* means "single stone") IC mounted in an 8 pin package. The *chip* is the monolithic silicon structure that provides the electrical and electronic functions. Some people use the word *chip* when referring to a complete IC package. The silicon chip in Fig. 2-62 is bonded to a plastic or ceramic base. Wire bonds form electrical connections from the silicon chip to the pins. A plastic or ceramic cap completes the assembly. The pins are identified by number. Count counterclockwise

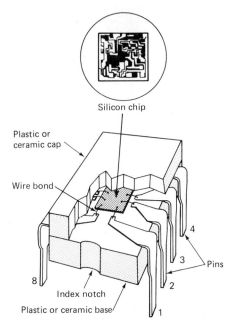

Fig. 2-62 Typical integrated circuit construction.

when viewing the package from the top. Pin 1 is the first one encountered after the index. The numbers in Fig. 2-62 represent pin numbers.

Figure 2-63 shows an abbreviated cross-sectional view of the chip. Now you can understand why it is considered a monolithic structure. Each component is an area buried in a single silicon structure and has been formed by the planar diffusion process. Note that P-type isolating wells are used to isolate one component function electrically from the next. For example, the NPN bipolar transistor at the left is diode-isolated from the P-channel resistor to its right. In normal operation, the PN isolation diodes are all reverse-biased, and electrical component integrity is maintained. The wells are produced during a manufacturing step called *isolation diffusion,* during which a P-type impurity such as boron is forced to penetrate the chip until it reaches the P-type substrate. This occurs not only at the sides, as shown in Fig. 2-63, but also at the front and rear of every component site. This leaves N-type islands that will become the collectors of transistors, the cathodes of diodes, the sites for P-channel resistors, one plate of a capacitor, and so on. The junction diode shown in the illustration may also be used as a capacitor if it is reverse-biased. Both NPN and PNP transistors are feasible, as are the various unipolar transistors discussed earlier in this chapter. Inductors are not feasible. After all of the components have been formed, evaporated aluminum is deposited onto the surface. The aluminum contacts selected areas of the chip through windows in the silicon dioxide layer. Then, the unwanted aluminum is etched away, leaving aluminum jumpers that interconnect the individual components to form a complete circuit.

The economy of the monolithic IC is now apparent. All of the needed components for a circuit are

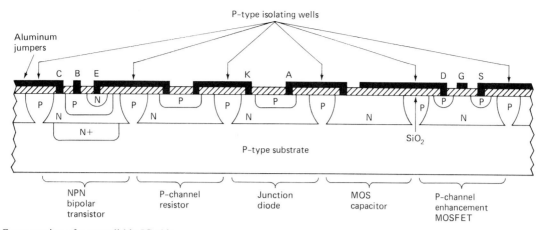

Fig. 2-63 Cross section of a monolithic IC chip.

formed at the same time by the planar diffusion process. Then aluminum interconnects these components to form circuits. Another economical feature is that monolithic ICs are batch-processed. Hundreds of them are processed at the same time on a silicon wafer that is typically several inches in diameter. The completed wafer, including aluminum jumpers, is about 400 micrometers (μm) [0.016 in.] thick. The wafer is scribed, and the individual chips are broken from it. Some complex chips contain thousands of individual components and are only 6250 μm (0.250 in.) by 6250 μm! This miracle of modern technology has allowed room-sized equipment to shrink to desktop size. The price has also decreased dramatically, with some computer-type circuits now costing tens or hundreds of dollars that formerly cost hundreds of thousands of dollars. It is easy to see why monolithic ICs have created a rapid expansion of electronics into every industrial sector. Very sophisticated control systems are now feasible, and they are relatively inexpensive, small, efficient, and reliable.

Most, but not all, ICs are monolithic. Another manufacturing process can combine monolithic ICs with larger component structures to permit larger signals to be controlled. Figure 2-64 shows the structure of a hybrid IC. It can be seen that several types of components are fixed to an insulating substrate of ceramic or glass. Hybrids are either of the thin-film or thick-film variety. A thin-film IC uses very thin films (about 0.3 μm) that are vacuum-deposited on a substrate. Resistors are usually formed by depositing tin oxide, nichrome, or tantalum strips; conductors are made by depositing gold nichrome. The other components, including one or more monolithic ICs, are in chip form and are fastened to the substrate with conductive epoxy. Thick-film ICs use screen-process printing to deposit resistive and conductive patterns onto the substrate. These patterns are much thicker than the vacuum-deposited ones. Hybrid ICs of both types offer several advantages over monolithic ICs. They can take advantage of separate power devices and therefore can handle larger signals

(up to several hundred watts). This makes them more flexible and provides a greater range of applications. They have a greater range of available capacitor and resistor values, and precision resistor values can be attained by trimming critical patterns with computer-controlled laser beams. Unfortunately, their higher cost makes circuit designers look to monolithics first, especially for high-volume applications.

Integrated circuits are also identified as being digital or linear. A *digital IC* works with only two circuit conditions: on or off. Chapter 11 deals with digital circuits. A linear IC works with an infinite number of possibilities. For example, the voltage in a linear control circuit might be 8.5 V, or 8.56 V, or 8.567 V, and so on. Several popular ICs are both; that is, they contain both linear and digital circuits.

Monolithic ICs are often further differentiated on the basis of the type of transistor that they use. Bipolar ICs favor NPN transistors since they are easier to fabricate in the chip and have higher performance than integrated PNP transistors. The MOS ICs use metallic oxide semiconductor field effect transistors. The PMOS ICs are based on P-channel transistors, and the NMOS ICs are based on N-channel devices. Also, some manufacturers "invent" terms such as *HMOS* to describe their particular innovative process. The term *HMOS* is used by Motorola, Inc., to identify their high-density NMOS devices. The complementary metallic oxide semiconductor (CMOS) ICs use both N- and P-channel transistors. Some ICs use both bipolar and unipolar transistors and are called *BI-FET* ICs. The field effect

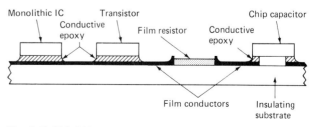

Fig. 2-64 Hybrid integrated circuit construction.

transistors are used in the integrated circuit's input circuits because of their advantages of high impedance, low noise, low leakage currents, and good temperature stability.

Finally, ICs may be identified according to their level of complexity. For example, in the digital world the number of logic gates is a way to categorize the complexity of an IC. A *logic gate* is a single decision-making element, with each gate potentially containing a dozen or so components. A *small-scale integration* (SSI) device will have up to 10 gates; a *medium-scale integration* (MSI) device has between 11 and 100 gates; a *large-scale integration* (LSI) device has between 101 and 1000 gates; and a *very-large-scale integration* (VLSI) device has over 1000 gates. Digital ICs are also identified as *transistor-transistor logic* (TTL), *emitter-coupled logic* (ECL), etc., integrated circuits. These designations will be explained in Chapter 11.

There are thousands and thousands of active IC part numbers, with new ones announced every month. This may be bewildering to a beginner. However, the industrial technician seldom needs to be concerned with the exact circuit located inside a particular IC. You will learn that some very common features which are important to the technician are pretty much the same for many ICs. Along these lines, let's take a look at a few very common IC characteristics. Figure 2-65 shows a totem-pole output stage that is very popular in digital ICs. The transistors form the totem-pole, which is capable of being driven high or low. When Q_1 is off there is no base current path for Q_3, and it is also off. When R_1 provides base current to Q_2 it is on. Therefore, if an external load is connected as shown, the IC will source current to the load. The typical TTL digital IC can source a few milliamperes before its output voltage drops too far below the threshold voltage.

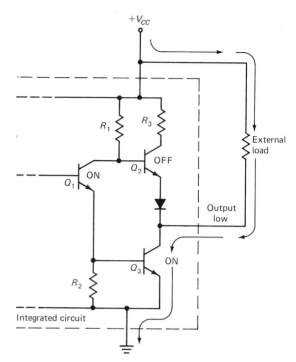

Fig. 2-66 Current sink mode for totem pole IC.

Now look at Fig. 2-66. Here, the totem-pole output is low; Q_1 is saturated on, and the drop across R_1 is large. The base voltage at Q_2 is too low for it to turn on, and Q_3 now has plenty of base current and is on. Note that the external load is now connected to the supply point, and the IC is said to be sinking load current. The typical TTL digital IC can sink several times more current than it can source. Compare Fig. 2-65 and Fig. 2-66 to clarify the difference between current sourcing and current sinking. Please note that either a high or a low output can produce load current, depending on the way the external load is connected. The totem-pole output stage can be redesigned with CMOS output transistors to allow both devices to be off at the same time. Integrated circuits with this feature are called *tri-state devices*. Their outputs can be high, low, or off (high-impedance).

Figure 2-67 shows another popular output stage. It is known as an *open collector IC* because there is no internal connection to the collector of the output transistor. Open collector devices offer the advantage of allowing the output collector circuit to operate at a different voltage level than the rest of the IC circuitry. This is convenient when logic voltages have to be translated from one level to another. Another advantage is that open collector outputs may be tied together without the danger of excessive current in the output circuit. Tying outputs together is usually avoided with totem-pole ICs since a high output would source excessive current to another output that was low. Note that an external pull-up resistor is required to develop an output swing in Fig. 2-67.

Figure 2-68 shows some package styles for ICs. The *dual-inline package* (DIP) is very popular for both digital and linear devices and is made with 14,

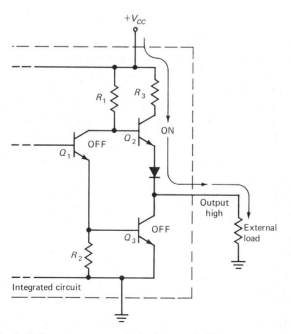

Fig. 2-65 Current source mode for totem pole IC.

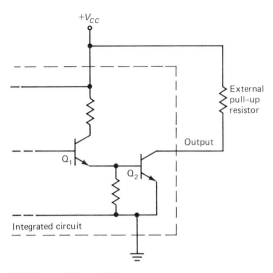

Fig. 2-67 Open collector IC.

16, 18, or 20 pins. The mini-DIP is an 8-pin dual-inline package. The LSI/VLSI package is available with 24, 28, 40, and, less commonly, 64 pins. It is often used for complex ICs such as microprocessors, memory devices, and programmable devices. It is wider than the DIP. The flat pack is used only in compact, low-profile applications and is less popular. The metal packages offer some advantages for heat dissipation and are popular for power applications such as amplifiers and voltage regulators. Hybrid ICs

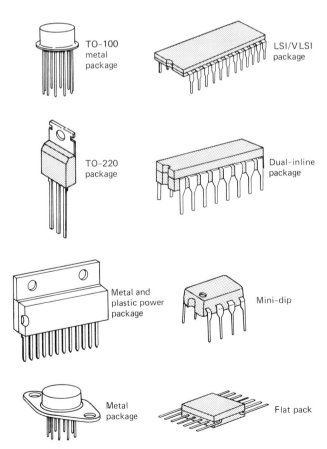

Fig. 2-68 Integrated circuit package styles.

may be housed in any of the packages shown in the left column of Fig. 2-68 or in a wide variety of custom packages.

REVIEW QUESTIONS

39. Integrated circuit pins are numbered by counting counterclockwise from the index when viewing the package from the _____.

40. In a monolithic IC, electrical integrity for each component is maintained by reverse-biased _____.

41. Refer to Fig. 2-65. Would the IC source any current if the external load resistor were connected to the supply rather than to ground?

2-8
TROUBLESHOOTING AND MAINTENANCE

Solid-state devices are usually very reliable. Some industrial equipment will operate for years without a single solid-state device failure. However, all of the devices covered in this chapter have two major enemies: heat and transients. In some cases, failures may occur frequently. The industrial technician must become proficient at locating defective devices and replacing them properly. A knowledge of circuit operation is usually mandatory when troubleshooting. This knowledge allows the fault or faults to be isolated. A piece of equipment may contain hundreds or even thousands of electronic components. Obviously, fault isolation is essential if the equipment is to be repaired in a reasonable length of time. Later chapters of this book explain the operation of many circuits and present specific troubleshooting information. This section will be limited to some general ideas concerning locating defective devices and replacing them.

Troubleshooting must always begin by verifying that the equipment is set up and properly connected. In some cases, there is nothing wrong with the equipment itself. A cable may have been pulled out, or a cable may be connected improperly. A control may be set wrong. Always check the obvious things first; it can save a tremendous amount of time.

When it is verified that the equipment is defective, refer to the manufacturer's service literature for the proper tear-down procedure. The literature may also contain specific troubleshooting procedures. Be certain that the power is off before removing panels, covers, or cabinets. Never pull or insert circuit boards with the power on. This can be a danger to you and to the equipment. Once access to the circuitry has been gained, it is time for a thorough visual inspection. Solid-state devices usually do not change in physical appearance when defective, but other components sometimes do. Be sure to inspect fans, air filters, and other cooling components since heat is a killer of electronic equipment.

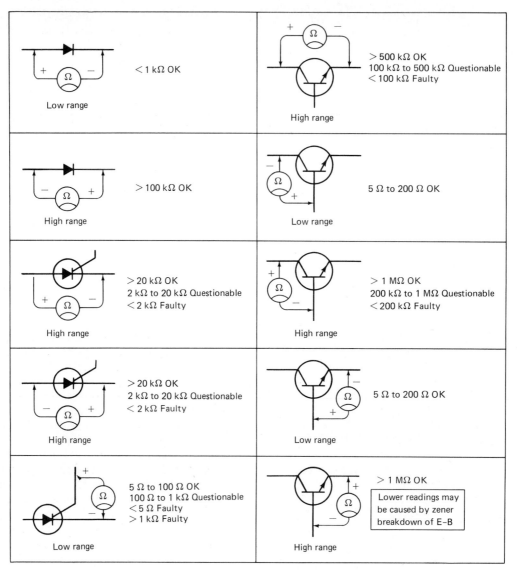

Fig. 2-69 Testing solid-state devices with an ohmmeter.

After any obvious problems have been corrected, it is time to decide how to isolate any remaining faults. Troubleshooting techniques vary from circuit to circuit and from technician to technician. They also depend upon the kinds of test equipment that are available. Whatever you choose, please remember to follow safe procedures. Some of the procedures that can be used to isolate defective solid-state devices include substitution, voltage analysis, waveform analysis, signal injection, and resistance analysis. Substitution is convenient when devices are in sockets. However, remember that a fault somewhere else in the circuit may have damaged the original part and is also likely to damage the substitute. Never change devices with the power on.

Voltage analysis involves comparing actual circuit voltages with voltages specified on a diagram by the equipment manufacturer to determine whether any are out of tolerance. Always verify supply voltages first. Power supply troubleshooting is covered in Chapter 5. A defective device may or may not upset

circuit voltages. Waveform analysis uses an oscilloscope to view signals into and out of circuits. If a circuit has the proper operating voltages and the correct input waveform, then a bad or missing output waveform is strong evidence that the circuit is defective. *Signal injection* is a procedure in which an input is stimulated with some external generator. It is often used in conjunction with waveform analysis to see whether the output responds as expected.

Resistance analysis is accomplished with the power off and often with the suspected device isolated from the rest of the circuit. Isolation may be necessary to prevent unwanted paths from producing abnormally low readings. Figure 2-69 shows the way that an ohmmeter can be used to test several devices. Verify your ohmmeter polarity and open circuit voltage. Some multimeters have reverse polarity on the Ohm's function, and others do not use an internal voltage large enough to forward bias junctions on some ranges. Always learn your equipment first. Ohmmeter tests are effective for detecting short-circuited junctions since a very low resistance will be measured in both directions. A very high reading in both directions indicates an open junction. The diode tests in Fig. 2-69 show the expected results for a good diode. It must be emphasized that the readings in this illustration are relative. For example, a diode that measures 200,000 Ω when reverse-biased may have excessive leakage for some circuits. Generally, more leakage is normal in large devices than in small devices. Leakage is also increased when the device is hot. Germanium devices exhibit quite a bit more leakage than silicon devices. You should also know that the forward resistance of any junction will measure differently on different ohmmeter ranges because the volt-ampere characteristic of a diode is nonlinear.

The transistor tests in Fig. 2-69 show the expected ranges of resistance for a silicon transistor. You will note that the collector-to-emitter reverse resistance is not expected to be as high as the collector-to-base reverse resistance. This is because the collector-base leakage is amplified beta times in the collector-emitter test. Your understanding of device theory will make ohmmeter testing more productive. Figure 2-69 also shows that the emitter-base reverse resistance may not test properly because the ohmmeter voltage may be high enough to cause zener breakdown. Most transistor emitter-base junctions show zener action at around 6 V, and many ohmmeters use 9 V or more on their high range. As you gain experience, you will learn the way to test quite a few solid-state devices with an ohmmeter. It is even possible to gate and latch some thyristors by placing the positive lead on the anode and the negative lead on the cathode, and then momentarily short-circuiting the gate to the anode. This will not work with large thyristors since the ohmmeter current, even on the lowest range, is less than the holding current. Field effect transistors are often very sensitive during out-of-circuit tests. The drain-to-source resistance may vary quite a bit as you bring a finger near the gate lead. Metallic oxide semiconductor field effect transistors are susceptible to damage by static discharges and should not be checked this way. Integrated circuits do not usually lend themselves to ohmmeter testing.

Repeated failure of the same device must be investigated. It is probably a thermal problem, a transient problem, a defective power supply, a defective load, or a design flaw. Check to be sure that all cooling components are installed and working properly. Inspect the mounting area on the heat sink. Burrs and peeled-over areas will reduce heat transfer. There should be no paint or heavy oxide on the mounting area. Galvanic action may occur in corrosive atmospheres between copper cases and aluminum heat sinks. Be sure that the correct thermal grease is applied. Everything must be installed correctly to prevent short circuits and excessive device temperature. The insulator may be of the beryllium oxide type. Be careful: Beryllium oxide parts must not be abraded or crushed because the dust is extremely dangerous if inhaled. Do not replace beryllium oxide washers with mica washers, since the heat transfer will be impaired. Do not over- or under-tighten stud mount devices. For example, the proper torques for DO-4 and DO-5 packages (Fig. 2-19) are 15 inch-pounds (in.·lb) and 30 in.·lb, respectively.

Repeated failures may also be caused by the load. Make sure the load is electrically normal. If it is a motor or solenoid, make sure that it is clean and properly lubricated and is operating freely. Also be sure that the mechanical load on the solenoid or motor is working smoothly and not binding. *Transients* are brief periods of overvoltage and may enter the equipment via the power lines. The industrial environment is often replete with transients since large inductive loads such as motors are constantly being turned on and off. It may be necessary to add transient suppression to the supply circuit. Transients are also generated when solid-state devices switch inductive loads within the equipment circuit. Check the suppression networks and devices across relay coils, solenoids, and motors. Suppression is covered in Chapter 4. Repeated failures due to design flaws can be checked by contacting the manufacturer. The engineering staff may supply a circuit modification or a substitute part with better ratings.

Parts Identification

Exact replacement parts are usually the best bet, especially for the technician with limited experience. Solid-state devices often have part numbers on their packages. The manufacturer's literature is also usually helpful for locating part numbers. The Joint Electronic Device Engineering Council (JEDEC) registers part numbers in this country. A registered part has been characterized to meet the specifications listed for that number. Registered solid-state parts (excluding ICs) have numbers prefixed with 1N, 2N, 3N, and 4N. This means that you can buy a 1N5000 rectifier, a 2N690 SCR, a 2N3055 bipolar transistor, a 3N128 field effect transistor, or a 4N32 opto coupler from any of several manufacturers and be reasonably assured that it will work as well as the original device. There are also registered JEDEC package numbers such as TO-3 that specify the physical parameters of devices.

Data manuals, cross-reference guides, and substitution guides are invaluable aids when trying to track down part numbers. These materials contain valuable information concerning physical characteristics, electrical characteristics, and lead identification drawings. You will find many good substitutions in these sources. In some cases the substitutions are not appropriate. It pays to check both electrical and physical parameters. Some companies build quite a few solid-state devices with their own part numbering system. These nonregistered devices can often be substituted for registered devices. An example is a Motorola MJ4502 power transistor, which can be substituted for a 2N5744. The guides usually list both registered and nonregistered device numbers. Some part numbers are proprietary and will not show up in the guides. It may be necessary to buy a replacement from the manufacturer of the equipment in these cases.

Integrated circuit part numbers are also referenced in some guides. The ICs have a part numbering system that can vary considerably from manufacturer to manufacturer. It pays to have a library of data manuals from the various companies that build solid-state devices. Supply catalogs are also helpful in many cases and can often be obtained just by asking for them. It pays to communicate with supply houses and parts jobbers to obtain valuable literature. Integrated circuit part numbers are usually a combination of a prefix, a part number, and a suffix. The *prefix* uses code letters to designate the type of circuit, the *part number* specifies the device type, and the *suffix* code specifies the package type and the temperature range. There are many variations of this basic system, and the manufacturer's data books are usually necessary to decipher all the information contained in the part number. Many IC makers also put date codes on their packages.

Handling Solid-State Devices

The final consideration in this chapter is the safe handling of solid-state devices. Many devices, especially the MOS types, are easily damaged by static discharge. It is a little disconcerting to realize that merely touching an expensive or hard-to-get device can destroy it. The human body can generate thousands of volts through simple movements such as walking, sliding in a chair, or sliding a sleeve across a bench top. These voltages are particularly high in low-humidity conditions. Some workers must wear a conductive wrist strap that is grounded through a high-value resistor to bleed off static charges. Note:

This is only practiced in an approved environment and with an approved grounding apparatus. The following guidelines are recommended to prevent static damage of solid-state devices:

1. Work on a metal surface. Plastic laminated table tops are a poor choice for a work surface since static build-up is likely.
2. Do not allow the relative humidity in the work area to go below 50 percent.
3. Do not handle devices any more than is required. They are shipped in protective carriers or pressed into conductive foam and should remain there until it is time to install them.
4. Immediately place removed parts into a protective carrier or conductive foam.
5. Touch the protective package to ground before removing the part.
6. Touch a grounded part of the equipment before removing or installing a part.
7. Use as little motion as possible. Remember, friction generates static electricity.
8. When instruments are connected to circuits, always connect the ground lead first.
9. Use only antistatic spray materials and static-controlled vacuum desoldering equipment.

REVIEW QUESTIONS

42. When testing with an ohmmeter, a reading of 0 Ω in both directions indicates a(n) _____ junction.

43. When testing with an ohmmeter, a reading of infinity ohms in both directions indicates a(n) _____ junction.

44. The number 1N4001 is an example of a(n) _____ part number.

CHAPTER REVIEW QUESTIONS

2-1. Electrons, in a P-type crystal, are considered _____ carriers.

2-2. Calculate the maximum reverse current for a 12-V, 10-W zener.

2-3. Would the zener in question 2-2 be safe when conducting maximum current at a temperature above 25°C?

2-4. Refer to Fig. 2-26 and calculate h_{FE} when V_{CE} is 100 V and the base current is 0.6 mA.

2-5. Calculate emitter current for the conditions of question 2-4.

2-6. Calculate the collector dissipation for the conditions of question 2-4. Is it within the safe thermal operating area?

2-7. What happens to transistor gain as temperature increases?

2-8. When a bipolar power transistor fails at a higher collector voltage and is within the safe thermal area, the failure mode is called _____.

2-9. How do the pulse-mode ratings of transistors compare with their dc ratings?

2-10. Suppose the thermal rating for a transistor is 200 W. Use Fig. 2-29 to determine the maximum thermal dissipation for an operating temperature of 80°C.

2-11. What are the three operating modes for a transistor?

2-12. A perfect switching transistor would show a collector-to-emitter drop of _____V at saturation.

2-13. Power FETs are normally _____ devices.

2-14. Refer to the 40-W transfer curve in Fig.

2-42. What is the change in drain current when the gate-to-source bias increases from 7 to 8 V?

2-15. What is the transconductance for the data given in question 2-14?

2-16. When a bipolar transistor heats, its gain increases, tending to make it conduct more and become even hotter. This is known as _____.

2-17. Refer to Fig. 2-49. What designates the minimum flow to keep the triac on?

2-18. Triac commutation is complicated by _____ loads since the internal recovery current can act to gate the device back on.

2-19. Refer to Fig. 2-59. What happens to the load power as the variable resistance is increased?

2-20. The circuit of Fig. 2-59 cannot achieve smooth control when being adjusted from a _____ power setting.

2-21. The name given to describe the effect of question 2-20 is _____.

2-22. The asymmetrical ac trigger device is designed to eliminate _____.

2-23. Refer to Fig. 2-66. Would the IC sink any current if the external load resistor were connected to ground rather than to the supply?

2-24. Refer to Fig. 2-67 and assume that the external resistor has been removed. With a 5-V supply, what output voltage swing will be developed as Q_1 is turned on and off?

2-25. The number MJ802 is an example of a(n) _____ part number.

2-26. Leakage currents are expected to be _____ in power devices.

ANSWERS TO REVIEW QUESTIONS

1. pure **2.** it is lower **3.** negative **4.** positive **5.** P-type **6.** minority **7.** reverse **8.** minority **9.** reverse
10. increases **11.** rectifier **12.** minority **13.** emitter, out **14.** controls **15.** reverse **16.** forward **17.** increases
from 5 to 85 mA **18.** 50 mA **19.** depletion **20.** current **21.** complements **22.** out **23.** insulate **24.** enhancement
25. off **26.** gate **27.** I_{G2} **28.** 2 **29.** 1 **30.** off **31.** ac **32.** increases **33.** 11.85 V **34.** 4.7 ms **35.** 213 Hz
36. it increases **37.** it would be off **38.** 2.26 V **39.** top **40.** diodes or junctions **41.** no **42.** short-circuited
43. open **44.** registered

3

INTRODUCTION TO MOTOR CONTROLS

This chapter will introduce direct current (dc) and alternating current (ac) motor controls. Of primary interest will be methods of starting and stopping motors and controlling their speed. Stepper motors, used where automatic or computer control is important, are covered, as is the brushless motor. Alternating current motors are very popular in both small and large sizes in a wide range of industrial applications. With the rapid introduction of solid-state control devices they are taking over a growing number of functions previously reserved for dc motors.

3-1
BRAKING DC MOTORS

There are several factors that must be considered in stopping a motor. When a motor is disconnected from the power source it will coast to a stop. Not all machines can be allowed to coast to a stop. When it is necessary to stop a motor quickly, it is accomplished by braking. Braking can be accomplished by several methods, each of which has advantages and disadvantages. There are three basic means of slowing down a motor: friction, dynamic action, and plugging.

Applications of braking vary greatly. For instance, a crane or hoist not only has to stop quickly but also must hold heavy loads. Other motors, such as those controlling machine tools, must stop quickly but do not have to hold a load. Safety braking may be required to protect an operator from injury. In this case, the fastest method is employed with little consideration to the potential damage to the motor or load. When dealing with equipment such as cranes and hoists, we must realize that the load has a tendency to turn the motor. This is known as an *overhauling load. Overhung loads* are those loads which are applied in a direction perpendicular to the axis of the shaft. These types of loads are created by supported weights such as those in hoists and elevators.

Friction brakes (also known as *magnetic* or *mechanical brakes*) have been used to stop motors for many years. Their application is similar to the braking of an automobile. The essential parts of the brake are the friction material, shoes, bands or disks, and operating devices. Most friction brakes are electrically released and spring-set, so they will be set in case of a power failure or interruption. Occasionally it is advantageous to have a brake that is electrically set. With this type of brake, it is possible to vary the applied torque by adjusting the voltage to the brake coil.

The *shoe brake* needs only a small movement to release it. For example, a solenoid may have to travel less than 0.10 inches (in.). This short stroke gives fast operation. A drum or wheel is driven by the motor shaft to provide a larger braking surface than that provided by using the motor's shaft alone. Braking torque is directly proportional to the surface area and the spring pressure. Spring pressure is adjustable on most friction brakes.

The *disk brake* is arranged for mounting directly to the motor end bell. The brake lining is a disk, which is supported by a hub keyed to the motor's shaft and rotates with the motor. When the brake is set, a spring pulls the stationary member into contact with the rotating disk. The simplicity of a single moving part, the *pressure plate,* makes for less mechanical maintenance. No linkages or levers are used, as they are in the shoe type.

The *band-type* brake has the friction material fastened to a band of steel which encircles the wheel and may cover as much as 90 percent of the wheel surface. The increased braking surface allows lower pressure per square inch of surface area, and a subsequent reduction in wear of the brake lining. This advantage is offset somewhat by the fact that the braking pressure is not equal over the whole band. The band brake also requires a longer stroke to release it.

The basic material used in all brake linings is asbestos. When servicing any brake parts, do not create dust by using a dry brush or compressed air; use a water-dampened cloth. The asbestos fibers may become airborne if dust is created during servicing.

Breathing dust containing asbestos fibers can cause serious bodily harm. The main disadvantage of any friction-type brake is that it requires more maintenance than other types. The more frequently a motor is stopped by the brake, the more maintenance will be required. The solenoids used to energize most friction brakes are available for either ac or dc operation.

Dynamic action, or *electrical, braking* is accomplished by changing the connections to a motor with or without the use of an auxiliary power source (depending on the type of motor to be braked). If a motor that is still running is reconnected as a generator, the result is dynamic braking. The motor acts as a loaded generator that develops a retarding torque, which stops the motor rapidly. The generator action converts the mechanical energy into electrical energy and dissipates this energy as heat in a resistor. Since dynamic braking is only present when the motor is rotating, a friction-type brake is required to hold any overhauling load after it stops.

The easiest type of motor to brake dynamically is the *shunt-wound* dc motor. While the machine operates as a motor, the counterelectromotive force (cemf) opposes the line voltage and limits the armature current to a value sufficient to provide the output torque requirements. Dynamic braking is accomplished by disconnecting the armature from the line (power source) and placing a current-limiting resistor across the armature terminals while the field remains energized. The armature will be rotating in the magnetic field and will continue to generate a cemf that is proportional to speed and the strength of the field. The armature current flowing through the limiting resistor will be opposite to that produced by the line. This reverse current in the armature will produce a torque opposite to the original motor action and cause the motor to slow down. As the forward speed is reduced, the generated voltage is also reduced. At the zero speed point, the generated voltage is also zero. The motor will stop at this point if no overhauling torque is available to continue the rotation. The connection for dynamic braking of a shunt-wound motor is shown in Fig. 3-1.

The braking resistor is usually selected so that the initial braking current is 150 percent of the normal current. Braking times of milliseconds are not un-

common on fractional horsepower motors. The field winding may or may not be disconnected from the line after the motor has stopped. When a field rheostat is employed, it is customarily short-circuited to increase the field to aid in the braking effect.

Dynamic braking may be used with a series motor, as shown in Fig. 3-2, but the connections are more complicated. If the motor were just disconnected from the line and shunted by a resistance (such as the shunt motor armature), no braking would be obtained. This is because the current would flow through the field in the wrong direction, thereby demagnetizing the field.

With the motor running and current flowing from L1 to L2, the cemf of the armature is in opposition, as shown by the arrow in Fig. 3-2. The field current is flowing from L1 to L2. For braking to occur, the field must be connected in the reverse direction so that the current flows through it in the same direction it does when the motor is across the line. When the switch is in position 2, the resistor is connected in series with the field and the armature, forming a closed loop. Notice that at this time the field connections are reversed so the current flow due to the armature emf is in the same direction as it is when connected as a motor. The energy is quickly dissipated in the resistor, and the armature stops rotating.

A combination of dynamic braking and friction braking can be used when a very large load must be stopped. Because the force of the load would wear out the friction brakes too quickly, dynamic braking can be used to slow down the load, and then the friction brakes can be applied to finally stop and hold the load if necessary.

Dynamic braking of a permanent magnet (PM) motor is accomplished the same way as it is with the shunt-wound motor: the armature is disconnected, and its terminals are shunted by a resistor. A very distinct difference should be noted: with the shunt-wound motor, the motor cannot be dynamically braked if there is a power failure. The field voltage and current must be available to generate the dynamic braking. With the PM motor, the power failure will not affect its braking ability because its field is not affected by any power failures. A normally

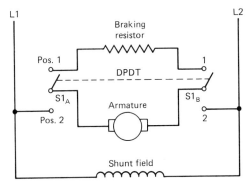

Fig. 3-1 Dynamic braking circuit of a shunt-wound motor.

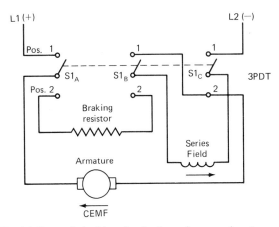

Fig. 3-2 Dynamic braking circuit of a series-wound motor.

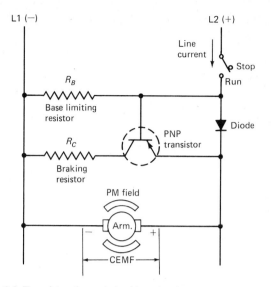

Fig. 3-3 Transistor dynamic braking circuit.

closed relay across the armature will automatically function in case of power failure and load the armature through its braking resistor. This inherent characteristic is very important and useful in many cases. For example it can be used on magnetic tape reel drives to prevent unwanted spillage of tape in the event of a loss of power. Figure 3-3 illustrates the utilization of solid-state electronic components to achieve dynamic braking of a PM motor. The diode drop biases the transistor off when the switch is in the run mode. When the switch is put into the stop position, the armature no longer draws current from the line. This is the *brake mode;* the transistor will conduct because of the polarity of the armature's cemf. The cemf turns on the transistor circuit and R_C acts as the armature load. This circuit will work just as well for a shunt-wound motor, but the field circuit must remain connected to the line until the motor stops.

A *compound-wound motor* is actually both a shunt- and a series-wound motor and can take advantage of either the shunt-braking circuit or the series-dissipating circuit or a combination of the two. However, the slower speed of the compound motor makes shunt-wound braking the preferred method.

Plugging is a way of braking a motor by reversing the power to the armature while the field remains connected as before. Plugging the motor allows for a very rapid and abrupt stop. Plugging can be used to brake a motor if the armature power is removed at the point where the motor speed drops to zero. Otherwise the motor will reverse. Plugging is more severe than the other braking methods mentioned because the voltage across the armature (in the shunt motor) and across the entire motor (in the series motor) is approximately twice its normal value at the instant that braking is initiated. The cemf voltage in the armature is additive to the line voltage (the armature leads have been reversed while the rotation

is still in the original direction) until the speed goes to zero. Under normal operating conditions, the cemf generated opposes the line voltage and thereby limits the armature current.

Plugging is not usually recommended because of the high armature currents drawn. Excessive armature heating and brush arcing will result, and brush life may be severely affected. Motors used for plugging are designed for this type of service. In some applications plugging is not permitted, and the motor controls are interlocked to enforce this condition.

Although manual and magnetic starters are used to reverse the direction of a motor, a special plugging switch is typically used in plugging applications. The plugging switch, sometimes called a *zero-speed switch,* is connected to the shaft of the motor or its load through a pulley or a shaft. The rotation of the switch shaft, at a given speed, causes a set of contacts to operate, either by a centrifugal mechanism or by a magnetic induction arrangement. The main function of the plugging switch is to prevent reversal once the countertorque action of plugging has brought the load to a standstill. Without this switch, the motor and load would stop and then run in the opposite direction.

The plugging switch will usually open one set of contacts and close another set of contacts as the shaft speed increases past a preset number of revolutions per minute (rpm). The continuous operating speed of the machine should be many times the speed needed to operate the switch contacts. This will ensure good contact holding force and reduce the possibility of chatter and subsequent false operation of the switch.

Figure 3-4 illustrates the wiring of a motor starter circuit that includes a plugging switch. The normally open (N.O.) contacts of the plugging switch (P) are wired in series with a set of interlock contacts to the reversing starter coil (R). Pushing the start button energizes the forward starter coil (F) and the motor starts forward. The forward coil is held in by the F_1 contacts across the start button now energized (closed). As the motor increases in speed the N.O. contacts on the plugging switch close at some preset rpm. The closing of these N.O. contacts will not energize the reverse coil (R) because of the F_2 con-

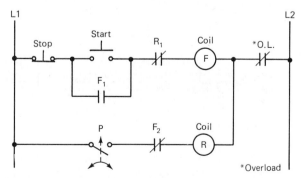

Fig. 3-4 Typical motor plugging circuit.

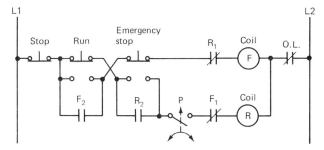

Fig. 3-5 Emergency stop plugging circuit.

tacts, which are now open. Pushing the stop button will drop out the forward coil (F) and cause the F_2 contacts to close, permitting the reversing coil (R) to energize through the still closed plugging switch. The motor connections are now reversed, and the motor's countertorque acts as a braking force. When the motor is stopped the plugging switch opens, disconnecting the reverse coil before the motor can physically reverse its direction.

Plugging is generally used for emergency stopping of a motor. The plugging circuit of Fig. 3-4 would be modified to include the emergency stop circuitry as in Fig. 3-5. The normal stop and run pushbuttons operate as in any standard starter circuit. The important difference is the addition of the emergency stop pushbutton. Pushing the emergency stop pushbutton will deenergize the forward starter (F) and simultaneously energize the reverse starter (R). The reverse starter is held in by the R_2 contacts through the plugging switch (P) and the now closed F_1 contacts of the forward starter coil (F). When the motor speed approaches zero speed, the plugging switch (P) will open to disconnect the starter before the motor reverses direction. There are numerous other types of plugging motors, which use timers, time delay relays, and other special devices.

REVIEW QUESTIONS

1. _____ braking is required to hold an overhauling load.

2. Friction brakes are electrically _____ and spring-set.

3. The _____ type friction brake uses a pressure plate and spring set.

4. The use of _____ in brake linings makes their maintenance a cautious and careful operation.

5. Name two other solid-state devices that could be used instead of the transistor in Fig. 3-3.

6. When plugging a series motor, the voltage across the motor is _____ that of normal operation.

7. Refer to Figs. 3-4 and 3-5. In which figure does the circuit show plugging accomplished at every actuation of the stop buttons?

3-2
SPEED CONTROL OF DC MOTORS

Industrial application of dc motors often requires, in addition to driving loads at a constant speed or torque, the ability to vary the speed of the motor. The speed of most dc motors can be easily adjusted from zero to speeds above the rated value. In addition to good speed control, the dc motor is well suited for applications requiring momentary high-torque outputs. A dc motor can produce torque three to five times its rated value for short durations. Because of their good speed control and high torque, they are used in many industrial applications and also in mining equipment. There are a variety of ways to control the speed of a dc motor, ranging from a simple series rheostat to a modern generation of solid-state devices.

There are three methods by which the speed of a shunt-wound motor can be controlled: field weakening, armature resistance control, and armature voltage control. The speed of a shunt-wound motor can be described by the following equation:

$$\text{rpm} = \frac{V_a - (I_a \times R_a)}{\phi_f}$$

The equation shows that the speed can be made to change by adjusting the variables, V_a, R_a, and ϕ_f (I_a changes proportionally with the load and is not considered a speed-control variable).

If the strength of the magnetic field of the shunt-wound motor is reduced, the motor will speed up. This speed-up occurs with the reduction in the field strength because less cemf is developed by the armature. The difference beween the line voltage and the new cemf produces an increase in armature current, resulting in an increase in output torque and speed. The torque is related to the flux field and armature current by the following equation:

$$T = K \times \phi_f \times I_a$$

The field can be weakened by connecting a rheostat in the field circuit as shown in Fig. 3-6(a). The rpm (speed) equation shows that weakening the field increases the rpm, provided the armature voltage is constant. Standard industrial motors will permit a speed increase of up to 400 percent. This method of speed control is considered efficient, because the power lost in the field rheostat is negligible; however, the field can only be weakened within certain limits. Weakening beyond a limit point can result in excessive speeds and instability. The armature may also overheat. The torque equation shows that a reduction of the field ϕ_f will require an increase in the armature current I_a to maintain a given torque. The torque-speed curves for different values of increasing resistance in the field are shown in Fig. 3-6(b). Field weakening will produce speeds above the normal rated speed. The motor can be overloaded easily because the rated torque drops as the speed in-

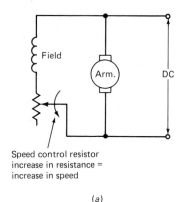

Speed control resistor
increase in resistance =
increase in speed

(a)

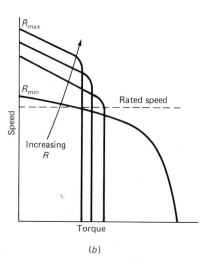

(b)

Fig. 3-6 Field weakening speed control of the shunt motor. (a) Schematic diagram. (b) Torque-speed curves.

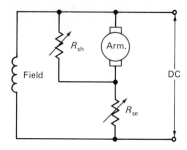

Fig. 3-7 Tandem (series and shunt) armature speed control.

creases, as shown in Fig. 3-6(b). Therefore, this type of control (field weakening) is limited to cases in which load conditions are very predictable and controlled.

A completely different set of characteristics occurs for a shunt-motor with a rheostat in the armature. In this case of armature resistance control, the field winding is kept at the rated or line voltage. Referring again to the rpm equation, if the armature voltage V_a is reduced (by increasing the resistance), the motor speed will decrease. Therefore, armature resistance control will lower the speed of the shunt-motor below its rated base speed. As indicated by the torque equation, an increase in load will result in an increase in the armature current and a subsequent increase in the voltage drop across the added series rheostat. If the motor is started with very little load and the load is increased, the speed will drop sharply. There will also be a corresponding I^2R (power) loss in the series rheostat.

The resistance methods of field weakening and armature resistance described are very simple and inexpensive and will provide control of a shunt-wound dc motor's speed above or below its rated value. These methods were and are still used in many applications in the industrial environment.

Both series and shunt rheostats may be employed in tandem to improve the speed regulation of the shunt-wound motor by making the operating speed less affected by the changes in load torque. The shunted armature shown in Fig. 3-7 has a variable resistance across (shunt) the armature and acts to increase the current through the series resistance and thereby reduce the difference between no-load and full-load current. The series resistance is used to control the armature voltage, as with the armature control method for the shunt motor.

The resistance methods of speed control just discussed are *open loop* methods (there is no feedback). These methods are still found in many industrial applications. Motors are frequently controlled by varying the supply voltage to the armature. Armature voltage control of shunt-wound motors can be open loop or closed loop (having feedback) to control the speed of the motor. Figure 3-8 shows open loop armature voltage control. This method of motor control has several advantages:

1. Wide speed-range control
2. Speed not appreciably affected by changing load
3. Less power wasted at low speed
4. Ease of interfacing with complex electronic control systems

The nonfeedback controller, shown in Fig. 3-8(a), produces the speed/torque curves of Fig. 3-8(b). Speed regulation as the motor load varies is essentially the inherent regulation characteristic of the motor as the curves illustrate. The motor will operate along the load line by jumping from one speed/torque curve to another. An infinite number of curves is made available by varying the armature voltage. Armature voltage control may be obtained from one of the simple control circuits shown in Fig. 3-9. These control amplifiers need only a simple reference voltage amplifier and a current (power) amplifier to provide infinite speed control. These circuits are also applicable to PM motors. Though effective, these methods are very inefficient, since substantial power is lost in the controlling device.

Open loop, or nonfeedback, control will be only as stable as the load and the individual components of the system. Figure 3-10 shows silicon controlled rectifier (SCR) motor speed circuits. The field cir-

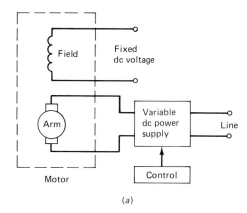

(a)

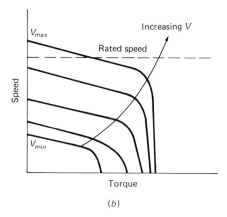

(b)

Fig. 3-8 Armature voltage control of speed. *(a)* Schematic diagram. *(b)* Torque-speed curves.

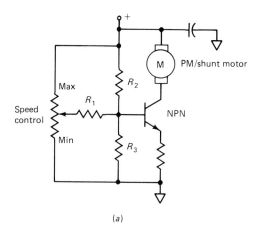

(a)

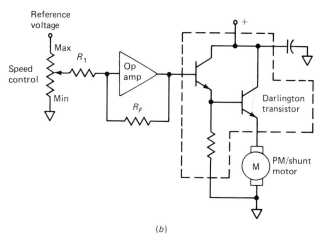

(b)

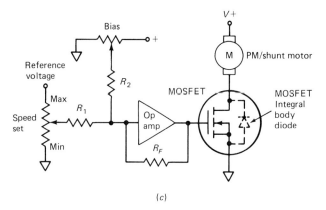

(c)

Fig. 3-9 Simple solid-state motor-speed controllers. *(a)* Single transistor speed control. *(b)* Op amp and Darlington power amplifier speed control. *(c)* Op amp with MOSFET power amplifier.

cuits have been omitted for simplicity. The motor may also be a PM motor, as previously mentioned. The circuit in Fig. 3-10(a) has a limited speed range but will maintain constant speed under varying load conditions. Though it is not obvious at first glance, this circuit includes a feedback element that enhances its control stability. This type of circuit makes use of the cemf of the armature, which is a function of the motor's speed and, therefore, can be used as an indication of speed changes as the load varies. When a load is applied to the motor, the motor speed will start to decrease and thus reduce the cemf induced in the rotating armature. With this reduced cemf, the SCR will fire earlier in the rectified cycle. The SCR control is less precise than a linear amplifier but offers high efficiency since the SCR is either on or off. The simplicity of the circuitry and low parts count make its use attractive.

The inclusion of the unijunction transistor (UJT) oscillator in Fig. 3-10(b) will give a wider range of speed control. It should be noted that in both these circuits, unfiltered direct current is used as the power source. This unfiltered source is necessary in order to commutate the SCR; otherwise it will remain on, and all control will be lost. In Fig. 3-10(b) the pulsating direct current also serves to turn off the UJT oscillator, so it starts a new charge cycle on each cycle of the pulsating supply voltage. Once the speed

control is set, the UJT firing point occurs at the same time, after V_{ac} goes to zero on each cycle. This delivers a constant power to the load. The circuit can turn the power completely off but cannot turn it completely on, as indicated by the waveforms in Fig. 3-11 (p. 49). Smooth variable control from 5 to 95 percent of available power can be obtained. This method of control is sometimes referred to as a variation of a pulse-width modulation (PWM) speed-control circuit.

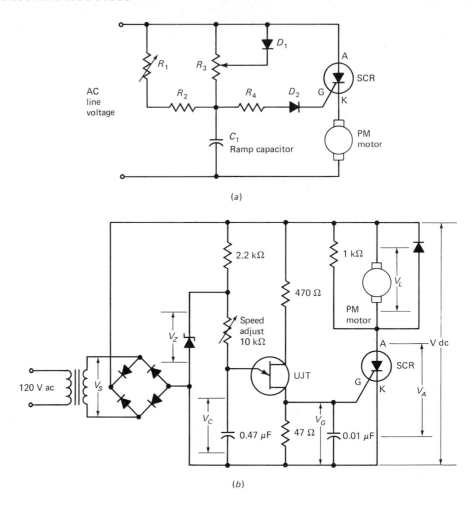

Fig. 3-10 Motor speed control using SCRs.

With the exception of the circuit in Fig. 3-10(a), all the circuits presented thus far have been open loop speed controllers. For applications demanding precise and constant velocity under varying load conditions, a closed loop system is needed. Block diagrams for closed loop controllers are shown in Fig. 3-12 (p. 50). The speed sensor (*tachometer*) generates a voltage proportional to motor speed. This voltage is fed back and compared to a reference voltage. Any error between the two voltages is amplified and corrects the speed of the motor. The extra element in this system that makes it work and closes the feedback loop is the speed sensor (tachometer, in this case). In many cases, a small PM motor is used as a tachometer. Most tachometers develop an output voltage proportional to the shaft speed. The polarity of the output voltage is dependent on the direction of rotation of the shaft. The output voltage from a tachometer may be expressed as

$$V_g = K \times \text{rpm}$$

where V_g is the tachometer output voltage and rpm is the armature shaft speed. A value of 1.0 V/500 rpm is typical. The output increases to 4 V if the tachometer is rotated at 2000 rpm. Alternatives to the generator-type tachometer include inductive pick-ups and optical pick-ups to provide pulses that are integrated (summed) for the speed feedback sig-nal. These pick-ups usually require amplification to be useful.

Figure 3-12(a) illustrates a *first quadrant regulation control circuit,* so named because the speed/torque curves are all in the first quadrant of the cartesian coordinate system. That is, they are positive when the shaft is rotating in the forward direction. The motor in Fig. 3-12(a) rotates in only one direction (it is *unidirectional*). Its speed will increase or decrease as the comparator turns on or off the input to the amplifier in response to the comparison of the speed (feedback) signal to that of the set point reference. The closed loop system in Fig. 3-12(b) is also referred to as a *servosystem,* which can also be used to control the position as well as the velocity of a system, with the addition of a position feedback element. These aspects will be covered in depth in Chapter 10. For now, it can be seen that by proper selection of the references, gains, motor, tachometer, and scaling components, a high-performance speed-control system is obtainable.

The series (universal) motor is designed to operate on either alternating or direct current and is capable of high speeds and starting torques. The speed of the series motor can be changed by changing the voltage across the motor. This is accomplished by three methods: series-resistance control, shunt-resistance control, and variable-voltage control. The

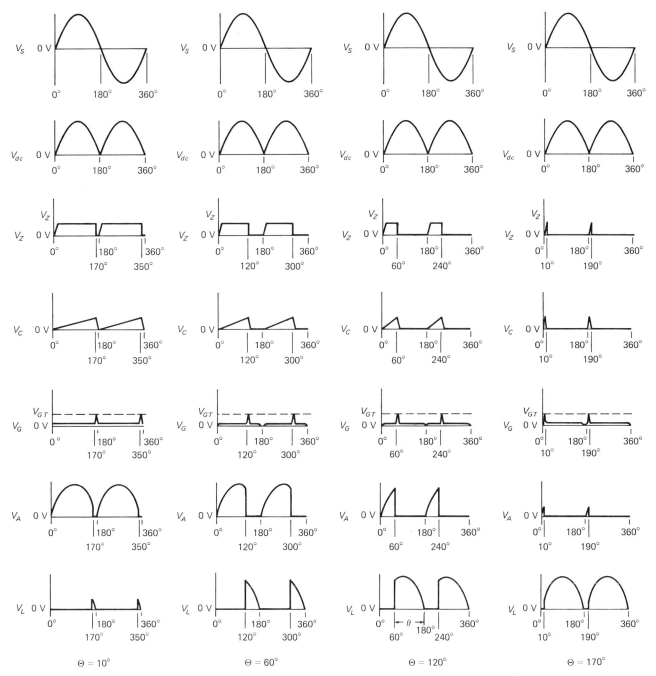

Fig. 3-11 Waveforms for SCR-UJT circuit in Fig. 3-10*b*).

speed/torque curve obtained by using a rheostat in series with the motor (armature and field) is illustrated in Fig. 3-13(*a*). Series-resistance control has good starting characteristics (high torque at low speed), but it is evident that the speed regulation of the motor declines with decreasing speed, thereby making good overall control of the motor's speed difficult. The series resistor produces a voltage drop in the circuit that is proportional to the current in the circuit. This voltage drop across the resistor will increase as the motor is loaded (motor current increases with an increase in load). It can be seen that the voltage across the motor will decrease with the increase in load, and the speed will drop rapidly with an increasing load when a series resistor is used.

Also, the higher the resistance, the greater the drop in speed as the load increases. A series resistor or rheostat will have its greatest effect on the starting torque of the motor. Maximum current flows when the motor is started and the resistive drop will limit the motor voltage to its lowest value. The series resistor usually will be adjusted for minimum resistance for starting and then increased as the motor gains speed. In theory, the motor can be adjusted to near standstill (complete stop). However, as a result of reduced inertia, *armature cogging* (the armature speeds up as it enters the flux field and slows down on exiting) is very pronounced at low speed; therefore, the lower limit must be set to a value at which cogging is avoided.

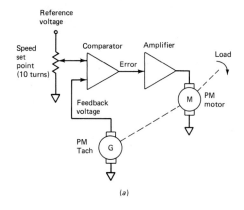

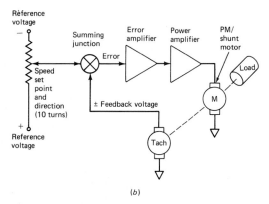

Fig. 3-12 Closed loop speed controllers. *(a)* Unidirectional feedback control. *(b)* Bidirectional feedback speed control.

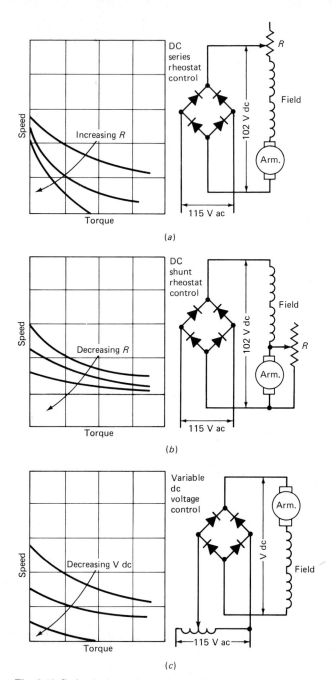

Fig. 3-13 Series (universal) speed control methods.

The *shunt-resistance control* of the series motor, along with its speed/torque curve, is shown in Fig. 3-13(*b*). The series motor can also be controlled by shunting an adjustable resistance across the armature as indicated. The speed control range is usually limited by this method because increased current must pass through the field coils, with a corresponding increase in heating. Although the speed range is limited, this method of control improves the speed regulation, as the curves indicate, while maintaining good starting torques. It is an excellent method for matching the speeds of motors operating in parallel.

Varying the voltage applied to the series motor typically produces a speed change of up to 7:1, depending on the individual motor. Figure 3-13(*c*) shows the circuit, along with the speed/torque curves obtained by *variable-voltage control*. It can be seen that the speed range increases, along with improved regulation and starting torque. The variable voltage can be obtained with the use of an autotransformer and a rectifier assembly. At this point the system may be considered simply as a blackbox: alternating current in, variable direct current out.

Since the series motor can run on alternating current as well as direct current, control for the motor can be half-wave or full-wave. The use of an SCR for control is shown in Fig. 3-14. This is a half-wave device with feedback. The circuit uses the cemf of the motor to vary the firing point, thereby maintain-

ing essentially constant speed control with varying torque requirements. The SCR conducts on the positive half cycles only; therefore the control will provide substantially less than full speed. However, special motors (designed to operate in this circuit) may be used for full-speed operation.

For 60-Hz ac operation, a simple full-wave control such as the one illustrated in Fig. 3-15 can be used. This is an open loop control (no feedback). Varying R_2 varies the time for capacitor C_1 to charge to the diac trigger voltage. When the diac triggers it will turn on the triac, as shown in the waveforms for V_{C1},

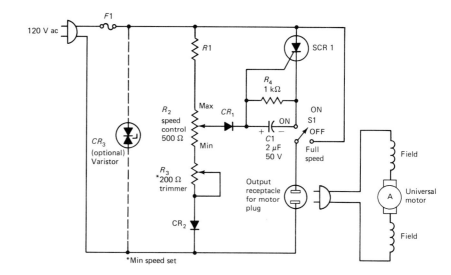

Fig. 3-14 SCR speed control of a universal motor.

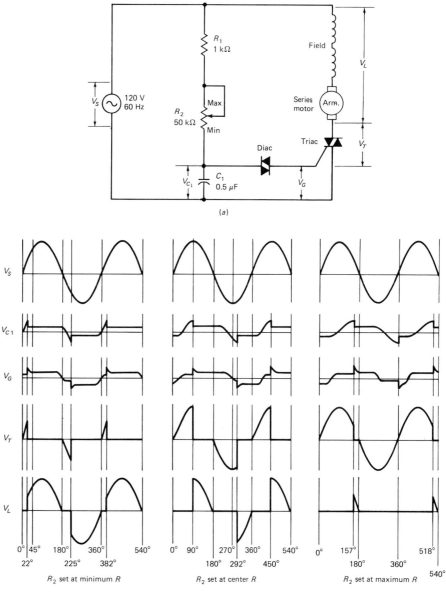

Fig. 3-15 Triac full-wave speed control. *(a)* Schematic diagram. *(b)* Waveforms.

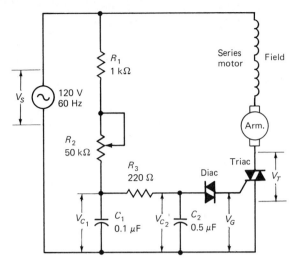

Fig. 3-16 Triac control with hysteresis compensation.

V_G, and V_T in Fig. 3-15(b). This will apply the supply voltage V_S, less the drop across the triac (V_T) across the load V_L. When the supply voltage goes through zero, the triac will turn off. It will remain off until capacitor C_1 again charges (in the reverse direction) to the diac trigger voltage. The waveforms show various settings of R_2 used to obtain early firing (22°) for maximum power to late firing (157°) for minimum power. This circuit does exhibit some dissymmetry of the load alternations. This is caused by circuit hysteresis due to capacitor C_1's retaining some charge at the polarity of the initial voltage applied across it. This dissymmetry is apparent by observing the V_L positive and negative waveforms of Fig. 3-15(b). The hysteresis can be reduced by using the circuit shown in Fig. 3-16. This circuit is also a popular arrangement for lamp dimmers.

REVIEW QUESTIONS

8. Field weakening of a shunt-wound motor will cause the armature speed to _____.

9. The field weakening method of speed control is reasonably efficient because the field current is _____ in comparison to the armature current.

10. A series rheostat in the armature of the shunt motor will cause its speed to _____ in comparison to base speed.

11. Resistance control of a motor is a form of _____ loop-type control.

12. _____ control of a shunt or PM motor is easily adapted to solid-state or computer-type devices.

13. The unfiltered (pulsating) voltage used in the UJT/SCR control is needed to _____ the SCR.

14. The _____ provides feedback for improved speed regulation in the SCR circuit of Fig. 3-10(a).

15. An increase in speed is obtained by _____ the firing angle of the SCR.

3-3
STEPPER MOTORS

The increasing trend toward digital control of machines and process functions has generated a demand for mechanical devices capable of delivering incremental motions of predictable accuracy. The stepper motor is often considered as a digital device which converts electric pulses into proportionate mechanical movement. Each revolution of the stepper motor's shaft is made up of a series of discrete individual steps. The motor usually provides for clockwise (cw) or counterclockwise (ccw) rotation. Therefore, the stepper motor is ideally suited for a wide variety of control and positioning applications in the industrial world. With the rapid growth of solid-state electronics and digital techniques, the stepper motor applications in peripherals, robotics, instrumentation controls, and machine tools have grown rapidly and continue to do so.

Conventional ac and dc motors have a free turning shaft. The stepper motor shaft rotation is incremental. The basic feature of the stepper motor is that upon being energized it will move and come to rest after some number of steps in strict accordance with the digital input commands provided. The stepping motor therefore allows control of the load's velocity, distance, and direction. The *repeatability* (ability to position through the same pattern of movements a number of times) is very good. The only system error introduced by the stepper motor is its single-step error, which is a small percentage of one step and is generally less than 5 percent (0.09°). Most significantly, this error is noncumulative, regardless of the distance traveled or the number of times repositioning takes place.

The stepping motor is an inherently reliable device with bearings being the only part subject to wear. In many applications, a stepping motor can replace shorter-lived, more maintenance-intensive devices such as brakes, clutches, and gears with an overall improvement in reliability.

Stepper motors are divided into three principal types or classes, each with distinct construction and performance characteristics: variable-reluctance (VR), permanent magnet (PM), and PM-hybrid.

A stepper motor's operation is based on the basic magnetic principle: like magnetic poles repel and unlike poles attract. If the stator windings in Fig. 3-17(a) are energized so that stator A is the north pole, stator B is the south pole, and the permanent magnet (PM) rotor is positioned as shown, a torque will be developed to position the rotor 180° from its indicated position. However, it would be impossible to determine the direction of rotation, and, in fact, the rotor may not move at all if the forces are perfectly balanced. If, as indicated in Fig. 3-17(b), two additional stator poles C and D are added and energized as shown, we are able to predict the direction of rotation of the rotor. As indicated in Fig. 3-17(b), the rotor's direction of rotation would be counter-

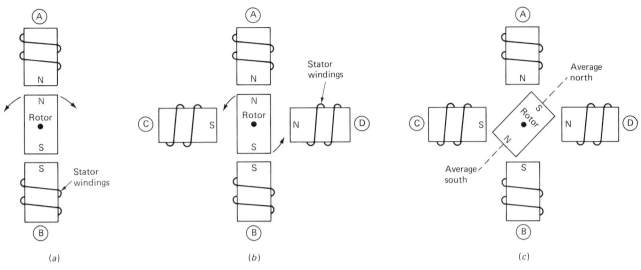

Fig. 3-17 Basic stepper motor rotation.

clockwise with the rotor aligning itself between the "average" south pole and the "average" north pole as in Fig. 3-17(c).

The distinguishing feature of the PM stepper motor is the incorporation of a permanent magnet, usually in the rotor assembly. To allow better step resolution, four more stator poles are added, and teeth are machined on each stator pole and also on the rotor. The number of teeth on the rotor and stator determines the step angle that will be obtained each time the polarity of one winding is changed. The stepper motor shaft responds with a specific angular increment each time the winding polarity is changed, moving to the average pole. This specific degree of shaft rotation or increment is known as the *step angle*. The PM stepping motor operates by means of the interactions between the rotor magnet biasing flux and the magnetic forces generated by the stator windings. If the pattern of winding energization is fixed, a series of stable equilibrium points is generated around the motor. If the windings are excited in a particular sequence, the rotor will follow the changing point of equilibrium and rotate in response to the changing pattern, as shown in Fig. 3-18.

By virtue of the rotor's permanent magnet, there is a detent torque developed in the motor even if the stator windings are not excited. The detent torque can be felt by turning a PM stepper by hand. A restoring torque is generated on the rotor whenever the rotor is moved from the position which has minimum reluctance (analogous to resistance in a dc circuit) for the permanent magnet flux. This torque is much lower than the normal energized torque and is typically only a small percentage of the maximum torque.

The variable-reluctance (VR) motor has a stator which has a number of wound poles. The rotor is a cylindrical, toothed unit whose teeth have a relationship to the stator poles and their teeth (the stator may not have teeth). The number of teeth will be

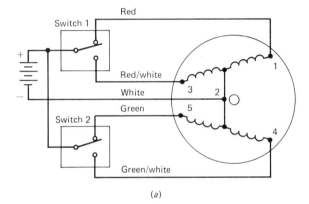

(a)

Step	Switch #1	Switch #2
1	1	5
2	1	4
3	3	4
4	3	5
1	1	5

*To reverse direction, read chart up from bottom.

(b)

Phase	1 Step	2 Step	3 Step	4 Step	5 Step
1	ON	ON	OFF	OFF	ON
3	OFF	OFF	ON	ON	OFF
5	ON	OFF	OFF	ON	ON
4	OFF	ON	ON	OFF	OFF

FWD ⟶

⟵ REV

(c)

Fig. 3-18 Four-step sequence of a permanent magnet stepper motor. *(a)* Schematic diagram. *(b)* Switching sequence. *(c)* Waveforms.

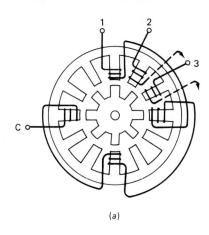

(a)

(b)

Common leads: yellow, green, blue

Fig. 3-19 Variable reluctance (VR) stepper motor. *(a)* Cross section of stator showing complete winding of one phase of a three-phase winding. *(b)* Three-phase wiring connection color code. *(c)* Three-phase, dual-excitation chart. *(d)* Stepping waveforms.

Phase Step	A Brown	B Red	C Orange
1	///	///	
2		///	///
3			///
4	///	///	

(c)

	Step 1	Step 2	Step 3	Step 4	Step 5
A	ON	OFF	ON	OFF	
B	ON	OFF	ON	OFF	
C	OFF	ON	OFF	ON	

(d)

determined by the step angle required. A typical VR motor is shown in Fig. 3-19. When a current flows through the proper windings, a torque is developed in such a way as to turn the rotor to a position of minimum magnetic path reluctance. This position will be statically stable in that external torque is required to move the rotor from this stable position. This particular position is not an absolute one. There are many stable positions in the average motor for any given stator energization pattern. When a different set of windings is energized, the stator field changes, causing the rotor to move to a new position. Proper selection of the energizing sequence of the windings allows the stable positions to be made to rotate smoothly around the stator poles, establishing the rotational speed and the direction of the rotor. When the energized pattern is fixed, the rotor position becomes fixed as well. Therefore, the shaft position is stepped by changing the pattern of winding energization. Figure 3-19(c) illustrates the standard excitation modes which produce a nominal step angle. The dual excitation (two windings always on) is chosen because of the higher torque available. Unlike the PM stepper, the VR stepper has very little residual magnetism, so there will be no force on the rotor (*detent torque*) when the stator is not energized. The *step angle* (determined by the number of stator and rotor teeth) varies from 7.5 to 30°. The VR steppers exhibit relatively low torque and inertial load capacity. They are, however, reasonably inexpensive and suitable for light computer and industrial instrument applications.

The PM-hybrid stepper rotor combines the rotor construction features of VR and PM types. A cross section of this stepper is shown in Fig. 3-20. The stator is of the wound type. Both the rotor and the wound stator are toothed. This construction gives the PM-hybrid higher torque capacity (50 to 2000 + ounce-inches [oz·in.]) with step accuracies of about ±3 percent and step angles that vary from 0.5 to 15°. PM-hybrid designs offer excellent speed capability: 1000 steps/s and higher can be obtained. Although their cost is relatively high, PM-hybrid designs deliver the best set of performance characteristics for many applications.

Steppers are popular because they can be used in an open loop mode while still offering many of the desirable features of the feedback-type system: feed them a defined number of pulses, and they will position within their step accuracy. The replacement of mechanical parts (which are susceptible to wear),

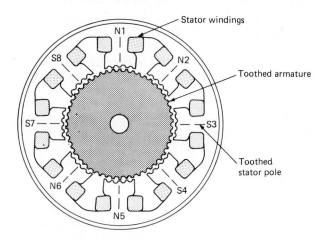

Fig. 3-20 Cross section of PM-hybrid motor with toothed armature and stator.

such as clutches and brakes, is eliminated because stepper motors provide a greater reliability and consistency, reasonable cost, and consistent performance. The stepper motor is an excellent positioning device. On the other hand, stepper motors are not very energy-efficient, but that is the price one must pay to obtain the unique characteristics of the stepper motor. Its limitations include the following: available torque is inversely proportional to speed; speed must increase gradually (if commanded to go from stop to full speed immediately, it will stall); and the stepper motor exhibits a low speed resonance point, where torque is reduced drastically.

Some stepper motors achieve torque amplification by means of a gear system integral with the motor. Methods to achieve torque amplification are the use of planetary gears and the use of a *flexing mechanical spline*. Both methods allow a reduction in speed and an increase in output torque by gearing down. For instance, a stepper that delivers 50 oz·in. of torque at 72 rpm will, with a 4:1 step-down gearing, deliver 200 oz·in. of output torque at a speed of 18 rpm.

To understand the stepping motor's unique characteristics thoroughly, we must examine the stepping motor logic sequencing. A simplified representation of a stepping motor is shown in Fig. 3-21. Initially, poles *A* and *B* are both energized with their north poles up, attracting the rotor's south pole to the position as shown in Fig. 3-21(*a*). Reversing the polarity of pole *A* (Fig. 3-21[*b*]) draws the rotor clockwise 90° to its new position: this is known as a *full step*. If pole *A* had been turned off, instead of being

reversed, the rotor would rotate 45° (clockwise) to line up with the field of pole *B*; this is known as a *half step*. The simple stepper motor in Fig. 3-21 would only have four full steps (90°) per revolution, or eight (45°) half steps. Actual stepper motors obtain small angle increments by using large numbers of poles, as shown in Fig. 3-20.

The most common stepper stator windings are center-tapped dual windings known as *bifilar windings*. Bifilar winding eliminates transformer coupling to adjacent windings. The use of bifilar windings on steppers also simplifies the required drive circuitry. Figure 3-22(*a*) shows a bifilar-wound stepper motor, its power supply, and the switching points. Only a single-polarity power supply is needed, whereas the motor of Fig. 3-21 would require a dual power supply for reversal of the poles. Only a single power supply is needed with the center-tapped windings.

The switching sequence shown in Fig. 3-22(*b*) is called a *four-step sequence* (full step). To reverse direction, read the sequencing chart upward from the bottom. Since current is maintained on the motor windings when the motor is not being stepped, a high holding torque results. As a rule of thumb you will find that the power supply is about five times the voltage rating of the motor. Series resistors are used in the common leads to limit the current and improve the inductive/resistive (*L/R*) time constant, for better performance.

Figure 3-22(*c*) illustrates the eight-step switching sequence, often called *electronic half stepping*. Under this condition, the rotor moves half its normal distance per step. For example: a 1.8°, 200 step per revolution motor would become a 0.9°, 400 step per revolution motor. Likewise a 0.72°, 500 step per revolution model moves in increments of 0.36° for 1000 steps per revolution. The advantages of operating in this mode include finer resolution and greater speed capability, but with less available torque.

The switching sequence for these motors was originally achieved by mechanical switches, mercury-wetted relays, or a commutator-brush arrangement. These were very awkward and expensive, besides being a maintenance headache. Electronic switching resolved this problem quite easily and efficiently. The circuits of Fig. 3-23 (p. 57) use solid-state devices to drive (switch) the stepper motor windings. These power drivers are used because most steppers require currents from hundreds of milliamperes up to amperes. A 200-oz·in. torque motor operating with 2.5-V windings will require approximately 2.5 A for each winding that is on. This amount of current switching or sinking is too large for logic circuitry to provide, and drivers are required. Metal-oxide semiconductor field effect transistors (MOSFETs) are taking over this type of switching application. An attractive feature of the MOSFET is the built-in diode. The *integral body diode* is so named because it is inherent in the silicon structure of the power MOSFET. When the transistor switches off, a large voltage due to the collapsing field is generated in the

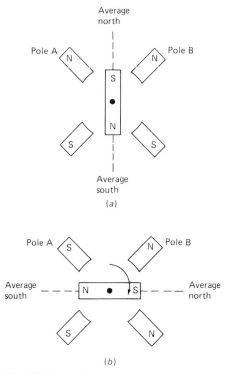

Fig. 3-21 Simplified stepping sequence.

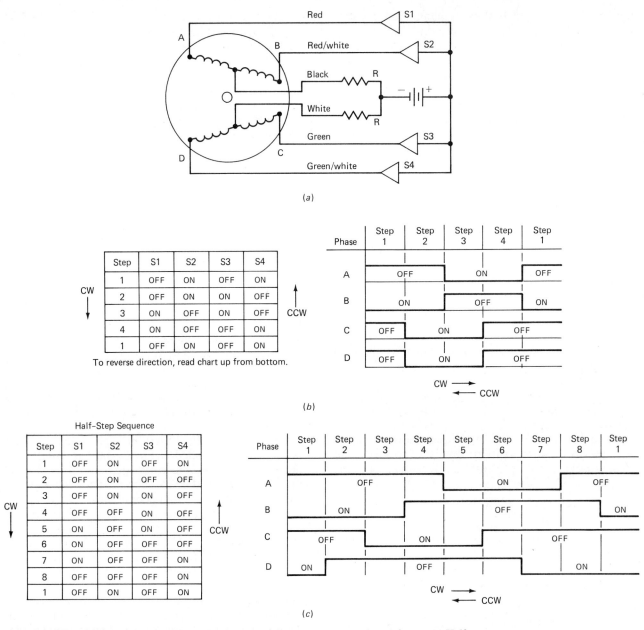

(a)

Step	S1	S2	S3	S4
1	OFF	ON	OFF	ON
2	OFF	ON	ON	OFF
3	ON	OFF	ON	OFF
4	ON	OFF	OFF	ON
1	OFF	ON	OFF	ON

CW ↓   CCW ↑

To reverse direction, read chart up from bottom.

Phase	Step 1	Step 2	Step 3	Step 4	Step 1
A	OFF		ON		OFF
B	ON		OFF		ON
C	OFF	ON		OFF	
D	OFF	ON		OFF	

CW ⟶
⟵ CCW

(b)

Half-Step Sequence

Step	S1	S2	S3	S4
1	OFF	ON	OFF	ON
2	OFF	ON	OFF	OFF
3	OFF	ON	ON	OFF
4	OFF	OFF	ON	OFF
5	ON	OFF	ON	OFF
6	ON	OFF	OFF	OFF
7	ON	OFF	OFF	ON
8	OFF	OFF	OFF	ON
1	OFF	ON	OFF	ON

CW ↓   CCW ↑

Phase	Step 1	Step 2	Step 3	Step 4	Step 5	Step 6	Step 7	Step 8	Step 1
A	OFF				ON		OFF		
B	ON			OFF			ON		
C	OFF		ON			OFF			
D	ON		OFF				ON		

CW ⟶
⟵ CCW

(c)

Fig. 3-22 Bifilar stepper motor. *(a)* Circuit diagram. *(b)* Full-step sequence and waveforms. *(c)* Half-step sequence and waveforms.

motor winding. The voltage forward-biases the body diode, and breakdown is avoided. The MOSFETs will also switch large currents faster than bipolar transistors.

The block diagrams for the logic circuitry required to obtain a full-step sequence for Fig. 3-22(b) is shown in Fig. 3-24. (Logic devices are covered in depth in Chapter 11.) Any type of logic family that supports the circuit requirements can be used. Figure 3-24(b) shows the typical circuitry required using transistor-transistor logic (TTL). The outputs are then applied to the drivers previously mentioned.

Figure 3-24(c) shows the simplicity of using a com-

plementary metallic oxide semiconductor (CMOS) circuit to obtain either the full- or half-step drive for the bifilar motor. It should be noted that the full-step control sequence is two on-time periods followed by two off-time periods. The rate of stepping is determined by the frequency of the applied clock, with each input pulse causing one step (or half step) in the stepper motor shaft. The half-step sequence pattern is three on-time periods followed by five off-time periods; like the full-step control it is very easily obtained by solid-state circuitry.

This type of control leaves a lot to be desired in some modern computerized industrial and robotic

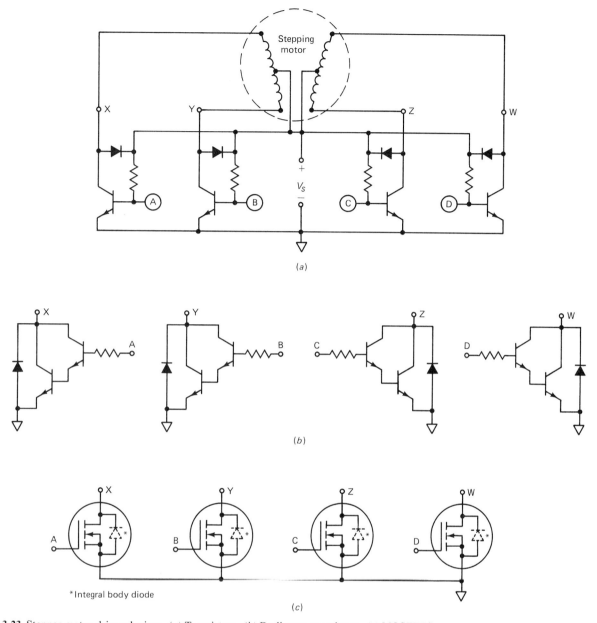

Fig. 3-23 Stepper motor driver devices. *(a)* Transistors. *(b)* Darlington transistors. *(c)* MOSFETS.

applications. Silicon monolithic integrated circuitry, as shown in Fig. 3-25, contains all the input stages, logic, and drivers necessary for steppers rated up to 500 mA/coil. The output can be applied to one of the previously discussed drivers if more power is required. The simplicity of one integrated circuit (IC) is very desirable and cost-effective.

Stepper motors come in a wide variety of sizes, types, and styles. The basic stepping principle for all is the same, and they fall into one of the three types discussed. They may have as little as two windings, or as many as ten phase windings (these take 2500 steps for one revolution).

REVIEW QUESTIONS

16. Compared to the free turning shaft of the dc motor, the stepper motor is _____.

17. The stepper motor rotor illustrated in Fig. 3-17(c) seeks the _____ pole.

18. The rotor of the PM and PM-hybrid stepper motor is made of what kind of material?

19. The variable-reluctance stepper has _____ residual magnetism.

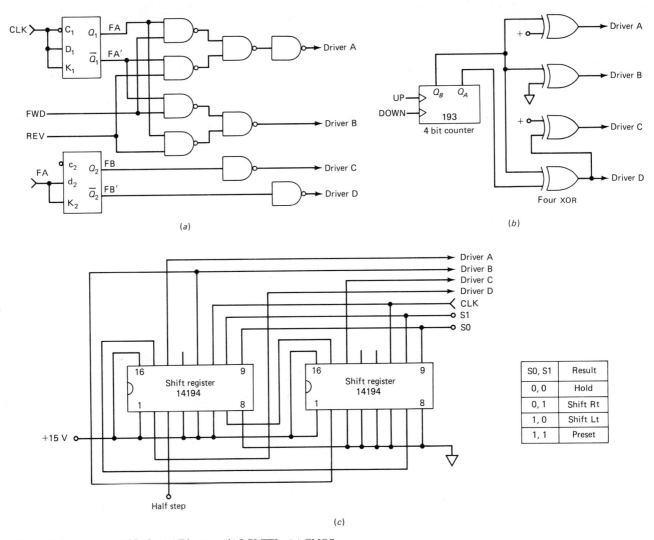

Fig. 3-24 Stepper control logic. *(a)* Discrete. *(b)* LSI TTL. *(c)* CMOS.

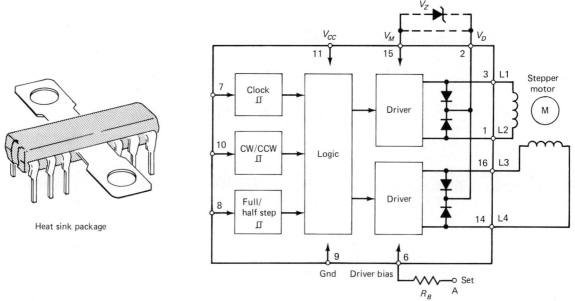

Fig. 3-25 Monolithic stepper motor driver with logic.

3-4
BRAKING AC MOTORS

Induction motors can be dynamically braked to an abrupt stop, just as dc motors can. They are dynamically braked by removing the ac power from the motor and substituting dc power. When this is done, the motor will act the same as the dc shunt motors previously described. The stator of the induction motor, with direct current applied, is similar to the field (stator) of a shunt motor, and the squirrel cage rotor is similar to a shorted armature when dynamically braked. The rotor acts like a dc generator with a shorted armature, with high circulating currents in the rotor bars. This rotational energy is dissipated in the form of heat (in the rotor) when the motor stops abruptly.

The source of direct current for braking may be large batteries or power supplies. The direct current may also be supplied by a charged capacitor. Figure 3-26(a–d) shows examples (simple circuits) of ac induction motor braking; Fig. 3-26(e) illustrates the *capacitor discharge method*. The capacitor is charged while the motor is running and discharges through the windings when the motor is braked. This arrangement eliminates the necessity for any external power source.

REVIEW QUESTIONS

20. Induction motors can be dynamically braked by applying _____ current to their windings.

21. Capacitor discharge braking requires _____ external power source.

3-5
BRUSHLESS MOTORS

The term *brushless* has been applied to a wide variety of electric rotating devices. Most earlier devices were used in applications in which unidirectional operation was acceptable and minimum electronics was a major consideration, as was elimination of the mechanical brush-commutator mechanism. The brushless dc motor uses ac stator voltages of two, three, four, or six phases. Typically, the brushless dc motor has the torque/speed characteristics of the conventional dc PM motor.

Development of low-cost position sensors (such as Hall effect integrated circuits and switching optical sensors), economies favoring the use of rare earth magnetic materials, and the availability of low-priced semiconductors permit a more competitive position for brushless dc motors. They are especially attrac-

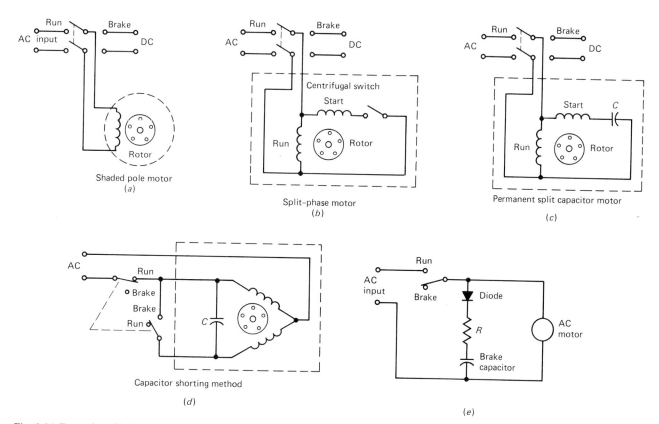

Fig. 3-26 Examples of induction motor braking.

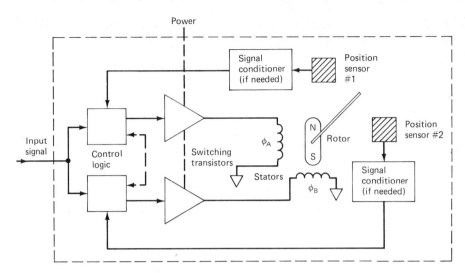

Fig. 3-27 Simplified block diagram of brushless dc motor.

tive in applications requiring tachometer feedback units. The brushless motor has high efficiency, long life, low noise, and low power consumption.

Electronically commutated brushless dc motors operate on the same motor action as the conventional dc motor, except that the supply current switching (in reference to rotor position) is accomplished by solid-state circuitry instead of the mechanical brush-commutator combination used in conventional dc motors. In brushless dc motors transistorized logic senses the position of the PM rotor and controls the distribution of current to the field windings. The field windings are energized in sequence to produce a revolving magnetic field. Rotor position is sensed by solid-state light emitters and sensors, Hall effect devices, or some other means. The sensor feedback signals return to the control unit, which feeds the drivers that turn on transistor switches, thereby delivering the sequentially rotating current to the field coils (stator). Figure 3-27 shows a brushless motor

in basic block form. It should be emphasized that the brushless dc motor is not a stepper motor and has smooth continuous shaft rotation and not the fixed-step detents found in the stepper motor. Because of the switching logic and circuitry associated with its control, it could be mistaken for a stepper motor.

The most commonly used methods for angular position sensing are Hall effect sensors and optical sensors. The *Hall effect sensor* detects the magnitude and polarity of a magnetic field. The signals are amplified and processed (within the IC) to form logic-compatible signals. A simplified block diagram of a Hall effect sensor is shown in Fig. 3-28. The Hall element has a constant current (I_H) passed through the element, which is usually indium. A magnetic field with a flux density (B) is applied at right angles to the element and causes the charge carriers to be redistributed within the element, causing a voltage V_H, the Hall voltage, to be induced in a direction perpendicular to the current and magnetic field. The

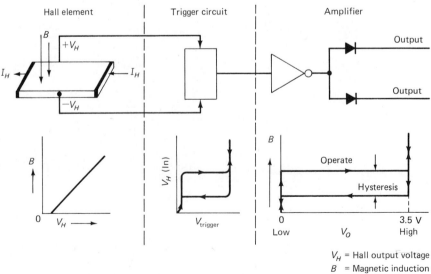

Fig. 3-28 Simplified block diagram of Hall effect sensor.

V_H = Hall output voltage
B = Magnetic induction
I_H = Hall current

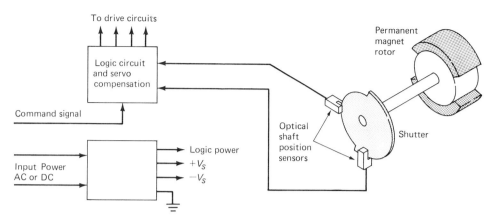

Fig. 3-29 Optical rotor position sensing.

Hall voltage is proportional to I (current) and B (flux density) and is expressed as

$$V_H = \frac{R \times I \times B}{D}$$

where

V_H = Hall voltage, V
B = Flux density, gauss (G)
I = Current, A
D = Element thickness, m
R = Hall constant

In Fig. 3-28, the Hall element exhibits a linear output, but the trigger circuit (comparator) transforms the Hall signal into an ON or OFF signal, and the amplifier shown is the interface to the desired control circuitry.

The other method of sensing rotor position for electronic commutation is the use of photodetectors. The basic concept is illustrated in Fig. 3-29. The stationary light source (solid-state or incandescent) emits a beam that is interrupted by a circular disk chopper mounted on the rotor shaft. The disk chopper will have some number of sectors removed, depending on the number of photodetector sensors and stator phases. A 60° sector would be used with six photodetectors, whereas a 180° sector would be used with two photodetectors (as shown in Fig. 3-29). The photodetectors (covered in Chapter 14) produce an output that is amplified and converted to a drive signal, as with the Hall effect device. These signals are then applied to the control circuitry to produce a rotating field. The electronics, light source, detectors, amplifiers, logic, and drivers are mounted integrally with the motor, as shown in Fig. 3-30. With the advances in solid-state technology, this type of control may eventually be reduced to a single integrated circuit.

The brushless motor stators can be connected in several variations: single-phase, two-phase, three-phase wye, three-phase delta, four-phase delta, and six-phase delta. A rotor having 4 poles and a six-phase delta-connected stator will commutate 12 times per revolution every 30° of the rotor shaft. This arrangement gives the highest power efficiency and peak torque. The major drawback is that it requires 12 switching transistors. By comparison, a four-phase, 4-pole configuration will yield the same peak torque output and requires only eight switching transistors. The three-phase wye and three-phase delta, 4-pole configuration can be full-wave excited by us-

Fig. 3-30 Typical brushless motor and control circuit board.

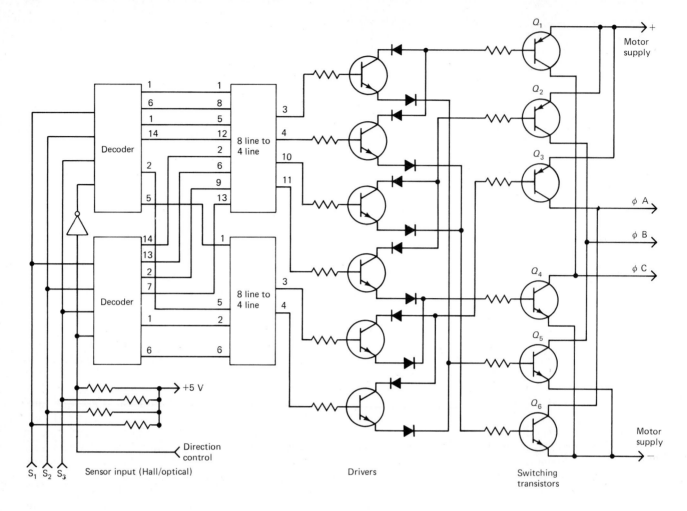

Fig. 3-31 Three-phase commutation circuitry.

ing six transistors with commutation 12 times per revolution.

A schematic for three-phase commutation circuitry is shown in Fig. 3-31. The input from the sensors (omitted for clarity) is connected to the decoders along with the direction signal to control the stator excitation and the direction of rotation.

The brushless dc motor is finding applications in servomechanisms, robotics, disk drives, and wherever direct drive motors are required. Speeds up to 25,000 rpm are not uncommon, and the nonarcing characteristics make it desirable for hazardous areas.

REVIEW QUESTIONS

22. The brushless motor exhibits torque/speed curves similar to those of the dc _____ motor.

23. The stator of the brushless dc motor has alternating or direct current applied to the _____ windings.

24. The brushless motor has a(n) _____ type rotor.

25. Two types of rotor position sensors are light-sensitive and _____ effect type.

26. The brushless motor, like the stepping motor, exhibits detent torque. (true or false)

3-6
SPEED CONTROL OF AC MOTORS

One of the principal characteristics of the ac induction motor is its ability to maintain a near-constant speed under normal load and supply voltage variations. All induction motors have a full speed lower than the synchronous speed. This reduction in speed is specified as *percent slip*. The synchronous speed N is a function of line frequency (f) and number of poles (P):

$$N = \frac{120f}{P}$$

This formula shows that the supply frequency and the number of poles are the factors that determine the speed of the motor. Unlike in the dc motor, the speed of an ac motor is not changed by varying the applied voltage. (One exception is the universal motor covered earlier in this chapter.) Reducing the applied voltage of a large motor in order to reduce its speed could damage the motor. This is due to the excess heat build-up inside the motor. Most ac motors are not designed to have their applied voltage vary more than 10 percent of the nameplate ratings.

Multispeed ac motors, designed to be operated at a constant frequency, are provided with stator windings that can be reconnected to provide a change in the number of poles and thus a change in the speed. These multispeed motors have only a few speeds (usually no more than four), and these speeds are widely separated from each other.

To change the speed of a split-phase motor, it is necessary to change the number of poles. Speed changing a capacitor-start motor is also done by changing the number of poles in the winding. A variable resistor or transformer can be used to vary the voltage across the winding of a permanent-split capacitor motor. These methods are shown in Fig. 3-32. The variable voltage transformer (autotransformer) is preferred to the series resistor. Figure 3-32(b) shows this method, which has the advantage of keeping a constant voltage on the motor whether under start or run conditions. There is also a large reduction in heat loss compared to the resistor method of Fig. 3-32(a). By reducing the voltage across the run winding only and maintaining full voltage across the capacitor (start) winding, more stable low-speed operation is attained. This method is illustrated in Fig. 3-32(c).

The *polyphase* (three-phase) *motor* utilizes two separate and independent windings for each pole. With this arrangement, any desired combination of two, three, or four different speeds is possible. Speed control of ac squirrel cage induction motors can be accomplished if the frequency of the applied voltage to the stator is varied to change the synchronous speed. The change in the synchronous speed (stator rotating field frequency) results in a change in the motor speed. There are two basic methods used to vary the frequency of the applied voltage to the ac motor: one uses an inverter, the other a converter. The inverter is used to change a dc voltage to an ac voltage whose output frequency can be varied. The converter will change the standard 60-Hz line frequency into almost any desired frequency. Both the inverter and converter use solid-state devices (mainly SCRs) for control. Prior to the use of solid-state devices, variable frequency was provided by variable-speed motor-alternator sets. The initial and maintenance costs are high for this approach in comparison to those of the solid-state devices. The solid-state designs are often referred to as *static frequency converters* (no moving parts). The operating frequencies are typically in the range of 10 to 200 Hz. The major applications for variable-frequency drives are textile machinery, machine tools, and steel and paper mills.

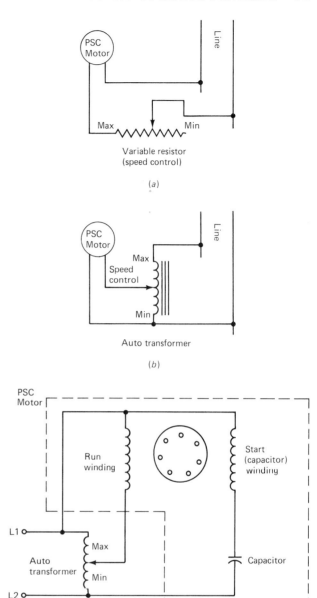

Fig. 3-32 Voltage control of permanent split capacitor motors.

The basic requirements for a variable-frequency inverter are shown in Fig. 3-33 in block form. This type of frequency conversion is also known as a *dc-link converter* because of the intermediate dc conversion that takes place. This method of conversion is covered in more detail in Chapter 8. At this point we will examine the system at the block level. The desired speed signal is applied to both the voltage-controlled oscillator and the pulse-width modulator. The three-ring counter (count 1,2,3, 1,2,3, as the name implies), via the logic circuits, controls the distribution of gate pulses to the six-step inverter. It contains six thyristors arranged for a three-phase

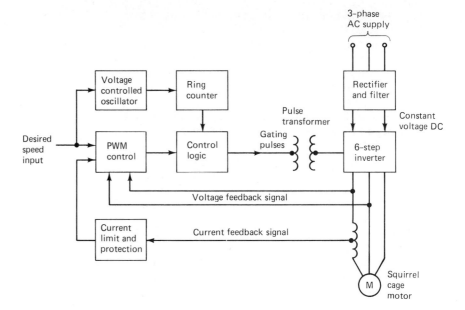

Fig. 3-33 Variable frequency converter block diagram.

full-wave bridge configuration. During each half cycle that the thyristors are gated on by the ring counter, pulse-width modulation ensures that the thyristors are switched on and off at the proper times to achieve voltage control within the inverter. Voltage feedback permits control of constant torque or horsepower as desired. At the same time, stator current control is obtained by means of current feedback to the pulse-width modulator control block.

The other method of static frequency conversion to obtain speed control of ac motors is the *ac-to-ac frequency converter,* which is commonly called a *cycloconverter.* Though inherently more efficient than the dc-link inverter, it has two major disadvantages. First, in order to keep the output harmonics at an acceptable level, it must be operated in a range from 0 Hz to one-third of the ac input source frequency. Second, to obtain bidirectional operation, a three-phase cycloconverter requires a minimum of 36 SCRs with complex control circuitry. The three-phase cycloconverter is presented in more detail in Chapter 8.

The easiest way to understand the principle of a cycloconverter is to study the seldom used single-phase to single-phase cycloconverter. The basic circuit shown in Fig. 3-34 has two two-pulse midpoint phase-controlled converters. One forms the positive group and the other the negative group, which is a dual-converter configuration. The output current of each group flows in only one direction (SCRs are unidirectional devices). To produce an alternating current in the load, the two groups, positive and negative, must be connected in inverse parallel as shown in Fig. 3-34. The positive group (SCR 1 and SCR 3) permit load current to flow only during the positive half of the cycle, when V_{AN} or V_{BN} is positive, and the negative group permits current flow during the negative half cycle, when V_{AN} or V_{BN} is negative.

The waveforms are shown in Fig. 3-35, and an inductive load is assumed. Voltages V_{AN} and V_{BN} are 180° out of phase (*antiphase*). By varying the firing points (t_1 to t_4) of the thyristors, the mean output voltage to the load can be varied as indicated. Varying the firing delay angles about 90° in a sinusoidal manner at the desired output frequency will result in a mean output load voltage that is sinusoidal at the desired frequency, as indicated in Fig. 3-35.

The ac shaded pole motor can be made to vary its speed by varying the voltage across its windings. This is an inefficient method and produces a lot of excess heat. For years these motors (used for fans) had their speed varied by inserting a choke in series with the main winding, which eliminated the IR (resistive) losses.

The best passive device, however, is an autotransformer. It has the advantage of maintaining the same voltage under high starting current. There is also

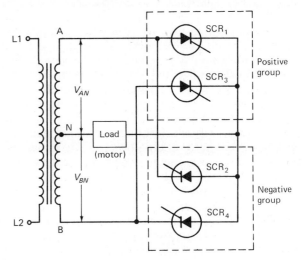

Fig. 3-34 Simplified single-phase cycloconverter.

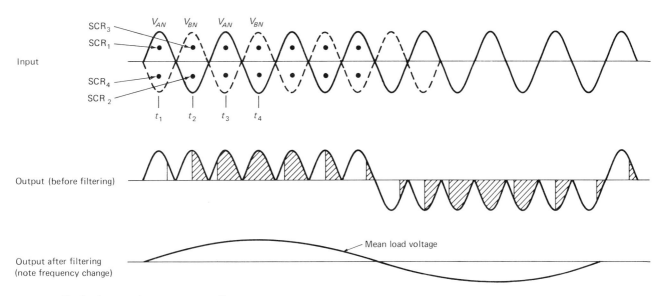

Fig. 3-35 Single-phase cycloconverter waveforms.

very little heat loss compared to the other methods. By reducing the voltage across the run winding (only) of the permanent-split capacitor motor (capacitor run), while maintaining full voltage across the start winding, the speed can be varied by varying the output voltage of an autotransformer.

Most modern variable ac voltage sources are derived from solid-state electronics, which is more cost-effective, reliable, and efficient than resistors, chokes, or transformers for varying ac voltages.

The ac power to a shaded-pole or permanent-capacitor motor is controlled by varying the rms value of the ac voltage using thyristors. The typical method used to vary the voltage is phase control. Unlike brush-type motors, induction motors give no convenient electrical indication of their mechanical speed. This means that direct speed feedback is not usually available. For many applications, such as fixed fan loads, direct voltage adjustment with no feedback (open loop) yields satisfactory performance. An example of this type of circuit is shown in Fig. 3-36(a). The single time constant circuit provides satisfactory proportional speed control for shaded-pole and permanent split-capacitor motors when the load is fixed. This type of circuit is best suited for applications that need speed control in the medium- to full-power range. It is especially useful for fan or blower control, in which a small change in motor speed produces a large change in air velocity. Caution must be exercised because the motor may stall if the speed is reduced below the motor's dropout speed. The single time constant circuit is employed because it cannot provide speed control from maximum to full off. Speed ratios as high as 3:1 can be obtained from the single time constant circuit used with these types of induction motors. As a result of the inductive load and because the triac turns off when the current reaches zero the source voltage may be at a value other than zero. This commutating

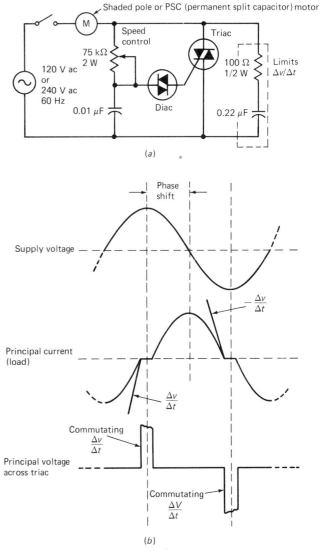

Fig. 3-36 Solid-state speed control. (a) Control circuit. (b) Voltage and current waveforms.

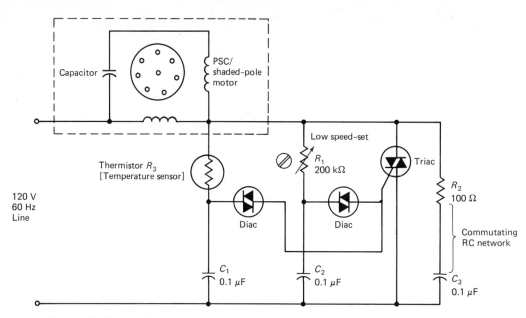

Fig. 3-37 Motor speed controlled by thermistor (temperature).

voltage of the motors may have a rate of rise $\Delta v/\Delta t$ which could retrigger the triac. The commutating $\Delta v/\Delta t$, shown in Fig. 3-36(b), can be limited by use of the RC network across the device as indicated in Fig. 3-36(a).

Figure 3-37 illustrates how this type of circuit could be easily modified to include a feedback signal (closed loop operation) by the addition of a temperature-sensitive resistor (*thermistor*) along with a capacitor and a diac. In this circuit, the thermistor R_3 acts in response to air temperature to control the power to the fan motor. Resistor R_1 and its phase-control network set the minimum fan (blower) speed to provide continuous air circulation and to maintain proper bearing lubrication.

REVIEW QUESTIONS

27. The induction motor speed equation indicates that the speed control terms are the _____ and the number of poles.

28. A split-phase motor with 4 poles runs at 1750 rpm. With 6 poles its approximate speed is _____ rpm.

29. The auxiliary winding in the capacitor-run motor is in series with the _____ winding at reduced speeds.

30. The use of a _____ is preferred to use of a variable resistor for voltage control of a shaded-pole motor.

31. Solid-state frequency changers are also called _____ converters.

32. Besides the inverter, the _____ is a solid-state frequency converter.

3-7 TROUBLESHOOTING AND MAINTENANCE

Preventive maintenance plays a key role in the economics of all industrial installations. In today's highly competitive industry, shutdown of equipment or processes can be very costly. Preventive maintenance, when properly performed, will help prevent troubles before they start. Keeping good records of all maintenance done will establish priorities and give notice of trends. Since motors constitute a major portion of the electrical load in any industrial situation, their maintenance is of extreme importance.

Direct Current Motors

The failure of a dc motor to start may be due to an open in the control circuit, such as a line fuse, low line voltage, or a frozen bearing (this condition will trip the fuses or breakers). Sparking at the brushes may show the need for reseating. The resistance method of controlling the speed of dc motors can also be a source of problems. Resistors can change value, usually because of overheating, so proper ventilation should be confirmed if this type of failure occurs. These resistors, except in the field weakening case, usually decrease the base speed of the motor. The wipers (sliding arm contacts) must be clean and have good contact pressure to ensure proper connections. Some adjustment arms have a small brush-like carbon contact that may need to be seated or replaced. Use the equipment manufacturer's maintenance guide as reference for any adjustments.

The solid-state devices in speed controls may fail by opening or by short-circuiting. The loss of all control in the SCR-type control can be due to a shorted SCR or a UJT that triggers it (if used). The opening of either device will stop the unit from operating or starting. After the power sources and supplies are verified, the solid-state devices can be checked (with power off) by using the methods described in Chapter 2.

The closed loop type of speed control relies on feedback of one sort or another. The tachometer signal, if missing (the line is either short-circuited or opened), would make the system become open loop. With no feedback available, the control would be lost, and the system would run away. In this case, the components and path of the feedback must be checked.

Reduction of control can be due to low voltages, low gain, poor connections, and changes in reference voltage or associated circuits. The mechanical coupling (linkage) in the feedback speed control must be maintained; otherwise, control can be erratic and unpredictable.

The universal motor will run slower on alternating than on direct current so if a rectifier shorts it may cause a drop in speed. The stepper motor will almost always be electronically controlled because of the switching sequence required by its windings. The resistance in series with the power supply affects the slewing (time constant) of the stepper motor. The inductive/resistive (L/R) time constant is inversely proportional to the resistance; therefore, a decrease in resistance will increase the slew speed and vice versa. A large decrease can also cause overheating due to the increase in the holding currents.

In many instances, the logic power is from another source or a separate regulator. These voltages must also be checked. The TTL-type logic families permit only a 250-mV deviation in the supply voltage. Excessive ripple in the power sources can cause false counting or gating of these control circuits and may be suspected in cases of erratic operation. Complementary metallic oxide semiconductor (CMOS) logic is less critical in this respect. Verify power supply parameters early in the troubleshooting process.

Output drivers can fail if they are open or short-circuited. Excessive heat is a major cause of premature failure. All heat sinks must be kept clean, and any air flow restrictions must be corrected. The diodes used for snubbing the inductance of the windings can fail, providing another source of problems. A snubber is a circuit or a component used to absorb the energy produced by the field collapsing in an inductive component.

Stepper Motors

Either a bad driver or its diode (if used) will abort the switching sequence for a stepper motor. In most cases, it will rock back and forth because of the loss of driving torque necessary to overcome the detent torque. In the case of the monolithic or integrated controller, after all input signals have been checked

along with the power supplies and the output signals, the chip may have to be replaced. Substitution is sometimes the only practical method available in these cases.

Do not dismantle stepper motors that use a PM rotor. A 5 to 15 percent loss of torque results. Consult the manufacturer's recommendations. A record of failures and maintenance procedures should be kept to alert you to all trends and causes, such as line surges, overheating, and so on. Be sure to replace all shields, fasteners, filters, and grounding straps before returning equipment to service.

Alternating Current Motors

Except small (fractional horsepower) motors, most ac motors are controlled and protected by one or more of the following devices: starters, breakers, fuses, contactors, and thermal overloads. The integrity of these components is the first thing to investigate. If overloads, breakers, or fuses are tripped or blown, you should investigate for overheating, overloading, or loose connections. Interaction with an operator who uses the equipment may yield important information as to the cause of the problem and should not be overlooked or ignored.

Brushless Motors

Modern brushless motors used in the industrial environment eliminate many of the mechanical problems of other types of motors. With no brush or commutation mechanisms, all problems involving the wear associated with these parts are eliminated. The rotors are permanent-magnet and should last indefinitely if not abused. The construction is of a modular form to assist in fast and efficient repair when required. The sensor unit has to be housed in the motor; the commutation logic may or may not be. These devices should be treated like any other semiconductors. The supply voltage to the sensors and logic circuits is critical and must be verified as being within manufacturer's specifications. The sensor and its logic circuits are modular; therefore, replacement with a spare unit is the best solution to a sensor-type problem. The modules can be best serviced at the bench level or returned to the manufacturer for service. As with all solid-state devices, heat is a prime problem. Ventilation systems must be kept clear and clean. A restricted air flow that goes uncorrected will lead to the eventual failure of the replacement part as it did with the original part.

The driver and switching transistors should be tested for short circuits or opens. Remember that there are two switching transistors per phase. Short-circuited transistors may point to overloads, stator coil shorts or grounds, or power supply problems. Power switching devices can be complex, so the manufacturer's manuals are your first source for troubleshooting information. The brushless motor is best serviced by viewing it as a system. The location of a faulty module is achieved by orderly investigation and the process of elimination.

Multispeed Motors

The troubleshooting and maintenance of static inverters are covered in Chapter 8. Speed controls using multispeed motors can encounter problems of loose connections, short circuits, or grounds to the stator coils, especially since multiple windings are involved. The normal testing for these failures is the same as for the other motors, but more elaborate procedures are necessitated by the complexity of the connections. Starters, contactors, and any timers that may be used are also suspect when multispeed motor connections are used. Loose connections are a very common occurrence and should always be checked along with the supply voltage.

REVIEW QUESTIONS

33. Changes in the _____ of the controls for dc motors will affect the speed.

34. If the power source is lost to a disk brake, will the brake remain on or off?

35. If the diode in Fig. 3-3 short-circuits, the transistor and the braking will be _____ all the time.

36. If the grounded end of the bias pots in Fig. 3-9 open, the motors will _____.

37. If C_2 in Fig. 3-16 opens, the hysteresis control will improve. (true or false)

38. Why are the diodes shown inside the symbols of the MOSFET drivers in Fig. 3-23(c)?

39. The brushless motor is usually serviced by replacing a suspect _____.

CHAPTER REVIEW QUESTIONS

3-1. To brake a shunt-wound motor dynamically, the resistor is placed across the field. (true or false)

3-2. Dynamic braking of a series-wound motor requires that the field, armature, and resistor be in a closed _____.

3-3. The _____ motor can be dynamically braked, but with loss of braking power.

3-4. The addition of a _____ in the feedback loop gives linear control of the motor's speed.

3-5. If a tachometer has a K of 20 mV/rpm, at 500 rpm V_g equals _____ V.

3-6. A unidirectional motor control is also called a _____ quadrant control.

3-7. Using resistance control near standstill in a series (universal) motor circuit can cause _____ of the armature.

3-8. _____ resistance control of the series (universal) motor is common when paralleling motors.

3-9. Voltage control of the series motor is the most common method used for speed control. (true or false)

3-10. Series voltage control using a triac is an example of _____ wave control.

3-11. Full-range speed control is usually accomplished with _____ wave-type control.

3-12. All the methods of speed control for the series (universal) motor cause the armature speed to _____.

3-13. The rotor and stator on the PM-hybrid stepper motor are _____ to give a high detent torque.

3-14. What method is used to increase torque in a stepper motor?

3-15. With an 8:1 step-down a 256-rpm stepper ratio, output would be _____.

3-16. Common center-tapped windings are also known as _____ windings.

3-17. The eight-step sequence of a four-winding stepper motor is known as a _____ step.

3-18. The _____ frequency is the prime speed-determining factor of a stepper motor.

3-19. A _____ degree sector would be used with four photodetectors on the rotor positioning mechanism in a brushless motor.

3-20. A three-phase delta stator would require _____ commutations per rotation.

3-21. The three-phase cycloconverter uses at least _____ SCRs.

3-22. The _____ is the solid-state device used for switching alternating current to shaded-pole motors.

3-23. If an induction motor is slowed too much, it may _____.

3-24. A 100 step per revolution stepper motor will take how many steps in the half-step mode for one complete revolution?

3-25. A gear reducer inside a stepper motor will do what to the motor's torque? Its speed?

3-26. Loss of a driver or its signal will cause most stepper motors to _____.

3-27. A three-phase brushless motor will have _____ switching transistors that require testing.

ANSWERS TO REVIEW QUESTIONS

1. friction **2.** released **3.** disk **4.** asbestos **5.** SCR, MOSFET **6.** twice **7.** Fig. 3-4 **8.** increase **9.** small **10.** decrease **11.** open **12.** voltage **13.** commutate **14.** cemf **15.** decrease **16.** locked/stiff **17.** average **18.** magnetic **19.** little **20.** dc **21.** no **22.** permanent magnet **23.** ac **24.** permanent magnet **25.** Hall **26.** false **27.** frequency **28.** 1150 **29.** run **30.** autotransformer **31.** static **32.** cycloconverter **33.** resistance **34.** on **35.** on **36.** rotate constantly **37.** false **38.** they are intrinsic **39.** module

4

BASIC CONTROL DEVICES

This chapter begins with some of the more simple devices of electronics such as switches and fuses, large and small. The wide range of timers from heater types through state of the art electronic types is thoroughly covered. Higher-level switching by relays, contactors, and starters for rotating equipment is presented with actual circuit implications. We will examine all their fundamentals, applications and maintenance requirements.

4-1
SWITCH BASICS

The electric *switch* is a device for making, breaking, or rerouting connections in an electrical circuit. This switching is accomplished by the opening or closing of two metal surfaces. Whichever type is used, the result will be the same. The perfect switch will have 0 Ω resistance when closed, for maximum transfer of power. Conversely, the open switch will have infinite resistance and be capable of withstanding high voltages without arcing. However, any switch tends to arc when it interrupts current flow. The greater the current switched, the hotter the arc will be and the greater the deterioration of the switch contacts. We will look at the types of switches common in today's electronic equipment. With the ever-decreasing size of electronic components it is only natural for switches now to be miniature and subminiature in size. But the current and voltage capacities will also follow the progression: the smaller the switch, the smaller the current and voltage capabilities.

Manual switches are the most widely used today, and each one of us will more than likely operate this type of switch sometime before the day's end. Figure 4-1 shows the basic construction for a manual-type switch. The operator shown in part 4 in Fig. 4-1 can be of a wide variety of styles. A few types of operators are the bat, flat, and baton types. This family of switches is classically referred to as *toggle-type switches*.

A simple single-pole, single-throw (SPST) switch has a single-pole, single-throw action and is capable of opening or closing a single electric circuit. A double-pole, single-throw (DPST) can open or close two

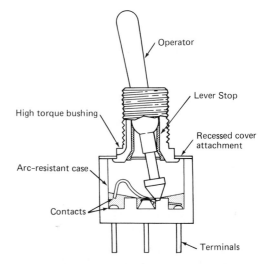

Fig. 4-1 Manual switch construction.

circuits. The DPST switch can be used to open circuits where two "hot" lines are involved, as in 240-V ac equipment. An SPDT has no center position; single-throw switches are either on or off. The three-position toggle switch has two extreme positions plus a center position. The center position is generally off, with the two extreme positions representing on conditions. As illustrated in Fig. 4-2 the B terminal is referred to as the "wiper," or "common" contact of the switch. The center position of the three-position switch is identified by the center OFF notation (c.o.). This notation separates this three-position type of single-pole, double-throw center off (SPDT [c.o.]) switch from a single-pole, double-throw (SPDT).

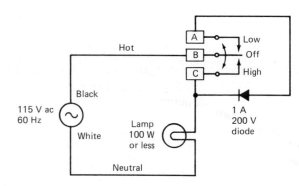

Fig. 4-2 Simple SPDT switch circuit.

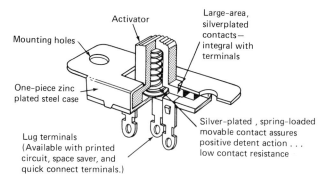

Fig. 4-3 Slide switch construction.

When an operator pushes the button, slide, or lever on a momentary-action switch, the switch contacts move to transfer the circuit to the second set of contacts. When the actuating force is removed, the contacts return to the original position. When an operator actuates a maintained-action switch, the contacts transfer to the second set of contacts. But no change takes place with the contacts when the operator removes the operating force. In some switches the actuator may stay there after removal of the force. Application of a second actuating force is required to return the actuator to its original position. This type of switch is known as *alternate* or *push-push action*.

The slide switch shown in Fig. 4-3 uses a simple slide action to produce the same connections as the toggle switch. Except for the different type of operator action, the switched poles accomplish the same results.

The dual-in-line package (DIP) switch is a miniature form of multipole rocker or slide switch and is made on standard IC socket centers of 0.300 in. The DIP switch can be soldered directly into the printed circuit board (PCB) or placed directly into an IC socket. Most switches are SPST and favor the 14- (seven switches) or 16- (eight switches) pin-package arrangement. Actuation is seldom changed and occurs mainly during installation, testing, and troubleshooting. A ballpoint pen makes a handy actuator for DIP switches. Figure 4-4 shows a 14-pin DIP switch with a typical circuit diagram.

The rotary switch is more indicative of electronic equipment applications. The basic switch structure is shown in Fig. 4-5. The rotary switch is generally a maintained-action switch, but special indexes can be spring-loaded for momentary action. Rotation can be *continuous* (turned through more than one complete circle) or limited to 360° or less. The rotary switch has a variety of positions. If positions are 10° apart, 36 positions are available. The most common angular differences between switch positions are 15, 30, 45, and 90°. One main advantage of the rotary switch is the sequencing ability of its contacts. The rotary switch can also have several banks of switch sections on one shaft, allowing contacts to change simultaneously and in sequence as needed for complex switching applications. Imagine operating an os-

cilloscope or multimeter without the aid of a rotary switch. Contacts can be short-circuiting or non-short-circuiting. Short-circuiting contacts are also called *make-before-break;* that is, when switching to the next position, the wiper contact will be made to the second position before breaking with the first set of contacts. The non-short-circuiting contacts are called *break-before-make contacts.* When switching from one position to the next, the wiper contacts will break with the first contact before making with the second contact. Applications of the short-circuiting-type contacts include avoiding momentary opening of the feedback elements in a system, dropping out a relay in a control application, and unloading an

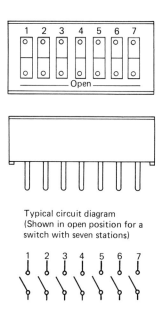

Typical circuit diagram
(Shown in open position for a switch with seven stations)

Fig. 4-4 Dip switch (14-pin).

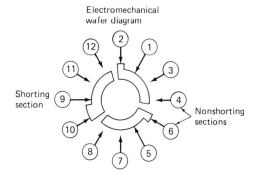

Wafer, 3-pole 3-circuit with 2 nonshorting and 1 shorting moving contacts

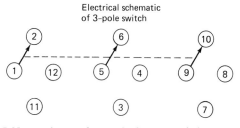

Fig. 4-5 Nomenclature of a standard rotary switch.

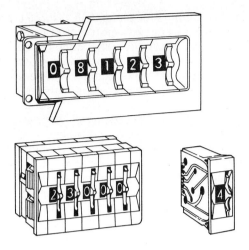

Table 1
10–Position Decimal

Dial Position	0	1	2	3	4	5	6	7	8	9
	Common to:									
0	•									
1		•								
2			•							
3				•						
4					•					
5						•				
6							•			
7								•		
8									•	
9										•

Table 2
8–Position Decimal

Dial Position	0	1	2	3	4	5	6	7
	Common to:							
0	•							
1		•						
2			•					
3				•				
4					•			
5						•		
6							•	
7								•

Table 3
10–Position BCD

Dial Position	1	2	4	8
	Common to:			
0				
1	•			
2		•		
3	•	•		
4			•	
5	•		•	
6		•	•	
7	•	•	•	
8				•
9	•			•

Table 4
10–Position BCD Complement

Dial Position	$\bar{1}$	$\bar{2}$	$\bar{4}$	$\bar{8}$
	Common to:			
0	•	•	•	•
1		•	•	•
2	•		•	•
3			•	•
4	•	•		•
5		•		•
6	•			•
7				•
8	•	•	•	
9		•	•	

Fig. 4-6 Thumbwheel switch and truth tables.

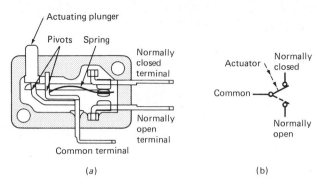

Fig. 4-7 Snap-action switch. *(a)* Operating principle. *(b)* Schematic representation.

ammeter circuit when switching ranges. Rotary switches are now being made for direct application on printed circuits. They may be smaller than the diameter of a dime, usually have only one deck, and switch low-level signals.

Thumbwheel switches have become more commonplace for today's numerical and computer-control applications. Their switching capability is limited to logic levels, with little or no current. Their specially made decks output binary coded decimal (BCD) (covered in Chapter 11), decimal, or hexadecimal codes (also covered in Chapter 11) necessary to input information for controllers. The switches may also be ganged, as shown in Fig. 4-6.

The *position-sensing switch* is typically a snap-acting switch. The switching function is performed by actuation of its input. In Fig. 4-7, the operating principle of the snap-action switch with three electrical contacts is shown in a cutaway view. This type of switch is often called a *microswitch*, after the company that is the world's largest manufacturer of snap-action switches. Figure 4-7 also shows the electrical symbology of this switch in its simplest form, SPDT.

With no force on the plunger, the spring holds the common contact tightly against the normally closed (N.C.) contact. As the plunger is depressed, the force holding the common and N.C. contacts goes through the gap and forces it firmly against the normally open (N.O.) contact. Upon removal of force on the plunger, the N.O. contact's force approaches zero, and the common contact snaps back to its rest, or N.C., position. A few of the actuating methods shown in Fig. 4-8 are the pin, leaf, lever, and plunger. The various methods will fit a variety of different applications.

Snap-action switches boast a low operating force (3 gram-inches [gm · in.] is common), yet ratings of 5 A, 250 V ac are not unusual. The DPDT is the maximum circuit configuration found, but tandem arrangements are possible. This switch is selected not only for its electrical rating, but also its mechanical characteristics, operating force, overtravel, and type of actuator. They are also available in enclosures that can be water- or oil-tight and even explosion-proof. Some environmental needs require these special housings, and nonsparking materials such as nylon may be used for the roller.

A *proximity switch* senses and indicates the presence or absence of an object without requiring physical contact. Simple proximity switches are made up of a sensing head, a receiver circuit for the sense information, and an output circuit including a relay or solid-state switch. The optoelectronic type is covered in Chapter 14.

The *radio frequency (RF) proximity switch* uses an RF oscillator circuit with its coil located in the sensing head. With no metal present, the oscillator output

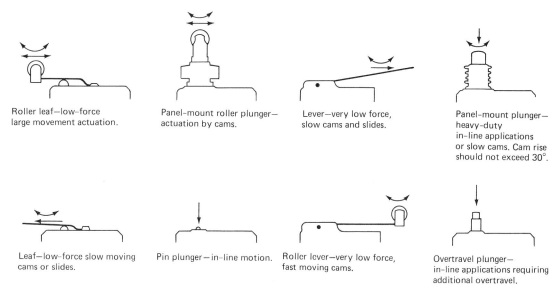

Roller leaf—low-force large movement actuation.

Panel-mount roller plunger—actuation by cams.

Lever—very low force, slow cams and slides.

Panel-mount plunger—heavy-duty in-line applications or slow cams. Cam rise should not exceed 30°.

Leaf—low-force slow moving cams or slides.

Pin plunger—in-line motion.

Roller lever—very low force, fast moving cams.

Overtravel plunger—in-line applications requiring additional overtravel.

Fig. 4-8 Basic switch actuators.

is normal or logically low. If any metal gets into the field of the sensing head the metal absorbs RF energy. The change in the oscillator is detected, and the logic level goes to the high state. The response can be very rapid as needed for high-speed counting applications. The RF proximity switch can also be used to sense the levels of acid or saline solutions. The operating frequencies are less than 3 megahertz (MHz) and are regulated by the Federal Communications Commission (FCC).

The *ultrasonic beam proximity switch* normally uses two heads: one for transmission and one to receive the signal. The transmitting and receiving heads are matched at a selected resonant frequency above the audio range. The detected output signal is amplified to energize a relay or set a logic device. Placing an object into the field of the beam between transmitter and receiver will lower the output signal amplitude and change the state of the logic or de-energize the relay. The ultrasonic proximity switch overcomes the metal and acidic requirements of the RF proximity switch.

The *Hall-type proximity switch* uses a solid-state structure and requires a magnetic field for actuation. This switch is covered in Chapter 3.

Keyboard switches are of prime importance for operator intervention into a numerical or computerized system. The membrane keyboard is simple and inexpensive and can be sealed. The base may be either the rigid or film type, as shown in Fig. 4-9. The top layer is a flexible polyester membrane, with movable screened contacts. The middle layer is the adhesive spacer that provides the gap between the movable contacts and the stationary switch circuits. The bottom layer is either a polyester film substrate or a PCB, depending on the type of construction preferred. A three-by-four keyboard matrix is popular for touch-tone operations and is indicated as *dual-tone-multifrequency* (DTMF) as required by the phone company.

The 16-position keyboard is often used for hexadecimal input of data or instructions to microprocessors, automatic controllers, and robots. Membrane keyboards are also available in an 8-by-12 matrix configuration and can be interconnected with an encoder (encoders are covered in Chapter 11) integrated circuit to produce the American Standard Code for Information Interchange (ASCII) (covered

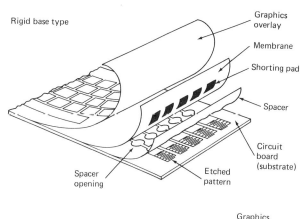

Rigid base type

Graphics overlay

Membrane

Shorting pad

Spacer

Circuit board (substrate)

Etched pattern

Spacer opening

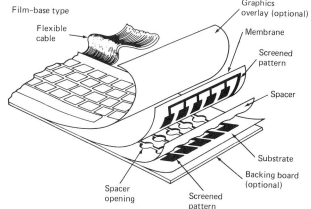

Film-base type

Flexible cable

Graphics overlay (optional)

Membrane

Screened pattern

Spacer

Substrate

Backing board (optional)

Spacer opening

Screened pattern

Fig. 4-9 Rigid and film membrane switches.

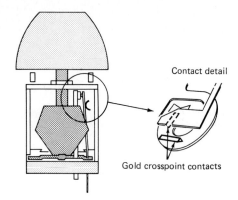

Fig. 4-10 Crosspoint keyboard module switch.

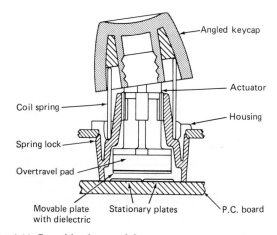

Fig. 4-11 Capacitive key module.

in Chapter 11) code needed to communicate with most computers today.

The *crosspoint,* or *hard contact,* keyboard is not temperature- or humidity-sensitive, is low in cost, and has a low standby power consumption. This type of contact is shown in Fig. 4-10. The switch matrix can be arranged in any of the styles previously mentioned. The contact material is precious metal to provide inertness to chemical action and provides a long life. The moving key action of this type of keyboard is often preferred by touch typists.

Some solid-state keyboards feature capacitive keyswitches to provide a keyboard with unique capabilities. They give high-speed data entry, including key-to-disk, key-to-tape, and other types. With a high-frequency clock (typically 666-kilohertz [kHz] at the input to the key matrix, depression of a key allows capacitive coupling of these pulses to the output of the matrix. Figure 4-11 shows the physical construction of a capacitive key switch. Looking closely at Fig. 4-11, you can see that no physical contact is required for this switch. This contactless style gives very high reliability.

REVIEW QUESTIONS

1. An ideal switch will have _____ ohms when closed and _____ ohms when open.

2. Can an SPDT be used to replace an SPST switch of the same voltage and current ratings?

3. The maximum number of switch positions for a 30° rotary switch is _____.

4. If you do not want to open a circuit momentarily when sequencing with a rotary deck, the switch should be of the _____ variety.

5. A _____ switch is commonly used to input BCD and hexadecimal codes.

6. The _____ action switch is commonly used for position sensing.

4-2
RELAYS

The *relay* is an electrically controlled device that opens and closes electric contacts affecting other devices in the same or another circuit. Basic single-pole relay action is shown in Fig. 4-12 with Fig. 4-12(*a*) being the unenergized state and Fig. 4-12(*b*) the energized state. Without going into the magnetics (to be covered later), when the switch is closed, a north and south pole are produced across the working gap by the solenoid winding of the coil, producing a magnetic flux as shown in Fig. 4-12(*b*). Note that the power source shown is direct current. Operation with alternating current will be discussed shortly. If the current in the coil is slowly increased from zero, a point will be reached where the armature snaps (closing the gap) and closes the normally open (N.O.) contact. This is defined as the *pick-up point* and is specified by current or voltage by the manufacturer. Conversely, if the current in the coil is gradually decreased, a point where the flux is too weak to hold the gap closed is reached and the armature snaps open. This is called the *drop-out current* (or voltage) and is usually considerably less than the pick-up current (or voltage). The only noticeable physical difference between ac and dc relays with the same ratings is the addition of a shaded pole at the end of the core. A shaded pole is a conducting ring that creates a magnetic field that lags the applied field, prohibiting the rapid collapse of the flux across the gap. If this were not the case, the relay would hum or chatter at a 60-Hz rate, and contact positioning would be indeterminate. The shading also causes ac relays to be inherently slower to release than dc relays of the same size. The delay is about 10 ms or longer to prevent contact release at each polarity reversal on 60-Hz operation.

The contacts in an electromagnetic relay make or break the connections in electric circuits. The various contact styles, materials, and ratings that are typical for most relays to be covered in this chapter are presented in Fig. 4-13. High-current contacts are usually of the *single-button* or *bifurcated* style. Noble metal alloys are used for the contact material to reduce oxidation that causes high resistance. The bifurcated contacts have two surfaces for less con-

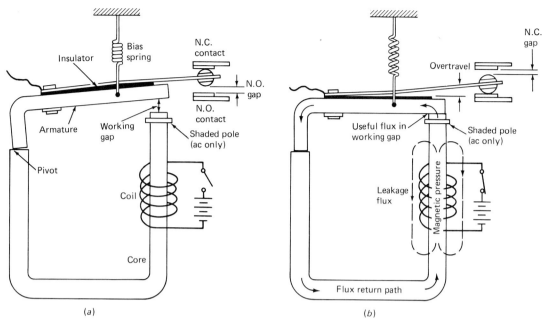

Fig. 4-12 Basic single-pole relay.

Contact style	Single button	Single button	Bifurcated	Single button	Bifurcated Low Level	Cross Bar Dry Circuit
Resistive load rating @ 28 V dc or 115 V ac	Typ. 7.5 A Max. 10 A Min. 0.500 A	Typ. 5 A Max. 7.5 A Min. 0.500 A	Typ. 5 A Max. 7.5 A Min. 0.500 A	Typ. 2.0 A Max. 3.0 A Min. 0.100 A	Typ. 100 mA Max. 2.0 A Min. 1.0 mA	Typ. 500 μA Max. 250 mA Min. dry circuit
Material	Silver–cadmium oxide	Silver–cadmium oxide	Silver–cadmium oxide	Fine silver and gold diffused	Fine silver and gold diffused	Gold/platinum/silver alloy
Inch dimensions	0.125 Dia.	0.100 Dia.	0.100 Dia.	0.078 Dia.	0.062 Dia.	0.017 Dia.
Contact data						

Fig. 4-13 Common relay contacts.

tact resistance at lower actuating force. The crossbar contacts are used for dry circuit (no or little current flow) and are made of gold to hinder any oxidation for low-level (millivolts or microvolts) switching circuits.

Contact arcing is more common in dc than in ac circuit interruptions. The ac circuits go through zero voltage at each half cycle and extinguish any arcing that occurs. Any metal transfer is generally eliminated, except for a roughening of the contact faces. The dc arcing or spark discharge is damaging and will cause metal to transfer from the negative contact to the positive contact, as shown in Fig. 4-14. The needle-type transfer which occurs at low voltage and low contact force is particularly objectionable, and the needlelike build-up may cause interlocking of the contacts. To prevent this, contact shunting or arc suppression may be used to eliminate or minimize contact arcing when switching inductive loads. The

type A circuit shown in Fig. 4-15 uses a capacitor connected across the contacts. When the contacts open, the inductive load generates a voltage due to its collapsing field. This voltage will cause the capacitor to charge, and arcing will be avoided. The addition of a resistor in the type B circuit limits the discharging current of the capacitor when the contacts reclose. The following formulas are applicable for determining R (ohms) and C (microfarads) for the RC suppression network across the contacts:

$$C = \frac{I^2}{10}$$

where

$$I = \text{circuit current, A}$$
$$C = \text{capacitance, } \mu F$$

$$R = \frac{0.1V}{I^X}$$

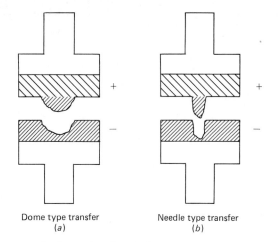

Dome type transfer
(a)

Needle type transfer
(b)

Fig. 4-14 Metal transfer of contacts.

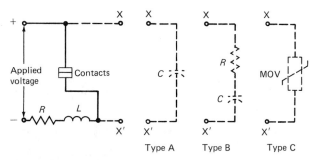

Type A   Type B   Type C

Fig. 4-15 Three types of arc suppression.

where

V = open circuit voltage (induced + supply)

$$X = 1 + \frac{50}{V}$$

Using 0.3 A (I) and 50 V (V), as an example:

$$C = \frac{0.09}{10} = 0.009 \quad \text{(use 0.01 μF)}$$

$$R = \frac{5}{0.09} = 55.6 \ \Omega \quad \text{(use 56 }\Omega\text{)}$$

The capacitor voltage rating should be greater than the open circuit voltage by at least 20 percent. The use of an oscilloscope will verify operation and permit trimming of the network. The type C network shown in Fig. 4-15 uses a metallic oxide varistor (MOV) for quenching transients (this device is discussed in Chapter 5).

Arcing also causes radio frequency interference (RFI) signals. Because RFI signals can cause problems with the operation of sensitive equipment and instruments, these suppression devices are very important for the role they play in reducing interference. The most effective point of installation is at the source or at the contacts.

Another problem with relays relates to the collapsing magnetic field of the relay coil, generating a transient or cemf voltage expressed as

$$V = \frac{-Ldi}{dt} \ \text{V}$$

where di is the change in current
dt is the change in time
L is the inductance, in henrys (H)

therefore $\dfrac{di}{dt} = \dfrac{\text{change in current, A}}{\text{change in time, s}}$

For example, if the coil inductance (L) is 3 H (typical) and the coil current decreases to zero in 5 ms (dt) when 500 mA (di) was flowing, the induced voltage from the preceding equation is

$$\begin{aligned} V &= \frac{-Ldi}{dt} \ \text{V} \\ &= \frac{-3 \times 0.5}{0.005} \\ &= \frac{-1.5}{0.5 \times 10^{-3}} \\ &= -300 \ \text{V} \end{aligned}$$

If unsuppressed, this voltage will destroy the transistor driver of Fig. 4-16. To minimize this transient, one of three methods shown in Fig. 4-16 can be used. The most effective may be a combination of the types shown. The simple diode solution does reduce transients, but it also delays the drop-out of the coil, and this delay may be undesirable. The diode becomes forward-biased as a result of the induced reverse

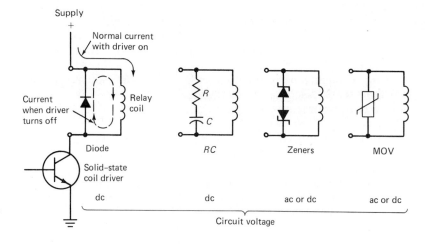

Supply
+

Normal current with driver on

Current when driver turns off

Relay coil

Diode

Solid–state coil driver

dc

R

C

RC

dc

Zeners

ac or dc

MOV

ac or dc

Circuit voltage

Fig. 4-16 Different types of coil transient suppression.

	Type A				Type B	
Form	Description	Symbol	Comments			
A	SPNO		Make when energized	A	Make SPST–NO	
B	SPNC		Break when energized	B	Break SPST–NC	
C	SPDT		Transfer when energized	C	Break–make (Transfer) SPDT	
D	SPDT (M–B)	CT	Make before break	D	Make–break (continuity transfer)	
E	SPDT (B–M–B)	CT	Break–make–break	E	Break– make– break	
F	SPST (M–M)		Make–make (sequential)	F	Make– make	
G	SPST (B–B)		Break–break (sequential)	G	Break– break	
H	SPDT (B–B–M)		Break–break–make	H	Break– break– make	
I	SPDT (M–B–M)	CT	Make–break–make	I	Make– break– make	
J	SPDT (M–M–B)	CT	Make–make–break	J	Make– make– break	
K	SPDTNO		Single-pole Double-throw– Center off	K	Center off SPDT	
L	SPDT (B–M–M)		Break–make– make	L	Break– make– make	
U	SPNODM		Make when energized. Includes pigtail.	U	Double make; contact on arm	
V	SPNCDB		Break when energized. Includes pigtail.	V	Double break; contact on arm	
W	SPDT (DB–DM)		Contacts transfer when energized. Includes pigtail.	W	Double break; double make; contact on arm	
X	SPNO (DM)		Make when energized	X	Double make	
Y	SPNC (DB)		Break when energized	Y	Double break	
Z	SPDT (DB–DM)		Transfer when energized	Z	Double make; double break SPDT–DB	

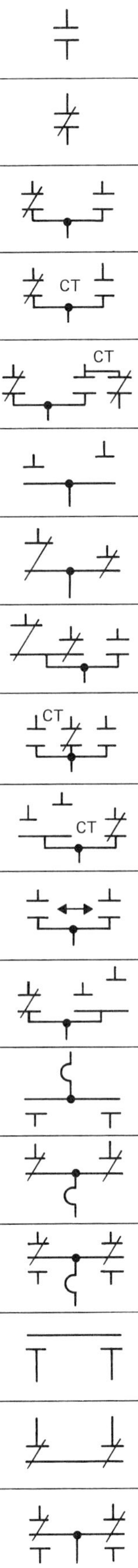

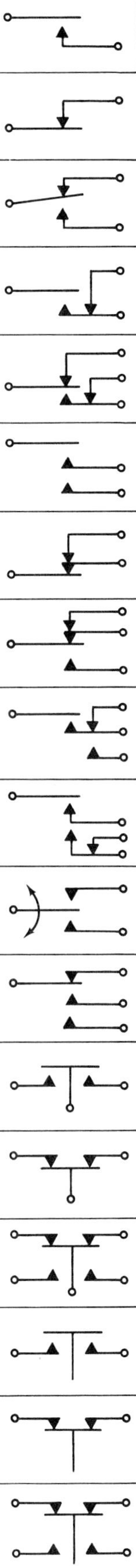

Fig. 4-17 Basic relay contact forms.

voltage of the coil as shown by the previous equation, and the diode remains on until the induced voltage drops to less than 0.6 V (for silicon diodes). The zener diodes and the MOV types shown in Fig. 4-16 preclude this delaying action, by stopping conduction at their higher breakdown potentials, which is far sooner than a diode will. Their breakdown voltage must be greater than the coil operating voltage, but less than the transistor's breakdown voltage.

The basic relay contact forms, of which 18 exist, are shown in Fig. 4-17 (p. 77). The illustration shows that there are two ways to represent relay contacts schematically. Type A and type B arc exactly the same electrically. The form identifications are commonly used by various federal, military, and industrial agencies, and familiarity with both types of symbols is necessary. Forms A, B, and C are most commonly encountered in typical relay circuits, as well as in the switches previously discussed as SPST and SPDT types, although form designation is more common for relay contacts.

Figure 4-18 shows contact action that can be observed by an oscilloscope and gives some standard terms used to describe relay action. The contact bounce on opening or closing can create arcing and associated problems. Bounce usually is due to improper armature seating, dirty or oily armatures, or excess ripple in the applied coil voltage.

Relays with misadjusted residual gap screws may suffer contact bounce as well. These screws are shown in Fig. 4-19. They are factory-set, and adjustment in the field should not be attempted. The contact chatter shown in Fig. 4-18 can lead to RFI and may indicate contact spring vibration or that the ac performance of the coil is inadequate.

The general purpose relay shown in Fig. 4-19 typifies a very broad category of relays that finds wide application. They can be designated as almost any type, depending upon the industry in which they are being used. A quick look through any manufacturer's relay catalog will verify this. A general purpose power relay uses the button-type contacts shown in Fig. 4-13 and is capable of switching up to 25 A. This type of power relay switching is common for power supplies, small motors, lights, and auxiliary circuits. A typical circuit for small motor reversing is shown in Fig. 4-20. The closure of S_1 in Fig. 4-20(a), or Fig. 4-20(b) will reverse the dc motor's direction of rotation. A very subtle situation can be observed by comparing these circuits. In Fig. 4-20(a), slow operation of either set of contacts (one contact is still made when the other transfers) will short-circuit the power source and will cause a catastrophic failure. This condition cannot occur in Fig. 4-20(b), and this is the preferred circuit.

There are over 20 types of primary relays used today in industry and their basic mechanical structure is similar to that shown in Fig. 4-19. A telephone-type relay (though not confined to the telephone industry) has a long coil compared to its diameter. Both ac and dc types are available, with a

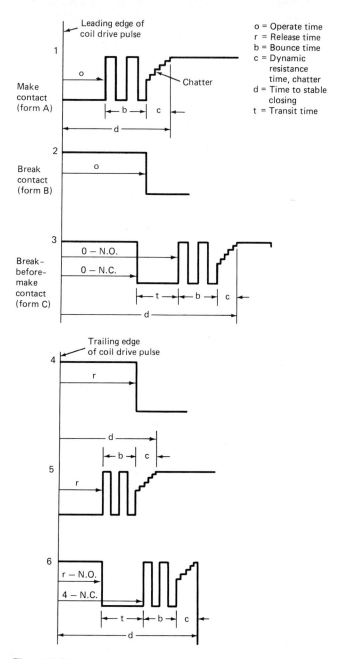

Fig. 4-18 Typical contact response as observed on an oscilloscope.

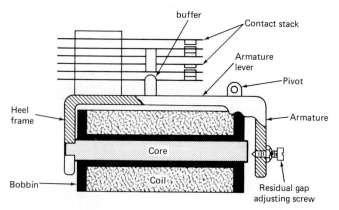

Fig. 4-19 General purpose relay.

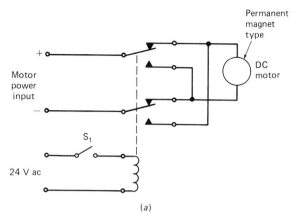
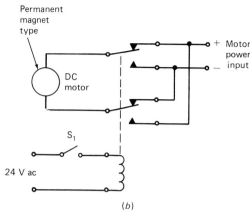

Fig. 4-20 Two different circuits for same function.

very wide choice of contact forms. A short time delay can be produced in a dc relay by placing one or more short-circuited turns around the core to produce an opposing flux on energization or deenergization. This short-circuited turn (or turns) is called a *slug*. It is most commonly used on telephone-type relays that have long coils.

The term *reed relay* includes dry reed relays and mercury-wetted contact relays, all of which use a hermetically sealed structure. The switch capsule of a typical dry relay (form A) is shown in Fig. 4-21. With the tremendous increase in low-level logic switching, computer applications, and communication applications, the dry reed relay is an important device.

The basic reed switch capsules consist of solid metallic contacts sealed in a glass envelope. These

flattened ferromagnetic reeds are sealed at each end of the capsule. The reeds are separated by an air gap and overlap inside the tube. The glass tube may contain an inert gas, such as nitrogen, or have a high vacuum for high-voltage switching. The contacts can withstand adverse conditions of humidity, salt spray, and high altitudes. When the capsule is surrounded by a electromagnetic coil of sufficient flux density or is exposed to a magnetic field, the extreme ends of the reeds assume opposite polarity, as shown in Fig. 4-21(a). The attraction forces of the opposing magnetic poles overcome the reeds' stiffness, causing them to move toward each other and close. Removal of the magnetic field will return the reeds to their open position. This action can be done at high speed and repeated millions of times. The capsule may contain form B or form C contacts (normally closed) with the normally closed side-biased magnetically with a small magnet, as shown in Fig. 4-21(b). The reeds may also be constructed to achieve a normally closed position mechanically. The first reed relays to appear had to be hand-wound, but today they are available with up to 4 poles and with standard coil voltages of 5, 6, 12, and 24 V. Electrostatic shielding is incorporated around the reed capsules to reduce pick-up and cross-talk. These relays are now available in DIP packages for standard IC sockets. High-voltage techniques permit some of them to withstand kilovolts and to have closed-contact resistances of 10 Ω or less.

The original mercury-wetted reed relays were formed with a movable reed having its base in a pool of mercury and the end arranged to move between two sets of stationary contacts. They had many disadvantages such as poor resistance to shock and vibration, and they had to be operated in a vertical position. Recent developments in mercury-wetted contacts have overcome these disadvantages. These relays offer, as their principal advantages, no contact bounce and low, stable contact resistance. Follow Fig. 4-22 as high-speed pictures 1 through 5 show the mercury-wetted contact action for an SPDT bridging (make-before-break) form C. The mercury is shown in black.

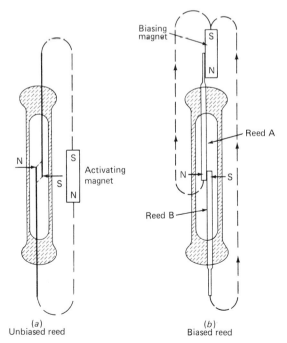

Fig. 4-21 Magnetic reed switches.

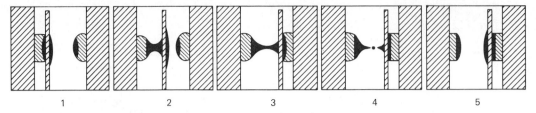

Fig. 4-22 Mercury wetted reed contact action.

The mercury plunger relay has contacts that are hermetically sealed in a glass or metal envelope inside a coil. Both normally open and normally closed contacts are available. These relays can handle up to 50 A. The sealed structure ensures that the contacts are free from dust and air-borne contaminants. This type of relay is position-sensitive and is not useful in vibration environments.

The mercury displacement relay (MDR) shown in Fig. 4-23 provides a self-renewing contact with a single moving part. There are no pins and hinges as in conventional relays, and the liquid mercury provides a new contact surface with every actuation. The MDR is very quiet, considering the current it will handle. (Note: If an MDR breaks, use extreme caution for clean-up because mercury vapor is highly toxic, and its disposal must follow accepted safe practices.)

Latching relays are available in two types: magnetic latching and mechanical latching. They may be of the power, telephone, or even the reed type. Magnetic latching usually employs permanent magnets to make the contacts bistable. Thus, after coil power is removed, the armature is magnetically held in the actuated position. Reset action is accomplished by applying either a voltage of proper polarity to a separate reset coil or a reverse voltage to the actuating coil. Mechanical latching relays are of two basic styles: mechanical reset and electric reset. The *mechanical reset* type employs a coil and armature with a reset mechanism. The mechanism locks the armature in the operated position after the coil has been deenergized. Reset is accomplished by manually tripping the locking mechanism. *Electric reset* types employ a second coil and armature to trip the latch mechanism for resetting to the original position. Some vendors call the latching relay an *impulse, alternating,* or *flip-flop* relay for obvious reasons. Solid-state relays will be covered in Chapter 14 on optoelectronics.

A few special relays, though infrequently encountered, are worth mentioning. The *RF,* or *coaxial,* relay is designed to attach directly to coaxial cables and match the cable impedance of 50 to 75 Ω. Radio frequency (RF) losses are minimized by the use of glass silicone and melamine for insulation.

Sensitive relays require only small amounts of power to operate (usually less than 100 mW). They can employ reeds or dry circuit cross-bar contacts. Low-voltage relays with coil resistances around 1kΩ can be directly driven by MOS, TTL, or HTL de-

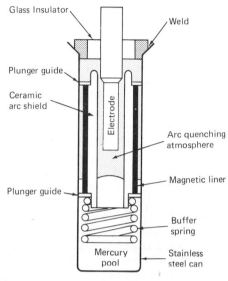

Fig. 4-23 Mercury displacement relay.

vices (covered in Chapter 11) without the need for buffers.

REVIEW QUESTIONS

7. Magnetic action of the armature closes the contacts in Fig. 4-12, and the _____ opens the contacts.

8. Is pole shading done on ac or dc relays?

9. Is arcing of contacts more predominant in ac or dc load circuits?

10. Metal transfer not only degrades contacts but can cause them to _____ closed.

11. Which coil suppressors shown in Fig. 4-16 should not be used in circuit Fig. 4-20(*a*) or 4-20(*b*)?

12. By magnetically biasing the reed in Fig. 4-21(*b*), form _____ contacts are realized.

4-3
TIMERS

We will now look at time delay relays (TDRs). Electrothermal or thermostatic delay relays rely on a heating element to provide a temperature differential for mechanical expansion to open or close a contact.

Originally built to a standard tube-type base, they are now available in miniature versions for PCB applications. Though in use for many decades, they still offer several advantages, including operation on alternating or direct current, small size and weight, simple construction, low cost, and long life. A big advantage or disadvantage is the recycle time. The cooling time can be five to ten times the actuating time, which renders them undesirable for use in some circuits. Delays of 2 to 300 s are very common, and some models are adjustable. Time delay relays can also serve as flashers. In a flasher service, the heater circuit is opened with the load circuit, and a 30 percent duty cycle is typical.

The electropneumatic form of TDR (also called the *dashpot time delay relay*) consists of a relay which has a plunger connected to a dashpot. The *dashpot* is a device that employs either a gas (such as air) or liquid to absorb energy and retard the moving parts so as to produce a time delay. When the coil is energized, the solenoid plunger pulls out the dashpot piston at an adjustable rate. At the end of its travel, the plunger completes the magnetic circuit and snaps the contacts closed. The contacts are held by full magnetic power until the coil is deenergized; then drop-out is instantaneous. The gas or liquid used for delaying is usually reused in a closed loop, thereby eliminating outside contaminants.

The addition of a copper collar slug on the telephone-type relay previously discussed delays the energization and sustains deenergization of this type of relay. For maximum pick-up delay, the slug is placed on the armature end of the core. Pull-in delays up to 150 ms and drop-out delays of 500 ms can be obtained. The electromechanical type of TDR is shown in its basic form in Fig. 4-24. A small synchronous ac motor drives a cam through an electromagnetic clutch and gear train. When the cam has rotated far enough, it operates a switch which controls the load. As long as the power-line frequency is constant, the time rotation of the cam will be dependent only on the power-line frequency (not voltage), thereby making the synchronous-motor TDR one of the most stable types. The delay range of the synchronous TDR is of the order of 50 to 1, for a given gear ratio. Delays of 10 s to several days can be obtained, with resolution on the order of 2 percent.

Adjustability is achieved by a mechanism that allows the user to rotate the switch with respect to the cam. Reset is accomplished by releasing the clutch, so a spring mechanism can return the cam to its initial position, and requires 100 ms or more. Some units have automatic reset and will recycle if a sustained circuit is used or wait for another start to begin another cycle. This type is known as the *interval timer*.

The *multicam timer* is an extension of the synchronous-motor TDR, with or without the clutch, but has no reset. A *cam* is a set of circular plates mounted on the shaft coupled to the motor by a gear assembly. Each cam is independently adjustable for

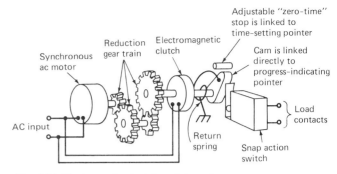

Fig. 4-24 Electromechanical time-delay relay.

2 to 98 percent of the total cycle time. Total revolution time is determined by motor speed and by the gear train assembly. A wide variety of models are available to suit a wide range of applications. A most common type with a push-to-start switch and a friction clutch is found in clothes dryers, dishwashers, and industrial applications such as chemical and other automatic processors that are cyclic in nature.

Electronic timers have been around for many years. Early designs used the vacuum tube, then the thyratron tube, for the active element used for timing. With the arrival of solid-state electronics (diodes, transistors, thyristors, and integrated circuits) vacuum and thyratron tubes have been mostly replaced in timing applications. Solid-state circuits have offered an unparalleled reliability, efficiency, and flexibility for use as control devices, in an ever-widening range of applications. The *RC*/threshold method is considered the oldest class for electronic timing. Figure 4-25 shows a typical threshold-type circuit using a unijunction transistor with an SCR. Although there are many variations of the circuit described here, the characteristics are attributable to all. With S_1 as shown, the capacitor voltage is at zero. Activating S_1 applies the reference voltage across the resistor and capacitor (R and C). The voltage across the capacitor rises to 63 percent of the reference voltage in $R \times C$ seconds, or one time

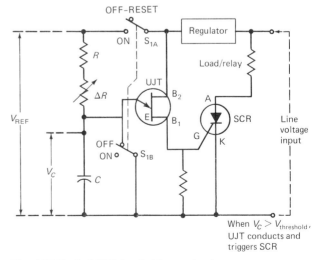

Fig. 4-25 Typical UJT threshold-type circuit.

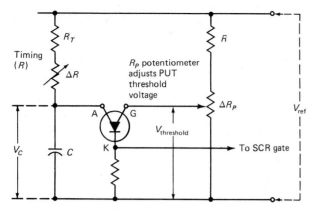

Fig. 4-26 PUT threshold timer circuit.

constant. This product is

$$T \text{ seconds} = R \text{ ohms} \times C \text{ farads}$$
$$1 \text{ s} = 1 \ \Omega \times 1 \text{ F}$$

If we wait longer, the voltage will continue to rise higher: to 86 percent at $2RC$ and to 95 percent at $3RC$. The curve of voltage against time is not linear but exponential. The larger the time constant, the longer it will take to reach the threshold voltage. Also, the closer the reference voltage is to the threshold voltage, the longer it takes to reach it. Let's look at some numbers. If the threshold voltage were 63 percent of the reference voltage, then one RC time constant would be required. If we use $R = 1 \text{ M}\Omega$ and $C = 1 \ \mu\text{F}$ then the time is

$$T = RC$$
$$= (1 \times 10^{6}) \times (1 \times 10^{-6})$$
$$= 1 \text{ s}$$

What if we want 60 s (1 min)? Then RC must be increased 60 times. A 60-MΩ resistor is not practical because of the loading of the UJT on the charging

circuit. If we need 1 h (60 min or 3600 s), we must increase C to 3600 μF. Such capacitors are large and expensive and have high leakage. We would have to use electrolytic-type capacitors, which vary in value and leakage with temperature changes. With aging, capacitors and resistors change value and drift. All in all, it is not unreasonable to expect a ''worst-case'' error of 30 percent in RC/threshold designs. The unijunction in Fig. 4-25 could be replaced by a programmable unijunction transistor (PUT), permitting a variation of the threshold voltage as shown in Fig. 4-26. About 90 percent of RC/threshold timers use either a PUT or UJT.

In the early 1970s, the integrated circuit (IC) timer was introduced; it still enjoys popularity among designers. This timer's success can be attributed to several inherent characteristics: versatility, stability, and low cost. The simplicity of the timer, along with its ability to produce long time delays, has lured designers from using mechanical timers and discrete circuits and into the ranks of IC timer users. The timer consists of two voltage comparators, a bistable flip-flop, a discharge transistor, and a resistor divider. To understand the basic concept, examine Fig. 4-27. This circuit shows the time delay mode, where the output changes state at some design time after a trigger signal is received. An internal resistive divider sets the two comparator levels. Since the resistors are of equal value, the threshold comparator is referenced at two-thirds of the supply voltage and the trigger comparator at one-third of the supply voltage. When the trigger voltage is moved below one-third of the supply, the trigger comparator changes state and sets the flip-flop, driving the output to a high state. An external capacitor can be connected to either the threshold comparator or the trigger comparator, depending on the desired mode of operation. Figure 4-27 shows the IC configured for time delay operation (note that the capacitor is connected to

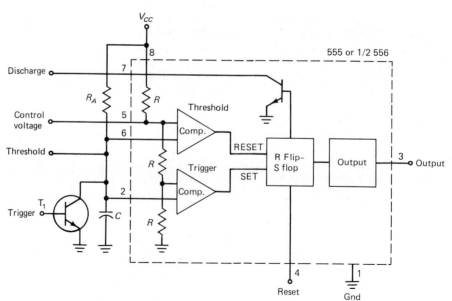

Fig. 4-27 Integrated circuit timer block diagram.

both comparators). The internal discharge transistor is not used in the time delay mode. The operation begins by turning on the external trigger transistor, which grounds the capacitor. The trigger comparator now sees a low state and sets the timer output high. When the transistor is turned off the capacitor begins charging; when it reaches the threshold (two-thirds V_{cc}) the threshold comparator trips and the output switches low and remains low until the external trigger transistor is turned on again. Thus time delay is accomplished by having the output go high, and, after a prescribed time interval, the output goes low.

A few details about a standard IC timer, the 555, are worth mentioning. Because of its circuitry, the 555 IC will only trigger on a negative-going edge of the input pulse. It is necessary that the trigger input be returned to a voltage greater than $1/3 V_{cc}$ or that ac coupling into the trigger input be used. The control voltage is brought in at the two-thirds point of the reference divider. Imposing an external voltage at this point can vary the timing. This feature makes it possible to use the timer as a voltage-controlled oscillator or pulse-width modulator. When this control feature is not used, the control voltage input should be bypassed to ground with a ceramic capacitor. The reset overrides all other functions of this timer.

With the arrival of large-scale ICs, mainly because of metallic oxide semiconductor (MOS) technology improvements, IC timers have been improved along with other integrated circuits. The programmable timer/counter of Fig. 4-28 is a typical example. The "programmable" feature increases its value for computer-controlled applications as well as others. The timing cycle is initiated by applying a positive-going trigger pulse to pin 11. This trigger activates the time-base oscillator, enables the counter section, and sets all counter outputs to the "low" state. The time-base oscillator generates pulses with a period equal to RC. These clock pulses are counted by the binary counter section. The timing cycle is completed when a positive-going reset pulse is applied to pin 10. The timing sequence of output waveforms at various circuit terminals after a trigger signal is shown in Fig. 4-28. When in the reset state, both the time-base and the counter sections are disabled, and all counter outputs are at "high" state. In most timing applications, one or more of the counter's outputs are connected back to the reset terminal (S_1 closed in Fig. 4-28). Connected this way, the circuit will start timing when a trigger is applied and will automatically reset itself to complete the timing cycle when a programmed count is complete. If none of the counter outputs is connected back to the reset terminal (S_1 open), the circuit will operate in its astable or free-running mode after an input trigger. The binary counter outputs (pins 1 to 8) are open collector type stages and can be tied together to a common pull-up resistor to form a *wired-or connection,* which simply means that the combined output will be low as long as any one of the outputs is low. This allows the time delay associated with each counter output to be summed by

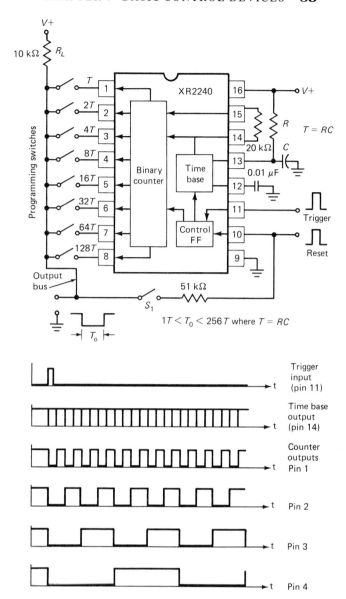

Fig. 4-28 Programmable timer/counter circuit block diagram and timing waveforms.

tying them together to a common bus, as shown in Fig. 4-28. For example, if pins 1, 5, and 6 were tied to the output bus, the total time delay would be $(1 + 16 + 32)T = 49T$. In this manner, one can program the timing cycle to be from 1 to $255T$, where $T = RC$.

When power is applied with no trigger or reset inputs, the circuit reverts to its "reset" state. The binary counter outputs are high or in a nonconducting state. Once triggered, the circuit is immune to additional trigger inputs, until the timing cycle is completed or a reset input is applied. During a timing cycle, the outputs change in accordance with the timing diagram of Fig. 4-28. The outputs can be used individually or can be "wired-or" as previously mentioned. The period T of the time-base oscillator can be modulated by applying a dc voltage to pin 13. The time-base oscillator can also be synchronized by applying a sync pulse to pin 12. The time-base period is determined by the external RC network connected to pin 13. The waveform at pin 13 is an exponential ramp with a period $T = RC$. The time-base output (pin 14) is an open collector stage and requires a 20 kΩ pull-up resistor to pin 15 for proper circuit operation. At reset, the time-base output is at high state, and subsequent to triggering, it produces a negative-going pulse train with a period $T = RC$. The time-base output is internally connected to the binary counter and in some applications may be used as an input for externally generated clock pulses.

REVIEW PROBLEMS

13. An electrothermal TDR operates on ac or dc voltage?

14. The dashpot of an electropneumatic relay speeds up actuation. (true or false)

15. Loss of liquid in the dashpot will _____ the delay.

16. To maximize the pick-up delay of a telephone-type relay, the slug is put on the _____ end.

17. What type of timer will wait to start another cycle?

18. What value of resistor is needed for an RC/threshold timer constant of 4 s with a 4 μF capacitor?

4-4
OVERCURRENT PROTECTION

Most electric and electronic circuits need some sort of safety valve for protection from dangerous overloads. The oldest and most common type of protector is the fuse. The common plug-type fuse was originated by Thomas Edison in 1890 and is available at current ratings up to 30 A at 125 V. The high currents that can be present during a short circuit can cause electrical damage to equipment, mechanical damage, and possibly an extremely hazardous fire risk. Therefore, it is extremely important not to bypass, short-circuit, or oversize the replacement of a blown fuse.

The fuse is a device which protects by melting (fusing), thereby opening the circuit in response to an overload or short-circuited current. The typical electronic fuse utilizes silver wire or some other metal link housed in a package that is often cylindrical. All fuses are temperature-sensitive devices and are rated for an ambient temperature of 25°C and must be derated as the temperature increases. This temperature requires good contact connections and air movement to achieve the predicted life.

Fuses have three important specifications—current rating, voltage rating, and fusing characteris-

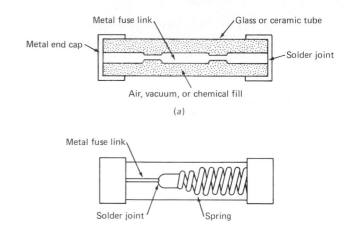

Fig. 4-29 Fuse construction. (*a*) Fast or medium acting fuse. (*b*) Slow action fuse.

tics—and are manufactured in a wide variety of styles and shapes. The glass and ceramic tube fuses are popular for protecting electronic equipment and are shown in Fig. 4-29.

The current rating of a fuse is expressed in RMS amperes or fractions of an ampere. Fuse ratings are established by a controlled set of tests referenced by the Underwriters Laboratories, Inc. It is extremely important to remember that fuses are sensitive to current, not voltage. A fuse will not blow (open) until its melting temperature is reached. The current rating is always stamped or otherwise indicated on the fuse body.

The voltage rating of a fuse should be equal to or greater than the circuit voltage. This rating is covered by National Electrical Code (NEC)® regulations and the Underwriters Laboratories (UL) to prevent fire risks. Ratings of 32, 125, and 250 V are common and adequately cover most electronic apparatus. The voltage ratings apply to an interrupting capacity of 10,000 A. This is a large safety factor for the type of short-circuit currents likely to occur in electronic equipment.

The *normal-blo* (standard) fuse is used for resistive or non-surge-type loads and should be replaced with the same type when being renewed. This type of fuse is mainly used for protection against short circuits. The chart in Fig. 4-30 shows the electrical characteristics of normal-blo fuses. Note that the fuses shown can withstand a 10 percent overload for at least 4 h. This style of fuse can have a glass or ceramic body, as shown in Fig. 4-29. Glass fuses give a better visual indication, but the ceramic fuses are more rugged to abuse and will not shatter on extreme overloads.

The second type of fusing characteristic is the *slo-blo*. This type may also be called *time delay* or *dual element*, depending on the manufacturer. These types provide a time delay for overload currents greater than 50 percent. They are useful in protecting surge-type loads such as motors and capacitors. Slow-acting fuses have two blowing mechanisms;

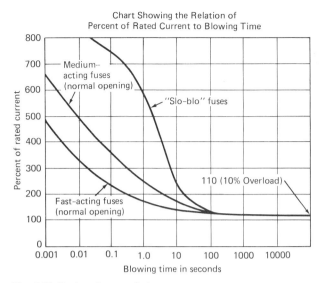

Fig. 4-30 Fusing characteristics.

Fig. 4-29(*b*) shows that a solder joint is under spring tension. A moderate overload will eventually melt the solder, and the joint will pull apart, interrupting the circuit. An extreme overload will melt the metal fuse link and provide a quick response. Compare the difference in Fig. 4-30 for the blowing time between a normal-blo and a slo-blo fuse for the same percentage of overload. For example, the blowing time at 500 percent is 0.01 s for the normal fuse and 2 s for the slo-blo type.

Semiconductor fuses are used to protect diodes, thyristors, and other semiconductors that have little thermal storage capacity and cannot handle heavy overloads for any length of time. This type of fuse is extremely fast-acting compared to normal-blo of the same ratings as indicated in Fig. 4-31, which covers currents up to 10,000 A. It is especially important to use an exact replacement when renewing semiconductor fuses. Note that the chart is a log-log scale so the numbers change rapidly along the time and current axes. The fuses used in some semiconductor circuits may be rated as high as 1000 A. Figure 4-32 shows the typical location of fuses in a solid-state power supply application. Fuses F_1 and F_2 placed in series with semiconductors and the supply lines are a protection against internal faults and overload conditions. Service interruptions cannot be tolerated in large rectifying equipment in some industrial applications such as electrochemical applications. The three-phase bridge of Fig. 4-32 is a usual method of including one or two redundant parallel paths (shown in dashed outline) to maintain uninterrupted operation. If a rectifier fails, the fuse opens, disconnecting the short-circuited rectifier without damage to the equipment or interruption of service. Fuse F_3 is protection against an external short circuit or excessive load currents at the output; F_2 protects against short circuits and overloads.

With international trade more commonplace today, a word about international fuses is appropriate at this

time. Most countries have certification by agencies similar to UL here in the United States. The following are acronyms for international, European, and Canadian agencies:

International Electrotechnical Commission (IEC)

Switzerland (SEV)

Verband (West German) (VDE)

Svenska (Sweden) (SEMKO)

Canada (CSA)

The European and Asian fuses are 5 by 20 mm and are not the same as the 1/4 by 1 1/4 in. 3AG fuses that are commonly used in the United States. These fuses are 6.3 by 32 mm. Trying to install a 3AG to renew a 5-mm and vice versa can cause a hazardous situation and should not be attempted. Also note that most foreign fuses are rated at 250 V because that is the primary operating voltage of their main line circuits, just as 120 V is the standard in this country. If foreign equipment is involved, observe fuse sizes and ratings carefully and judiciously.

Circuit breakers are automatic overcurrent protectors designed to open under overload conditions. We will cover the general purpose types applicable to electronic equipment. Circuit breakers are mainly limited to three poles. The simple thermal circuit breaker shown in Fig. 4-33 responds to temperature changes of its internal bimetallic element. This element is in series; if the load current increases to

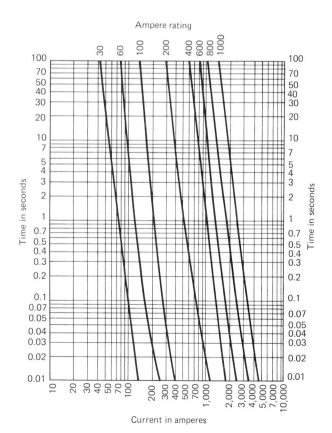

Fig. 4-31 Fuse characteristic curves.

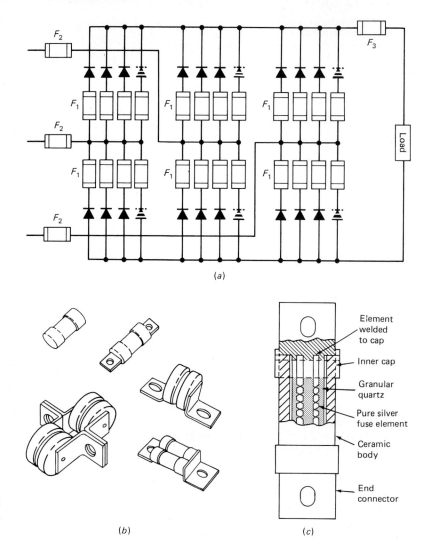

(a)

(b)

(c)

Element welded to cap

Inner cap

Granular quartz

Pure silver fuse element

Ceramic body

End connector

Fig. 4-32 Semiconductor fuses. (a) Placement. (b) Package. (c) Cross-section.

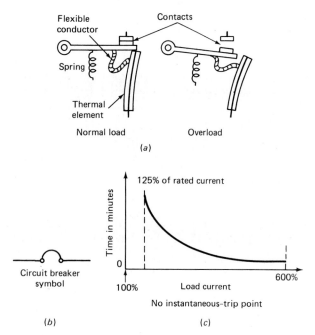

Flexible conductor

Contacts

Spring

Thermal element

Normal load   Overload

(a)

Circuit breaker symbol

(b)

125% of rated current

Time in minutes

0

100%   Load current   600%

No instantaneous-trip point

(c)

Fig. 4-33 Thermal breaker. (a) Action. (b) Schematic symbol. (c) Characteristic curve.

about 125 percent of the breaker's rating, it bends and the contacts open. To provide additional protection, some types of thermal breakers contain a built-in trip-free mechanism which allows the contacts to open, even if the breaker's lever is manually held reset. The performance curve in Fig. 4-33(c) shows that the thermal breaker has no true instantaneous tripping point. This limitation precludes the simple thermal breaker from most applications.

The *thermal-magnetic circuit breaker* and its performance curve are shown in Fig. 4-34. This breaker operates by using the same principle as the simple thermal breaker for moderate overloads. On extreme overloads, the magnetic feature comes into play and takes over to assist in tripping the mechanism. The magnetic field created by the fault current attracts the magnetic plate and opens the contacts. This action provides the thermal-magnetic breaker with an instantaneous trip capability for extreme overloads. The thermal-magnetic types are therefore more desirable for most applications.

The *magnetic-hydraulic circuit breaker* consists of a solenoid with a *dashpot* (piston in fluid). The solenoid coil is wound so that the spring-loaded core

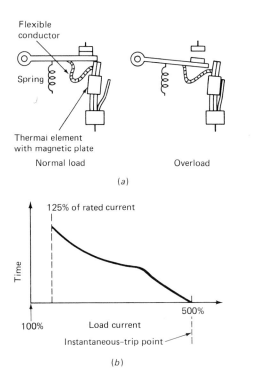

Fig. 4-34 Thermal magnetic breaker. (a) Action. (b) Characteristic curve.

does not move at normal currents. On heavy overloads (usually ten times the rating), the strength of the magnetic field attracts the armature to a cylinder cap to open the contacts instantaneously. At small, sustained overloads the field of the solenoid pulls the iron plunger into the core. As the core moves into the solenoid, it reduces the gap of the magnetic circuit and in turn increases the magnetic flux. The speed of movement of the piston is determined by the viscosity of the silicone fluid in the dashpot. This style of breaker has a time delay for moderate overloads. If the overload is removed before the piston gets to the top, the spring pushes the piston back to the bottom before the armature can trip the contacts. Since breakers are basically mechanical devices, it is a common practice to manually trip them occasionally to ensure their proper operation.

REVIEW QUESTIONS

19. What other fuse characteristic is important in addition to the current and voltage ratings?

20. Fuse current rating is always given in _____ amperes.

21. Fuses are _____ sensitive devices and therefore require good solid connections.

22. What type of fuse is used for motors or capacitive loads?

23. What is the main criterion of semiconductor fuse usage?

24. What is the major limitation of the simple thermal breaker?

4-5
SOLENOIDS, CONTACTORS, AND STARTERS

A *solenoid* is an electrically energized coil in which the turns of the windings are insulated from each other even if bare wire is used. They are the basis for all electromagnets. We are interested in the solenoid with a moving core called a *plunger*. This device will convert electrical energy into mechanical motion, which may be either linear or rotary. The basic linear solenoids are shown in Fig. 4-35. The plunger is the movable bar of high-permeability steel or soft iron and may be laminated for ac operation to reduce eddy current loss. Referring to Fig. 4-35(a), when the coil is energized the plunger becomes magnetized and mutual attraction takes place between the coil and plunger. The traditional solenoid is the iron-clad or box style shown in Fig. 4-35(b). The solenoid and plunger are provided with an iron or steel jacket, which is usually laminated for ac operation. The effect of the frame is to increase the magnetic and mechanical forces slightly at the initial position and much more so at or near the final position of the plunger. In the magnetic-cushion type, Fig. 4-35(b), the plunger cannot strike the opposite end of the open frame as it passes into the coil, thereby eliminating the hammer blow effect which occurs in iron-clad solenoids. This effect is utilized in electric hammers, but in other applications the constant hammering is detrimental to the solenoid and the noise is objectionable. This effect occurs on the solenoids shown in Fig. 4-35(c) and 4-35(d). In Fig. 4-35(c), the iron-clad solenoid has its plunger travel limited by the closed frame. The type illustrated in Fig. 4-35(d) is known as the *stopped iron-clad solenoid* because of the iron or steel plug or stop extending downward toward the plunger to increase the pull at or near the final position and also to increase the seated pull.

One of the important factors considered in selecting a solenoid is the work requirement for the solenoid. The force required by the load must not exceed the force developed by the solenoid during any portion of its travel; if it does, the plunger will not pull

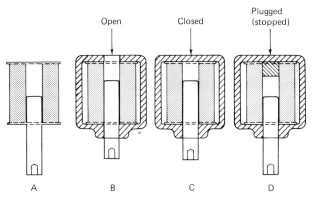

Fig. 4-35 Solenoids and plungers.

Volts—60 Hz	120
Duty	Cont.
Coil resistance (Ohms @ 25°C)	88
Watts seated	9.5
Amps seated	0.24
Amps ⅛″	0.72
Amps ⅜″	1.0
Amps ⅝″	1.22

Fig. 4.36 Sample of solenoid specifications.

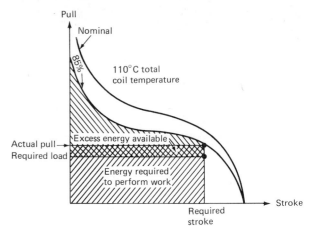

Fig. 4-37 Typical ac solenoid pull-stroke curve.

in or seat. With an ac solenoid, this could burn out the coil. The coil reactance will not reach its intended maximum because the core (plunger) does not come into position. Excess current will flow in the coil. Never select or replace with an overrated solenoid because it will develop more force than is required by the load. The extra energy must be dissipated by the plunger or field piece on impact, causing premature failure. Proper alignment of plunger and load is also important. Loading which is not centered along the line of plunger travel should be avoided or corrected. If misalignment occurs, plunger wear increases, and with ac solenoids, seating will be impaired. Where an ac laminated plunger is directly linked to a mechanical load, jamming of the plunger could burn out the coil.

With ac solenoids, an inrush current occurs until the plunger is seated. The table in Fig 4-36 gives typical values for an ac solenoid. As can be seen, the longer the stroke the higher the inrush current, and the inrush continues until the plunger is seated. The operating limit of any solenoid is determined by the temperature rating of the insulating materials used in its coil. As long as this maximum is not exceeded, the solenoid's pull characteristics will not be materially degraded. With modern insulation materials, this maximum temperature is usually 110°C. As the coil temperature increases, pull decreases. A widely used rule of thumb is to select a solenoid which will deliver a slightly greater force at 110°C than the load requires for the anticipated coil voltage as shown in Fig. 4-37.

The *duty cycle percentage* is defined as the ratio of solenoid coil ON time to the period of one cycle (ON plus OFF time). For example, if a solenoid is energized for 1 min and deenergized for 4 min the duty cycle will be 20 percent, as follows:

$$\text{Duty cycle, \%} = \frac{\text{ON time} \times 100}{\text{ON time} + \text{OFF time}}$$
$$= \frac{1 \times 100}{1 + 4} = 20\%$$

A majority of intermittent-duty cycles fall into three categories: 50, 25, and 10 percent. Most solenoids operated for up to 30 s ON time fall into the 50 percent category. The 30-s to 3-min range falls into the 25 percent category, and the 3- to 5-min maximum ON time would most likely fall in the 10 percent category.

The continuous-duty solenoid may remain permanently energized at rated voltage without exceeding its maximum temperature rise. A continuous-duty dc solenoid may be subject to continuous ON-OFF cycles without overheating; not so with the ac solenoid because of its inrush current. As Fig. 4-36 shows, high coil current occurs until the plunger seats. The longer the stroke the higher the inrush current. In continuous-duty ac solenoids, subject to ON-OFF cycling, repetitive surges could cause overheating. A reduction in duty cycle or the use of a dc solenoid is necessary. The intermittent-duty solenoid is designed to be energized for a maximum of 5 min without excessive overheating. After 5 min it must be permitted to cool before resuming operation. In some cases thermal cut-out devices are incorporated as a fail-safe measure for excessively high ambient temperatures or overloads which could cause the plunger to stall.

The speed of operation is a function of the applied load and power. An ac solenoid's operating speed will vary, depending on the point of the applied voltage cycle. If a consistent operating speed is required, a dc solenoid should be selected.

Push-type solenoids are usually modified pull types and can use alternating or direct current. They have a nonmagnetic pusher rod projecting through their stops with the plunger protruding out the other end.

We must look at the rotary solenoid with its unique style of operation. Smooth rotary motion is achieved by the use of ball bearings and a varying pitch ball race. The solenoid can turn, step, index, lock, punch, or lift in milliseconds. These units can have right- or left-hand strokes and can weigh up to 5 lb and have starting torques as high as 100 lb · in. for a 25° stroke. In ordinary electromagnets, the magnetic pull increases sharply as the air gap closes, but in the rotary solenoid, this is compensated for by the compound angle of incline of the ball races. The incline of the ball races is steep at the beginning of the rotary stroke and gradually decreases as the balls approach the deep end of the ball races. This arrangement amplifies torque at the start of the rotary stroke, where it is usually needed.

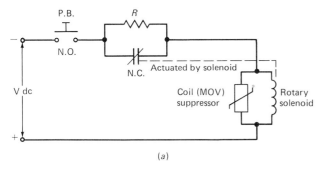

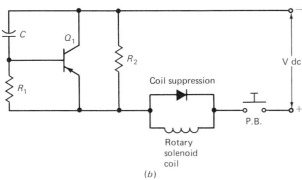

Fig. 4-38 Hold-in current-reduction circuits.

TABLE 4-1 CONTRACTOR RATINGS

Motor Rating* 600 V	Resistive Rating* 600 V	Horsepower* 120 V	
30 A	40 A	1.5	(1φ2P)
		3	(3φ3P)
40 A	50 A	2	(1φ2P)
		5	(3φ2P)

* At 25°C, 60 Hz.

Contactors are, in general, constructed in the same way as relays. Normally, they are supplied with two, three, or four N.O. contacts. They are used for switching power to control motor loads in refrigerating, air conditioning, heating, lighting, and ventilating systems. Applications also include controlling heating elements and primary power to an assembly such as a copy machine or a computer. Ratings are generally listed by continuous amperes or horsepower. A typical listing will look like the one shown in Table 4-1. The symbol 1φ2P denotes single phase with 2 poles and 3φ3P denotes three-phase with 3 poles. Contacts are usually field-replaceable. As an option, an auxiliary snap-action switch is available for low-power switching or coil interlock and latching. With the use of a control transformer, the contactor coil may be operated at a lower voltage, as shown in Fig. 4-39. The addition of the control transformer is desirable in some instances to keep high voltages off the switches and lights, and away from the operators. This simple circuit provides a means of energizing the contactor. Depressing the momentary start button completes the circuit; the contactor energizes, the light (R) lights, and the associated power is applied to the load. The contact's ICR (usually auxiliary) will now close, holding the circuit on when the momentary start button is released. This condition will persist until the operator intervenes by momentarily pushing the stop button. In some cases, a limit or thermal switch may be put in series with the stop button or hold-in contacts (ICR) for protection. One disadvantage of the second arrangement is that an operator can hold the start button in and override the interlocks if they are wired in series with the hold-in contacts.

Contactors can be ac- or dc-operated, that is, for the line or the actuation coils. In dc operation, a magnetic blowout coil wound on a steel core is mounted between pole pieces. The blowout coil is connected in series with the contactor and carries the dc load current with the contactor closed. The current sets up a magnetic field through the core and the pole pieces. When the contacts open, the arc is magnetically forced up and away from the contacts. This lengthening and subsequent extinguishing of the arc is rapid and greatly reduces wear and contact burning. *Contact wear allowance* is the total thickness of material that may be worn away before the contact surfaces become ineffective. The contacts

As for all the other electromagnetic devices discussed, the temperature is a very important factor. As the temperature of the coil rises, so does its resistance. Increased resistance reduces current flow with a constant voltage source, consequently decreasing the output torque. Heat can be dissipated by controlling the air flow, by mounting the solenoid on a large surface (heat sinking), or by resorting to duty cycle limiting, which is how they are rated by the manufacturers. One way to decrease the temperature rise in a rotary solenoid is to use a hold-in circuit to reduce current to a point at which torque is sufficient to maintain the solenoid in the energized position. One common method to reduce coil current during hold-in is a normally closed (N.C.) switch in parallel with a hold-in resistor, as shown in Fig. 4-38(a). When the pushbutton (PB) closes the circuit, full voltage is impressed across the solenoid coil, resulting from the bypassing of the resistor by the N.C. switch. As the rotary solenoid approaches the end of its stroke, a mechanical connection opens the N.C. contacts. This reduces the solenoid voltage and lowers power dissipation. Figure 4-38(b) shows another technique for reducing hold-in current. When the PB switch is first closed, current flows through the base-emitter junction while charging capacitor C to the input voltage. This base current turns on the emitter-collector circuit and allows full power to be impressed across the solenoid. The hold-in continuous current is supplied through R_2 thus limiting dissipation in the solenoid. When switch PB is released, the solenoid is released and the capacitor is discharged by R_1 and R_2 to prepare the circuit for the next activation cycle.

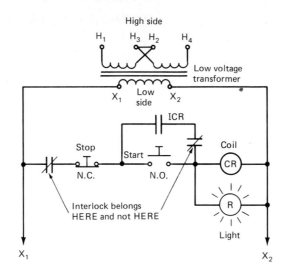

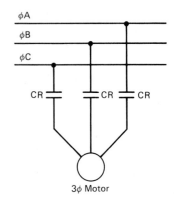

Fig. 4-39 Simple pushbutton contactor circuit.

should be renewed when worn below the amount specified by the manufacturer.

Starters can be of the manual or magnetic type. *Manual contacts* are closed by operator force. Today's machines often require the convenience and safety of remote control. The most common manual type employed today is the drum-type switch. A *drum switch* consists of stationary contact fingers held by spring pressure against contacts on the periphery of a rotating cylinder or sector. Their use is in applications in which the starter is operated frequently, as in a machine shop by a machinist. A reversing drum switch's internal connections are shown in Fig. 4-40(*a*). A stop can be inserted or removed by installing the proper face plates for reversing or ON-OFF operation. By connecting, as shown in Fig. 4-40(*b*), a three-phase, three-wire motor can be reversed. Because no starting windings, capacitors, or centrifugal switches are involved, reversing can be almost instantaneous. When used for trolley- and crane-type operation, dynamic breaking may be included in the starter switch.

The *magnetic starter* is very similar to the magnetic contactor in design and operation. Both have the feature of operating contacts when the coil is energized. The important difference between contactors and starters is the use of overload heater elements in the starter. There are motor starting switches, for controlling small ac or dc motors up to 2 hp, that are equipped with their own overload protective devices. It is not always possible to control the load that is applied to a motor. As a result, the motor may overheat, and serious damage can occur. For this reason, overload heater elements are added in the motor starter. The same current that is causing the motor to overheat is also going through the thermal element. We will now and in the future refer to these elements as *thermal overload relays,* as they are designated by industrial convention. They are available in the bimetallic type and the fusible-alloy type. The *bimetallic* type has two heaters in series with the circuit to be protected, and above these

heaters are two bimetallic strips. Heating will activate the bimetallic strips, release, and cause the contacts to open. The *fusible-alloy* type has two heaters, each surrounding a thermal element consisting of a small tube, inside of which is a loose-fitting shaft. The tube and shaft are joined together by a low-melting alloy. On overload, the increased current melts this alloy, allows the shaft to turn, and open the contacts. All overload relays should be *trip-free*. This means that they cannot be held reset, causing damage to the motor. If an overload trips, the cause should be investigated and removed before the reset is actuated to put the starter back into operation. The overloads are selected to have an ultimate tripping current of 15 percent overload on the motor. Figure 4-41 shows a typical three-phase motor starter circuit diagram with thermal overloads included. Pushing the start button energizes the CR coil and closes all CR contacts. The motor runs until the stop button is pushed or until an overload activates the thermal elements in line 1 or line 3, opening one of the overload (O.L.) switches in series with the CR coil. If it is the fusible type, it must be replaced

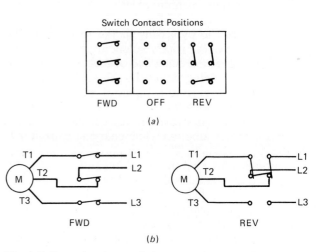

Fig. 4-40 Reversing drum switch.

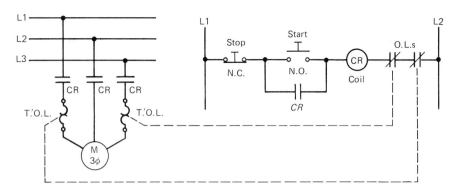

Fig. 4-41 Starter circuit for three-phase motor.

before resetting can be done. The bimetallic type can be reset after it cools and proper investigation of the cause has taken place. A commercial full-voltage starter is shown in Fig. 4-42.

Reduced-voltage starters are divided into several types:

1. Autotransformer
2. Primary resistor
3. Wye-delta (or star-delta)
4. Part winding

Why use a reduced-voltage starter? In connecting a motor directly across the line, the in-rush (starting) current may be on the order of five to ten times the full load current of the motor. This high current often will cause large line disturbances and excess stress on components, connections, and wiring. The basic principle of the reduced-voltage starter is to apply some percentage of the line voltage to start the motor. After the motor starts to rotate, it is switched or incremented to full line voltage. In the *autotransformer* type, starting steps are usually 50, 65, or 80 percent of full voltage. It should be noted that at reduced voltage, the output torque available is also reduced. This effect is important when the motor is starting under a heavy load.

For example,

50 percent voltage tap, the torque is 25 percent.

65 percent voltage tap, the torque is 42 percent.

80 percent voltage tap, the torque is 64 percent.

(Note: Torque varies as the square of the voltage.) The circuit diagram in Fig. 4-43, shows the contacts of a reduced voltage autotransformer starter. The two autotransformers are connected in open delta to provide reduced-voltage starting. The 50 percent taps have been selected for 25 percent starting torque. Five start contacts S and three run contacts R are required. The overloads have been omitted for clarity but would be included and are always necessary. A timing relay is operated by the starter contactor. The motor accelerates at reduced voltage through the S contacts; after a few predetermined seconds, the timer contacts close, deenergizing the starter (S) contactor and energizing the running (R) contactor. This also disconnects the autotransformer and puts full line voltage on the motor.

Figure 4-44 illustrates a *resistor-type starter*. Similar to the autotransformer type previously mentioned, it has the same sequence. After a preset time the run contactor short-circuits the line resistors and applies full line voltage to the motor. This type of starting gives smooth acceleration, because the mo-

Fig. 4-42 Full-voltage magnetic starter *(courtesy General Electric Co.).*

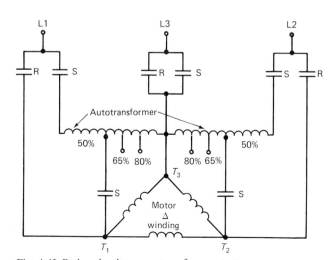

Fig. 4-43 Reduced-voltage autotransformer starter.

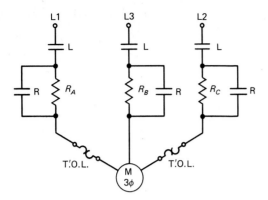

Fig. 4-44 Resistor starter.

tor is never disconnected from the line, but is very inefficient as a result of the high power loss in the resistors during starting.

When using the *wye-delta* (or *star-delta*) *reduced-voltage types,* all three-phase motor windings (six connections, T_1-T_6) must be available for connecting to the starter as indicated in Fig. 4-45. Figure 4-45(*b*) shows the equivalent start ciruit. This circuit is achieved by closing the S contacts and the LC contacts. After a preset delay, the S contacts open and the D contacts close, and the LC contacts remain closed. Figure 4-45(*c*) shows the equivalent run circuit. The S and D contacts are mechanically interlocked to prevent both from ever activating at the same time. This type of starting requires no extra components as the previously discussed types do, but the motor must be made for this type of starting. That is, all winding connections must be available.

The *part winding starter* also requires that the motor have all windings available (six or nine connections). There are six basic arrangements for this type

of starting, and it is best understood by referring to the diagrams and table of Fig. 4-46, in which the run (R) contacts and start (S) contacts are shown. The common two-step circuit which is illustrated provides a starting current equal to about 65 percent of the motor's normal locked rotor current for a starting torque of about 45 percent. The equivalent circuit connections for the table of Fig. 4-46 are shown. The starter is designed so that when the start contacts are energized the part winding of the motor is connected to the input lines. After about 4 s the second set of windings is connected, and the motor develops its normal torque.

The last type of starter we will investigate is the *reversing starter.* The reversing starter is, in effect, two starters, of equal size, used for a given motor application. To reverse direction of any three-phase motor any two line connections are interchanged, as with the drum switch of Fig. 4-40. It is a problem to connect the two starters to the motor properly so that the line feed from one starter is isolated from the other. Either mechanical or electrical or both types of interlocks are employed to prevent both starters from closing their line contacts at the same time. Figure 4-47 (p. 94) shows a commercial reversing magnetic starter. Note in Fig. 4-48 (p. 94) that only one set of overloads is needed to protect both forward and reverse connections. When the forward (F) button is actuated, the normally closed F contacts in the reverse circuit open, disabling the reverse start circuit until the stop button is actuated. And the converse is true for reverse (R) motion; note also the R contacts in the forward circuit. The interlocking described is the electrical type. Catastrophic failure would occur if both sets of contacts were allowed to close at the same time. As a double precaution, a mechanically interlocking type of reversing starter is

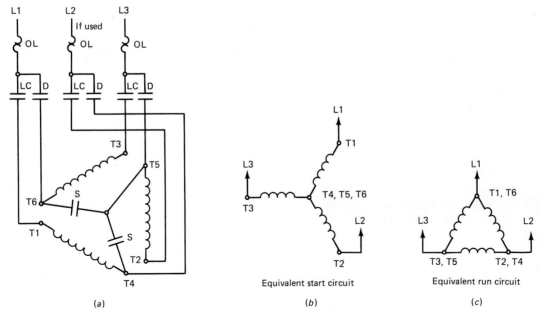

Fig. 4-45 Wye-delta starter. (*a*) Starter circuit. (*b*) Wye connection. (*c*) Delta connection.

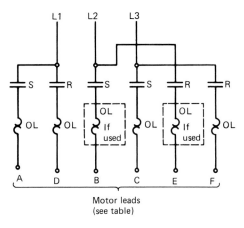

Motor Lead Connections

Part Winding Schemes		Lettered Terminals in Panel					
		A	B	C	D	E	F
1/2 Y or Δ	6 leads	T7	T2	T3	T1	T8	T9
1/2 Y	9 leads ○	T7	T2	T3	T1	T8	T9
1/2 Δ	9 leads □	T1	T8	T3	T6	T2	T9
2/3 Y or Δ	6 leads	T9	T8	T1	T3	T2	T7
2/3 Y	9 leads ○	T9	T8	T1	T3	T2	T7
2/3 Δ	9 leads □	T1	T4	T9	T6	T2	T3

○ Connect terminals 4, 5, and 6 together at motor terminal box.

□ Connect terminals 4 and 8, 5 and 9, 6 and 7 together in three separate pairs at terminal box.

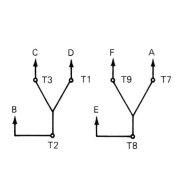

6 Leads-wye
AA

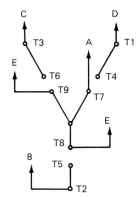

○ Tie T4, T5, and T6 together at box.
9 Leads-wye
BB

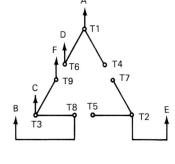

□ Tie T4 and T8, T5 and T9, T6 and T7, together in three separate pairs.
9 Leads-delta
CC

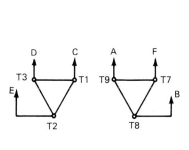

6 Leads-delta
DD

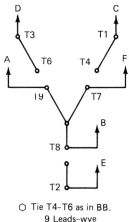

○ Tie T4-T6 as in BB.
9 Leads-wye
EE

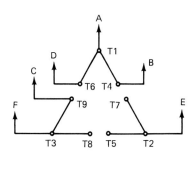

□ Connect pairs as in CC.
9 Leads-delta
FF

Fig. 4-46 Part winding starter.

available and is preferred by many designers and technicians.

REVIEW QUESTIONS

25. The "stopped iron-clad solenoid" shown in Fig. 4-35(d) is designed for increased mechanical _____.

26. If misaligned with its load, a solenoid will not pull in or _____.

27. A jammed plunger in an ac solenoid can _____ the coil.

28. What is the maximum ON time for a 33 percent duty cycle if the OFF time is 2 min?

Fig. 4-47 Magnetic reversing starter *(courtesy General Electric Co.).*

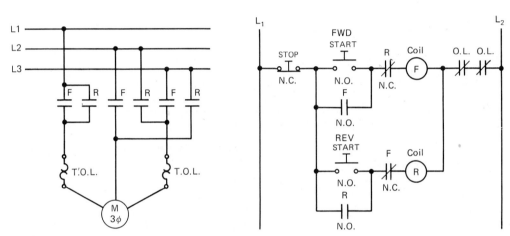

Fig. 4-48 Reversing starter.

4-6 TROUBLESHOOTING AND MAINTENANCE

Basic controls are the primary controls for most industrial equipment. Because most of these devices are line-operated, extreme care and safe troubleshooting techniques are necessary. Because of the nature of these controls the analysis of any problem has to be broken down as to whether it is electrical or mechanical. In some cases, problems may be both. The line voltage may be 110, 220, or 440 V so it is recommended that battery-operated test equipment be used. A buddy system is desirable when controls are remote from the device they are controlling. The best sequence of effective troubleshooting starts with a careful analysis of the problem. If resistance measurements are to be made or if it is necessary to disconnect leads, you must lock out circuits and tag them OFF for maintenance as a standard safety practice. Occasionally, a problem may be due to personnel. As long as people operate controls that operate machines and processes, problems that are not within the normal realm of troubleshooting techniques can arise. They may be due to a mis-

understanding or a lack of knowledge of the process. Whatever the problem, it should be recognized and handled carefully, efficiently, and as diplomatically as possible.

Visual inspection of all external controls, lights, fuses, breakers, and connections is the first order of business. Any unusual odors or signs of burning or overheating should be investigated immediately, especially where solenoids, transformers, and relay coils are involved. A blown fuse or tripped circuit breaker indicates the need to investigate, not merely to replace or reset the protection device. Fuses are temperature-sensitive, and therefore poor or loose connections of their holders can cause a failure not related to the existing circuitry. This is also true for circuit breakers and especially thermal overloads used for motor starters. Some overload units may be on the same frame as the motor or processor and thereby subject to constant vibration. This vibration can cause connections to loosen and produce intermittent connections or overheating in high-current circuits. Troubleshooting can be done with the power on (*hot*) or with the power off (*dead*). In the OFF circuit, the ohmmeter is one of the best ways to check continuity of a circuit. Parallel or branch circuits may have to be disconnected to eliminate alternate paths. Most toggle, slide, and snap-action types of switches can be checked this way with good results.

Rotary switches may make at some positions and not at others, so careful testing is necessary. Cleaning can be done with a sharpened pencil eraser or commercial spray cleaners. Be sure to follow the directions on the can. Caution must be observed because many contact cleaners are not safe to use on plastics and should be avoided. Many of today's switches have plastic parts, and instead of a simple cleaning job, a major replacement may be required if the wrong cleaner is used. Most thumbwheel-type switches can be checked with the aid of a schematic or a table such as the one shown in Fig. 4-6.

Keyboard types are best serviced with the use of an oscilloscope since a serial-type code may be produced on actuation. Some keyboards require whole blocks of key replacements, although others allow individual keys to be replaced. The service literature is the best aid in such circumstances.

Electropneumatic TDRs may develop leaks or bind internally and usually require replacement. A shortened time-out can be due to a loss of fluid or gas, as an increase in timing may be caused by dirt or grime on moving parts. If at all possible, remove the unit, if cleaning is indicated, to a vented hood, using the manufacturer's recommendations on cleaning.

Electromechanical TDRs are very susceptible to mechanical wear and fatigue of movable parts and need periodic lubrication. Some have covers to prevent dust, dirt, or vapors from contaminating the enclosure and should be replaced immediately during maintenance.

Most solid-state timers are virtually maintenance-free as far as their mechanical aspects are concerned. But semiconductors do fail and must be replaced. If the frequency of failures increases, the power supply should be investigated. Chapter 5 discusses power supply troubleshooting. Line transients (brief overvoltage condition) are another common cause of solid-state failures and should not be overlooked.

Relays should not hum or chatter when energized. In the case of dc types, the power supply should be checked first. The armature and gap should be free of all dirt and grime. Oil should never be on any part of a relay's magnetic circuit because it attracts and holds dust and dirt, which hinder the magnetic action of the armature.

Most contacts are plated with special materials and should not be cleaned with abrasive-type tools, such as files or emery cloth. Excessive pitting, as in Fig. 4-14, requires replacement of the relay or its contacts. Pitting may be due to failure of the suppression network or a load change from the original design, in which case a different suppression network may be required.

Reed relays are maintenance-free but do fail and need replacement periodically. They are magnetic devices and must not be subject to large stray magnetic fields; they have a low current capacity and will not tolerate overloading. Low supply voltage has a great effect on all relays and should be the first priority when doing checks with power on.

Solenoids are typically on-off, in-out, etc. devices. Any deviation such as noncompletion of a full stroke or sluggish operation should be investigated. Many solenoids are attached to valves that control air, water, oil, or other industrial fluids or gases. They can appear almost any place in a control system and are usually remotely controlled. Alternating current solenoids may hum slightly; however, an excessive hum usually indicates that the plunger is not seated properly. Remember the coil may overheat, as shown in Fig. 4-36, as the result of excessive current flow. Cleanliness when dealing with any electromechanical device cannot be overemphasized. Also remember to check the voltage on the coil because it may indicate a loose or intermittent connection and possibly a low voltage, which can cause poor actuation and result in eventual failure of the unit. A check of the dc resistance of coils is a good practice when looking for an open coil. Remember that ac coils rely on inductive reactance so their dc resistance will be less than that of an equivalent dc type. A current probe or clamp-on ammeter will give the best results to verify the manufacturer's indicated ratings.

Rotary solenoids are intermittent devices; if failures occur too frequently, the current limiting or timing network may be defective. For instance, if the capacitor (C) in Fig. 4-38(b) is leaky, the transistor (Q_1) may shunt the limiting resistor (R_2) and cause the coil to overheat, resulting in premature failure. Or an increase in the resistance of R_2 may cause the rotary solenoid not to hold; a simple ohmmeter test is best to verify this problem.

Common problems encountered with contactors and starters are loose connections and grounds. Most wiring to these devices is through conduits. Nicks and scrapes on wire insulation that occur during installation may produce problems at a later date. Grounding on one lead of a three-phase motor, as in Fig. 4-39, may reduce its speed or torque or even change its rotation. When voltage checking, phase-to-phase voltages, as well as phase-to-neutral measurements, are necessary. Contacts used in starters and contactors are available from the manufacturer and should be replaced when excessive wear is determined. These units should not hum, as discussed in connection with solenoids, and often indicates improper seating. Along with checking the coil voltages, look for loose parts or connections and dirt. Dirt or grime in the armature gap will inhibit proper seating and lead to excessive contact wear due to low pressure. Looking at closed contacts will not provide the necessary information, since contact damage will not be apparent. As mentioned about relays, most contacts are plated and should not be filed. When contacts are pitted or worn, contact replacement is the best solution and may prevent a failure at a future date.

Thermal overloads are temperature-sensitive devices, just as fuses are. Therefore, cleanliness, proper ventilation, and good solid connections are very important for efficient operation. Frequent tripping of an overload may be the first indication of motor damage, a load increase, or some sort of binding. A current measurement of the motor's line current under load will be in order if no short circuits in the wiring or motor are indicated.

Fuses and circuit breakers play an important role in the safe operation of electrical equipment. A violent rupture of a fuse or instaneous tripping of a breaker usually indicates a dead short or breakdown. Checking for short circuits with an ohmmeter is the best procedure before reenergizing the system. If a breakdown is suspected, a Megger (a high-resistance/high-voltage ohmmeter) test is one of the best but should be used only if the suspected device can be isolated from the circuit. Never overrate or override a fuse or circuit breaker. Doing so may result in a fire or other severe equipment damage. The most important aspect of good troubleshooting and maintenance is logical and sensible application of your knowledge.

REVIEW QUESTIONS

29. Basic controls besides having electrical problems may also have _____ problems.

30. A _____ inspection of all external controls is the first approach to solving a problem.

31. In Fig. 4-27, if C were short-circuited what symptom would result?

32. Excessive hum on an ac relay or contactor may indicate what?

33. Filing pitted relay contacts is a good maintenance practice. (true or false)

34. Overheating in solenoids is the most common cause of their failures. (true or false)

CHAPTER REVIEW QUESTIONS

4-1. What type of switch matrix is used for the DTMF coding? Hexadecimal?

4-2. Name a contactless type of keyboard switch.

4-3. Calculate the R and C needed across contacts breaking 1 A at 25 V dc.

4-4. Draw a circuit that makes form C relay contacts by using form A and form B contacts.

4-5. Would you expect more or less contact bounce and chatter from a mercury-type relay?

4-6. What type of relay uses a separate reset coil?

4-7. Caution during cleanup of a broken _____ relay is a necessary safety practice.

4-8. Relay drop-out current is usually considerably _____ than relay pick-up current.

4-9. If the capacitor voltage of Fig. 4-27 is zero, the output is _____.

4-10. If pins 5, 6, and 7 are connected to the output bus on Fig. 4-28, and RC equals 10 ms, T equals _____.

4-11. To operate in the astable mode, S_1 of Fig. 4-28 is left _____.

4-12. What controls the delay of the magnetic-hydraulic breaker?

4-13. What can happen to a breaker that has not been tripped for a long period?

4-14. If Q_1 short-circuits in Fig. 4-38, the rotary solenoid will _____.

4-15. If you were to put a pilot lamp into the circuit of Fig. 4-41 to indicate control, should it be connected across start, stop, O.L.s, or the coil?

4-16. If the overloads trip in Fig. 4-48 in the forward direction, will the motor stop or reverse direction?

4-17. Increasing a solenoid's stroke will help it to seat better. (true or false)

4-18. Refer to Fig. 4-39. If the 1CR contact failed to close, what would be the symptom?

4-19. What is the purpose of the N.C. R and F contacts in Fig. 4-48?

4-20. Refer to Fig. 4-32. If any of the F_1 fuses opens, a _____ may be short-circuited.

4-21. What is the primary difference between a 3AG and a European fuse?

ANSWERS TO REVIEW QUESTIONS

1. zero, infinite **2.** yes **3.** 12 **4.** short-circuiting **5.** thumbwheel **6.** snap-action **7.** spring action **8.** ac relays
9. dc loads **10.** stick or lock **11.** single diode, ac power **12.** B **13.** both ac and dc **14.** false, slows actuation
15. shorten **16.** armature **17.** interval **18.** 1.0 MΩ **19.** fusing characteristics **20.** RMS **21.** temperature
22. slo-blo or dual element **23.** fast-acting **24.** no instantaneous trip point **25.** pull **26.** seat **27.** overheat/burnout
28. 1 min **29.** mechanical **30.** visual **31.** cannot trigger, inoperative **32.** dirty magnetic circuit
33. false **34.** true

POWER SOURCES

This chapter deals with one of the more fundamental units in industrial electronics and robotics. Power sources are critical to any industrial system. A malfunctioning source is almost sure to prevent normal system operation. Practically all of the electronic circuits used in industrial systems require direct current (dc) for energization. This chapter examines modern dc sources, their principles and characteristics, and troubleshooting techniques.

5-1
BATTERIES

Cells and batteries have long been important items in industrial systems. Lately, they are increasing in importance because of two major trends. The first is the increasing amount and diversity of portable equipment. Modern components and manufacturing techniques have made it possible to create mobile and portable equipment that formerly would not have been feasible because of size and weight restrictions. This portable equipment must often operate independently of fixed power circuits. The second major trend is uninterruptible circuits and equipment. These devices must continue to operate, or at least retain important information, in the event of a loss of primary power. Such devices are often said to be *battery-backed-up* and are increasing as more digital and computer systems are applied in the industrial environment. A good example is an industrial control system that stores important data in memory circuits as it operates. These memory circuits may be backed up by battery power in case of a failure of the main alternating current (ac) supply. In this case, the control system can quickly resume proper operation when the main supply is restored. Without backup, restoration to normal operation could be very complicated because the system would have no way of determining various data parameters and other control conditions at the time of the interruption.

A battery is made up of series, parallel, or series-parallel combinations of electrochemical cells. Series combinations yield more voltage than a single cell can provide. Parallel combinations yield more ampere-hour (A · h) capacity than a single cell can provide, and series-parallel can increase both voltage and ampere-hour capacity. Cells are divided into two broad categories: primary and secondary. A *primary cell* invokes an irreversible chemical change in the cell structure upon discharge. This means that primary cells cannot be recharged. They are replaced with new cells (or new batteries) when they become discharged. A *secondary cell* or battery also produces chemical change in the cell structure when it is discharging, but the chemical change is reversible. Secondary sources can be restored to a full-charge, or near full-charge, condition by passing a charging current of the proper magnitude and duration through the source in a direction opposite to the discharge current.

Carbon Zinc Cell

Cells and batteries are usually identified by the major materials used to build them. The carbon-zinc cell or battery is a very inexpensive and therefore popular primary type. They are often called *dry cells* since the electrolyte is in paste form to prevent the leaks and spills often associated with *wet cells,* in which the electrolyte is in liquid form. In a carbon-zinc cell (Fig. 5-1) zinc acts as the anode, the electrolyte is ammonium chloride, and a carbon rod serves as the positive contact. The ammonium chloride splits into positive ammonium ions and negative chlorine ions. The zinc anode dissolves in the electrolyte and gives off positive zinc ions, leaving an excess of electrons on the anode and the negative terminal shown in Fig. 5-1. The zinc ions combine with the chlorine ions and form neutral zinc chloride. The ammonium ions are repelled by the zinc ions going into solution and migrate to the carbon electrode, where they pick up an electron and split into ammonium and hydrogen gases. This produces a deficiency of electrons on the carbon electrode and positive terminal shown in Fig. 5-1. When an external load is connected, conventional current flows from the positive terminal, through the load, and into the negative terminal of the cell.

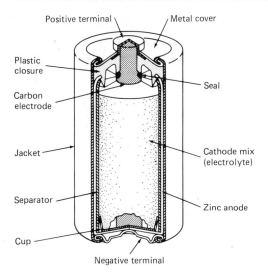

Fig. 5-1 Construction of a carbon-zinc cell.

During high discharge rates, the chemical action of the carbon-zinc cell can cause the cell capacity to fall off. The hydrogen gas that is formed collects on the surface of the carbon rod. Hydrogen gas acts as an insulator, and the resistance of the cell increases. In this condition the cell is said to be *polarized* and cannot deliver adequate current flow. Manganese dioxide is added to the electrolyte to act as a depolarizer. It provides oxygen that combines with the hydrogen gas to form water. However, the depolarizer will not keep up with the hydrogen production at high discharge rates. Therefore, carbon-zinc cells may have to be rested to realize their expected ampere-hour capacity when they are used in high-current applications.

Carbon-zinc cells should not be stored for long periods of time. Storage time is referred to as *shelf life* and should not exceed 1 year. The electrolyte tends to dry out, raising the internal cell resistance. Another problem is that impurities in the zinc anode will set up many *local cells,* and the resulting local action deteriorates and depletes the cell. Shelf life can be improved somewhat by storing carbon-zinc cells and batteries at reduced temperatures.

Alkaline Cell

The alkaline-manganese cell (usually called *alkaline cell* or *alkaline battery*) is constructed inside-out when compared to the carbon-zinc cell and offers considerable improvements. It is also a primary type and cannot be recharged. A zinc rod in the center of the cell serves as the anode, and the electrolyte is an alkaline solution of potassium hydroxide and zinc oxide. The outer, positive electrode is cylindrical manganese dioxide, which provides faster depolarization because of the large surface area of the cylinder. As a result, alkaline cells and batteries are much better suited to high-current applications. They do not require rest periods and can last several times

as long as equivalent carbon-zinc types. They generate 1.5 V per cell (as does carbon-zinc) and are built with a button top that contacts the positive electrode and a bottom plate that contacts the zinc rod. This makes their external appearance and polarity compatible with those of carbon-zinc cells. They are more expensive to buy, but they may be more economical in the long run.

EXAMPLE

A size D carbon-zinc cell will provide 50 mA for 15 h. At this point, the cell voltage will have dropped to 1.3 V, which is usually considered the discharged condition. A size D alkaline cell will provide 50 mA for 56 h before reaching the 1.3-V cutoff. What is the capacity of the two cells?

SOLUTION

The ampere-hour capacity for the two types can be calculated as follows:

$$50 \text{ mA} \times 15 \text{ h} = 750 \text{ mA} \cdot \text{h}$$
$$= 0.75 \text{ A} \cdot \text{h}$$
$$50 \text{ mA} \times 56 \text{ h} = 2800 \text{ mA} \cdot \text{h}$$
$$= 2.8 \text{ A} \cdot \text{h}$$

It is clear that the alkaline cell provides almost four times the ampere-hour capacity provided by the carbon-zinc cell. Alkaline cells also enjoy a longer shelf life (about 3 years) and are less likely to leak and damage expensive circuits and equipment.

Lithium Cell

Lithium cells are available in several types. The construction of a lithium thionyl chloride cell that boasts the highest energy density available among primary sources is shown in Fig. 5-2. The densities are as high as 420 W · h/Kg and 800 mW · h/cm^3. Lithium cells and batteries can be considered permanent components in some electronic systems. Because of their extremely low self-discharge characteristics and their

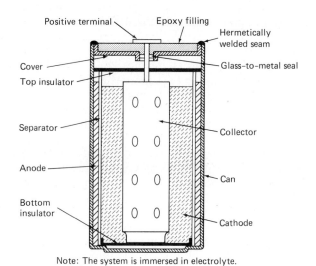

Note: The system is immersed in electrolyte.

Fig. 5-2 Construction of a lithium thionyl chloride cell.

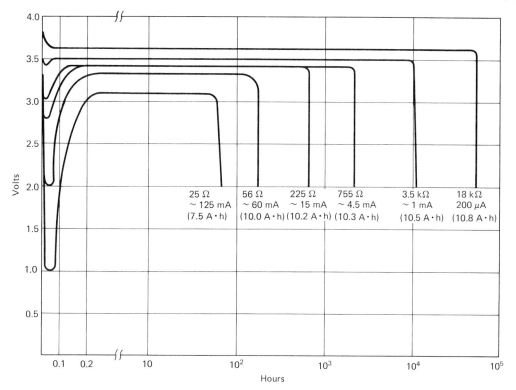

Typical discharge 25°C after 1 year storage

Fig. 5-3 Lithium thionyl chloride cell discharge characteristics.

hermetically sealed construction, their projected shelf life is greater than 10 years when they are stored at room temperature. They develop an open circuit potential of 3.68 V and a nominal working 3.4 V.

Figure 5-3 shows the discharge characteristics for a size D lithium thionyl chloride cell. Notice the unusual voltage response for the first 0.1 h of operation. With a 25-Ω load (approximately 125 mA), the cell output drops to 1 V and then recovers to 3.1 V after 0.3 h. This is called the *transition period* and may be an unacceptable characteristic because the transition minimum voltage may not be adequate to operate some systems. This type of cell is not generally used in high-current applications for this reason. Note, from Fig. 5-3, that the transition minimum voltage is about 3.4 V for a load of 3500 Ω (approximately 1 mA). Also note that the cell will deliver 10.5 A · h of energy when loaded this way. This is 14 times the energy that can be realized from the same size carbon-zinc cell! Figure 5-3 also shows that the plateau voltage is extremely flat after the transition period and extends to over 50,000 h for a load current of 200 microamperes (μA). Fifty thousand hours equates to 5.71 years. It should be clear why lithium batteries and cells are attractive sources for low-current applications such as memory backup and as voltage references.

Figure 5-4 shows a typical backup circuit for a complementary metallic oxide semiconductor random access memory (CMOS RAM). When the main 5-V supply is available, diode D_1 is forward-biased,

and current is supplied to the memory circuit as shown. Diode D_2 is now reverse-biased, and its leakage is normally in the nanoampere (nA) range and can be ignored. If the main 5-V supply fails, D_1 is reverse-biased, and D_2 becomes forward-biased. A backup current flows as shown in Fig. 5-4. The backup voltage will be approximately 3 V since D_2 will drop 0.6 V when forward-biased. A value of 3 V is adequate for data retention in the CMOS RAM. The resistor is optional and is included to prevent a high charging current through the lithium cell in the event that D_2 short-circuits.

Another variation is the lithium-iodine cell. Its construction can be seen in Fig. 5-5. It uses a lithium anode and an iodine cathode. It is designed to be permanently soldered onto a printed circuit board (PCB). The iodine version does not have the energy density of the type previously discussed. However,

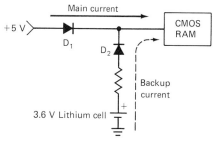

Fig. 5-4 Power back-up for memory chip.

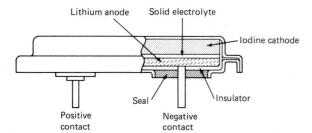

Fig. 5-5 Construction of a lithium-iodine cell.

it is considered a quasisecondary source in that it is capable of some recharging under conditions of low reverse currents. The charging current is typically limited to 1 μA. The circuit of Fig. 5-4 can be modified for a trickle charge of 1 μA by shunting D_2 with a high-value resistor. Since lithium-iodine cells develop an open-circuit voltage of about 2.8 V and since the drop across D_1 can be assumed to equal 0.6 V, the calculation for the resistor is straightforward:

$$R = \frac{V}{I}$$
$$= \frac{5 - 0.6 - 2.8}{1 \ \mu A}$$
$$= 1.6 \ M\Omega$$

Figure 5-6 shows that the storage time for a lithium-iodine cell is 100 years at room temperature! Other primary cells with industrial applications include the lithium manganese dioxide, mercury, and silver oxide cells.

Nickel-Cadmium Cell

Secondary cells are often used in those cases in which more current is required than in memory backup and reference applications. The discussion will be limited to two designs: nickel-cadmium cells and gelled electrolyte types.

Nickel-cadmium cells and batteries are often referred to as *ni-cads*. A typical ni-cad cell is shown in Fig. 5-7. It uses a potassium hydroxide electrolyte, a cadmium and iron oxide negative electrode, and a nickel hydroxide and graphite positive electrode. They develop an almost constant 1.25 V per cell over 90 percent of their discharge cycle. The discharge characteristics are shown in Fig. 5-8. Note that the discharge capacity can range from 90 to 120 percent,

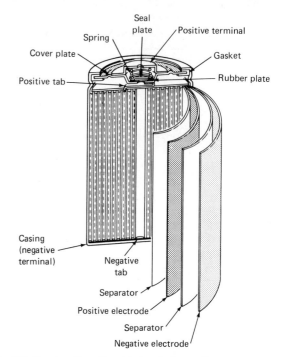

Fig. 5-7 Construction of a nickel cadmium cell.

depending on the rate. A 0.1C rate indicates that the current will be 0.1 times the ampere-hour rating. A size D ni-cad is rated at 4 A · h and will actually deliver 4.8 A · h (120 percent) when it supplies a current of 400 mA (0.1C). The cutoff voltage is usually considered to be 1 V per cell. Ni-cads feature a low internal impedance and are well suited to high-current applications, including pulse applications, where rather high currents must be supplied for short periods. Due to their ability to supply high currents, they must not be short-circuited because cell damage or circuit damage may result. They are usually rated for at least 500 charge-discharge cycles.

Ni-cads will retain 50 percent of their capacity when stored for 3 months at 20°C. They may be stored in either a charged or discharged condition. Several cycles of charge-discharge may be required after an extended storage period to restore normal capacity. They may also require several deep cycles

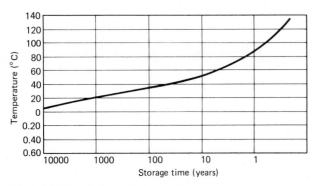

Fig. 5-6 Lithium-iodine cell storage time.

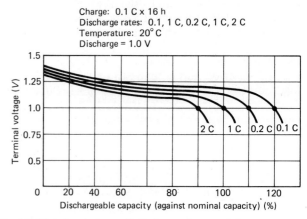

Fig. 5-8 Nickel cadmium cell discharge characteristics.

(drained to 1.0 V per cell and then completely recharged) after an extended number of shallow cycles. A *shallow cycle* is a situation in which the cell is recharged after it has delivered only a portion of its rated capacity. Too many shallow cycles may cause a "memory" effect, and the cell will tend to deliver only that portion of its capacity that it remembers delivering. This effect, although interesting, is often overrated, and ni-cad batteries containing three or more series cells are more often damaged by overly deep discharging. The individual cells are seldom matched in capacity and the weakest cell will eventually drop to 0 V and then become a load. When this occurs, its polarity will reverse and permanent damage often results.

Ni-cads can also be damaged by overcharging and by rapid charging. For this reason, a constant voltage source is not recommended. You may recall that a constant voltage source is characterized by a low internal resistance. Constant current is the preferred charging mode, and a constant current source is characterized by a high internal resistance. A voltage source can approximate a current source simply by addition of a series resistor.

EXAMPLE

It is desired to charge a 12-V ni-cad battery from a 24-V source. Also assume that the charging rate is to be the standard *C/10 rate,* at which the charging current is found by dividing the ampere-hour capacity (C) by 10. Find the value of the series resistor required.

SOLUTION

A C/10 rate will be 0.4 A for a 4-A · h battery. The resistor will have to drop 12 V (24 − 12) at a current of 0.4 A:

$$R = \frac{V}{I}$$
$$= \frac{12}{0.4}$$
$$= 30 \ \Omega$$

The C/10 rate will require at least 14 h for full charge. The battery will not completely recharge in 10 h since no secondary source is 100 percent efficient. More charging energy must be delivered to the battery than its rated discharge energy. A C/10 rate is considered very safe since the battery may be left on charge indefinitely with no resulting damage due to overcharge. Faster rates are possible, such as a C/4 rate, which will recharge the battery in 5 h. However, it must be noted that a C/4 rate must not be applied to a ni-cad for more than 6 h; if it is, the cell temperature will begin to rise and permanent damage will result.

Special quick-charge ni-cads are available to achieve a full charge in a minimum of 4 h. Special charging circuits are often used with these types to provide automatic cutback to a C/10 rate when full charge is reached. The cutback is usually triggered by a circuit that monitors cell voltage and cell temperature.

Gel Cell

Gelled-electrolyte cells (usually referred to as *gel-cells*) are close cousins to the lead-acid cells used in automobiles. They enjoy a high output of 2.2 V per cell, and their cutoff point is 1.75 V per cell. They can be totally discharged without any danger of damage due to cell reversal. They use lead-calcium electrodes and are completely sealed to operate in any position. However, they do require a one-way safety vent in the event that cell pressure would exceed some safe limit. They are capable of several hundred cycles and should not be charged indefinitely, even at a C/10 rate. They also should not be stored for extended periods in a discharged condition. When this occurs, permanent sulfation of the electrodes will reduce capacity and eventually lead to cell failure. Gel-cells, like ni-cads, have a low internal resistance and should not be short-circuited. They serve well in high-current applications, although the delivered capacity drops in these situations, and rest periods may be required. The major drawback of gel-cells is their tendency to release hydrogen gas. This is especially prevalent at high charging rates. Hydrogen gas is extremely explosive, and this characteristic precludes the use of gel-cells in some industrial environments.

REVIEW QUESTIONS

1. Which broad category of cells undergoes an irreversible chemical change during discharge?

2. Carbon-zinc cells may require rest periods when used at high discharge rates because of cell _____.

3. Calculate how long an alkaline D cell will supply 80 mA before reaching its cutoff voltage.

4. Which type of cell, under a condition of high initial discharge, produces a minimum transition voltage before reaching its plateau voltage?

5. It is desired to trickle-charge a 2.8-V lithium-iodine cell at 2 μA from a 10-V supply. Using Fig. 5-4 as a guide, calculate the value for the resistor that would be connected across D_2.

6. You wish to charge a 6-V, 2-A · h ni-cad battery at the C/10 rate. A 12-V constant-voltage source is available. Calculate the series resistor.

5-2
RECTIFICATION

Rectification is the process of changing alternating to direct current. Solid-state diodes have proved to be very efficient rectifiers. Figure 5-9 shows a half-wave rectifier circuit that uses a diode (D) to change alternating current to direct current. An alternating voltage appears across the secondary of the transformer and forward-biases the diode every half cycle. Current flows through the diode and the load resistor only half the time and only in the direction shown.

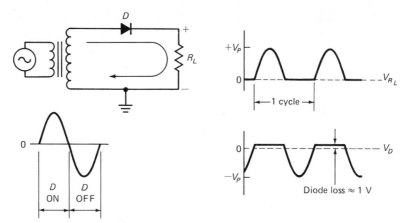

Fig. 5-9 Half-wave rectifier circuit.

Note the waveforms across the load (V_{R_L}) show in Fig. 5-9. This waveform is called *half-wave pulsating direct current* and shows one load pulse for every cycle of the input. The waveform across the diode (V_D) shows the reverse voltage peaks when the diode is off. It also shows the *diode loss,* which is the voltage drop across the diode when it is conducting. For silicon rectifiers, this loss ranges from 0.7 to 1 V and is directly related to load current.

The polarity of the rectified direct current can be predicted by assigning a current direction through the load and then the drop across the load. Figure 5-9 shows conventional current flowing down through the resistor; the resistor polarity is therefore positive at the top. Another way to predict polarity is to observe that the cathode lead of the diode contacts the positive end of the load. This is always the case in any rectifier circuit and is a good way to predict load polarity. If the diode is reversed, the anode lead will contact the top of the load, thus making it negative. Since the bottom of the load resistor is grounded and since ground is the normal reference point, Fig. 5-9 can be considered a positive supply as drawn and a negative supply if the diode is reversed.

Half-wave rectifiers are not time-efficient since they load the source only on every other alternation. The extreme pulsation of load voltage and current is also considered a disadvantage. The pulsations can

be removed by filtering, but the process often causes high-heating (root mean square [rms]) currents to flow in the transformer secondary and in the rectifier. Also, the unidirectional current flow in the secondary biases the transformer core with a dc flux, and a larger and heavier core may be required to avoid saturation. Because of their disadvantages, half-wave rectifiers are generally limited to low-power applications. They are more attractive for rectifying high-frequency square waves, for example, in switch-mode power supplies, which are covered in section 5-7.

A full-wave rectifier circuit using a center-tapped transformer is illustrated in Fig. 5-10. It uses two diodes and produces two load pulses for every cycle of the input. The load waveform is called *full-wave pulsating direct current.* When the top of the secondary is positive, D_1 is forward-biased, and load current flows as shown by the solid arrow in Fig. 5-10. Notice that only the top half of the transformer secondary is conducting, since D_2 is now reverse-biased. On the next alternation, D_2 is forward-biased; load current flows as shown by the broken arrow, and only the bottom half of the secondary is conducting. In Fig. 5-10, V_P will be equal to half the peak secondary voltage since only half of the transformer winding conducts at any given time. Note that the top of the load resistor is positive and is in contact with the cathodes of the rectifiers.

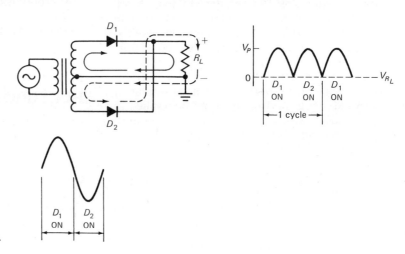

Fig. 5-10 Full-wave (center-tap) rectifier circuit.

The bridge rectifier in Fig. 5-11 is full-wave and uses four diodes. It eliminates the need for a center tap. When the top of the secondary is positive, D_1 and D_2 are forward-biased, and current flows as shown by the solid arrow. When the bottom of the secondary is positive, D_3 and D_4 are forward-biased, and current flows as shown by the broken arrow. Once again it is seen that the positive end of the load is in contact with the cathodes of the rectifiers. The load waveform in Fig. 5-11 is drawn as negative-going. Compare it to the load waveform shown in Fig. 5-10. The reason for the difference is that Fig. 5-11 is a negative supply. The positive end of the load is grounded. The circuit can be changed to a positive supply by reversing all four diodes or by changing the ground connection to the negative end of the load. The waveform shown in Fig. 5-11 is what would be seen on an oscilloscope if all conditions were normal. *Normal* means that the instrument ground is connected to the circuit ground, and a negative voltage causes a downward deflection on the oscilloscope screen.

The center-tap and bridge circuits both seem to be capable of the same performance. However, there are several differences. The bridge circuit uses two diodes conducting in series. The diode losses will therefore be twice as great. A typical diode loss in solid-state rectifier circuits is 1 V per diode. Therefore, 2 V will be lost in the bridge circuit and only 1 V in the center-tap circuit. This is significant in low-voltage supplies, and the center-tap circuit may be preferred in those applications. One advantage of the bridge circuit is that it requires only half as many secondary turns to develop any given voltage. Using half as many turns results in a less bulky transformer even though larger wire will be required because the entire secondary must conduct on both alternations. Another advantage of the bridge circuit is that the rectifier diodes are subjected to half the peak inverse voltage (PIV) when compared to the center-tap circuit. Because of these differences, the bridge circuit is generally preferred at higher voltages and the center-tap circuit at lower voltages. However, in practice, both circuits will be found operating over a broad range of voltages.

Figure 5-12 summarizes five important rectifier circuits. It lists their average dc output voltage (V_O), PIV per diode, rms ripple voltage, and ripple frequency and shows their output waveforms. Notice that the average dc output voltage is lowest for the half-wave circuit and is only 45 percent of the rms input voltage. Also notice that the ripple voltage is the highest, at 54 percent of the rms input voltage. This makes the percentage of ac ripple very high for the half-wave supply. The percentage of ripple in the output voltage is given by

This high ripple percentage emphasizes one of the disadvantages of the half-wave rectifier circuit.

Figure 5-12 also shows two three-phase rectifier circuits. These circuits are very popular in industrial equipment and have several advantages over the single-phase circuits already discussed. Note the low rms ripple voltages for the three-phase circuits. For example, in the case of the three-phase bridge circuit:

$$\text{ripple, \%} = \frac{0.057}{1.35} \times 100$$
$$= 4.22\%$$

This low percentage of ripple shows that the output of three-phase rectifiers is a much more pure form of direct current than is provided by single-phase circuits. Figure 5-13 shows why. The waveform for the line-to-neutral circuit is shown in Fig. 5-13(a). Three sine waves are drawn 120 electrical degrees apart. The negative alternations are shown as broken lines and are eliminated in the line-to-neutral circuit. The resulting ripple is formed by the positive alternations only. Figure 5-13(b) shows that the negative alternations are folded up to the positive part of the graph by the three-phase bridge circuit, resulting in a very small ripple voltage.

Two popular multiphase rectifier circuits are shown in Fig. 5-14. The six-phase star circuit (Fig. 5-14[a]) has a small (4.22 percent) ripple content and a ripple frequency equal to six times the line frequency. The graph shows six output pulses for one cycle of input. A six-phase bridge would achieve even less ripple content and a ripple frequency equal to 12 times the input frequency. Figure 5-14(b) is a three-phase double wye with an interphase transformer. Again, the ripple percentage is only 4.22 percent, and the ripple frequency is six times the line frequency.

Multiphase rectifiers have low ripple percentages and high ripple frequencies. These are both advantages. Another advantage, especially in high-current supplies, is the low rectifier-cell ratios found in multiphase supplies. The *cell ratio* is the comparison of rectifier current to load current. For example, in a single-phase half-wave circuit the cell ratio is 1.00 since the single diode must conduct the entire load current. A single-phase full-wave circuit shows a cell ratio of 0.5 because any diode is on half the time. Three-phase circuits have cell ratios of only 0.333, and the six-phase circuit shown in Fig. 5-14(a) shows a cell ratio of only 0.167. Small cell ratios relax the current ratings for individual diodes used in rectifier service.

Rectifier diodes must be derated according to case temperature and service. The curves in Fig. 5-15 show that these diodes must be derated at temperatures above 150°C. The diodes must also be derated

$$\text{ripple, \%} = \frac{\text{Ripple voltage as fraction of input voltage}}{\text{average dc output as fraction of input voltage}} \times 100$$
$$= \frac{0.54}{0.45} \times 100$$
$$= 120\%$$

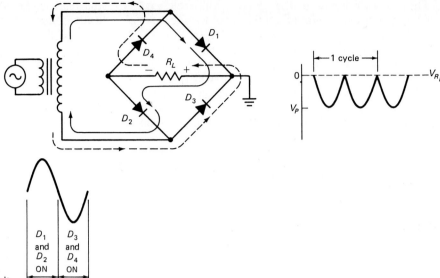

Fig. 5-11 Full-wave (bridge) rectifier circuit.

Schematic	Name	V_O (peak)	V_O dc	PIV per Diode	RMS Ripple Voltage	Ripple Frequency	Output Waveform
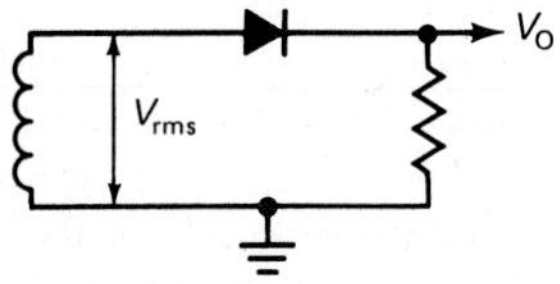	Half-wave	$1.41\ V_{rms}$	$0.45\ V_{rms}$	$1.41\ V_{rms}$	$0.54\ V_{rms}$	$1f_L^*$	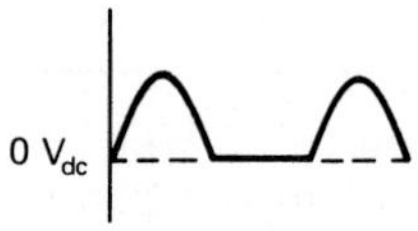
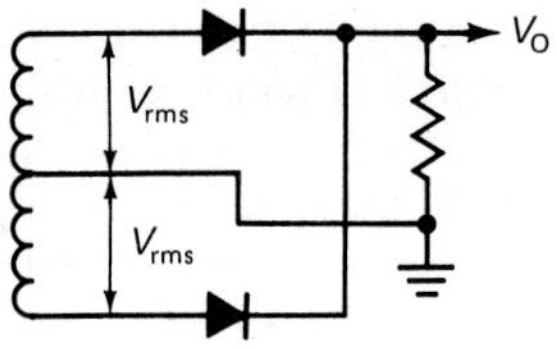	Full-wave	$1.41\ V_{rms}$	$0.90\ V_{rms}$	$2.82\ V_{rms}$	$0.43\ V_{rms}$	$2f_L$	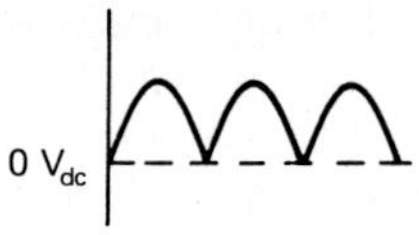
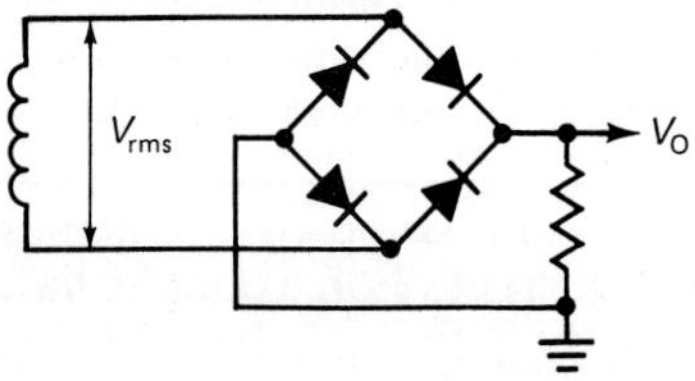	Bridge (full-wave)	$1.41\ V_{rms}$	$0.90\ V_{rms}$	$1.41\ V_{rms}$	$0.43\ V_{rms}$	$2f_L$	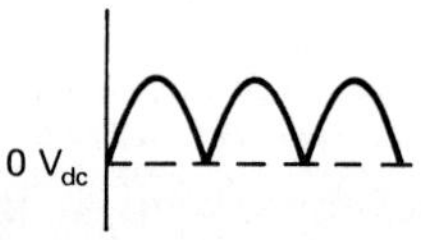
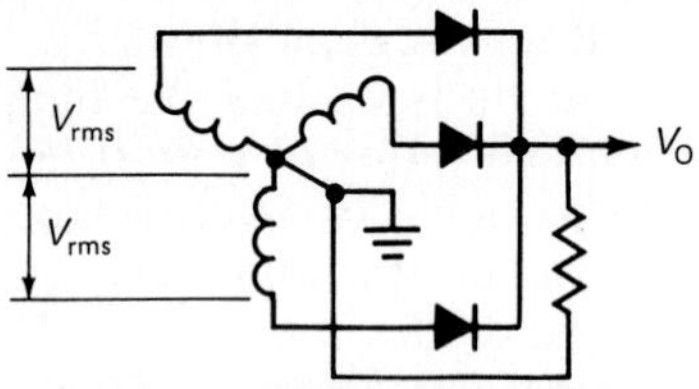	Three-phase wye line to neutral (half-wave)	$1.41\ V_{rms}$	$1.17\ V_{rms}$	$2.45\ V_{rms}$	$0.21\ V_{rms}$	$3f_L$	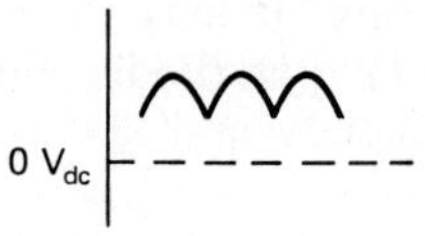
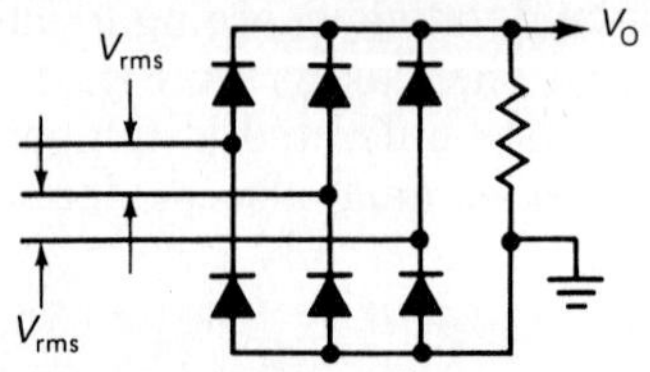	(Y or Δ) Three-phase bridge line to line (full-wave)	$1.41\ V_{rms}$	$1.35\ V_{rms}$	$2.45\ V_{rms}$	$0.057\ V_{rms}$	$6f_L$	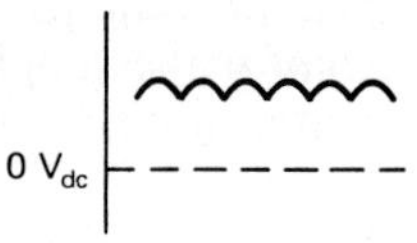

*f_L = Line frequency

Fig. 5-12 Summary of rectifier circuits.

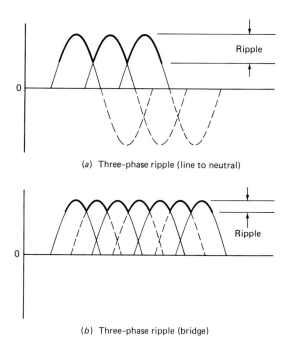

(a) Three-phase ripple (line to neutral)

(b) Three-phase ripple (bridge)

Fig. 5-13 Three-phase ripple voltage.

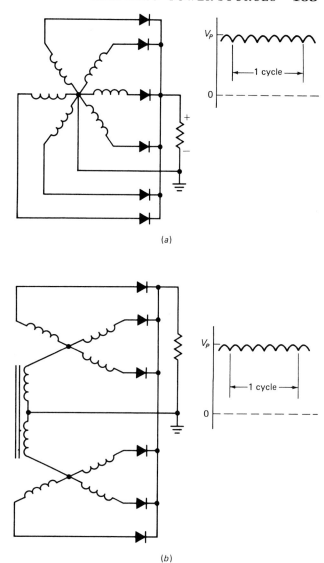

(a)

(b)

Fig. 5-14 Multiphase rectifier circuits. (a) Six-phase star circuit. (b) Three-phase double-wye and interphase circuit.

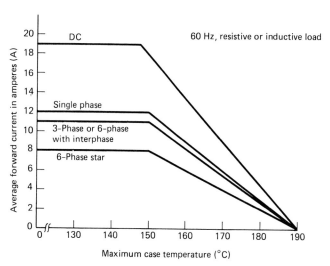

Fig. 5-15 Diode derating according to temperature and service.

according to the type of circuit in which they are to be used. This is due to the heating effect of pulse current. Power dissipation varies as the square of the current ($P = I^2R$), and the high-current amplitude associated with pulse waveforms creates more heat loss in diodes than is indicated by the average current. A diode is capable of its greatest forward current at direct current (no pulsation) and its least forward current in six-phase star service (the narrowest pulse). The curves show that the same diode can support 12 A in single-phase service and only 8 A in six-phase service up to 150°C. However, do not forget the cell ratios. A single phase full-wave circuit shows a cell ratio of 0.5, and the six-phase star shows a cell ratio of 0.167. This means that six diodes of the type shown in Fig. 5-15 could provide a dc load current of up to 47.9 A [(1/0.167) × 8)] in six-phase star service, and two diodes could supply up to 24 A [(1/0.5) × 12] in single-phase, full-wave service. In actual practice, they would not be called upon to deliver their maximum currents as most circuit designers derate manufacturer's specifications to approximately 70 percent for good reliability.

You may be curious about the dc curve shown in Fig. 5-15. Since we have investigated the use of diodes to rectify (change alternating to direct current), the purpose of this curve may not be readily apparent. In addition to rectification, diodes can be used to isolate one circuit from another. Two dc power supplies can be used in parallel to supply current to a load when one supply cannot provide adequate current. Diodes in the output lines to the load prevent either supply from delivering a current to the other supply. This is an example of diode isolation and indicates one example where diode dc ratings would be appropriate.

The two most obvious rectifier diode specifications are their average forward current rating and their PIV rating. Others include the following:

1. Physical characteristics (mounting details, etc.)
2. Thermal characteristics
3. Power dissipation
4. Recovery time
5. Transient voltage rating
6. Avalanche rating
7. Current surge rating

Recovery time is a measure of how quickly a diode can stop conducting when suddenly reverse-biased. It takes time to sweep the carriers from the junction region, and high-frequency rectifiers must have a rapid recovery time or circuit efficiency will suffer and diode heating will be a problem. The *transient voltage* rating is a measure of the amount of nonrepetitive voltage stress a diode can withstand. It is especially useful in industrial environments in which large inductive loads are being switched. The *avalanche characteristic* occurs in controlled-avalanche rectifiers, in which excess voltages can be expected to produce predictable (controlled) results. Noncontrolled avalanche often results in the rectifier's being destroyed. Finally, *current surge ratings* are nonrepetitive current demands and are especially important when capacitor-input filters are used, as discussed in the next section.

REVIEW QUESTIONS

7. Refer to Fig. 5-9. Assume no diode loss and an rms secondary output of 50 V. Calculate the average dc voltage across the resistor. Hint: Refer to Fig. 5-12.

8. Again use Fig. 5-9 and the same assumptions as in question 7. Calculate the ac ripple voltage across the resistor.

9. Is the circuit of Fig. 5-10 a positive supply or a negative supply with respect to ground?

10. Refer to Fig. 5-10. Assume no diode loss and an rms secondary output of 30 V across the entire winding. Calculate the average dc voltage across the resistor.

11. Again use Fig. 5-10 and the same assumptions as in question 10. Calculate the ripple voltage across the resistor and the percentage of ripple.

12. Refer to Fig. 5-11. Assume a per-diode loss of 1 V and a 10-V (rms) secondary. Calculate the average dc output voltage.

13. Why is the diode in Fig. 5-15 rated up to 19 A in dc service and only up to 8 A in six-phase star service?

5-3 FILTERING

Except for applications such as welding, electroplating, battery charging, and motoring, the output of single-phase rectifiers contains too much ac ripple. A filter circuit will be needed to smooth the pulsating waveform and make it more like pure direct current (a straight line). Figure 5-16 shows some power supply filter circuits. They are used between the rectifier output and the load. They are examples of low-pass filters since they are designed to pass direct current (0 Hz) and to block the ac ripple (some multiple of the line frequency).

Figure 5-16 shows that power supply filters may be divided into two broad categories: capacitor input and choke (inductive) input. The capacitive input types are shown in Fig. 5-16(a). The rectifier output would be connected at the left of the circuits drawn. The left represents the input side of the filter. Notice that the filter input contains a shunt capacitor. Also notice that in Fig. 5-16(b) the inductive input filter types show a series coil (choke) at the filter input. Supply performance is affected by whether the input component is a capacitor or a choke.

Capacitor input filters draw large current pulses from the rectifiers and from the transformer secondary. They produce a high load voltage when little current is drawn and much less voltage when full load is reached. This makes their voltage regulation poor. Voltage regulation is calculated as:

$$\text{Percent regulation} = \frac{\Delta V}{V_{FL}} \times 100$$

where ΔV = change in voltage from no-load to full-load

V_{FL} = full-load voltage

EXAMPLE

Suppose the output of a power supply drops from 13 V at no-load to 11 V at full-load. Find the percent regulation.

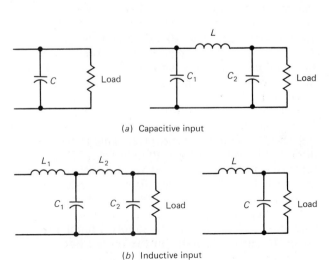

(a) Capacitive input

(b) Inductive input

Fig. 5-16 Power supply filter circuits.

SOLUTION

$$\text{Percent regulation} = \frac{\Delta V}{V_{FL}} = \frac{13 - 11}{11}$$

$$= \frac{2}{11} \times 100$$

$$= 18.2\%$$

Choke input filters lengthen the rectifier and transformer conduction time. This decreases the heating effect in these components for any given value of load current. They also show a lower output voltage and better voltage regulation when compared to the capacitor input types. In spite of these facts, the single-capacitor filter, shown in Fig. 5-16, has become very popular. The weight, size, and cost of inductors have eliminated them from most solid-state power supply designs. It is now possible to build supplies with adequate characteristics without using filter inductors because of the excellent current ratings of solid-state diodes, the improvements in electrolytic capacitors, and the performance of modern voltage-regulator circuits. Inductors are more attractive at high frequencies where far less core is required. Switch-mode power supplies use frequencies in the tens and hundreds of kilohertz, and filter inductors are employed in these designs. This topic is covered in section 5-7.

The full-wave, center-tapped circuit in Fig. 5-17(a) has a single capacitor filter in parallel with the load. The voltage waveform across this parallel combination (Fig. 5-17[b]) shows that the capacitor charges to the peak value of the ac input (1.41 × V_{rms}). If the capacitor is large enough, it will hold the load

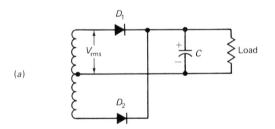

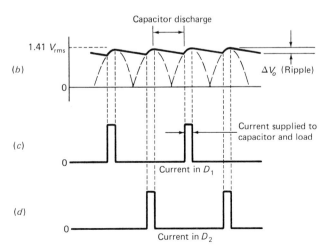

Fig. 5-17 Full-wave rectifier with a capacitive input filter. (a) Rectifier circuit. (b) Voltage across the capacitor. (c) Current in diode D_1. (d) Current in diode D_2.

voltage near this peak value over the period of the half cycle until the next rectifier pulse comes along. The capacitor has eliminated most of the ac ripple across the load. The remaining ripple is designated as ΔV_o and resembles a sawtooth waveform. The current waveforms (Fig. 5-17[c] and [d]) indicate that the rectifiers conduct for only short periods of time. Of course, the transformer secondary current is also of a pulse nature. The amplitudes of the current pulses are about five times greater than the load current in a typical power supply of this type. The heating effect is much greater for pulse-type waveforms, and the components must be derated when capacitor input filters are used. For example, the transformer in this circuit should not be expected to deliver more than 80 percent of its rated current.

The current pulses can be understood by examining the waveforms in Fig. 5-17. The stored voltage across the capacitor keeps both diodes reverse-biased most of the time. Neither diode will begin conducting until the peak secondary voltage exceeds the capacitor voltage by about 0.6 V. This 0.6-V difference is required to collapse the diode depletion region. At the time of turn-on, the diode supplies charging current to the capacitor and some current to the load. The diode continues to conduct until the secondary voltage reaches peak. It should be clear that the diode current is relatively short in duration under these conditions and that extra stress is placed on some power supply components. Extra voltage stress is also created in three-phase rectifier circuits. The PIV is increased from 2.45 × V_{rms} to 2.82 × V_{rms} with a capacitive input filter.

Selecting the capacitance for a simple filter as shown in Fig. 5-17 depends on three factors: ripple frequency, load current, and allowable ripple voltage. It is accomplished by

$$C = \frac{I}{\Delta V} \times T$$

where C = capacitance, F
I = load current, A
ΔV = peak-to-peak ripple, V
T = ripple period, s

EXAMPLE

Suppose a capacitor is needed for a 10-A power supply. Also assume that the supply is full-wave, runs from the 60-Hz single-phase ac line, and may have 1-V peak-to-peak ripple. Find the value of the capacitor.

SOLUTION

The ripple frequency is twice the line frequency in this case, and the period may be found by

$$T = \frac{1}{F}$$

$$= \frac{1}{120}$$

$$= 8.33 \text{ ms}$$

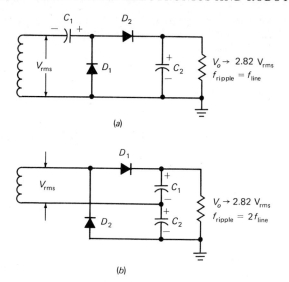

Fig. 5-18 Voltage multipliers. (*a*) Half-wave doubler. (*b*) Full-wave doubler.

Now the capacitance can be found by

$$C = \frac{10}{1} \times 8.33 \times 10^{-3}$$
$$= 0.0833 \text{ F}$$
$$= 83,300 \text{ } \mu\text{F}$$

There are two other important capacitor ratings. The first is the voltage rating. For safety, the voltage rating must be somewhat greater than the peak secondary voltage. The second rating is the current capability of the capacitor. The capacitive ripple current (for an input capacitor) is approximately 2.5 times the load current. This amounts to 25 A for the previous example. This ripple current heats the capacitor. Heating shortens the life of the capacitor, and this factor must be taken into account for good reliability.

Voltage Multipliers

Capacitive filters also lead the way to voltage multipliers. Figure 5-18(*a*) is a half-wave doubler. Assume the first alternation makes the top of the secondary negative. This will forward-bias D_1, and C_1 will be charged to the peak secondary voltage. The next alternation will forward-bias D_2, and the stored charge across C_1 will series add with the secondary voltage. Also, C_2 will be charged to twice the peak secondary voltage. The load will see a peak voltage of $2 \times 1.41 = 2.82$ V_{rms}. The ripple frequency is equal to the line frequency, since C_2 and the load receive a line pulse once per cycle. The full-wave doubler in Fig. 5-18(*b*) provides two line pulses per cycle. Capacitor C_1 is charged through D_1 to the peak secondary voltage, and C_2 is charged through D_2 to the peak secondary voltage. The capacitors are in series, and the load sees twice the peak secondary voltage. Both capacitors are in the load circuit, and both alternations pulse the load; thus full-wave operation is realized and the ripple frequency is twice the line frequency.

The half-wave doubler in Fig. 5-18(*a*) is safer in some applications because it allows one side of the ac source to be grounded. Neither side of the ac source may be grounded in the full-wave doubler circuit. Grounding the chassis is possible in line-operated equipment (no on-board isolation transformer) with the half-wave doubler and can prevent a "hot" chassis and ground loops. This topic is covered in more detail in a later section of this chapter.

Voltage doublers can provide a dc voltage that is nearly three (2.82) times the ac input. Their output drops rapidly under load, however, and double the rms input is normal under working conditions. They are noted for poor voltage regulation. Figure 5-19 shows that high-order voltage multiplication is also possible; C_1 is charged through D_1 to the peak voltage V. Then C_2 is charged through D_2 with C_1 series aiding the secondary voltage. Now C_3 is charged through D_3, and capacitors C_1 and C_2 series aid the secondary. Thus, a voltage equal to three times V is available at the cathode of D_3. If the circuit shown in Fig. 5-19 is to be used to supply any of the evenly multiplied voltages ($2V$, $4V$, $6V$), the ground will have to be moved to the top of the secondary. In theory, any multiplication factor is possible. In practice, circuit efficiency limits high-order multiplication to very-low-current applications such as in cathode-ray tube supplies.

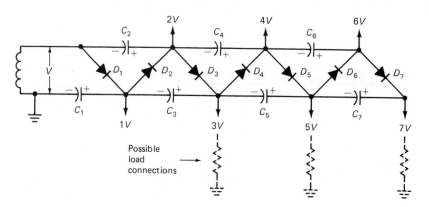

Fig. 5-19 High-order voltage multiplier.

REVIEW QUESTIONS

14. A power supply develops 48 V at no-load and 40 V at full-load. Calculate its regulation.

15. Which broad category of power supply filter develops a high no-load output voltage?

16. Which broad category of power supply filter is noted for poor voltage regulation?

17. Which broad category of power supply filter causes high peak rectifier current?

18. Refer to Fig. 5-17. Assume no diode loss, a very small load current, and a $30\text{-}V_{\text{rms}}$ secondary. Calculate the dc load voltage.

19. What will happen to the dc load voltage calculated in question 18 if the load current increases?

20. What will happen to the percentage of ripple in the circuit of question 18 if the load current increases?

5-4
VOLTAGE REGULATION

Voltage regulation is the ability of a power supply to hold its output potential constant under conditions of changing line voltage, temperature, and load current. Some industrial circuits are critical and demand voltage regulation of less than 1 percent. These circuits will require power supplies with voltage regulators.

Line voltage varies with demand. A *brownout* is a condition in which the demand is so high that the power company is forced to reduce the line voltage intentionally to protect its equipment. Industrial customers are not protected from brownouts and may even experience a greater reduction in line voltage than residential consumers.

When line voltage is abnormally low, the output from a power supply will also tend to be low. The ability of a power supply to hold a constant output over a range of line input voltage is termed *line regulation*. One way to achieve line regulation is to use a ferroresonant power transformer. These transformers are designed with separate core windows for the primary and secondary windings. They have two magnetic circuits: the main flux path and the shunt flux path. The main flux path couples the primary and secondary circuits. The secondary circuit is tuned to resonate at the line frequency by connecting a capacitor across a part of or all of the secondary. The value of this capacitor is several microfarads in 60-Hz supplies. The Q of the tuned circuit is high enough to cause large circulating currents that saturate the core in the main flux path. *Saturation* is a decrease or increase in magnetizing force that is not accompanied by a corresponding change in flux density. With the main flux path saturated, line voltage fluctuations will not change the main flux density, and the secondary voltage will remain constant. Of course, if the primary voltage drops too low, the core

can come out of saturation, and the secondary voltage will drop.

The shunt magnetic flux path in a ferroresonant transformer is prevented from saturating by air gaps placed in its magnetic circuit. Air has a much higher reluctance than transformer iron, and this characteristic limits the flux in this part of the magnetic circuit. If a greater demand is placed on the secondary circuit, the Q of the secondary tuned circuit is reduced, and therefore the circulating currents are also decreased. Since the shunt path is linear (nonsaturated), it can respond to this change, and a decrease in flux results. With fewer lines of force in the shunt circuit, the main flux density can increase; more energy is transferred from primary to secondary to make up for the increased demand on the secondary circuit. Thus, the ferroresonant transformer also regulates for load changes. Unfortunately, the size, weight, and cost of these transformers eliminate them from many designs, but they are counted among the most reliable voltage regulators available.

A more common way to achieve line and load regulation is to use a separate regulator circuit after the power supply filter. In Fig. 5-20, a simple regulator uses a zener diode (D) in shunt with the load. As long as the zener operates in its constant voltage region, the load voltage will also remain constant. Zener diodes can generate noise (especially when operating near the knee) so that a capacitor is often included in this circuit to bypass the noise from the load. This circuit will operate properly over a range of load and line conditions. If the unregulated input voltage drops too low, the zener will stop conducting and regulation will be lost. If the load demand goes too high, the drop across series resistor R will increase to the point where the zener stops conducting,

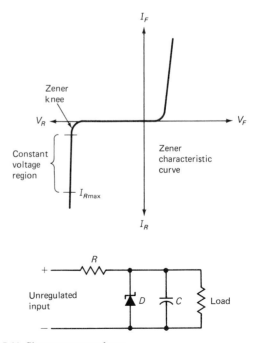

Fig. 5-20 Shunt zener regulator.

and once again regulation will be lost. Another problem is that the safe zener dissipation may be exceeded if the load demand drops to zero or if the unregulated input voltage goes too high.

The regulator circuit of Fig. 5-20 is limited to low-power applications. A few simple calculations will illustrate why.

EXAMPLE

Suppose a regulated 15 V at 2 A is required and the unregulated input is 20 V.

SOLUTION

The zener current is often set at half the load current in this design so it should be 1 A. Ohm's law is used to calculate the required series resistor R. It will conduct the load current plus the zener current $(2 + 1 = 3$ A$)$ and will drop the difference between the input voltage and the load voltage $(20 - 15 = 5$ V$)$:

$$R = \frac{V}{I}$$
$$= \frac{5}{3}$$
$$= 1.67 \ \Omega$$

The zener dissipation will be equal to its voltage drop times its current. However, if there is any possibility that the load current can drop to zero the diode will have to conduct all of the current:

$$P = V \times I$$
$$= 15 \times 3$$
$$= 45 \ W$$

A 45-W zener diode is unacceptable because of its cost, and the efficiency of the circuit is poor when the load current is low. This is why the shunt zener regulator is limited to low-power applications.

Figure 5-21 shows positive and negative regulators that are an improvement over the shunt circuit. In Fig. 5-21(a) a zener diode and an NPN transistor form a positive voltage regulator. The transistor is called a *series pass transistor* in this application: It passes the load current from collector to emitter. The zener is used to regulate the base voltage of the transistor. Regulating the base voltage also regulates the emitter voltage if we can assume a constant base-emitter drop in the transistor. If the emitter voltage is regulated, then the load voltage is also regulated in Fig. 5-21. The current gain of the transistor greatly relaxes the zener dissipation. If we assume a gain of 50 from the base to the emitter and a load current of 2 A, the base current will be 40 mA (2/50). Now the zener current can be set to half this value, or 20 mA. If we again assume a 20-V unregulated input and a 15-V zener, resistor R is now

$$R = \frac{5}{0.06}$$
$$= 83.3 \ \Omega$$

Notice that the resistor is much larger than in the shunt regulator. Once again, the worst-case zener

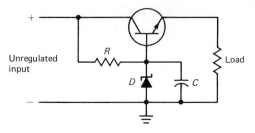

(a) Positive regulator circuit

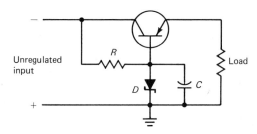

(b) Negative regulator circuit

Fig. 5-21 Series pass circuits.

dissipation occurs if the load current drops to zero. In the case of Fig. 5-21, with no-load current there can be no emitter current and no base current. The zener will now conduct 60 mA for a dissipation of

$$P = 15 \times 0.06$$
$$= 0.9 \ W$$

Now this is far more acceptable. The cost of the zener is reasonable, and circuit efficiency is much better at low load currents. The circuit can also be arranged for regulating negative voltages. This is shown in Fig. 5-21(b). Note that the zener is reversed and that the series pass transistor is a *PNP* device.

The shunt circuit of Fig. 5-20 and the series pass circuits of Fig. 5-21 will not develop the same load voltages with a 15-V zener diode. The series pass circuit will have a lower output since the pass transistor will show some drop from base to emitter. This is usually 0.7 V in a silicon transistor that is conducting moderate currents. Therefore, the load voltage can be expected to be 0.7 V less than the zener voltage in Fig. 5-21. However, at high load currents the base-to-emitter drop is going to be greater. A series-pass transistor that is conducting 5 A will show a base-emitter voltage closer to 1.2 V, and the output voltage will then be 1.2 V less than the zener voltage. So the problem with the circuits shown in Fig. 5-21 is that the no-load to full-load voltage will drop a half volt or more in a 5-A supply. This amount of voltage change may not be acceptable in some applications.

A feedback regulator with better performance is shown in Fig. 5-22. The output is sampled by the voltage divider R_2 and R_3, and some portion of the load voltage is fed back to the base of Q_2, which acts as an error amplifier. The feedback voltage is compared to the zener voltage, which acts as a reference. Any error between the two voltages is amplified and used to reduce the error. Suppose the load voltage

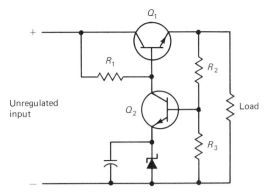

Fig. 5-22 Feedback regulator.

in Fig. 5-22 is 15 V and that $R_2 = R_3$. They will divide the 15 V in half, and therefore the base voltage of Q_2 will be $+7.5$ V with respect to ground. If the zener is a 6.8-V unit, the emitter of Q_2 will be $+6.8$ V. The base-emitter bias of Q_2 will therefore be 0.7 V $(7.5 - 6.8)$, and there will be moderate emitter and collector current in Q_2. Resistor R_1 conducts two currents: the base current for Q_1 and the collector current for Q_2. Now, suppose the load demands more current. As always, this will tend to make the output voltage drop. This drop will reduce the base voltage of Q_2. The emitter voltage is constant because of the zener, and now there will be less voltage from base to emitter in Q_2. As V_{be} decreases, so must base current I_b. As I_b decreases, so must collector current I_c. Finally, with Q_2 demanding less collector current, R_1 can supply more current to the base of Q_1 and turn it on harder. When a series-pass transistor is turned on harder it has less resistance and will drop less voltage from collector to emitter. In Fig. 5-22, if V_{ce} in Q_1 drops, more of the unregulated input voltage must drop across the load. Thus, the load voltage has been stabilized by the feedback.

The application of feedback achieves good voltage regulation. It also provides ripple rejection since ripple is also an error in output voltage. Feedback is the basis for most regulator designs. It is based on sampling the output and comparing it to some reference. Any change in output, including ripple, is amplified by an error amplifier. The error amplifier then controls a series-pass transistor to eliminate most of the error. If the output is low, the pass transistor is turned on harder. If the output is high, the pass transistor is turned on less. Using feedback, it is possible to build regulated supplies that show no significant change in output over the full range of load currents and over some range of the unregulated input. It should be emphasized that in circuits such as that in Fig. 5-22 the unregulated input must be at least 2 V greater than the regulated load voltage. Otherwise, the pass transistor will saturate, and control will be lost.

The regulation and ripple rejection of a feedback regulator is related to the gain of the error amplifier. A high-gain error amplifier will provide very good regulation and ripple rejection. However, the accuracy of the output can be no better than the accuracy

of the reference voltage. Some power supplies use an integrated circuit zener diode in place of an ordinary zener diode. These integrated circuits are also called *reference diodes* and provide a much more accurate reference voltage. For example, an ordinary zener 6.9-V shunt regulator will show a 17-mV change in output when its input changes 10 percent. An equivalent IC zener, the LM129, will show only a 180-μV change under the same conditions. The IC references are also more temperature-stable than ordinary zener references. The schematic symbol used to represent a reference diode or IC zener is the same as the ordinary zener symbol. The package may be a small type such as the TO-92 with two leads. Thus, it is possible to misidentify an IC reference and replace it with an ordinary zener. This will cause considerable degradation in the accuracy of the output voltage.

Integrated circuit voltage regulators that contain the reference circuit, the error amplifier, the pass transistor, and protection circuits all in one package are available. Figure 5-23 shows how easy these devices can be to apply. In the schematic diagram (Fig. 5-23[c]) three connections are made between the unregulated source and the load. The capacitors may not be required in some applications. Capacitor C_1 is needed only if the regulator is located some distance from the main filter capacitor. This is the case with on-card regulators. An on-card system uses a separate regulator on each printed circuit card in lieu of one large main regulator. Capacitor C_2 may be required to improve the response of the regulator to transients. Three terminal regulators are available in small packages (such as the TO-92) and in larger packages such as the TO-3 and TO-220, as shown in Fig. 5-23(a) and Fig. 5-23(b). Grounding the case or

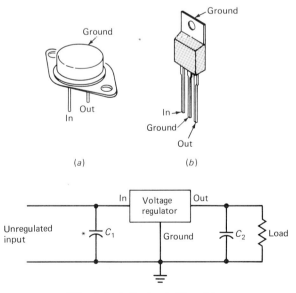

Fig. 5-23 Integrated voltage regulator. (a) TO-3 package. (b) TO-220 package. (c) Regulator circuit.

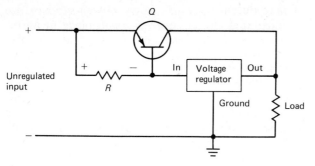

Fig. 5-24 Current boost circuit.

tab of the package is shown only as a general practice since these connections are inputs on some devices. The small package regulators have ratings of around 100 mA, the TO-220 devices as high as 3 A, and some TO-3 devices as high as 10 A. They are available with several fixed voltages such as in the 78XX series. The 7805 is a fixed 5-V regulator, the 7812 regulates at 12 V, the 7815 at 15 V, and so on. They are available as negative regulators in the 79XX series (for example, the 7905 is a negative 5-V regulator). Many integrated voltage regulators are adjustable, or can be configured for adjustment, and some are designed for switch-mode service, as will be seen in section 5-7.

The current of an integrated circuit voltage regulator can be boosted by adding an external PNP pass transistor to a fixed positive regulator as in the circuit of Fig. 5-24. The integrated circuit will supply all the load current up to the point where the current in resistor R is enough to cause a voltage drop of 0.6 V. At this point, the transistor will be turned on, and it too will help support the load current. If R is 4.7 Ω, the turn-on current will be

$$I = \frac{0.6}{4.7}$$
$$= 128 \text{ mA}$$

As the load demands more and more current, the drop across R will continue to increase and turn the pass transistor on harder. Figure 5-24 shows that the polarity of the drop across R is correct for forward-biasing the base-emitter junction of the PNP transistor. A similar circuit arrangement is possible with a negative regulator and an NPN pass transistor.

Fixed integrated voltage regulators can also be used to supply variable output voltages. In this service, the ground terminal is connected to some reference voltage rather than directly to ground. Figure 5-25 shows a dual-complementary supply with adjustable output voltages. The supply is *dual-complementary* since it provides both positive and negative voltages with respect to ground. The rectifier circuit is a combination of two full-wave, center-tapped supplies. This arrangement is also called a *center-tapped bridge rectifier*. Capacitors C_1 and C_2 filter the positive and negative voltages. The 7805 is a fixed positive 5-V integrated regulator, and the 7905 is a fixed negative 5-V regulator. However, the supply is adjustable over a range of 5 to 20 V positive and negative with respect to ground. Notice that the ground terminals of the fixed regulators are driven by triangular symbols. The triangle is commonly used in schematics and block diagrams to represent an amplifier. The inputs of the amplifiers in Fig. 5-25 are at the right and are marked with minus (−) and plus (+) symbols. The minus input is called the *inverting input*. Any positive-going signal applied here will cause the output to go in a negative direction, and any negative-going input will drive the output in a positive direction. Notice that the wiper arm of R_4 supplies the inverting input of the top amplifier. If the wiper arm is moved toward R_5, more of the negative output voltage will be applied to the top circuit. This negative-going signal is inverted by the top amplifier, and the ground lead of the 7805 is driven in a positive direction. The 7805 will develop an output that is 5 V plus the positive voltage sup-

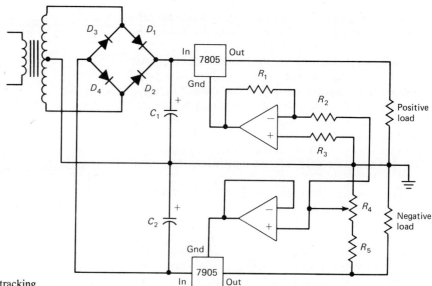

Fig. 5-25 Dual complementary supply with tracking.

plied to its ground lead. This makes the positive supply adjustable by changing the setting of R_4.

What happens to the negative voltage in Fig. 5-25 when R_4 is adjusted? The wiper arm of R_4 also supplies a signal to the bottom amplifier. In this case the signal is supplied to the plus input, which is a noninverting input. When the wiper arm is moved toward R_5 the resultant negative-going signal is amplified and drives the ground lead of the 7905 in a negative direction. The negative output voltage will be -5 V plus the negative voltage supplied to its ground lead. This makes the negative supply voltage adjustable by changing the setting of R_4.

Both outputs shown in Fig. 5-25 are adjustable with R_4. When the wiper arm of R_4 moves toward R_5, the positive output goes more positive and the negative output goes more negative. One control adjusts both supplies. This is known as a *tracking supply*. The positive output voltage tracks the negative output voltage. Supplies can be single or can have more than one output. If they have two outputs, they can be complementary (one positive and one negative). Supplies can be fixed or adjustable. Finally, supplies that have more than one output can be independent or tracking.

REVIEW QUESTIONS

21. Refer to Fig. 5-20. Assume 5 V across the load, a load current of 100 mA, an unregulated input of 8 V, and a zener current of half the load current. Calculate the resistance of R and its power dissipation.

22. Use the data of question 21 and calculate the zener dissipation with the load connected and with the load disconnected.

23. Use the data of question 21. At what load current will the circuit stop regulating? What happens to the zener current at this point? What happens to the output voltage if the load current increases even more?

24. Use the data of question 21. At what unregulated input voltage will the circuit stop regulating? (Hint: Try calculating the load resistance and draw an equivalent circuit.)

25. Refer to Fig. 5-21(b). Assume an unregulated input of 10 V, a current gain from base to emitter of 80, a 5.7-V zener, and a load current of 100 mA. Calculate a value for R that will set the zener current to half the base current.

26. Use the data of question 25. What load voltage can be expected?

5-5
CURRENT REGULATION

Most power supplies operate in the constant-voltage mode. They maintain a fixed load voltage over a range of load currents. A constant-current supply will maintain a fixed load current over a range of load

resistances. Constant-current supplies are useful for charging batteries, for supplying bias currents in reference circuits, for energizing electromagnets, and for performing various control applications. Many constant-voltage supplies also have a constant-current mode. This mode is useful to protect the supply and the circuits it energizes in the event of a fault such as a short circuit. A supply that reacts to an overload by changing from constant-voltage to constant-current operation is said to have *automatic crossover*.

Figure 5-26 shows three constant-current circuits. Each will supply a fixed current over a range of load resistance. At some high value of load resistance, the current will fall off as the load voltage cannot exceed the input voltage. The output voltage is called the *compliance voltage,* and the *compliance range* predicts the values of load resistance that will receive a constant current. Figure 5-26(a) is a field effect transistor (FET); the load current will be equal to its I_{DSS} rating. The FETs (constant-current diodes) may be used to supply a constant bias current to zeners or IC zeners for a more stable reference voltage. A transistor circuit in which the drop across the emitter resistor is zener-regulated is depicted in Fig. 5-26(b). The constant voltage across R_1 predicts a constant

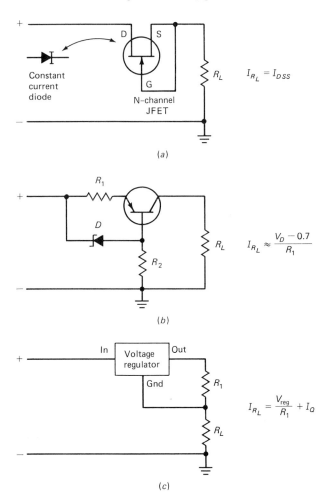

Fig. 5-26 Constant-current circuits.

current in R_1 and also in the emitter and collector circuits of the transistor. Figure 5-26(c) shows how an IC voltage regulator can be used to provide a constant current. The load current will be equal to the regulator output voltage divided by R_1 plus the quiescent drain current of the IC.

Figure 5-27 shows a constant-voltage power supply with current limiting. It uses an error amplifier to compare a reference voltage against a sample of the load voltage. Any change in load voltage will cause the amplifier to bias the pass transistor so as to reduce the error. Resistor R_1, diode D, and capacitor C form a stable reference voltage in Fig. 5-27. This reference voltage is fed to the noninverting input of the amplifier. The inverting input of the amplifier is connected to the junction of R_4 and R_5, which divide the load voltage. The amplifier output drives Q_2, which is direct-coupled to Q_1, the pass transistor. The direct connection from the emitter of Q_2 to the base of Q_1 produces a high overall current gain, and the two transistors are usually called a *Darlington pair*. Transistor Q_3 is off when the supply is operating in the constant voltage mode and has no effect on circuit performance. If the load should demand less current, the load voltage will tend to increase. The divided voltage at the inverting input of the error amplifier will also increase. When a positive-going signal appears at the inverting input, the output goes less positive. This means less drive to the Darlington pair, and the pass transistor now drops more voltage. This action decreases the load voltage, and much of the original error is eliminated.

The constant-voltage mode of Fig. 5-27 will range from zero load current to that value of load current which causes enough drop across R_3 to turn on Q_3. Resistor R_3 is in series with the load and can be considered a current-sensing resistor. When the load current goes high enough to drop about 0.6 V across R_3, Q_3 will turn on, assuming it is a silicon transistor. Notice that the polarity across R_3 forward-biases the

base-emitter junction of Q_3. When Q_3 comes on, it loads the output of the error amplifier. Current that normally was supplied through R_2 to the base of Q_2 is now supplied to Q_3. This reduces the drive to the Darlington pair, and the output voltage drops. The load resistance can continue to decrease toward short-circuit conditions, and the load current stays reasonably constant. Power supplies that cross over from constant-voltage to constant-current operation at some value of load current employ conventional current limiting.

Conventional current limiting applied to the current boost circuit studied earlier is shown in Fig. 5-28. Most integrated circuit voltage regulators have internal current limiting and protect themselves from overloads. When they are current-boosted, the internal current limiting will not protect the pass transistor. Figure 5-28 shows how two components can be added to provide current limiting for pass transistor Q_1. Resistor R_2 is the current-sensing resistor. It senses that portion of the load current supplied by pass transistor Q_1. When Q_1's current goes high enough, the drop across R_2 reaches 0.6 V and turns on Q_2. When Q_2 is on, it acts in parallel with R_1 to decrease its effective resistance. Less resistance means less voltage drop across R_1 and less bias for pass transistor Q_1. Transistor Q_1 will now drop more voltage from its collector-to-emitter terminal, and the load voltage will begin to drop. The supply has crossed over from constant-voltage to constant-current operation. The maximum load current in this circuit will be equal to the limiting current of the integrated circuit plus the limiting current of the pass transistor.

Conventional current limiting may not always protect the pass transistor and other components from damage. A sustained short circuit will create high-power dissipation in the pass transistor. For example, a 2N3055 pass transistor is rated at 15 A and a collector dissipation of 115 W. If it is operated in a 5-A 12-V power supply with conventional current limiting, it may seem that it should be safe under all load conditions. However, it can be destroyed by excessive collector dissipation. Suppose it is operated with an unregulated input of 18 V. If the load is

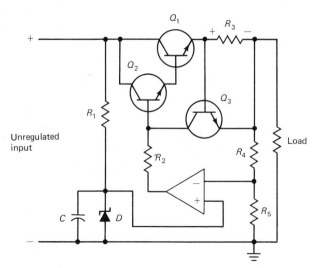

Fig. 5-27 Conventional current limiting.

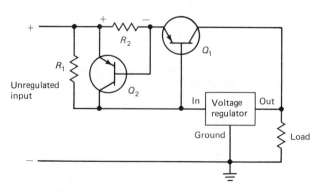

Fig. 5-28 Conventional current limiting added to a current boost circuit.

a short circuit, all of the input will drop across the pass transistor. The transistor dissipation will be

$$P_c = V_{ce} \times I_c$$
$$= 18 \times 5$$
$$= 90 \text{ W}$$

It still seems safe since 90 W is less than its rated 115 W. However, its rated dissipation is for a case temperature of 25°C (77°F). When a transistor is dissipating 90 W it becomes very hot. Figure 5-29 shows the derating curve for the transistor in question. Notice that the curve shows 90 W of dissipation to be maximum at a temperature of 63°C. The transistor may be destroyed if the short circuit lasts long enough for the case to exceed this temperature.

Another form of current limiting may be more desirable, especially when long-term overloads are expected. Foldback current limiting provides better protection for the pass transistor and other circuit components. Figure 5-30 compares the graphs for two types of current limiting. Notice that in Fig. 5-30(a) the supply operates at a constant output of 12 V up to load currents just over 5 A. This is the constant-voltage region. As the load resistance drops, the current demand goes beyond 5 A, and the output voltage starts to drop. It continues to drop until it reaches 0 V at short-circuit conditions. There is little current change from the beginning of limiting to the short-circuit condition. In foldback limiting (Fig. 5-30[b]) the constant-voltage region is the same. As the load demands more than 5 A, the current starts to fold back (decrease). At short-circuit conditions, the current is 2 A. The dissipation in the pass transistor can now be calculated by using the same conditions as before but with a foldback current of 2 A:

$$P_c = 18 \times 2 = 36 \text{ W}$$

This is a much more reasonable dissipation. According to Fig. 5-29 the transistor is safe up to a case temperature of 140°C. A moderate heat sink will be able to accomplish this, and the pass transistor will be safe even if subjected to a prolonged short circuit.

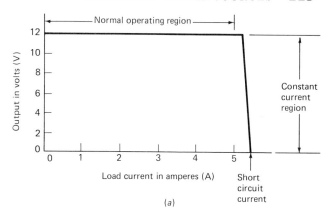

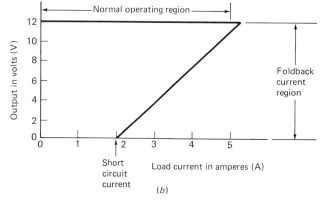

Fig. 5-30 Conventional current limiting compared to foldback current limiting. (a) Conventional current limiting. (b) Foldback current limiting.

Figure 5-31 shows a circuit that employs foldback current limiting. Transistor Q_3 is off under normal load conditions. Components R_1, D, and C develop a stable reference voltage for the noninverting input of the error amplifier. Resistors R_6 and R_7 divide the

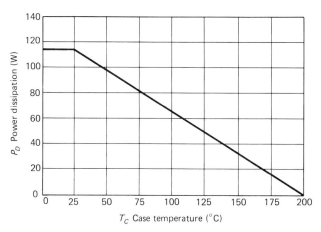

Fig. 5-29 Power temperature derating curve for a 2N3055 transistor.

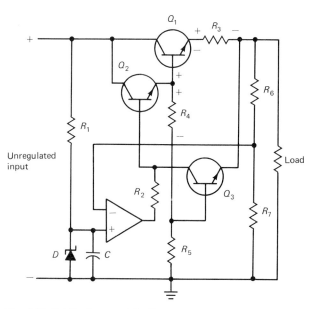

Fig. 5-31 Foldback current limiting.

load voltage for the inverting input. Any error produces more or less drive to the Darlington pair, Q_2 and Q_1, and the supply operates as a constant voltage source. If the load demand continues to increase, the drop across R_3 (the current-sensing resistor) and V_{be} of Q_1 will increase. Note that the polarity of these two drops provides forward bias for the base-emitter circuit of Q_3. Also notice that the drop across R_4 provides a reverse bias for Q_3; Q_3 will not come on until the two forward drops can overcome the drop across R_4 and the required junction voltage of Q_3. When Q_3 comes on, the limiting action begins. Now Q_3 can divert drive current from the pass circuit. This causes the output voltage to begin dropping. It also reduces the voltage across R_4 and R_5 since the base voltage of Q_1 tracks the output. With less drop across R_4, Q_3 turns on harder since the R_4 drop is a reverse-bias source for Q_3. The output is now folding back since Q_3 is conducting more than it did before with a given voltage across the current-sensing resistor. If the load becomes a short circuit, the current in R_4 approaches zero and no longer acts to produce any reverse bias for Q_3. Now, only a fraction of the rated supply current is required in R_3 to keep Q_3 on.

Current-limiting circuits of each type can "latch-up" under certain conditions, although the foldback type is more likely to do so. *Latch-up* occurs when the supply is turned on and the output fails to come up to its full value. It latches at some point on its limiting curve. Some loads present a low resistance at turn-on; for example, an incandescent lamp shows a very low resistance when cold. This type of load can latch the supply at some point on its current-limiting curve. With the current limited, the lamp will never reach its normal operating temperature or its normal operating resistance, and the output will remain latched at some voltage lower than normal. Dual-complementary power supplies may also latch up if one polarity supplies a bias to a circuit energized by the other polarity. At turn-on, the bias is missing, the circuit draws excessive current, and the supply goes into limiting. If the bias side uses voltage from the limiting side, the bias may never reach a normal level, and the limiting action will continue indefinitely.

REVIEW QUESTIONS

27. Refer to Fig. 5-27. Assume R_3 is a 0.15-Ω resistor and Q_3 is a silicon transistor. At what load current will the limiting action begin?

28. Use the same conditions as in question 27. If the output is 12 V, over what range of load resistance will the supply act as a voltage source?

29. Use the same conditions as in question 28. Over what range of load resistance will the supply act as a current source?

30. Refer to Fig. 5-28. Assume that the integrated circuit regulator is internally limited to 1.5 A, that Q_2 is a silicon transistor, and that R_2 is

a 0.07-Ω resistor. Calculate the maximum load current.

31. Use the information provided by Fig. 5-29 and Fig. 5-30(b). Calculate the maximum safe case temperature if the unregulated input is 15 V and the load is a short circuit.

32. Use Fig. 5-30(a) and predict the load current for a 2-Ω load resistor. (Hint: Use Ohm's law and a graphical approach that satisfies a slope of 2 Ω.)

5-6 PROTECTION DEVICES AND CIRCUITS

The current regulation circuits discussed in the previous section provide protection for the power supply and load circuits for certain kinds of faults. This section examines some other devices and circuits that offer additional protection. Fuses and circuit breakers are covered in Chapter 4. You may wish to review the discussion of these devices since they are commonly used to protect power supplies and other industrial circuits.

Line transients are common in the industrial environment. They are caused by lightning, switching of large inductive loads such as motors, and malfunctions in other parts of the industrial plant. They can cause all sorts of damage to wiring and equipment, and semiconductors are especially susceptible. Studies show that one 5000-V transient can be expected per year for every 120-V circuit in this country; lower voltage transients can also be expected even more frequently. Fuses and circuit breakers are too slow-acting to prevent transient damage.

A varistor (Fig. 5-32) can be used to absorb transient energy safely and to protect sensitive diodes, transistors, and integrated circuits. *Varistors* are voltage-dependent resistors. When the line voltage is normal, a varistor has a very high resistance and draws very little current from the line. When a tran-

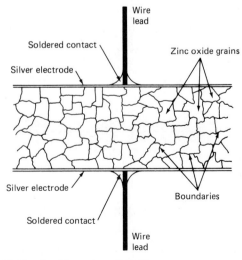

Fig. 5-32 Metal oxide varistor (MOV) structure.

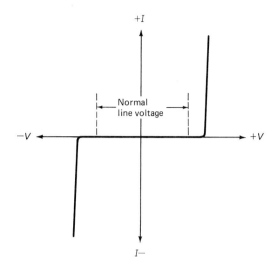

Fig. 5-33 MOV volt-ampere characteristic curve.

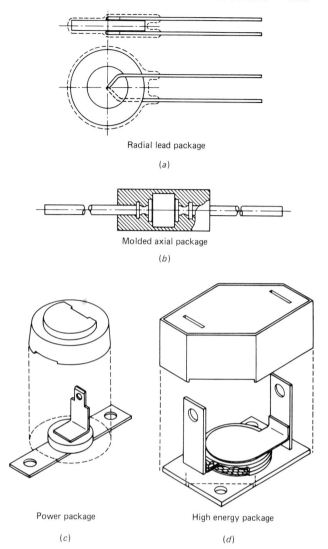

Fig. 5-34 MOV package styles (*Courtesy General Electric Co.*).

sient comes along, the resistance of the varistor drops sharply. This drop will cause high current in the varistor, and the transient will be safely absorbed. The wafer of zinc oxide in Fig. 5-32 contains boundaries between the grains. These boundaries act as semiconductor junctions, and each requires about 3 V to become forward-biased. The boundaries act in series, and the breakdown characteristics may be controlled by the thickness of the wafer. The metallic oxide varistor (MOV) structure produces a rather high capacitance. This characteristic is usually of no consequence in power line applications.

Figure 5-33 shows the characteristic volt-ampere curve for an MOV. There is no current flow over the normal ac line swing. However, if the line swings positive enough, the positive current in the MOV will increase sharply. The graph shows the same action for positive- and negative-going transients.

Four common packages for MOV devices are shown in Fig. 5-34. The axial devices (Fig. 5-34[*b*]) can absorb 2 joules (J) of transient energy at currents up to 100 A. The high-energy package (Fig. 5-34[*d*]) is rated up to 6500 J and 50,000 A. Fortunately, transients are usually very short in duration.

EXAMPLE

Suppose a transient reaches 5000 V, lasts 10 μs, and causes a current flow of 100 A in an MOV device. What is the absorbed transient energy?

SOLUTION

$$\text{Energy} = V \times I \times t$$
$$= 5000 \times 100 \times 10 \times 10^{-6}$$
$$= 5 \text{ J}$$

Five joules is too much for the axial package but could be easily handled by the larger packages shown in Fig. 5-34.

Figure 5-35 shows a typical application of an MOV in a power supply circuit. The varistor is connected in parallel with the transformer primary. Normally, it shows a very high resistance and draws almost no current from the ac line. In the event of a transient,

it drops in resistance and absorbs much of the transient energy. It protects the transformer, rectifier, filter, regulator, and load. Depending on the amplitude and duration of the transient, the fuse in Fig. 5-35 may not blow. The MOVs act in nanoseconds and are therefore 100,000 times faster than fuses.

Another type of overvoltage situation can develop in power supply circuits. Many circuits use series pass transistors. These are hardworking devices and are therefore failure prone. Unfortunately, the most common failure mode in a series-pass transistor is an emitter-to-collector short. This fault places the entire unregulated supply voltage across the load. Since the load may contain many sensitive devices such as integrated circuits, extensive damage can result in a piece of equipment if a series-pass transistor short-circuits.

Figure 5-36 shows a crowbar circuit added to a high-current power supply to prevent circuit damage in the event of a regulator failure. The crowbar circuit is made up of D_1, R_9, C_5, and the SCR. With normal load voltage, the zener (D_1) is off. If the load

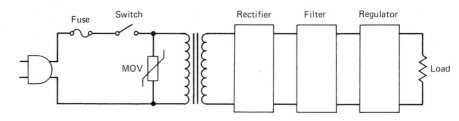

Fig. 5-35 MOV protected power supply.

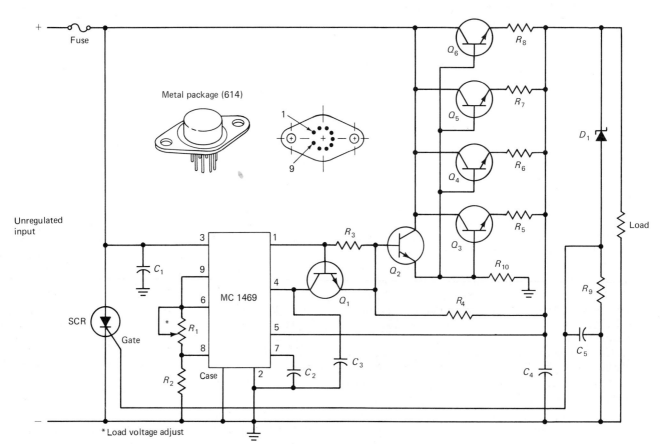

Fig. 5-36 High current supply with crowbar overvoltage protection.

voltage goes too high, the zener breaks over and current flows in R_9. The drop across R_9 gates the SCR, and it turns on. Notice that the SCR sits across the unregulated input and therefore quickly blows the fuse. Capacitor C_5 prevents noise from false-gating the SCR. The simple zener gating circuit does not provide a precise crowbar action. Some circuits are more elaborate. Overvoltage-sensing integrated circuits that provide accurate trip points and programmable delay characteristics are available.

The rest of Fig. 5-36 works as the series-pass circuits already discussed. However, there are a few details worth mentioning. There are four pass transistors operating in parallel. Transistors Q_3 to Q_6 share the load current and give this regulator high current capacity. Resistors R_5 to R_8 help balance the transistor currents. If one transistor has a higher current gain than the others, it will conduct more than its share of the load current. This will make it run hotter than the other three transistors. Transistor

gain increases with temperature rise. As it heats, it will continue to conduct more current and become even hotter. This situation is called *thermal runaway* and can lead to the destruction of the transistor. The emitter resistors prevent thermal runaway in this circuit by dropping more voltage with an increase in current. This drop subtracts from the base-emitter bias and decreases the transistor current. Typically, these resistors (R_5 to R_8) are 0.1 Ω. Transistor Q_2 in Fig. 5-36 is a driver for the pass transistors. The MC1469 integrated regulator cannot supply enough drive current for four pass transistors, and Q_2 provides the needed current gain. Transistor Q_1 senses the drop across R_3 and, along with the internal circuits of the MC1469, provides conventional current limiting. Notice that the MC1469 is housed in a 614 metal package with 9 pins.

Thermal protection involves sensing the temperature of some component (usually the pass transistor) and shutting the supply down at some critical temperature. Many of the integrated circuit regulators, such as the 78XX series, employ this technique. A thermal shutdown transistor is located close to the series-pass transistor in the circuit. The shutdown transistor is biased to 0.4 V and remains off at reasonable temperatures; if the shutdown transistor gets too hot (150°C or so), 0.4 V is enough to turn it on. When it comes on, it removes the base drive from the pass transistor, and the output is turned off. The shutdown temperature is several degrees above the temperature at which the regulator will turn back on. The difference between these two temperatures, known as *hysteresis*, prevents the regulator from rapidly *oscillating* (switching back and forth) between on and off conditions.

REVIEW QUESTIONS

33. Is the relationship between voltage and resistance in a metal oxide varistor (MOV) direct or inverse?

34. Is the relationship between the duration of a transient and its energy content direct or inverse?

35. Calculate the energy in joules contained in a 2000-V, 350-A transient that lasts 20 μs.

5-7
SWITCH-MODE SUPPLIES

Almost all of the voltage regulators studied up to this point have used a series-pass transistor. These regulators can provide excellent performance but are relatively inefficient. The pass transistors consume a significant part of the total energy as they conduct the full load current and drop part of the unregulated input voltage. They run hot and require substantial heat sinking in many cases. A design that eliminates series-pass transistors is illustrated in Fig. 5-37. It is not a switch-mode supply but is included in this section because it introduces an important concept.

The bridge section of Fig. 5-37 uses two rectifiers and two silicon-controlled rectifiers. The SCR control circuit gates the bridge into conduction early or later in the half cycle, depending upon demand and line voltage. The waveforms show how the average dc output of the bridge is related to SCR gating time. Early gating produces a high dc average, and late gating produces a low dc average. An error amplifier compares a reference voltage with a sample of the load voltage. Any difference will change the gating of the SCRs and reduce the error.

The SCR-regulated supply is more efficient than any of the series-pass circuits. When an SCR is off, it blocks current. Zero current ensures zero power dissipation in the SCR. When an SCR is on, it drops very little voltage, thus ensuring little power dissipation in the SCR. The point is that the SCR does not dissipate very much power because it is a switching device. It is either off or on. On the other hand, a pass transistor is somewhere between off and on. It operates in the linear region; it acts as a resistor

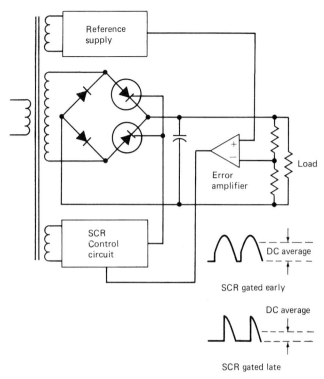

Fig. 5-37 SCR regulated power supply.

and therefore consumes energy. In switch-mode power supplies the control transistors are not operated in the linear region. This gives this type of supply an efficiency advantage similar to that achieved by the SCR supply and also allows it to be operated at frequencies far beyond the line frequency. The higher frequencies allow a significant reduction in the size and weight of transformers, inductors, and filter capacitors. Switching supplies are about one-third the size and weight of equivalent linear types.

Switch-mode operation can achieve regulation by using pulse-width modulation. Figure 5-38(a) represents a high-duty cycle since the width of the positive-going pulse is a large percentage of the cycle. In this case it is about 75 percent. A square wave has a duty cycle of 50 percent since the positive pulse is equal to one-half cycle. Assuming an operating fre-

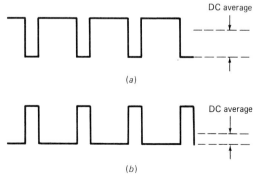

Fig. 5-38 Pulse width modulation. (a) High duty cycle. (b) Low duty cycle.

quency of 40 kHz, the calculation would appear as follows:

$$\text{Duty cycle, \%} = \frac{t_{\text{high}}}{T} \times 100$$
$$= \frac{18.75 \times 10^{-6}}{25 \times 10^{-6}} \times 100$$
$$= 75\%$$

With rectangular waveforms, the dc average is equal to the dc peak times the duty cycle. Assuming peak values of 100 V, the waveform in Fig. 5-38(a) would average to 75 V and the waveform in Fig. 5-38(b) to 25 V because its duty cycle is 25 percent.

Figure 5-39 shows three switch-mode regulator configurations. In each case, the switching transistor is driven by a pulse width modulator (PWM). The dc load voltage is controlled in these circuits by controlling the duty cycle of the rectangular waveform supplied to the base of the switching transistor.

In the circuit of Fig. 5-39(a) the load voltage is less than the input voltage. When the output of the PWM goes positive, the transistor turns on. Load current is supplied through the transistor and the inductor. When the output of the PWM goes negative, the transistor turns off. Load current continues to flow since the field of the inductor collapses and forward-biases the diode. The filter capacitor also helps to maintain load current while the transistor is off. The load receives relatively pure direct current. The step-down configuration is much more efficient than its series-pass equivalent, especially in those cases in which the load voltage is considerably less than the input voltage.

Not only is the switching mode much more efficient, it also allows regulator configurations not possible with the series-pass arrangement. In Fig. 5-39(b) the load voltage is greater than the input voltage. When the PWM goes positive it turns the transistor on and current flows through the transistor and the inductor. When it goes negative, the transistor turns off, and the field in the inductor collapses and generates a voltage that adds in series with the input voltage. The diode is now forward-biased, load current flows, and the filter capacitor is charged. When the transistor turns on again, the diode prevents it from discharging the filter capacitor. The load voltage can be opposite in polarity to the input voltage as shown in Fig. 5-39(c). These circuits are handy in cases such as a positive ground system in which one or two negative voltages are required. When the transistor turns off, the inductor discharges through the diode and the load. Since the direction of discharge current must be the same as the direction of charge current, demonstrating that the load polarity is opposite to the source polarity is easy.

Regulation in a switch-mode supply is a matter of comparing the load voltage with a reference voltage and using any error to correct the duty cycle of the output of the PWM. Figure 5-40 shows a step-down regulator based on an integrated circuit specially designed for switch-mode power supplies. The PWM is contained in the chip and consists of an oscillator, AND gate, and a latch. The oscillator produces a rectangular waveform, and its frequency of operation is controlled by external capacitor C_1. The AND gate is a circuit that turns on and passes a signal to the latch only when both of its inputs are positive. The *latch* is a storage circuit that remembers a high signal that was applied to its S input and supplies that high signal to its Q output until it is reset at its R input. Transistors Q_1 and Q_2 form a Darlington switch that will be turned on when the Q output of the latch is high. Voltage regulation is achieved in Fig. 5-40 by comparing an internally developed reference voltage with a sample of the load voltage. Any error is compensated for by pulse-width modulation. For example, if the load demands more current, the voltage tends to drop. Dividers R_3 and R_2 will then send a negative-going signal to the error amplifier. It is inverted, and the output of the amplifier goes positive (high), enabling the AND gate. The positive-going oscillator will then set the latch and turn on the Darlington pair. The latch is reset at its R input by the oscillator signal. The duty cycle is a function of the error amplifier output. For example, if the amplifier never goes high, the duty cycle will be 0 percent since the AND gate will never be enabled. On the other hand, if the amplifier output is continuously high, the latch will be immediately set after it is reset and the duty cycle will approach 100 percent. Normal operation falls in between these two extremes.

Figure 5-40 also provides for current limiting. Resistor R_1 senses the current to the Darlington pair.

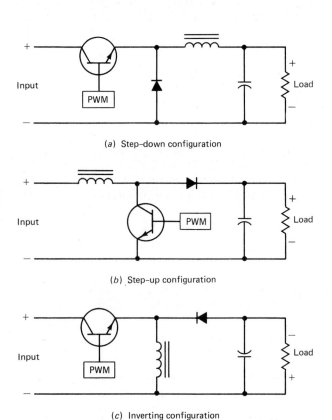

(a) Step-down configuration

(b) Step-up configuration

(c) Inverting configuration

Fig. 5-39 Switch-mode regulator configurations.

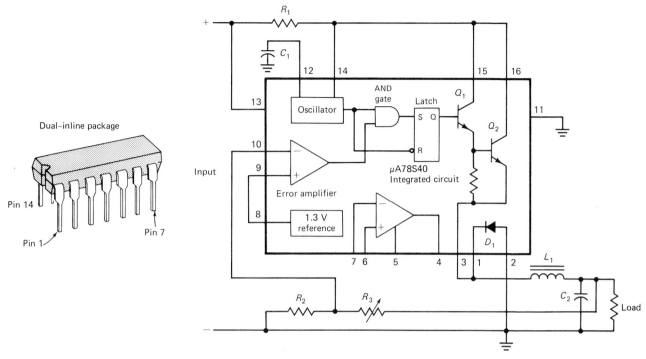

Fig. 5-40 Integrated circuit step-down regulator.

Dual-inline package

Pin 14

Pin 1

Pin 7

When the drop across R_1 reaches about 0.3 V the oscillator waveform is reduced in duty cycle. Now it is not possible for the Darlington switch to be driven with a duty cycle as high as before, and the output is limited. As shown, the circuit can safely deliver about 1 A of load current. It is possible to use pin 3 of the integrated circuit to drive an external switching transistor for more load current. It is also possible to use an external diode and therefore not use pins 1 and 2 on the device. Pins 4 through 7 access an additional amplifier that can be used for temperature control or some other power supply feature. The integrated circuit can be used in any of the three configurations shown in Fig. 5-39.

Converters are another category of switch-mode supply. A *converter* is a circuit that changes direct current to alternating current and then back to direct current again. This makes it possible to transform one dc voltage to another dc voltage. It also permits transformer isolation at a considerable savings in size and weight because the frequency of operation is much higher than the normal line frequency. Figure 5-41 shows a width-controlled converter with voltage regulation. Transistors Q_1 and Q_2 are driven with out-of-phase rectangular waves. Note that they will alternately conduct and allow primary current in T_1. The load voltage is sampled by divider R_1-R_2 and compared with a reference. Any error is used to modulate the width of the pulses supplied to the base circuits. Increased load demand will be compensated for by increasing the pulse width and the ON time for both transistors. This will increase the energy delivered to the transformer and compensate for the increased load demand.

Diodes D_1 and D_2 in Fig. 5-41 rectify the high-frequency alternating current. Inductor L_1 and capacitor C_1 form a choke-input filter. There are periods of time when both transistors are off, and L_1 will then discharge to maintain the load current. Diode D_3 is forward-biased by L_1 at those times and completes the discharge circuit. The circuit will function without D_3, but then the discharge current is forced to flow through the rectifiers and the secondary of the transformer. This condition is not desirable because it increases the dissipation and lowers circuit efficiency.

Switch-mode regulators and converters are quite a bit more efficient than series-pass arrangements. Better efficiency translates to reduced heat sink requirements, and a smaller supply results. The high frequency of operation reduces the size of filter components and the size of the cores in the inductive components. All in all, it seems that they have every advantage. There is one disadvantage, however. They are noisy. They operate with rectangular waveforms. *Rectangular waveforms* are composed of a fundamental frequency plus a series of harmonically related frequencies. The third harmonic is three times the fundamental; the fifth is five times the fundamental; and so on. Harmonics add to the noise in the output of the supply. They also create another problem, called *electromagnetic interference* (EMI). The energy content of the harmonics falls off at the higher frequencies, and the majority of the noise ranges from 10 to 500 kHz. There is often enough high-frequency energy to cause radiated interference with nearby equipment. Shielding and output filters are required to control radiated EMI.

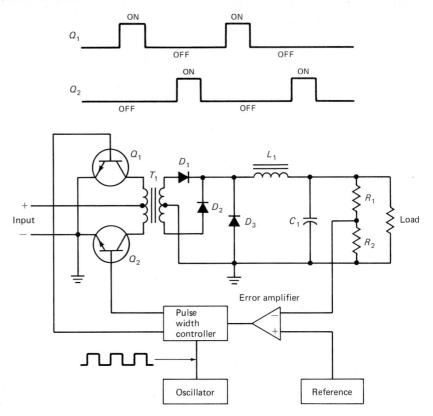

Fig. 5-41 Width-controlled converter/regulator.

Switchers also may create problems with conducted EMI. The location of switching transistors directly off the power line and their high peak currents can create significant line noise. Switchers can place noise on both sides of the line referenced to ground (*common mode noise*) and can also place noise on one side of the line referenced to the other (*differential mode noise*). Line filters are required to attenuate these forms of line noise.

Figure 5-42 shows a frequency-controlled sine wave converter. Sine waves solve some of the noise and EMI problems because they have no harmonic content. The circuit uses power FETs to switch the direct current produced by D_1-D_4. The resulting square wave is converted to a sine wave by a resonant tank circuit formed by L_1 and C_3. These two components are tightly coupled by transformer T_1 and behave as though they were in parallel. The high-frequency sine waves are changed to direct current by Schottky rectifiers D_5 and D_6 and L_2 and C_4 form a choke-input filter.

The circuit of Fig. 5-42 uses frequency modulation rather than pulse-width modulation. The duty cycle of the drive signals supplied to the FETs is fixed at 50 percent. Any change in load voltage will shift the frequency of the drive signal. This is accomplished in the voltage-controlled oscillator (VCO). The tank circuit has some natural resonant frequency established by the inductance of L_1 and the capacitance of C_3. Suppose the VCO is generating a signal above the resonant frequency. A tank circuit shows maximum voltage when driven at its resonant frequency.

The circuit Q is such that an octave-frequency increase translates to a 12-decibel (dB) drop in tank voltage. An octave increase means double and a 12-dB drop means one-fourth. Therefore, if the tank voltage were 20 V at a VCO frequency of 150 kHz it would drop to 5 V at a frequency of 300 kHz. Any error in load voltage is corrected by shifting the VCO in the proper direction. Figure 5-42 operates above resonance, and an increased load demand will lower the VCO frequency closer to resonance which will increase the tank voltage. Decreased load demand will raise the VCO frequency further away from resonance and decrease the tank voltage. The slope of the control curve is 12 dB per octave.

The power FETs and Schottky rectifiers in Fig. 5-42 permit good circuit efficiency into the hundreds of kilohertz. This is not possible with bipolar transistors and ordinary rectifiers because of their poor switching performance at high frequencies. They cannot turn off fast enough, because of carrier storage. It takes time to sweep all carriers from their junctions so conduction does not cease immediately when forward bias is removed. Field effect transistors are unipolar and do not exhibit the carrier storage associated with bipolar devices. The Schottky diodes use metal on one side of the junction and doped silicon on the other. This construction technique eliminates the depletion region and the storage problem. Power FETs are also more rugged, in that they do not suffer from *secondary breakdown,* which is a phenomenon suffered in bipolar devices when the crystal develops hot spots. The hot spots are

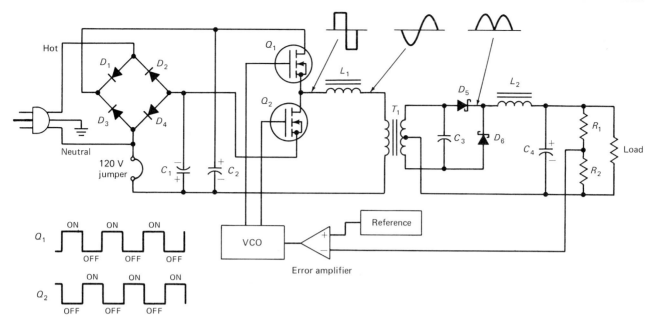

Fig. 5-42 Frequency-controlled sine wave converter.

caused by current crowding associated with the intense fields generated in power devices and can lead to breakdown even though the transistor is operating in its safe area. Primary breakdown is caused by operation outside the safe area.

Diodes D_1 to D_4 and capacitors C_1 and C_2 in Fig. 5-42 make up a bridge-doubler circuit. The circuit acts as a doubler with the 120-V jumper installed; the dc voltage supplied to the switching transistors would be around 240 V with a 120-V ac input. With the jumper removed, the circuit acts as a bridge rectifier and supplies about 240 V to the transistors with a 240-V ac input. Thus, the circuit can be configured for either of two line voltages with the jumper. Line isolation is achieved in T_1. Since the frequency of operation is so high, this transformer is tiny compared to an equivalent line frequency device.

REVIEW QUESTIONS

36. Calculate the efficiency of a series-pass regulator that supplies 5 A at 12 V from an unregulated input of 18 V. (Hint: *efficiency* is found by dividing useful output by total input.)

37. If a switch-mode supply operates at 30 kHz, calculate the duty cycle if the ON time is equal to 12 μs.

38. Assume that you are observing the drive signal to the base of a switching transistor in a power supply. What can you expect to see happen if the load current is suddenly decreased?

39. Calculate the average voltage of a rectangular waveform with a peak value of 25 V and a duty cycle of 85 percent.

40. Can the integrated circuit shown in Fig. 5-40

be used to develop a negative voltage in a positive ground system?

41. What will happen to the load voltage in Fig. 5-40 if R_3 is adjusted for less resistance?

42. Which component in Fig. 5-41 is included to reduce the dissipation in the transformer and the rectifiers?

43. What do schematic symbols D_5 and D_6 in Fig. 5-42 depict?

5-8 TROUBLESHOOTING AND MAINTENANCE

Power sources are considered the heart of electronic equipment. They must function properly for the equipment to work as designed. Technicians must be familiar with power supply operation and know how to verify correct performance. They must be able to diagnose malfunctions and replace components to restore proper and safe operation. Because the power supply can affect all parts of the system, verification of supply voltages, currents, and waveforms must precede any troubleshooting efforts in other parts of the system. More than one technician has spent valuable time troubleshooting a normal circuit that is acting abnormally because of a faulty supply. It is usually easy to verify proper operation in a power supply, and this must be done early in the troubleshooting process.

Safety is the main consideration when troubleshooting. Severe electrical shock, burns, fires, equipment damage, and losses in production time are some of the penalties for improper work procedures.

Grounded test equipment may lead to ground loops. Some parts of the ac line and some parts of the power supply circuitry can be hot with respect to an earth ground. The act of plugging in a piece of test equipment normally connects its case and any ground leads to an earth ground. Therefore, if any of the ground leads is brought into contact with a hot circuit, a ground loop results and very high fault currents will flow. Or the technician's body can become part of the loop if the case of the test equipment and a hot circuit are touched at the same time.

It may be necessary to use battery-operated test equipment when working on some industrial circuits. This permits the instrument case and test leads to remain "floating" with respect to ground. Floating measurements require special safety procedures. For example, the case of a piece of test equipment that is being used in a floating measurement may be hundreds of volts with respect to ground. Touching the case may cause a severe shock if the technician's body has any conductivity to ground. Special test equipment, insulating mats, clothing, and other protective gear are musts when making floating measurements.

Isolation transformers can be used in some cases to prevent ground loops. The power supply can be isolated from the ac line by energizing it from the transformer secondary. This eliminates ground loops but does not eliminate other shock hazards. No matter what the working conditions are, safe practices must be followed. Safe workers have an orderly way of working, regardless of the conditions at the time. They regard all circuits as potentially dangerous and are not lulled into sloppy work habits by terms such as *low voltage*. For example, Fig. 5-42 shows a power supply design that could be used in an industrial control computer to supply 5 V to the logic circuits. What could be safer than 5 V? Well, the power supply does develop 5 V for the load but works initially at 240 V dc in the switching section! Direct current of 240 V dc is potentially very dangerous and must be treated with respect. Figure 5-42 also points out that line isolation is achieved in only part of the power supply circuit. The bridge-doubler runs directly off the ac line, and a ground loop can result during work on this part of the circuit.

The industrial technician must also be aware that the metal case of certain power supply components may be at a high potential with respect to the common ground. For example, refer to the voltage-doubler circuits shown in Fig. 5-18. Depending on capacitor construction, C_1 could be a shock hazard in these circuits. In the half-wave doubler, the case of C_1 would be at the full secondary voltage referenced to ground. In the full-wave doubler, the case of C_1 would be at half the load voltage referenced to ground.

Power supplies are potential shock hazards even when turned off. Today, the large capacitive filter is the most common way to achieve pure direct current.

Capacitors can store quite a bit of energy and may store it for long periods of time. Safety demands that the capacitors be discharged before touching parts of the circuit or making some measurements. Short-circuiting capacitors by a hand tool or test lead is not advisable in high-energy supplies. The energy delivered by some capacitors when short-circuited is enough to vaporize an alligator clip! Special discharge rods with internal current-limiting resistors are used to bleed off capacitors in high-energy equipment.

Troubleshooting is a logical procedure. It begins by carefully observing the symptoms and all operating conditions. The logical technician verifies all control settings, external connections and cabling, and power to the unit before tearing it down. Once the technician is sure that there are no obvious external reasons for malfunction, then it is time to power down. In the industrial environment, it is often necessary to lock circuits off and tag them. The tag warns that maintenance is underway. Tear-down procedures are often presented in the equipment manufacturer's literature. So are important safety precautions. Find all relevant literature and use it!

A visual inspection of the inside of the equipment follows tear-down. Look for obvious problems such as foreign objects, leaking batteries and capacitors, broken wires, cracked circuit boards, burned components, components not seated properly in their sockets, circuit boards not seated properly in their connectors, loose connectors, and dirt. The industrial environment is often very dirty, and you may find that the equipment is loaded with grime and dust. This kind of build-up is often conductive and must be cleaned. Be sure to wear a respirator when cleaning up since the dust may be hazardous. Do not forget to check ventilation systems and to clean any air filters. Any liquid material inside the equipment may present a more difficult problem. Try to find out where it came from and what it is.

It may be necessary to remove circuit boards and wash them, in some cases. Check the manufacturer's recommendations before proceeding. Some may recommend a solution of 90 percent ethyl alcohol and 10 percent water. If corrosive materials such as chlorides are on the board, a solution of water and bicarbonate of soda may be used to clean and neutralize the board. The boards must be thoroughly rinsed with deionized water and dried before installation.

Battery and cell maintenance is usually straightforward. A voltage check will verify whether the unit must be replaced. However, do not make the mistake of measuring open-circuit voltage. A weak battery will often show a normal voltage until loaded. Energize the load circuit or select a resistor of the proper resistance and power ratings and make a test under normal load conditions. Some equipment may have a battery test switch for this purpose. If a secondary battery is below the cut-off voltage, verify proper operation of the charging circuit. Keep a maintenance log and always date the replacement of any cell or battery. Many technicians also tag the equipment to indicate the date that the battery was replaced. These habits save time and money. Finally, leaking cells and batteries must be replaced because electrolyte material is highly corrosive and conductive.

The most common failure mode for solid-state rectifiers is a short circuit. Make sure the power is off and that the filters are discharged and then run an ohmmeter check when you suspect a bad rectifier. Use a low range and check for different readings as the ohmmeter polarity is reversed. Do not expect to see infinite reverse resistance when running in-circuit checks. For example, refer to Fig. 5-10. Suppose the ohmmeter positive lead is on the cathode of D_2 and the negative lead is on the anode. This reverse-biases D_2 and D_1, but ohmmeter current will flow through the bottom half of the secondary and the load. If the diode is good, the reading will be a function of the load and secondary resistance. The forward resistance of a solid-state diode is also important. Most ohmmeters do not turn the diode on very hard, and a good diode will show a resistance considerably greater than 0 Ω. A short-circuited diode will show 0 Ω in both directions. Removal of at least one diode lead from the circuit will allow conclusive tests. Silicon rectifiers normally have a reverse resistance higher than the top range of the ohmmeter. Replace any leaky units. Don't forget that at least 0.6 V is necessary to turn on a silicon diode. Some ohmmeters have a special low-voltage ohms function, which cannot be used when testing diodes. Finally, if a rectifier is rated at more than 1000 V it may be a series combination of diodes and cannot be tested with an ordinary ohmmeter. It will test open (infinite resistance) in both directions.

Filter capacitors, especially the electrolytic type, are failure-prone. They may develop excessive leakage and may even short-circuit. They can also dry out and lose much of their capacity. Finally, they can develop a high series resistance which limits their ability to deliver load current. In-circuit testing can be used to find a short-circuited capacitor. Observe polarity and, as in the case of rectifier testing, be aware of other paths for the ohmmeter current. Remove at least one capacitor lead for more conclusive testing. A momentary low resistance followed by increasing resistance is to be expected when ohmmeter-testing large capacitors. Large electrolytics always show some leakage, and the ohmmeter will not reach an infinite reading on its highest range. Testing for excessive leakage is best done at the rated voltage. Also, testing for capacity and series resistance demands a capacitor tester. It may be most effective to try a new capacitor when symptoms such as excessive ripple point to the filter. Electrolytics have a shelf life and the technician should be aware that a "new" capacitor can be defective, especially if it has spent 10 years in storage.

As mentioned before, troubleshooting is a logical process. Good troubleshooters use analysis to limit the possibilities. They can take a set of symptoms and zero in on a set of possible causes. They understand circuit laws and know how circuits operate normally. This knowledge leads them to the answers they are looking for. For example, look at Fig. 5-20. Suppose that the load voltage is zero. First verify that there is some input to the circuit. It is not productive to troubleshoot a regulator circuit until it is verified that the input is normal or at least low. Zero input usually points to a defect in a circuit before the regulator. Low input may point to an overload condition. Suppose the unregulated input in Fig. 5-20 is low or normal. What kinds of faults could cause the load voltage to be zero? Resistor R may be open, or the zener, capacitor, or load may be short-circuited. If resistor R is open, it will be cold. If there is a short circuit, it will probably be hot. In fact, it may smell and look burned. Use all of your senses but be careful what you touch. You could be burned or shocked.

Suppose the load voltage in Fig. 5-20 is too high. This changes the analysis. Now, the only probable fault is an open zener diode. It is unlikely that R is short-circuited because resistors seldom short-circuit; if it did short-circuit, it would probably destroy the diode.

We are dealing with voltage analysis. We are measuring circuit voltages and analyzing possible causes for improper readings. Learn how to take measurements safely when the equipment is on because much troubleshooting must be done this way. Watch where you put your arm, wrist, hand, and fingers. Use insulated probes and never have more than one hand in the circuit at one time. A forearm-to-finger shock can be bad enough, but a hand-to-hand shock can be lethal.

After the voltage analysis, it may be time to go back to resistance analysis. If the load voltage is zero in Fig. 5-20, we know that any of three short circuits and one open circuit are among the possibilities. Power down, discharge the capacitors, and use the ohmmeter to find the problem. Start with an in-circuit check of R. Try both polarities even though it is not a diode. Sometimes you can avoid forward-biasing junctions in other parts of the circuit with this technique. The highest reading that you can obtain is closest to the correct reading. If R measures too high, you have found the trouble or part of it. There still could be a short circuit that caused excess current in R and burned it out. Use the ohmmeter to check; if a short circuit is found, leads will have to be disconnected since the diode, capacitor, and load are connected in parallel.

Look at the series-pass circuit in Fig. 5-21(a). Assume a normal unregulated input and zero load voltage. What kinds of component failures are possible? The transistor could be open, resistor R could be open, and the diode or the capacitor could be short-circuited. A short-circuited load is another possibility, but it would probably cause the unregulated input voltage to be lower than normal, and the pass transistor would be hot because of the high current flow. Suppose the load voltage in Fig. 5-21(a) were too high. This might be caused by an open diode or a short-circuited pass transistor. A quick voltage reading from ground to the base of the transistor will eliminate one or the other. A high voltage here points to an open zener, and a normal reading indicates that the transistor has short-circuited.

Low output voltage often points to an overload. If a regulator uses current limiting, the output voltage will drop below normal with excessive load current. Even without current limiting, overloads always can be expected to cause voltages to drop below normal. Overloads also make circuits run hot. However, do not jump to conclusions, because some components are quite safe at temperatures that will burn your finger. Sometimes it is necessary to use current analysis. Most technicians avoid this since the circuit has to be broken to insert the ammeter. If a series resistor of known value is available, measure the drop across it and calculate the current with Ohm's law. Don't forget the possibility of latch-up (discussed in the section on current limiting). In some cases, it may be necessary to remove some or all of the load from the supply to determine whether normal operation can be restored.

Another cause of voltage error is a fault in the reference supply. This is easy to verify with a voltage check. Don't forget to check the divider that samples the output voltage. Any problem here will send the wrong voltage to the error amplifier and cause an output error. A current-limit circuit can also cause voltage error if it is defective. Suppose R_3 in Fig. 5-27 increases in value. It will cause the circuit to go into current limiting at less than maximum load current, and the load voltage will be below normal.

Repeated pass transistor failures may indicate a defect in the current-limit circuit. Suppose Q_1 in Fig. 5-31 is replaced and normal operation is restored. It will be wise to check the supply for foldback; otherwise the replacement transistor may fail in a short period of time. If Q_3 is open, the supply will not limit, and the pass transistor will not be protected from overloads. Sometimes the difference between a good technician and a poor technician is that the good technician repairs the equipment once, and the poor technician repairs it once a week.

Blown fuses with glass tubes should be visually inspected. It is easy to tell the difference between a moderate overload and a severe overload. A severe overload often covers the inside of the glass with spattered metal and black smoke. These kinds of clues are valuable. For example, suppose the circuit of Fig. 5-35 is dead. Inspection of the fuse reveals a severe overload. If lightning made the lights flicker when the equipment went dead, it is easy to piece together what happened. If there was no transient, then it is likely that there is a dead short circuit in the equipment, and it will be wise to look for it before energizing the equipment again. A severely blown fuse in Fig. 5-36 indicates that the crowbar circuit was tripped. The pass transistors and Q_2 should be checked for emitter-to-collector short circuits before turning the supply on. If the transistors are good, the integrated circuit may be defective, divider R_1-R_2 may be wrong, D_1 may be short-circuited, the SCR may be short-circuited, or noise may have caused false gating of the SCR.

Switch-mode supplies can exhibit some additional symptoms when compared with linear supplies. They can make sounds such as clicking, chirping, and squealing. They are designed to operate above the limit of human hearing, but defects and overloads may cause the frequency to drop. Overloads can also cause a switcher to shut down and then restart repeatedly. This can create audible clicking and chirping. Always investigate the possibility of an overload when unusual sounds are heard.

An oscilloscope is the preferred instrument for analyzing switch-mode supplies. Rectangular waveforms can cause misleading readings in a voltmeter unless the meter is capable of true rms performance at the frequency of operation. Most meters are not true rms indicators, and some true rms meters are not accurate at switch-mode frequencies. It is also good practice to check the output of linear supplies for ripple and noise with an oscilloscope. The frequency of operation also must be verified in switch-mode supplies. This can be done with sufficient accuracy on a good oscilloscope by measuring the period of the waveform and then deriving the frequency.

The last step is the replacement of defective components. It is mandatory to use exact replacements or substitutions with equal or better specifications. However, this can be a trap for the beginning technician. Replacing a 1-W resistor with a 2-W resistor can cause failure of a more expensive component if the original design intended that the resistor would increase in value under overload conditions. Substituting a larger-capacity filter may cause excessive current peaks in the rectifiers and transformer. Generally, it is safest for the beginner to match replacements to the original parts as closely as possible; this means according to component type, as well. It is poor practice to replace a film resistor with a carbon composition resistor, for example. Such substitutions can even lead to fires. The physical size is also important. Make sure the replacement has the proper lead arrangement and that it will fit in the space available. Make your work look like the factory wiring as much as possible. Replace all components, even seemingly unimportant ones such as ferrite beads. Do not leave long leads on components. For example, unless bypass capacitors such as C_1, C_2, and C_3 in Fig. 5-36 have short leads, the integrated circuit may become unstable.

Finally, it must be emphasized that certain components are special and have rather critical specifications. Rectifiers D_5 and D_6 in Fig. 5-42 must be of the Schottky type. Ordinary rectifiers will not work in this circuit. They will overheat and be destroyed and may cause additional circuit damage. Capacitor C_3 is another example of a special component. It is a special four-lead capacitor designed for low effective series resistance (ESR). An ordinary capacitor has a much higher ESR and would be overheated and destroyed in a circuit of this type. The inductors in switchers and sine wave converters are also special. They are quite often *cup-core* types, where the cores surround the windings or toroid types, where the windings are placed on doughnut-shaped cores.

REVIEW QUESTIONS

44. What must the technician guard against when working on "hot" circuits with grounded test equipment?

45. Which part of a power supply may store a charge for some time?

46. Refer to Fig. 5-11. A technician connects the positive ohmmeter lead to the cathode of D_2 and the negative lead to its anode. Is the technician measuring the forward resistance of D_2? Is the reverse resistance of D_2 being measured?

47. What is being measured in question 46?

48. Refer to Fig. 5-22. Transistor Q_2 is open. What is the symptom?

CHAPTER REVIEW QUESTIONS

5-1. If a 6-V, 2-A · h ni-cad battery is completely discharged, what is the minimum time required to completely restore it at a C/10 charging rate? How long can it be left on charge before it is damaged?

5-2. Why is it poor practice to store discharged gel-cells for an extended period?

5-3. Assume a line frequency of 60 Hz. What is the ripple frequency in a line-to-neutral three-phase wye rectifier?

5-4. What is the rectifier cell ratio for a single-phase, full-wave bridge circuit?

5-5. Refer to Fig. 5-17. Calculate the capacitor value required to keep the ripple voltage at 1 V peak-to-peak. Assume a load current of 10 A and a line frequency of 40 kHz.

5-6. What can you conclude from question 5-5 regarding the size of filter components required in high-frequency switch-mode power supplies?

5-7. Refer to Fig. 5-18(*b*). Assume no diode loss, a light load, and a 120-V rms secondary. What is the dc load voltage?

5-8. Refer to Fig. 5-22. What can be expected to happen to the output voltage if resistor R_2 opens?

5-9. Refer to Fig. 5-24. Assume a pass transistor with a V_{be} of 1.5 V at a collector current of 4 A. Select a value for R that will set the integrated circuit-regulator current at 1 A when the load current is 5 A.

5-10. Use Fig. 5-30(*b*) and predict the load current for a 2-Ω load resistor.

5-11. An MOV is rated at 5 J and 200 A. Assume maximum current flow and calculate the maximum voltage that it can safely withstand for 5 μs.

5-12. Suppose the circuit of Fig. 5-35 is designed to operate on common 120 Vac. Why would it not be possible to use an MOV designed to break over at 150 Vdc?

5-13. Refer to Fig. 5-36. Which component sets the crowbar trip point?

5-14. Refer to Fig. 5-36. Potentiometer R_1 is used to adjust the load voltage. Why must it be adjusted carefully in a circuit of this type?

5-15. Calculate the frequency of the seventh harmonic in a switch-mode supply that operates at 35 kHz.

5-16. What two undesired effects are caused by harmonic energy in switch-mode supplies?

5-17. How much harmonic energy can be found in a sine wave?

5-18. Name two problems or limitations associated with bipolar power transistors that are not associated with power FETs.

5-19. Refer to Fig. 5-22. If R_1 is open, what is the symptom?

5-20. Refer to Fig. 5-27. If R_1 is open, what is the symptom?

5-21. Refer to Fig. 5-36. Transistor Q_2 has a collector-to-emitter short circuit. What is the symptom?

5-22. Refer to Fig. 5-36. Transistor Q_2 is open. What is the symptom?

5-23. Refer to Fig. 5-40. Resistor R_1 has increased in value. What is the symptom?

5-24. Refer to Fig. 5-22. What can be expected to happen to the load voltage if the series-pass transistor develops a collector-to-emitter short?

ANSWERS TO REVIEW QUESTIONS

1. primary **2.** polarization **3.** 35 h **4.** lithium thionyl chloride **5.** 3.3 MΩ **6.** 30 Ω **7.** 22.5 V **8.** 27 V
9. positive **10.** 13.5 V **11.** 6.45 V; 47.8 percent **12.** 7 V **13.** pulse waveforms show a greater heating effect
14. 20 percent **15.** capacitor input **16.** capacitor input **17.** capacitor input **18.** 42.3 V **19.** it will decrease
20. it will increase **21.** 20 Ω; 0.45 W **22.** 0.25 W; 0.75 W **23.** 150 mA; drops to zero; drops below 5 V **24.** 7 V
25. 2.29 kΩ **26.** 5 V **27.** 4 A **28.** from 3 to infinity Ω **29.** from 0 to 3 Ω **30.** 10.1 A **31.** 150°C **32.** 5.3 A
33. inverse **34.** direct **35.** 14 **36.** 0.667 or 66.7 percent **37.** 36 percent **38.** the duty cycle will decrease
39. 21.3 V **40.** yes **41.** it will go down **42.** D_3 **43.** Schottky rectifiers **44.** ground loops and shock **45.** filter
capacitor **46.** no; no **47.** the load plus the forward resistance of D_3 (depending on primary resistance, D_1 may also
be turned on) **48.** high output and no voltage regulation

6

AMPLIFIERS

This chapter covers electronic amplifiers. An electronic amplifier is a circuit using active devices such as transistors and ICs that allows an input to control a power source to produce some useful output. Industrial systems may also use hydraulic, pneumatic, or magnetic amplifiers.

6-1
SIMPLE LINEAR TYPES

A *linear amplifier* is one whose output signal is a replica of the input signal. For example, if the input signal is a sinusoid, then the output signal from the amplifier will also be a sinusoid. The purpose of a linear amplifier is to increase the level of the signal. In an industrial control system, the output from a temperature sensor may vary only several millivolts over its entire operating range. Small signals such as this must be amplified to be useful. Amplifiers are often identified according to the power level that they produce. Amplifiers that produce signals at significant levels of voltage or current are usually called *power amplifiers*. Amplifiers that work at small voltage and current levels are usually called *voltage amplifiers* or *small-signal amplifiers*.

The amount of gain that an amplifier produces is often measured in decibels (dB). The dB power gain of an amplifier is evaluated by

$$dB = 10 \times \log \frac{P_{\text{out}}}{P_{\text{in}}}$$

For example, if an amplifier develops a 100-W output signal when driven by a 1-W input signal, then the power ratio is 100. The common logarithm of 100 is 2; therefore the power gain of the amplifier can be stated as 20 dB. Small-signal amplifiers are usually evaluated in terms of their voltage gain. Since power varies as the square of the voltage, the dB gain is evaluated as follows:

$$dB = 20 \times \log \frac{V_{\text{out}}}{V_{\text{in}}}$$

Suppose an amplifier develops a 5-V output signal when driven by a 100-mV input signal. The voltage ratio is 50, the common logarithm of 50 is 1.7, and the gain of the amplifier is 34 dB. Since the dB system is based on power ratios, it must be adapted to work with voltage ratios by doubling the logarithm, which is equivalent to squaring the voltage ratio. However, there is an implicit assumption: that the input impedance of the amplifier is equal to the output impedance of the amplifier. This rule is commonly violated, and the dB voltage gain of a circuit is often evaluated with the preceding equation even though the impedances are not equal.

Figure 6-1 shows a simple common emitter amplifier circuit. The name *common emitter* is used since the input signal is supplied to the base circuit, the output signal is taken from the collector circuit, and the emitter terminal is grounded and is therefore common to both the input and output circuits. This circuit is useful for amplifying small signals since it can be expected to show a voltage gain of about 100 times (40 dB, ignoring the impedances). It can be viewed as a voltage amplifier even though it is based on a bipolar transistor, which is a current amplifier. This idea was discussed in Chapter 2.

The input signal is capacitively coupled in Fig. 6-1 to the base of the transistor. As the input signal goes in a positive direction it will aid the supply voltage (V_{cc}) and increase the base current. Since the transistor is controlled by base current, the collector current will also increase but many times more since there is quite a bit of current gain from the base to the collector. The increase in collector current will cause a greater drop across the collector load resistor (R_L), and therefore less voltage will drop across the transistor from collector to ground. It can be seen that the output signal goes in a negative direction (less positive) when the input signal goes in a positive direction. Thus, the output signal is phase-inverted

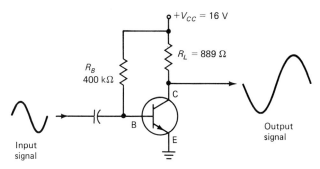

Fig. 6-1 Common emitter voltage amplifier.

129

180° when compared to the input signal. As the input signal goes in a negative direction, the base current decreases, the collector current decreases, and the drop across the transistor increases, making the collector terminal go in a positive direction.

Figure 6-2 shows a graphic presentation of the amplifier performance. It consists of a collector family of characteristic curves upon which a load line has been added to show how the collector load and supply voltage will interact with the transistor to develop an output signal. Notice that one end of the load line terminates on the horizontal axis at a value of 16 V. This is the cutoff voltage and is equal to the supply voltage in Fig. 6-1. Also notice that the other end of the load line terminates on the vertical axis at a value of 18 mA. This is called the *saturation current* and represents the maximum flow with a 16-V supply and a load resistor of 889 Ω. Use Ohm's law to verify this current. An amplifier can operate anywhere along its load line between saturation and cutoff. Small-signal amplifiers usually operate near the center for best linearity and maximum signal output swing. The amplifier of Fig. 6-1 will operate where the 40-μA base curve intersects the load line due to the 16-V supply and the 400,000-Ω base-limiting resistor R_B. Use Ohm's law to verify this current.

The operating point is also called the quiescent point. It is marked with a Q in Fig. 6-2. The *quiescent value* is the steady-state or resting condition of the amplifier. Quiescent values can be measured with no input signal applied. The quiescent collector-to-emitter voltage can be found by projecting down from

the Q point in Fig. 6-2 and is seen to be a little less than 8 V. The quiescent collector current is found by projecting to the left and is a little over 9 mA. What happens when an input signal is applied? An ac input signal will drive the amplifier above and below its Q point. Figure 6-2 shows the amplifier being driven with an input signal of 40 μA peak-to-peak (from 60 to 20 μA). By projecting this swing down, the output signal is shown to be about 8 V peak-to-peak. The voltage gain can be calculated if the input resistance of the base-emitter circuit is known. This resistance is approximately 2000 Ω in a circuit of this type. We can now use Ohm's law to calculate the input signal voltage required to develop a signal current of 40 μA peak-to-peak:

$$V = 40 \times 10^{-6} \times 2000 = 0.08 \text{ V}$$

The voltage gain of the amplifier is 8 divided by 0.08, or 100 times. It should now be clear how bipolar transistors can be viewed as voltage amplifiers even though they are inherently current-amplifying devices.

Figure 6-2 is also useful to explain clipping in a linear amplifier. If the input signal is too large, the output signal will be clipped. The limits at which the clipping will occur are saturation and cutoff. The amplifier will clip if the output tries to swing more than about 15 V peak-to-peak. The output signal will no longer be a good reproduction of the input. A severely clipped sine wave, for example, looks more like a square wave. This is not desirable in linear amplifiers. It is avoided by operating the amplifiers near the center of their load lines and by not allowing the input signals to become too large.

Figure 6-1 can also be evaluated with a few simple calculations. We know how to use Ohm's law to calculate the base current at 40 μA. If h_{FE} is known, the collector current is found by:

$$I_C = h_{FE} \times I_B = 230 \times 40 \times 10^{-6} = 9.2 \text{ mA}$$

You can verify h_{FE} and the quiescent collector current from Fig. 6-2. Chapter 2 covers how to calculate h_{FE} from the curves if you have forgotten. The h_{FE} varies quite a bit from transistor to transistor. Suppose the circuit is constructed with another transistor which has an h_{FE} of only 50. In this case, the collector current will be only 2 mA. The drop across the load resistor will be 2 mA × 889 Ω, or 1.78 V. This means that the drop across the transistor will be 16 − 1.78, or 14.22 V. The transistor will be operating near cutoff, which is not a good arrangement. The output signal can now be no larger than 3.56 V peak-to-peak before clipping starts to appear on the positive peaks. It should be clear that the circuit of Fig. 6-1 is not practical because it is too sensitive to h_{FE}. It is also temperature-sensitive, and its Q point will move toward saturation as it gets warmer.

Figure 6-3(*a*) shows an improved common emitter amplifier. This circuit is not nearly so sensitive to h_{FE} and temperature. It is improved by the addition of two resistors: one in the base circuit and one in

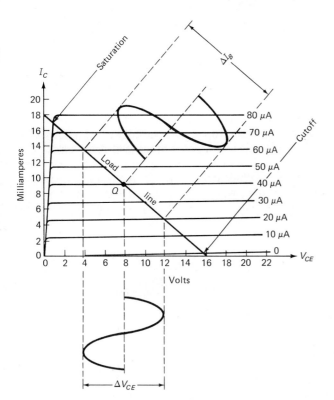

Fig. 6-2 Amplifier load line.

the emitter circuit. Resistors R_{B1} and R_{B2} form a voltage divider to set the base voltage at some fraction of the supply voltage. The divider current is much greater than the base current (typically 20 times as much) and therefore behaves essentially as an unloaded divider to provide a stable base voltage. Placing different transistors into the circuit will not change the base voltage or current appreciably. Resistor R_E is added in the emitter circuit to provide negative feedback. If a transistor tends to conduct more current, the drop across the emitter resistor will increase, thus subtracting from the forward bias for the base-emitter junction. This condition tends to decrease the device current. The negative feedback stabilizes the operating point. The negative feedback increases the input impedance of the amplifier by an amount equal to h_{FE} times R_E. The negative feedback also affects the signal performance of the amplifier. The voltage gain is much lower and is approximately equal to R_L divided by R_E. However, the gain can be restored for ac signals by eliminating the ac negative feedback. This is done by adding an emitter bypass

capacitor (shown as a phantom component in Fig. 6-3[a]). The capacitor is selected to have a reactance of about one-tenth the value of R_E at the lowest frequency of operation.

Figure 6-3(b) shows the PNP version of the improved amplifier circuit. It has the same ac characteristics as the NPN version but uses a negative collector supply. Compare the two versions and note that the dc currents are reversed.

The circuits discussed thus far have shown capacitors for coupling signals and for eliminating ac negative feedback. These techniques are not useful in dc amplifiers. Direct coupling must be used in dc amplifiers. One very popular direct-coupled arrangement is the Darlington amplifier shown in Fig. 6-4(a). The emitter of Q_1 is directly connected to the base of Q_2. This circuit provides a very high gain from the input to the output. The input impedance is also increased by this arrangement. If each transistor has an h_{FE} of 200, the overall current gain will be approximately equal to h_{FE1} times h_{FE2}, or 40,000. In practice it is difficult to achieve this much gain, however. The first transistor must be operated at a very low level to avoid driving the second transistor into saturation. Transistor gain tends to drop off at low current levels. In spite of this, the circuit still provides considerable current and voltage gain. When two transistors are connected in the Darlington arrangement, they essentially behave as a single transistor with super gain. For this reason, they are avail-

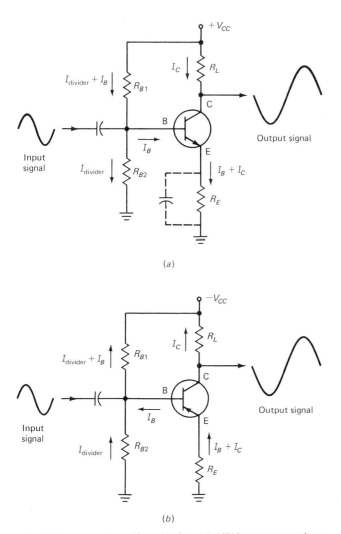

(a)

(b)

Fig. 6-3 Improved amplfier circuits. (a) NPN common emitter amplifier. (b) PNP common emitter amplifier.

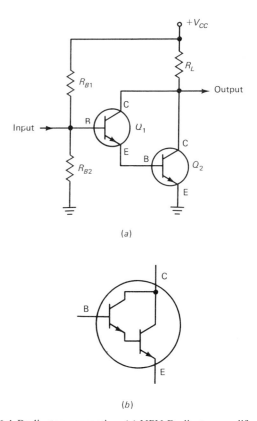

(a)

(b)

Fig. 6-4 Darlington connection. (a) NPN Darlington amplifier. (b) NPN Darlington transistor.

TABLE 6-1 AMPLIFIER CONFIGURATION SUMMARY

	Common Emitter	Common Base	Common Collector (Emitter Follower)
Input signal applied to	Base	Emitter	Base
Output taken from	Collector	Collector	Emitter
Current gain	High 50 to 100	Low ≈ 1	High 50 to 100
Voltage gain	High ≈ 100	High ≈ 100	Low ≈ 1
Input impedance	Medium $\approx 1200\ \Omega$	Low $\approx 50\ \Omega$	High $\approx 100\ k\Omega$
Output impedance	Medium $\approx 50\ k\Omega$	High $\approx 1\ M\Omega$	Low $\approx 50\ \Omega$
Application	General purpose	Low to high impedance transformer	Buffer amplifier

able packaged as a single device with one base, one emitter, and one collector lead (Fig. 6-4[*b*]). In addition, PNP Darlingtons are available.

The common emitter configuration is the most popular way to arrange an amplifier circuit. However, two other configurations exist, as shown in Fig. 6-5. The common base amplifier in Fig. 6-5(*a*) is so named because the base is common to both the input and output circuits. This configuration has a low input impedance since the input signal is applied to the emitter terminal, which is a high-current point. It is therefore limited to amplifying signal sources that have a low characteristic impedance. The output impedance is high. The output signal is not phase-inverted as it is in the common emitter amplifier. It

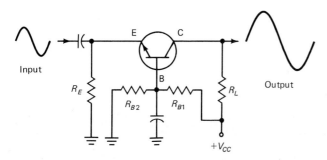

(*a*) Common base amplifier

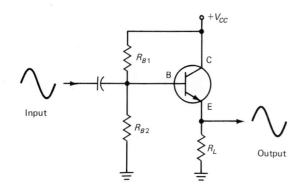

(*b*) Common collector amplifier

Fig. 6-5 Other amplifier configurations.

provides no current gain but does provide voltage and power gain. The common collector amplifier configuration is shown in Fig. 6-5(*b*). It uses the emitter terminal as the output and shows no voltage gain; there is actually a slight loss in signal voltage. The output is in phase with the input, and the circuit is usually called an *emitter follower*. Even though it has no voltage gain it is still very useful, especially for eliminating loading effects on a signal source. It has a reasonably high input impedance and a low output impedance and is often used as a buffer amplifier. It does provide current gain and power gain. The circuit of Fig. 6-4(*a*) can be converted to a Darlington emitter follower by moving the load resistor to the emitter circuit of Q_2 to provide an even higher input impedance. The common emitter amplifier is the only configuration that provides both voltage and current gain. It also provides the highest power gain and is best suited to most applications. Table 6-1 provides a summary of the amplifier configurations.

Other transistor types can be used to build amplifiers. For example, PNP devices can be substituted in the circuits shown by using a negative collector supply. Metallic oxide semiconductor field effect transistors (MOSFETs) can be used, although the gate-biasing techniques may be different. For those applications where a signal source has a high impedance MOSFETs are attractive. Since they are voltage-controlled transistors, they do not load the signal source as much as biopolar transistors do. They are often used in the first stage of a multistage amplifier. Bipolars will be used in the subsequent stages since they are less expensive and provide better gain. Power MOSFETs are attractive in some applications and are replacing power bipolars. Some of the industrial applications are in dc motor control, ac motor control, power supplies, induction heating, and high-frequency welding.

REVIEW QUESTIONS

1. Calculate the dB power gain for an amplifier with an input signal of 1 mW and an output signal of 5W.

2. Calculate the dB gain of an amplifier with an input signal of 0.5 V peak-to-peak and an output signal of 20 V peak-to-peak (assume the input impedance is equal to the output impedance).

3. A transistor amplifier is driven at its base terminal, and the output signal is taken from its collector terminal. What is the configuration of this amplifier?

4. What is the phase relationship between the input and the output voltages for question 3?

5. Refer to Fig. 6-1. If the drop across the collector load resistor is 12 V, what is the voltage from the collector terminal to ground?

6. Suppose the base bias resistor in Fig. 6-1 opens (infinite resistance). At which end of the load line will the amplifier operate?

7. What will be the collector voltage for question 6?

6-2
DIFFERENTIAL AMPLIFIERS

A *differential amplifier* (diff amp) responds to the difference between two input signals. Figure 6-6 shows a simple example. The circuit has two inputs and two outputs. It is possible to drive such an amplifier at one, or at both, of its inputs. It is also possible to take the output signal from one, or both, of its outputs. In a *single-ended output* the signal is being taken from one of the outputs (referenced to ground). In a *differential output* the signal is being taken from both outputs (not referenced to ground). The amplifier is very flexible in this respect since it offers four combinations of input/output conditions. Differential amplifiers are important in their own right. They are very flexible and stable and capable of rejecting some types of noise. They are even more important as an integral part of operational amplifiers, which are covered in the next section of this chapter.

Direct coupling is the method used to provide dc gain. When quite a bit of gain is needed, several stages must be arranged, with the output of the first amplifier directly coupled to the input of the second amplifier and so on. When this multistage arrangement is used with the simple designs of the preceding section, drift becomes a problem. Transistors are temperature-sensitive. When many stages are direct-coupled, even a small temperature change in the first stage can drive the last stage into saturation or cutoff. Temperature compensation and negative feedback can be used to stabilize the operating point, but the circuit tends to become expensive and complicated. A differential amplifier is an attractive choice because its design inherently cancels drift due to temperature change (refer to Fig. 6-6). If the two transistors are closely matched and maintained at the same temperature, any temperature change will affect both outputs by the same amount. We will see that such changes are canceled in the differential output. Matching transistors is an expensive proposition in discrete circuitry, but fortunately it is a byproduct of monolithic circuit construction. Since all of the devices are formed at the same time, they have well-matched characteristics. They also track well in temperature since they exist together in a single monolithic structure. Thanks to integrated circuit technology, differential and operational amplifiers have found wide application in industrial circuitry.

Suppose the amplifier of Fig. 6-6 is driven with a sinusoid at its B input only. As you would expect, the input signal will cause changes in the base current, which, in turn, will create a collector signal at output B. We can also expect that output B will be phase-inverted. Now look at Fig. 6-7. It shows that both outputs are active. Why is this so, since only input B is being driven? The reason is that both sides of the differential amplifier share emitter resistor R_E. When input B goes positive, the base, collector, and emitter currents of Q_B all increase. This causes an increase in the current through R_E which increases the voltage drop across it. This drop acts as a positive-going signal fed to the emitter of Q_A, which responds as a common base amplifier, and it produces a positive-going output at A (Fig. 6-7). Thus, both outputs are active even though only one input is driven. The same results can be obtained by driving only input A, except that output A will be out of phase and output B will be in phase with the input. Driving only one of the inputs creates a difference signal to which this type of amplifier responds.

Figure 6-7 also shows that the differential output $(A - B)$ has twice the peak-to-peak swing when compared to either single-ended output. For example, if output A has swung 2 V positive, then output B has swung 2 V negative, and the difference is $+2 - (-2) = +4$ V. What happens if both inputs are driven with the same signal? Of course, if they are driven the same, then the difference is 0 and there is no difference to amplify. Any signal applied to both inputs with exactly the same phase and voltage is

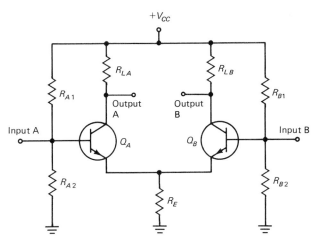

Fig. 6-6 Differential amplifier.

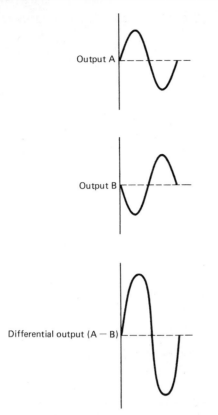

Fig. 6-7 Outputs with differential input.

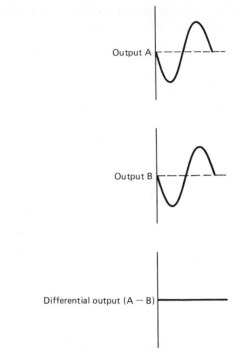

Fig. 6-8 Outputs with common mode input.

known as a *common mode signal.* Figure 6-8 shows what happens. When input A and input B are driven by a common mode signal, the outputs are in phase. If output A has swung 2 V negative, then output B has also swung 2 V negative, and the difference is $-2 - (-2) = 0$ V. So when the output is taken differentially, any common mode signal is canceled. It will be shown a little later that it is also possible to cancel common mode signals at the single-ended outputs by adding a current source to the emitter circuit of the diff amp.

The ability of differential amplifiers to cancel common mode signals is an important one. Many signals have *common mode noise.* A prime example is in the case of floating measurements, where the oscilloscope probe cannot be grounded because of the presence of ac voltages (reference to ground) at both points in a piece of equipment across which a waveform must be measured. Figure 6-9 shows how a differential amplifier can be used to make such a measurement. Notice that input A is a higher-frequency wave that is riding on a lower-frequency signal. The lower-frequency signal is undesired and is known as *noise,* or *hum.* Input B is about the same, but note that the phase of the higher-frequency signal is inverted. Now look at the output. The low-frequency noise, or hum, has been canceled, and the high-frequency signal has been amplified because it appeared as a differential signal. A differential amplifier can also eliminate high-frequency noise if it appears as a common mode signal.

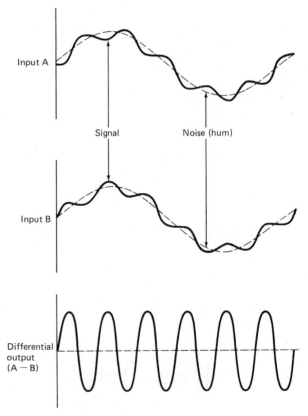

Fig. 6-9 Elimination of common mode noise.

A differential amplifier must be perfectly balanced if it is to cancel the common mode signal completely. Perfect balance can never be achieved. For this reason, a way of measuring performance is required.

The figure of merit is known as the *common mode rejection ratio* (CMRR). It is found by

$$\text{CMRR} = 20 \times \log \frac{\text{Gain (differential)}}{\text{Gain (common mode)}}$$

For example, suppose an amplifier shows a voltage gain of 300 for a differential signal and a voltage gain of 0.01 for a common mode signal. The gain ratio will be 30,000, the common logarithm is 4.5, and the CMRR is 90 dB. Note that the amplifier actually decreases the common mode signal to only 1 percent of its original value. This is the opposite of amplification and is called *attenuation*.

Figure 6-10 shows an important improvement to the differential amplifier. It is powered by a dual supply. Compare this with Fig. 6-6 and verify that the voltage divider bias has been eliminated. The negative emitter supply (V_{EE}) allows the bases of both transistors to be operated at dc ground potential. Base current is very small, and the drop across R_A and R_B is negligible. This is usually desired in a dc amplifier since the signal source is often referenced to ground. The original circuit (Fig. 6-6) is awkward to use since the inputs have a dc offset with respect to ground.

Figure 6-11 shows another improvement for the differential amplifier circuit. The emitter resistor has been replaced with a constant current source. This particular example uses a 5.7-V zener biased by the negative supply. The zener drop is applied across the base-emitter (B-E) junction of transistor Q_C, and the 1000-Ω resistor. Subtracting 0.7 V for the B-E junction leaves 5 V across the resistor. Ohm's law sets the current through the resistor at 5 mA. Since the emitter and collector currents are almost equal, Q_C will conduct a total of 5 mA for both transistors in the diff amp. Assuming balance, each transistor will have 2.5 mA of emitter current. Now assume a common mode signal that is going positive. Ordinarily, both transistor currents would increase, but in this case they cannot. The constant current source supplies a total of 5 mA regardless of the change in

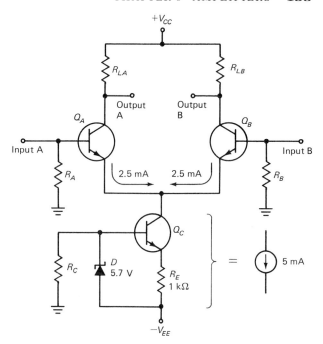

Fig. 6-11 Diff amp with constant current source.

collector-emitter resistance in the differential pair. The same thing is true if a common mode signal is negative-going and attempts to decrease the current in both transistors. The common mode signal will not appear at output A or output B. In other words, adding the constant current source provides good common mode rejection for the single-ended outputs. The previous circuits provided good common mode rejection only in the differential output. A differential input in Fig. 6-11 will unbalance the transistor currents. For example, a positive-going signal applied to one input might cause that transistor current to increase to 3 mA, in which case the other transistor current will decrease to 2 mA. The current changes will cause both differential and single-ended outputs to appear. The constant current source has a second advantage. It raises the input impedance of the differential amplifier to the megohm region. The input impedance is increased by an amount equal to h_{FE} times the impedance of the current source, which is characteristically very high.

REVIEW QUESTIONS

8. Refer to Fig. 6-6. Assume that only input A is driven with a signal. Will a signal appear at output B? If so, what is its phase relationship to the input signal?

9. The circuit of Fig. 6-6 is driven with a common mode signal. Will there by any single-ended output? Assuming balance, will there be any differential output?

10. Refer to Fig. 6-9. Suppose the low-frequency (noise) component of input signal B is out of phase

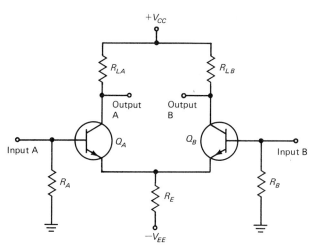

Fig. 6-10 Dual-supply differential amplifier.

with the same component at input A. Would the amplifier be able to attenuate the noise?

11. What is the CMRR (in decibels) of an amplifier that provides a voltage gain of 100 for a differential signal and a voltage gain of 0.01 for a common mode signal?

6-3
OPERATIONAL AMPLIFIERS

The original operational amplifiers (often called op amps) were based on vacuum tubes and were used to do mathematical operations in analog computers. They were large, expensive, power-hungry, and subject to drift. The digital computer has replaced the analog computer, and solid-state devices have replaced vacuum tubes. Modern operational amplifiers, thanks to integrated circuit technology, are small, inexpensive, power-efficient, and much more stable. They have found a wide range of applications in industrial circuitry even though their first application, the analog computer, has vanished. They are direct-coupled high-gain amplifiers and are usually powered by a dual supply. Dual-supply operation allows them to conveniently amplify signals near ground potential and allows their output to swing above and below ground potential.

The major sections of a modern IC operational amplifier are shown in Fig. 6-12. There are two inputs to the first stage, which is a differential amplifier. The outputs of most op amps are single-ended. One of the differential inputs is in phase with the output and is called the *noninverting input*; it is marked plus (+). The other input is out of phase with the single-ended output, is called the *inverting input,* and is marked minus (−). An intermediate voltage amplifier follows the differential input amplifier to provide high gain. An output amplifier is the third major stage and provides a low output impedance so that the op amp can drive most loads. In addition to the terminals shown in Fig. 6-12, an op amp may have offset null

terminals, frequency-compensation terminals, or gain-control terminals. The standard schematic symbol for an operational amplifier is a triangle with + and − inputs and a single output. The power supply and other connections may be omitted from some schematics for simplicity, as they are for many of the circuits shown in this book.

Figure 6-13 shows the equivalent circuit for a very popular op amp, the 741. Since it is a monolithic IC, studying the circuit in detail is not necessary. It is shown here to demonstrate how complex and expensive it would be in discrete form and to illustrate a few important concepts. Transistors Q_1 and Q_2 are the input devices and are part of a differential amplifier which acts as a voltage-to-current converter. Emitter resistors R_1 and R_2 are brought out to offset null terminals for the purpose of externally balancing the diff amp. The current signal from the differential input is sent to the second stage, consisting mainly of Q_{16} and Q_{17}, which acts as current-to-voltage converter. Capacitor C_1, a 30-pF frequency-compensation capacitor, is connected across the second stage to roll off (decrease) its frequency response 20 dB for every decade increase in frequency. A decade equals 10; thus a frequency change from 100 to 1000 Hz represents a decade increase. This gain roll-off is characteristic of most modern op amps with internal frequency compensation. It is used to ensure that the amplifier will remain stable with all feedback configurations. An unstable amplifier does not respond as planned and is useless. As we will see, op amps are almost always used with feedback. The last detail we will look at is the output circuit, which consists mainly of transistors Q_{14} and Q_{20}, which are a complementary pair. These two devices act as emitter followers to give the op amp a low-impedance output so that it can drive loads down to around 2000 Ω.

Figure 6-14 shows some popular op amp packages. The dual-in-line style is widely applied and can house one, two, or four amplifiers. The metal package is hermetically sealed and rated to operate over a slightly wider temperature range but is more expen-

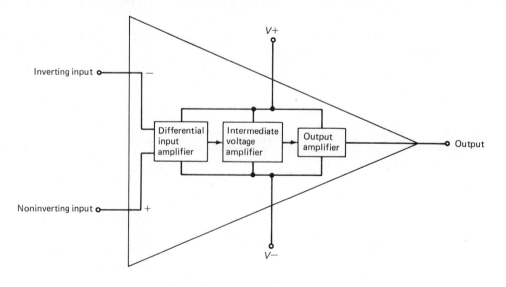

Fig. 6-12 Major sections of an operational amplifier.

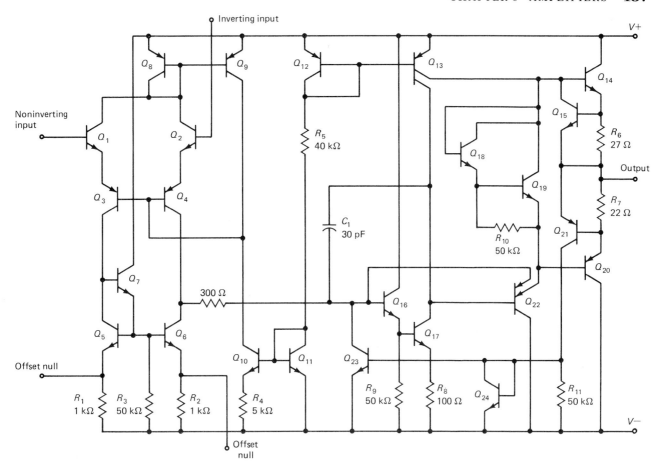

Fig. 6-13 741 operational amplifier equivalent circuit.

sive. Note that the metal case is internally connected to the $V-$ supply. The quad op amp package eliminates the offset null terminals since too many pins would be required. Many applications for operational amplifiers do not require the offset null function, and the quad package may be a good choice in those cases. The dual package uses a separate $V+$ pin for each op amp. Both pins must be energized when both amplifiers are in use.

Operational amplifiers are widely applied because they approach ideal amplifiers, especially for dc and low-frequency signals. Let's see how some specifications for common op amps compare to the ideal. The ideal amplifier has an infinite input impedance so that it can be connected to any signal source with no loading effects. An ordinary op amp, such as the 741, approaches the ideal with an input impedance of 6 MΩ. Super Beta op amps offer input impedances 10 times higher, and BI-FET operational amplifiers are available with FETs in the input amplifier and boast an input impedance of 10^{12} Ω. Another ideal amplifier characteristic is infinite gain. Common op amps provide over 100 dB of gain at low frequencies. Premium devices, called *instrumentation amplifiers,* provide 130 dB of gain at low frequencies. The ideal amplifier would infinitely reject common mode signals. Operational amplifier CMRR ranges from 90 dB

for standard types to 130 dB for instrumentation amplifiers. The ideal amplifier has zero output impedance and is capable of driving any load. Common operational amplifiers, such as the 741, can supply at least 5 mA to a 2000-Ω load. Monolithic power devices are capable of up to 1 A of output current, and hybrids with even higher ratings can be found. The ideal amplifier also has infinite bandwidth, meaning it can amplify any signal frequency. In this area the typical op amp falls considerably short of the ideal, with useful gain extending only into the tens of kilohertz region. Wide-band devices are available and extend performance to the tens of megahertz. Many, many applications do not require wide-band performance, and the typical device is all that is required. Modern op amps approach the ideal closely enough to make them very attractive for many applications.

Figure 6-15(*a*) shows the gain versus frequency response for a 741 op amp. The gain is in excess of 100 dB for low frequencies. Around 7 Hz, the gain begins to drop at a rate of 20 dB per decade as a result of the internal frequency compensation capacitor as previously mentioned. Some op-amps without internal frequency compensation are available but are not very popular since they must be externally compensated with resistors and capacitors. This fea-

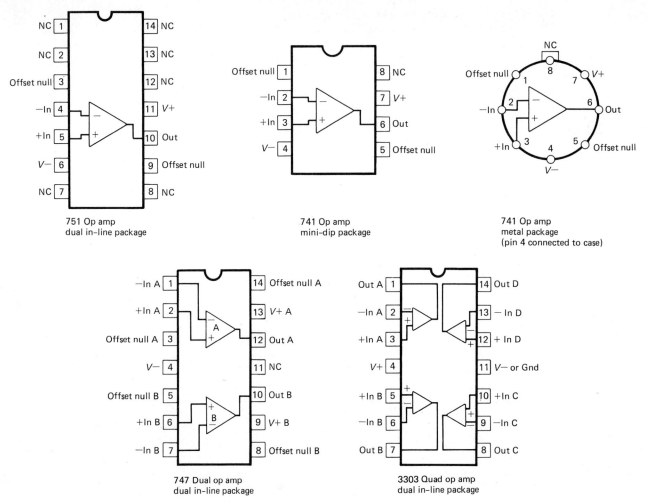

Fig. 6-14 Typical operational amplifier packages (top view).

ture complicates circuit design. Internally compensated op amps have a constant gain-bandwidth product. Referring again to Fig. 6-15(a), the gain is 1 at a frequency of 1 MHz for a gain bandwidth product of 1 MHz. At a frequency of 1 kHz, the gain is 1000, and the gain bandwidth product is still 1 MHz. As we will see later, it is possible to use negative feedback and attain constant gain over a reasonable frequency range. The term *open loop* means there is no feedback. Figure 6-15(b) shows the amplifier phase response. Plots of this type are used to ensure stability in amplifier designs. The phase plot shows a phase angle of −45° at the cutoff frequency (10 Hz). The phase plot is approximately 0° a decade below the cutoff frequency and −90° a decade above the cutoff frequency. Note that the phase response is limited to no more than −90° over the frequency range where the gain is greater than unity. This helps ensure stability in the op amp.

Amplifier slew rate can also affect high-frequency performance. It is a measure of how quickly the output can change in response to a step change at the input. The typical op amp slews at 0.7 V/μs. Figure 6-16 shows how this can affect the output signal, which is inverted in this case. When the square wave input goes positive, the output goes negative but can only slew (change) so fast. Later, the square wave goes negative, and the output begins to slew positive. It is clear that the output waveform is not square. This is known as *slew rate distortion*. The power bandwidth (BW_P) of an amplifier refers to its ability to produce maximum voltage swing at higher frequencies. At some point, the output will be slew-rate-limited and will begin to degrade. This point is defined by

$$BW_P = \frac{\text{Slew rate}}{6.28 \times V_P}$$

If an op amp is powered by a +15-V and −15-V dual supply, its output can swing within about 1 V of the positive and negative rails (28 V peak-to-peak). Its peak swing will be half that value, or 14 V. Assuming a slew rate of 0.7 V/μs, the power bandwidth of the amplifier will be

$$BW_P = \frac{0.7 \text{ V/μs}}{6.28 \times 14 \text{ V}}$$

$$= \frac{1}{6.28 \times 14 \text{ V}} \times \frac{0.7 \text{ V}}{\text{μs}}$$

$$= 7.962 \text{ kHz}$$

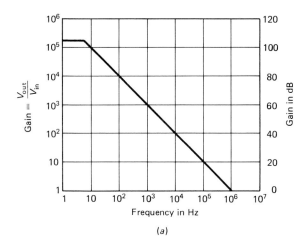

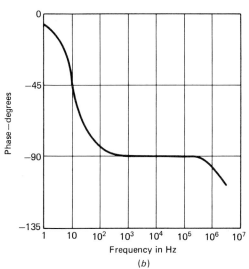

Fig. 6-15 Open loop performance curves for the 741 op amp. (*a*) Gain versus frequency. (*b*) Phase versus frequency.

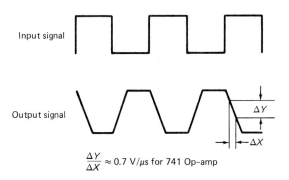

$$\frac{\Delta Y}{\Delta X} \approx 0.7 \text{ V}/\mu s \text{ for 741 Op-amp}$$

Fig. 6-16 The effect of op amp slew rate on the output signal.

High-speed operational amplifiers are manufactured with slew rates of 100 V/μs. They have gain-bandwidth products over 50 MHz and are useful in high-frequency applications.

A perfectly balanced op amp will produce 0 V output for 0-V differential input. Practical amplifiers show a slight dc voltage at the output due mainly to slight imbalances in the differential input stage. A

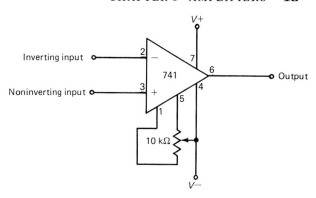

Fig. 6-17 Voltage offset null circuit.

slight differential input voltage of 1 mV or so will typically be required to zero the output. This error is called the *input offset voltage* and can be corrected in critical applications with the voltage offset null circuit shown in Fig. 6-17. A 10-kΩ potentiometer is connected across the offset null terminals 1 and 5. The wiper arm is connected to the negative supply. The potentiometer is adjusted for 0 V at pin 6 when the differential input is 0. You may wish to refer to Fig. 6-13 to verify how the offset null circuit trims the emitter resistors in the differential input amplifier to achieve balance. The offset null connection is not required in all applications, and pins 1 and 5 are left floating in those cases. The pin numbers used in Fig. 6-17 are for the mini-dip and metal packages shown in Fig. 6-14.

REVIEW QUESTIONS

12. Which op amp input is in phase with the output, and how is it usually marked on a schematic diagram?

13. If the gain of an internally compensated op amp is 80 dB at 100 Hz, what will it be at 10 kHz?

14. What is the input impedance of an ideal amplifier? Why?

15. What is the output impedance of an ideal amplifier? Why?

16. Refer to Fig. 6-15. What is the gain bandwidth product at 100 Hz? Is it constant at other frequencies?

17. Calculate the power bandwidth for an op amp with a slew rate of 100 V/μs and an output signal of 20 V peak-to-peak (10-V peak).

18. If the amplifier of question 17 has a gain bandwidth product of 65 MHz, what limits its high-frequency performance for large output signals?

6-4
OP AMP APPLICATIONS

Practical operational amplifiers approach the ideal amplifier in several important ways. The approach is close enough to allow simple and straightforward

hniques to be used with good results. The
are based on several assumptions. First,
input current in an ideal amplifier since it
e impedance. Second, the output imped-
ro. Third, the gain is infinite and reduces
ntial input to zero when negative feedback
Refer to Fig. 6-18. Resistor R_2 provides
feedback because it connects the output
he inverting input. With infinite gain, it will
ossible to make the voltage at the inverting
all different from the voltage at the nonin-
input, which is at ground. Suppose, for ex-
that the input terminal to the left of R_1 is
positive by a signal. The inverting input will
y to go positive, but the gain of the amplifier
mmediately produce a negative-going output
feeds back through R_2 to cancel the positive
ge and keep the inverting input at ground poten-
It is not possible to have any voltage difference
oss the + and − inputs since the feedback acts
cancel it. The noninverting input of Fig. 6-18 is
ounded through R_3. We have assumed no input
urrent, and there is no voltage drop across R_3. Both
nputs will remain at ground potential even when the
amplifier is driven with a signal. The feedback keeps
the inverting input at ground potential, and that terminal is considered to be a *virtual ground*.

The voltage gain for the inverting amplifier of Fig.
6-18 is easy to derive. The virtual ground sets the
current in R_1 at V_{in}/R_1 and the current in R_2 at
V_{out}/R_2. Because there is no amplifier input current,
these currents are equal:

$$\frac{-V_{in}}{R_1} = \frac{V_{out}}{R_2}$$

The input voltage is indicated as negative because it
is inverted from the output voltage. Solving for V_{out}
gives

$$V_{out} = -V_{in} \times \frac{R_2}{R_1}$$

The voltage gain for the inverting amplifier in Fig.
6-18 is equal to the feedback resistor value divided
by the input resistor value. For example, if R_2
is 100 kΩ and R_1 is 1 kΩ, the voltage gain will be
−100. Again, the minus sign indicates the phase inversion between the input and the output. Now refer
to Fig. 6-19. The open loop gain of the typical op
amp is several hundred thousand. By using negative
feedback, the gain is reduced to an absolute value of

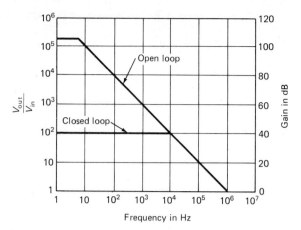

Fig. 6-19 Closed loop frequency response.

100 (the absolute value ignores the minus sign). This
value is called the *closed loop gain*. One of the advantages of closing the loop and reducing the gain is
increased bandwidth. Figure 6-19 shows that the
closed loop gain is constant to a frequency of
10 kHz, giving a bandwidth of 10 kHz. The open
loop bandwidth is only 7 Hz. This greatly improved
bandwidth makes the amplifier more useful for many
applications.

It is easy to predict the bandwidth when the amplifier is operated with negative feedback. Calculate
the gain by using the ratio of the feedback resistor
to the input resistor. Draw a line from the calculated
gain value on the vertical axis until the open loop
gain plot is intersected. This intersection is called the
corner frequency and represents the closed loop
bandwidth of the amplifier. The plot in Fig. 6-19 does
not show it, but the gain is down 3 dB at the corner
frequency. For the example given, the gain will be
37 dB (40 dB − 3 dB) at a frequency of 10 kHz. The
−3-dB point is the standard limit when specifying
amplifier bandwidth and is also known as the *cutoff
frequency*. Note that the amplifier does not abruptly
stop working beyond this frequency, but its gain
drops at a rate of 20 dB per decade for signals higher
in frequency.

The input impedance of the inverting amplifier in
Fig. 6-18 is equal to the input resistor R_1 because of
the virtual ground at the inverting input. In our example, the signal source would see a load of 1000 Ω.
The output impedance of the amplifier is equal to the
inherent output impedance divided by the loop gain.
Loop gain is equal to the open loop gain divided by
the closed loop gain. The open loop gain is 200,000,
and the closed loop gain is 100 for our example,
yielding a loop gain of 200,000/100 = 2000. The inherent output impedance is 75 Ω, and the closed loop
output impedance is equal to 75/2000, or 0.04 Ω. This
indicates that adding negative feedback also decreases the output impedance of the amplifier. The
low output impedance is advantageous because the
circuit can deliver a signal to almost any load, provided that the current capabilities of the op amp are
not exceeded. Resistor R_3 in Fig. 6-18 may be replaced by a direct connection to ground with only

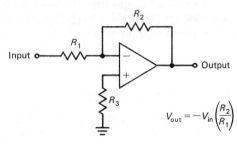

Fig. 6-18 Inverting amplifier.

$$V_{out} = -V_{in}\left(\frac{R_2}{R_1}\right)$$

minor impact on circuit performance. When used, it is usually selected to have a value equal to the parallel resistance of R_1 and R_2 (990 Ω in our example). The idea is to produce identical voltage drops due to the input bias current at both amplifier inputs. Even though we have assumed zero input current, real op amps do have a small input current. Balance reduces offset error. However, the input current is only 30 nA in a typical op amp, and the imbalance created by directly grounding the noninverting input can be ignored in most applications.

Thus far we have seen that it is easy to use an op amp as an inverting amplifier. The merits of negative feedback for increasing bandwidth and decreasing output impedance have been established. Now let's look at some other applications. Figure 6-20 shows a voltage follower which is a noninverting amplifier since the input signal is applied to the noninverting input. The voltage gain of the circuit is 1. This circuit is useful even though it has no voltage gain. It has a very high input impedance, which is equal to the inherent input impedance of the op amp. Since this impedance is 6 MΩ in a standard op amp, the voltage follower makes an excellent isolation amplifier. The inverting amplifier that we looked at previously exhibits a much lower input impedance due to the virtual ground created by negative feedback. A high-impedance signal source will suffer loading effects when connected to such an amplifier. Therefore, it is sometimes necessary to isolate the signal source by connecting it to a voltage follower and then connecting the voltage follower output to the next stage.

A noninverting amplifier with gain is shown in Fig. 6-21(a). The gain for this circuit is derived by starting at the inverting input. Notice that the voltage there is a function of the output voltage and the divider network formed by R_1 and R_2:

$$V_{\text{inv}} = V_{\text{out}} \times \frac{R_1}{R_1 + R_2}$$

Once again we make the assumption that both amplifier inputs are at the same potential; therefore,

$$V_{\text{in}} = V_{\text{out}} \times \frac{R_1}{R_1 + R_2}$$

Solving for V_{out} gives

$$V_{\text{out}} = V_{\text{in}} \times \frac{R_1 + R_2}{R_1}$$

Suppose R_1 in Fig. 6-21(a) is a 1-kΩ resistor and R_2 is a 22-kΩ resistor. The voltage gain will be $(22 + 1)/1$, or 23. The bandwidth will be 40 kHz and

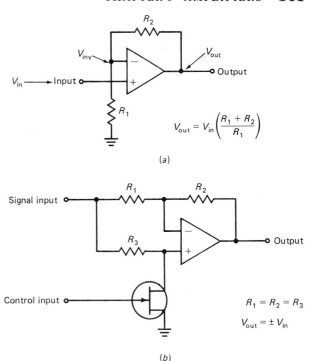

$$V_{\text{out}} = V_{\text{in}}\left(\frac{R_1 + R_2}{R_1}\right)$$

(a)

$R_1 = R_2 = R_3$

$V_{\text{out}} = \pm V_{\text{in}}$

(b)

Fig. 6-21 Amplifiers. (a) Noninverting. (b) Switchable.

can be verified by using Fig. 6-19. The input impedance is equal to the input impedance of the op amp, which is high. The output impedance will be reduced by a factor equal to the loop gain and will therefore be quite low.

Figure 6-21(b) shows a switchable amplifier that will invert or not invert the input signal, depending on the control signal applied to the gate of the FET. Suppose the control signal is 0 V. The FET will be on, grounding the noninverting input of the op amp and the right end of R_3. The amplifier will function as an inverting amplifier with a gain of 1 since $R_1 = R_2$. If the control input is made negative enough to cut off the FET, the amplifier will switch to the noninverting mode since the input signal now also drives the + input of the op amp. Once again, we can assume no differential input due to the large gain of the op amp, and therefore the signal voltage at the left end of R_1 will be equal to the signal voltage at the right of R_1. There will be no signal current in R_1, and it appears as an infinite resistance as far as the signal circuit is concerned. Now, look at the gain equation in Fig. 6-21(a) to determine that the voltage gain approaches 1 as R_1 approaches infinity. Therefore, the voltage gain of the switchable amplifier is -1 or $+1$, depending on the control signal.

Operational amplifiers are also capable of subtraction and addition. Figure 6-22 shows an amplifier that subtracts one input signal from another and amplifies the difference. Input 1 is applied to the inverting terminal and is subtracted from input 2, which is applied to the noninverting terminal. A summing amplifier is shown in Fig. 6-23. It can be used to add two or more ac or dc input signals. All inputs are

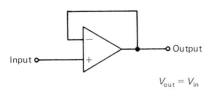

$V_{\text{out}} = V_{\text{in}}$

Fig. 6-20 Voltage follower.

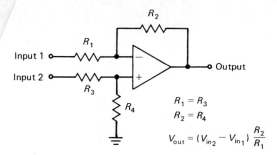

Fig. 6-22 Subtracting (differential) amplifier.

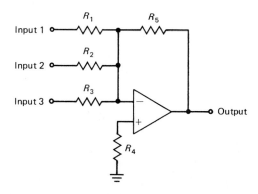

$$V_{out} = \left[-V_{in_1}\left(\frac{R_5}{R_1}\right)\right] + \left[-V_{in_2}\left(\frac{R_5}{R_2}\right)\right] + \left[-V_{in_3}\left(\frac{R_5}{R_3}\right)\right]$$

Fig. 6-23 Summing amplifier.

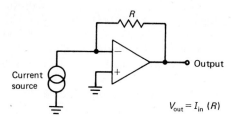

Fig. 6-24 Current-to-voltage converter.

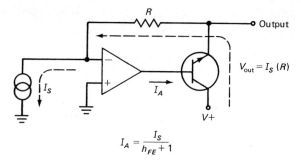

Fig. 6-25 Current-to-voltage converter with current boost.

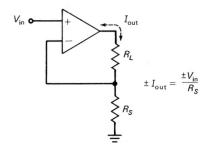

Fig. 6-26 Voltage-to-current converter.

applied to the inverting terminal, and the output will be the inverted sum. If $R_1 = R_2 = R_3$, then the output will be proportional to the nonweighted sum of the inputs. A weighted sum can be obtained by varying the value of the input resistors. Since the inputs are summed at the inverting terminal, which is a virtual ground, there is no interaction among the inputs. A signal at any input will not cause any effects at the other inputs.

An operational amplifier used as a current-to-voltage converter is shown in Fig. 6-24. The output voltage is equal to the input current times the feedback resistor. Since the inverting terminal is a virtual ground, this circuit places a short circuit across the current source. This is a valuable asset for some signal sources since they must be short-circuited for accurate measurements. Other approaches, such as using a current meter or a load resistor, may adversely affect measurement accuracy because of the

loading effect of the meter or the load resistor. The internal impedance of the meter adds to the effective series resistance of the current source (assuming that it is not an ideal current source) and unloads the circuit. The unloading causes the measured current to be less than the true short-circuit current. The current-to-voltage converter circuit eliminates unloading effects for improved accuracy. Figure 6-25 shows a current-to-voltage converter circuit with current boost. The typical op amp can supply only milliamperes. This circuit extends the current capacity into the ampere range by using the current gain of an external NPN transistor. The current gain from the base terminal to the emitter terminal is $h_{FE} + 1$, and the required operational amplifier output current is reduced by this factor.

Figure 6-26 shows an op amp voltage-to-current converter. This circuit is also called a *transconductance amplifier*. Resistors R_L and R_S divide the output voltage for the inverting input. Once again, we can assume that there will be no difference between the inverting terminal voltage and the noninverting terminal voltage; therefore,

$$V_{out} = \frac{V_{in}(R_S + R_L)}{R_S}$$

The output current is predicted by Ohm's law:

$$I_{out} = \frac{V_{out}}{R_S + R_L}$$

Combining, we obtain:

$$I_{out} = \frac{V_{in}}{R_S}$$

This equation shows that the current in the load resistor (R_L) is independent of the value of the load resistor. It is a function of the input voltage and the value of the sense resistor (R_S). By making the input voltage constant, this circuit can serve as a constant current source or sink.

REVIEW QUESTIONS

19. Refer to Fig. 6-18. Calculate the voltage gain if $R_1 = 10$ kΩ and $R_2 = 100$ kΩ. What is the input impedance of the amplifier?

20. Use Fig. 6-19 to find the bandwidth of the amplifier in question 19.

21. Refer to Fig. 6-21. Calculate the gain if $R_1 = 1$ kΩ and $R_2 = 100$ kΩ. What is the input impedance of this circuit if the op amp is a standard type?

22. Refer to Fig. 6-22 and assume that all resistors are 10 kΩ in value. Calculate the output voltage if input 1 = -1 V and input 2 = 2 V.

23. Refer to Fig. 6-23. Assume that $R_1 = R_2 = R_3 = 10$ kΩ and that $R_5 = 47$ kΩ. Calculate the output voltage if input 1 = $+0.5$ V, input 2 = $+0.7$ V, and input 3 = $+1$ V.

24. Are the inputs in question 23 weighted or nonweighted? How could the circuit be changed to achieve the other condition?

6-5
NONLINEAR APPLICATIONS

In a *nonlinear circuit* the output signal is not a replica of the input signal. For example, refer to Fig. 6-27. The input signal is a sine wave, and the output signal is a rectangular wave. This circuit is called a *comparator* and can be used to convert other waveforms to rectangular. It is also used to compare a signal to a reference voltage and change output states when the reference threshold is crossed. Since the op amp is running open loop, the gain is very high, and only a few millivolts difference between the input voltage and the reference voltage (V_{REF} in Fig. 6-27) will drive the output to positive or negative saturation. Operational amplifiers can be used as comparators, but their slew rate may limit performance for many applications. Integrated circuit comparators that have optimized characteristics such as fast switching, wide supply range, and high output current are available.

Figure 6-27 is an inverting comparator because the input is applied to the inverting input of the op amp or comparator IC. When the input signal goes more positive than V_{REF}, the output goes to negative saturation. When the input is less than V_{REF}, the output goes to positive saturation. The output waveform shows that the saturation voltages are a little less than the supply voltages. Some comparator ICs, such

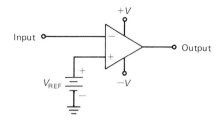
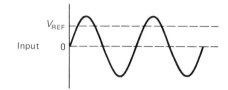
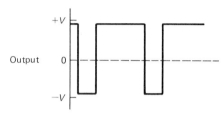

Fig. 6-27 Inverting comparator.

as the type 311, will require an external pull-up resistor on the output. This is convenient when an analog signal must be converted for another circuit where the voltage levels are different. For example, the comparator IC can be connected to a 15-V dual supply, and its output can be pulled up to a separate $+5$-V supply. The output will swing between 0 and $+5$ V. This swing is compatible with many digital circuits.

The reference voltage in Fig. 6-27 can be changed for different results. If the noninverting input of the IC is grounded, the reference voltage will be zero. This will change the duty cycle of the output waveform to 50 percent with the input waveform that is shown because the output will switch at the zero crossings. If the reference voltage is made greater than the peak value of the input, the output will remain at the positive saturation point. By making the reference voltage adjustable, the circuit could be used to produce a rectangular output with varying duty cycle. The circuit can also be reconfigured for noninverting operation, as shown in Fig. 6-28. In this example, the reference voltage is negative. When the input signal goes more negative than the reference, the output is driven to negative saturation. When the input is more positive than $-V_{\text{REF}}$ the output is at positive saturation.

Figure 6-29 shows a *window comparator*. This circuit is used to determine whether a voltage or signal is within or without a given range called the *window*. It uses two op amps or comparators and two reference voltages. The waveforms show that the output signal is 0 as long as the input is between $+4.5$ and $+5.5$ V. If the signal is outside this window, the

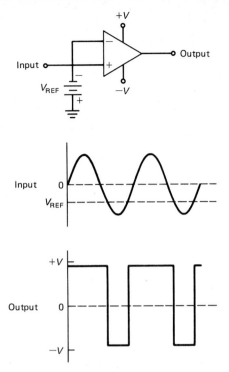

Fig. 6-28 Noninverting comparator.

output is at positive saturation. Suppose the signal is in the window. The top IC will be at negative saturation, since its noninverting input is negative (less than +5.5 V). The bottom IC is also at negative saturation because its inverting input is positive (more than +4.5 V). Both diodes are reverse-biased (off), and the combined output is zero. Now, suppose

the input goes more positive than 5.5 V. The top IC goes to positive saturation, D_1 is forward-biased (on), D_2 is still off, and the combined output goes to positive saturation less the drop across D_1. If the input goes below +4.5 V, the bottom IC goes to positive saturation as its inverting input will be negative with respect to its noninverting input. This will forward-bias D_2, and the combined output will be at positive saturation less the drop across D_2. Window comparators can be used to monitor a critical voltage such as a power supply to determine whether or not it is in tolerance.

The waveforms in Fig. 6-30 show the operation of a Schmitt trigger. At first glance this circuit appears to be accomplishing the same result as the inverting comparator circuit since the input waveform is sinusoidal and the output is rectangular. However, there is a difference because positive feedback is used. Note that R_1 connects the IC output back to the noninverting input. This is positive feedback and produces two switching points: the upper threshold point (UTP) and the lower threshold point (LTP). These points are set by the supply voltages and resistors R_1 and R_2. For example, if the IC is energized by a 15-V dual suppy the output will saturate at near +14 V or −14 V. If we assume that R_1 is 22 kΩ and that R_2 is 1 kΩ, we have the information needed to calculate the two threshold points:

$$\text{UTP} = \frac{R_2}{R_1 + R_2} \times (+V_{\text{SAT}})$$
$$= \frac{1 \text{ k}\Omega}{23 \text{ k}\Omega} \times (+14 \text{ V})$$
$$= +0.61 \text{ V}$$

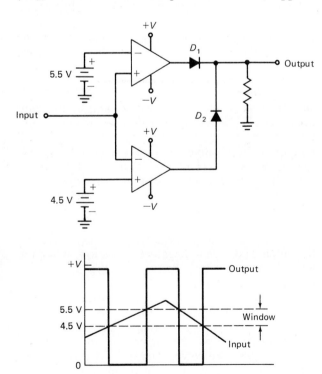

Fig. 6-29 Window comparator.

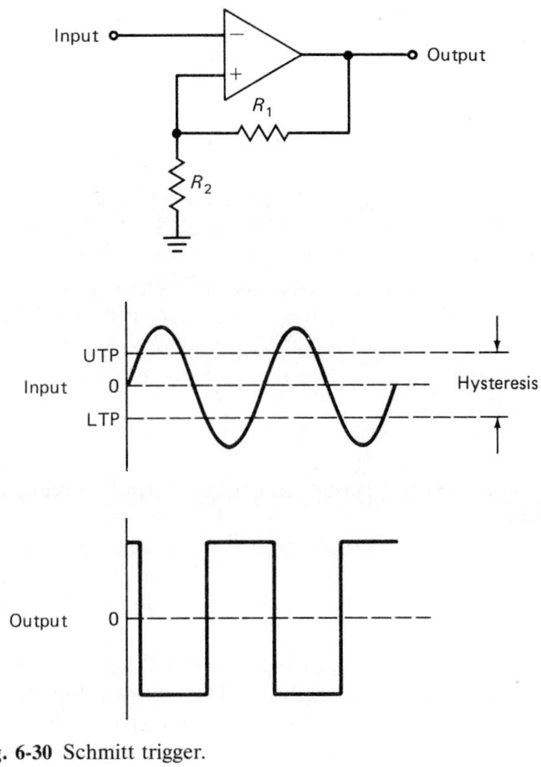

Fig. 6-30 Schmitt trigger.

The LTP is determined by using $-V_{SAT}$ in the preceding equation and calculates to -0.61 V in this example. The waveforms of Fig. 6-30 show how the threshold points work. At the beginning of the input cycle, the output is at positive saturation. Therefore, the voltage divider sets the noninverting input at the UTP. When the input signal exceeds this point, the IC output goes to negative saturation, and the voltage divider now sets the noninverting input at the LTP. The output does not switch again until the input signal goes more negative than the LTP.

The difference between the two threshold points is called *hysteresis*. For our example, the hysteresis will be equal to $+0.61 - (-0.61) = 1.22$ V. Hysteresis is desirable when signals are noisy. Refer to Fig. 6-31, which shows the noisy signal performance for a comparator circuit and for a Schmitt trigger. The output waveform in Fig. 6-31(a) has extra transitions. They are caused by noise when the average value of the input signal is near the reference voltage. The noise adds and subtracts from the input signal, and the instantaneous value goes above and below

the reference value several times. The output frequency is higher than the input frequency. The Schmitt trigger waveforms in Fig. 6-31(b) show no frequency distortion. As long as the hysteresis is greater than the peak-to-peak noise amplitude, extra transitions are eliminated. The hysteresis must be less than the peak-to-peak signal, however, or the output will not switch at all.

Figure 6-32 shows an operational amplifier differentiator. A *differentiator* is a circuit that responds to the rate of change of the input signal. It is essentially a high-pass filter, meaning that it produces more output for signals at higher frequencies. Its instantaneous output is proportional to the instantaneous rate of change at its input (or to the slope of the input waveform):

$$V_{out} = -RC \times \frac{\Delta V_{in}}{\Delta T}$$

Assuming an input signal that is changing at a rate of 100 V/s and the circuit values shown in Fig. 6-32:

$$V_{out} = -1 \times 10^4 \times 1 \times 10^{-6} \times \frac{100 \text{ V}}{1}$$
$$= -1 \text{ V}$$

The illustration shows the input and output waveforms. The simple differentiator circuit suffers from high-frequency noise since the gain goes up as frequency does. A practical differentiator will often use a resistor in series with the input to limit the high-frequency gain.

The opposite of differentiation is *integration*. Figure 6-33 shows an op amp integrator. It is essentially a low-pass filter and produces more output for signals at lower frequencies. When an input waveform steps from 0 to -1 V, the output ramps positive at a rate of 100 V/s. The integrator output is proportional to

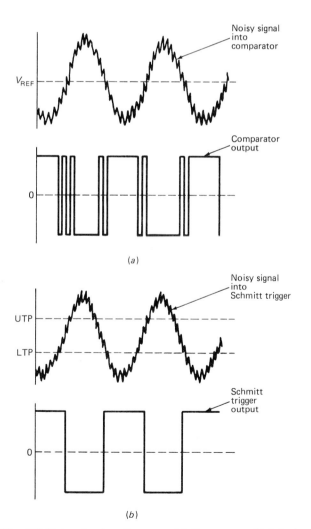

Fig. 6-31 Noisy signal performance. (a) Comparator. (b) Schmitt trigger.

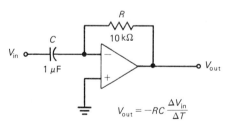
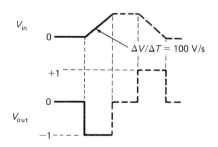

Fig. 6-32 Differentiator.

the product of the amplitude and duration of the input signal (the area under the curve). The output voltage is given by

$$V_{out} = -V_{in} \times \frac{1}{RC} \times T$$

By dropping the T term in the preceding equation, the output is expressed as a rate of change in volts per second. With the input and circuit values shown in Fig. 6-33, the rate of change will be

$$V_{out} = -(-1\text{ V}) \times \frac{1}{1 \times 10^4 \times 1 \times 10^{-6}}$$
$$= 100\text{ V/s}$$

The simple integrator circuit suffers from drift. Any dc offset voltage at the input will cause the integrator to drift and eventually saturate. Practical circuits usually use a high value of resistor in parallel with the feedback capacitor to reduce this drift.

Comparing the waveforms of Fig. 6-32 and Fig. 6-33 shows the opposite natures of differentiation and integration. If the differentiated function is fed to an integrator, its output will be the same as the original function. Figure 6-34 shows circuit performance with other waveforms. The differentiator output shown in Fig. 6-34(a) is an inverted cosine wave with a sine wave input. A cosine wave is 90° out of phase with a sine wave. The total phase shift shown is 270° since the output is differentiated and inverted (90 + 180). Note that the output is zero when the input is peaking, because the instantaneous rate of change of a sine wave at peak is zero. The maximum rate of change occurs during the zero crossing at the output peaks at this time. The integrator waveforms of Fig. 6-34(b) show that an inverted cosine at the input produces a sine wave at the output. Once again it is demonstrated that when a signal is differentiated and

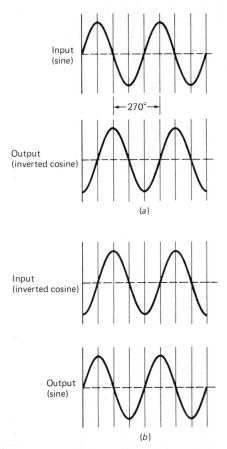

Fig. 6-34 Output waveforms. (a) Differentiator. (b) Integrator.

then integrated, the original signal will appear at the output of the integrator.

Integrators and comparators can be used together to achieve some other valuable functions. Figure 6-35 shows a voltage-to-frequency converter. The positive input voltage is applied to the integrator,

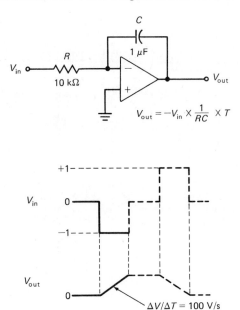

Fig. 6-33 Integrator.

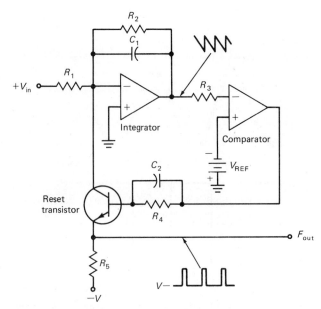

Fig. 6-35 Voltage-to-frequency converter.

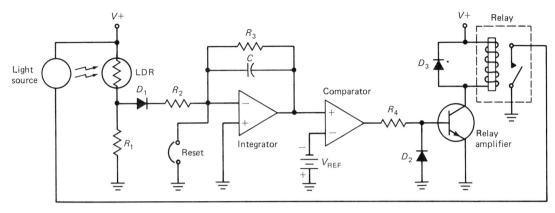

Fig. 6-36 Light integrator.

which begins ramping negative at a rate proportional to V_{in}. When the integrator ramp exceeds the value of V_{REF}, the comparator output switches to positive saturation and turns on the reset transistor. With the transistor saturated (low resistance), the integrator output is rapidly driven positive because $V-$ is applied to the input. The comparator then saturates negative, and the reset transistor turns off so the next integration cycle can begin. The output is a series of positive-going pulses, and the pulse frequency is proportional to V_{in}. For example, if the value of V_{in} is doubled the integrator will ramp negative at twice the rate, and the reset pulses will appear in half the time. The output frequency will double. Capacitor C_2 is a speed-up capacitor. It decreases the switching time of the reset transistor. The voltage-to-frequency converter can also be called an *analog-to-digital converter*. The input voltage is analog, and the output is a digital pulse train.

Figure 6-36 shows a light integrator. The light-dependent resistor (LDR) shows a decrease in resistance when it is illuminated. It acts as part of a voltage divider along with R_1. The divided positive voltage is applied to the integrator, which ramps negative at a rate that is proportional to the light intensity. When the ramp exceeds $-V_{REF}$, the comparator output saturates negative. This turns off the relay amplifier, and the contacts open and turn off the light source. Pushing the reset button will discharge the

integrator and begin another cycle. The value of integration in this application is that the circuit ensures the proper light dosage regardless of supply variations, light source variations, or light transmission path changes. The integrator/comparator combination will not turn the light source off until the proper sum has been acquired. This type of a circuit can be used in various photochemical processes and is far superior to a simple timer, which is subject to errors when light intensity fluctuates. Diode D_1 prevents the divider from discharging the integrator when the light source fluctuates or is momentarily interrupted. Diode D_2 clamps the base voltage of the transistor when the comparator output is negative. Diode D_3 protects the transistor from relay coil transients.

Ordinary diode circuits cannot be used to rectify signals in the millivolt range because several tenths of a volt are required to turn diodes on. A precision rectifier circuit that overcomes this limitation is shown in Fig. 6-37. The op amp on the left, along with its diodes and resistors, forms a precision half-wave rectifier. When the input signal goes positive, the output goes negative, and D_1 turns on to conduct the feedback current. Diode D_2 is reverse-biased, and the left end of R_3 remains at zero as long as the input is positive. When the input goes negative, the output of the left op amp goes positive. Now D_1 is off, and D_2 is on and supplies the feedback current through R_2. Resistor R_1 = resistor R_2 and the signal

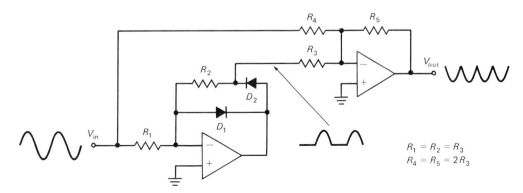

Fig. 6-37 Precision full-wave rectifier.

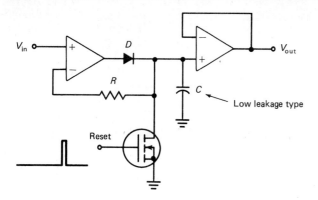

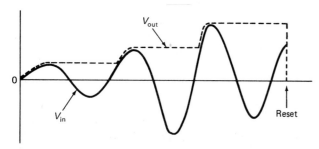

Fig. 6-38 Positive peak detector.

at the cathode of D_2 is an inverted replica of the negative portion of the input signal. If a precision half-wave rectifier is all that is required, the other op amp and its associated parts can be eliminated. Full-wave operation is realized by combining the half-wave signal with the original signal in the second op amp, which serves as a weighted adder. Its feedback resistor is twice the value of R_3, and the half-wave signal receives twice the gain that V_{in} receives.

Figure 6-38 shows a positive peak detector circuit. It "remembers" the greatest positive input value until it is reset. The op amp at the left serves as a noninverting amplifier. The diode in series with its output allows the capacitor to charge positive with respect to ground. If the input signal goes less positive, the diode turns off and prevents the capacitor from being discharged. The op amp at the right acts as a voltage follower. Its high input impedance allows the capacitor to retain its charge for long periods of time. The circuit is reset at the end of the sampling interval by applying a positive pulse to the enhancement mode MOSFET.

REVIEW QUESTIONS

25. Refer to Fig. 6-27. What will happen to the duty cycle of the output if V_{REF} is reversed?

26. What will have to be added to Fig. 6-27 if the IC is a type 311 comparator?

27. Refer to Fig. 6-29. Will the circuit work as a window comparator if the reference supplies are reversed?

28. Refer to Fig. 6-30. Assume that $R_1 = 4.7$ kΩ, $R_2 = 1$ kΩ, and V_{SAT} is equal to plus and minus 11 V. Calculate UTP and LTP.

29. What output waveform would you expect from the circuit in Fig. 6-32 with a triangular input wave?

30. Refer to Fig. 6-32. If the input is going negative at a rate of 50 V/s, what is the output voltage?

31. Refer to Fig. 6-33. If the input wave is square, what will the output be?

6-6
SPECIAL FUNCTIONS

It is often desirable to have a differential amplifier that has high gain, high input impedance, and a high CMRR. Using an operational amplifier produces conflicting design requirements in these cases since the gain is decreased by making the input resistors high in value. Conversely, making the input resistors low (for good gain) lowers the input impedance. You may wish to refer again to Fig. 6-22 to verify this point. It is also difficult to change gain because resistor ratios must be closely matched (within 0.1 percent). These op amp limitations have led to the development of a special circuit or device called the *instrumentation amplifier* (IA).

Figure 6-39(*a*) shows how an IA can be "built up" by using three op amps. The input section uses two devices, and the signal is fed to the noninverting input in each case. This meets the requirement of

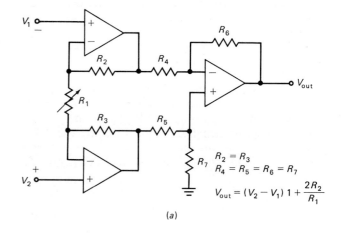

$$R_2 = R_3$$
$$R_4 = R_5 = R_6 = R_7$$
$$V_{out} = (V_2 - V_1)\left(1 + \frac{2R_2}{R_1}\right)$$

(*a*)

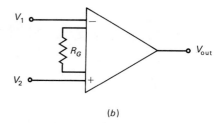

(*b*)

Fig. 6-39 Instrumentation amplifier. (*a*) Three op amp. (*b*) Schematic symbol.

high input impedance. The output section uses a single amplifier to combine the two signals differentially. The bottom input (V_2) is noninverting because it drives the noninverting input of the output amplifier. The top input (V_1) is inverting because it drives the inverting input of the output amplifier. Note that the gain is set by a single resistor in the sense that R_1 can be changed without the need to change R_2 or R_3. In fact, R_1 can be made adjustable, a great convenience for some applications. Changing the gain has no effect on the input impedance of the IA.

Instrumentation amplifiers are also available in a single package. They can be monolithic or hybrid integrated circuits. Figure 6-39(b) shows the schematic symbol commonly used for a single-package device. Resistor R_G sets the gain of the IA along with an internal precision feedback network. The National LH0038 is one example of a hybrid IA and is available in a 16-pin dual-inline package. It is shown in Fig. 6-40 in a bridge amplifier application. It features a low input offset voltage of 25 μV and a low offset drift of no more than 0.25 μV/°C. It has a CMRR of 120 dB and can amplify low-level signals in the presence of high common mode noise. The internal gain setting resistors are precision thin-film types and exhibit excellent thermal tracking. The voltage gain is set by using pins 5 through 10 on the IC package, and a range of 100 to 2000 is available. With pin 7 jumpered to pin 10 as shown in Fig. 6-40, the voltage gain is 1000.

The bridge circuit is a very popular arrangement for accurate measurements using certain types of transducers such as strain gauges. Chapter 9 will treat transducers in detail, but a brief discussion here is appropriate to demonstrate the need for precision amplification. A *strain gauge* is a wire device that is used to measure strain (elongation or compression) of some member of a physical system. If a strain stretches the wire in the gauge, it causes its resistance to increase. The bridge circuit of Fig. 6-40 is an excellent arrangement for a strain gauge since the bridge balance will be affected if the gauge is one element of the circuit, such as R_4. Also, a second identical gauge can be used in the R_3 leg of the bridge. If the two gauges track thermally, then tem-

perature effects are canceled since bridge balance will not be affected when both elements change by the same amount. The thermal (compensation) gauge will be mounted in such a way that it will not react to the axis of strain. The output of the bridge will therefore be a function of strain only.

Ideally, the amplifier that follows a precision bridge circuit should not degrade the accuracy of the signals. It should also reject common mode hum and noise picked up on the bridge elements and the interconnecting cable. The IA of Fig. 6-40 has the high performance and thermal stability required. The guard output (pin 11) is maintained at the common mode voltage and greatly reduces noise pickup since it also maintains the shield at the common mode voltage. The output sense and ground sense terminals are utilized in those applications in which errors in load voltage can occur. For example, the output (pin 1) may be used to drive a buffer amplifier for more current capacity. The load voltage at the output of the buffer amplifier will show some error due to drops in the buffer amplifier. The error is eliminated by connecting the ground sense and output sense directly to the load.

Operational amplifiers are normally energized from a dual supply. In some cases this arrangement is not convenient. Figure 6-41 shows that it is possible to "float" an op amp across a single supply. Note that the $-V$ terminal is grounded and that a voltage divider "floats" the noninverting input terminal at half the supply voltage. The dc output voltage of the op amp will therefore be at one-half the supply voltage. This arrangement is convenient for ac amplifiers, and a coupling capacitor can be used in series with R_1 to block any dc component in the input signal. The blocking capacitor will be required if the signal source is at dc ground because the output will be driven to positive saturation without it. It may also be necessary to use a bypass capacitor across R_3 to eliminate any ac noise at the noninverting input.

A special type of amplifier called a *Norton op amp* offers biasing advantages for single-supply operation. It is a current-differencing amplifier and uses a current mirror instead of the typical differential pair to achieve the noninverting input function. Figure 6-42 shows a partial schematic for a Norton amplifier. The common emitter input amplifier uses a current source

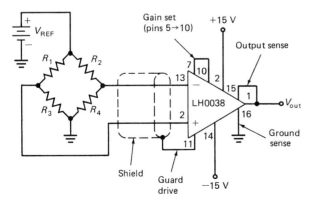

Fig. 6-40 Bridge amplifier.

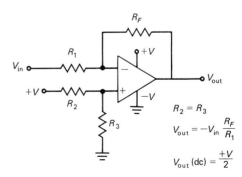

Fig. 6-41 Single-supply inverting amplifier.

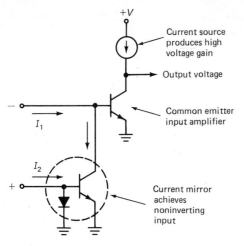

Fig. 6-42 Partial schematic of Norton amplifier.

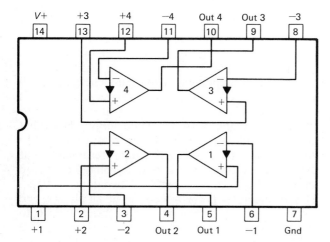

Fig. 6-43 Pin configuration of LM3900.

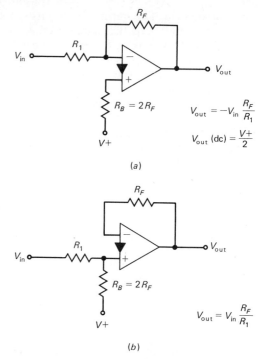

$$V_{out} = -V_{in}\frac{R_F}{R_1}$$

$$V_{out}\ (dc) = \frac{V+}{2}$$

(a)

$$V_{out} = V_{in}\frac{R_F}{R_1}$$

(b)

Fig. 6-44 Norton amplifier circuits. (a) Inverting. (b) Noninverting.

as a load for high-voltage gain. A current source has a very high internal impedance; any change in collector current will therefore produce a large swing in collector voltage. The noninverting input function is achieved with a *current mirror circuit.* Any current flowing into the mirror (I_2) will have the opposite effect on the output voltage when compared to any current flowing into the inverting input (I_1). For example, if I_2 increases, the mirror transistor will turn on harder and remove some of I_1 from the base circuit. This has the same effect as decreasing I_1.

The pinout for an LM3900 Norton op amp is illustrated in Fig. 6-43. It contains four separate amplifiers. It is designed for single-supply operation, pin 7 is grounded, and the positive supply is applied to pin 14. The supply range is 4 to 36 V dc. The arrow on each amplifier symbol denotes that it is a Norton type. Figure 6-44 shows two single-supply circuits utilizing the Norton op amp. The noninverting input is biased with a single resistor, R_B. When this resistor is twice the value of the feedback resistor, the dc output voltage is equal to half of the supply voltage.

Filters are circuits used to remove unwanted frequency components from a signal. They are useful for improving signal-to-noise ratio and for rejecting undesired signals. Practical filters can be built by using only passive devices such as resistors, capacitors, and inductors. Inductors are expensive components and are often physically large in low-frequency designs. It is possible to eliminate the need for inductors by using active devices such as op amps. Filters designed this way are known as *active filters.* They can save space and weight and usually cost less than passive designs. They also eliminate filter loss and are easier to apply since they have a high input impedance and a low output impedance.

Figure 6-45 shows the circuit and frequency response plot for an active low-pass filter. It is a Butterworth design and provides an attenuation of 40 dB per decade for signals beyond the cutoff frequency f_C. The plot shows that the gain is -3 dB at the cutoff frequency. Two or more filter sections can be cascaded for improved attenuation of out-of-band signals. Cascading will affect the cutoff frequency, however. If two filters with the same cutoff frequency are cascaded, the overall gain will become -6 dB at the original cutoff frequency, and the gain roll-off will be 80 dB per decade above the cutoff frequency. The cutoff frequency is always specified at the -3-dB point and will now occur at a slightly lower frequency. The circuit is easy to design. Suppose a cutoff frequency of 1 kHz is required. The design procedure can begin with a selection of resistor or capacitor values depending on which is more convenient. If a 0.005-μF capacitor is desirable for

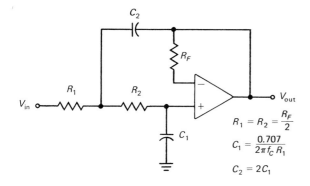

$$R_1 = R_2 = \frac{R_F}{2}$$

$$C_1 = \frac{0.707}{2\pi f_C R_1}$$

$$C_2 = 2C_1$$

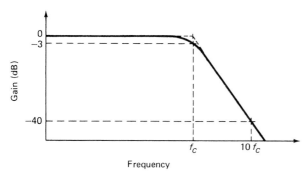

Fig. 6-45 Unity gain low-pass Butterworth filter.

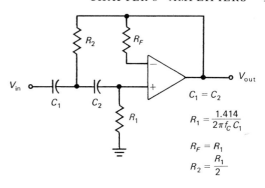

$$C_1 = C_2$$

$$R_1 = \frac{1.414}{2\pi f_C C_1}$$

$$R_F = R_1$$

$$R_2 = \frac{R_1}{2}$$

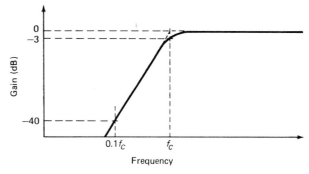

Fig. 6-46 Unity gain high-pass Butterworth filter.

C_1, the equation from Fig. 6-45 can be rearranged to solve for R_1:

$$R_1 = \frac{0.707}{6.28 \times 1000 \times 0.005 \times 10^{-6}}$$
$$= 22.5 \text{ k}\Omega$$

The feedback resistor will have to be twice that value, or 45 kΩ, and C_2 will have to be twice C_1, or 0.01 μF.

The circuit for a unity gain high-pass Butterworth filter is shown in Fig. 6-46. It produces an attenuation of 40 dB per decade for signals lower than the cutoff frequency. It is useful for the reduction of low-frequency noise on a signal. Additional sections may be cascaded to achieve better attenuation for out-of-band signals. Cascading will shift the cutoff frequency upward a little. The design process is straightforward and usually begins with a selection of a capacitor value. Suppose the cutoff frequency is 1 kHz, and 0.005-μF capacitors are available:

$$R_1 = \frac{1.414}{6.28 \times 1000 \times 0.005 \times 10^{-6}}$$
$$= 45 \text{ k}\Omega$$

The feedback resistor will be the same value, and R_2 will be half this value, or 22.5 kΩ.

Figure 6-47 illustrates the design of a multiple-feedback bandpass filter. This type of filter rejects frequencies above and below the pass band. It is used to select one frequency, called the *resonant frequency* (f_R), out of a range of frequencies. The figure of merit of a bandpass filter is its Q, which is

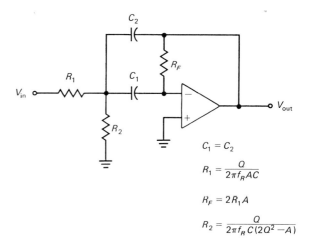

$$C_1 = C_2$$

$$R_1 = \frac{Q}{2\pi f_R A C}$$

$$R_F = 2R_1 A$$

$$R_2 = \frac{Q}{2\pi f_R C (2Q^2 - A)}$$

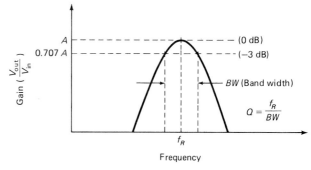

Fig. 6-47 Multiple feedback bandpass filter.

equal to f_R divided by the bandwidth. The *bandwidth* is the difference between the upper and lower cutoff points. As always, the response is −3 dB at the cutoff frequencies. The higher the Q, the more narrow the bandwidth of the filter. This particular circuit is useful for Q values up to about 20. Higher Qs are obtainable by cascading filter sections. The design process is based on the resonant frequency, the Q value, and the desired gain at resonance (A).

EXAMPLE

A filter that resonates at 1 kHz with a bandwidth of 100 Hz is required. The necessary Q is 1000/100, or 10. If the desired voltage gain at 1 kHz is 10, and 0.01-μF capacitors are selected, find the values of R_1 and R_2.

SOLUTION

The value of R_1 can be calculated first:

$$R_1 = \frac{10}{6.28 \times 1000 \times 10 \times 0.01 \times 10^{-6}}$$
$$= 15.9 \text{ k}\Omega$$

The feedback resistor is next:

$$R_F = 2 \times 15.9 \times 10^3 \times 10$$
$$= 318 \text{ k}\Omega$$

Finally, R_2:

$$R_2 = \frac{10}{6.28 \times 1000 \times .01 \times 10^{-6} \times (200 - 10)}$$
$$= 838 \ \Omega$$

Figure 6-48 depicts the circuit for a *bandstop filter*, which is used to remove one interfering frequency. It can be used to eliminate 60-Hz hum from a signal. It provides an attenuation of approximately 40 dB at the resonant frequency and a gain of 1 for signals above and below the stopband. It is designed for a given value of resonant frequency and Q. The design process begins with the selection of a convenient value of capacitor. Suppose the selected capacitor value is 0.01 μF, the resonant frequency is 1 kHz, and the stopband is 100 Hz wide. The Q is 1000/100 = 10, and the feedback resistor is found by the following equation:

$$R_F = \frac{Q}{\pi \, f_R \, C_1}$$
$$R_F = \frac{10}{3.14 \times 1000 \times 0.01 \times 10^{-6}}$$
$$= 318 \text{ k}\Omega$$

R_1 is found next:

$$R_1 = \frac{R_F}{4Q^2}$$
$$R_1 = \frac{318 \times 10^3}{4 \times 10^2}$$
$$= 796 \ \Omega$$

The value of R_2 will also be 796 Ω, and R_3 is

$$R_3 = 2 \, Q^2 \, R_1$$
$$R_3 = 2 \times 10^2 \times 796$$
$$= 159 \text{ k}\Omega$$

The next special function we will look at is the four-quadrant multiplier. Its transfer characteristic is

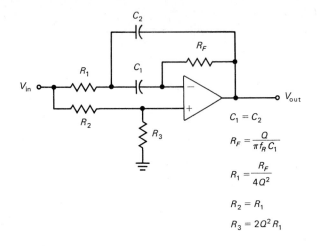

$$C_1 = C_2$$
$$R_F = \frac{Q}{\pi f_R C_1}$$
$$R_1 = \frac{R_F}{4Q^2}$$
$$R_2 = R_1$$
$$R_3 = 2Q^2 R_1$$

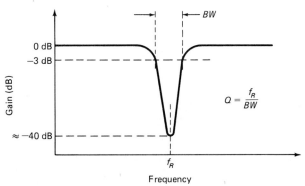

$$Q = \frac{f_R}{BW}$$

Fig. 6-48 Unity gain bandstop filter.

shown in Fig. 6-49(*a*), and its schematic symbol is shown in Fig. 6-49(*b*). Its output is the linear product of its two input voltages. Quadrant I operation results when both the Y and the X inputs are positive. Quadrant II results when X is negative and Y is positive. When both inputs are negative, the circuit operates in quadrant III. Quadrant IV means that X is positive and Y is negative. The output voltage is algebraically correct. For example, it is positive when both inputs are positive or both inputs are negative. It is negative if one of the inputs is negative. To avoid saturation with large inputs, the output is divided by 10. Therefore, with X = 10 V and Y = 10 V, the output will be equal to 10 $\times$ 10 = 100 divided by 10, or 10 V.

In addition to multiplying two signals, the four-quadrant multiplier may also be used to square a single voltage. This operation is illustrated in Fig. 6-50(*a*). Note that the X and Y inputs are tied together. The output signal is equal to one-tenth of the input signal squared. Note that the output of a squaring circuit can never be negative. If the input signal is a sine wave, the output signal is also a sine wave but at twice the frequency. The output sine wave for the frequency doubler also has a dc component. If this feature is undesirable, it can be removed with a coupling capacitor. Figure 6-50(*b*) shows a divide circuit. The output of the multiplier is summed with V_Z at the inverting terminal of an op amp. Voltage V_X is applied to the X input of the multiplier, and the

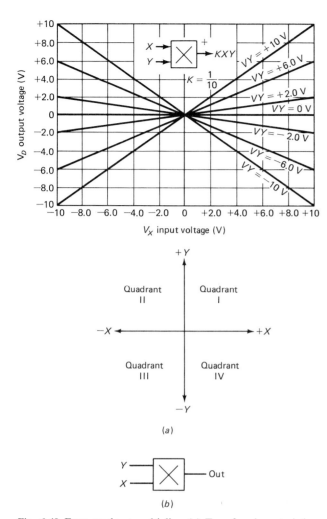

Fig. 6-49 Four quadrant multiplier. (*a*) Transfer characteristics. (*b*) Schematic symbol.

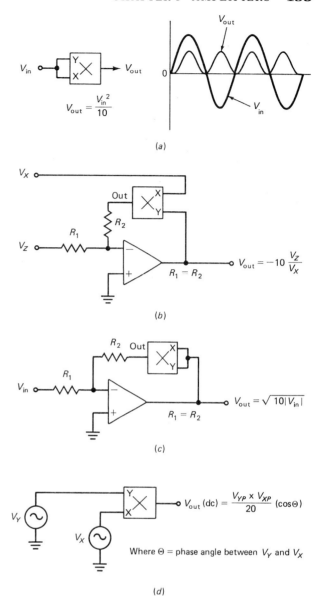

Fig. 6-50 Four quadrant multiplier applications. (*a*) Squaring circuit (frequency doubler). (*b*) Divide circuit. (*c*) Square root circuit. (*d*) Phase angle circuit.

op amp output is applied to the *Y* input. The overall output is inverted and is equal to ten times of V_Z divided by V_X. Figure 6-50(*c*) shows a square root circuit. The circuit works for negative input voltages only, and the output is equal to the square root of 10 times the absolute value of V_{in}. The last multiplier application is shown in Fig. 6-50(*d*). It produces a dc output voltage proportional to the peak value of two input waveforms and the cosine of the phase angle between them. For example, if one waveform is a sine and the other is a cosine, the phase angle between them will be 90°. The cosine of 90° is 0, and the dc output will be zero. If both input waveforms have a peak value of 10 V and happen to be in phase, the dc output voltage will be equal to 5 V because the cosine of 0° is 1.

Some industrial applications are hazardous and require an isolation amplifier to protect equipment from high-voltage surges and ground loops. Isolation amplifiers are also required to measure low-level signals in the presence of high common mode voltages. Malfunctions may cause power line voltages to be imposed on low-voltage signal lines. The isolation am-

plifier must be able to withstand such a mishap and protect expensive equipment on the other end. There are two commonly used isolator designs. In Fig. 6-51 the block diagram for an Analog Devices 284J, which is of the modulation type, is shown. High-frequency transformers couple energy into the input section. Here, a current-limited converter rectifies and filters the high frequency to provide direct current for the input circuits. Since transformer coupling is used, the input is guarded (isolated) from the primary supply, the output, and ground. The signal to be amplified is applied to a phase modulator, which is also fed by the high-frequency signal. The resulting phase-modulated signal is transformer-coupled to a phase demodulator in the output section. The demodulator recovers the original signal, and a low-

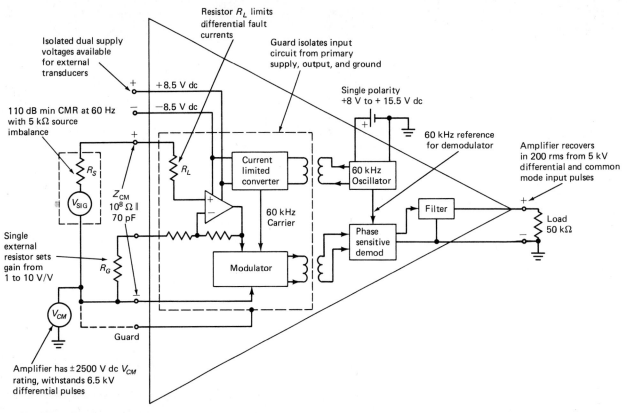

Fig. 6-51 Block diagram of 284J isolation amplifier (*Analog Devices*).

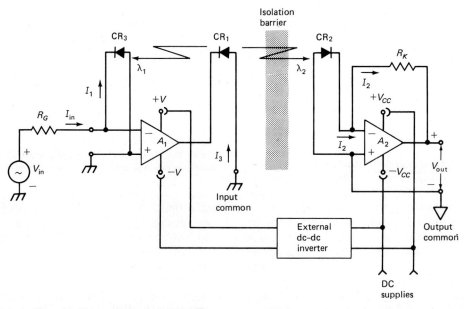

Fig. 6-52 Block diagram of an optically coupled isolation amplifier.

pass filter removes the high-frequency component. The other type of isolation amplifier is optically coupled and is shown in Fig. 6-52. It uses photodiodes and photodetectors to communicate across the isolation barrier. The optically coupled isolator is less expensive and smaller and has a wider bandwidth than the modulation type. However, it requires an external dc-to-dc converter for operation. The optically coupled types also suffer from gain inaccuracy and are more susceptible to temperature effects.

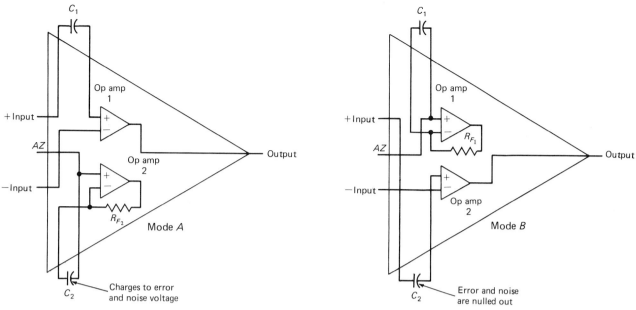

Fig. 6-53 The two half cycles of operation for a commutating autozero op amp.

Their signal-to-noise ratio is also poorer than that of the modulation types.

Temperature drift is the major enemy of high-gain dc-coupled amplifiers. An early solution was to use a *chopper amplifier.* These circuits chopped the dc signal to produce an ac signal. The ac signal was then amplified in a high-gain ac amplifier, where drift was not a problem. Finally, the ac signal was converted back to dc and filtered. Chopper stabilization has now progressed to a related but different approach called *commutating auto-zero* (CAZ) op amps. One example is the Intersil ICL7600 CAZ op amp. It features a low-input offset voltage of 2 μV. Its long-

term drift is only 0.2 μV/year. It also features an input offset temperature drift of only 0.005 V/°C. It is based on two internal op amps that are connected so that when one amplifier is processing a signal the other is maintained in the auto-zero mode. Since the device auto-zeros its internal offset errors, it is extremely stable with temperature and time. Figure 6-53 shows its operating principles. The voltage at the *auto-zero* (AZ) input is the voltage to which each op amp will be zeroed. In mode A, op amp number 2 operates in unity gain and charges external capacitor C_2 to a voltage equal to the dc offset of op amp number 2 and the instantaneous low-frequency noise voltage. Later, the internal switches connect the device in the configuration of mode B. Operational amplifier number 2 now has capacitor C_2 connected in series with its noninverting input. Since it was previously charged to its offset and noise voltage, the offset and noise are nulled out.

The last special function to be covered here is the *sample and hold amplifier* shown in Fig. 6-54. It is used in those cases in which circuits or measuring devices cannot accept varying signals. It works by pulsing the sample input, which turns on the enhancement mode FET. With the FET on, the input voltage can charge or discharge the capacitor. The acquisition time can be made short by using a capacitor of reasonable value and a low impedance op amp and FET switch. After the acquisition time, the FET is turned off. The capacitor now holds the value of V_{in}. The hold time can be made long by using a high-input-impedance op amp at the output. The droop voltage is usually on the order of several millivolts per second and is caused by capacitor leakage, FET leakage, and the input current of the second amplifier. The sample and hold circuit is available on a single chip.

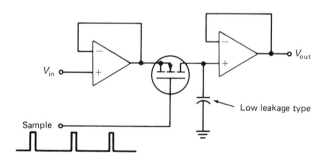

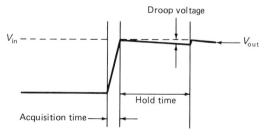

Fig. 6-54 Sample and hold amplifier.

REVIEW QUESTIONS

32. Refer to Fig. 6-39(a). What is the differential gain of the amplifier if $R_2 = 100$ kΩ and $R_1 = 1$ kΩ?

33. What sets the input impedance of the IA in question 32?

34. Refer to Fig. 6-40. Assume no offset error and a balanced bridge. What is the output voltage?

35. Refer to Fig. 6-40. Will the output voltage change if R_3 and R_4 change by the same amount?

36. Refer to Fig. 6-40. Resistors R_3 and R_4 are identical strain gauges, one of which does not react to elongation. What is the function of the nonreacting gauge?

37. Refer to Fig. 6-41. What will the output voltage be if R_3 opens?

38. Why are the biasing resistors in Fig. 6-44 made twice the value of the feedback resistors?

39. Use Fig. 6-45 and design a filter that cuts off at 100 Hz by using a value of 0.1 μF for C_1.

40. What is the gain of the filter in question 39 at 20 Hz? At 100 Hz? At 1 kHz?

6-7
OSCILLATORS

An *oscillator* is a circuit that creates an ac signal or changes direct to alternating current. Oscillators are often based on amplifiers. An amplifier can be made to oscillate by adding positive (in-phase) feedback. If the gain of the amplifier is greater than the loss in the feedback network, the circuit will oscillate. Figure 6-55 shows a Wien bridge oscillator. The *Wien bridge* is made up of resistors R_1 and R_2 and capacitors C_1 and C_2. The bridge produces a zero phase shift at the resonant frequency. Frequencies higher than resonance produce a lagging response through the bridge, and frequencies lower than resonance produce a leading response. Therefore, only the resonant frequency arrives at the noninverting input of the op amp with the correct phase to sustain oscillations.

A Wien bridge oscillator is capable of very-low-distortion sinusoidal output because the bridge acts as a filter and allows in-phase feedback to occur for only a single frequency. This is an important feature in low-distortion sinusoidal oscillators because a pure sine wave has energy content at only one frequency. However, there will be considerable distortion in the circuit if the gain of the amplifier is more than that required to sustain oscillations. The circuit of Fig. 6-55 keeps the gain low by the voltage divider action of R_3, C_3, and the FET. The FET is biased by rectifying and filtering the output signal with the diodes and C_4. If the gain goes too high, the output signal becomes larger, more of it is rectified, and the

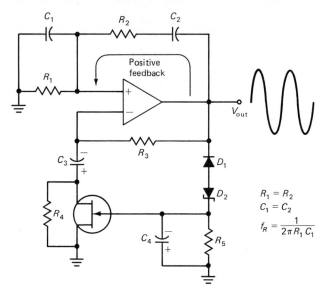

Fig. 6-55 Wien bridge oscillator.

$R_1 = R_2$
$C_1 = C_2$
$f_R = \dfrac{1}{2\pi R_1 C_1}$

negative gate voltage at the FET increases. This increases the drain-to-source resistance of the FET, and the negative feedback voltage at the inverting input is now larger. This decreases op amp gain, and the output is stabilized. Ideally, the automatic gain portion of the oscillator should keep the op amp gain just large enough to sustain oscillations.

The initial study of oscillators is sometimes confusing since it is difficult to understand how the circuit starts oscillating in the first place. When an oscillator is first energized, there is no strong signal at any particular frequency, but there is wide-band noise. This noise comes from the amplifying components and any parts connected to the input of the amplifier. The amplifier provides gain and wide-band noise is present at the output. The feedback network is frequency-sensitive and ensures that only one frequency arrives at the amplifier input with the correct phase (positive) relationship. This signal receives the most gain and soon appears at the output. The oscillator of Fig. 6-55 starts very quickly because the op amp gain is maximum when the circuit is first turned on.

Figure 6-56 shows another type of oscillator, known as a *multivibrator*. Resistors R_2 and R_3 provide positive feedback and upper and lower threshold points, as was discussed earlier for the Schmitt trigger. When the output is at positive saturation, current flows through R_1 to charge the capacitor. As long as the capacitor voltage is less than the UTP, the output remains at positive saturation. Eventually, the capacitor does charge to the UTP, the inverting input goes positive with respect to the noninverting input, and the output is driven to negative saturation. The capacitor now must charge in the opposite direction until the LTP is crossed. The multivibrator is not a single-frequency oscillator in the same sense as the Wien bridge circuit. It is in a category of *relaxation*

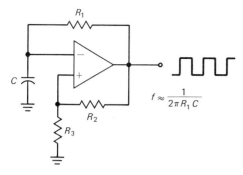

Fig. 6-56 Astable multivibrator.

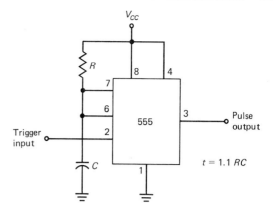

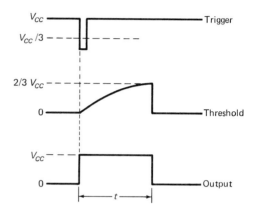

Fig. 6-58 Integrated circuit timer monostable (one-shot) multivibrator.

oscillators, which are governed by RC time constants. The output frequency is a square wave at some fundamental frequency that is determined by the time constant of R_1 and C. The output also contains considerable energy at the odd harmonics of the fundamental. For example, if the fundamental frequency is 100 Hz, there will be energy at 300 Hz, 500 Hz, 700 Hz, and so on.

The 555 IC timer can also be used as an astable multivibrator. This circuit is shown in Fig. 6-57 (the 555 timer is discussed in Chapter 4). The resistors can be adjusted for an entire range of duty cycles in the output waveform. A diode will be required for duty cycles of 50 percent and less. At the moment of power on, the capacitor is discharged, holding the trigger low. This triggers the timer and establishes the capacitor charge path through R_A and R_B, assuming no diode. When the capacitor voltage reaches the threshold of $\frac{2}{3}$ V_{CC}, the output goes low, and the discharge transistor turns on. The capacitor discharges through R_B. When the capacitor voltage drops to $\frac{1}{3}$ V_{CC}, the trigger comparator trips and retriggers the timer. Adding the diode allows the charge path to be R_A and the diode to provide a duty cycle range of 5 to 95 percent by adjusting the ratio of the timing resistors. Resistor R_B should be no smaller than 3000 Ω to ensure reliable starting of the circuit.

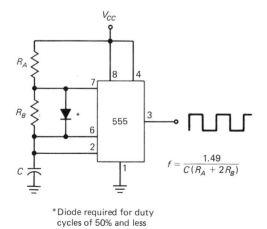

*Diode required for duty
cycles of 50% and less

Fig. 6-57 Integrated circuit timer astable multivibrator.

The IC timer may also be used in the *monostable,* or one-shot, mode to provide a timed output pulse in response to an input trigger pulse, as shown in Fig. 6-58. Prior to the trigger input pulse, the discharge transistor in the IC is on, thus short-circuiting the timing capacitor (C) to ground. When a trigger pulse drives pin 2 below $V_{CC}/3$, the device triggers, turning the discharge transistor off. The output goes high and remains high until the timing capacitor charges to $\frac{2}{3}$ V_{CC} through the timing resistor (R). When the $\frac{2}{3}$ V_{CC} threshold is reached, the comparator trips, in turning driving the output low and turning the discharge transistor on. The transistor quickly discharges the timing capacitor. The cycle is now complete, and the timer is ready for another input pulse. The width of the output pulse is equal to $1.1RC$. If the timing resistor is 4.7 kΩ and the timing capacitor is 0.02 μF, the pulse width will be 1.1 × 4.7 × 10^3 × 0.02 × 10^{-6} = 103.4 μs.

Figure 6-59 shows an oscillator circuit that produces both a square and a triangular waveform output. Assume that the threshold detector output is initially at negative saturation. This will cause the integrator to begin ramping in a positive direction. It will continue to ramp until its output crosses the lower threshold point of the detector. Its output will

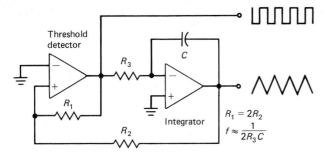

Fig. 6-59 Square and triangle wave generator.

$$R_1 = 2R_2$$
$$f \approx \frac{1}{2R_3 C}$$

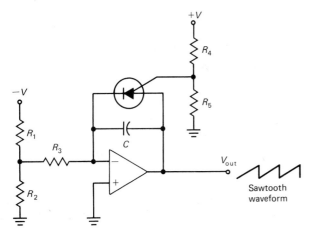

Fig. 6-60 Sawtooth oscillator.

now snap to positive saturation. The integrator will now begin ramping negative until the lower threshold point is crossed. The detector reassumes positive saturation, and another cycle begins.

A *sawtooth oscillator* is shown in Fig. 6-60. Voltage divider R_1 and R_2 apply some fraction of the negative supply to the integrator. It begins ramping positive at a rate that is determined by the magnitude of the input voltage, R_3, and the capacitor. Eventually, the anode-to-cathode voltage of the programmable unijunction transistor (PUT) will reach the breakover voltage. The PUT will turn on and quickly discharge the capacitor to about 1 V. Then the integrator will begin ramping again. The firing point of the PUT is adjusted with the positive divider made up of resistors R_4 and R_5. The output frequency is partly controlled by the gate voltage of the PUT because if it is adjusted to fire earlier, the positive ramp will be terminated sooner.

REVIEW QUESTIONS

41. What two conditions must be met for an amplifier to oscillate?

42. What is the function of the FET in Fig. 6-55?

43. The oscillator of Fig. 6-56 produces energy at a fundamental frequency and at _____ harmonic frequencies.

44. Refer to Fig. 6-57. Which resistor(s) determine(s) the length of time the output will be in the high state when the diode is used? When the diode is not used?

45. Refer to Fig. 6-60. What determines the output frequency in addition to R_3, C, R_4, and R_5?

6-8
TROUBLESHOOTING AND MAINTENANCE

Amplifier troubleshooting should begin with a thorough visual inspection. Do not forget to check cables, sockets, and connectors. Edge connectors on plug-in circuit boards are notorious for causing problems. Power down and remove the boards. Brighten the contact surfaces on the board by using an ordinary pencil eraser. Do not use highly abrasive materials. Check devices in sockets because vibration can cause them to work loose. Reseat any that look suspicious. If the pins on ICs look black, that is a clue that some corrosion has taken place. Use an IC puller or a small screwdriver and work the ICs partly out of their sockets and then firmly reseat them. This procedure produces a wiping action that will usually restore the conductivity of the contacts to an acceptable level. Of course, severe corrosion or sockets that have lost their tension will probably dictate replacement.

After you have verified that everything looks good, it is time to verify supply voltages. A circuit usually cannot work as intended if its supply voltage or voltages are out of tolerance. Too many technicians waste time troubleshooting circuits that do not have anything wrong with them. A complete power supply check may include an oscilloscope test. A dc voltmeter reading can be in the normal range when there is excess ac ripple on the supply line. This check is especially important if one of the symptoms is erratic output or ac hum in the output of the amplifier.

After the supply has been verified, it is time to identify the symptoms. Is the amplifier "dead" (no output at all)? Is the output weak? Is the amplifier unstable (acting as an oscillator)? Is the output of the amplifier distorted? Is there noise or hum on the output signal? Does the dc output drift? Is the output latched at the positive or negative rail?

Amplifier troubleshooting is usually done while working "live." If the amplifier is part of a control system with motors, hydraulic, or pneumatic actuators, you must be very careful that a circuit disturbance will not produce a dangerous mechanical output. It may be necessary to deactivate part of the system while working. However, if the system produces feedback signals, then deactivation may introduce its own set of symptoms. Refer to the manufacturer's recommendations when servicing systems of this type.

A dead amplifier is often the easiest to troubleshoot. If it is a multistage arrangement, you should view it as a signal chain. It must be verified one link at a time. Signal tracing with an oscilloscope is a useful technique. Start with the first stage and verify the presence of an input signal. Move to the output of the first stage and check again. If you find a stage where there is an input signal and no output signal, the defective stage has been isolated. It is then usually possible to take a few dc readings to isolate the defective component. For example, in a common emitter amplifier such as the one shown in Fig. 6-3, you may find no base voltage on the transistor. This situation could be caused by an open bias resistor or a short-circuited coupling capacitor. Another possibility is normal base voltage and a collector voltage equal to the supply. If the collector voltage is supposed to be about half the supply, then the transistor or the emitter resistor is open. The kinds of problems that upset dc terminal voltages in amplifier circuits (assuming normal supplies) include short-circuited or leaky coupling capacitors, short-circuited or leaky bypass capacitors, open or out of tolerance resistors, open solid-state junctions, and short-circuited solid-state junctions. Your knowledge of circuit principles and device behavior will lead you in the right direction.

If the circuit is a dc amplifier, the oscilloscope tracing technique may be less useful. It may be more desirable to use a digital voltmeter to verify the dc signal at various points in the signal chain. It is also sometimes possible to utilize circuit adjustment as a way to verify proper circuit response. For example, the circuit may have an offset null adjust potentiometer. The potentiometer setting can be changed slightly while monitoring the dc output voltage to determine whether the response is as expected. Do not use this technique until you carefully note the original setting of the potentiometer. If it is a multiturn potentiometer, carefully count the turns so you can accurately reset the circuit when you are done testing. Also make sure that the manufacturer's literature contains the adjustment procedure so you can restore the circuit to original specifications. If you find that it is possible to restore a normal dc output by making a large change in some adjustment point, you should not consider this a "fix" until you have investigated further. This condition often indicates that some component has drifted out of tolerance. It is probably going to continue drifting with time, and your fix will be temporary at best.

A weak output indicates that the gain of one of the stages is below normal or that the input signal is out of tolerance. Once again, tracing with an oscilloscope or digital voltmeter will usually isolate the stage. Low gain in a stage may be caused by a defective bypass or coupling capacitor, a bias resistor, a decoupling resistor, or a solid-state device. Even though the power supply voltages have been checked, it is possible that one stage will operate at a voltage lower than normal. Decoupling resistors are sometimes used to isolate a stage from the power bus. If a decoupling resistor increases in value, the supply voltage for that one stage will drop, and the gain will often be less than normal. When measuring gain, remember that follower-type amplifiers are expected to produce unity gain only.

An unstable output may be caused by undesired oscillations. Remember that any amplifier can become an oscillator if there is in-phase feedback and more gain than loss. Decoupling resistors are used in conjunction with bypass capacitors to prevent amplifier coupling through the power bus. An open bypass capacitor may cause ac feedback and oscillations. Another source of instability is phase error that causes feedback to become positive at some frequency. Stray capacitances in a circuit, along with resistances, form a group of RC lag networks that do not show on schematic diagrams. Negative feedback can become positive feedback if the total lag reaches 180°. As the frequency goes higher and higher, the individual lags can sum to produce enough phase shift to make the feedback positive. If there is enough gain at this frequency, oscillations will occur. This is why op amps are often internally compensated for a gain roll-off of 20 dB per decade. This compensation ensures that the gain will be too low for oscillations to occur at the frequency where phase shift makes the feedback become positive. If an op amp is unstable and is externally compensated, the compensation components should be checked. If the gain is too high, it may also cause oscillation for the reasons already discussed.

Distortion in an amplifier is often caused by a bias error. For example, if an amplifier is not operating near the center of its load line, then clipping of either the negative-going or the positive-going part of the signal may result. It is usually easy to trace distortion to one stage by using an oscilloscope. In some cases, it is also helpful to use a function generator to feed a known test signal through the system. Most function generators are capable of producing sine, square, and triangle waveforms. The triangle waveform is usually the best choice when looking for distortion in a linear amplifier. The sharp peaks make it easy to spot any tendency toward clipping or compression, and the straight sides make any nonlinearity apparent. Once the defective stage is isolated, dc voltage checks will often reveal which component has failed. If the dc readings are correct, a defective solid-state device should be suspected.

Noise and hum are problems that can indicate power supply difficulties, open bypass capacitors, broken ground wires and shields, or defective devices. If the noise or hum is a common mode signal that the amplifier should reject, check the op amps and IAs. Circuit imbalance is also a possible culprit when the CMRR is not what it should be. For example, the IA shown in Fig. 6-39(a) will not exhibit good common mode rejection if one of the four resistors (R_4 through R_7) in the output stage changes value. It may be possible to adjust the CMRR in some amplifiers. Once again, if a significant change in a potentiometer must be made to restore proper operation it may be an indication that some component is drifting. Another noise source is a defective component in the input stage of a high-gain amplifier. A resistor, transistor, diode, IC, or some other component may become a noise generator. Remember that some components such as zener diodes normally generate significant noise and must be bypassed to confine the noise. Check the appropriate bypass capacitors. A transducer (or some other signal source) connected to the amplifier input may also be defective and noisy. Check for dirty and corroded connections. Finally, the amplifier may be responding to a strong electromagnetic field from some RFI source.

Drift is often a problem in high-gain dc amplifiers. Some drift is normal. If the drift is causing problems, it must be investigated. It is usually worst when it occurs in an early stage because it is amplified by all succeeding stages. Often a pattern to the drift can provide a valuable clue. For example, the drift may always be in the same direction as the amplifier warms up. This usually indicates that some component or device is changing with temperature. You can use a source of heat or cold to help isolate a sensitive part. Some cold spray materials build up damaging static charges and attack plastics, so be careful to use an approved type. Heat sources include soldering pencils, lamps, and heat guns. Be careful not to overheat circuit boards, components, and plastic parts.

Latch-up is a problem encountered with some operational amplifiers. The output gets "stuck" at the positive rail or at the negative rail. If the inputs seem normal, check the feedback circuit components. If no problems can be found there, investigate the dual supply on power up by using a dual-trace oscilloscope. If the plus and minus voltages are not applied at the same time, latch-up may result.

REVIEW QUESTIONS

46. Refer to Fig. 6-3. Suppose R_{B2} opens up. The amplifier operating point will move toward _____.

47. Will the dc collector voltage measure normal, high, or low for the fault discussed in question 46?

48. Refer to Fig. 6-4(a). Will the dc output voltage measure normal, high, or low if the emitter of Q_1 opens up?

49. Refer to Fig. 6-5(b). Should the normal voltage gain of this amplifier be high, moderate, or near unity?

50. Refer to Fig. 6-10. What dc output voltage can you expect at A and B if the $-V_{EE}$ supply fails and goes to 0 V?

51. Refer to Fig. 6-40. What dc condition will exist at V_{out} if R_2 opens?

CHAPTER REVIEW QUESTIONS

6-1. Refer to Fig. 6-2 and assume a quiescent base current of 80 μA. Where will the amplifier operate?

6-2. What will the quiescent collector-to-emitter voltage be for question 6-1?

6-3. Refer to Fig. 6-3. Which two components establish the base voltage?

6-4. What type of coupling is required for dc amplification?

6-5. What is the function of the emitter bypass capacitor in Fig. 6-3?

6-6. What is the overall current gain in a Darlington pair in which the first transistor has an h_{FE} of 50 and the second an h_{FE} of 200?

6-7. What is the phase relationship between input and output for common base and emitter follower amplifiers?

6-8. Refer to Fig. 6-11. Assuming balance and a common mode input, what will the differential output be? What will the single-ended output be?

6-9. A particular op amp requires a 1.5-mV input differential to produce 0 output. What is this error called?

6-10. What terminals are provided on some op amp packages to eliminate the error of question 6-9?

6-11. Refer to Fig. 6-23. Input 1 is driven 0.2 V negative. What effects will be noticed at the other two inputs?

6-12. Calculate the output voltage for Fig. 6-25 if the feedback resistor is 27 Ω and the current source is conducting 0.5 A.

6-13. Refer to Fig. 6-26. If $V_{in} = 0.01$ V, $R_L = 10,000$ Ω, and $R_s = 10$ Ω, how much current will flow in the load resistor? How much current will flow if the load resistor is changed to 4.7 kΩ?

6-14. Refer to Fig. 6-33. If the input is $+0.2$ V, what are the polarity and rate of change at the output?

6-15. What is the instantaneous output voltage in question 6-14 after the input has been applied for 100 ms?

6-16. Refer to Fig. 6-35. What will happen to the output frequency if the reference voltage is made more negative?

6-17. What is the function of R_3 in Fig. 6-36?

6-18. Refer to Fig. 6-36. What will happen to the ON time of the light source if the reference voltage is made more negative?

6-19. Suppose the output waveform in Fig. 6-37 shows every other negative peak at a greater amplitude. Could this be caused by R_3 being out of tolerance?

6-20. Refer to Fig. 6-38. How would the circuit act if the diode were short-circuited?

6-21. What is the bandwidth of a bandpass filter having a Q of 20 and a resonant frequency of 100 Hz?

6-22. Two unity gain highpass filters are cascaded. Assuming that both have a cutoff frequency of 100 Hz, what is the overall response at 100 Hz? Is the cascaded cutoff frequency higher, lower, or the same?

6-23. What is the output voltage for the multiplier in Fig. 6-49 if $V_X = -5$ V and $V_Y = +6$ V? In which quadrant is the multiplier operating?

6-24. Calculate the dc output voltage for Fig. 6-50(a) if the input signal is a 6.32-V peak sinusoid. (Hint: Refer also to Fig. 6-50[d].)

6-25. Calculate the output voltage for Fig. 6-50(c) if the input voltage is -5 V.

6-26. What are the two types of isolation amplifiers?

6-27. Why are chopper-type amplifiers used?

6-28. What is the function of the external capacitors in the CAZ amplifier shown in Fig. 6-53?

6-29. Refer to Fig. 6-41. What dc condition will exist at V_{out} if R_3 opens?

6-30. Refer to Fig. 6-55. What ac output condition will exist if C_3 opens?

ANSWERS TO REVIEW QUESTIONS

1. 37 dB **2.** 32 dB **3.** common emitter **4.** 180° **5.** 4 V **6.** cutoff **7.** 16 V **8.** yes; 0° **9.** yes; no **10.** no **11.** 80 dB **12.** noninverting; + **13.** 40 dB **14.** infinity; to avoid loading effects **15.** zero; so it can drive any load **16.** 1 MHz; yes **17.** 1.59 MHz **18.** slew rate **19.** 10; 10 kΩ **20.** 100 kHz **21.** 101; 6 MΩ **22.** +3 V **23.** −10.34 V **24.** nonweighted; change any of the input resistors **25.** it will decrease **26.** an output pull-up resistor **27.** no (the output is at positive saturation) **28.** +1.93 V; −1.93 V **29.** square wave **30.** +0.5 V **31.** triangle **32.** 201 **33.** the intrinsic impedance of the op amps **34.** zero **35.** no **36.** thermal compensation **37.** positive saturation **38.** to set V_{out} at half the supply **39.** $R_1 = 11.3$ kΩ; $R_F = 22.5$ kΩ; $C_2 = 0.2$ μF **40.** 0 dB; −3 dB; −40 dB **41.** in-phase feedback and more gain than loss **42.** to control gain and minimize distortion **43.** odd **44.** R_A; R_A and R_B **45.** R_1 and R_2 **46.** saturation **47.** low **48.** high **49.** near unity **50.** equal to V_{CC} **51.** +14 V (positive saturation)

MAGNETIC DEVICES

This chapter deals with some important magnetic devices used in modern industry. These devices are for the most part passive and contain no moving parts. They are relatively maintenance-free and reliable. They will last indefinitely if not abused by overloading or overheating, provided they are adequately protected for the environment in which they are designed to operate. Industry relies heavily on these devices for control, transfer, and conditioning of ac power. In this chapter emphasis is on 60-Hz power, though other frequencies are mentioned where appropriate.

7-1
SIGNAL AND SPECIAL PURPOSE TRANSFORMERS

An impedance relationship exists across a transformer that is not equal to the turns ratio. The following analysis will show what is true for all power and signal transformers. Since the voltage ratio of a transformer is equal to the turns ratio,

$$\frac{V_p}{V_s} = \frac{N_p}{N_s}$$

Also, the current ratio is equal to the reciprocal of the turns ratio:

$$\frac{I_p}{I_s} = \frac{N_s}{N_p}$$

Then, by dividing one equation by the other we obtain,

$$\frac{\dfrac{V_p}{I_p}}{\dfrac{V_s}{I_s}} = \frac{N_p{}^2}{N_s{}^2}$$

Since V_p/I_p is equal to Z_p, the primary impedance, and V_s/I_s is equal to Z_s, the impedance of the secondary, by substitution we obtain

$$\frac{Z_p}{Z_s} = \left(\frac{N_p}{N_s}\right)^2$$

Therefore, the ratio of the impedances across a transformer varies as the *square* of the turns ratio.

Rearranging the terms,

$$Z_p = Z_s \times \left(\frac{N_p}{N_s}\right)^2$$

and

$$Z_s = Z_p \times \left(\frac{N_s}{N_p}\right)^2$$

If the transformer shown in Fig. 7-1 has a step-up ratio of 10 to 250 and draws 100 mA at 6 V in the primary circuit, the primary impedance is

$$Z_p = \frac{V_p}{I_p} = \frac{6}{0.1} = 60 \ \Omega$$

The impedance of the secondary is

$$Z_s = 60 \times \left(\frac{250}{10}\right)^2$$
$$= 60 \times 625 = 37{,}500 \ \Omega$$

For a step-down transformer with a turns ratio of 15 to 1 supplying an impedance of 2 Ω, the impedance of the primary is

$$Z_p = 2 \times \left(\frac{15}{1}\right)^2$$
$$= 2 \times 225$$
$$= 450 \ \Omega$$

The value of a signal transformer is in its ability to transform a low impedance to a high impedance or vice versa. No external power is required to accomplish this feat. The isolation characteristics are also very beneficial for the elimination of ground loops that can override a small signal. Signal transformers have well-defined bandwidths. Most iron core types are limited to audio frequencies. With ultrasonic circuits of 20 kHz or more, ferrite or air cores are used to eliminate the hysteresis and eddy current losses that increase with frequency.

Maximum transfer of power from source to load is important in any circuit and especially so when moderate to high power levels are involved. The maximum transfer of power occurs when the impedance of the source is equal to the impedance of the load. The simple dc circuit of Fig. 7-2 illustrates this principle. A 10-V battery with an internal resistance R_B of 1 Ω feeds a variable load resistance R_L. When R_L is 4 Ω, the total circuit series resistance is 5 Ω ($R_B + R_L$), the current is 2 A, and there is an 8-V drop across the load. The power absorbed by the load is 8 V times 2 A or 16 W. The power dissipated

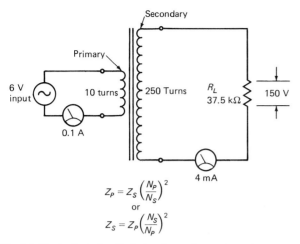

$$Z_P = Z_S \left(\frac{N_P}{N_S}\right)^2$$
or
$$Z_S = Z_P \left(\frac{N_S}{N_P}\right)^2$$

Fig. 7-1 Equivalent circuit of signal transformer impedance.

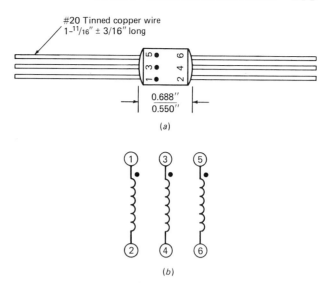

Fig. 7-3 Thyristor pulse transformer. (a) Physical dimensions. (b) Schematic symbol.

in the battery as heat is R_B times I^2, or 4 W. Analysis of the chart in Fig. 7-2(b) indicates that the maximum power absorbed by the load is 25 W. Further analysis shows that this maximum power occurs with R_B and R_L equal to 1 Ω. Thus, *the greatest power is delivered to the load when the impedance of the load is equal to the impedance of the source.*

Pulse transformers are often used to couple a trigger pulse to a thyristor in order to obtain electrical isolation between two circuits. The transformers usually used for thyristor control can be arranged with a 1:1 ratio (two windings), a 1:1:1 ratio (three windings), or a 1:1:1:1 ratio (four windings). Figure 7-3 shows the physical dimensions, along with the schematic representation for a typical unit. Note the black dot that appears at each winding. This is a

phasing (polarity) reference dot. The dots show that the indicated terminals have the same polarity with respect to each other.

For pulse triggering a thyristor, a minimum amount of energy is required, and a minimum trigger pulse width is also necessary. A typical gate current to gate pulse width plot is shown in Fig. 7-4, along with the

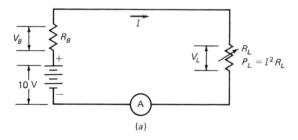

(a)

$V(V)$	R_B	R_L	R_T	$I(A)$	V_B	V_L	P_L (W)
10	1	4	5	2	2	8	16
10	1	3	4	2.5	2.5	7.5	18.75
10	1	2	3	3.33	3.33	6.67	22.2
10	1	1	2	5	5	5	25
10	1	.5	1.5	6.67	6.67	3.33	22.22
10	1	.33	1.33	7.5	7.5	2.5	18.75
10	1	.25	1.25	8	8	2	16

(b)

Fig. 7-2 Power transfer and impedance. (a) Circuit. (b) Table of impedances.

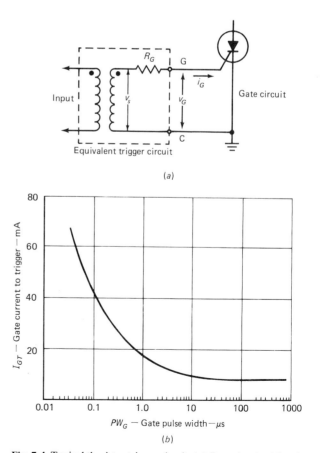

Fig. 7-4 Typical thyristor trigger circuit. (a) Gate circuit with pulse transformer. (b) Gate current to gate pulse width curve.

equivalent trigger circuit. The transformer shown in Fig. 7-3(*a*) has been specifically designed for triggering thyristors. The prime requirement of a trigger pulse transformer is efficiency. The simplest test is to use the desired trigger pulse generator to directly drive a 20-Ω resistor and then to drive the same resistor through the pulse transformer. If the pulse waveforms across the resistor are the same under both conditions, the transformer is considered to be perfect.

A similar type of pulse transformer is used to drive MOSFETs in switching and flyback converters. The windings are interleaved for lowest practical leakage inductance. When driven by these transformers, drain to source rise times of 25 nanoseconds (ns) are

achieved, and useful frequencies up to 200 kHz are obtainable. Ferrite cores are commonly used at these frequencies. These devices are for pulse service and will not work as input or output transformers, which are primarily low-frequency devices.

Bucking and *boosting transformers* (sometimes called *corrector transformers*) are used to provide an economical and convenient means of boosting or bucking voltage on single- and three-phase circuits. These transformers have series-multiple 12/24- or 16/32-V secondary windings suitable for a wide variety of applications. They are connected as auto-transformers in single-phase or three-phase circuits, for boosting or bucking voltages. Each *boost or buck* (B-B) transformer will have eight leads brought out

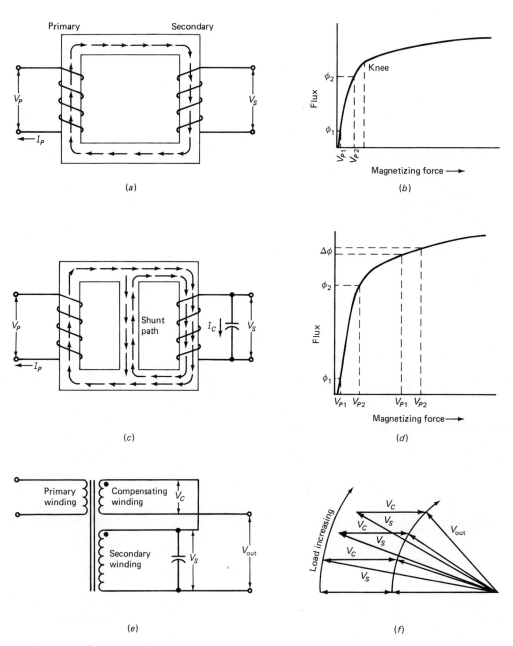

Fig. 7-5 Constant voltage transformer operation.

for wiring. In accordance with American National Standards Institute (ANSI) standards, the four leads on the high-voltage winding will be marked H1, H2, H3, and H4. The low-voltage leads will be marked X1, X2, X3, and X4. A few points of interest should be noted. The *open-delta* connection requires only two B-B transformers to transform three-phase voltages. The open-delta does not provide a neutral and therefore is used only on three-phase applications in which the neutral (if one exists) is not connected to the load. The geometric neutrals of the input and output voltages on the open-delta connection are not the same. Thus, circulating currents will result if the neutral of the line is connected directly to the load; therefore, the neutral must not be connected for this type of operation.

As an example of the use of a B-B transformer, suppose we have a 60-Hz, single-phase, 12-kilovolt-ampere (kVA) load that is rated at 230 V. However, the only line available is 208 V, 60 Hz. The solution would be to select a B-B transformer rated 12/24 V with a kVA rating of 10 percent of 12 kVA or 1.2 kVA. The primary currents cancel, so only the low-voltage winding conducts full load current.

For many types of industrial applications a simple, reliable source of constant potential is a requisite for successful operation. The static-magnetic regulator, with its simplicity, ruggedness, and reliability, is commonly employed. These units have many generic names: *ferromagnetic, flux-former,* etc. Ferroresonant constant-voltage transformers are discussed in Chapter 5. The basic operating principle is the same; only the name is different, as they are all constant-voltage transformers.

The basic principles of constant-voltage transformer operation are illustrated in Fig. 7-5. Figure 7-5(*a*) reviews how the conventional transformer primary voltage (V_p) sets up magnetizing current (I_p) and a resultant flux (arrows in core) that links the secondary winding to induce a voltage (V_s). When operating below the knee of saturation (Fig. 7-5[*b*]), flux ϕ_1-ϕ_2, V_s is proportional to the primary voltage V_p. Above the knee, changes in V_p have little effect on V_s. With a magnetic shunt added (Fig. 7-5[*c*]) and a capacitor across the secondary, the capacitor current I_c generates additional flux in the secondary leg through the shunt path. The total flux in the secondary leg is now above the knee (Fig. 7-5[*d*]), so changes in V_p have considerably reduced effect on the secondary flux and on V_s as compared to Fig. 7-5(*b*). To minimize changes of V_s with V_p further, a compensating winding can be added over the primary winding and is connected in series opposition (Fig. 7-5[*e*]) to the secondary to subtract from V_s. Figure 7-5(*f*) shows a locus of points, illustrating that with increasing load current (V_p constant), the compensating winding subtracts less. This means that as the load current is increased, the compensating winding bucks out less of the secondary voltage. This effect offsets the drop in secondary voltage normally associated with increasing load.

REVIEW QUESTIONS

1. What is the primary impedance of a transformer with 300 primary turns and 20 secondary turns if the secondary load is 40 Ω?

2. The _____ characteristics of the signal transformer are useful for the elimination of ground loops.

3. Signal transformers are usually employed for frequencies below _____.

4. The maximum transfer of power is obtained when the source impedance is _____ to the load impedance.

7-2
MAGNETIC AMPLIFIERS

The *magnetic amplifier* is a device used to reproduce an applied input signal at an increased amplitude. All amplifiers do this, but the method of amplification is unique for magnetic amplifiers. The increases in amplitude that occur in magnetic amplifier circuits are produced by the variations in magnetism and inductance within the unit. The magnetic amplifier is known for its dependability, ruggedness, high efficiency, and ability to withstand high temperatures and other severe environments.

They are used as industrial regulators, relays, starters, amplifiers, servosystems, and converters, and they are employed in some computer applications. Since a good knowledge of magnetism is essential to understanding the magnetic amplifier, a brief review of basic magnetic theory is in order.

The region around a bar magnet where its influence is felt is called a *magnetic field*. This field can be viewed as a pattern of lines arranged in an orderly fashion emanating from the poles of the magnet. A similar pattern of magnetic lines will exist around a current-carrying coil. The strength of the field is called *magnetomotive force* (mmf) measured as *NI*, proportional to the ampere turns:

$$\text{MMF} = NI$$

where

$$\begin{aligned} \text{MMF} &= \text{ampere turns} \\ N &= \text{number of turns in the coil} \\ I &= \text{current flow in the coil, A} \end{aligned}$$

Another parameter to be considered is the field intensity (*H*), sometimes called *magnetizing force*. It is the magnetomotive force per unit length:

$$H = \frac{NI}{l}$$

where

$$\begin{aligned} H &= \text{ampere turns per meter} \\ NI &= \text{ampere turns} \\ l &= \text{length, m} \end{aligned}$$

Magnetic flux (Φ) is similar to the current in an electric circuit and comprises the total number of lines of force existing in the magnetic field. The unit of flux is the weber (Wb). A weber is composed of 10^8 lines. *Flux density* (B) is the means of measuring the amount of flux lines per unit area. In the SI (the international system of measurements), the tesla (T) is the unit of flux density. The flux density is 1 T when there is 1 Wb/m², expressed as

$$B = \frac{\Phi}{A}$$

where

$$B = \text{flux density, T}$$
$$\Phi = \text{flux, Wb}$$
$$A = \text{area, m}^2$$

Permeability (μ) is a comparative factor depicting the ease by which a material can conduct magnetic flux as compared to the ease of which a vacuum conducts flux. The permeability formula is

$$\mu = \frac{\Delta B}{\Delta H}$$

where

$$\mu = \text{permeability}$$
$$\Delta B = \text{change in flux density}$$
$$\Delta H = \text{change in field intensity}$$

It should be noted that in a vacuum (or air) flux density equals field intensity at all times. Therefore, the μ (permeability) of a vacuum equals 1. Materials with a permeability of less than 1 are called *diamagnetic,* and those with a μ slightly greater than 1 are called *paramagnetic*. Materials such as iron, cobalt, and nickel with a μ much greater than 1 are referred to as *ferromagnetic*.

Reluctance is the opposition to magnetic flux offered by a magnetic material. The equation for reluctance is

$$\mathscr{R} = \frac{l}{\mu A}$$

where

$$\mathscr{R} = \text{reluctance}$$
$$l = \text{length, m}$$
$$\mu = \text{permeability}$$
$$A = \text{area, m}^2$$

Reluctance is analogous to the resistance of an electric circuit. Up to this point, we have discussed electromagnetic circuits having fixed or slowly changing direct currents and constant or slowly changing magnetomotive forces applied. Circuits with alternating currents will now be considered.

When an alternating current flows through an air-core coil, as shown in Fig. 7-6(a), the flux density (B) increases and decreases in phase with the force (H) that produces it.

The *B-H* curve of an electromagnetic coil with a ferrous core is shown in Fig. 7-6(b). The curve forms

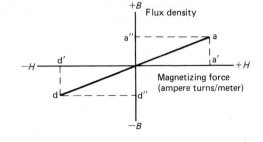

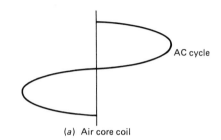

(a) Air core coil

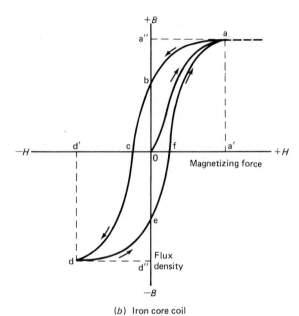

(b) Iron core coil

Fig. 7-6 *B-H* curves.

a loop as a result of the residual magnetism. The ability of a core to remain magnetized after the force is removed is called *retentivity*. The greater the retentivity, the greater the residual magnetism will be. It is important to note that in Fig. 7-6(b) if the force is increased beyond point a', the flux density increases very little (as indicated by the dashed line) since B has reached saturation. The permeability (μ) is very low at this time since the change in flux density (Δ B) is very small. The inductance of the coil is also very low at this time. It should be noted that the permeability (μ) is a direct quantity in the equation for inductance. This, of course, also affects the coil's reactance, since

$$X_L = 2\pi f L$$

where

$$f = \text{frequency in Hz}$$
$$L = \text{inductance in henrys}$$

The reactance of a coil is very low if its core is allowed to saturate and is very high during any time when flux density (B) changes rapidly in response to field-intensity change. Hysteresis losses depend on several factors. Among them are the core type, the temperature of the core, and the frequency of the applied voltage.

Figure 7-7 compares hysteresis loops for high, medium, and low retentivities. A core with high retentivity (Fig. 7-7[a]) has a wide hysteresis loop. Use of such a core results in high hysteresis loss. The use of a low-retentivity core results in a narrow loop (Fig. 7-7[b]) and low hysteresis loss. If a rapidly varying force is applied, the hysteresis loop will widen. The higher the frequency, the greater the amount of lag and the wider the loop. The use of ferromagnetic cores becomes impractical when the frequency goes too high.

The terms *magnetic amplifier* and *saturable reactor* are frequently used interchangeably. This practice, however, is erroneous because a magnetic amplifier is a device consisting of a combination of saturable reactors, rectifiers, resistors, and transformers. Even though the saturable reactor is the main component in all magnetic amplifiers, the term (*saturable reactor*) applies to only one part, the reactor.

Amplification occurs when a low-level signal controls a relatively large amount of power. In a saturable reactor, a control signal can determine the permeability of the core. It has been shown that a change in μ will change the inductance and thereby vary the inductive reactance (X_L) of the coil. The resistance is low compared to the inductive reactance, so the impedance of the coil is due mainly to X_L. It was also shown that when a core saturates, its μ is very low, and the impedance of the coil in this case approaches the coil's dc resistance. The ideal saturable reactor core would have the highly rectangular *B-H* curve, as shown in Fig. 7-8. Many materials approach this ideal curve.

When faithful reproduction of a waveform is needed, a core of nickel-iron alloy provides a linear

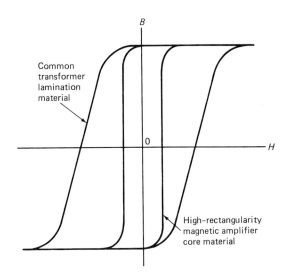

Fig. 7-8 Typical and ideal core materials.

relationship between B and H. Most saturable reactor cores use nickel-iron alloys. These are subdivided into (1) high-permeability (mumetal or Permalloy) and (2) grain-oriented alloys (orthonol, deltamax, and Permirson). High-permeability cores are used in low-level input amplifiers. The grain-oriented materials are used in cores of high-level output stages.

Two desirable features of the saturable rector core are (1) thin laminations for reduced eddy current loss and (2) gapless construction to minimize flux leakage. *Toroidal,* or circular, cores will have much less flux leakage than square or rectangular cores. Flux lines follow smooth curved paths, not sharp corners, as illustrated in Fig. 7-9.

The μ of the core in a saturable reactor is controlled by sending a dc current through a control winding. The flux produced when current flows through the dc control winding is not confined to the core. Some flux lines flow outside the core or through the insulation on the core as shown in Fig. 7-10. Consequently, the entire flux that is produced by the current in the control coil (N_C) does not pass through the load coil (N_L). The amount of leakage increases as the core approaches saturation. This leakage can be minimized by winding the control coil on the same leg of the core as the load winding (N_L), as shown in Fig. 7-11(a). Another method is the use of dual

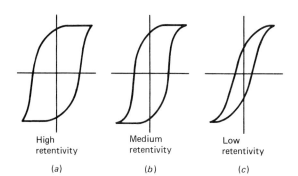

Fig. 7-7 A comparison of hysteresis loops.

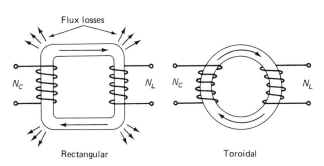

Fig. 7-9 Comparison of a rectangular and a toroidal core.

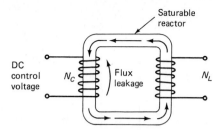

Fig. 7-10 Flux leakages in a saturable reactor core.

coils on two cores, as shown in Fig. 7-11(b). Both of these methods require a high amount of insulation between the coils.

Another method is the *three-legged core,* which requires less insulation. The coils are wound separately, as shown in Fig. 7-11(c). Reactors are classified as to whether they have a single, double, or three-legged core or according to their construction as follows:

1. Rectangular cores
   a. Stacked
   b. Spiral- or tape-wound

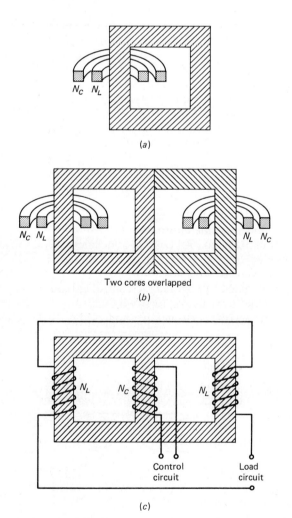

Fig. 7-11 Saturable reactor core types. (a) Single ring. (b) Twin ring. (c) Three-legged.

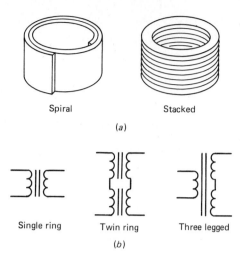

Fig. 7-12 Core types. (a) Construction. (b) Schematic representation.

2. Toroidal cores
   a. Stacked
   b. Spiral- or tape-wound

Toroidal cores are more common than rectangular cores in reactors. Figure 7-12(a) shows the spiral and stacked cores; Fig. 7-12(b) schematically illustrates the single-ring, double (twin ring), and three-legged cores. Figure 7-13(a) shows a saturable reactor con-

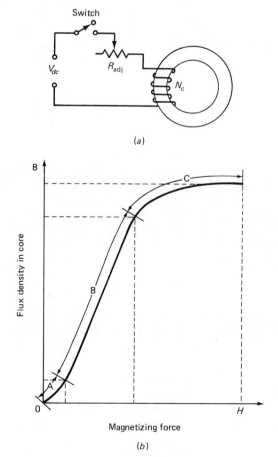

Fig. 7-13 Saturable reactor. (a) Circuit. (b) B-H curve.

trol circuit. It contains a ferromagnetic core with a control winding (N_C), an adjustable series resistor, a switch, and the control voltage V_{dc}.

With the switch closed, the current flowing in the control circuit depends on the amount of series resistance R_{adj}. Therefore, the magnetizing force (H) applied to the core is controlled by adjusting R_{adj}. The *B-H* curve for this core is illustrated in Fig. 7-13(*b*). As the control current is increased from zero, the flux density (*B*) increases slowly through region *A* of the *B-H* curve. The core permeability is relatively low in this region.

As the control current (magnetizing force *H*) is increased into region B, the flux density increases very rapidly. The core permeability in this area is high, because a small change in *H* results in a large change in B. Further increases in the control current will drive the core into saturation, area C. In area C, any increase in *H* results in only a very small change in B; therefore the permeability is very low. Figure 7-14 shows the addition of the load circuit to complete the saturable reactor circuit.

The load circuit consists of the load winding and a load resistor in series with an ac voltage source. The voltage developed across the load resistor is determined by the current flowing in the load circuit. This current is determined by the inductive reactance, X_L, of the load winding and the resistance of the load resistor. The inductance of a coil is directly proportional to its permeability, μ, and can be expressed as

$$L = \frac{\mu N^2 A}{l}$$

where

L = inductance, henrys (H)
μ = core permeability, *B/H*
N = number of turns
A = cross-sectional area, m^2
l = magnetic path length of coil, m

The inductive reactance of the coil is directly proportional to its inductance expressed by $X_L = 2\pi f L$.

It has been shown that the permeability of the core can be controlled by varying the control current, thereby changing the operating point on the *B-H* curve. If the control current is set so that the operating point is in the B area of the *B-H* curve of Fig. 7-13(*b*), the permeability of the core is very high. This results in a low current flow in the load resistor,

and a low voltage drop occurs across the load resistor.

By increasing the control current to the point at which the core is driven into saturation, the permeability is now reduced to a low value. The inductance and inductive reactance of the load winding will also be very low. This will now permit a high load current, and most of the ac source voltage will appear across the load resistor. Accordingly, a saturable reactor uses a small control voltage variation to change or control a large ac load voltage over a wide range.

The basic circuit illustrated in Fig. 7-14 is inefficient because of the ac in the load winding. This current causes two detrimental effects:

1. Voltage is induced in the control winding from the load winding by transformer action.
2. During one half cycle, the load winding current flow will produce a flux which opposes that produced by the control winding. Power from the control winding is needed to return the core flux to its normal operating point. The problem of the opposing flux is eliminated by using a rectifier in series with the load winding. The load current becomes unidirectional, and if the rectifier (diode) is connected properly, the flux fields never oppose each other. The addition of a rectifier converts the saturable reactor into a magnetic amplifier.

Another way of reducing these effects is to use a nonpolarized three-legged core, as shown in Fig. 7-15(*a*). The control winding is wound on the center leg, one load winding on one outside leg, and the other on the opposite leg, so the flux fields cancel in the center leg, as seen in Fig. 7-15(*a*). The schematic

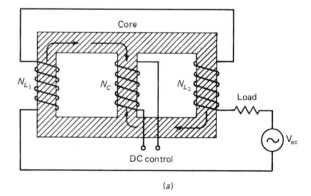

(*a*)

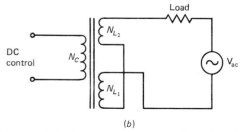

(*b*)

Fig. 7-15 Nonpolarized saturable reactor. (*a*) Three-legged core. (*b*) Circuit.

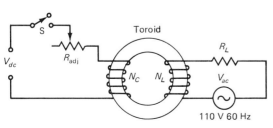

Fig. 7-14 Saturable reactor with load circuit.

circuit representation of this configuration is shown in Fig. 7-15(*b*). It can be seen that the two load windings, being in opposition, induce no voltage into the control winding. When the load voltage changes (reverses), the flux fields still oppose each other in the center leg. Also it can be seen that one flux field aids the control flux, and the other opposes it. This characteristic allows a control voltage of either polarity to be used and is the reason for the term *non-polarized three-legged core.*

In some instances it is necessary to use a three-legged core with the load winding wound so the flux will aid in the center leg, as shown in Fig. 7-16(*a*). The schematic circuit representation of this core is illustrated in Fig. 7-16(*b*). The load winding flux fields now aid each other in the center leg. These fields may aid the control field, thereby helping to saturate the core, or they may oppose the control field, reducing the flux density. The polarity of the applied voltage of either the control coil or load coils will determine flux density. The core is now said to be *polarized.*

The addition of a rectifier in the circuit of Fig. 7-17 converts the reactor circuit to a magnetic amplifier circuit. The saturable reactor is used with a solid-state diode (rectifier) to produce a controlled dc voltage across the load resistance R_L. The reactor used is a three-legged core. The two load coils oppose, as was previously indicated, and this reactor is nonpolarized. The current and the resultant voltage across the load R_L will depend upon the level of the dc control voltage.

The operation of the circuit with the control voltage, across N_C, at zero will be considered first. When

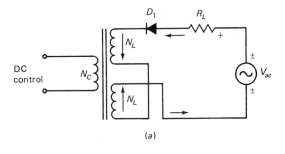

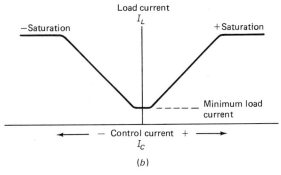

Fig. 7-17 Basic magnetic amplifier. *(a)* Nonpolarized. *(b)* Transfer curve.

the ac supply voltage is positive, current will flow through R_L and D_1, and through the load windings (N_L) back to the source. With no direct current to saturate the control core, a high permeability exists, which results in a high inductance and a high inductive reactance. The current through the load circuit will be small, and consequently the voltage across R_L will be low.

When a dc potential is applied across the control winding, current will now flow and magnetize the core. If the current is of sufficient magnitude the core becomes saturated, and μ decreases drastically. This decrease will make the inductance of the load windings approach zero, and the load current will be high. Therefore, the voltage across R_L will be high. Increasing the current beyond the saturation point produces an insignificant increase in the load current.

The dynamic characteristic curve can be plotted to illustrate the effects of varying the control current in a magnetic amplifier. This plot is shown in Fig. 7-17(*b*). Note that the direction of the control current is unimportant in a nonpolarized magnetic amplifier. The load current in Fig. 7-17(*b*) never goes to zero, but only to some minimum value. This minimum value of load current, called the *no signal* or *quiescent current,* appears when the control current is zero. The control signal is analogous to the input signal between the base and emitter of a transistor. Amplification is achieved in the transistor circuit because a small input current in the base-to-emitter circuit produces a relatively large change in the collector current. In the magnetic amplifier, small changes in the control winding current cause large changes in load current.

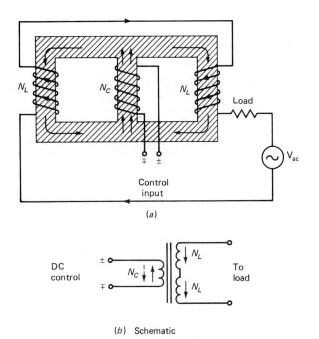

Fig. 7-16 Polarized saturable reactor. *(a)* Three-legged core. *(b)* Circuit.

The load current of the magnetic amplifier is also dependent on the size of the load resistance R_L. Figure 7-18 illustrates the output curves for load current I_L for two different values of load resistances. The curve shows that the load current is greater at saturation in the magnetic amplifier that has a lower value load resistor. It should be noted that operating on the steep portion of the curve (from point a to saturation) will produce a large change in the load current; consequently, amplification is achieved.

As with other types of amplifiers, feedback may be added to magnetic amplifiers. Positive feedback is used to increase the gain, and negative feedback is used to limit the gain or to improve the stability or linearity. Magnetic amplifiers can be classified as follows:

1. Without feedback
2. With external feedback
3. With internal feedback

A magnetic amplifier with external feedback may have positive or negative feedback supplied by means of an external, inductively coupled winding, with the load current flowing through it. Usually, the rectified load current flows through a feedback winding.

If this field in the feedback winding aids the control winding field, the feedback is positive. If the two fields (control and feedback) oppose each other, the feedback is negative. The effect of positive external feedback on the magnetic amplifier is shown in Fig. 7-19(a). A comparison of the curves shown illustrates that positive feedback produces a nonsymmetrical curve, whereas the nonfeedback curve is symmetrical. It should also be noted that the load current does not reach its minimum when the control current is zero (as in the nonfeedback case) but reaches its minimum when the control current is negative.

Figure 7-19(b) shows the way that the slope of the feedback curve is determined by the percentage of

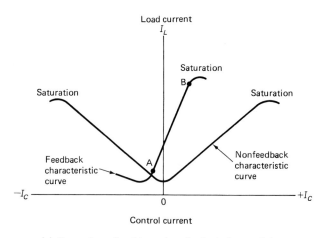

(a) Comparison of positive and nonfeedback characteristics

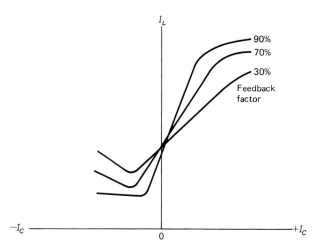

(b) Transfer characteristics of varying amounts of positive feedback

Fig. 7-19 Magnetic amplifier feedback and nonfeedback transfer curves.

feedback. If the positive feedback is increased above 100 percent, the amplifier will become unstable and will probably lock up into maximum conduction even with very small control current. The feedback factor is usually kept below 85 percent to maintain good stability.

Figure 7-20 shows the schematic symbols of a reactor using a feedback winding. The terms N_C, N_F, and N_L refer to the control, feedback, and load windings, respectively. These designations will be used throughout the remainder of this section. Figure 7-21 shows a simple positive feedback magnetic amplifier with a feedback winding (N_F) added to each reactor.

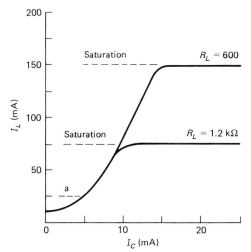

Fig. 7-18 Magnetic amplifier load curves.

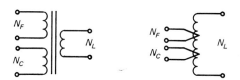

Fig. 7-20 Magnetic amplifier with external feedback.

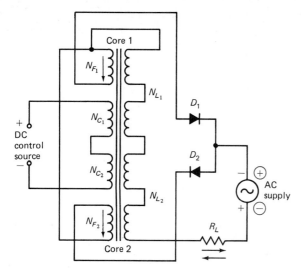

Fig. 7-21 Magnetic amplifier with feedback circuit.

To follow the circuit of Fig. 7-21, it is assumed that the ac supply is on the half cycle having the circled polarity. Current will flow from the positive end of the supply through D_2, N_{F_2}, series load windings N_{L_1} and N_{L_2} of both reactors, and load R_L back to the supply. On the other half cycle, current flows through the load R_L, N_{L_2}, N_{L_1}, the feedback winding N_{F_1}, and back to the supply through D_1. Note that the current that flows in the feedback windings will aid the flux set up in core 1 and core 2 by the dc control voltage. This causes a higher load current to flow and causes the output load voltage to increase.

If the control polarity is reversed, negative feedback occurs. The control winding flux will now oppose the flux of the feedback winding. Under this condition, an increase in the control current will increase the load winding reactance until the point where the control flux and feedback flux are equal is reached. If the control flux overrides the feedback flux, the load winding reactance will slowly decrease, causing the load current to increase. The input impedance of a magnetic amplifier is not affected by external feedback. In contrast, in other electronic amplifiers, the feedback partly determines the input impedance. Since the feedback current of a magnetic amplifier flows through an isolated winding, the input impedance is determined only by the winding resistance and any additional series resistance used to increase the response time. The time constant (T) of an inductive circuit is $T = L/R$; T is in seconds, L is in henrys, and R is in ohms.

It was mentioned previously that increased current will flow through the load winding of a magnetic amplifier using positive external feedback. This tends to make the quiescent load current too high for some applications. Quiescent current reduction may be accomplished by the addition of a bias winding, (N_B), to the core of each reactor, as shown in Fig. 7-22. The bias winding reduces the load current to an acceptable value when the control current is equal to

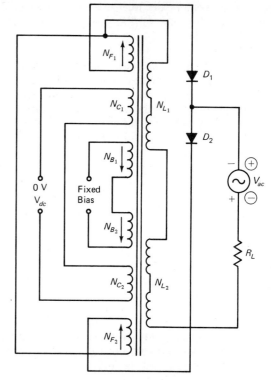

(a) Feedback and bias windings

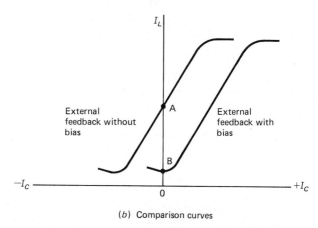

(b) Comparison curves

Fig. 7-22 Magnetic amplifier with external feedback and bias windings.

zero. Figure 7-22(b) compares the characteristic curves with and without bias. Operating point A of Fig. 7-22(b) shows a high quiescent current with no bias. This large quiescent current is due to the flux in the core caused by the current through the feedback windings as indicated earlier.

Operating point B shows that the quiescent current is minimum in the magnetic amplifier using feedback with a bias winding. Figure 7-22(a) shows that the flux from the bias winding opposes and cancels the flux produced by the feedback winding when the control field is at zero. This effect provides minimum flux in the reactor core and means that the permeability will be high. The inductance and inductive reactance of the load windings are maximum, and the load current, I_L, is at minimum with no control current. But once control current is applied the feedback current will increase and overcome the bias. By varying the bias current amplitude, the operating point may be adjusted to any desired value, as shown in Fig. 7-23.

Feedback in a magnetic amplifier can be accomplished by having the load winding produce the feedback directly. This is called *internal, intrinsic, electric,* or *self-saturated feedback.* This type of circuit feedback has a (rectifier) diode in series with the load so that direct current will flow through the load windings. If a three-legged core is used, the load windings are wound on the center leg of the core and are connected in series aiding. An internal feedback magnetic amplifier is one that uses a self-saturation (auto-excited) circuit because of its nature of operation. Comparison of the characteristics of internal and external feedback amplifiers yields small differences between the two. Self-saturation has the advantage of eliminating the feedback windings and also eliminates the additional resistance in series with the load. On the other hand, use of external feedback allows easy adjustment of amplifier gain.

A basic self-saturating magnetic amplifier is shown in Fig. 7-24(*a*). The determining factor in this self-saturating magnetic amplifier is the rectifier (diode). This configuration is a polarized magnetic amplifier; therefore, it responds differently if the control winding is reversed in polarity. The full wave output is more common and will be analyzed rather than the half-wave output of Fig. 7-24(*a*).

To produce a full sine wave output requires a two-reactor magnetic amplifier circuit, as shown in Fig. 7-24(*b*). The control input voltage for this circuit is also direct current. With no dc control voltage present and the ac supply being that of the circled polarities indicated in Fig. 7-24(*b*), the current will flow from the ac supply through R_L, D_1, N_{L_1}, and back to the source. On the other half cycle (uncircled polarities) current will flow through, N_{L_2}, D_2, R_L, and back to the source. The load windings N_{L_1} and N_{L_2} of Fig. 7-24(*b*) are wound series aiding on the center leg of their respective cores.

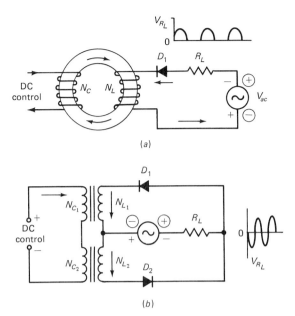

Fig. 7-24 Self-saturating magnetic amplifier. (*a*) Half wave. (*b*) Full wave.

Note that the voltage drop across the load R_L will reverse on each half cycle, producing the alternating voltage (V_{R_L}) shown in Fig. 7-24(*b*).

The polarity and magnitude of the input direct current to the control windings will determine the alternating current developed across the load. With the dc control polarity as shown, the flux produced in the control winding aids the load-winding flux. This will decrease the reactance of the load winding, and the load current will increase, producing a greater voltage drop across R_L. Increasing the load current increases the flux developed in the core. The load winding reactance decreases further, and the load current will increase. Therefore it can be seen that if the load flux aids the control flux, the internal feedback is positive.

When the control voltage is reversed, the control flux will oppose that produced by the load current. This causes the reactance of the load winding to be high, resulting in a low output voltage across R_L. A point where the control flux and the load flux are equal and cancel each other can be reached. The reactance will be maximum at this point, and the load voltage will be minimum. Any further increase in the control voltage will cause the flux in the core to increase, thus decreasing the reactance of the load coils. The load current will increase only a little since the increased load flux opposes the control flux and reduces the overall flux in the core. The internal feedback is now negative.

The transfer characteristic curve for Fig. 7-24(*b*) (with positive feedback) is shown in Fig. 7-25(*a*). The quiescent current (point A) is due to the high level of load flux in the core. The control flux aids the load flux, and the gain is high as a result of the positive feedback. Negative control flux opposes

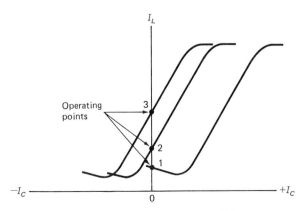

Fig. 7-23 Operating points for various values of bias.

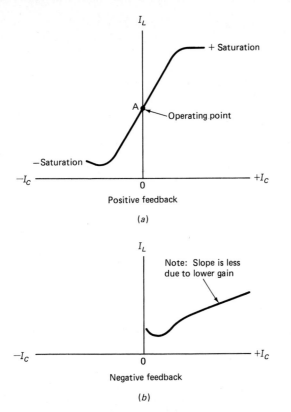

Fig. 7-25 Transfer curve of an internal feedback magnetic amplifier.

the load flux, and the resulting negative feedback results in a reduction of gain. The addition of a bias winding can be used to set the operating point, as discussed previously.

A typical application of a magnetic amplifier is illustrated in Fig. 7-26. Many servosystems (covered in Chapter 10) use a two-phase induction motor to position the load. The direction of rotation in such a motor is dependent upon the phase of the ac signal, and the speed is proportional to the signal amplitude. The circuit of Fig. 7-26 will produce the required output to drive such a servomotor. The bias windings (N_{B_1} and N_{B_2}) minimize the quiescent load-winding current. With no control signal applied, the reactance of N_{L_1} and N_{L_2} are high and equal, and the small

current that flows through the field windings of the servomotor produces opposing effects. Thus the rotor will remain stationary.

The control windings of the two reactors of Fig. 7-26 are connected in series opposition, so that a given polarity of control current will increase the amount of core flux of one reactor, and the opposite polarity of control current will increase the amount of core flux in the other reactor. If the circled polarity of control voltage is applied, reactor SX$_1$ approaches saturation, and the reactance of N_{L_1} decreases. Since this reactor controls the current through field 1 of the servomotor, the decrease in N_{L_1} reactance will allow more current through field 1. The reactance of N_{L_2} remains high so that very little current flows through field 2. A resultant torque is generated within the motor, causing the rotor to turn the output shaft in a particular direction.

If the control winding is driven by the uncircled polarity, the reactor SX$_2$ will approach saturation. The reactance of N_{L_2} will decrease, allowing more current of flow through field 2 of the servomotor. Winding N_{L_1} has a high reactance, and the current through field 1 is low. A resultant torque is produced, causing the rotor to turn in a direction opposite of that previously considered. The speed of the servo is approximately proportional to the magnitude of the control (error) signal.

REVIEW QUESTIONS

5. Materials with a permeability much greater than 1 are called _____.

6. Air core coils have a higher permeability than iron core coils. (true or false)

7. Referring to Fig. 7-6(b), the inductance is minimum (μ is low) at what points?

8. The higher frequencies of operation will produce a _____ hysteresis loop.

9. A saturable reactor may operate in either the A or B regions of Fig. 7-13(b). (true or false)

10. The saturation current for a load of 800 Ω on Fig. 7-18 would be _____ than 0.1 A.

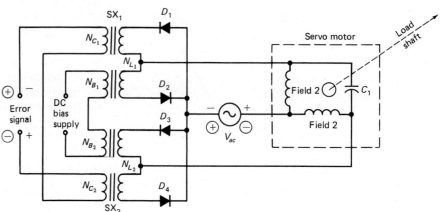

Fig. 7-26 Application of magnetic amplifier servomotor.

7-3
TROUBLESHOOTING AND MAINTENANCE

In general, magnetic devices are fairly hardy and less sensitive to abuse than solid-state devices. However, excessive current produces excessive heat, which can cause insulation failure and short-circuited turns within the inductor. As always, cooling and ventilation must be adequate and not restricted in any way. A visual inspection may show signs of overheating. A distinct ''burned'' odor is another clue. Some units have a built-in thermal switch to protect them. Tripping of this device can be an indication of future or current problems and should be thoroughly investigated. An inductor may fail by short-circuiting to its core (frame). This is easily checked with a high-voltage ohmmeter (megger). Most inductors have breakdown voltages greater than 500 V. Unless a very low resistance short circuit has occurred, an ohmmeter with a 10- to 30-V source is not adequate for most insulation tests for breakdowns.

The dc resistance of a high-current inductor is on the order of milliohms and would require a Wheatstone bridge or milliohmmeter to check its value adequately. If an inductance bridge is available, measurement of the inductance will also verify the integrity of the coil. In a switching power supply or in a line filter, a test frequency close to the device's operating frequency will yield a more meaningful value when using an ac bridge or impedance-measuring equipment. Do not interrupt a choke-type circuit when current is flowing. The induced voltage can be many times the applied voltage and may cause arcing and a breakdown of the insulation. Also, it creates a shock hazard, and the arc may cause eye damage. The arc caused by interrupting a 5-A current flow in a several-henry inductor is in excess of 1000 °F.

An abnormally high output voltage from a dc supply using a choke input filter can occur if the load is below the minimum for the critical inductance value used. Excessive load current can saturate the core and lower the inductance, causing abnormally high ripple.

Signal transformers should be tested for short circuits to the core or frame as well as between the windings. Since most units are low-power, the use of a signal or function generator is best suited for dynamic testing along with an oscilloscope or wideband ac voltmeter. The pulse- or trigger-type transformer can be tested, as mentioned in Sec. 7-1. The insulation resistance of these devices is suspect in many cases and is best checked with the aid of a high-voltage ohmmeter. If the thyristor it triggered had a fault (anode-to-gate short circuit) the trigger transformer should be checked for signs of overheating (discoloration or a burnt odor). A complete test may save a future return call for a marginal system.

The saturable reactor and the magnetic amplifier require no preventive maintenance other than cleaning and inspecting any associated cooling systems. The windings should all be well insulated from each other and the core. This is easily verified with a megger. The diode (rectifier), if used, can be checked in the usual manner. The MOSFETs may find applications in magnetic circuits, especially in their constant-current mode (gate tied to source). An oscilloscope is an invaluable tool to troubleshoot these circuits. Cases of saturation or distortion can only be detected this way. Caution: The alternating current used may or may not be referenced to ground or neutral. The use of an isolation transformer may or may not be a safe way to prevent ground loops through the test equipment. Floating measurements are covered in Chapter 1. Changes in the bias source can shift the quiescent point and cause excessive no-signal load current and possible overheating. The loss of positive feedback (if external) will give a reduction in gain or sensitivity, whereas loss of negative feedback can give such an increase in gain as to lock up (latch) the system, which will become immune to all external signals. A comparison of each winding for double-reactor systems is very beneficial, and these values should be recorded for future reference.

The devices covered in this chapter are ac devices, and in most cases (except extreme short circuits, etc.) the dc resistance is not a good measure of the integrity of the device, especially where turns ratios are involved. Short circuits to the core (frame) are common, especially if overheating had occurred some time in the past. A high-voltage ohmmeter is a very useful tool for these types of checks.

REVIEW QUESTIONS

11. Suspected electric leakage of an inductor is best tested by using a _____.

12. A Wheatstone bridge can be used to check the dc resistance of an inductor. (true or false)

13. Interrupting the current through an inductor may create a dangerous _____.

14. A load waveform that appears as half-wave direct current in Fig. 7-24(b) is due to D_1 being open or short-circuited?

CHAPTER REVIEW PROBLEMS

7-1. The polarity of transformer windings may be indicated with phasing _____.

7-2. The constant-voltage transformer uses a magnetic _____, along with a compensating winding to stabilize the voltage.

7-3. One method used to reduce leakage flux is the use of _____ ring cores.

7-4. A basic magnetic amplifier is a saturable reactor with a _____ added.

7-5. The minimum value of load current is called the _____ or quiescent current.

7-6. Positive feedback will _____ the gain of a magnetic amplifier.

7-7. Positive feedback is kept below 85 percent to protect a magnetic amplifier from _____.

7-8. A _____ winding is used to minimize the quiescent current of the magnetic amplifier.

7-9. Negative feedback increases the input impedance of the magnetic amplifier. (true or false).

7-10. The two-reactor _____ circuit will provide a full sine wave to the load.

7-11. The servomotor magnetic amplifier typically has how many reactors?

7-12. An increase in the ripple from a filter may be due to excessive current causing it to _____.

7-13. Trigger transformers are best tested by using _____.

7-14. The _____ are usually suspect in a magnetic amplifier failure since they are the least reliable part of the system.

7-15. Loss of negative feedback can make the reactor _____.

7-16. Excessive quiescent current may indicate a loss of the _____.

ANSWERS TO REVIEW QUESTIONS

1. 9000 Ω **2.** isolation **3.** 20 kHz **4.** equal **5.** ferromagnetic **6.** false; opposite is true **7.** a or d **8.** wider
9. true **10.** greater **11.** megger **12.** true **13.** arc **14.** open

OPEN-LOOP MOTOR CONTROL

This chapter treats open-loop motor control and power conversion, including inverters, choppers, and cycloconverters. It also covers phase control of dc motors and static conversion as applied in ac motor circuits. The improved reliability and increased ratings of solid-state devices have made their presence commonplace in industry. They also interface well with computer-type controls; that interface is the major thrust of automation technology.

8-1
DC MOTOR PHASE CONTROL

There are two basic categories of motor control: open-loop and closed-loop. A *closed-loop system* senses the motor output and uses this information to correct the drive to the motor to eliminate error. Closed-loop systems are covered in Chapter 10. In some cases, where the load on the motor is reasonably constant, open loop motor control is adequate. The half-wave drive circuit discussed in Chapter 4 is an example of an open loop control system.

For many years, dc motors were controlled by *thyratrons,* vacuum tubes containing an inert gas, a heated cathode, an anode (plate), and a control grid located between the cathode and the anode. These tubes can control loads from a few milliamperes to several amperes. A simplified thyratron motor control circuit is shown in Fig. 8-1(a). The thyratron's plate-to-cathode resistance is very low when the tube is in the conducting state. The grid of the thyratron controls the point at which the tube fires or the gas in the tube ionizes. Once ionization takes place, the grid loses control and cannot stop plate current flow.

The tube can only be extinguished by lowering the plate voltage below the ionization level (usually 15 V). The thyratron is either full on or full off, just like a switch. The device is operationally comparable to the solid-state silicon controlled rectifier (SCR). Motor performance is controlled by shifting the firing point of the control device, whether it is a thyratron or an SCR. This is known as *phase control* and is illustrated in Fig. 8-1(b). When the dc bias and ac bias add together to equal the critical grid voltage, the thyratron fires. The other thyratron fires on the next half cycle. The conduction period can be lengthened for increased motor output or shortened for decreased motor output.

A phase-sensitive control circuit capable of supplying reversible half-wave power for a permanent magnet or shunt dc motor is shown in Fig. 8-2. The circuit is called a *balanced bridge reversing drive circuit.* It consists of two half-wave circuits back to back (SCR$_1$, D_1 and SCR$_2$, D_2) triggered by the unijunction transistor Q_1, on either the positive or negative half cycle of the applied line voltage. Which half-wave circuit fires will depend upon the direction of the imbalance of the reference bridge from the value of R_1 (the sensing element), which can be a thermistor, photodiode, potentiometer, or the output from a control-type amplifier. In some cases, an opto-isolation amplifier may be employed to avoid the direct connection between the sensor and the bridge circuit.

Resistor R_5 is set so the dc bias on the emitter of the UJT, Q_1, is just below the peak point which would trigger Q_1. Zener D_4 sets the dc voltage across R_5 and C_1 (the RC time constant for Q_1). With R_1 and R_2 equal, the bridge circuit (R_1, R_3, and R_2, R_4) will be balanced, and the UJT (Q_1) will not trigger; therefore no output will appear across the load (armature). If R_1 increases in resistance it will unbalance the bridge circuit. Capacitor C_1 will be charged

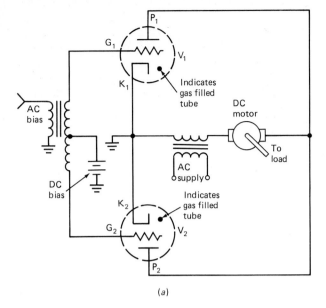

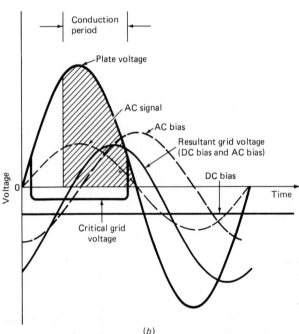

Fig. 8-1 Thyratron circuit and phase control signals. (*a*) Motor control circuit. (*b*) Voltage relationships with one thyratron.

to a greater potential on one line alternation, raising the emitter voltage of the UJT and causing it to trigger for one half cycle of the ac line voltage. When Q_1 triggers, one of the SCRs is forward-biased so it will turn on for the remaining half of the cycle. If R_1 were to decrease in resistance a similar action would occur, but with the other SCR triggering on the opposite half of the cycle. This reverses the polarity across the load, so now the motor will reverse its direction from the previous imbalance condition. Resistor R_2 is used to match the quiescent resistance of sensor R_1 at its null (balanced) position. No feed-

back is employed, so this is an open loop control. Any changes in the load (torque, speed, etc.) will not automatically be corrected.

If a series motor is used, a circuit similar to that of Fig. 8-3 may be employed. This circuit uses a triac in lieu of SCRs. In this circuit, the triac is triggered on either the positive or negative half cycle. The bridge rectifier (D_3—D_6) provides a dc voltage for the armature circuit. Thus, the armature current is always in the same direction. However, the field current is reversed, depending on the triac's triggering polarity. This circuit employs controls for gain, balance, and *dead band* (that range of input voltage to which a control does not respond). The circuit also provides for an analog input control signal, from a sensor, transducer, or process controller.

Figure 8-4(*a*) illustrates the internal block diagram for a silicon integrated circuit designed for phase control from ac mains with resistive or inductive loads. The IC has a voltage regulator and a voltage monitor circuit to reset timing functions and inhibit triac firing pulses when a power-up occurs. A ramp generator is provided to control motor acceleration and can be controlled by the speed program input, pin 5. Charging currents for the ramp generator are set by an external resistor for a slow ramp and internally for a fast ramp. A frequency-to-voltage converter is included to enable a rate generator (tachometer) to be used for motor speed sensing.

The control amplifier in Fig. 8-4(*a*) has differential inputs that compare the ramp voltage against the actual speed voltage (if used in a closed loop system). Synchronization of the triac pulse is achieved by delaying the pulse with reference to the zero voltage points of the line voltage. Inductive loads such as motors produce a phase lag of the load current. Under high-speed or heavy-load conditions, the triac must be fired after the load current from the previous cycle has ceased. The current synchronization input (pin 1) performs this task by ensuring that there is a voltage across the triac before a trigger pulse is permitted (when the triac is conducting only a small voltage drop appears across it). The gating pulse width is dependent upon an external capacitor connected to pin 14, which also delays the pulse from the zero voltage point.

Figure 8-4(*b*) shows a permanent magnet motor whose armature is driven from an ac supply by means of a bridge rectifier. When driving highly inductive loads of this type with a phase control circuit, the triac cannot be gated until a quarter cycle after the line zero crossing because of the electromotive force (emf) generated by the armature. Resistor R_{11} senses the motor current, and the developed voltage signal is applied to pin 3 of the IC. The motor can be rated up to 220 V, and currents up to 30 A are possible. Resistor VR_1 is used to adjust the speed of the motor, which could range up to 10,000 rpm. The circuit provides unidirectional motor operation which is adequate for many industrial applications.

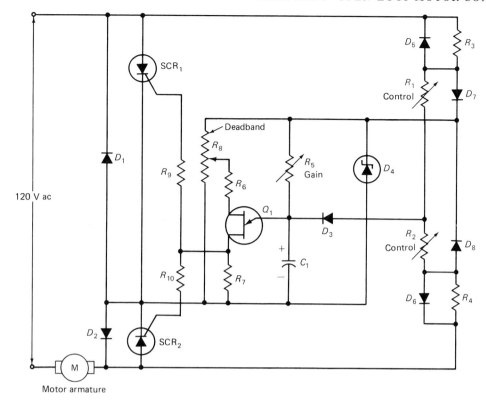

Fig. 8-2 Balanced bridge (reversing) drive for PM or shunt motors.

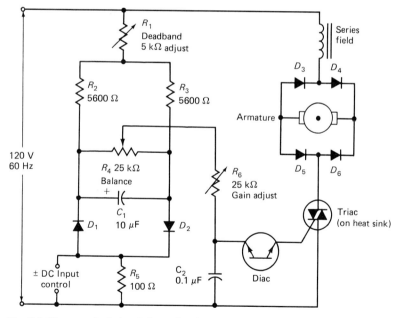

Fig. 8-3 Phase control circuit for series dc motor.

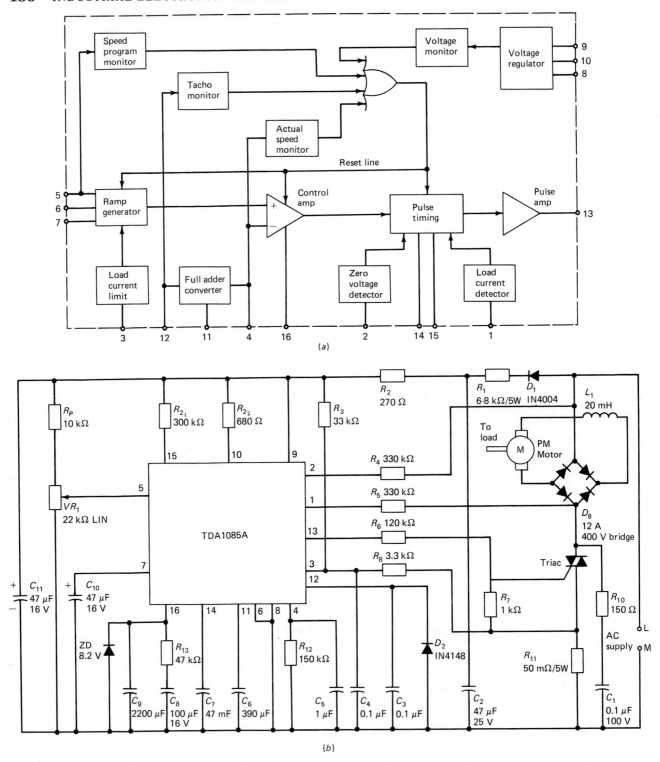

Fig. 8-4 Integrated circuit for phase control.

REVIEW QUESTIONS

1. The thyratron is a linear control device. (true or false)

2. The thyratron tube is analogous to the solid-state _____.

3. Phase control of a load is accomplished by shifting the _____ point of the thyratron or SCR.

4. In Fig. 8-2, _____ is used to balance the sensor (R_1) resistance at null.

5. In Fig. 8-2, which solid-state device produces the gating pulses for the SCRs?

6. The range of input at which a control circuit produces no output is known as its _____.

8-2
DC-DC CHOPPER CONTROL

The dc counterpart to ac phase control is the dc-dc chopper (converter). Choppers control the average load voltage by switching a fixed dc source. The switching may be accomplished by bipolar transistors, thyristors, or MOSFETs. The basic circuits for these devices are shown in Fig. 8-5(*a*). Figure 8-5(*b*) shows a typical chopper waveform. The average load voltage can be controlled by several methods. There

are three techniques commonly used to vary the ratio of the switch ON time to the switch OFF time (duty cycle) of the "chopped" output waveform. The duty cycle can be adjusted by:

1. Varying the ON time (pulse width modulation).
2. Varying the OFF time (pulse rate modulation).
3. Varying both times, which is a combination of pulse width and pulse rate modulation.

Figure 8-6 shows chopper action for pulse width modulation, pulse rate modulation, and a combination of both. Note that the duty cycle controls the average voltage. Choppers can be used in variable-speed dc drives to supply the armature voltage for speed control of separately excited dc motors. They can also be used to provide a variable-supply voltage for series dc motor-speed control. The chopper offers the advantage of higher efficiency than that of traditional electromechanical devices. The improved efficiency is due to the elimination of the wasted energy when using starting and control resistances for these motors.

The thyristor makes an ideal switch for chopper applications; however, a method of turning off the thyristor must be incorporated. The turn-off methods (commutation) are discussed in section 8-3. Figure 8-7 shows a diagram of chopper control of a vehicle motor. Contacts S_2, S_3, S_4, and S_5 are field-reversing relay contacts. With S_2 and S_5 closed, the vehicular direction is forward. With S_3 and S_4 closed, the direction is reversed. This type of chopper arrangement has a typical duty cycle from 20 to 80 percent. At the stopped condition, all four relay contacts (S_2, S_3, S_4, S_5) are open. With S_2 and S_5 closed and the chopper at low speed (rate), the motor voltage averages about 20 percent of the battery voltage. This voltage may be increased to 80 percent of the battery voltage to provide greater motor output by increasing the duty cycle of the chopper. When the 80 percent point is reached, contact S_1 is closed to apply full

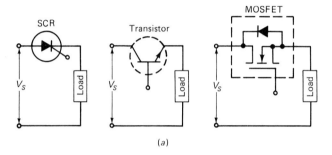

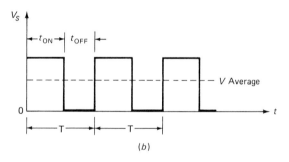

Fig. 8-5 The dc-dc basic chopper circuits and waveform.

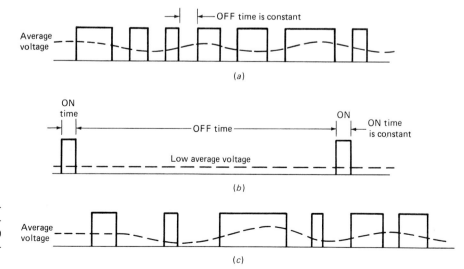

Fig. 8-6 Chopper pulse width modulation waveforms. (*a*) Pulse width modulation. (*b*) Pulse rate modulation. (*c*) Combination rate and width modulation.

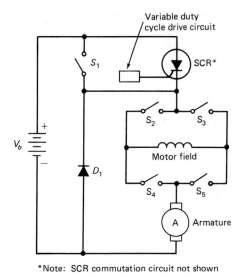

*Note: SCR commutation circuit not shown

Fig. 8-7 Basic vehicle control circuit.

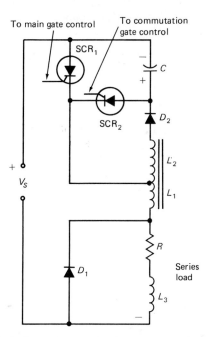

Fig. 8-8 Basic Jones chopper circuit.

battery voltage to the motor, and maximum torque will be obtained. Diode D is the freewheeling diode. Its purpose is to protect the motor from high voltage transients that can be produced when the thyristor is turned off. The chopper may use a variable-frequency constant-pulse width (rate modulation) system or a pulse width modulation control if so desired.

Choppers are an energy-efficient method for controlling a series dc motor that normally operates from a dc power source supplied by a third rail (as with electric trains) or an overhead conductor (trolley) or a battery bank, as with electric fork lifts. One basic problem encountered in chopper control is the maximum armature current that can be commutated by the thyristor. As the motor size (horsepower rating) increases, so will the locked rotor armature requirements. Another problem encountered when using thyristors in a chopper circuit is to achieve commutation, without which the control would be lost. Commutation is more or less automatic in ac power systems since the thyristors will turn off at the line zero crossings. Commutation in dc systems requires extra circuitry.

One circuit that achieves dc commutation is shown in Fig. 8-8. The Jones circuit controls the mean load voltage by varying the ratio of the ON time to the OFF time (*pulse width modulation*). To understand the operation of the Jones circuit better, six working circuits illustrating the various phases of operation are shown in Fig. 8-9. These phases represent time intervals from t_0 through t_5. Single-pole, single-throw (SPST) switches replace SCRs to indicate their state so as to simplify the analysis. The cycle starts in Fig. 8-9(a) by triggering SCR_1 at t_0. The lower plate of capacitor C will start to charge positively. By time t_1, the lower plate of the capacitor has resonantly charged, via L_2, to its peak positive voltage as shown in Fig. 8-9(b). This peak positive voltage is equal but

opposite to the peak negative voltage found on the capacitor just before SCR is triggered. Turning on SCR_1 has the effect of reversing the voltage across the commutating capacitor. This voltage is held on the capacitor by the charging diode D_2. Energy is now available to commutate SCR_1, which is conducting load (motor) current through L_1 as shown in Fig. 8-9(b). Inductors L_1 and L_2 are closely coupled and form an autotransformer. When SCR_1 is on, the load current flows through L_1 and induces a positive-going voltage at the top of L_2 that charges the commutating capacitor.

At time t_2 in Fig. 8-9(c), SCR_2 is triggered on, and the capacitor is now effectively connected across SCR_1. This reverse-biases the thyristor, and SCR_1 is commutated (turned off). The load current is now being supplied by V_B and the capacitor. Figure 8-9(d) shows that the capacitor is now charged to an opposite polarity as the load current continues to flow. As the capacitor voltage reaches the supply voltage, the charging current decreases and eventually goes below the holding current for SCR_2. Thyristor SCR_2 turns off at time t_4. Now, both thyristors are off, and the motor current is decreasing. Diode D_1 is forward-biased by the emf generated by the collapsing motor field. This allows the energy stored in the motor to be dissipated. At t_5, all currents have ceased, and the circuit is ready for a gate pulse to SCR_1 to begin another cycle.

The Jones circuit provides an efficient control of motor speed by varying the duty cycle. To increase motor speed, the time between SCR_1's gating pulse and SCR_2's gating pulse will be increased. This allows motor current to flow for a greater period of time and raises the average motor voltage. The discussion has been limited to basic circuit operation.

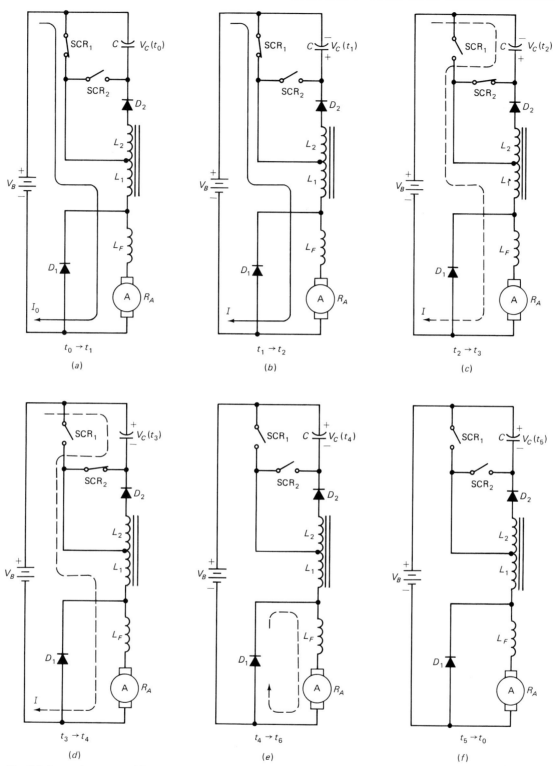

Fig. 8-9 Jones chopper working sequences.

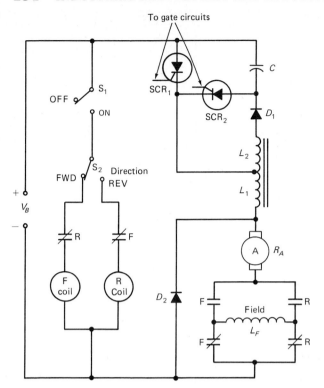

Fig. 8-10 Jones circuit modified for forward and reverse.

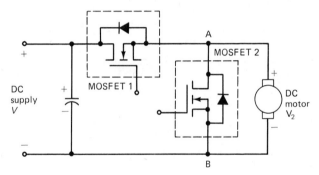

Fig. 8-11 Two-quadrant chopper using MOSFETs.

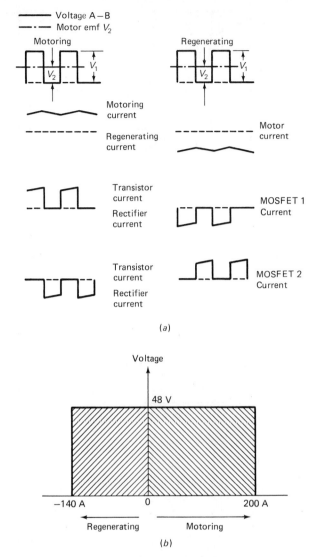

Fig. 8-12 Chopper waveforms and quadrant of operation.

Some applications may include acceleration control, braking, current limiting, and other features. The circuit can also be modified for forward and reverse operation, as shown in Fig. 8-10.

Modern power MOSFETs may be rated for continuous current as high as 40 A and may be connected in parallel for higher currents. They are attractive candidates for controlling electric motors at currents up to several hundreds of amperes. Figure 8-11 shows the basic circuit for a dc-to-dc chopper that provides continuous speed control in the motoring mode of operation (with the motor receiving power from the dc source). It also allows the motor to return regenerative energy to the dc source. This is known as a *two-quadrant chopper circuit*. Waveforms that describe the operation are shown in Fig. 8-12(*a*), and Fig. 8-12(*b*) defines the two operating quadrants of the circuit. When in the motoring mode, MOSFET 1 is switched on and off, at an appropriate repetition

rate, providing control of the average voltage applied to the motor. At this time MOSFET 2 is off, and its diode acts as a conventional freewheeling diode to conduct the freewheeling motor current when MOSFET 1 is off. When the motor acts as a generator and returns energy to the dc source, MOSFET 2 is chopped on and off to control the current fed back from the motor to the source. In this mode, MOSFET 1 is off, but its internal diode is used to carry the motor current to the dc source during the intervals when MOSFET 2 is off. In order for a motor to regenerate, it is necessary for it to have either a shunt or separately excited field. A series-connected field is not feasible because of the reversing of connections required and is not practical.

Figure 8-13 shows a partial schematic for a 48-V, 200-A two-quadrant chopper based on power MOSFETs. This unit employs ten MOSFETs connected in parallel for the motoring switch and five MOSFETs connected in parallel for the regenerating switch. Continued improvements in power MOS-

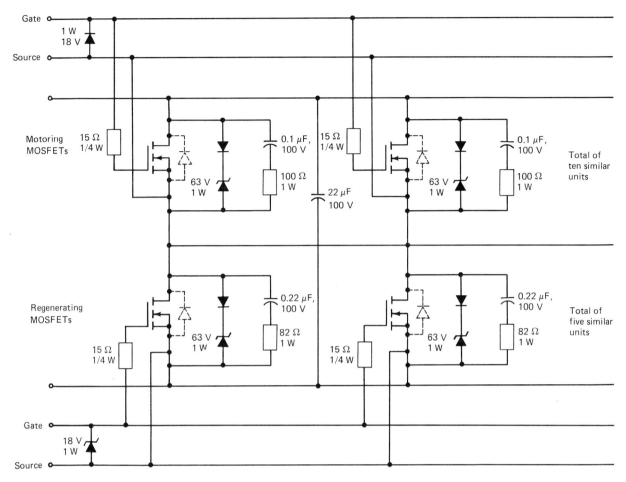

Fig. 8-13 Power circuit schematic.

FETs are expected to make motor control circuits of this type economically and technically superior to chopper systems employing transistors or thyristors.

REVIEW QUESTIONS

7. Three common solid-state choppers are the transistor, the MOSFET, and the _____.

8. Chopper circuits control the average load voltage by one of three _____ methods.

9. The freewheeling diode in the Jones circuit of Fig. 8-8 is _____.

10. In Fig. 8-7 S_1 is used for dynamic braking of the motor. (true or false)

11. A basic problem when using a thyristor in a chopper is being able to _____ the device.

12. In Fig. 8-8 L_1 and L_2 form an _____.

8-3
COMMUTATION CIRCUITS

The gate has no more control over a thyristor once the device is triggered on for any current exceeding the latching current. External means must be applied to stop the flow of current through the device. The two basic methods used for turning thyristors off are:

1. Current commutation
2. Forced commutation

Current commutation may be achieved by opening or closing a switch. Figure 8-14(a) shows a series switch; Fig. 8-14(b) shows a shunt switch. Commutation is achieved by opening the series switch or by closing the shunt switch. Each case of switch operation would produce a high value of the rate of change of voltage across the SCR. Mechanical switches are seldom used for current commutation. There are some static switching circuits, but this mode of commutation is generally not employed in industrial circuits.

Forced commutation employs a momentary reverse bias to turn off a thyristor. The reverse bias must be applied for a period longer than the device's turn-off time. Also, the rate of rise of the reapplied voltage must not exceed the critical value. With inductive loads, the stored energy of the collapsing field must be diverted from the thyristor by a freewheeling diode or some other method. The six

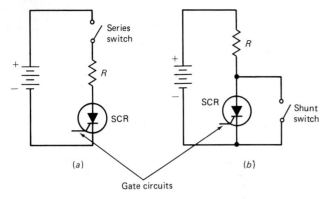

Fig. 8-14 Current commutation.

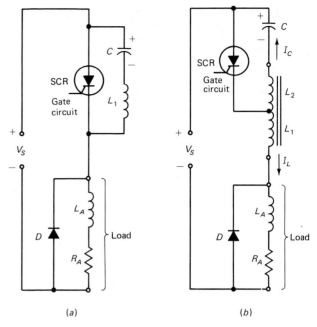

Fig. 8-16 Parallel capacitor-inductor/Morgan commutation circuits.

classes of forced commutation are given various names by different manufacturers in their literature. Once the circuits are compared, the exact method of commutation may be determined. Understanding of the commutation method used is the important point, not its name or class.

The simplest form of forced commutation is self-commutation and employs a series capacitor for resonating the load, as shown in Fig. 8-15. When the SCR is turned on, the capacitor is charged to the source voltage through the thyristor. The current will decay below the holding current as the capacitor voltage approaches the supply voltage V_s. With an underdamped resonating inductive load, the voltage on the capacitor will reverse and exceed the applied voltage V_s to assist in the thyristor turn-off. The load forms part of the tuned circuit. This method of commutation is sometimes employed in inverter circuits. In circuits with unidirectional load current, there must be a method to discharge the capacitor. Figure 8-15(b) shows the use of a parallel resistor; Fig. 8-15(c) shows the preferred method, using a second SCR in parallel with the capacitor.

Another method of forced commutation is the LC self-commutated or parallel capacitor-inductor

circuit shown in Fig. 8-16(a). The previously discussed series-capacitor (resonant load) circuit has limited control range, and load variations affect its operation. Placing the underdamped LC circuit in parallel with the SCR, as shown in Fig. 8-16(a), eliminates this problem. When the power is turned on, the capacitor will charge up to source voltage V_s through inductor L and the load. When the SCR is turned on, current will flow through the load and the capacitor, and inductor L_1 will begin an oscillatory action. After one-half cycle, the stored energy will reverse-bias the SCR and reduce its current to a value less than the holding current, and the SCR will then turn off. The capacitor will start to recharge to

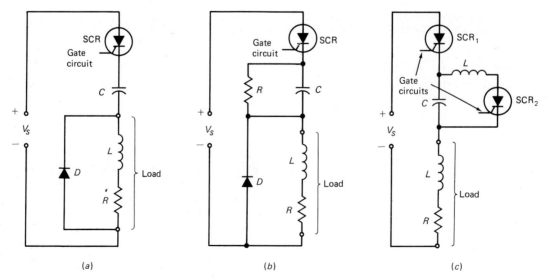

Fig. 8-15 Series capacitor commutation. (*a*) Basic circuit. (*b* and *c*) Improved circuits.

its original polarity through the inductor and the load. The circuit has several restrictions. The mode of operation can only be pulse rate modulation. The SCR can only be fired after the capacitor has fully recharged, or the circuit will fail to commutate. The ON time t_{ON} is limited to $\pi\sqrt{LC}$, and t_{OFF} is load-dependent. These restrictions of t_{ON} and t_{OFF} limit the range of load voltage control.

An improved version of this circuit is shown in Fig. 8-16(b). The circuit, known as the *Morgan circuit,* utilizes the properties of a saturable reactor. When the reactor is unsaturated, its inductance will be high. When the reactor is saturated, its inductance will be low. This point is covered in Chapter 7. Before the SCR is turned on, the capacitor is charged to V (supply) as shown. With no current flow through the tapped saturable reactor, the core is unsaturated, and the reactor is in the high-inductance state. When the SCR is triggered, the capacitor current (I_C) and the load current (I_L) flow in opposing directions through the reactor. The reactor tends to keep the rate of rise equal for both currents, and this rate of rise is determined mainly by the load inductance. As soon as full load current is flowing, capacitor current (I_C) will decrease, and load current will remain constant. The voltage induced across L_2 will initially be very small and reversed in polarity. This voltage will build in value as the charging current decreases. The period of oscillation is long, since the reactor core is unsaturated (high L). As the capacitor-charging current I_C goes to zero, the reactor current approaches I_L, and the reactor core saturates. Now the L_2 inductance will be very small. The period of oscillation is thus short, and the stored energy in L_2 reverse-biases the SCR and turns it off. The remaining charge on the capacitor is then dissipated in the load and is ready to recharge to the supply voltage V_S. As I_L approaches zero, the reactor core unsaturates, and the cycle can begin again when the SCR is turned on.

A parallel-capacitor commutation circuit is shown in Fig. 8-17. When SCR_1 is turned on, the capacitor is charged to the supply V through R_A, with the right-hand plate positive as shown. Commutation is initi-

ated when SCR_2 is turned on, applying the capacitor across SCR_1. The discharge current of the capacitor opposes the load current in SCR_1, and the SCR commutates.

An alternative parallel-capacitor configuration, in which the capacitor-charging current flows through the load, is shown in Fig. 8-18. The capacitor is charged with the polarity indicated to the supply potential by turning on SCR_2. When the capacitor is fully charged SCR_2 will automatically commutate off. When SCR_1 is gated on, V_S supplies load current I_L. Now SCR_1 provides a discharge path for the capacitor through D_1 and L_1. The capacitor voltage will reverse as L_1 maintains current flow after the first resonant period. Now D_1 prevents the charge from being drained off the capacitor. Commutation is initiated by turning SCR_2 on. This reverse-biases SCR_1 and reduces the current below the holding value. When SCR_1 is commutated, the capacitor recharges to its original condition through SCR_2 and the load.

Another class of commutation depends on the commutation energy being supplied from an external source. There are several common configurations that may be used; two of them are shown in Fig. 8-19. In Fig. 8-19(a) SCR_1 is commutated off by means of an auxiliary transistor switch Q_1. The thyristor is assumed to be initially on when turn-off is desired. A signal applied to the base of Q_1 turns it on and reverse-biases the SCR. The SCR is now commutated off. The drive signal to the base of Q_1 must be of sufficient duration to ensure thyristor turn-off and of sufficient amplitude to place Q_1 in saturation. If Q_1 comes out of saturation before the thyristor turn-off is complete, commutation failure results. Therefore, Q_1 and its drive circuitry are selected to satisfy worst-case conditions for SCR turn-off under the heaviest load conditions. Figure 8-19(b) utilizes a pulse transformer with a square-loop *B-H* core to achieve thyristor turn-off. When the SCR is conducting, the core of the pulse transformer is saturated (low impedance) by the load current. When the time comes to turn off the SCR, the first step is to unsaturate the pulse transformer. The application

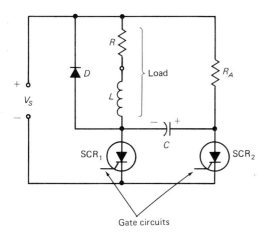

Fig. 8-17 Parallel-capacitor commutation circuit.

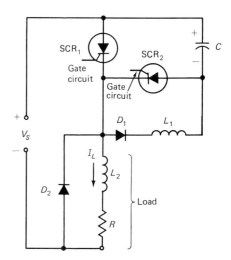

Fig. 8-18 Parallel-capacitor commutation through the load.

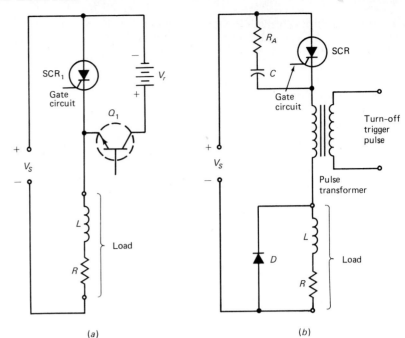

Fig. 8-19 External commutation techniques. *(a)* *(b)*

of a drive pulse to the primary reverses the flux in the transformer for several microseconds. A voltage pulse is developed across the pulse transformer secondary, which reverse-biases the SCR and turns it off. If an inductive load is present, a freewheeling diode is connected across the load to prevent damage to the SCR from voltage spikes developed by the inductive load discharging. External pulse commutation circuits allow both pulse width and pulse rate modulation techniques to be used. Note that the commutation is independent of the load current and the supply voltage source.

If the supply is an alternating voltage, as shown in Fig. 8-2, load current will flow only during the positive half cycle. During the negative half cycle the SCR will turn off because of the negative polarity across it. The only constraint is that the half cycle must be longer than the turn-off time of the SCR.

REVIEW QUESTIONS

13. Two methods of thyristor commutation are _____ and forced commutation.

14. Forced commutation requires that the _____ bias be applied for a longer time than the thyristor's turn-off time.

15. Resonating commutation uses an over-damped *LC* circuit. (true or false)

16. In most circuits that use a capacitor for commutation, it will be necessary to charge the capacitor to _____ polarity.

17. The Morgan commutation circuit is identified by its use of a _____ reactor.

18. In the external transistor commutation cir-

cuit, the transistor must be driven into _____ to turn off the thyristor.

8-4
STATIC FREQUENCY CONVERTERS FOR AC MOTOR CONTROL

The ac motor, especially the squirrel cage induction type, possesses many virtues in comparison to dc motors. These include significantly lower cost, weight, and inertia; higher efficiency; and fewer maintenance requirements. They are also capable of operating in dirty and explosive environments because they do not have a commutator and brushes.

In spite of the many virtues of the ac motor, the cost of converters and circuit complexity were main factors preventing the widespread application of ac drives. However, as the ratings of solid-state devices continuously improve and as their cost decreases, variable-speed ac drives are increasing in popularity. Integrated circuit technology is also assisting in this changeover as complex circuits are reduced to the chip level.

The speed of an induction motor is determined by the synchronous speed and by the slip of the rotor. The synchronous speed is dictated by the supply frequency and the number of poles. Slip can be controlled by regulating the voltage or current supplied to the motor. There are several methods for controlling the speed of induction motors:

1. Variable-voltage constant-frequency or stator voltage control

2. Variable-voltage variable-frequency control

3. Variable-current variable-frequency control

4. Regulation of the amount of slip

The term *inverter* normally refers to equipment used for transforming direct to alternating current. A cycloconverter is used for transforming a higher-frequency alternating current to a lower-frequency without any intermediate dc link.

The most commonly used ac drive system is the variable-frequency dc-link inverter. Polyphase induction motors or synchronous motors are employed because their operating characteristics are retained over the range of the inverter, which is typically from 10 to 200 Hz. Variable-frequency drives are found in machine tools and textile, paper, and steel mill equipment. In most variable-frequency drives, a constant voltage per hertz is maintained up to the rated frequency of the motor, and then the stator voltage is maintained at its rated value as the frequency is increased. Failure to maintain a constant volts/hertz ratio affects the torque output and can cause an increase in stator current and may overheat the motor.

Figure 8-20 shows a simple two-transistor inverter. The circuit uses a square wave fed through transformer T_2 to drive the bases of Q_1 and Q_2. Each transistor is alternately switched into saturation. When Q_1 is on, Q_2 is off, and the current flows through the top half of the primary of T_1. Later, Q_1 is turned off and Q_2 is turned on. The current now flows in the bottom half of the T_1's secondary. This induces a square wave in the secondary of T_1. This square wave output is filtered (L_1, C_1) to make it more sinusoidal before it is applied to the load.

A MOSFET three-phase full-wave inverter is illustrated in Fig. 8-21. This type of inverter is suitable to drive a three-phase induction motor and is capable of a variable-frequency output. This type of inverter is often referred to as a *two-on inverter* since two MOSFETs are always on. Figure 8-22 shows simplified schematics and the firing sequence required for three-phase outputs. The output waveforms are shown in Fig. 8-23. Figure 8-22(*a*) shows the state of the MOSFETs for 0° to 60°. Transistor Q_2 connects the positive end of supply V to motor terminal *A* while Q_3 connects the negative end of V to the *B* terminal. Therefore, from 0° to 60° the voltage across terminals A and B (V_{AB}) is +V. The voltage from terminals A to C (V_{AC}) and from C to B (V_{CB}) is +1/2V since the two windings are in series across V_{AB}. The waveforms show V_{CA} and V_{BC}, the inverse of V_{AC} and V_{CB}. Hence, V_{CA} and V_{BC} are equal to −1/2V for the 0° to 60° timing sequence.

Figure 8-22(*b*) shows Q_2 and Q_5 on, which makes V_{AC} equal to +V (or V_{CA} equal to −V). Both V_{AB} and V_{BC} are +1/2V for 60° to 120° as shown in Fig. 8-23 (page 192). By further analysis of Fig. 8-22(*c,d,e,f*), the three-phase composite waveforms shown in Fig. 8-23 are produced.

A similar type of inverter can be obtained by using thyristors in the three-phase configuration shown in Fig. 8-24(*a*) on page 193. The commutation and firing circuits have been omitted to simplify the explanation. The thyristors are fired in a sequence to produce positive-phase sequence output voltages V_{AB}, V_{BC}, and V_{CA} at the output terminals *A*, *B*, and *C*.

From Figure 8-24(*b*) it can be seen that SCR₁ con-

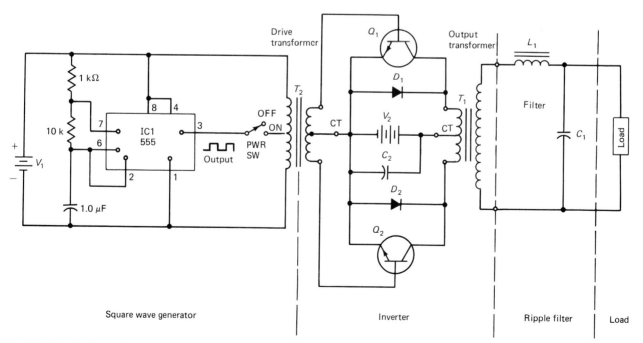

Fig. 8-20 Simple single-phase inverter.

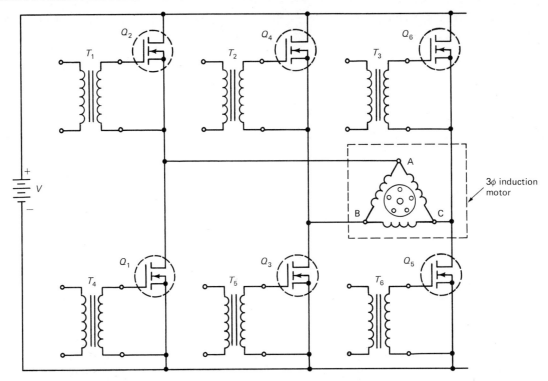

Fig. 8-21 MOSFET three-phase inverter for an induction motor.

ducts from 0 to 180°, and SCR_4 conducts from 180 to 360°. This condition relies on SCR_1 being commutated prior to SCR_4 firing; if there is any time delay in the turn-off SCR_1, a short circuit or "shoot through" can exist between the positive and negative supply rails through SCR_1 and SCR_4. The associated waveforms along with the gating pulses are shown in Fig. 8-24(*b*).

Figure 8-25(*a*) on page 194 shows a wye-connected load for the six-step three-phase inverter shown in Fig. 8-24(*a*). Figure 8-25(*b*) shows the load connections for the 0 to 60° time interval, with SCRs 1, 5, and 6 conducting. Two windings of the wye load are always in parallel, and one is in series with the parallel pair. Thus, as a voltage divider, the parallel pair drops $1/3V$ while $2/3V$ is dropped by the single winding. This assumes that the wye load is balanced. The output voltages for the load V_{AN}, V_{BN}, and V_{CN} are similar to those of Fig. 8-23. One important difference, since the load is wye-connected, is that the peak voltage (across any winding) to the neutral is $2/3V$ instead of the full V output as realized by a delta load.

The trigger circuit for the inverters is often obtained from a ring counter. A three-state ring counter is shown in Fig. 8-26 (page 195). The output frequency is determined by the applied input square wave. This input waveform is generated by a separate oscillator. Varying the frequency of the oscillator will vary the speed of a three-phase induction motor connected as a load. Two three-state ring counters

may be used to trigger a six-step inverter. When the power is applied through S_2, capacitors C_1 to C_6 charge to V through T_1, T_2, or T_3 and the associated resistors. The input square wave is differentiated by C_C and R_5 to produce the spikes shown in Fig. 8-26. The spikes would trigger all three SCRs through diodes D_1 to D_3, but the positive charge on C_1 to C_3 reverse-biases the diodes and keeps them off. When the start switch (S_1) is closed, C_3 is discharged by R_6 and the other resistors in the discharge circuit. This discharge allows a positive spike to be coupled through D_1 to C_1 to fire SCR_1 and produce a trigger pulse at the secondary of T_1. This in turn discharges C_4 through SCR_1 and R_4. Capacitor C_2 is also discharged through R_2 and the 1-kΩ resistor. With C_2 discharged, D_2 is no longer reverse-biased. Now the next positive spike from the differentiator will fire SCR_2. The firing of SCR_2 through R_4 and the subsequent discharge of C_5 through R_4 produce a positive voltage pulse at the cathode of SCR_1. This pulse momentarily reverse-biases SCR_1, which commutates. The circuit operation continues as the SCRs fire in a sequence of 1–2–3–1–2, and so on.

The voltage waveforms for two-ring counters are shown in Fig. 8-27 (page 195). The (*a*) group was just discussed; the (*b*) waveforms are for a second three-state ring counter whose input is delayed by one-half period to produce the remaining input triggers for the inverter switches.

The output of these types of inverters has a high harmonic content, especially when they operate over

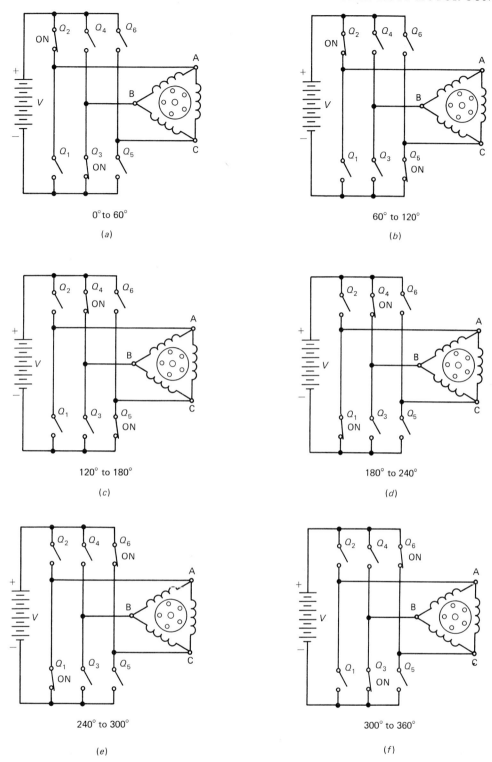

Fig. 8-22 Simplified firing sequence of Fig. 8-21.

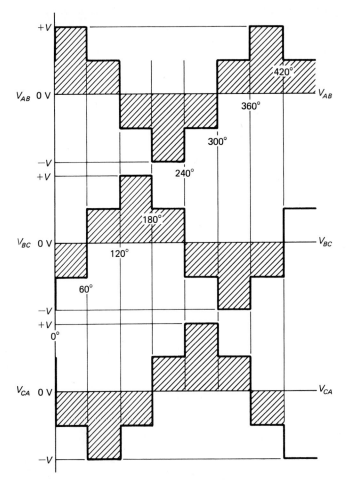

Fig. 8-23 Three-phase inverter output waveforms.

a wide frequency range such as 10 to 200 Hz. Output filtering is not practical because of the wide range of frequencies. There are several other methods that may be used to minimize the output harmonic content. One way is to use pulse width modulation techniques within the inverter. Another method is to combine a number of square-wave inverters with each one phase-shifted and fired at the desired output frequency. This method is known as *harmonic neutralization*. Figure 8-28 (page 196) shows how a three-phase inverter can be constructed by using three single-phase bridge inverters. This arrangement eliminates the third harmonics from the inverter's output.

The single-phase cycloconverter is covered in Chapter 3. To review, the cycloconverter can produce a variable-frequency output by using phase-controlled converters. They normally operate at a frequency of one-third the supply frequency or less. The major advantages for using the cycloconverter are (1) elimination of the intermediate dc link, thereby improving the overall efficiency; (2) voltage control accomplished within the converter; (3) achievement of line commutation, eliminating any forced external commutation circuitry.

The output of the single-phase, two-pulse midpoint converter (covered in Chapter 3) has a high ripple content. Increasing the number of pulses will reduce the ripple content in the load waveform. The pulse number can be increased by using converters as shown in Fig. 8-29 (page 196), arranged to form three dual converters. As shown in Fig. 8-29(*b*), this configuration uses 18 thyristors and permits only unidirectional operation. Another increase in the pulse number can be obtained by using six six-pulse midpoint converters with interphase reactors as shown in Fig. 8-30 (page 197). This unit requires 36 thyristors and also provides unidirectional operation only. Circulating currents are reduced by the interphase reactor, which presents its full reactance to the passage of circulating currents but only a quarter of its reactance to the load current.

The cycloconverters discussed up to this point are for continuous variable-frequency applications. In some applications the output frequency is a fixed percentage (less than 100) of the supply frequency. In these cases an *envelope cycloconverter,* a less complex circuit, may be employed. Figure 8-31 (page 197) shows a block diagram with logic control of the

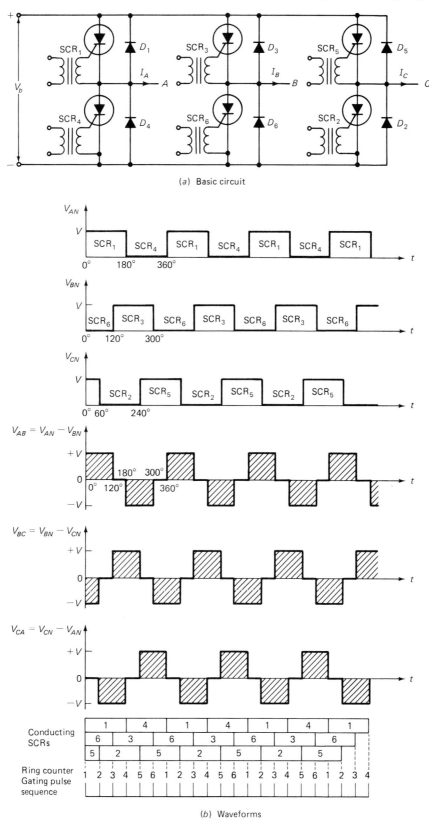

(a) Basic circuit

(b) Waveforms

Fig. 8-24 Three-phase six-step inverter.

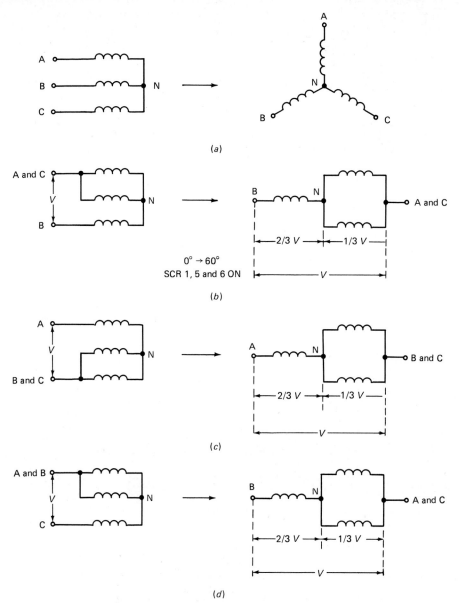

Fig. 8-25 Line to neutral voltages for wye-connected load.

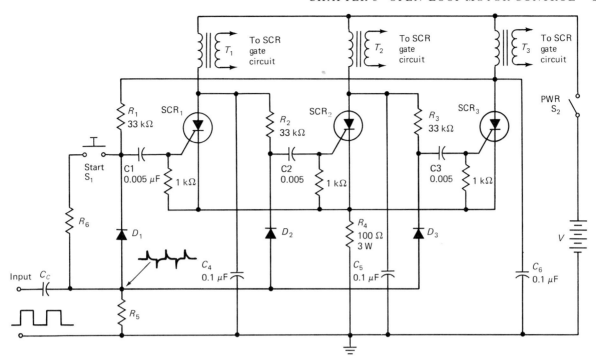

Fig. 8-26 Three-stage ring counter.

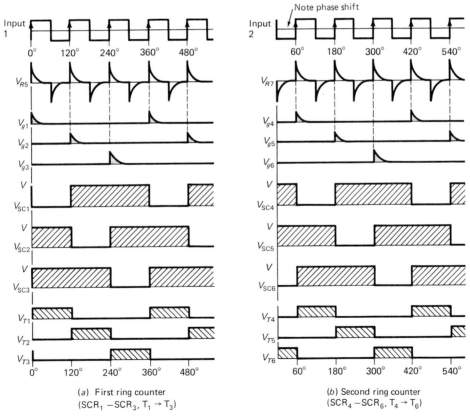

(a) First ring counter
(SCR$_1$ —SCR$_3$, T$_1$ → T$_3$)

(b) Second ring counter
(SCR$_4$ —SCR$_6$, T$_4$ → T$_6$)

Fig. 8-27 Waveform from ring counter circuit.

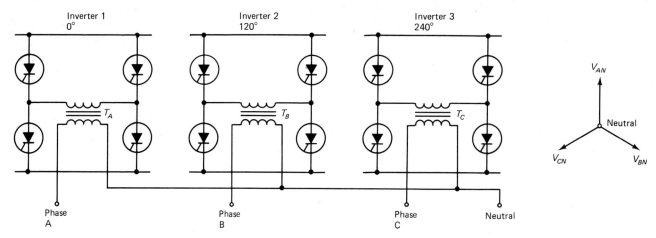

Fig. 8-28 Three single-phase inverters forming a six-step inverter.

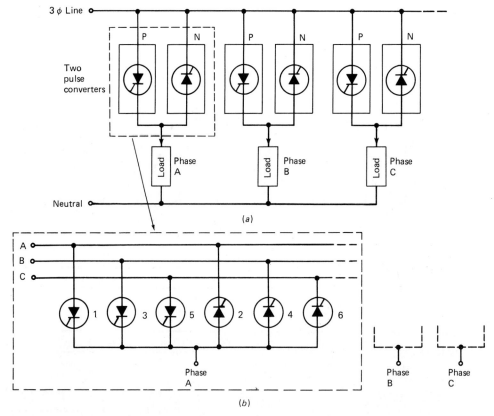

Fig. 8-29 Three two-pulse converters combined. (*a*) Block diagram. (*b*) Schematic diagram.

thyristor gates to obtain 6:1 reductions of the supply frequency. Using a three-phase configuration greatly reduces the harmonic content. The output of the six-pulse configuration is fed into a wye-double-wye transformer with four different output voltages. This arrangement produces a composite waveform that is a good approximation of a sine wave, as shown in Fig. 8-32. The waveform shows the 20-Hz envelope obtained from the 60-Hz supply. A basic disadvantage of the envelope cycloconverter is it operates only at a fixed number of frequencies when controlled by a counter with a variable number of states.

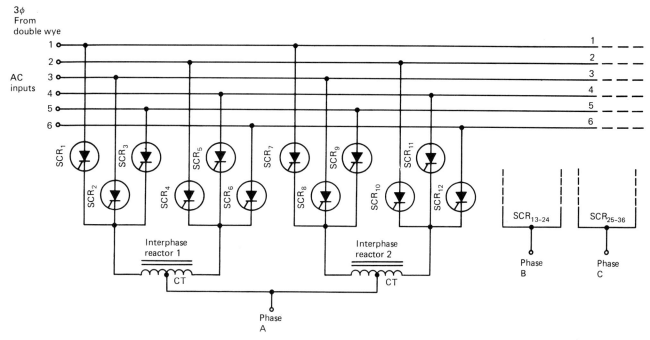

Fig. 8-30 Three-phase six-pulse cycloconverter.

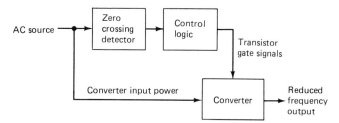

Fig. 8-31 Formation of envelope cycloconverter output.

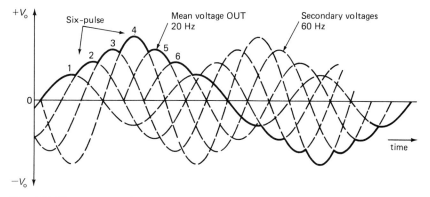

Fig. 8-32 Mean output voltages of synchronous envelope cycloconverter.

REVIEW QUESTIONS

19. The speed of induction motors is related to the line _____.

20. The _____ is used to change direct current to alternating current.

21. A cycloconverter converts a low supply frequency to a higher frequency. (true or false)

22. The cycloconverter changes frequency without any intermediate _____ link as required with an inverter circuitry.

23. Inverters typically operate from _____ Hz to 200 Hz.

24. In Fig. 8-24, shoot-through occurs if there is a delay in the _____ of the thyristor.

8-5
TROUBLESHOOTING AND MAINTENANCE

The devices used in the static converters covered in this chapter are also discussed in other chapters. One exception is the thyratron tube, whose appearance in modern industrial equipment is very unlikely. However, not all equipment is modern, and the technician must be aware that the tubes are gas-filled and exhibit a steady bluish glow when operating. In most cases, the tubes are balanced and share the load equally. If one tube is much brighter than another, a pair should be reversed to see whether the symptom follows the tube reversal. If so, the replacement of groups of tubes is recommended in this circumstance. If the problem does not follow the tube reversal, a circuit problem is indicated, so voltage and resistance checks are in order. Use the manufacturer's maintenance manual for reference. The tubes also have filaments which may be separately controlled or powered. The loss of filament power (usually less than 12 V_{ac}) will prevent tube conduction.

If possible, gather information from the operators of the equipment. This information may be useful in deciding whether a thermal problem, noise, regulation, or worn controls may exist. As with solid-state devices, heat dissipation is very important. Dirty or restricted fans and heat sinks should be checked first whenever any thermal problems exist.

Since phase-control circuits are line-synchronized, an oscilloscope triggered at the line frequency can be used for localizing any trouble. Silicon controlled rectifiers must hold off the line voltages and in most cases are in bridge-type configurations. This type of arrangement gives one a comparison device to use for checking the operation of all other devices. In the case of small motors, a power resistor or a lamp bank can be substituted for the motor to prevent overheating or damage due to latched-up conditions. If control is lost in one direction of a bidirectional control, the thyristor is usually suspect. An open device will prohibit movement in one direction, although a leaky or short-circuited device will keep the unit locked in that direction.

Most chopper circuits rely on self-commutation. In many circuits such as the Jones chopper, the capacitor is part of a resonant circuit needed for commutation, so its stability and leakage are important factors. If the capacitor is leaky, the SCR can latch up, and all control is lost. Any significant change in its value can shift the period of the resonant ringing and may not produce a voltage of sufficient amplitude to commutate the SCR. A leaky or short-circuited diode will disrupt all commutation. In the case of a freewheeling diode, a short-circuited device will stop the motor and severely overload the line and the control devices. A blown fuse or a tripped breaker can be expected in these cases. Parallel MOSFETs are employed in many applications and must be separated, at least at the drain or source lead, to check for short circuits. Remember that many devices have an integral (internal) diode that can influence the testing. In some cases, the devices can be first tested as a group. If the group test failed, the units would have to be separated to locate the one that was short-circuited. No commutation is needed for MOSFETs, so if latch-up occurs it usually indicates a short-circuited device.

Single-phase inverters must operate at their specified frequency, and any large deviation can cause a loss in efficiency. Overheating in the load can result from excess harmonic content if the inverter's filter fails to function properly. The ac current capability of any inductor or capacitor in a filter must be observed when replacements are in order. A set of manufacturer's spares is recommended in all cases. Inverters can usually be isolated into individual blocks for troubleshooting purposes. Isolation at the component level involves voltage or resistance checks. Various dc power supplies are employed for biasing and power conversion. Verification of all supply voltages must occur early in the fault analysis procedure.

Ring counters rely on the symmetry of parts (resistors, solid-state devices, capacitors, etc). Any large change in value or leakage will stop the counter or prohibit its starting. If it is stopped, one can detect the stuck stage and check the associated components. An oscilloscope can be used to trace trigger waveforms up to the gates of the thyristors or MOSFETs. If any SCR has failed, its shunt diode should also be checked. An open diode is often the cause of an SCR failure when inductive loads are involved.

In the case of large cycloconverters, the thyristors may be individually fused, and the fuses should be checked first. If a fuse is blown, the SCR should be verified before returning the unit to service. The logic used in developing the pulses can be tested to ensure that the timing and amplitude of the trigger pulses are within the manufacturer's specifications. Some units may have a test position and test points available for these purposes. In all cases, the manufacturer's maintenance manual should be consulted before any servicing is attempted.

REVIEW QUESTIONS

25. Thyratrons exhibit a _____ glow when ionized.

26. Before servicing open loop controls, a technician should obtain information from the equipment _____.

27. Latch-up in an SCR can be due to loss of commutation. (true or false)

28. Commutation in the Jones or Morgan circuit relies on energy stored in a _____.

29. Individual MOSFETs can be tested while connected in parallel in a circuit. (true or false)

30. If a capacitor opens in an inverter's filter, the load may _____.

CHAPTER REVIEW QUESTIONS

8-1. In Fig. 8-3, reversing the control voltage will reverse the motor's direction. (true or false)

8-2. In Fig. 8-4(*b*), which component senses motor current to provide overload protection?

8-3. Varying the OFF time of the waveform is called *pulse* _____ *modulation*.

8-4. Which solid-state switches are used in the Jones circuit?

8-5. The capacitor in the Jones circuit is usually an electrolytic type. (true or false)

8-6. Two-quadrant choppers are well suited for series-connected motors. (true or false)

8-7. When used with an alternating supply voltage, commutation is frequency-independent. (true or false)

8-8. To keep the output torque constant, an inverter must keep the _____ ratio constant.

8-9. Failure to keep the previous ratio constant may cause the motor to _____.

8-10. The two-on inverter refers to two phases being on at any time. (true or false)

8-11. The wye load voltage referenced to the neutral is _____ of the delta voltage for the two-on six-step inverter.

8-12. A method to gate inverters sequentially is to use a _____ counter.

8-13. The SCRs in a ring counter are commutated by positive pulses at their cathodes. (true or false)

8-14. Cycloconverters usually produce an output at _____ of the supply frequency or less.

8-15. Circulating currents are reduced in multiple cycloconverters by using a _____ reactor.

8-16. The _____ cycloconverter combines the output of multiple secondaries to produce the sine wave.

8-17. Thyristors in the cycloconverter may be individually _____ for overcurrent protection.

8-18. The _____ maintenance manuals should first be checked before attempting any servicing.

8-19. In most cases, the oscilloscope can be triggered from the line or _____ input of a cycloconverter.

8-20. Verification of all _____ voltages is an important early step in servicing any inverter.

ANSWERS TO REVIEW QUESTIONS

1. false **2.** SCR **3.** firing **4.** R_2 **5.** Q_1; the UJT **6.** deadband **7.** thyristor **8.** modulation **9.** D_1 **10.** false
11. commutate **12.** autotransformer **13.** current **14.** reverse **15.** false **16.** reverse/opposite **17.** saturable
18. saturation **19.** frequency **20.** inverter **21.** false **22.** dc **23.** 10 **24.** commutation **25.** bluish **26.** operator
27. true **28.** capacitor **29.** false **30.** overheat

9

INPUT TRANSDUCERS

Control is based on information. Industrial automation systems must extract information from the physical process that is being controlled. Input transducers convert physical parameters into electrical signals that correspond to what is happening. In the broadest sense, a transducer *is any device that receives energy from one system and retransmits it, usually in another form, to another system. Thus, an electric motor can be viewed as a transducer. The word* sensor *is more restrictive. It refers to that part of a transducer that responds to the quantity being measured. This chapter will use the terms* transducer *and* sensor *to describe components and devices used to measure physical conditions.*

9-1
POSITION AND DISPLACEMENT

Displacement is the difference between the position of some object and a reference point. Displacement can be *linear* (straight-line) or *rotary* (angular). Potentiometric transducers can be used to measure both linear and angular displacement. Potentiometers are very common transducers in the industrial environment. Figure 9-1 shows a linear displacement potentiometer. A resistance element is shown at the top of the diagram. This element can be formed by winding resistance wire on a form or by depositing resistance material. The wiper contact moves along the resistance element in response to motion applied to the input shaft. If a voltage is applied across terminals A and B, then some portion of that voltage will

appear across A and C. Most potentiometric transducers are nominally linear. If the input shaft is at its mechanical center position, half the applied voltage will appear at C. If it is at its far left position, the output voltage will be zero. If it is one-quarter from its far left position, the output voltage will be one-fourth the applied voltage. In other words, there is a linear relationship between the shaft position and the output signal.

The actual performance of a displacement potentiometer will deviate from nominal linearity. Manufacturers rate them according to worst-case deviation. A 1 percent linearity rating or better is typical for transducer service. Thus, a 100-Ω transducer may have an error of plus or minus 1 Ω at its worst-case position. Potentiometer resolution is another source of error. Suppose a 100-Ω resistance element is made up of 200 turns of wire. This means that each turn represents 0.5 Ω of resistance. As the wiper moves, the resistance across the wiper contact and either end contact changes in half-ohm steps. This is the smallest change that the transducer can resolve. It is also expressed in percentage form; the resolution would be 0.5 percent (0.5 Ω/100 Ω × 100) in this example. For best accuracy, the percentages of linearity and resolution should be as small as possible.

An angular displacement potentiometer is shown in Fig. 9-2. The input shaft turns, and the wiper contact moves with it. As before, if a voltage is placed across the resistance element, the voltage at the wiper contact will be a function of shaft position or shaft displacement. The preceding discussion of linearity and resolution also applies. The angular potentiometer is usually limited to about 320° of rotation. However, gearing can be used to make a transducer with greater mechanical range. Other arrangements to provide greater mechanical range include worm drives for the wiping contact and assemblies with more than one wiper.

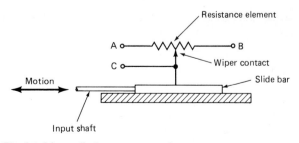

Fig. 9-1 Linear displacement potentiometer.

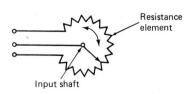

Fig. 9-2 Angular displacement potentiometer.

200

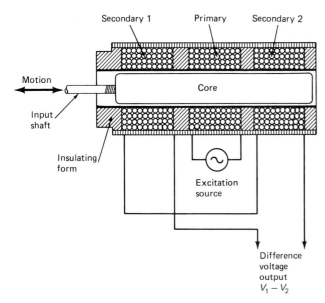

Fig. 9-3 Linear variable differential transformer (LVDT) construction.

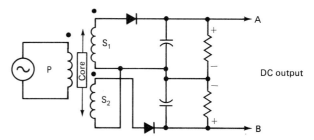

Fig. 9-5 Converting LVDT output to dc.

Potentiometric transducers are relatively inexpensive and easy to apply. However, they have some limitations. For example, attempts to make the resolution very high usually result in poorer linearity. They are temperature-sensitive, a characteristic that also affects their accuracy. Potentiometers are considered to be low- to medium-accuracy transducers. The wiper contact is another limiting factor, being subject to wear and dirt and potentially producing electrical noise.

The *linear variable differential transformer* (LVDT) is more costly but outperforms the potentiometric transducer. Its construction is shown in Fig. 9-3. It consists of a primary, two secondaries, and a

movable core. The primary is excited by an ac source. When the core is in its exact center location, the amplitude of the voltage induced into secondary 1 will be the same as the voltage induced into secondary 2. The secondaries are connected to phase cancel, and the output voltage will be 0 at that point. Figure 9-4 illustrates what happens to the output voltage as the core is moved to the left (Fig. 9-4[a]) and to the right (Fig. 9-4[b]). Note that the magnitude of the output voltage is a linear function of core position and that the phase is determined by the side of the null position on which the core is located.

Figure 9-5 shows a simple circuit for converting the ac output of an LVDT to dc. With the core centered, both S_1 and S_2 produce equal amplitudes. Both half-wave rectifiers produce equal dc voltage drops across the two resistors. The polarities are opposing, so the output voltage from A to B is equal to zero. If the core is moved up, S_1 produces more voltage than S_2. The drop across the top resistor is now greater, and A becomes positive with respect to B. If the core is moved down, B becomes positive with respect to A. Although this simple arrangement works, higher performance is available with a more elaborate detector circuit. The last section of this chapter deals with transducer signal conditioning in more detail.

A *rotary variable differential transformer* is shown in Fig. 9-6. This transducer allows the measurement of angular displacement up to about 90°. This range may be extended with gearing.

Linear variable differential transformer accuracies are very good. Typical linearities are between 0.25 and 0.05 percent of full range. In general, even better linearity can be obtained by operating the LVDT over something less than its maximum range. The typical industrial LVDT has a total range of approximately plus and minus 2½ cm. Resolution is excellent, with typical specifications near 0.000001 cm.

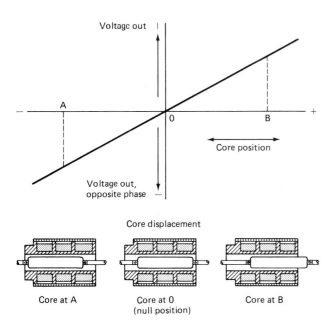

Fig. 9-4 Linear variable differential transformer (LVDT) output phase and voltage versus core position.

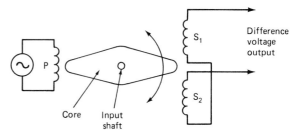

Fig. 9-6 Rotary variable differential transformer.

The excitation frequency for an LVDT varies, depending on its design and the application. For example, if the transducer must accurately track rapidly changing displacement, the higher frequencies are advantageous. Typical values range from 50 Hz to 30 kHz. The voltage applied to the primary is usually around 10 V. Reliability of the LVDT is superb, with ratings of millions of hours mean time before failure (MTBF).

Some displacement measurements involve very small movements. Strain gage transducers lend themselves to these applications. They also are noted for low cost and ease of use and, in some cases, can simply be epoxied to the physical member under measurement. Strain gages are based on the principle that the resistance of a conductor is directly proportional to its resistivity and length and inversely proportional to its cross-sectional area. If a conductor is stretched, its length will increase, thus increasing its resistance. At the same time, its cross-sectional area decreases, also increasing its resistance.

Figure 9-7(a) shows a bonded wire strain gage. It consists of several loops of fine wire bonded to a paper or plastic backing. Note that the sensitivity axis is parallel to the long portions of the wire runs. If the gage is slightly elongated along this axis, maximum resistance increase will result because the greatest total length of wire is involved. In Fig. 9-7(b) the foil-type gage is depicted. These gages are made with a printed-circuit-type process using conductive alloys rolled to a thin foil. A grid configuration is used for the strain-sensitive element to allow higher values of gage resistance while maintaining short gage lengths. Gage resistance varies from 30 to

3000 Ω with 120- and 350-Ω values being rather common. Gage lengths vary from 0.02 to 10 cm (0.008 to 4 in.). The sensitivity axis for the foil-type gage is parallel to the long runs of foil, and the alignment marks are used to accurately install the gage along the proper axis.

The typical industrial strain gage has a gage factor of 2. *Gage factor* is the resistance change ratio divided by the length change ratio:

$$\text{GF (gage factor)} = \frac{\Delta R/R}{\Delta L/L}$$

Suppose a 4-cm strain gage has a nominal resistance of 350 Ω and a gage factor of 2. How much will the resistance increase if the gage is elongated by 0.02 cm. First, find the ratio of length change by dividing 0.02 by 4. This yields 0.005. Then, multiply this ratio by the GF and the nominal resistance to find the change in resistance: $0.005 \times 2 \times 350 \ \Omega = 3.5 \ \Omega$. Thus, the gage will increase from 350 to 353.5 Ω when elongated by 0.02 cm.

It is obvious that the resistance change is small in strain gages. How can such a small change be converted into a useful signal? The Wheatstone bridge circuit is well known for its accuracy and sensitivity in measurement applications. Figure 9-8 shows a bridge with a strain gage serving as one of the bridge elements. If all four bridge elements equal 350 Ω, the bridge is balanced and the output voltage is zero. What happens to the output if the resistance of the gage increases to 353.5 Ω? By using the equation and input voltage shown in Fig. 9-8, you should calculate an output voltage of approximately 0.03 V. The bridge output will normally be applied to the input of an instrumentation amplifier with a gain of 100 (or more). The final output would be nearly 3 V for our example.

Another feature of the bridge circuit is that it can be accurately balanced to produce zero output for zero strain. This process is called *nulling* the bridge. In Fig. 9-8 R_2 could be replaced by a fixed resistor in series with an adjustable resistor. The adjustable resistor would allow a change of a few percentage points in that leg of the bridge. This would facilitate accurate nulling of the bridge. Finally, the bridge circuit makes temperature compensation easy. Strain

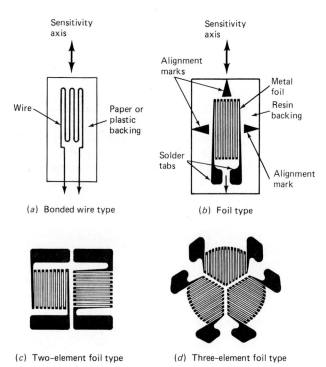

(a) Bonded wire type   (b) Foil type

(c) Two-element foil type   (d) Three-element foil type

Fig. 9-7 Strain gages.

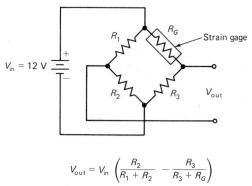

$$V_{out} = V_{in} \left(\frac{R_2}{R_1 + R_2} - \frac{R_3}{R_3 + R_G} \right)$$

Fig. 9-8 Wheatstone bridge circuit.

gages react to temperature as well as to strain. Refer again to Fig. 9-7(*c*). The two-element strain gage is often used where temperature compensation is required. With proper alignment, only one of the gages will react to strain. However, both will react to temperature and presumably by the same amount. If the two-element gage is applied to a bridge circuit such as the one shown in Fig. 9-8, one will act as a strain sensor, and the other will provide temperature compensation. For example, the gage sensitive to strain would serve as R_G, and the compensation gage would serve as R_1. The bridge equation shows that any resistance change that affects R_G and R_1 by an equal amount will produce no change in V_{out}.

Temperature compensation with a two-element gage does create some error. All gages have some sensitivity to strain perpendicular to their longitudinal axis. This is called *transverse sensitivity*. It is minimized by the gage manufacturer by placing extra material in the end loops of the conductors and by keeping the grid lines close together. The manufacturer will specify the *transverse sensitivity factor,* which is the ratio of transverse GF to longitudinal GF. The gain of the instrumentation amplifier can be adjusted to compensate for the characteristic that the temperature compensation gage also increases slightly in resistance when the assembly is elongated.

Certain semiconductor materials exhibit a characteristic known as *piezoresistance,* which is a change in resistance with strain. Semiconductor strain gages capitalize on this effect to provide very sensitive strain transducers. They have gage factors higher than those of the bonded wire or foil types. The semiconductor gage factors range from 45 to 175. This high sensitivity permits them to be used without amplifiers in some applications. Their resistance change is less linear over large ranges, however.

Figure 9-9 shows another type of displacement transducer. Capacitance is directly related to plate area and inversely related to the plate spacing. As the metal tube moves to the right, the distance between plates decreases, increasing the capacitance. The capacitive transducer can be placed into an ac bridge to provide an ac output voltage that is a function of linear displacement. Or the capacitor can be part of a tuned circuit for an oscillator. This arrangement will produce a frequency change with any change in position. Figure 9-10 shows an angular displacement capacitor. Capacitance will increase as the moving plate covers more of the fixed plate.

None of the transducers shown in this section lend

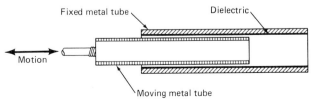

Fig. 9-9 Linear displacement capacitor.

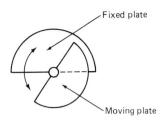

Fig. 9-10 Angular displacement capacitor.

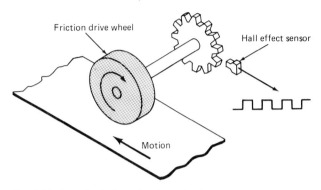

Fig. 9-11 A method of sensing large displacements.

themselves to the measurement of very large displacements. There are various ways to accomplish this, and Fig. 9-11 shows one example. As the material moves beneath the friction wheel, it causes it to turn. The wheel is shaft-coupled to a toothed wheel made of ferrous material. As the toothed wheel turns, it alternately provides a high and then a low reluctance path for the Hall-effect sensor. The output of the sensor is in the form of pulses. These pulses can be accumulated by counting circuits to measure extremely large displacements.

REVIEW QUESTIONS

1. Refer to Fig. 9-1. Assume that 12 V is across terminals A and B and that the slide bar is positioned at three-fourths of its right-most displacement. What voltage drop should appear from A to C?

2. Assume that the transducer in question 1 has a linearity of 1 percent. What range of output voltages would be considered normal for the data given?

3. A potentiometric transducer has a total resistance of 500 Ω, and the smallest resistance change it can produce is 1 Ω. What is its percentage of resolution?

4. Refer to Fig. 9-5. Assume that the core is 1 cm above its null position and that the drop across the top resistor is 6 V and the drop across the bottom resistor 4 V. What is the output voltage? Is A positive or negative with respect to B?

5. Assuming perfect linearity, what would happen in question 4 if the core were moved to a position 1 cm below the null position?

6. Refer to Fig. 9-8. All bridge elements are equal at 350 Ω. How much output will be produced if R_1 and R_G both increase by 1 Ω?

7. Refer to Fig. 9-7(*d*). How many sensitive axes are there?

9-2
VELOCITY AND ACCELERATION

Speed is the rate of change in displacement. It may be measured in meters per second, kilometers per hour, or centimeters per minute. *Velocity* is a measure of speed and direction. It is a vector quantity. Industrial measurements most often deal with fixed directions. A ram on a machine, for example, can only travel back and forth along a fixed path. Since the direction is established, the term *velocity* rather than speed is used to describe the rate of ram travel. One ram direction would yield a positive velocity and the opposite direction a negative velocity. Rotating machine parts also usually travel in a fixed path. Again, one direction would be measured in terms of positive velocity and the other in negative velocity. *Angular velocity* is the rate of change in angular displacement. The most common unit of angular displacement is the revolution (360°) and angular velocity is usually measured in revolutions per minute (rpm). It may also be measured in radians per second (rad/s). There are 6.28 (2 × π) radians in 1 revolution. Linear velocity is often converted to angular velocity by an arrangement such as that shown in Fig. 9-11. Velocity information can be extracted from this arrangement by timing the sensor pulses.

High velocities are ordinarily measured by timing the period required for an object to travel from one fixed point to another. Optical detectors and timing circuits are used to make the measurement. For example, an object passes one point, where it interrupts a light beam and the timing circuit starts. Later, a second point is passed, where a second light beam is interrupted and the timing circuit stops. Very high accuracy can be achieved with such an arrangement if the distance between the two points is accurately measured and if the timing circuit is precise. Modern counting circuits can easily resolve millionths of a second. Angular velocity can also be measured by optical techniques. Figure 9-12 shows an optical tachometer that can be used to measure shaft, gear, or pulley velocity. A contrasting stripe of paint or tape is applied to the rotating part. The tachometer has a light source and lens assembly. The light reflected back into the photo detector is alternately brighter and darker because of the contrasting stripe. The detector produces one pulse for every shaft revolution. Counting and timing circuits convert the pulses to angular velocity.

Direct current tachometers are also popular for measuring angular velocity. They are usually equipped with a permanent magnet field, a rotating armature circuit, and a commutator and brush assem-

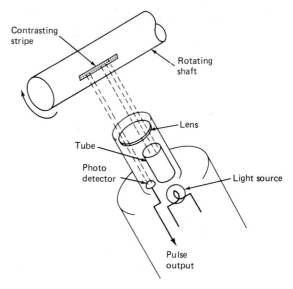

Fig. 9-12 Optical tachometer.

bly. Ignoring loading effects, the output of a dc tachometer is proportional to the flux density of the field, the length of the armature circuit, and the angular velocity of the shaft. Since flux and the armature circuit are fixed for any given tachometer, the manufacturers rate them with an output constant of volts per rpm. Typical dc tachometers produce outputs from 0.01 to 0.02 V/rpm. They produce a polarity reversal when rotated backward. This is a necessary feature when the tachometer must also provide direction information to the system.

Direct current tachometers, because they have brushes, suffer from noise and maintenance problems. Alternating current tachometers have been developed to eliminate these problems. They employ a permanent magnet rotor and a polyphase stator. The output is ac, with both frequency and voltage proportional to the angular velocity of the input shaft. A frequency-to-voltage converter circuit is sometimes used to convert the output to a dc signal. Or the stator output can be rectified to produce a dc signal. Neither technique will produce a polarity change if the input shaft changes direction, however.

The *drag-cup tachometer* works on an induction principle. It is also called an *ac induction tachometer.* Figure 9-13 shows its construction. One set of stator coils is connected to an external excitation source. The other stator coils are positioned 90° from the excited coils and form the output circuit. The drag cup is made of copper or aluminum and is connected to the input shaft. The side view shows that there is also a laminated inner core. The drag cup has a clearance gap between this core and the stator poles. The theory of operation is shown in Fig. 9-14. With no rotation, the output is zero. An eddy current flux is set up in the drag cup, but it is at 90° to the output coils. If the shaft is turned, a second flux appears in the drag cup. It is caused by an armature reaction current flowing in the drag cup and is at

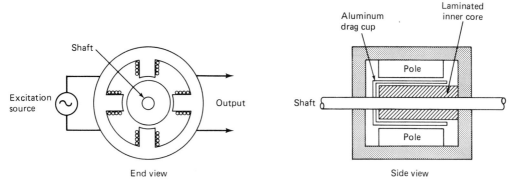

Fig. 9-13 Drag-cup tachometer.

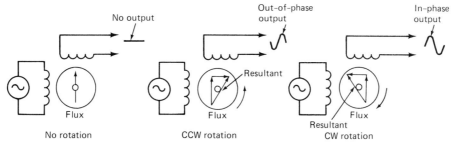

Fig. 9-14 Drag-cup tachometer operation.

right angles to the exciter coils. The resultant flux produces an output. The faster the shaft turns, the greater the reaction flux and the better the angle of the resultant for output. Therefore, the output is proportional to shaft velocity. If the shaft is reversed, the reaction flux reverses, the resultant angle shifts, and the phase of the output changes. With the appropriate phase-detector circuitry, the drag-cup tachometer can provide shaft direction information.

A linear tachometer produces an output that varies smoothly and continuously with angular velocity. For cxample, if the output is 10 V at 1000 rpm, it will ideally be 10.5 V at 1050 rpm and 11 V at 1100 rpm. Digital tachometers produce a fixed number of output pulses for every shaft revolution. The optical tachometer presented earlier produced one pulse per revolution. In Fig. 9-15 three pulses are produced for every revolution because there are three slots in the rotating disk. It is possible to sense direction with this type of tachometer by using two sets of staggered, overlapping slots and a dual LED/phototransistor assembly. The overlap provides a point where both beams will be on. Then one beam will be interrupted first, depending on the direction of rotation. Magnetic digital tachometers are also popular. These operate on the Hall-effect principle or the variable-reluctance principle.

Acceleration is the rate of change in velocity. An object that is accelerating is increasing its velocity with time. In *deceleration,* or negative acceleration, the object is losing velocity with time. It is measured in units of displacement per time per time. For example, gravity will cause a falling object to accelerate

at 981 cm/s/s. This is often written as 981 cm/s^2 and is known as the gravitational constant *g*. A signal proportional to acceleration (or deceleration) can be obtained by differentiating the output of a velocity transducer. Likewise, the output of an accelerometer can be integrated to provide velocity information. Differentiators and integrators were covered in Chapter 6. Or acceleration can be measured by another indirect means. Newton's law gives us $F = ma$ (force equals mass times acceleration). If the mass is known, it is possible to measure a displacement produced by the force and derive the acceleration. Fig. 9-16 shows the basic *accelerometer,* in

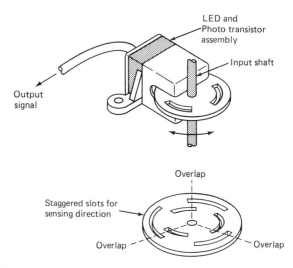

Fig. 9-15 Optical tachometer.

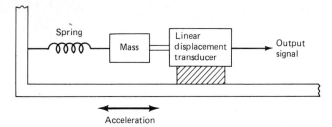

Fig. 9-16 Basic accelerometer.

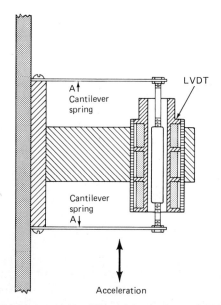

Fig. 9-17 Linear variable differential transformer (LVDT) accelerometer.

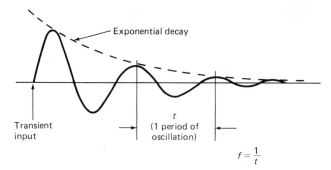

Fig. 9-18 Damped sine wave.

which a linear displacement transducer is mechanically coupled to a spring-loaded mass. Acceleration along the axis shown will produce a reaction force on the mass. Spring tension will allow more or less displacement of the mass, depending on the amount of force created by the acceleration. The output signal from the displacement transducer will be proportional to acceleration.

An LVDT accelerometer is shown in Fig. 9-17. The LVDT core and spring assembly form the mass. The cantilever springs resist core motion. Core displacement will be proportional to acceleration. Deceleration will cause an opposite core displacement and can therefore also be measured. If an accelerometer such as the one in Fig. 9-17 is set into motion and then suddenly stopped, it will oscillate at its natural resonant frequency. In fact, any transient mechanical input will cause oscillations in a spring-mass system. The oscillations are typically around 60 Hz for an LVDT-type accelerometer.

Figure 9-18 shows the damped sine wave produced by a spring-mass system. The natural resonant frequency is the reciprocal of one period of oscillation. The system continues to oscillate, but friction eventually brings it to a rest. This friction is called the *damping coefficient*. The greater the damping coef-

ficient, the more quickly the system will recover from transients. Some accelerometers use a viscous material such as oil to increase the damping coefficient. However, too much damping will make the response sluggish, and the transducer will not accurately track rapid changes in acceleration. Resonant frequency and the damping coefficient are very important specifications in an accelerometer. It has been found that when acceleration is changing with time (which is normally the case) an accurate response cannot be obtained from an accelerometer unless its resonant frequency is at least several times the frequency of the acceleration change. It is also important that the mass of the accelerometer not appreciably change the mechanical response of the system to which it is connected.

Figure 9-19 shows a piezoelectric accelerometer that lends itself to the very rapid accelerations involved in measuring shocks and vibrations. These

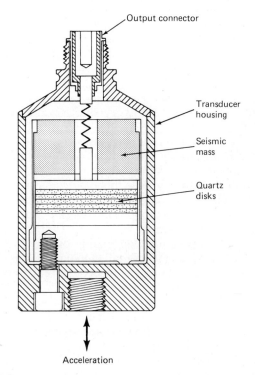

Fig. 9-19 Piezoelectric accelerometer.

transducers have natural resonant frequencies up to 40 kHz or so. The surfaces of piezoelectric materials, such as quartz, become charged when the materials are mechanically stressed. The seismic mass is supported by the transducer housing through the quartz disks. Preloading allows the accelerometer to respond to both negative and positive acceleration. The mass can be made high for large-transducer sensitivity or low for high-resonant frequency. A typical high-sensitivity unit will have a range of plus or minus $100g$ (g is the earth's gravitational constant given earlier) and a resonant frequency of 2 kHz. A low-sensitivity unit has a typical range of $-20,000g$ to $+50,000g$ with a resonant frequency of 40 kHz.

Angular acceleration is measured in revolutions per time per time (rev/t^2) or radians per time per time (rad/t^2). Angular acceleration is measured indirectly with an angular displacement transducer. The transducer output goes to a computer, where timing and position information can be combined to calculate acceleration and deceleration. Sophisticated machine tools and robots achieve very precise control of motion and position by using these techniques.

REVIEW QUESTIONS

8. What happens to the output of a dc tachometer when its shaft is turned in the opposite direction?

9. Why might it be said that ac tachometers are angular speed indicators rather than angular velocity indicators?

10. What is the advantage of the drag-cup tachometer over the ac tachometer?

11. When the output of a velocity transducer is differentiated, a signal proportional to _____ is the result.

12. The basic accelerometer in Fig. 9-16 measures _____ directly. The calibrated spring allows this to be interpreted as a _____ measurement, and the calibrated mass allows the output to be expressed in units of _____.

13. Refer to Fig. 9-18. The period of one oscillation cycle is 1.5 ms. What is the resonant frequency?

14. What would a larger damping coefficient achieve in Fig. 9-18?

9-3
FORCE AND FLOW

Force is measured in newtons (N); 1 N is equal to 0.225 lb. Small forces may be measured in dynes (dyn); 1 N is equal to 100,000 dyn. Force transducers are often based on displacement principles. For example, refer to Fig. 9-20, which shows a force-measuring device based on a compression spring and an

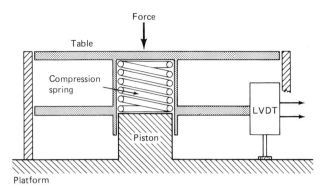

Fig. 9-20 Measuring force.

LVDT. When a force is applied to the table, the spring compresses in proportion to the amount of force. The body of the LVDT moves down with the table. The core of the LVDT is fixed to the platform. The greater the force, the more the relative displacement of the LVDT core. Devices such as the one shown in Fig. 9-20 are often called *load cells*.

Figure 9-21 shows how force can be converted to strain and a corresponding resistance change for measurement purposes. The cantilever beam assumes a semicircular shape because of the applied force. The top surface of the beam elongates, and the bottom surface compresses. The equation shows that the stress at a given point is directly proportional to the force magnitude and force distance and is indirectly proportional to the beam width and the square of the beam thickness. With all distance measurements in meters and the force in newtons, the stress will be found in units of newtons per meter. It is also necessary to know the modulus of elasticity for the beam material in newtons/square meter. This allows the strain in meters per meter to be calculated. Once this is accomplished, the resulting increase in the resistance of the strain gage can be calculated.

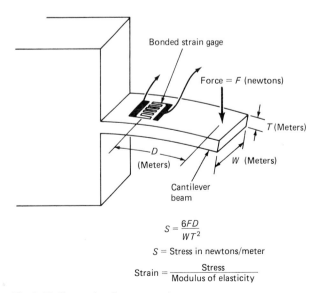

$$S = \frac{6FD}{WT^2}$$

S = Stress in newtons/meter

$$\text{Strain} = \frac{\text{Stress}}{\text{Modulus of elasticity}}$$

Fig. 9-21 Converting force to strain.

EXAMPLE

A beam is 2 cm wide and 0.25 cm thick. With a force of 100 N, calculate the stress on the beam at a point 10 cm from the point where the force is applied.

SOLUTION

$$S = \frac{6\,FD}{WT^2}$$

where S = stress
F = force
D = distance between strain gage and point of application of force
W = width
T = thickness

$$= \frac{6(100)(0.10)}{(0.02)(0.0025)^2}$$
$$= 4.8 \times 10^8 \ \text{N/m}$$

If the beam is made of steel with a modulus of elasticity of 2×10^{11} N/m², the strain can be found by

$$\text{Strain} = \frac{S}{\text{modulus of elasticity}}$$
$$\text{Strain} = \frac{4.8 \times 10^8 \ \text{N/m}}{2 \times 10^{11} \ \text{N/m}^2}$$
$$= 2.4 \times 10^{-3} \ \text{m/m}$$

Now, if we assume that a 120-Ω strain gage with a gage factor of 2 is mounted at the strain point, the increase in resistance due to the 100-N force may be found by

$$\Delta R = \text{GF} \times \text{strain} \times \text{resistance}$$
$$= 2 \times 2.4 \times 10^{-3} \times 120$$
$$= 0.576 \ \Omega$$

A tension load cell is illustrated in Fig. 9-22. It can be used to measure the force required to pick up heavy loads in industry. The metal proving ring will elongate along the tension axis. Four strain gages are placed around the proving ring to sense the changes. The four gages are electrically arranged in the bridge configuration. Tension will slightly distort the proving ring and elongate gages R_2 and R_4, while R_1 and

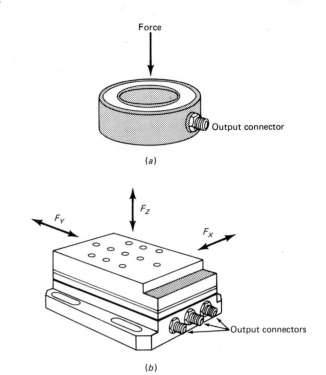

Fig. 9-23 Piezoelectric force transducers. *(a)* Load washer. *(b)* Three-component dynamometer.

R_3 will be slightly compressed. The bridge is arranged to produce the maximum change in output voltage for these changes. Temperature effects are minimized by the bridge arrangement since all four gages tend to track thermally.

Two examples of piezoelectric force transducers are shown in Fig. 9-23. The load washer type shown in Fig. 9-23(*a*) is designed to measure axial forces. It is preloaded when manufactured and can measure both tensile and compressive forces. Load washer force transducers are available with ratings from 7 kN to 1 MN. The three-component dynamometer type shown in Fig. 9-23(*b*) measures three orthogonal (right-angle) components of force. It has a natural resonant frequency of about 4 kHz and measures a

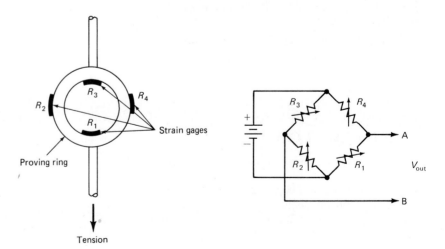

Fig. 9-22 Tension load cell.

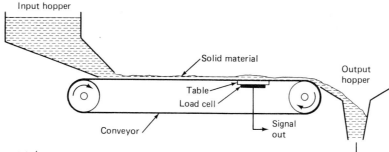

Fig. 9-24 Solid flow measurement.

maximum force of plus or minus 5 kN along any of the three axes. Typically, the work piece is mounted on the dynamometer to provide signals proportional to the forces involved in grinding and milling operations. Or the tool can be mounted on the dynamometer to measure the cutting forces produced in turning operations. Piezoelectric transducers do not have a dc frequency response and are not useful for monitoring steady-state conditions.

Force transducers can be applied in the measurement of solid flow. The units may be in kilograms/minute or in some other form; Fig. 9-24 shows one arrangement. A load cell is placed under a conveyor table to measure the force exerted by the flowing material. Another transducer, not shown, provides a signal proportional to conveyor velocity. A circuit or a computer is used to multiply the two transducer signals because the flow is directly proportional to the product of the weight and the velocity. Another technique that is used is to measure the sag of the conveyer with a displacement transducer such as an LVDT. The sag will be directly related to the weight of the material being carried. Again, when these data are combined with the velocity, flow is the result.

Fluid (liquid and gas) flow is measured in other ways. One popular way is to use some type of a flow restriction to create a differential pressure. *Bernoulli's principle* states that as the velocity of a fluid increases, its pressure decreases. Likewise, as the velocity of a fluid decreases, its pressure increases. When a fluid flows through a restriction, its velocity must increase and its pressure must drop. This is the principle behind many flowmeters. Figure 9-25 shows a venturi differential pressure flowmeter. The *venturi* is that part of the pipe where the passage necks down. This reduced area forces the fluid to increase

in velocity through the venturi. Note that two pressure tubes connect bellows to two points: P_1 and P_2. Point P_1 is at a point before the venturi, and P_2 is in the venturi. When fluid is flowing, P_1 will be greater than P_2, and the pressure difference is proportional to the flow. The bellows extend in proportion to pressure. When P_1 is greater than P_2, the core in the LVDT will move to the right in Fig. 9-25. If the flow is increased, the core will move farther to the right. If the flow stops, there will be no Bernoulli effect and the core will be centered.

Some alternatives to the venturi tube are illustrated in Fig. 9-26. The orifice plate provides the necessary restriction to flow. This construction is less expensive than the venturi type. However, it sets up flow turbulence in the pipe and is less accurate. The nozzle type is a compromise between the venturi-tube and orifice-plate flowmeters. It produces less turbulence and subsequent pressure loss than the orifice type but is not quite so efficient as the venturi tube. The Dall tube has the least turbulence and insertion loss. However, it cannot be used

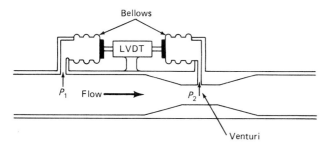

Fig. 9-25 Differential pressure flowmeter.

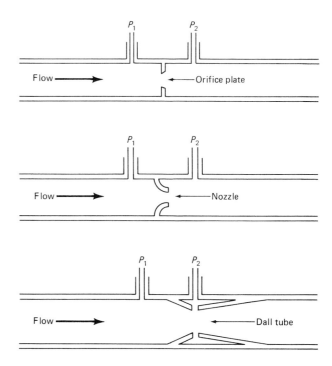

Fig. 9-26 Other differential pressure flowmeters.

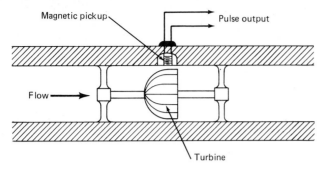

Fig. 9-27 Turbine flow measurement

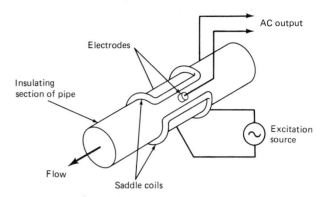

Fig. 9-29 Magnetic flowmeter.

with slurries, as the venturi tube can. A *slurry* is a suspension of solid particles in a liquid carrier such as water.

The turbine flowmeter shown in Fig. 9-27 is a very linear and accurate flow transducer. The fluid turns the turbine as it moves past the blades. The faster the flow, the greater the speed of turbine rotation. The blades of the turbine are magnetized, and pulses are induced in the magnetic pick-up coil located in the wall of the tube. These transducers are expensive and cannot be used with slurries.

Positive displacement flowmeters divide the flow into measured units and are noted for good accuracy. The *nutating disk flowmeter* is shown in Fig. 9-28. Fluid enters the left side of the housing. It then works its way into the left chamber of the inner housing, where it applies pressure to the disk. The disk wobbles (nutates) and releases measured amounts of flow into the right chamber and on to the exit pipe. As the disk nutates, it turns the top shaft and drives a pulse generator located in the top housing. The pulse rate is proportional to flow.

All of the flowmeters presented to this point offer some restriction to flow. Only one of them, the venturi type, lends itself to measuring slurry flow. Figure 9-29 shows a magnetic flowmeter that can be used with electrically conducting fluids and slurries and offers no restriction to flow. It is based on the principle of voltage induction in a conductor that is mov-

ing through a magnetic field. The faster the flow, the greater the induced voltage at the electrodes. The saddle coils provide an ac field. Direct current field flowmeters are also used but tend to cause polarization at the electrodes when used with fluids with low conductivity. Alternating current excitation also has limitations caused by dielectric losses and direct pick-up at the electrodes. In either case, the output signals are usually in the microvolt range and require quite a bit of amplification. Magnetic flowmeters are typically the most expensive to apply.

Other flowmeters use moving vanes or metering floats to respond to flow. They may only provide a visual indication or may be coupled to a potentiometer or an LVDT. Phototransistors have also been used to provide an electrical interface. Finally, temperature-sensing devices have also been used to measure flow. If the fluid is colder than a sensor, then the flow will remove heat from the sensor. The section on temperature transducers includes a simple circuit that detects air flow in this manner.

REVIEW QUESTIONS

15. Refer to Fig. 9-21. For what purpose might a second, identical gage be mounted next to the first gage, but with its axis of sensitivity arranged perpendicular to the first gage?

16. Refer to Fig. 9-21. Suppose the beam is 3 cm wide and 0.15 cm thick, and that a force of 5 N is applied 8 cm from the gage. What is the stress at the gage?

17. Refer to question 16. Assume that the beam is made from aluminum with a modulus of elasticity of 6×10^{10} N/m^2. What is the strain at the gage?

18. Refer to questions 16 and 17. The gage factor is 2, and the gage resistance is 120 Ω. Calculate the increase in resistance due to the 5-N force.

19. Refer to Fig. 9-22. Assume a null bridge with zero tension. What will the polarity of A with reference to B be when the proving ring is loaded?

20. What other type of transducer or information is required in Fig. 9-24 to calculate flow?

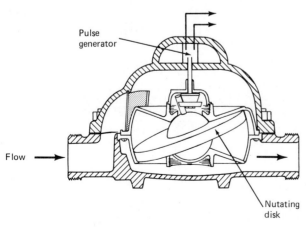

Fig. 9-28 Nutating disk flowmeter.

21. Refer to Fig. 9-25. The direction of flow is as shown. Which pressure is greater?

9-4
PRESSURE AND LEVEL

Pressure is defined as force per unit area. Unfortunately, there are many ways to measure pressure, and many units have evolved. The international system of units has come to the rescue and established the *pascal* (Pa) as the standard unit of pressure; 1 Pa of pressure is defined as a force of 1 N applied over an area of 1 m^2. However, standards are sometimes adopted slowly in industry, and the technician who works with pressure transducers will find several measurement units with which to cope. Industrial pressures are often measured in pounds per square inch (psi). Pascals can be obtained by multiplying psi by 6.8948×10^3.

There are three different reference conditions for pressure measurements. Gage pressure is referenced to atmospheric pressure, which is 14.70 psi (101 kPa) at sea level. Gage pressure will change with altitude. Absolute pressure is referenced to a perfect vacuum and does not change with altitude. Gage pressure can be obtained by subtracting the ambient atmospheric value from the absolute value. Likewise, absolute pressure can be found by adding the ambient pressure to the gage pressure. Differential pressure is referenced to an arbitrary value. A *g*, *a*, or *d* may be suffixed to a pressure measurement to clearly denote gage pressure, absolute pressure, or differential pressure.

The most common way to measure pressure is to use a force-summing device to convert the pressure into a displacement. Any of the displacement transducers already discussed can then provide the output signal. At very high pressures, the force-summing device may be eliminated and the pressure directly applied to a sensor based on the piezoresistive or piezoelectric effects. Figure 9-30 shows some examples of Bourdon tubes that are used to convert pressure into a proportional change in displacement. In Fig. 9-30(*a*) the cross section of a Bourdon tube is shown. Bourdon tubes are made of metal, such as steel, phosphor bronze, or brass. Figure 9-30(*b*) and 9-30(*c*) shows the circular and spiral shapes often used for these tubes. When pressure is applied, the tube tends to straighten, and a displacement results. The helical Bourdon tube shown in Fig. 9-30(*d*) produces a rotary motion when pressure is applied. Figure 9-31 shows a circular Bourdon tube coupled to an LVDT. When pressure is applied, some of the spring tension is overcome, and the core moves up.

Other force-summing devices used in pressure transducers include the bellows type and the diaphragm type. The *bellows type,* shown in Fig. 9-32(*a*), is a pressure cylinder with a thin corrugated metal wall. The bellows expand with pressure, producing a displacement to the right. Bellows can be

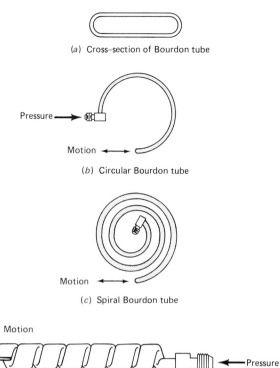

(*a*) Cross-section of Bourdon tube

(*b*) Circular Bourdon tube

(*c*) Spiral Bourdon tube

(*d*) Helical Bourdon tube

Fig. 9-30 Bourdon tubes.

made more sensitive than Bourdon tubes and are used at lower pressures. An auxiliary spring can be added to allow the bellows to be used at higher pressures. A pair of bellows makes a good arrangement for differential pressure measurements. (This is illustrated in Fig. 9-25.) A diaphragm force-summing device is shown in Fig. 9-32(*b*). The *diaphragm* is a flexible plate and can be made from rubber, neoprene, metal, or corrugated metal. A pair of dia-

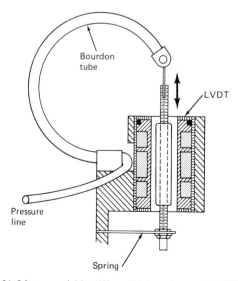

Fig. 9-31 Linear variable differential transformer (LVDT) pressure transducer.

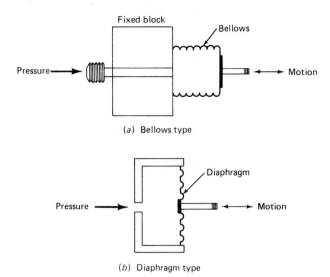

Fig. 9-32 Bellows and diaphragm pressure transducers.

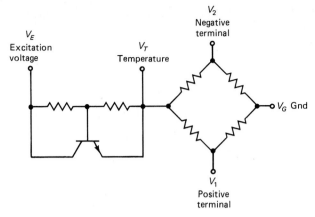

Fig. 9-34 Schematic diagram of LX04XXA monolithic pressure transducer (National Semiconductor).

phragms can be employed for differential pressure measurements.

Figure 9-33 shows a *strain gage pressure transducer*. The gage is bonded to a diaphragm. Pressure flexes the diaphragm, stretching the gage, and its resistance increases. The reference side of the diaphragm may be a perfect vacuum. If it is, the measured pressure will be absolute. Or a second pressure port may enter the reference side. If this is done, the output will be referenced to the second pressure and will be called *arbitrary* or a *differential pressure*. Finally, the reference side of the diaphragm can be vented to the atmosphere. Pressures measured this way are known as *gage pressures*.

Figure 9-34 shows the schematic diagram for a National Semiconductor LX04XXA monolithic pressure transducer. These units are actually piezoresistive integrated circuits. They provide an output voltage proportional to applied pressure. They also supply a separate temperature-dependent output that can be used to temperature compensate the transducer. Model LX0420A is rated to 100 pounds per square inch absolute (psia) and provides an output sensitivity of 0.2 to 0.8 mV/psi. It has a natural resonant frequency of 100 kHz. Other models are rated to 1000 psia and 3000 psia.

A *piezoelectric pressure transducer* uses a dia-

phragm to transmit pressure to a quartz element. An output voltage proportional to pressure is the result. Piezoelectric transducers are noted for high-frequency performance and can be used to monitor rapidly changing pressures. They are not dc sensors and cannot be used for static pressure measurements.

Sudden pressure spikes can be damaging to some transducers. These transients are caused by pumps and valves and by resonance effects. Snubbers are sometimes used with pressure transducers to prevent transient damage. A *snubber* amounts to a restriction in a pressure port that prevents sharp increases or decreases in the pressure applied to the transducer. Snubbers are effective in preventing transducer damage, but they do limit the high-frequency performance of the measuring system.

Liquid level measurements can be accomplished indirectly with pressure transducers. Figure 9-35 shows such an application. The pressure at the transducer is proportional to the liquid density, the level above the transducer, and gravity. Note that the top of the tank is open (vented to the atmosphere). Suppose a tank contains water which has a density of

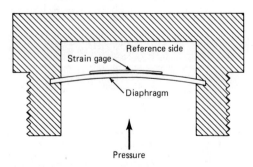

Fig. 9-33 Strain gage pressure transducer.

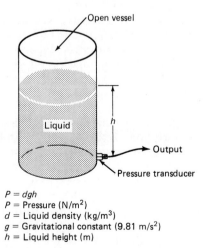

$P = dgh$
P = Pressure (N/m^2)
d = Liquid density (kg/m^3)
g = Gravitational constant (9.81 m/s^2)
h = Liquid height (m)

Fig. 9-35 Level measurement by static pressure.

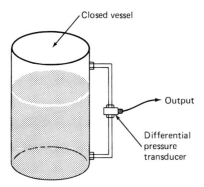

Fig. 9-36 Level measurement by differential pressure.

1000 kg/m^3 and that the level is 10 m above the pressure sensor. The pressure P at the sensor is found by

$$
\begin{aligned}
P &= dgh \\
&= 1000 \times 9.81 \times 10 \\
&= 98{,}100 \text{ N/m}^2 \\
P &= 98.1 \text{ kPa (14.22 psi)}
\end{aligned}
$$

You might wonder why the pressure is indicated in units of newtons per square meter. Recall that 1 N is defined as the force required to accelerate a 1-kg mass 1 m/s^2.

If the vessel holding the liquid is closed, a different technique is used to measure level. The height of the liquid above the pressure transducer is proportional to the difference between the pressure at the top of the tank and the static pressure at the transducer. A differential pressure transducer applied to a closed vessel is illustrated in Fig. 9-36.

Figure 9-37 shows a tank supported by load cells. The force on the load cells is proportional to the level in the tank. A summing amplifier may be used to add the signals from the individual load cells. Figure 9-38 illustrates a capacitive probe inscrtcd into the tank. The liquid is an insulator with a dielectric constant that is different from the constant for the air or gas above the liquid. This produces a capacitance change in the probe as the level changes. The capacitive reactance of the probe also changes;

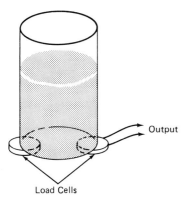

Fig. 9-37 Using load cells to measure level.

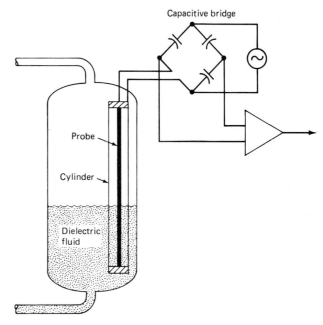

Fig. 9-38 Capacitive level measurement.

the ac bridge produces an output signal that is proportional to level. In addition to load cells and capacitive probes, there are various techniques used to convert level to displacement. Floats, levers, cables, and pulley systems are employed, and the resulting displacement can be changed into an electrical signal with potentiometric transducers or LVDTs.

REVIEW QUESTIONS

22. A force of 1 N applied to an area of 1 m^2 is defined as a pressure of 1 _____.

23. The three ways to reference pressure are absolute, gage, and _____.

24. List three force-summing devices used in pressure-measuring transducers.

25. Refer to Fig. 9-33. The strain gage environment is a perfect vacuum. What type of pressure does the transducer measure?

26. Refer to Fig. 9-35. The tank contains water. The transducer registers a static pressure of 4 psi. How far above the transducer is the top of the water?

27. The top of a tank is sealed. What type of pressure measurement will be appropriate for evaluating the liquid level in the tank?

9-5
TEMPERATURE

There are three temperature scales widely used in industry: Fahrenheit, Celsius, and Kelvin. The *Fahrenheit* scale is the oldest and dates back to the early 1700s. It originally used the freezing point of water and the temperature of human blood as its two reference points. About 40 years later, Celsius proposed that the melting point of ice and the boiling point of water be used as reference points. His system became known as the *centigrade scale,* and in 1948 the name was officially changed to the *Celsius* scale. Lord Kelvin first proposed the concept of absolute zero in the early 1800s. His scale uses 0° to represent absolute zero. The conversions for the three scales are as follows:

$$C = 5/9(F - 32)$$
$$F = 9/5C + 32$$
$$K = C + 273.15$$

where *C, F,* and *K* are the Celsius, Fahrenheit, and Kelvin temperatures, respectively. The *Rankine scale* also finds some application in industry. It is the Fahrenheit equivalent of the Kelvin scale. Degrees Rankine may be found by adding 459.67 to the Fahrenheit value.

Temperature can be sensed in many ways. A metal tube can be filled with liquid and connected to a Bourdon tube. The liquid will expand as heat is applied, and the Bourdon tube will provide a displacement proportional to temperature. Bimetallic strips can also be used to provide a displacement that is proportional to temperature. A displacement transducer can be added to the Bourdon tube or bimetallic strip to provide an electrical output. However, it is usually easier to use a sensor that directly converts temperature into an electrical signal. Figure 9-39 shows the four common temperature sensors. The *thermocouple* produces an output voltage directly related to temperature. The *resistance temperature detector* (RTD) shows an increase in resistance with temperature. The *thermistor* has an opposite response, in that its resistance decreases with a temperature increase. Finally, the *integrated circuit sensor* produces a voltage or a current signal that increases with temperature increases. This section will deal with these four devices.

Some industrial thermocouples are shown in Fig. 9-40. They are based on the junction of two dissimilar metals. When the junction is heated, a voltage is generated; this is known as the *Seebeck effect.* The Seebeck voltage is linearly proportional for small changes in temperature. Various combinations of metals are used in thermocouples. Thermocouples are sometimes connected in series to provide higher output and better sensitivity. The series arrangement is known as a *thermopile.* Type E thermocouple units use chromel alloy as the positive electrode and constantan alloy as the negative electrode. Type S thermocouples produce the least output voltage but can be used over the greatest temperature range.

Figure 9-41 shows a type T thermocouple, which uses copper and constantan. Copper is an element and constantan is an alloy of nickel and copper. The copper side is positive with respect to the constantan side. Assuming that copper wires will be used to connect the thermocouple to the next circuit, a second copper-constantan junction is unavoidable as the illustration shows. This second junction is called the *reference junction.* It generates a Seebeck voltage that opposes the voltage generated by the sensing junction. If both junctions are at the same tempera-

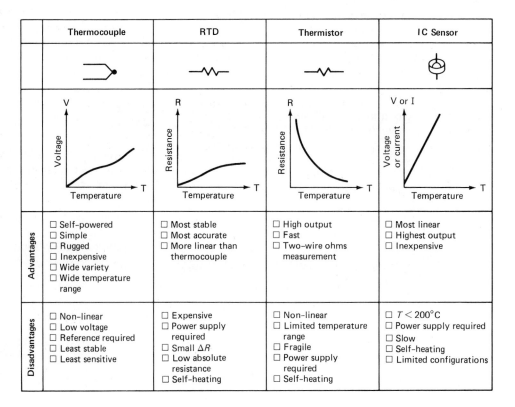

	Thermocouple	RTD	Thermistor	IC Sensor
	V / Voltage vs Temperature	R / Resistance vs Temperature	R / Resistance vs Temperature	V or I / Voltage or current vs Temperature
Advantages	□ Self-powered □ Simple □ Rugged □ Inexpensive □ Wide variety □ Wide temperature range	□ Most stable □ Most accurate □ More linear than thermocouple	□ High output □ Fast □ Two-wire ohms measurement	□ Most linear □ Highest output □ Inexpensive
Disadvantages	□ Non-linear □ Low voltage □ Reference required □ Least stable □ Least sensitive	□ Expensive □ Power supply required □ Small ΔR □ Low absolute resistance □ Self-heating	□ Non-linear □ Limited temperature range □ Fragile □ Power supply required □ Self-heating	□ $T < 200°C$ □ Power supply required □ Slow □ Self-heating □ Limited configurations

Fig. 9-39 Four common temperature sensors (Omega Engineering).

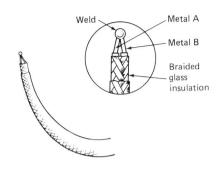

Fig. 9-40 illustration labels: Weld, Metal A, Metal B, Braided glass insulation

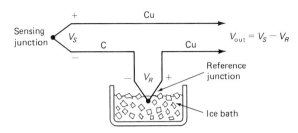

Fig. 9-42 Cold-junction compensation.

Fe = Iron

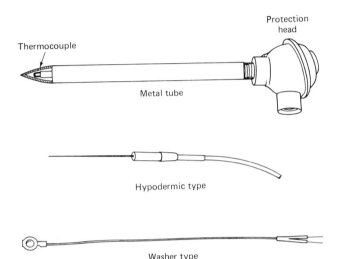

Fig. 9-40 Industrial thermocouples.

Fig. 9-43 Four-junction circuit.

ture, V_{out} will be zero. If the sensing junction is at a higher temperature, V_{out} will be proportional to the differences between the two junction temperatures. The problem is that the temperature cannot be derived directly from the output voltage alone. It is subject to an error caused by the voltage produced by the reference junction.

One solution to the problem is shown in Fig. 9-42. The reference junction is placed in an ice bath to keep it at a known temperature. This process is known as *cold junction compensation*. The reference junction is maintained at 0°C, and the reference voltage is now predictable from the calibration curve of the type T thermocouple. The reference voltage is subtracted from V_{out}, and the temperature of the

sensing junction is found from the calibration curve. When copper is not one of the thermocouple metals, the four-junction circuit of Fig. 9-43 results. The type J thermocouple uses iron and constantan. When it is connected to copper wires, two iron-copper junctions result. These junctions present no additional compensation problems, however, because of the isothermal block. This block is made of a material that is a poor conductor of electricity but a good conductor of heat. Both iron-copper junctions will therefore be at the same temperature and generate the same Seebeck voltage. Note that these two voltages will cancel. Also notice that cold-junction compensation is used at the reference junction.

It is obvious that ice baths are not the most convenient way to compensate the reference junction. This technique is used in the calibration laboratory. The industrial environment demands a different approach; Fig. 9-44 shows one possibility. The isothermal block contains two reference junctions and a thermistor. The resistance of the thermistor is a function of temperature. A circuit is used to sense

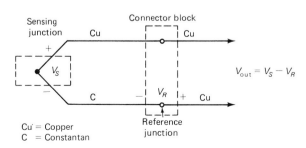

Cu = Copper
C = Constantan

Fig. 9-41 Thermocouple structure.

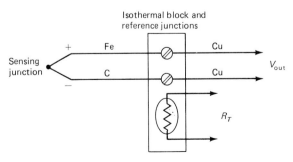

Fig. 9-44 Sensing the reference junctions.

this resistance and to compensate for the voltage introduced by the two reference junctions. This arrangement is sometimes called *electronic ice point reference*. If the sensor is interfaced to a computer, the reference temperature will be converted to a reference voltage and then subtracted from V_{out}. This process is known as *software compensation*. The question now is why bother with any of these processes since the thermistor appears to be capable of sensing absolute temperature with no compensation problems? Thermocouples are useful over a much wider temperature range than the other three sensors. They can be optimized for various atmospheres and are rugged and inexpensive. They lend themselves to monitoring a large number of locations. An isothermal block with one temperature sensor can provide compensation for several units. The term *zone block* is often used in this application. A scanner circuit using reed relays selects one junction from the zone block at a time. If software compensation is used, the individual thermocouples do not have to be of the same type. Different correction voltages for the various metals will be stored in computer memory. The outstanding advantages of thermocouples outweigh their disadvantages for many industrial applications.

Metals exhibit a *positive temperature coefficient;* their resistance increases with temperature. This effect is exploited in resistance temperature detectors (RTDs). They are usually made from platinum, which can maintain its stability at high temperatures. Figure 9-45 shows several styles of platinum RTDs. The glass-encapsulated type is bifilar-wound with platinum wire on a glass or ceramic bobbin. This type of

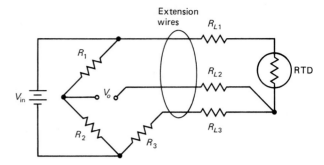

Fig. 9-46 Three-wire bridge circuit.

winding reduces magnetic pick-up, and a sensor less susceptible to electrical noise results. The assembly is then sealed with molten glass. The film types are manufactured with a platinum film on an alumina substrate. If the film is screen-deposited, it is a *thick-film* type. If it is vapor-deposited, a *thin-film* type results. The film types are less costly, yet are as accurate as the wire-wound types. They can be made very small, making their response time faster as a result of the low thermal mass.

Platinum RTDs are available from 10 Ω to several thousand ohms. The most popular value is 100 Ω at 0°C. Platinum's temperature coefficient is +0.00385. Thus, the typical RTD will increase its resistance by 0.385 Ω/°C. This small change in resistance demands the accuracy and sensitivity of a bridge circuit. The sensor is usually mounted away from the bridge so that the bridge resistors are not subjected to a temperature that would cause them to drift or to be damaged. The extension wires are a source of error because of their resistance. They also exhibit a positive temperature coefficient. The effects of the extension wires can be minimized with the three-wire bridge circuit of Fig. 9-46. The resistance of the extension wires is represented by R_{L1} and R_{L2}. If the wires are matched in length and material, their effects are canceled because each is in an opposite leg of the bridge. Lead R_{L3} is a sensor lead and carries little or no current. Therefore, its resistance is not a source of error. Unfortunately, the three-wire bridge circuit creates a nonlinear relationship between resistance change of the RTD and output voltage. For this reason, the four-lead circuit of Fig. 9-47 is pre-

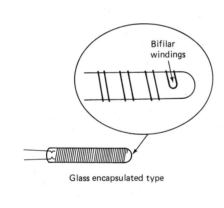

Glass encapsulated type

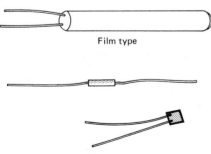

Film type

Miniature thin–film type

Fig. 9-45 Resistance temperature detectors (RTDs).

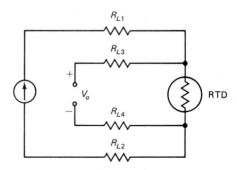

Fig. 9-47 Resistance temperature detector four-lead circuit with current source.

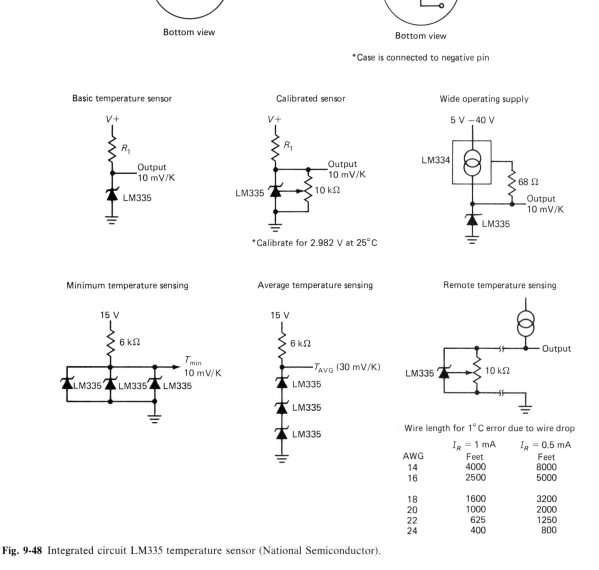

Fig. 9-48 Integrated circuit LM335 temperature sensor (National Semiconductor).

ferred for those applications requiring greatest accuracy. A constant current source supplies the RTD. The resistances of R_{L1} and R_{L2} are no longer a factor since the current will be held constant. Leads R_{L3} and R_{L4} are the sense leads. Their resistance is also not a factor, assuming that V_O is applied to a very high input impedance circuit such as an instrumentation amplifier.

A *thermistor* is a negative-coefficient sensor made from semiconducting material. Oxides of titanium, iron, and nickel are among the materials used. They are very sensitive, with temperature coefficients that range from -2 to -6 percent/°C. As such, thermistors are capable of detecting minute changes in temperature. Their resistance at 25°C ranges from 100 Ω

to 100 kΩ, with 5000 Ω being a very common value. Their linearity is the poorest of those of all temperature sensors, and they are susceptible to permanent decalibration if exposed to high temperatures. The normal limit is 200°C, but they are also subject to decalibration if operated below, but near, their upper limit for extended periods of time. They can be built as small as 0.1 mm (about 0.005 in.), and their small thermal mass provides a very fast response time. They are available in a wide variety of shapes and sizes. Thermistors are much more delicate than thermocouples and RTDs.

The higher resistance of thermistors makes them less error-prone than RTDs. Lead resistance is not nearly so significant, and a simple two-wire extended

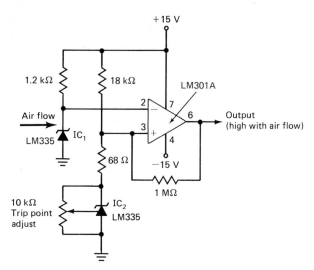

Fig. 9-49 Air flow detector.

bridge connection is usually adequate. However, they are subject to self-heating error. The bridge voltage is usually reduced to a low value to minimize this effect, or a pulsed supply may be used.

Integrated-circuit (IC) temperature sensors eliminate the linearity errors associated with thermistors. However, as semiconductor devices, they exhibit the other limitations. They are available in both voltage- and current-output configurations. Figure 9-48 shows the National Semiconductor LM335 IC temperature sensor. It provides a proportional output of 10 mV/°K. It operates as a two-terminal zener. Its dynamic impedance is less than 1 Ω, and it operates over a current range of 400 μA to 5 mA with virtually no change in performance. When calibrated at 25°C, it typically shows less than 1°C error over a 100° range. Its usable range is −10 to +100°C, and an LM135 is also available with a range of −55 to +150°C.

Figure 9-48 (p. 217) shows how to apply and calibrate the IC temperature sensor. The minimum temperature-sensing circuit is feasible because of the low dynamic impedance of the sensors. The coolest sensor will set the output voltage. The average circuit simply adds the individual output voltages. A simple potentiometer circuit provides one-point calibration. Single-point calibration works because the output of the sensor is proportional to absolute temperature with an extrapolated output of 0 V at 0 K. Errors in output voltage versus temperature are only slope errors. Thus, a slope calibration at one temperature corrects all temperatures.

Self-heating errors can be reduced by operating the IC at the minimum current suitable for the application. Sufficient current must be available to drive the sensor and the calibration pot at the maximum operating temperature. Self-heating can be exploited in some applications. Figure 9-49 shows a detector in which an air flow is directed onto an LM335 sensor. If the flow stops, the sensor temperature increases as a result of self-heating, and its output will go in a positive direction. This voltage is applied to the inverting input of the comparator. The LM301A comparator output will go negative when the threshold set by the voltage divider at its noninverting input is crossed. The second LM335 sensor provides trip point adjustment and adjusts the positive threshold for changes in ambient temperature. It also must be exposed to the air flow. Hysteresis is provided by the 1-MΩ feedback resistor. This circuit does not provide a linear measure of air flow but provides a negative-going output if the air flow stops or falls below the trip point value.

REVIEW QUESTIONS

28. Refer to Fig. 9-43. Ideally, how much error is introduced by the two iron-copper junctions? Why?

29. Refer to Fig. 9-44. The output of the thermistor goes to a computer, which uses stored information to correct for the reference temperature. This technique is known as _____ compensation.

30. Examine Fig. 9-46. Why is R_{L3} not significant?

31. Refer to Figs. 9-46 and 9-47. Which circuit is more accurate?

32. Refer to the calibrated sensor circuit shown in Fig. 9-48. What is its output at 100°C?

33. Refer to Fig. 9-49. The ambient air temperature increases. What happens to the threshold voltage at the positive input of the comparator? What is the net effect of this?

34. Refer to Fig. 9-49. What circuit feature reduces the possibility of multiple output pulses from the comparator as the air flow approaches the trip point?

9-6
MISCELLANEOUS MEASUREMENTS

Humidity is the moisture content of air. *Relative humidity* is the ratio of water vapor pressure in the atmosphere to that of the saturated water vapor pressure of the atmosphere at the same temperature. It is usually expressed in percentage form: 0 percent means there is no water vapor at all in the air, and 100 percent indicates that the air is holding all the water vapor that it can at that temperature. When the relative humidity is 100 percent, any drop in air temperature will initiate condensation of some of the water vapor. The temperature at which this occurs is known as the *dew point*. Relative humidity may also be used to express moisture content in artificial environments, as well as in gas or gaseous mixtures. Relative humidity affects electromagnetic propagation, ballistics, aerodynamics, and many industrial processes.

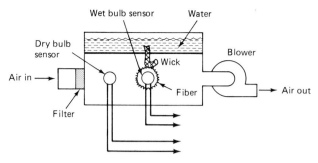

Fig. 9-50 Psychrometer.

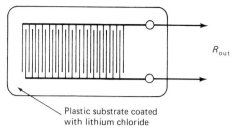

Fig. 9-51 Resistive humidity sensor.

Relative humidity may be measured with a *psychrometer,* as shown in Fig. 9-50. Air, or gas, is drawn into a chamber containing two temperature sensors. Thermistors or IC sensors are suitable for this application. One sensor is dry and measures the air temperature. The other sensor is encased in wet fiber. A wick and a reservoir maintain the wet condition. Water will evaporate from the wet fiber and cool the sensor. With low relative humidity, the water will evaporate quickly, and the wet sensor will achieve a temperature substantially lower than that of the dry sensor. High ambient air temperature also speeds the evaporation and increases the temperature difference. High relative humidity slows the evaporation, producing a lower temperature difference. Table 9-1 shows the relationship between temperature difference, air temperature, and relative humidity. More detailed data will be required for accurate measurements than are shown in this table. The data can be stored in a computer, and, with the proper sensor interface, an automatic measuring system can be realized.

The psychrometer is not a particularly convenient instrument. Hygrometers provide a more simple alternative for measuring relative humidity. Hygroscopic materials that readily absorb moisture from the air are available. For example, human hair is hygroscopic and can be used as a relative humidity sensor. A hair increases in length by about 3 percent when the relative humidity changes from 0 to 100 percent. The hair can be placed under tension, and an LVDT can be used to translate its length into an

electrical signal. Resistance hygrometers also use hygroscopic materials to sense relative humidity. Figure 9-51 shows an example. The plastic substrate holds interlaced foil electrodes and is coated with lithium chloride. As the lithium chloride absorbs moisture from the air, the resistance between the electrodes drops. The performance of a resistance hygrometer is illustrated by Fig. 9-52. The resistance ranges from 80 kΩ at 100 percent to over 300 MΩ at 20 percent relative humidity.

Robotics has increased the interest in *proximity sensors.* Such sensors can help a robot find an object and are also useful for detecting obstructions and human personnel that have entered the work envelope. Figure 9-53 shows a simplified block diagram for an ultrasonic proximity sensor. A burst generator produces seven cycles at a 30-kHz frequency. This frequency is above the human range of hearing and is therefore considered ultrasonic. The burst rate is 2 Hz. The bursts are amplified and applied to an output transducer, where they become ultrasound. The sound waves travel at approximately 340 m/s. The reflected waves arrive at an input transducer and are changed back into an electrical signal. The time (t) between the transmit burst and the receive burst can be used to calculate distance. For example, if the time is 30 ms the total distance traveled is 340 $\times$ 0.03 = 10.2 m. The object that caused the reflection is half that distance away: 5.1 m (16.7 ft). Sound velocity changes with the temperature of the atmosphere. It travels at 331 m/s at 0°C and at 386 m/s at 100°C. A temperature sensor can be used to correct for this effect.

TABLE 9-1 RELATIVE HUMIDITY LOOK-UP TABLE

	Dry Bulb-Wet Bulb Difference, °C					
	0.56	2.78	5.56	8.33	11.1	13.9
4.4	92	60	—	—	—	—
10.0	93	68	38	12	—	—
15.6	94	73	49	26	6	—
21.1	95	77	55	37	20	3
26.7	96	79	61	44	29	16
32.2	96	81	65	50	36	24
37.8	96	83	68	54	42	31

Dry Bulb, °C

Relative Humidity, %

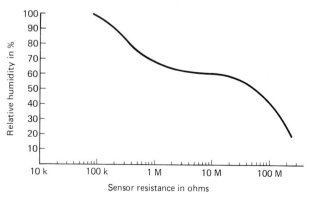

Fig. 9-52 Relative humidity versus resistance.

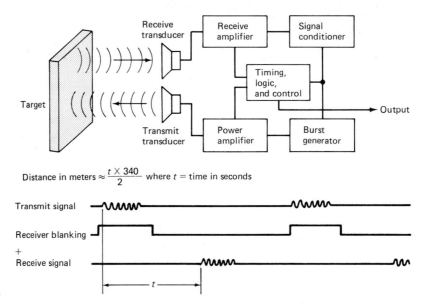

Distance in meters $\approx \dfrac{t \times 340}{2}$ where t = time in seconds

Transmit signal

Receiver blanking

+
Receive signal

Fig. 9-53 Ultrasonic proximity sensor.

Some of the waveforms in an ultrasonic proximity sensor are shown in Fig. 9-53. A receiver blanking pulse turns the receiver off during the time the burst is applied to the output transducer. This pulse protects the receiver circuits from overload. It is possible to use a single transducer for both transmit and receive. A single piezoelectric element can be switched between the receiver input and the power amplifier output. It will be switched to the output only during the time of the transmit burst. The gain of the amplifier may be controlled by a ramp waveform not shown in the illustration. The strength of the received signal falls off as the inverse square of the total distance traveled. Therefore, for measuring distant objects, high receiver gain is needed. However, with high receiver gain, extraneous reflections and other sounds may give a false indication. The solution is to ramp the gain up as time increases. The receiver gain is set low for the time immediately following the transmit burst. As time increases, the receiver gain is also increased to compensate for the path loss.

Pulsed infrared systems are also finding increased application. They use light-emitting diode transmitters that operate below the frequency range of the visible spectrum. Phototransistors with high infrared sensitivity are used to receive the reflected signals. These devices are described in the discussion of optoelectronics. Infrared systems make excellent motion detectors and are used in industrial robotic, security, and safety installations.

The sensing of near ferrous objects can be based on magnetic principles. Figure 9-54 shows a *reluctance proximity sensor.* The assembly uses a permanent magnet and a core with a coil wound on it. The magnet produces a flux that surrounds the turns of the coil. There is no coil output since the flux is static. When a ferrous object enters the field of the magnet, flux distortion is produced and a cutting action results. The graph shows the coil voltage with an object approaching the sensor field and then leaving the sensor field.

Sensing of ferrous or nonferrous objects can be accomplished with the *eddy current killed oscillator (ECKO)* sensor shown in Fig. 9-55. An oscillator provides an ac signal for a coil which in turn generates an electromagnetic field. When a target is intercepted by this field, eddy currents are induced in the metal. These currents represent a circuit loss, and the amplitude of the oscillations decreases. In fact, if the coupling between the coil and the target is tight and if the target is made from a metal with large eddy

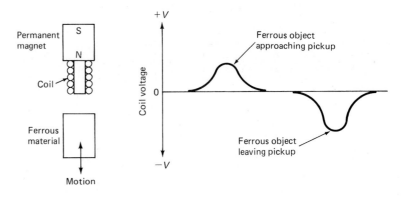

Fig. 9-54 Reluctance pick-up proximity sensor.

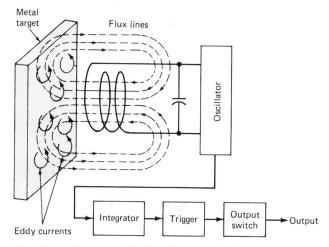

Metal target

Flux lines

Oscillator

Eddy currents

Integrator → Trigger → Output switch → Output

Fig. 9-55 Eddy current killed oscillator proximity sensor.

current losses, the oscillator will stop working. This effect established the name *killed oscillator.* It is not necessary to kill the oscillator to trigger the output, however. The integrator output can trip the trigger before oscillations cease altogether.

Figure 9-56 shows a tilt transducer. *Tilt* is any departure from the horizontal and is usually measured in degrees or radians. The tilt cell contains an electrolyte that conducts electricity. There are also three electrodes sealed in the tilt cell. When the cell is horizontal, the bubble is centered as shown, and electrodes A and B have the same exposure to the electrolyte. Any tilt will cause the bubble to move left or right. Suppose the seating plane tilts clockwise. This tilt causes the bubble to move to the left, decreasing the electrolyte contact on electrode A and increasing the contact area on electrode B. The resistance from the main electrode increases to A and decreases to B. The bridge circuit takes advantage of this dual effect, and an output voltage proportional to tilt results.

Atomic radiation must be sensed in some industrial environments for the protection of human personnel. Radiation may also be used in some measuring applications such as *thickness gaging,* in which a material placed between a radioactive source and a sensor absorbs an amount of energy proportional to its

density and thickness. If the density is a known constant, the thickness of a material can be gaged by the output of a radioactive sensor.

Ionization transducers are used to measure atomic radiation. They are often tubes filled with inert or organic gas. Atomic particles enter the tube and collide with the gas molecules, creating free electrons and positive ions. The tube also contains electrodes. A voltage across the electrodes produces current pulses due to the free electrons and ions. Sensitivity may be improved by operating some tube types at a higher voltage. This higher voltage will accelerate any dislodged electrons to create more collisions, and an avalanche will result. Therefore, the current pulse will be larger. Geiger-Muller tubes are operated at very high potentials. They avalanche over the entire electrode area, and a single event becomes self-perpetuating. They require some means to stop the action after each event. Organic materials are used inside the tube to provide a self-quenching action after each discharge.

Solid-state radiation detectors are based on a reverse-biased junction. Normally, little or no current flows. Atomic particles entering the depletion region raise the energy level of electrons from the valence band to the conduction band. In addition to free electrons, holes are created at these sites. The holes and electrons serve as carriers to support the flow of current, and output pulses result.

Some substances absorb atomic energy and reemit the energy as photons. The flashes of light that result can be coupled by fiber-optic cable to a photomultiplier tube. The high gain of the photomultiplier tube produces a very sensitive radiation detector. These detectors are known as *scintillation counters.*

Not all thickness-gaging systems use radiation. Other techniques include variable reluctance, variable inductance, and variable capacitance. The inductance and reluctance types work with metallic materials, and the capacitance sensor is suited to gaging nonmetallic materials.

REVIEW QUESTIONS

35. What is the name of the instrument that uses wet and dry sensors to measure relative humidity?

36. What types of humidity sensors use hygroscopic materials?

37. Refer to Fig. 9-53. If $t = 20$ ms, how far away is the target?

38. Refer to Fig. 9-53. The burst rate is 2 Hz. What is the greatest distance that can be measured?

39. Reluctance proximity sensors are useful for detecting _____ objects.

40. The ECKO proximity sensor can detect _____ and _____ objects.

41. Refer to Fig. 9-56. The bridge output is zero when the seating plane is horizontal. What is the

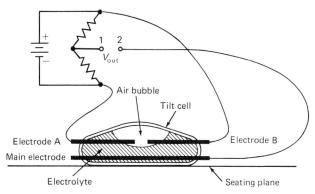

V_{out}

1 2

Air bubble

Tilt cell

Electrode A

Main electrode

Electrode B

Electrolyte

Seating plane

Fig. 9-56 Tilt transducer.

polarity of terminal 1 with respect to terminal 2 if a counterclockwise tilt is introduced? A clockwise tilt?

9-7
SIGNAL CONDITIONING

Signal conditioning provides any required gain, isolation, noise rejection, offsetting, or linearization of the output of a transducer. Gain is required in those cases in which the output is too small to be directly useful in a measurement or control system. The required gain is usually supplied by operational amplifiers or instrumentation amplifiers. These circuits are discussed in Chapter 6. The required gain may also be supplied by special signal-conditioning devices, several of which are covered in this section. Isolation is required in those applications in which ground loops must be eliminated. Noise rejection includes common mode nulling and low-pass filtering.

Offsetting is required when the level of a signal must be shifted by some predictable amount. For example, an application may require the measurement of small changes about some large initial value. Offsetting may also be required to remove an error, as in thermocouple cold-junction compensation. Another example is the conversion of one measurement scale to another, such as gage pressure to absolute pressure or degrees Celsius to degrees Kelvin. Finally, offsetting also includes the conversion of a voltage signal to a current signal for transmission purposes. This topic will be covered later in this section.

Linearization may be accomplished by digital or analog methods. The digital approach lends itself to applications in which the transducer output is applied to the input of a computer. The computer may linearize the readings by performing mathematical operations on them. Another computer approach is to convert each digitized value to a corresponding corrected value by using readings stored in computer memory. The linearized values are stored in a look-up table, such as Table 9-1. Analog linearization uses amplifiers and other circuits that have a nonlinear response that is complementary to the characteristic curve of the transducer. For example, an amplifier with a logarithmic response can be used to linearize a sensor with an exponential output.

In addition to transducer nonlinearity, circuit nonlinearities can also introduce significant errors in some cases. Let's review the basic instrumentation bridge circuit shown in Fig. 9-57. When $R_1/R_4 = R_2/R_3$ the bridge is at null and $V_O = 0$. If the ratio R_2/R_3 is fixed at K, then a null condition guarantees that $R_1 = KR_4$. If R_1 is unknown, it can be measured by nulling the bridge by adjusting R_4. If R_4 is calibrated, the unknown is found by multiplying by K. Null-type measurements are typically used in feedback systems.

Figure 9-57 shows a different set of equations because transducer measurements involve the devia-

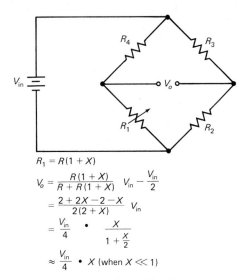

$$R_1 = R(1 + X)$$
$$V_o = \frac{R(1 + X)}{R + R(1 + X)} V_{in} - \frac{V_{in}}{2}$$
$$= \frac{2 + 2X - 2 - X}{2(2 + X)} V_{in}$$
$$= \frac{V_{in}}{4} \cdot \frac{X}{1 + \frac{X}{2}}$$
$$\approx \frac{V_{in}}{4} \cdot X \quad (\text{when } X \ll 1)$$

Fig. 9-57 Typical instrumentation bridge circuit.

tion of one or more of the bridge elements. The equations show that the output voltage is not a linear function of the resistance change in R_1. Assume that all of the bridge elements are nominally equal and that R_1 is variable with a fractional deviation of X. Note that the final equation shows that the output voltage will be approximately equal to $V_{in}/4 \times X$ for very small values of X. As X gets larger, the nonlinearity becomes more evident. As a demonstration, the following values of V_O were calculated for $V_{in} = 10$ V:

$X = 0.05$	$V_O = 0.12195$ V	(2.44% error)
$X = 0.10$	$V_O = 0.23810$ V	(4.76% error)
$X = 0.20$	$V_O = 0.45455$ V	(9.09% error)
$X = 0.30$	$V_O = 0.65217$ V	(13.0% error)

Notice that the nonlinearity becomes more significant with large values of X. For this reason, the simple bridge circuit is most accurate when the sensor shows only small changes in resistance.

Figure 9-58 shows a linear bridge circuit. This ar-

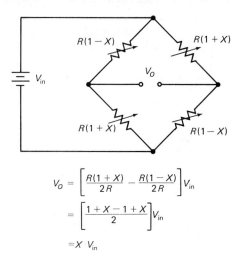

$$V_O = \left[\frac{R(1 + X)}{2R} - \frac{R(1 - X)}{2R} \right] V_{in}$$
$$= \left[\frac{1 + X - 1 + X}{2} \right] V_{in}$$
$$= X \, V_{in}$$

Fig. 9-58 Linear bridge.

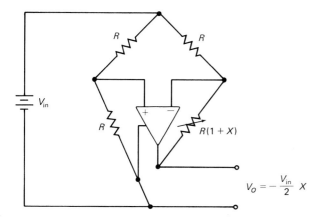

Fig. 9-59 Active bridge.

$$V_O = -\frac{V_{in}}{2} X$$

sistance ratios of the top elements to the bottom bridge elements large. This process, however, decreases the bridge sensitivity. Figure 9-59 shows an active bridge with good linearity and sensitivity. An operational amplifier enforces a null condition even for large values of X. It adds a variable voltage in series with the sensor to maintain the null. This voltage is also the output voltage and is inherently linear with changes in X and twice the magnitude of the basic circuit of Fig. 9-57.

The bridge excitation voltage (or current) is another source of error. It must be well regulated for high accuracy. Figure 9-60 shows the application of a 2B31 transducer signal conditioning module manufactured by Analog Devices. It solves many of the problems associated with accurate measurements. This module also provides gain and stable bridge excitation for measurements using resistive sensors such as RTDs or strain gage sensors. It features a high CMRR (140 dB) and active low-pass filtering for a low-noise output. Pin 29 is available for offsetting the output over a plus and minus 10 V range. If no offsetting is required, pin 29 is grounded. The gain can be set from 0 to 66 dB by varying the resistor across pins 10 and 11.

Linearization is provided by feeding back a small percentage of the amplifier output to modulate the bridge excitation voltage. The sense of the feedback is determined by whether the nonlinearity is concave

rangement might represent two identical, two-element strain gages mounted on opposite sides of a cantilever beam. As the beam stretches on one side, it compresses on the other side. This would increase the resistance of two bridge elements and decrease the resistance of the other two bridge elements. The output of this bridge is linear and is four times the magnitude of a bridge where only one element changes resistance.

Unfortunately, the circuit of Fig. 9-58 does not lend itself to all measurement applications. Another technique that improves linearity is to make the re-

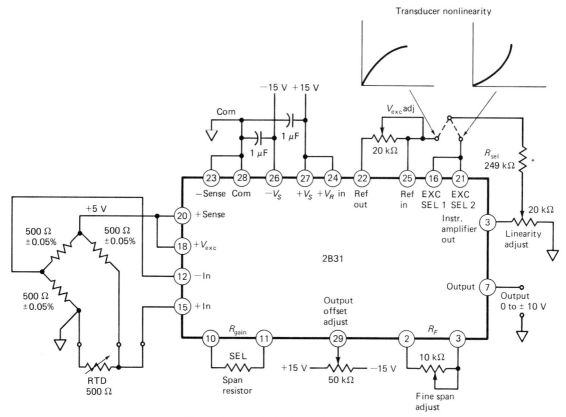

Fig. 9-60 2B31 Linearizing circuit (Analog Devices).

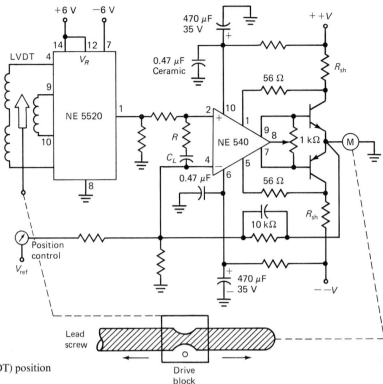

Fig. 9-61 Linear variable differential transformer (LVDT) position servo (Signetics).

upward or downward. Figure 9-60 shows the appropriate jumper connection for each case. The magnitude of the correction is set by R_{sel}, and the linearity pot is used for final trimming. The circuit can be adjusted by using a precision resistance decade in place of the sensor. The offset is adjusted at the low end of the measurement range. The fine span is adjusted at one-third range, and the linearity is adjusted at the top end of the range. There is some interaction, and two or three trials are required for best performance.

An LVDT signal conditioner manufactured by Signetics includes an amplitude-stabilized oscillator to drive the primary of the transducer. The oscillator has a range of 1 kHz to 20 kHz. The IC also contains a synchronous demodulator to convert the amplitude and the phase of the LVDT secondary signal to a dc voltage proportional to position. The synchronous demodulator compares the phase of the secondary signal with that of the primary signal. The IC also has an auxiliary amplifier to provide gain and filtering. It will operate from a single or a dual supply. The internal oscillator generates a triangle wave, which is converted to a sine wave and applied to two driver amplifiers in the IC. Their output appears at two different pins in phase opposition. One pin provides the reference signal for the synchronous demodulator. An LVDT secondary signal is applied to a second demodulator input. The demodulator output is in the form of a bipolar full-wave rectified signal. The active low-pass filter formed with the external resistors and capacitors and the internal auxiliary

amplifier removes the carrier component (ripple) and provides gain. The dc output signal is equal to half the reference voltage when the LVDT core is at null. Core motion results in a dc shift that is proportional to displacement.

Figure 9-61 shows the Signetics LVDT conditioner used in a position servo circuit. The output of the NE5520 conditioner drives the noninverting input of the NE540, which in turn drives a complementary pair of transistors. The permanent magnet dc motor will change direction as the output signal reverses polarity with respect to ground. The motor mechan-

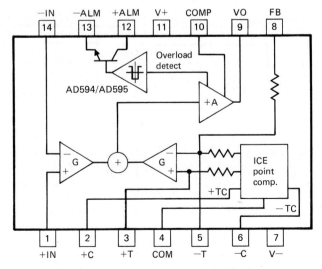

Fig. 9-62 Thermocouple amplifier (Analog Devices).

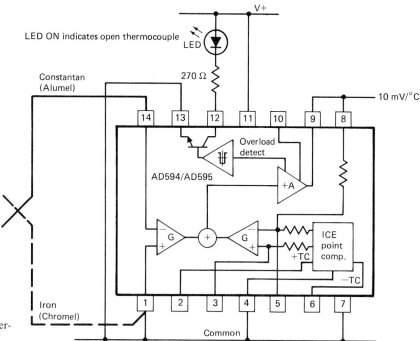

Fig. 9-63 Single supply circuit with open thermocouple indicator (Analog Devices).

ically drives a lead screw, and the drive block is mechanically coupled to the core of the LVDT position sensor. The inverting input of the NE540 is established by V_{ref} and the position control. Any discrepancy between the setting of the position control and the actual position sensed by the LVDT will result in a differential signal at the NE540. The motor will run in a direction to reduce the discrepancy. Servos are covered in detail in Chapter 10.

The Analog Devices AD594/AD595 monolithic thermocouple amplifier is shown in Fig. 9-62. The 594 is precalibrated by laser wafer trimming to match the characteristics of type J sensors, and the 595 matches type K thermocouples. The chip contains a complete instrumentation amplifier and a cold-junction compensator. It also includes a thermocouple failure alarm circuit that activates if one or both thermocouple leads open. Figure 9-63 shows the device interfaced to a thermocouple. The thermocouple leads are soldered directly to pins 1 and 14, producing copper-constantan (or copper-alumel) and copper-iron (or copper-chromel) reference junctions. These junctions are isothermal with the IC itself, and the ice point compensator offsets their Seebeck voltages. Special solders are recommended for these connections. They should be composed of noncorrosive rosin flux and an alloy of 95 percent tin with 5 percent antimony, 95 percent tin with 5 percent silver, or 90 percent tin with 10 percent lead. Ordinary electronic solder is usually 60 percent tin and 40 percent lead.

Many transducer signals in industry are converted to current signals for transmission to a distant controller or computer. The standards are 4 to 20 mA and 10 to 50 mA, with the 4 to 20 range being the most common. Current transmission offers several advantages. The signal is not affected by noise, drops in the line, stray thermocouples, contact resistance, or contact emf. Only two wires are needed, and an open circuit fault is easily detected by a 0-mA signal condition. The National Semiconductor LH0045 is a two-wire transmitter. It is a linear IC that converts the voltage signal from a sensor or a bridge to current. A single twisted pair of wires is all that is required for both signal output and supply circuits. The device contains an internal reference to power the bridge, an input amplifier, and an output current source. The output is adjustable to meet either industry current standard. It interfaces easily with thermocouples, strain gages, RTDs, and thermistors.

REVIEW QUESTIONS

42. Refer to Fig. 9-57. The bridge is at null when $V_O = $ _____.

43. Refer to Fig. 9-57. Calculate the actual value of V_O when $X = 0.03$ for $V_{in} = 5$ V. Calculate the nominal value by using the approximation. Calculate the percentage of error.

44. Refer to Fig. 9-57. What happens to the nonlinearity as X increases?

45. Refer to Fig. 9-58. Calculate V_O for $V_{in} = 5$ V and $X = 0.1$. For $X = 0.2$. Is V_O a linear function of X?

46. Refer to Fig. 9-59. Calculate V_O for $V_{in} = 5$ V and $X = 0.07$. Is V_O a linear function of X?

CHAPTER REVIEW QUESTIONS

9-1. Stretching a conductor will _____ its resistance.

9-2. An industrial strain gage has a GF of 2 and a resistance of 350 Ω. If its normal length is 2 cm, what will its resistance be if it is stretched to 2.02 cm?

9-3. What output voltage would be produced by the data in question 9-2 if the gage were in a Wheatstone bridge circuit with three fixed 350-Ω resistors?

9-4. Assume that a two-element strain gage is used for temperature compensation. What would have to be done to the gain of the instrumentation amplifier to compensate for the transverse sensitivity factor?

9-5. Why might dual-staggered slots be used in an optical tachometer?

9-6. When the output of an accelerometer is integrated, a signal proportional to _____ results.

9-7. The LVDT accelerometers do not lend themselves to vibration analysis because of their rather _____ resonant frequency.

9-8. The principle that relates fluid flow rate and pressure is attributed to _____.

9-9. Identify a transducer that divides flow into measured units.

9-10. Which temperature sensor converts heat to a voltage?

9-11. Which temperature sensor has the greatest temperature range?

9-12. The signal generated by a thermocouple is due to the _____ effect.

9-13. Which temperature sensor requires compensation?

9-14. A series arrangement of thermocouples is known as a _____.

9-15. Low thermal mass is desired in a sensor requiring a _____ response time.

9-16. Why do wire-type RTDs use bifilar windings?

9-17. A platinum RTD is rated at 100 Ω at 0°C. Assuming a positive temperature coefficient of 0.00385, what is its resistance at 50°C?

9-18. Low-resistance sensors, such as RTDs, require three or four wire connections to eliminate the error due to _____ resistance.

9-19. Radiation detectors that use a gas-filled tube are based on the _____ principle.

9-20. Atomic radiation entering a semiconductor depletion region can cause the junction resistance to _____.

9-21. What type of radiation detector utilizes light?

9-22. Linearization based on look-up tables is a(n) _____ technique.

9-23. Refer to Fig. 9-63. Why must the thermocouple leads be soldered at the IC pins and not to copper extension wires?

9-24. Why do the current transmission standards use 4 or 10 mA to represent zero rather than 0 mA?

ANSWERS TO REVIEW QUESTIONS

1. 9 V **2.** 8.88 to 9.12 V **3.** 0.2 percent **4.** 2 V, positive **5.** A would be 2 V negative with respect to B **6.** none **7.** 3 **8.** the polarity reverses **9.** they cannot indicate shaft direction **10.** it provides direction information **11.** acceleration **12.** displacement; force; acceleration **13.** 667 Hz **14.** fewer cycles of oscillation (more rapid decay) **15.** for temperature compensation **16.** 3.56×10^7 N/m **17.** 5.93×10^{-4} m/m **18.** 0.142 Ω **19.** negative **20.** velocity **21.** P_1 **22.** pascal **23.** differential **24.** Bourdon tube; bellows; diaphragm **25.** absolute **26.** 2.81 m **27.** differential **28.** none; the voltages cancel **29.** software **30.** it is a sense lead (little or no current) **31.** Fig. 9-47 **32.** 3.732 V **33.** it increases; IC_1 will have to become warmer to trip the comparator **34.** hysteresis **35.** psychrometer **36.** hygrometers **37.** 3.4 m **38.** 85 m (in practice it is much less) **39.** ferrous **40.** ferrous; nonferrous **41.** positive; negative **42.** zero **43.** 0.03695 V; 0.03750 V; 1.47 percent **44.** it increases **45.** 0.5 V; 1.0 V; yes **46.** −0.175 V; yes

10

SERVOMECHANISMS

This chapter discusses servomechanisms and the components that make up servosystems. Some components are electrical, some electronic, and some mechanical. Servomechanisms are systems that position an object by comparing position feedback signals with command signals. Systems that use feedback are closed-loop systems; those that do not use feedback are open-loop systems. Feedback can be continuous or discontinuous. This chapter will only deal with continuous types of control systems.

10-1
POTENTIOMETERS AND ENCODERS

Precision potentiometers are simple rotary devices for obtaining shaft position information. The most straightforward application is the conversion of mechanical position to a voltage. Basically, a precision potentiometer consists of a resistive element with a movable arm, or *slider,* in contact with the element. As the arm (slider) rotates, the resistance varies between the end of the resistive element and the slider, indicating shaft position. The resistive element can be made of wire, conductive film, or a cermet element.

Potentiometers used for servomechanisms are generally about ⅞ to 3⁵⁄₁₆ in. (22.2 to 84.125 mm) in diameter. The early models were mostly of the wirewound type. Current technology provides other choices such as conductive plastic, which offers a better temperature stability, longer life, and lower sensitivity to the environment. Potentiometers can be excited with alternating or direct current. Single-turn potentiometers have a rotation that is usually limited to 350°. Some models have continuous rotation with no internal stops. Potentiometers may be ganged so that a single shaft will rotate several sliders. *Multiturn potentiometers* are limited to 3, 5, 10, 15, 25, or 40 revolutions before hitting internal stops. Figure 10-1 shows the internal construction of a ten-turn potentiometer. The winding (resistance element) is in the form of a helix, and the contact assembly is

such that the slider travels a helical path while making contact with the resistance element.

A linear potentiometer produces a resistance change that is linearly related to its shaft position. A position of one-half rotation will produce 50 percent of maximum resistance, and a position of three-fourths rotation will produce 75 percent of maximum resistance, and so on. *Linearity* is specified as the deviation (in percentage of the total resistance) of the actual resistance at any point from the expected resistance. This is called *normal,* or independent, linearity. A standard value of linearity is 0.1 percent with 0.01 percent types available. If a load resistance is placed between the slider and one end of a linear potentiometer, as shown in Fig. 10-2, the potentiometer will no longer be linear. The magnitude of loading error is a function of the ratio of the slider load to the total resistance of the potentiometer. This error varies inversely with the load ratio. That is, a small load ratio will produce a large error.

There are several other important characteristics of potentiometers, such as resolution, noise, and mechanical tolerances. *Resolution* in a potentiometer is the minimum change of resistance output expressed

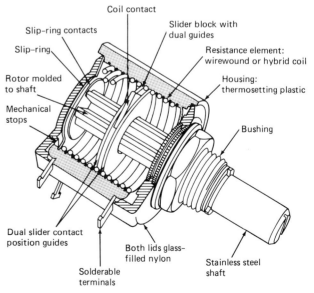

Fig. 10-1 Multiturn potentiometer.

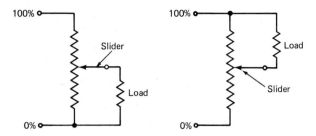

Fig. 10-2 Potentiometer loading.

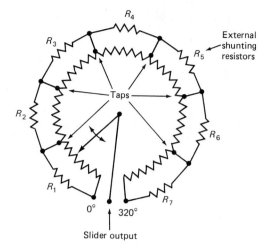

Fig. 10-3 Tapped potentiometer.

as a percentage of its total resistance. It is dependent on the number of turns of wire per inch on the winding and the arc diameter of the slider.

A nonwire potentiometer is stepless and has essentially infinite resolution. The typical resolution of a wire-wound potentiometer is 0.05 percent. Many high-gain servomechanisms have a tendency to "hunt" between the turns of wire on the potentiometer, seeking a voltage that does not exist in the surface. Plastic conductive and cermet units have all but eliminated this problem. A multiturn wire-wound potentiometer may also be employed because of its better resolution.

Noise in a potentiometer appears as spurious unwanted voltages. If a wire-wound potentiometer is excited with direct current, for example, the finite resolution of the potentiometer will cause a ripple voltage to appear at the slider as the shaft is rotated. In some systems this noise can cause problems, especially if the potentiometer's slider is worn and the noise becomes excessive. There will also be an increase in the hunting of the system. This noise is all but eliminated with the nonwire potentiometer. Mechanical tolerances of potentiometers are important in servomechanism applications. They are manufactured with accurate external surfaces to allow interchangeable units and to permit low backlash (see Chapter 10) and matching coupling surfaces.

Why use wire-wound potentiometers? Two reasons are apparent. First, the wire-wound potentiometer can be made with very low values, such as 10 Ω. Values less than 1000 Ω are hard to obtain with non-wire-wound types. The second reason is that nonlinear functions are needed in many servomechanism systems, and it is all but impossible to duplicate the specifications obtainable from wire-wound units. Nonlinear potentiometers can be made by winding the element on a mandrel which has the slope of the function. Examples of mathematical functions that are available include the tangent, secant, cosecant, square root, and inverse functions. The use of taps on a linear winding is another method used to obtain nonlinear functions. Sections of the resistance between taps can be loaded with external resistors to warp the resistance to obtain a given function. From 4 to 10 taps are normally used, as shown in Fig. 10-3.

A relatively new potentiometer that solves numerous linear control applications is the *rectilinear po-*

tentiometer. This unit has strokes (travels) from ½ to 6 in. (12.7 to 152 mm), with an independent linearity of 0.1 percent.

Optical encoders may also be used to provide position feedback. The primary parts of an optical encoder are shown in Fig. 10-4. Two types of rotary optical encoders available are *incremental* (outputs a fixed number of N pulses per revolution) and *absolute* (the output is a unique code for each angular position). *Incremental optical encoders* generally provide two signals which are in quadrature (90° phase difference). The phase difference provides directional information. The output waveforms may be square wave or sine wave, as illustrated in Fig. 10-5. The marker pulse is used to provide index information and is typically 180 electrical degrees. Not all optical encoders provide a marker pulse.

Figure 10-6 shows a 28-mm two-channel incremental optical encoder. The emitter end plate contains two light-emitting diode (LED) light sources (with molded lenses) to form a parallel beam for each channel. The code wheel is a metal disk which, in this case, has 500 equally spaced slits around its circumference. An aperture with a matching pattern is positioned on the stationary phase plate. The light beam

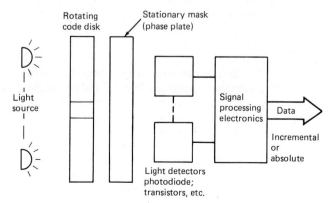

Fig. 10-4 Primary elements of an optical encoder.

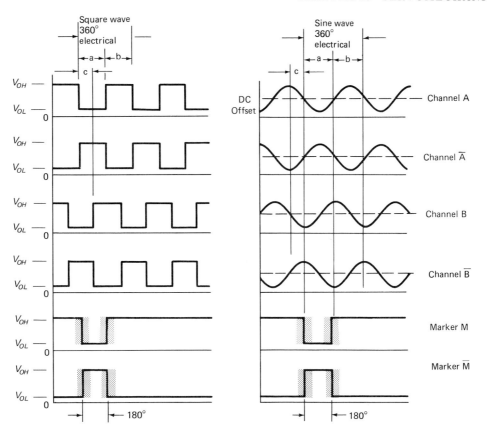

Shaded areas represent the locus of leading and trailing edges of marker pulse.

Fig. 10-5 Incremental encoder output waveforms.

is transmitted only when the slits in the code wheel and the aperture line up; therefore, during a complete shaft revolution, there will be 500 alternating light and dark periods. A molded lens beneath the phase plate aperture collects the modulated light into the photodiode detector. Each channel consists of an integrated circuit with two photodiodes and amplifiers, a comparator, and output circuitry as shown in Fig. 10-7. The two quadrature output signals are indicated at their respective channels. The direction of rotation is determined by observing which of the channels is the leading waveform.

Figure 10-8(*a*) shows a circuit approach to interface to a microprocessor. The logic gates and microprocessors are discussed in the next two chapters. An encoder used to provide position information for a shearing process is shown in Fig. 10-8(*b*). The material moves the measuring wheel, and the encoder provides pulses to the computer. When the desired count is reached, the shear solenoid is activated.

The *absolute optical encoder* makes absolute position information available in the form of binary, gray, or binary-coded decimal (BCD) formats, depending on the coding pattern of the optical encoder's disk. These codes are covered in Chapter 11. The advantage of the absolute encoder is that it can maintain position information even during a power

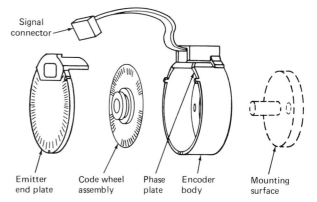

Fig. 10-6 Optical encoder kit.

failure, whereas the incremental type of encoder must be rotated to the index marker for initialization once power is restored. In the absolute encoder, there are signal output lines (one for each bit position) going to the digital processor. An optical encoder disk is illustrated in Fig. 10-9(*a*). The parallel digital data (four lines in this case) require conditioning and eventually are interfaced to a computer.

Many industrial processes and computer controls also require linear motion inputs. They may be from *X-Y* tables, plotters, quality control (QC) equipment,

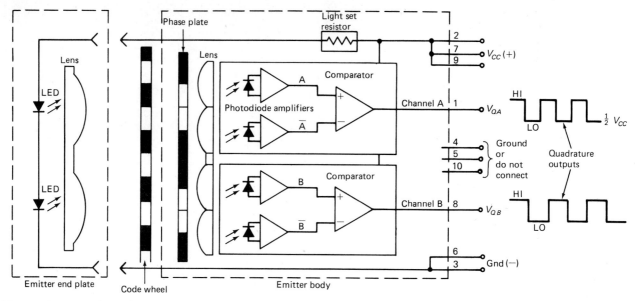

Fig. 10-7 Incremental encoder block schematic.

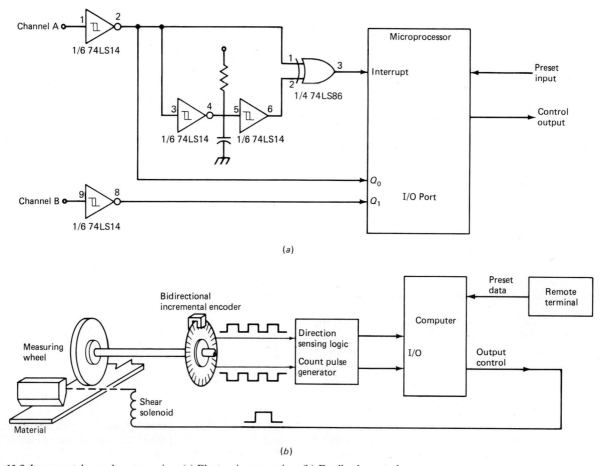

(a)

(b)

Fig. 10-8 Incremental encoder processing. (a) Electronic processing. (b) Feedback control.

machine tools, etc. These units have a similar type of optoelectronics and also produce quadrature signals with an optional reference just as the incremental encoders do. High-resolution units using special signal electronics provide resolutions as fine as 0.5 μm (2000 pulses/mm) at 300 kHz. A wide variety of linear encoder lengths are also available. Velocity information can also be derived from the incremental signals.

Position sensing varies from application to appli-

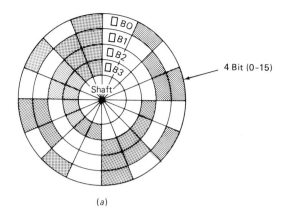

(a)

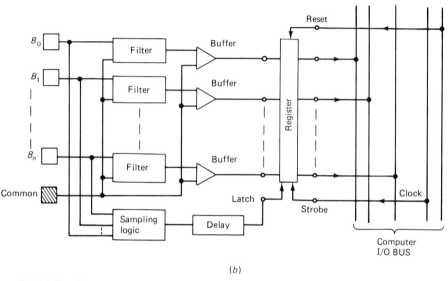

(b)

Fig. 10-9 Absolute encoder. (a) Simple encoder disk. (b) Typical encoder interface.

cation. Potentiometers are used for systems that have low resolution requirements. In some cases, a potentiometer is used as a coarse position sensor, and an optical encoder is used for the high-resolution positioning. Areas with heavy contamination (such as oil and dust) are a problem for optical encoders. The problem can be solved with special enclosures.

REVIEW QUESTIONS

1. List three common types of potentiometers.

2. The movable contact on a potentiometer is called the *wiper* or _____.

3. Suppose a 10 kΩ linear potentiometer has a total travel of 350.° What is its nominal resistance from the wiper contact to the far-end contact when it is rotated 270°?

4. The deviation of a potentiometer from its nominal straight-line resistance is rated as a percentage and is known as its _____ specification.

5. A noisy or dirty wiper on a potentiometer can cause a feedback system to _____.

6. Taps may be used to linearize a potentiometer. (true or false)

7. The two outputs from an incremental encoder are _____ degrees apart.

10-2
SYNCHROS AND RESOLVERS

The term *synchro* is a generic name for a family of inductive devices which can be connected in various ways to form shaft angle measurements. All of these devices work on essentially the same principle, which is that of a rotating transformer. A synchro looks like an ac motor and consists of a rotor and a stator. Synchros vary in diameter from 0.5 to 3.7 in. (12.7 to 94 mm).

Internally, most synchros are similar in construction. They have a rotor with one or three windings (depending on the synchro type) capable of revolving inside a fixed stator. There are two common types of rotors: the salient pole and the wound rotor. During one complete cycle, the magnetic polarity of the

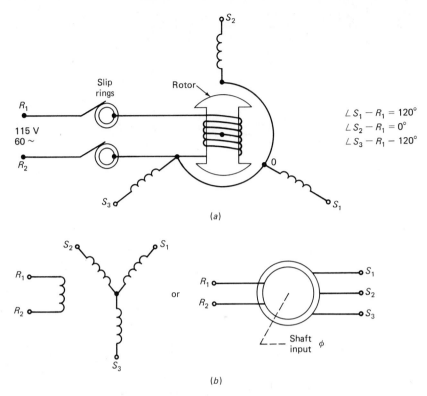

$$\angle S_1 - R_1 = 120°$$
$$\angle S_2 - R_1 = 0°$$
$$\angle S_3 - R_1 - 120°$$

Fig. 10-10 Synchro. (a) Electric circuit. (b) Schematic symbol.

TABLE 10-1 SYNCHRO TYPES

Functional Classification	Military Abbreviations	Input	Output
Torque transmitter	TX	Rotor positioned mechanically or manually by information to be transmitted	Electric output from stator identifying rotor position supplied to torque receiver, torque differential transmitter, or torque differential receiver
Control transmitter	CX	Same as TX	Electric output same as TX but supplied only to control transformer or control differential transmitter
Torque differential transmitter	TDX	TX output applied to stator; rotor positioned according to amount; data from TX must be modified	Electric output from rotor (representing angle equal to algebraic sum or difference or rotor position angle and angular data from TX) supplied to torque receivers, another TDX, or a torque differential receiver
Control differential transmitter	CDX	Same as TDX but data usually supplied by CX	Same as TDX but supplied only to control transformer or another CDX
Torque receiver	TR	Electrical angular position data from TX or TDX supplied to stator	Rotor assumes position determined by electric input supplied
Torque differential receiver	TDR	Electrical data supplied from two TDXs, two TXs, or one TX and one TDX (one connected to rotor, one to stator)	Rotor assumes position equal to algebraic sum or difference of two angular inputs
Control transformer	CT	Electrical data from CX or CDX applied to stator; rotor positioned mechanically or manually	Electric output from rotor (proportional to sine of the difference between rotor angular position and electric input angle)

rotor changes from zero to maximum in one direction, back to zero, then to maximum in the opposite direction, and then back to zero. The wound rotor is used in most synchro control transformers. It often consists of three coils arranged so that their axes are displaced from each other by 120°. One end of each coil terminates at one of the three slip rings on the shaft, and the other ends are connected together.

The stator of a synchro is a cylindrical structure of slotted laminations with three Y connected coils wound with their axes 120° apart. The stator windings are not connected directly to the ac power source. Their excitation is supplied by the ac magnetic field of the rotor.

The synchro may be viewed as a variable coupling transformer. The rotor is energized by an ac voltage, and the coupling between the rotor and the stator windings varies as a trigonometric or linear function of the rotor position. Figure 10-10 shows a synchro schematic diagram. Synchro systems consist of two or more interconnected synchros. Units are grouped together according to their intended function. The seven common types are listed in Table 10-1.

The conventional *synchro transmitter* (TX) uses a salient pole rotor with skewed slots. When an ac excitation voltage is applied to the rotor, the resultant current produces a magnetic field and by transformer action induces voltages in the stator coils. The effective voltage induced in any stator coil depends upon the angular position of the coil's axis with respect to the rotor axis. When the maximum coil voltage is known, the induced voltage at any angular displacement can be determined. Figure 10-11 shows the voltages induced in one stator coil as the rotor is turned to different positions.

The turns ratio between the rotor and stator is such that when single-phase 115-V excitation is applied to the rotor, the highest value of effective voltage induced in any one coil will be 52 V. Because the common connection between the stator coils is not accessible, it is only possible to measure the stator coil-to-coil effective voltage. Figure 10-12 shows how these voltages vary as the rotor is turned. Values are shown above the line when the terminal-to-terminal voltage is in phase with the R_1 to R_2 voltage and below the line when the voltage is 180° out of phase with the R_1 and R_2 voltage. Therefore, negative values indicate a phase reversal. As an example, when the shaft (Fig. 10-12) is turned 30° from the reference (zero degree) position, the S_1 to S_3 voltage will be about 45 V and in phase with the R_1 to R_2 voltage. The S_1 to S_2 voltage will be about 90 V and is 180° out of phase with the R_1 to R_2 voltage. Although the curves of Fig. 10-12 resemble time graphs of ac volt-

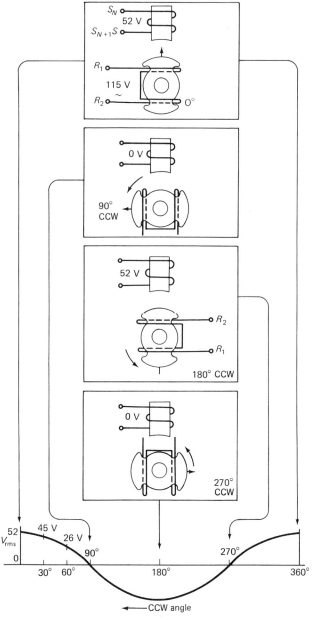

Fig. 10-11 Curve of stator voltage versus rotor position.

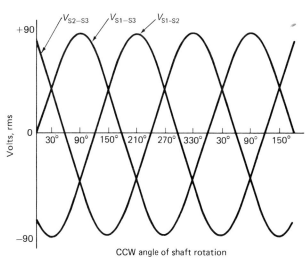

Fig. 10-12 Waveform of synchro stator output voltages versus shaft rotation.

ages, they show only the variations in the effective voltage amplitude and phase as a function of the mechanical rotor position.

It should be noted that the synchro is *not* a three-phase machine or generator. In a three-phase machine, there are three voltages *equal* in magnitude, displaced from each other by 120 electrical degrees. With the synchro, which is a single-phase device, the three stator voltages vary in magnitude, and one stator coil is in phase or 180° out of phase with another coil, as illustrated in Fig. 10-12.

In general, if the rotor of a synchro is excited by 60-Hz or 400-Hz ac (called the *reference voltage*), the voltage induced in any stator winding will be proportional to the cosine of the angle between the rotor coil axis and the stator axis, as was indicated in Fig. 10-11. The voltages induced across any pair of stator terminals (S_1 to S_2, S_1 to S_3, S_2 to S_3) will be the sum or difference, depending on the phase, of the voltages across the coils measured.

For example, if a reference voltage $V \sin (\omega t)$ excites the rotor of a synchro (R_1 to R_2), the stator terminals will have a voltage of the following form:

$$V (S_1 \text{ to } S_3) = V \sin (\omega t) \sin \theta$$
$$V (S_1 \text{ to } S_2) = V \sin (\omega t) \sin (\theta + 120)$$
$$V (S_2 \text{ to } S_3) = V \sin (\omega t) \sin (\theta + 240)$$

where θ = synchro shaft angle.

Note: The expression $V \sin (\omega t)$ predicts the instantaneous voltage of a sine wave at time (t) where V represents the maximum voltage and $\omega = 2\pi f$. The rms stator voltages are given by:

$$V_{rms} (S_1 \text{ to } S_3) = 0.707\ V \sin \theta$$
$$V_{rms} (S_1 \text{ to } S_2) = 0.707\ V \sin (\theta + 120)$$
$$V_{rms} (S_2 \text{ to } S_3) = 0.707\ V \sin (\theta + 240)$$

These voltages are known as the *synchro format voltages* and will be referred to as such from now on. Synchros are divided into two basic types, torque synchros and control synchros. *Torque synchros* are required if it is necessary to transmit angular displacement information from a shaft of one synchro to the shaft of another synchro without using any additional amplifiers or gearing. The two most common torque synchros connected in a repeater system are the *torque transmitter* (TX) and the *torque receiver* (TR). Figure 10-13 shows a TX and TR connected as a repeater system. In the repeater system, the rotor of the transmitter (TX) is excited with a reference voltage to produce the synchro format voltages on output terminals S_1, S_2, and S_3. The stator voltages induced in the receiver (TR) stator coils as a result of its rotor excitation will be equal to the voltages induced by the transmitter stator current. In this balanced condition, shown in Fig. 10-13, there is no current flow in the stator coils or in the stator interconnections. The total current drawn is that used by the excitation of the two rotors. Therefore, the transmitter will supply current only when the receiver rotor is out of alignment with the transmitter rotor. These repeater systems are accurate to ±1° and are used in systems in which a rotating device's output is required to position a remote pointer. Today, most of these units have been replaced by a synchro-to-digital converter driving an LED display of the digital readout of angular position. There also exists a digital-to-synchro transmitter to convert digital data to synchro format voltages to drive a remote electromechanical pointer. A mix of these devices is not uncommon. Digital synchro converters will be covered shortly.

Two other members of the torque synchro family are worth mentioning. The first device is the *differential synchro transmitter* (TDX). The stator is quite similar to that of the transmitter and receiver just discussed. It has three sets of coils wound around the stator frame to produce poles 120° apart. The rotor is quite different from those of conventional synchro units. Electrically, it has three sets of coils

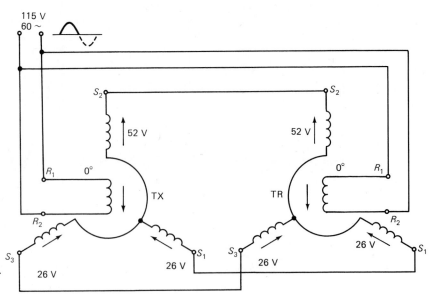

Fig. 10-13 Synchro transmitter and receiver pair with both rotors at 0°.

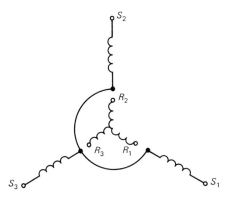

Fig. 10-14 Schematic diagram of differential synchro.

wound in slots equally spaced on the rotor and connected to produce poles 120° apart. The schematic of the differential transmitter is shown in Fig. 10-14.

The TDX usually obtains its input from a torque transmitter (TX) and produces electrical synchro format signals. The power comes from the synchro stator outputs (there is no reference applied to the TDX). The TDX may be connected to add or subtract two inputs. Figure 10-15(a) shows a connection for subtraction. If a mechanical input of 75° is applied to the TX and its output signals go to the TDX stator, the TDX subtracts its own mechanical input (30°) and transmits the results to the TR, which indicates the system's mechanical output by position of its

rotor (45°), as indicated in Fig. 10-15(a). In some cases, the system is set up for addition. This is done by reversing the S_1 and S_3 leads from the TX to the TDX stator, and from the TDX rotor to the TR stators. This will result in the behavior shown in Fig. 10-15(b). With the same mechanical inputs of 75° and 30°, the receiver will provide an output equal to the sum of signals by turning to 105°. A wiring schematic for the subtraction system is shown in Fig. 10-16. The torque synchro system is suitable only for very light loads and is never really accurate. In addition the torque system places a drag on the associated equipment it is measuring.

When larger amounts of power and more accuracy are required, torque synchros give way to the *control synchros*. These devices are used for providing and handling control signals to a servo power amplifier when more power and accurate angular displacement of a large load are required. The control synchros are not designed to handle any mechanical load. The two most common control synchros are the *control transmitter* (CX) and the *control transformer* (CT). The CT develops an ac rotor output voltage that is proportional to the relative shaft angles between the synchro transmitter and the control transformer. The devices are normally connected as shown in Fig. 10-17. The output of the CX (transmitter) is fed to the stators of the CT (transformer). The CT is a high-impedance version of the torque receiver with its

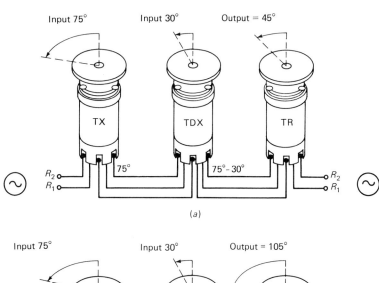

(a)

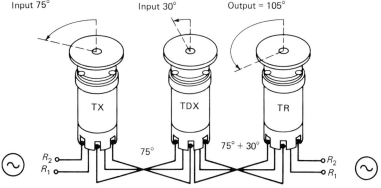

Fig. 10-15 Applications of the differential synchros. (a) Subtraction with TDX. (b) Addition with TDX.

(b)

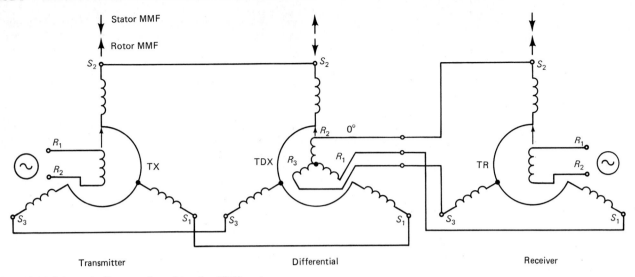

Fig. 10-16 Schematic diagram of a subtraction TDX system.

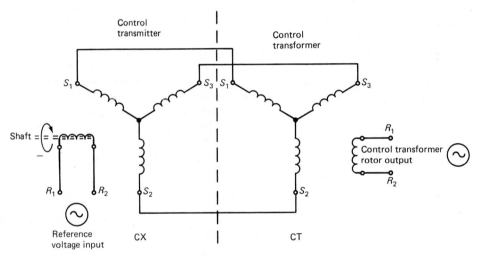

Fig. 10-17 Synchro control system (chain).

rotor aligned at 90° from that of a TR. In a control system (chain), when the shaft angle of the CX equals that of the CT shaft angle, a null (minimum) voltage will appear on the rotor terminals, R_1 and R_2, of the CT. Any variation from this null will produce a signal in the CT rotor whose phase will depend on in which direction it is moved off null. Figure 10-18 shows the output of a CT rotor as it travels near alignment (null) with the transmitter rotor. Typically, for a 115-V CT the null voltage would be about 30 mV rms.

A simple closed loop servo system using a CX and CT control system is shown in Fig. 10-19. When the shaft of the CX is turned to some angle, the S_1, S_2, and S_3 outputs provide the synchro format voltages previously mentioned. These voltages are transmitted to the CT stators S_1, S_2, and S_3. If the CT is not at the input angle θ, a voltage will be produced at the output of the CT rotor winding. This signal (error)

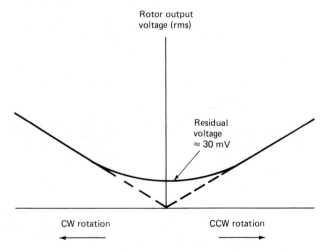

Fig. 10-18 CT rotor null voltage near alignment.

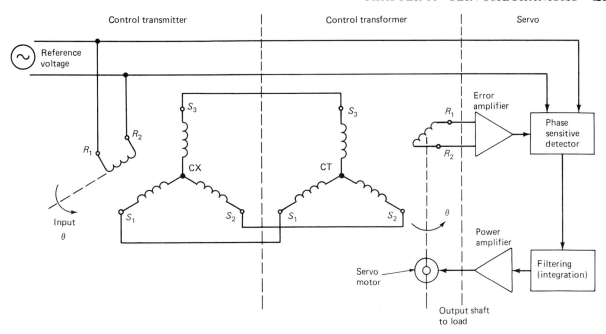

Fig. 10-19 Simple servo system using control synchros.

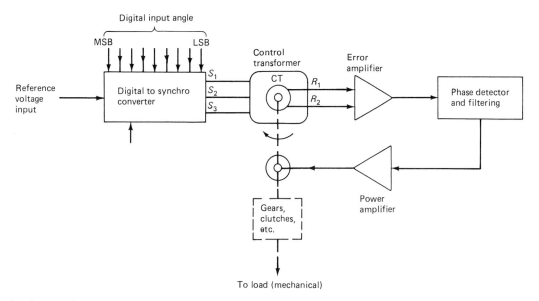

Fig. 10-20 Digital-to-synchro converter.

is amplified, phase-detected and fed to a servo amplifier to cause a servomotor to position a load and the CT shaft to a position where the CT rotor output is minimum (null). The direction in which the motor turns toward angle θ is determined by the phase of the CT rotor signal with respect to the reference voltage.

In some later applications, a digital-to-synchro converter (a solid-state CX) can be used, with the input angle in digital form as shown in Fig. 10-20. There also exists a *control differential transmitter* (CDX), which is the control equivalent of the TDX previously discussed and is used to add or subtract an additional shaft angle.

The resolver is basically a trigonometric function generator that resolves for an angle θ, the hypotenuse, or sides of a right triangle. Although it resembles a synchro device outwardly, internally the resolver is quite different. There are a wide variety of winding and ratio configurations available. The most common has two isolated primary (rotor) windings at right angles to each other. The two windings of the stator are isolated and are also placed 90° apart. A *resolver* is illustrated in Fig. 10-21. It solves the unknowns of a right triangle. If θ is the shaft input angle, one can determine B and A directly from a resolver by applying voltage C as an input to the stator, positioning the rotor to angle θ, and reading A and B as outputs

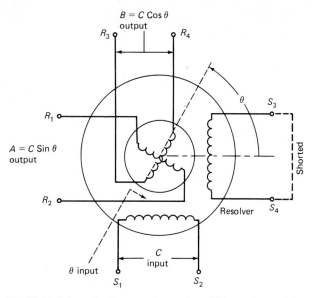

Fig. 10-21 Schematic diagram of a resolver. Either rotor or stator can be the primary.

from the rotor windings. To be more mathematically correct, if the stator is excited with $V \sin \omega t$, the resolver format voltages will be

$$V_{R1\ R2} = V \sin \omega t \sin \theta$$
$$V_{R4-R3} = V \sin \omega t \cos \theta$$

For example,

$$V_{R1-R2} = 100 \times 0.500 = 50 \text{ V}$$
$$V_{R3-R4} = 100 \times 0.866 = 86.6 \text{ V}$$

where

$$\text{resolver shaft angle} = 30°$$
$$V \sin \omega t = 100 \text{ V}$$
$$\sin \theta = \sin 30° = 0.500$$
$$\cos \theta = \cos 30° = 0.866$$

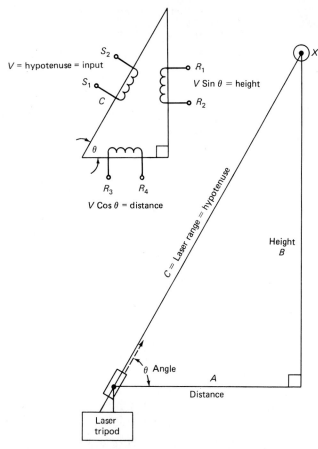

Fig. 10-22 Example of computing resolver for determining height.

The unused stator rotor is usually short-circuited. These voltages represent the rectangular or cartesian coordinates of the point.

Figure 10-22 is an example of polar-to-cartesian conversion used to determine the precise height of

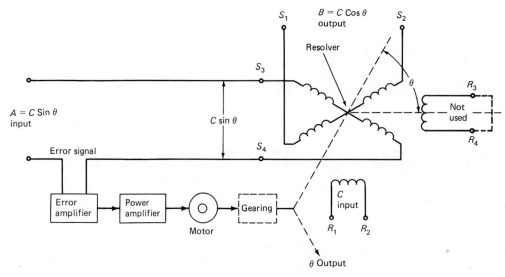

Fig. 10-23 Computation using a resolver.

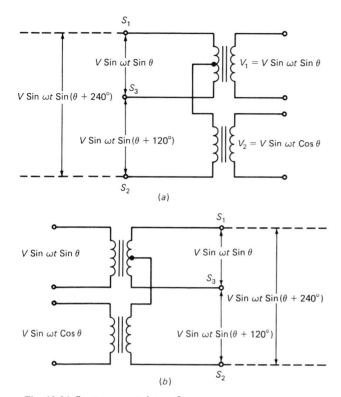

Fig. 10-24 Scott connected transformers.

use signals in the resolver format. Figure 10-24 shows two common Scott transformer connections. In Fig. 10-24(a) the synchro-to-resolver connection and the resolver output format voltages obtained are shown. In Fig. 10-24(b) the resolver-to-synchro Scott connected transformers, which are simply the inverse of the synchro-to-resolver case, are illustrated.

REVIEW QUESTIONS

8. The stator windings S_1, S_2, and S_3 of a synchro are connected to a three-phase power source. (true or false)

9. Referring to Fig. 10-11, the voltage at the 180° point should be _____.

10. In Fig. 10-12, the voltage of S_1, S_2 at 270° is out of phase with R_1—R_2 voltage. (true or false)

11. Referring to Fig. 10-13, if TX is moved clockwise, TR will move _____.

12. If the differential synchro system of Fig. 10-15(b) has an input of 130°, the receiver will be at _____.

13. A _____ transformer is used to input angular signals into a servo power system.

14. The control transformer error signal is composed of a voltage and _____ signals.

10-3
SERVOMOTORS AND RATE GENERATORS

The requirements for most small servomechanisms are met by ac two-phase induction motors. Their mechanical output power varies from 0.5 to 100 W. Above 10 W, most two-phase servomotors are cooled by a separate motor-driven blower included in the same housing with the servomotor. A 10-W frame will deliver about 25 W output with the added blower cooling. Direct current servomotors vary in size from 1/20 hp to many horsepower and are generally used in large power servomechanisms.

A typical connection for a two-phase motor is shown in Fig. 10-25. A voltage V_M is applied to the main (fixed or reference) winding. Voltage V_C is supplied from the controller, which is usually an amplifier. The magnitude of V_C is a function of the degree of action required of the motor. The windings are usually identical and equally rated. The voltages V_M and V_C must be in synchronism and are derived from the same ac source. They must also be in time quadrature, which may be produced by introducing a 90° phase shift in the amplifier or by connecting a suitable capacitor in series with the main (reference) phase V_M. When V_C has a voltage value leading V_M by approximately 90°, rotation in one direction is obtained; when V_C lags V_M, rotation in the other direction will occur. Since torque is a function of both V_M and V_C, changing the magnitude of V_C changes the developed torque of the motor. Some

an object by using an adjustable laser. The output voltage $VR_1 - VR_2$ is proportional to the height, and output $VR_3 - VR_4$ is proportional to the ground distance when a voltage V (equal to the laser range) is applied to $S_{1-}S_2$, and the resolver is rotated to an angle θ. A servo system to perform this function is shown in Fig. 10-23. The servomechanism is satisfied only when the resolver angle is equal to θ, which occurs when the input to the error amplifier is at a null. One output is $B = \cos\theta$, which is the voltage proportional to the height of the Fig. 10-22, when a voltage C (equivalent to the laser range) is applied to the stator as shown.

By using all four windings of the resolver, a two-dimensional space problem can be solved. If the stators are excited with a voltage representing X and Y, respectively, when the shaft is positioned to an angle ψ, the voltages produced at the rotor terminals will be

$$V_{R1-R3} = X \cos\Psi + Y \sin\Psi$$
$$V_{R4-R2} = Y \cos\Psi - X \sin\Psi$$

The applications of these mathematical functions are commonly used in guidance and robot control systems.

It is possible to convert synchro input signals into resolver format signals and to convert resolver format signals into synchro format signals. These conversions are generally done by using Scott connected or Scott T transformers. In many cases this is or was done to upgrade a system to use synchro-to-digital (SDC) or digital-to-synchro converters (DSC), which

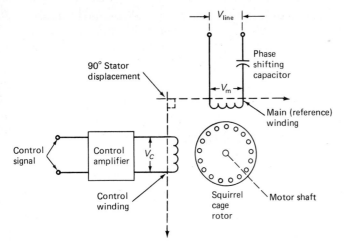

Fig. 10-25 Conventional two-phase servomotor circuit.

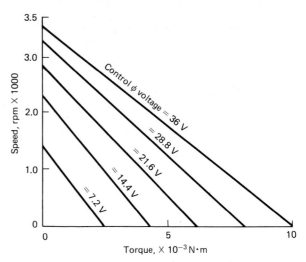

Fig. 10-27 Torque-speed characteristics of a typical servomotor.

servomotors are designed with a center-tapped winding to be fed from a push-pull output amplifier as shown in Fig. 10-26.

The speed-torque characteristic curves for a typical ac servomotor are shown in Fig. 10-27. They are typical of servomotor characteristics. The torque is large at zero speed to aid in servo static sensitivity, to give internal damping for the servomechanism, and to prevent single phasing (the tendency of the rotor to continue to rotate when one winding is opened and the other winding remains excited) of the servomotor. Two-phase servomotors are inherently high-speed, low-torque devices and are geared down to drive the load.

Although two-phase servomotors are available in a wide variety of configurations, the most popular type has a squirrel cage rotor with a low ratio of rotor-to-fame diameter and high rotor resistance. This type gives the best overall performance and is efficient for converting input watts to shaft torque. There are other motor configurations that are used in specific applications. These include drag-cup motors and solid iron rotors.

The drag-cup motor shown in Fig. 10-28 is constructed with its rotor made of copper, aluminum, or an alloy. For a given size and weight as compared to the squirrel cage motor, it generally has lower torque. The heavy iron laminations are stationary, only the lightweight cup rotates, and the inertia is very small. The solid iron rotor core is used to carry both flux and induced rotor currents. This motor was designed for operation from the output of vacuum tube amplifiers and is rarely encountered today. The two-phase induction motor consists of two input windings (coils in slots of a laminated-iron structure) spaced 90 electrical degrees apart. Under a balanced condition, the windings are excited with equal voltages, 90° apart in time phase. The motor currents therefore generate magnetic fields in the air gap which are also in space and time quadrature. As with the induction motors, the rotating speed N_S is

$$N_S = \frac{120f}{p}$$

where f = frequency of line
p = number of poles

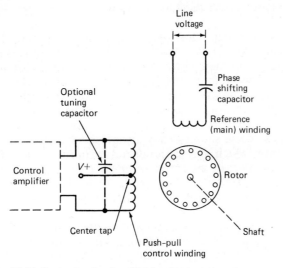

Fig. 10-26 Center-tap (push-pull) two-phase servomotor.

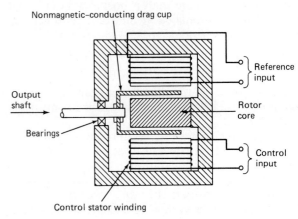

Fig. 10-28 Drag-cup two-phase servomotor.

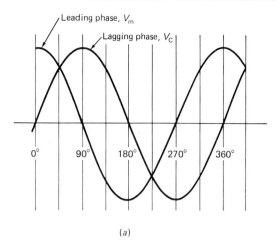

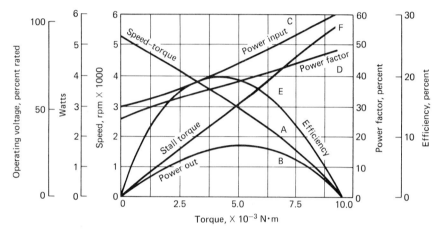

Fig. 10-29 Waveforms and characteristic curves for two-phase servomotor.

(b)

A 4-pole, 60-Hz winding causes the resultant field to rotate at 1800 rpm, and a 4-pole 400-Hz winding rotates at 12,000 rpm. Figure 10-29(a) shows the two-phase stator currents which develop the rotating field in a 2-pole machine. This rotating flux field induces a voltage in the rotor conductors with a magnitude proportional to the relative speed. The rotor voltages in turn cause currents; as a result a torque is developed by interaction of the current-carrying conductors and the rotating field. This drags the rotor along after the synchronous field of the stator. Since the rotor must overcome friction, it cannot reach synchronous speed. The difference between actual and synchronous speed is known as *slip*:

$$\text{Slip} = \frac{\text{synchronous} - \text{actual speed}}{\text{synchronous speed}}$$

In servomotors, the no-load speed is approximately five-sixths synchronous speed, corresponding to a slip of one-sixth, or 16.7 percent.

Figure 10-29(b) presents the characteristics of a typical servomotor from no load to the stalled condition. Curve A shows the variation of torque for rated phase voltage and varying load. As previously shown in Fig. 10-27, it is nearly linear. Curve B shows the power output and has a parabolic shape which peaks near one-half of the no-load speed. Curves C and D show the power input and the power factor, respectively. Curve E shows that the efficiency peaks near one-half of the no-load speed. Curve F plots the variations of stall torque as a function of control voltage with the fixed-phase voltage constant. This curve is linear and is a measure of the motor's stiffness. A servomotor should develop torque with a minimum amount of input wattage.

In ac servomotors, as mentioned earlier, there has to be a phase shift of the voltage on the main (reference) winding with respect to the control winding. There is no simple method of maintaining this phase shift for all motor speeds. Some designs use the two-capacitor method shown in Fig. 10-30, which gives good results on small servomotors.

A dc motor can be controlled by varying either the field current or the armature current. The types of dc servomotors are the series motor, the shunt motor, and the permanent magnet (PM) motor. These motors offer higher efficiency than an ac motor of the same size, but radio frequency interference (RFI) is a problem in some applications.

Most of the dc servomotors used in low-power applications are of the PM type. The ease of con-

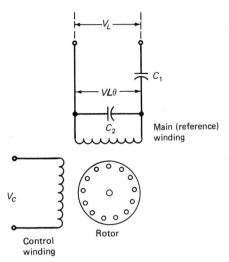

Fig. 10-30 Two-capacitor method of phase shifting main excitation.

trollable speed, along with the linear torque-speed control curve, makes the PM motor ideal for servomechanism applications. The characteristic curves are shown in Fig. 10-31. The speed-torque curve is quite similar to that of the ac servomotor presented earlier in this section. These motors are available in 6-, 12-, and 24-V models, making them applicable to solid-state circuitry. By comparison, the dc motor has some advantages over the ac motor. The dc motor inertia is greater than that of the ac motor (inertia is covered in the next section). This greater inertia is due to the wound armature and commutator, which produce a heavier rotor. The dc motor does not require any standby power; however, the ac servomotor continuously draws power for its main (reference) winding.

Figure 10-32 shows a modern dc servomotor that is only one-third as thick as a conventional motor. Such motors are popular in compact systems, such as numerical control and robotics systems, and in automatic control machines. At 24 V and 3.8 A, it develops 1.5 kilogram-centimeters (kg · cm) of

torque. Larger units with up to 4.5 kW of output power are available.

The dc servomotor in some modern servomechanisms may be one of the brushless types (covered previously) which lend themselves to easy computer control. Along the same lines, the stepping motor has become a valuable type to be used as a servo-type motor. This type of application will be covered in a later section.

In the past, large dc servo systems employed constant armature current, with the control signal applied to the field winding. These systems operated with vacuum tubes, thyratrons, or amplidynes providing the field drive. The power required for field control is only a fraction of the power required for armature control. Power amplification of the order of 20,000:1 can easily be obtained.

A *rate generator* (tachometer) is an electromechanical device resembling a small motor, which produces an output voltage that is proportional to its shaft speed that can be read out or used for closed loop speed control or stabilization. The dc rate generator is usually separately excited (shunt-wound) or is a permanent magnet generator. Though any dc generator can be used with a calibrated meter to indicate speed, more precise units are required for control-system applications. Special consideration is given to certain electrical and mechanical characteristics. The electrical output signal should be a noise-free voltage that varies linearly with speed. Mechanically, the unit should run quietly and smoothly, with low drag and low inertia. The relationship between voltage and speed can be expressed mathematically as

$$V = KN$$

where N = rate generator's speed
K = constant of proportionality between the voltage and speed

Most dc rate generators have an output sensitivity of about ½ V/100 rpm. Other important factors that differ from those of a standard dc machine are func-

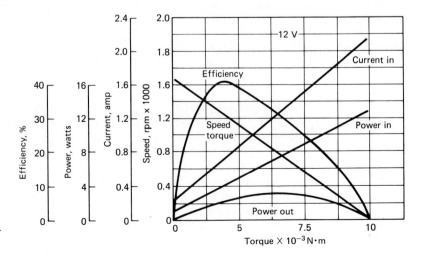

Fig. 10-31 Characteristic curves for a dc servomotor.

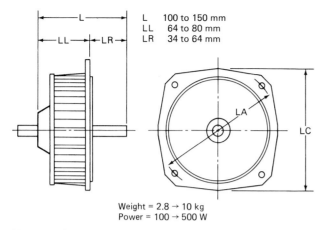

L   100 to 150 mm
LL   64 to 80 mm
LR   34 to 64 mm

Weight = 2.8 → 10 kg
Power = 100 → 500 W

Fig. 10-32 Space saver dc servomotor.

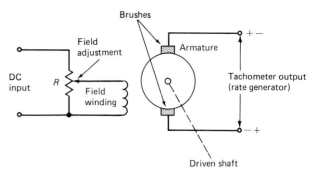

Fig. 10-33 Separately excited dc rate generator.

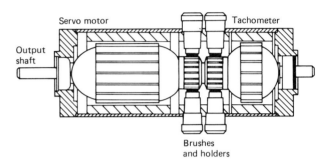

Fig. 10-34 Servomotor and rate generator on common shaft.

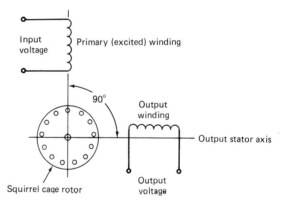

Fig. 10-35 Squirrel cage induction rate generator.

tions of the mechanical and electrical quality. Cogging is prevented by skewing the armature slots, and friction is reduced by limiting brush pressure and by using precision high-quality bearings.

Some of the advantages of the dc rate generator are the following:

1. Freedom from waveform and phase-shift problems
2. Absence of any residual zero-speed voltage (present in ac units)
3. High output gradients: up to 20 V/1000 rpm
4. Easier temperature compensation than in an ac rate generator

Some of the disadvantages of using dc rate generators are the following:

1. Brush problems: contact, vibration, arcing, and breaking contact. The position of the brushes must be exactly at neutral for reversible units.
2. Noise generation: filtering of high-frequency brush commutation noise is necessary.
3. Output ripple is undesirable and must be attenuated in some systems.
4. Brush friction and hysteresis effects require a higher driving torque.

The selection of a separately excited or a permanent magnet rate generator is based on factors such as line-voltage variations, ambient temperature variations, and the possiblity of demagnetization of the PM type. The output of the separately excited rate generator can be line-voltage-compensated, as indicated in Fig. 10-33, whereas the output of a PM type can only be attenuated.

Since a majority of rate generators (tachometers) are closely connected to the servomotor, many devices combine the motor and the tachometer on the same shaft. An example is shown in Fig. 10-34.

The ac rate generator must produce a sinusoidal signal output of *constant frequency* whose amplitude is proportional to its speed. The amplitude of the

voltage from an ac generator is proportional to its speed, but the frequency of the ac signal also varies with the speed. An ac rate generator used for servomechanisms is shown in Fig. 10-35. The unit has two stator windings: an excitation (reference) winding and an output winding. These windings are placed in the stator so that they are 90 electrical degrees apart. Because these coils are at right angles to each other, no output voltage is induced when the rotor is stationary. When the rotor turns, the flux produced by the eddy currents in the rotor is along the secondary axis and produces a voltage at the reference frequency in the output winding. The magnitude of the output voltage is proportional to the rotational speed. The direction of shaft rotation is indicated by the phase of the output voltage (compared to the reference voltage). If the output voltage is in phase with the reference, the direction is said to be *positive*.

If the output is 180° out of phase, the direction is said to be *negative*. The sensitivity of ac rate generators ranges from 1/10 to 1 V/100 rpm, with output impedances of 100 to 1000 Ω. The output voltage is mathematically expressed as it is for the dc rate generator.

There are three·types of errors associated with the induction-type ac rate generator: residual voltage at zero speed, nonlinearity, and voltage and phase errors at low speed. The zero speed output has a detrimental effect on a servo system's performance. The fixed and variable residual components as a function of rotor position are shown in Fig. 10-36(*a*). These residual voltages are minimized by the manufacturer by using precision machining techniques. The residual output voltage must be considered when replacements are required. Figure 10-36(*b*) shows a simple compensation network that may be employed to cancel the fixed component of the residual voltage.

If an ac rate generator is directly coupled to the servomotor shaft it adds directly to the inertia. For this reason, the drag-cup rate generator finds wide application. The drag-cup rate generator consists of the same type of stator (two-winding) as the two-phase induction rate generator just discussed. Its rotor consists of a thin nonmagnetic conducting material of aluminum or copper. Drag-cup rotors yield

maximum uniformity. This is accomplished by various manufacturing techniques and by having matched temperature coefficients for all mating parts.

No output voltage is induced when the drag-cup is stationary. Upon rotation, the eddy currents induced in the rotor cup distort the path of the flux so that a voltage proportional to shaft speed appears at the output winding. A notch or some other dissymmetry may be added to the rotor to cancel the inherent output dissymmetry.

A wide variety of incremental encoders are available and are being integrated into servomechanisms as rate generators. These devices are discussed in Chapters 3, 13, and 14. In many cases their usage is dictated by the environment of the application.

REVIEW QUESTIONS

15. The phase relationship between the reference voltage and the control voltage in a two-phase servomotor is _____ degrees.

16. The reference voltage is usually shifted by use of a _____.

17. Single phasing of a servomotor is a method of braking. (true or false)

18. The rotor of a two-phase servomotor is either a squirrel cage or a _____.

19. A 2-pole, 400-Hz motor will rotate at _____ rpm.

20. If the motor in question 19 rotates at 20,000 rpm, the slip is _____ percent.

21. The resistor in Fig. 10-36(*b*) is used for phase-shifting purposes. (true or false)

10-4
MECHANICAL COMPONENTS

A servomechanism will usually contain mechanical components such as gears, couplings, bearings, limit stops, and clutches. These parts are manufactured to tight tolerances for ease of assembly and for conformance with performance specifications.

Couplings are used to connect the ends of two shafts together so they always rotate at the same speed with the same angular position. Most couplings fall into one of the four following categories:

1. Rigid (sleeve) coupling
2. Flexible three-piece (Oldham)
3. Flexible-bellows, spring, one piece
4. Flexible-sleeve, one piece
5. Universal joints

The *rigid (sleeve) coupling,* as the name implies, is a one-piece coupling that rigidly couples two shafts together. Each end has a set screw to secure the sleeve to each shaft. One type requires that both

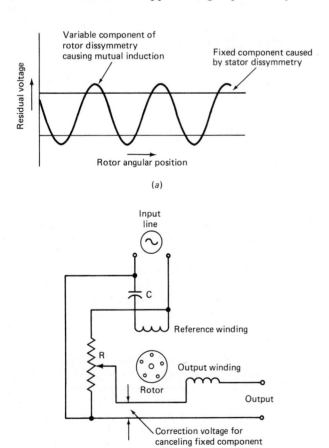

Fig. 10-36 Fixed and variable components of residual voltage and canceling circuit.

shafts be of exactly the same diameter; no misalignment is allowed. Another type of adapter coupling allows mating of two different diameter shafts. In specifying couplings, the letters *OD* stand for outside diameter, and the letter *B* indicates the bore (inside diameter). Dimensions are specified in inches and millimeters by most manufacturers. Mixing of sizes, that is, inches with millimeters, though it may seem tolerable, is not good practice. The set screws usually require use of an allen wrench for any adjustment. These will also be in inch or metric sizes and will require the appropriate tool. Some very-high-speed high-torque applications will use a recessed inner set screw with a second set screw on top.

A *flexible three-piece coupling* is also known as an *Oldham-type coupling*. This coupling allows for a slight angular or lateral misalignment of the shafts being coupled. It consists of two end hubs with machined surfaces to receive the center interlocking floating member. The floating (center) member, called the *torque disk,* is usually not metallic, but a plastic material such as delrin. The machined parts must be handled with care, and distortion will cause backlash in the coupling. *Backlash* is the play or lost motion that occurs between two loosely fitting parts.

The *bellows* or *spring coupling* consists of two hubs connected by a flexible metal bellows or a spring. It also allows for shaft misalignment. However, flexing of the bellows can cause metal fatigue, so these couplings are found only in low-torque applications. The spring coupling is preferred for high-speed applications and acts as a shock absorber.

A *flexible one-piece coupling* consists of a flexible element that may be polyurethane, rubber, or neoprene. This coupling can accommodate shafts that are out of alignment by as much as 1 in. (25.4 mm). They are quiet-running and absorb end play.

Universal joints can be either single or double types, as indicated in Fig. 10-37. The *single joint* will operate at angles up to 30°, and the *double joint* can couple shaft angles approaching 90°. Figure 10-37(*c*) shows a typical position for each type of universal joint.

The *clutch* can be thought of as a special type of coupling device. The most commonly used clutch in servomechanisms is the *slip clutch*. The clutch slips when the shafts reach a torque limit. Without the clutch, the servo may stall, potentially resulting in damage to the gearing or motor if the rotation energy is not dissipated in the clutch. Some models can be adjusted at the installation; others are fixed and are not field-adjustable. Occasionally, an electrically operated magnetic clutch is employed. For example, it could be used to disconnect a hand crank so it does not turn during normal system operation.

Bearings are an important part of any servomechanism. Sleeve bearings are usually oil-impregnated and are not commonly found in precision servomechanisms. For the most part, *ball bearings* are used because of their low-friction and low-wearing characteristics. Two common types of ball bearing

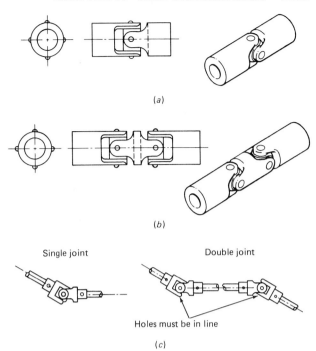

Fig. 10-37 Universal joints. (a) Single. (b) Double. (c) Typical applications.

mounts are the plain and flanged types. The *flange* type is popular because of the ease of snapping it in and out of place in its housing.

Some servomechanisms employ *limit stops* as a mechanical safety feature to prevent a shaft from rotating past a particular point. They are often used in conjunction with clutches to avoid damage. Some limit stops allow many revolutions before stopping by using a traveling nut on a screw thread. The threaded shaft is stopped when the traveling nut contacts either stop plate. In some cases, rubber or springs are used to absorb the shock when the stop is reached.

Gearing is required in a servomechanism to convert the high-speed low-torque power from the servomotor to a lower-speed higher-torque power to the controlled shaft. Many types of gears are available, including spur, helical, worm, bevel, internal pinion, and gear racks. Of all the gears mentioned, the *spur gear* is the most commonly employed. A *worm* is a gear with teeth in the form of screwthreads. Figure 10-38(*a*) shows a worm gear and a spur gear that can be mated to obtain a right-angle drive. Because of their high friction, right-angle drives are rarely found in servomechanisms. *Bevel gears* are conical in form and operate at intersecting axes, usually at right angles (Fig. 10-38[*b*]). *Internal gears* are usually limited to planetary drives and must mesh with an external gear. Figure 10-38(*c*) is an illustration of an external gear. A *helical gear* is cylindrical and has either right-hand or left-hand teeth, as shown in Fig. 10-38(*d*). They may be operated on parallel or crossed axes. Crossed helical gears are sometimes called *spiral gears*. When two gears run together, the one with

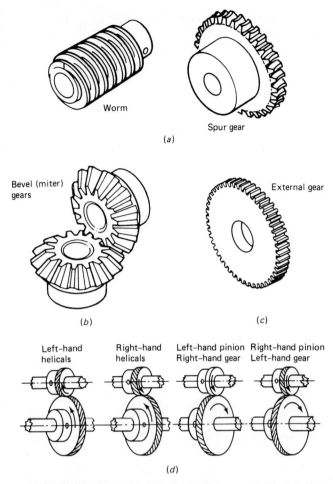

Fig. 10-38 Typical servo gears. (a) Worm gears. (b) Bevel (miter) gears. (c) External gear. (d) Helical gears.

the larger number of teeth may be called the *gear,* and the one with the smaller number of teeth may be called the *pinion.* A *rack* is a gear with teeth spaced along a straight line.

Uniform clearance of gears is critical in determining backlash between two gears. Backlash is proportional to the difference between the tooth space and the mating gear's tooth thickness. Increasing clearance will increase the difference and the backlash. Wear also increases the backlash, and so does loss of lubrication.

Differential gearing is used to mechanically add or

subtract the angular position of two shafts. Differentials are usually made from bevel gears, as indicated in Fig. 10-39. Differentials are available with speed ratios from 1:1 to 3000:1.

Some dynamic characteristics of gearing are important in servomechanism applications. Assume a pinion of diameter d_1 is driving gear d_2. An external torque T_1 is applied to the pinion shaft, causing a rotation with an angular velocity ω_1. The torque on the shaft is equal to the force developed at a point on the circumference of the gear times half the gear diameter:

$$T_1 = F\frac{d_1}{2}$$

Therefore,

$$F = \frac{2T_1}{d_1}$$

Similarly, with the second gear:

$$T_2 = F\frac{d_2}{2}$$

For meshed gears, the forces are equal and substituting for F gives

$$T_2 = T_1\frac{d_2}{d_1}$$

For proper gear meshing to take place, all teeth must be the same size, and the number of teeth on each gear is proportional to the pitch diameter. The *gear ratio* is the ratio of the number of teeth on the second gear to the number of teeth on the first gear. This leads to

$$\frac{d_2}{d_1} = \frac{N_2}{N_1}$$

Combining with the torque equation,

$$\frac{T_2}{T_1} = \frac{d_2}{d_1} = \frac{N_2}{N_1}$$

These equations show that the ratio of the torque is proportional to the ratio of the diameters, which is also proportional to the gear ratio. It can be shown that

$$\frac{\alpha_2}{\alpha_1} = \frac{\omega_2}{\omega_1} = \frac{\theta_2}{\theta_1} = \frac{N_1}{N_2}$$

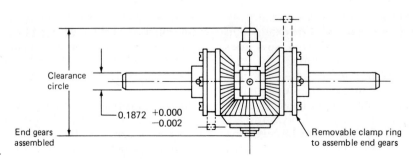

Fig. 10-39 Differential.

where α_1 and α_2 = angular acceleration of gears 1 and 2, respectively

θ_1 and θ_2 = angular position of gear 1 and gear 2

ω_1 and ω_2 = angular velocities of gear 1 and gear 2

Hence, the ratio of accelerations, positions, or angular velocities is inversely proportional to the ratio of the number of gear teeth. If the moment of inertia of the load is defined as

$$J = \frac{1}{2} Mr^2$$

where J = inertia, kilogram-meter^2 (kg · m^2)
M = mass, kg
r = radius of gear, m

The angular acceleration of a gear is equal to torque divided by inertia:

$$\alpha_1 = \frac{T_1}{J_1}$$

and

$$\alpha_2 = \frac{T_2}{J_2}$$

Substituting, we obtain

$$\frac{\alpha_2}{\alpha_1} = \frac{T_2 J_1}{T_1 J_2} = \frac{N_2 J_1}{N_1 J_2} = \frac{N_1}{N_2}$$

which results in

$$J_1 = J_2 \left(\frac{N_1}{N_2}\right)^2$$

Therefore, using a step-down gear box reduces the apparent inertia affecting the servomotor by a factor equal to the square of the gear ratio.

For example, if a rotating load has an inertia of 250 kg · m^2, through a 40:1 gear ratio, what is the resulting inertia on the motor?

$$J_m = J_l \left(\frac{1}{40}\right)^2$$

$$J_m = \frac{250 \text{ kg} \cdot \text{m}^2}{40^2} = 0.156 \text{ kg} \cdot \text{m}^2$$

where J_m = inertia of motor
J_l = inertia of load

The gearing also provides an improvement in load torque as compared to motor torque. The torque and inertia advantages are the main reason for the use of gear boxes in a servomechanism. However, do not forget that these advantages are obtained at the sacrifice of speed.

REVIEW QUESTIONS

22. The B diameter on a coupling signifies the _____ diameter.

23. The bellows coupling is only used in high-torque applications. (true or false)

24. Large angular or lateral alignment corrections are obtained by using _____.

25. The mechanism that allows slip to occur when torque limits are reached is called a _____.

26. Torque multiplication is obtained at the expense of _____.

10-5
AMPLIFIERS AND FEEDBACK

A system can be represented by a combination of blocks. Each block may have a single line input and a single line output. Each block may represent a single function. For example, the block shown in Fig. 10-40(a) represents an amplifier with a voltage gain of 250 times. With a 1-mV signal in, the amplifier will produce a 250-mV signal at its output. Figure 10-40(b) is a more general form with the input voltage, output voltage, and amplifier gain given as V_1, V_2, and G_1, respectively. It can be said that G_1 operates on the input V_1 to give V_2. If V_2 is divided by V_1, this result will be equal to the gain (G_1), provided that the amplifier stays within its linear range. This function can be expressed as

$$\frac{V_2}{V_1} = G_1 = \text{a constant}$$

Figure 10-41(a) shows a second amplifier connected in cascade with the first amplifier. The entire system can also be represented in reduced form, as shown in Fig. 10-41(b). It is important to note that G_1 is multiplied by G_2, which operates on V_1 to obtain V_3.

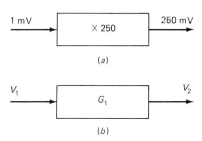

Fig. 10-40 Block representations of an amplifier.

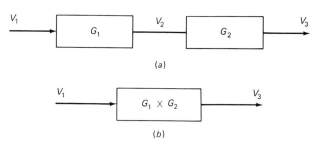

Fig. 10-41 Cascaded blocks.

EXAMPLE

If $V_1 = 1$ mV, $G_1 = 50$, and $G_2 = 50$, find V_3.

SOLUTION

$$V_3 = V_1 \times G_1 \times G_2$$
$$= 1 \text{ mV} \times 50 \times 50$$
$$= 2500 \text{ mV} = 2.5 \text{ V}$$

If G_2 were a potentiometer (voltage divider), its gain would be less than 1 (unity). At 50 percent rotation its gain (G_2) would be ½. The preceding total gain would now be 25 for the $G_1 \times G_2$ product. Since the gain of G_2 is ⅟₁₀₀ of the first example, the output will now be 25 mV.

The *summing junction* is used where signals are added or subtracted. In a block diagram, a circle is used as a summing junction as shown in Fig. 10-42. The Greek letter Σ may be used inside the circle to signify that a summing operation is to be performed. Summing junctions may perform addition or subtraction of two or more variables as shown in Fig. 10-42(*a* to *c*).

Figure 10-43(*a*) shows a gain block cascaded with a summing junction. The overall function can be found by

$$V_w = V_x - V_y + V_z$$
$$V_2 = V_w \times G_1$$
$$= G_1 (V_x - V_y + V_z)$$

if

$$G_1 = 20$$
$$V_x = -1 \text{ V}$$
$$V_y = -2 \text{ V}$$
$$V_z = +2 \text{ V}$$
$$V_w = -1 - (-2) + 2 = (-1 + 2 + 2) \text{ V}$$
$$= (3) \text{ V}$$
$$V_2 = G_1 \times V_w$$
$$= 20 \times 3$$
$$= 60 \text{ V}$$

An alternate representation is shown in Fig. 10-43(*b*). It illustrates that a junction can be viewed as several junctions. This may lead to simplified analysis in some cases.

Summing junctions are often based on operational amplifiers (op amps). Sum and difference amplifiers were covered in Chapter 6. Operational amplifiers may be used as noninverting or inverting amplifiers, depending on whether the signal is applied to the plus input or to the minus input. Figure 10-44 shows a summer (adder) circuit and a subtractor (difference) circuit. Other variations that are commonly used in feedback systems are shown in Fig. 10-45. Note that the summer (adder) is now of the noninverting type and that the difference amplifier output is now $V_2 - V_1$ as compared to $V_1 - V_2$ for the circuit of Fig. 10-44. These circuits are but a few examples of those available to fit the numerous mathematical needs of servomechanisms.

Servo amplifiers can be divided into two types. Most of the gain is contained in an early amplifier

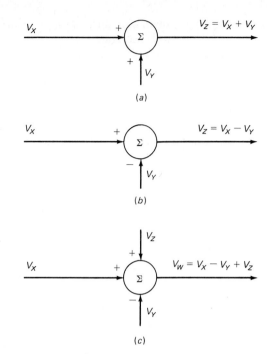

Fig. 10-42 Summing junctions.

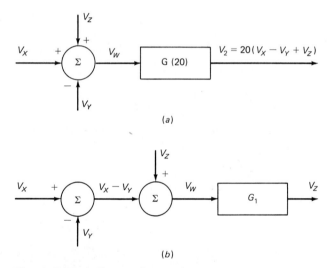

Fig. 10-43 Block diagram alternate forms.

often called the *preamplifier.* The power gain is produced in the final amplifier. The power amplifier must supply the required load voltage and current, which can be substantial, as in the case of a large servomotor.

Power amplifiers usually provide a frequency response from dc to 1000 Hz. A total servo amplifier (package) may appear as a solid-state unit. Figure 10-46 shows what is contained within such units. The amplifier gain can be adjusted as needed by resistor selection. The frequency response is adjusted during installation by selecting and connecting frequency-compensation components. In some low-power applications, the power amplifier stage may be a com-

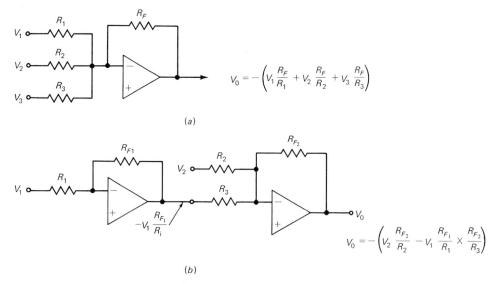

$$V_0 = -\left(V_1 \frac{R_F}{R_1} + V_2 \frac{R_F}{R_2} + V_3 \frac{R_F}{R_3}\right)$$

(a)

$$V_0 = -\left(V_2 \frac{R_{F_2}}{R_2} - V_1 \frac{R_{F_1}}{R_1} \times \frac{R_{F_2}}{R_3}\right)$$

(b)

Fig. 10-44 Op Amp circuits. (a) Summer (adder). (b) Subtractor.

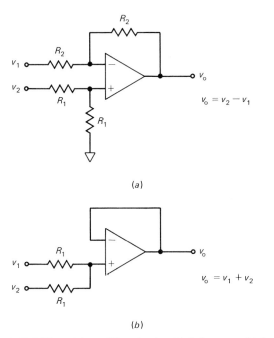

$v_0 = v_2 - v_1$

(a)

$v_0 = v_1 + v_2$

(b)

Fig. 10-45 Differential amplifier circuits. (a) Subtractor. (b) Summer (adder).

plementary pair of transistors connected directly to the output of the preamplifer, as discussed in Chapter 6.

Servo amplifiers are available with output power ratings up to 5 kW. In those cases in which maintaining low-voltage offsets with time and temperature is essential or when external set adjustments are not practical, a chopper-stabilized amplifier is used to achieve drifts as low as 0.1 μV/°C. The chopper amplifier is a high-gain feedback amplifier, containing a MOSFET chopping transistor. The chopper converts the difference between the dc or low-frequency input voltage and the feedback voltage to a high-frequency square wave and amplifies it with no drift. The high-frequency square wave is then rectified and filtered to produce an output waveform that is an amplified version of the input.

Figure 10-47 presents a detailed block diagram of a chopper-stabilized amplifier. In a system diagram, it may be simplied and represented by a single OP-AMP symbol. The chopper-stabilized amplifier is best understood by looking at the waveforms in Fig. 10-47. The input signal, V_{in}, is split into two components by high-pass network $C_1 - R_1$ and by the low-pass network $R_2 C_2$. The low-frequency signals are applied to the chopper, along with a square wave that gates the chopper transistor off and on at a high frequency rate. The ac amplifier can be designed for high gain and drift will not be a problem since its frequency response does not extend down to dc (it will utilize coupling capacitors between stages). The amplified output is then peak-detected to recover the low-frequency and dc components. Capacitor C_3 filters the detected signal to remove any component of the chopper signal. The output signal V_{out} is formed by summing the detected and filtered signal with the high-frequency signal.

In cases in which overloads are unavoidable or may create very long recovery time, an overload circuit may be incorporated into the amplifier. Such a circuit is illustrated in Fig. 10-48. This overload recovery circuit will prevent the amplifier circuitry from saturating. The input circuit is protected from large signals by a diode clipper. The feedback circuit will drop in impedance with a large swing in V_{out} due to the zener's becoming forward-biased. This will lower the amplifier gain and allow the amplifier to recover in 1 μs. Without this overload protection, it may take up to 10 s to recover.

A *closed-loop system* is one in which the output of a process affects the input. Adjustments are made by the control system until the difference between

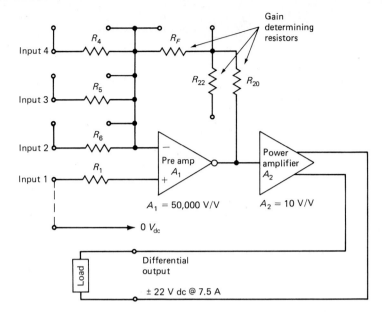

Fig. 10-46 Servo amplifier block diagram.

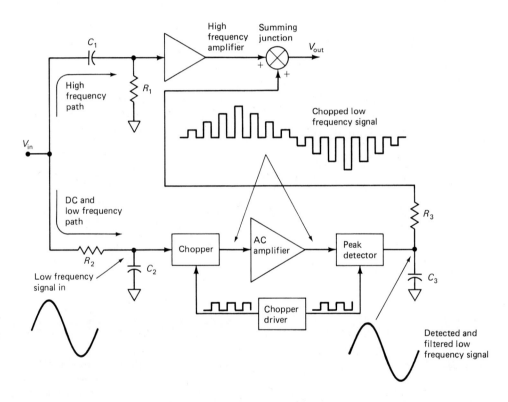

Fig. 10-47 Chopper-stabilized amplifier block diagram.

the desired and actual output is as small as practical. In other words, the controlled parameter (output), whether position, angle, or speed, is sampled and fed back to the input, where it can be compared with the desired condition. Look at Fig. 10-49, in which a human operator is trying to maintain a speed of a trolley at 50 mph (80 kph). The operator observes the speedometer and decides (compares) whether to increase or decrease the speed control, depending upon whether the indicator is above or below the desired 50 mph (80 kph). Note the two signal paths: (1) a forward path from the operator handle control-

ling the speed (speed is the output), (2) a backward (feedback) path via the speedometer to the operator, serving as the comparator to the speed controller. The input to the speed-control handle is the difference between that indicated by the speedometer and the desired speed computed by the operator. It would be almost impossible to maintain a constant speed without feedback. If we had no way of knowing the speed we would have to guess.

Block diagram representation is a technique commonly used in control system analysis. Consider the illustration in Fig. 10-50(a), which shows a simple

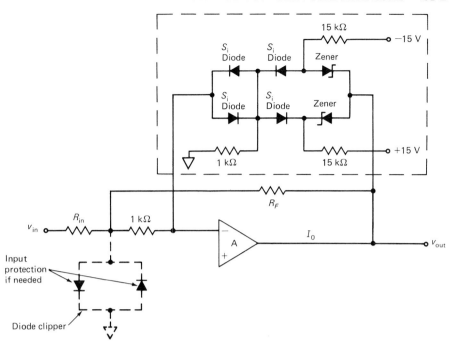

Fig. 10-48 Overload recovery circuit.

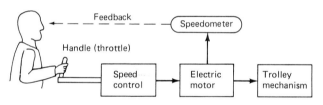

Fig. 10-49 Simple closed loop feedback system.

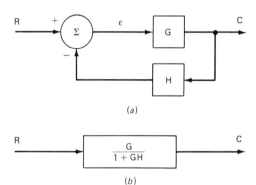

(a)

(b)

Fig. 10-50 Servo system block reduction. (a) Simple system. (b) Block reduction.

form of feedback control system. The block diagram can be reduced by block diagram algebra into a form that is commonly used and should be remembered. From Fig. 10-50(a)

$$\epsilon G = C$$

also,

$$R - CH = \epsilon$$

Substituting for ϵ yields

$$(R - CH)G = C$$

or,

$$RG - CGH = C$$

Combining

$$RG = C + CGH$$

then,

$$RG = C(1 + GH)$$

The control ratio is defined as the output C over the command signal R and is equal to

$$\frac{C}{R} = \frac{G}{1 + GH}$$

Figure 10-50(a) can now be reduced to the single block shown in Fig. 10-50(b) having the same input and output as the original.

The unity feedback system shown in Fig. 10-51(a) can also be reduced to one block and the control ratio C/R derived. The feedback loop can be considered to have a value of $H = 1$. All the output C is fed back to the summing junction. The reduction will therefore be the same as previously shown in Fig. 10-50(b):

$$\frac{C}{R} = \frac{G}{1 + G \times 1}$$
$$\frac{C}{R} = \frac{G}{1 + G}$$

The reduction is shown in block form in Fig. 10-51(b).

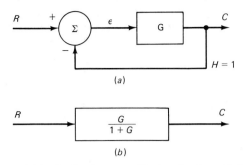

Fig. 10-51 Unity feedback system block reduction. (a) System. (b) Block reduction.

There are five important characteristics for a closed loop system:

1. Accuracy
2. Sensitivity
3. Resolution
4. Linearity
5. Frequency response

Together, all five characteristics provide a complete description of the system.

The *accuracy* is an indication of how closely a system meets the desired control parameter. If the normal speed of a system is 77 kph and the actual varies between 70 and 84 kph, the system is said to be accurate to ±7 kph. Accuracy may also be defined as a percentage of deviation. Instead of ±7 kph we could specify ±9 percent deviation. The percentage of error may be based on a maximum value. For example, if the maximum value were 210 kph, the maximum error of ±7 kph would become

$$\text{max error, }\% = \frac{7 \times 100}{210} = 3.33\%$$

which is less than a third of the nominal percentage error.

The *sensitivity* of a control system specifies the level of input required to obtain a desired output. A speed-control system may have a sensitivity of 1000 rpm/V; a temperature-control system may have a 100°/V sensitivity. Therefore, sensitivity has various interpretations, depending on the specific system used as reference.

Resolution is defined as the smallest quantity recognizable by the system. Resolution can be specified in percentage of maximum or in absolute units. If a thermometer has 1° markings, 1° is the smallest recognizable quantity.

Linearity is defined as the amount by which a relationship between two quantities deviates from a straight line. Linearity is usually expressed as a percentage of a value or as a percentage of maximum. Operation within a linear working range is obtained only if the system signals are restricted in magnitude to avoid nonlinear regions of the components. For

example, if an amplifier is driven into saturation by a large input signal, linearity is lost.

Amplifier frequency response was covered in Chapter 6. The same definition applies to a control system. The output of the system may be plotted on a frequency-response curve and the −3-dB points obtained. The *response time* of a system is usually used, rather than the frequency response, in control systems. This is the time a system takes to respond to an input signal. By definition it is the time needed for the output to change from 10 to 90 percent of its final value when a step signal is applied to the input as shown in Fig. 10-52(a).

This is referred to as *rise time* and it is also the response time of a servomechanism. In Fig. 10-52(b), curve (a) shows an underdamped response. The output overshoots the value dictated by the systems input, then undershoots this value, and finally settles to a value close to the input value. This type of response has an oscillating or ringing effect. In the overdamped response (b) the output does not overshoot the desired value but takes a very long time to reach its final value. The third response (c) is that of a critically damped system, in which the output reaches its final value in the minimum possible time without overshooting the desired final value.

Both the mechanical and electrical components of a control system have response-time characteristics, and both determine the overall system performance. After a transient period, a final steady-state value is

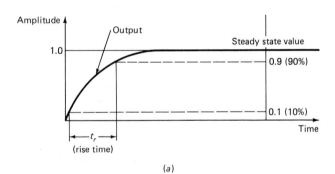

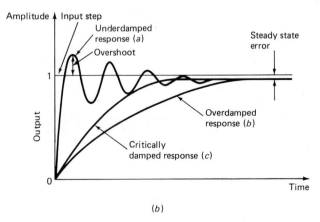

Fig. 10-52 Transient response curves. (a) Rise time (response time). (b) Response curves.

obtained. The difference between the final steady-state output reached and the value called for by the input is called the *steady-state error*. The steady-state error is shown in Fig. 10-52(*b*). The gain of the system can be increased to reduce the steady-state error. However, with increased gain, a situation in which the oscillations obtain a fixed amplitude and the system never settles or reaches steady state may arise. This is an *oscillatory system,* and it is said to be unstable. Open-loop systems are never unstable, but when feedback is introduced, the control system can become unstable. Any feedback system may become unstable if the feedback is in phase with the input. Amplifiers and feedback networks exhibit phase errors that are especially pronounced at the frequency limits. It is possible for negative feedback to become positive feedback at some frequency extreme. For this reason servomechanisms tend to become unstable when the gain is increased in an effort to reduce the steady-state error. In order for oscillations to occur, the magnitude of G must be greater than 1 when the phase error is $-180°$. Various techniques, such as Bode and Nyquist plots are used to ensure stability when the system is designed. It is important to understand that system stability can be lost if gain or phase-shift networks are altered.

There are three types of servomechanisms:

1. Type 0: A constant input signal (*x*) will result in a constant position at the controlled output (*y*).
2. Type 1: A constant input signal (*x*) results in a constant velocity at the output (*y*).
3. Type 2: A constant input signal (*x*) results in a constant acceleration at the output (*y*).

The control system type is determined mathematically by examining the *G* and *H* factors of the loop which determine its transfer function. A mathematical analysis of a control system's transfer function yields two key factors. The first is the steady-state response of the system to three types of inputs. The second is the steady-state error, which is either zero, finite and constant, or infinite.

Figure 10-53(*a*) shows a *type 0 servomechanism* (often called a *position* or *follow-up system*). The output shaft (either driven directly or through a gear box) is to follow the angular setting of the input potentiometer. If a step input is applied to the type 0 system, the steady-state error, E_{ss}, for a step (also called a *position* or *setpoint*) with a value of *P* is

$$E_{ss} = \frac{P}{1 + K}$$

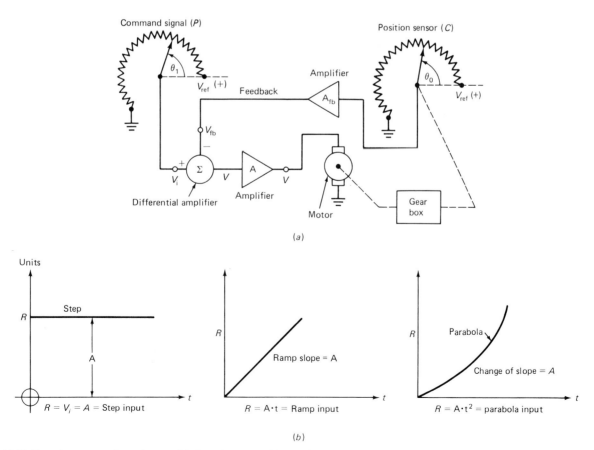

Fig. 10-53 Type 0 system and test inputs. (a) Type 0 system. (b) Test inputs.

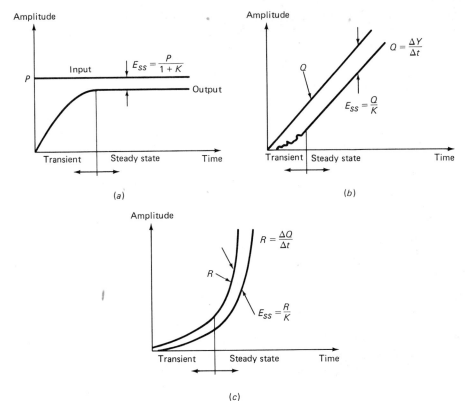

Fig. 10-54 Steady-state error for different system types. (a) Type 0 system. (b) Type 1 system. (c) Type 2 system.

The larger the value of K (gain) the smaller the error, but as mentioned earlier, large values of gain can make a system unstable. This response for a type 0 system is shown in Fig. 10-54(a). If a ramp or acceleration (parabolic) input is applied to a type 0 system, the output cannot follow it, and the steady-state error increases with time and approaches a value of infinity. These inputs are shown in Fig. 10-53(b).

The *type 1 control system*, which is known as a *rate (velocity) servo*, is shown in Fig. 10-55. As defined earlier, the output shaft will run at a constant velocity (speed) for a constant input. The steady-state error of the type 1 system to the step input is zero, which is the desired circumstance. The steady-state error of a type 1 system due to a ramp input Q is shown in Fig. 10-54(b). If the system gain is K, the steady-state error is

$$E_{ss} = \frac{Q}{K}$$

As with the type 0 system, increasing K will decrease the steady-state error. A type 1 system cannot follow an acceleration (parabolic) input, and the steady-state error for this type of input diverges (increases) as time increases.

A *type 2 system* has a steady-state error of zero for both position and velocity inputs. If the input is an acceleration of value R, then the steady-state error E_{ss} is as shown in Fig. 10-54(c) and is equal to

$$E_{ss} = \frac{R}{K}$$

The type 2 control system is seldom used industrially and is more commonly used in missile and guidance systems. The steady-state errors for the three types of systems are summarized in Table 10-2.

If a control system is found to be marginally or inherently unstable, a compensating network may be added to improve the system's gain and phase margin. These networks may be based on integrators or differentiators, and their applications in servo damping are shown in Fig. 10-56. These stabilizing networks are usually found either preceding or incorporated within the amplifier (gain stage) of the system. They are critical for proper system response

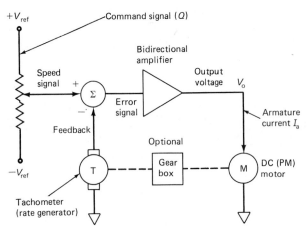

Fig. 10-55 Type 1 (rate/velocity) system.

TABLE 10-2 STEADY-STATE ERRORS FOR VARIOUS INPUTS TO THREE TYPES OF SERVOMECHANISMS

Input Type	System Type		
	0	**1**	**2**
Position, P	$\dfrac{P}{1 + K}$	0	0
Velocity, Q	Infinity	$\dfrac{Q}{K}$	0
Acceleration, R	Infinity	Infinity	$\dfrac{R}{K}$

and should not be modified; if they are modified, oscillations may occur.

Steady-state error can be reduced by increasing loop gain, but stability can be lost if the gain is too high. Compensating networks improve the phase margin and allow greater gain for improved performance. Integral damping is a related technique to reduce steady-state error. Integrators are presented in Chapter 6. The output of an integrator will ramp in response to a steady signal applied to its input. In the case of a type 0 servo, as in Fig. 10-56(a), suppose a step command signal is applied to position the output to a new location. An immediate and relatively large error signal results, driving the input of the summing amplifier. The summing amplifier output drives the controller in a direction that eliminates the error. When the step signal is first applied, the integrator output is zero, and the system responds as if the integrator were not in the circuit. As time passes, the error signal decreases, but the integrator output increases. Without the integrator, a small residual error signal would be present when the controller finally stopped. However, with the integrator the error signal will eventually be removed because the integrator output continues to ramp as long as any residual error voltage remains at its input.

Integral damping may also be added to a type 1 servo. Suppose there is a sudden change in the command set point. The immediate reponse will be a relatively large error signal that will be amplified and will act on the output device, thus reducing the error to a small value as the output speed comes close to matching the command signal. However, because of friction and loading torque on the output, there will be some steady-state error and some residual error voltage. If this error voltage is integrated and applied to the input of the amplifier, it will eventually be removed as the integrator output continues to ramp in response to the steady-state error signal.

The underdamped response that was shown in Fig. 10-52 is undesirable. Because of inertia (both mechanical and electrical) the controlled parameter overshoots the set point value, then undershoots it, and so on. Gain can be reduced to control overshoot but at the expense of response time and accuracy. Derivative damping is a technique based on a second

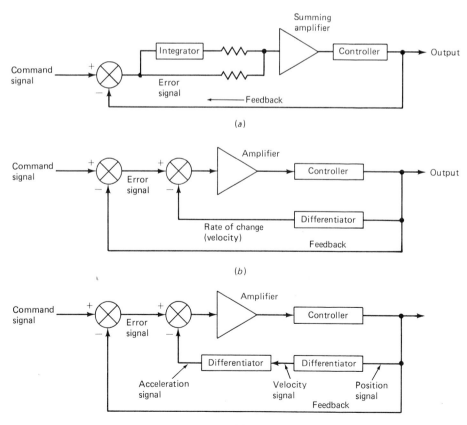

Fig. 10-56 Compensation/damping techniques. (a) Integral damping. (b) Derivative damping. (c) Second-derivative damping.

feedback loop that contains a differentiator. (Differentiators are covered in Chapter 6.) The output of a differentiator is proportional to the rate of change at its input. When derivative damping is added to a servomechanism, overshoot and undershoot can be reduced without sacrificing loop gain. Suppose a change in command signal is applied to the damped system shown in Fig. 10-56(b). Because of system inertia, the immediate response is slow; the output of the differentiator is also small, so it has little impact on the input to the amplifier. As the system continues to respond, the error signal begins to decrease, and the rate of change is now increasing, so the output of the differentiator is increasing. Eventually, the error signal and the rate signal (velocity) will be equal, resulting in no input to the amplifier. However, the system continues to respond because of its inertia. Now, the differentiator output is greater than the error signal which reverses the polarity of the amplifier output. The controller responds by reversing, and the overshoot is reduced (damped).

Everything in electronic systems has an analog in mechanical systems, and the reverse is also true. For example, mechanical damping (the analog of the electronic damping just discussed) is also possible. Friction forces that oppose oscillatory motion can be developed. Simple brakes are usually not used because they degrade response time and accuracy. Rate brakes that respond to the rate of change are used. These are usually called *viscous damping systems* and include devices such as fluid turbines, magnetic particle brakes, and eddy current types.

Derivative damping can cause an error in type 0 servomechanisms if there is any output from the differentiator under steady-state conditions. The first derivative of motion is velocity, and the second derivative of motion is acceleration [Fig. 10-56(c)]. Therefore, a second-derivative damping system provides damping correction only when the output is accelerating or decelerating.

Servomechanism characteristics can also be controlled by digital techniques. With the advent of the microprocessor, there is an increased tendency to calculate the correct input to the servomotor at any given time to provide the desired response. Digital techniques and microprocessors are covered in the next two chapters.

REVIEW PROBLEMS

27. If in Fig. 10-42(c) V_x is $+3$, V_y is $+3$, and V_z is -4, then V_w is _____.

28. If V_w of question 27 is fed into a block with G_1 of 0.25 V/V the output is _____ V.

29. In Fig. 10-44(b), V_1 is $+4$, V_2 is -2, and R_F equals R_1 equals R_2 equals R_3, V_{out} equals _____ V.

30. The amplifier that drives a servomotor is a _____ amplifier.

31. For special high-stability cases, a _____ amplifier may be employed.

32. To prevent lock-up, an amplifier may incorporate an _____ circuit.

33. A type 1 system will have an error equal to _____ for a ramp input as shown in Fig. 10-54.

10-6
ROTATING AMPLIFIERS

When the control of a large power load is required, the choice is usually limited to hydraulic systems or electrical systems involving motors, mostly dc. If a dc motor is chosen for the output, a rotating power amplifier may be used to excite it. Rotating amplifiers are being replaced with solid-state devices as the power capabilities of these devices continues to improve. Most new systems employ solid-state amplifiers. But rotating amplifiers are still in use at many industrial sites, and familiarity with their characteristics is still important.

Rotating amplifiers fall into one of three basic categories. They are the Ward Leonard system, the Regulex or Rototrol generator, and the Amplidyne generator. Each system has a generator driven by an ac induction motor at a constant speed, 3600 rpm for small to medium systems and 1800 rpm for large systems. They can be considered as amplifiers because a small change in field current will produce a large change in armature current.

The Ward Leonard system employs a simple motor-generator set. The circuit shown in Fig. 10-57 is a Ward-Leonard system in its simplest form. The dc motor in the circuit is fed directly from a dc generator, which is operated at a constant speed. The generator may be viewed as a power amplifier since the power required to excite the field is much lower than the power output from the armature. The dc field to the generator (F_g), is adjustable in magnitude and polarity by means of the field rheostat and the reversing switch. The motor armature is supplied by the generator, which provides a smoothly varying voltage from zero to some full-load value. The dc motor field (F_m) is fed from a constant source derived from the ac line. The generator is driven by an ac induction motor of constant speed. This motor may be either single- or three-phase, depending upon the size of the application. The Ward Leonard system allows a small variation in field current to provide a smooth, reversible, flexible, and stable control of a large dc motor. Such systems are employed in hoists and elevators and in large machining centers.

The system shown in Fig. 10-57 is open loop and only applicable where an operator must have complete control at all times. This system can be modified for closed loop operation by providing a feedback signal to be compared to the control input. The feedback can be proportional to rate (type 1) or to position (type 0). Figure 10-58(a) shows how the system can be connected with tachometer feedback to provide motor-speed regulation. The motor speed is measured by the output from the dc tachometer (rate

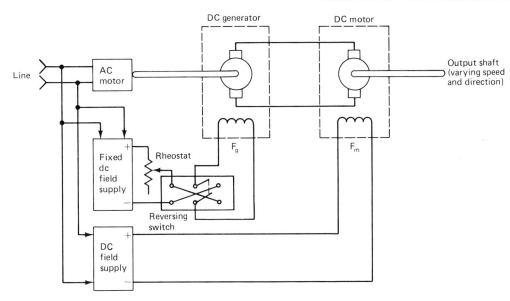

Fig. 10-57 Simple Ward-Leonard system.

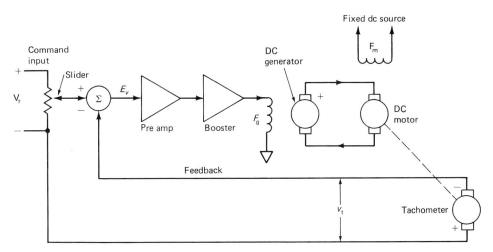

Fig. 10-58 Ward-Leonard rate (velocity) feedback system.

generator) coupled to the motor shaft. The tachometer voltage V_t is compared with the reference voltage V_r. The difference is voltage-amplified by the preamplifier and then power-amplified by the booster to drive the field of the generator.

The Ward Leonard system can also be used to control the angular position of an output shaft at a high power level by following an input signal at a lower power level. This type of system is shown in Fig. 10-59. The output angle θ_o is converted into an electrical signal, which is compared with the command signal. Any error is amplified and applied to the field circuit of the dc generator. The slider/wiper of the position-sensing potentiometer is driven by the output shaft through a gear box. Some systems use synchros and control transformers in place of potentiometers for input and feedback devices.

The Regulex generator and the Rototrol generator are trade names of Allis-Chalmers and Westing-

house, respectively. They are employed in systems rated at 5 kW and above. Both are basically dc generators which employ self-excitation as a method of increasing amplification. The Rototrol generator commonly uses a series field for self-excitation. The Regulex generator, in most applications, uses a shunt field. The typical magnetization curve of the dc shunt generator shown in Fig. 10-60(b) is shown in Fig. 10-60(a). The simple shunt generator is driven at a constant speed with the field switch (S) open. A small residual magnetism is assumed to generate some value of emf at zero excitation. The straight line (θ_a) is called the *field-resistance line* and is a plot of

$$V_f = R_f \cdot I_f$$

The slope of the field-resistance line is determined by the field rheostat. This slope (O_a) is less than the air-gap line (a straight-line projection of the magnetization curve through the origin). Point (a) is the

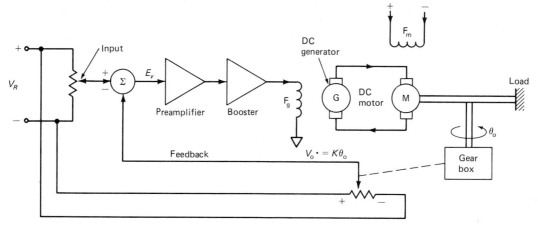

Fig. 10-59 Ward-Leonard position control system.

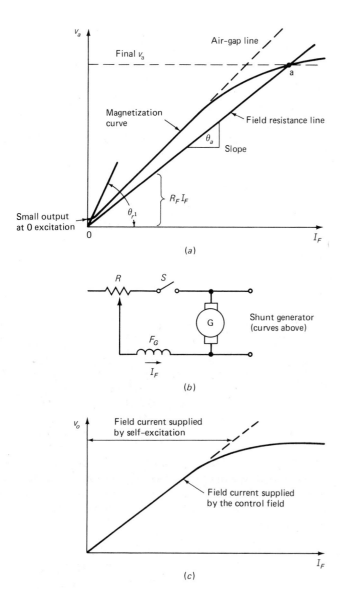

Fig. 10-60 Shunt generator and Regulex-generator curves.

intersection of the field-resistance line with the generator's magnetization curve. If the switch (S) is closed, the residual output will start a build-up of the shunt-field current. If the connections are such that positive feedback results, the generator voltage builds up until limited by magnetic saturation at point (a). The generated voltage will just satisfy the field current required to sustain itself.

If the field resistance is large, the resulting field-resistance line will be small ($\theta,'$ in Fig. 10-60[a]), and very little output voltage build-up will occur. The value of R_f corresponding to the slope of the air-gap line is called the *critical field resistance*. An adjustment to this value is referred to as *tuning*. Suppose the resistance of the self-excited field is increased until it has the slope shown in Fig. 10-60(c), and the necessary additional field current required is supplied by an auxiliary control field. Up to the saturation of the iron, the output will be proportional to the applied control current, which is only a small percentage of the total field required.

The self-excited generators have a number of field windings. Up to eight separate windings can be put on a machine. For a given power output, these added windings will increase the physical size of the generator compared to that of a conventional model. Figure 10-61 shows two basic self-exciting generators. Both have critical values of resistance, and both can be tuned. The main differences show up in the magnitudes of field currents involved. The shunt generator will have a small field current flowing in a high-impedance winding of many turns. The series generator shows a large current flowing in a low-impedance winding of few turns. Whether shunt or series, the field is *tuned* to the air-gap line, and operation is on the linear part of the magnetization curve.

When a change in the output voltage is necessitated by a change in the load requirements, the control fields are used to establish the new operating point. The power required by the control winding is very small because it only has to initiate change or

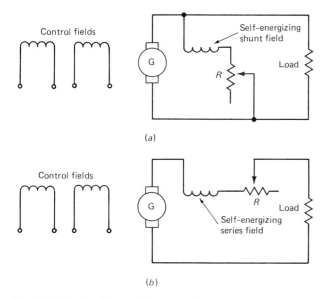

(a)

(b)

Fig. 10-61 Basic self-energizing generators.

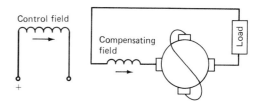

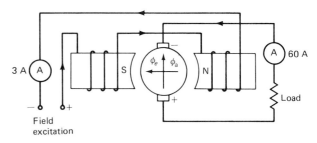

Fig. 10-63 Schematic diagram of an amplidyne.

Fig. 10-64 Magnetic fields and currents in a conventional dc generator.

stabilize steady operating conditions. Figure 10-62 shows a voltage control system with an exciter (E), control winding (C), and a tuned shunt-field winding (S_f). The control field voltage V_c is obtained by feedback comparison of the generator voltage V_g with the reference voltage (V_r). The exciter is driven at a constant speed. At steady-state conditions, the shunt field provides all the required excitation, and error voltage V_c is zero. Any tendency to drift from this point results in a correction voltage V_c across the control field. Suppose a load change decreases the output voltage. This error voltage $V_c = V_r - V_g$ is applied to the control field, and the exciter output builds up. This output will continue to change as long as the error exists. The final operating point will shift to the point at which the self-energizing shunt field again alone supplies all the excitation required by the new load requirements. At this new operating point, the steady-state error V_c is again zero. The series-excited generator will act in essentially the same way as the shunt generator just discussed.

The amplidyne motor generator consists of a constant-speed ac drive motor and a two-stage electromechanical power amplifier contained in a single housing. As the symbol in Fig. 10-63 shows, the

amplidyne generator has two sets of quadrature brushes with one set shorted. The drive motor, usually a squirrel cage type, has its rotor shaft coupled to the armature of the generator section. Since this motor drive is similar to previously discussed systems, it need not be covered again. The amplidyne is radically different from the conventional generator because of the unusual method employed to obtain high-power amplification.

Figure 10-64 shows a dc generator with a 60-A load on the armature. To meet this demand, the armature must have induced in it sufficient voltage to force the required current to the load. Therefore, the armature conductors must cut a magnetic field of a certain flux density to provide the required output. A field current of 3 A may be necessary in this case. The generator can now be considered as a current amplifier with a gain of 20.

Figure 10-65 is the same as Fig. 10-64, except that the load has been removed and the armature leads short-circuited. Since the load resistance is gone, the only significant opposition to current flow is the resistance of the armature windings. This condition would produce abnormally high armature current and

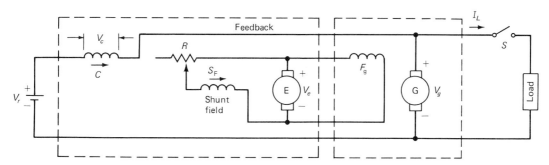

Fig. 10-62 Voltage-regulating system with tuned shunt-field exciter.

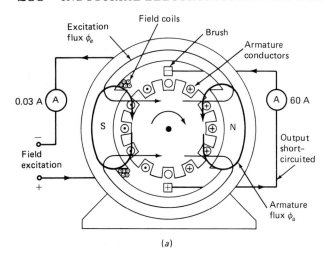

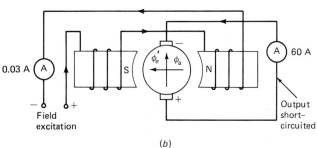

Fig. 10-65 Magnetic fields and currents in a short-circuited dc generator.

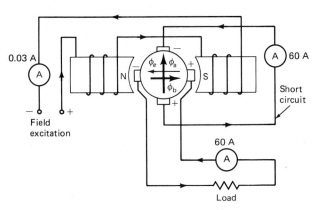

Fig. 10-66 Short-circuited generator supplied with additional brushes.

quickly lead to a burned-out armature. However, one way to reduce the enormous current is to reduce the excitation flux to a much lower level. It is possible to reduce the short-circuit current in the armature to 60 A by reducing the flux. Since the armature handled a 60-A load before, a short circuit of the same current will not cause any damage. A reduction in the field excitation current will weaken the flux to the proper level, in this case, perhaps 0.03 A (30 mA). It can be seen that 0.03 A now controls a short-circuited current of 60 A. Before, 3 A with a load applied was required. The generator gain has increased to 2000. The problem as to how this increased power gain can be put to use now arises. Obviously, the load cannot be put in series with the short circuit, since this would just be a return to the original circuit. The short circuit must remain. It can be seen in Fig. 10-65(b) that two flux circuits exist: ϕ_e, a weak excitation flux, and ϕ_a, a strong armature flux due to the 60 A. The cross section of Fig. 10-65(a) shows that the armature conductors are evenly spaced around the core. They will cut across the heavy armature flux, ϕ_a, at the same rate that they will cut the excitation flux ϕ_e. The maximum voltage induced in the conductors as they cut the armature flux will be at right angles to the voltage induced by the excitation flux. To take advantage of this second voltage, a second set of brushes, shown in Fig. 10-66, is added to the commutator at right

angles to the short-circuited brushes and connected to the load. The voltage developed across the second set of brushes is sufficient to supply a 60-A current to the load.

Another problem arises, as can be seen in Fig. 10-66. As the armature current in the short-circuited section creates a flux at 90° to the excitation flux, so will the load current set up a flux at 90° to the armature flux. This new reaction flux, ϕ_b, is 180° from the original excitation flux, ϕ_e. The reaction flux is much stronger than the excitation flux, and because it opposes it, the excitation flux no longer has control of the output. To overcome this condition, a compensating winding is placed on the pole pieces and is connected in series with the load. The number of turns is adjusted so that the compensating flux ϕ_c will exactly cancel the load armature reaction flux for all values of load current within the operating range. The equivalent circuit is shown in Fig. 10-67.

Since any residual magnetism along the axis of the control field would considerably affect the Amplidyne output, it is necessary to demagnetize the core material. A small ac generator is used to eliminate any residual magnetism. This generator has a permanent magnet attached to the end of the armature.

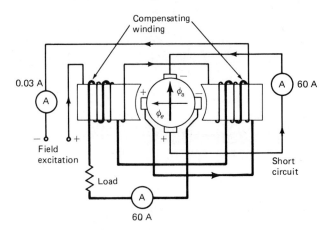

Fig. 10-67 Amplidyne generator equivalent circuit, showing magnetic fields.

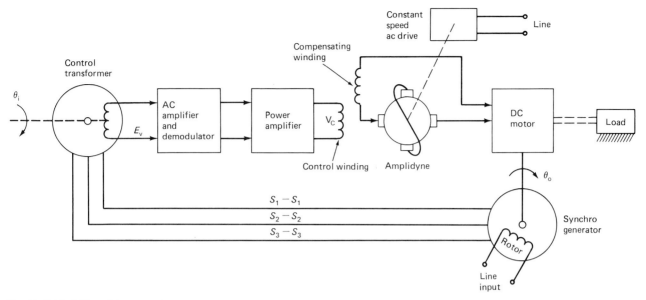

Fig. 10-68 Hydrid position control system.

The magnet revolves within a separate field winding and generates a small ac voltage, which is applied to the two sets of opposed windings on the field pole pieces. They are sometimes called *killer windings*. The generated alternating current neutralizes any residual magnetism when the control field is zero.

The power gain obtainable from an Amplidyne varies from 700 to 100,000. For example, typical gains are 5000 for a 500-W unit and 25,000 for an 8-kW unit. Figure 10-68 shows a type 0 position control system utilizing an Amplidyne with synchros as the feedback control elements.

REVIEW QUESTIONS

34. The Ward Leonard system is always an open loop system. (true or false)

35. The Ward Leonard system in Fig. 10-59 is _____-directional.

36. The Regulex generator typically uses a _____ field for self-excitation.

37. The adjustment of the field resistance in a Regulex generator is called _____.

38. In Fig. 10-62, voltage V_c is also called the _____ voltage.

10-7
TROUBLESHOOTING AND MAINTENANCE

Servomechanisms are much like other electronic systems in that the fundamental measurements of voltage, current, and resistance are of primary importance. The technician must test and analyze the system to ensure that it is performing properly and within specified tolerances. The best way to maintain a system is to be aware of its performance, so that if there is a tendency toward error it can be discovered and corrected before it becomes detrimental.

During their useful life, potentiometers will undergo certain changes in their characteristics. The two most common characteristics that change are linearity and noise. Linearity is checked with a suitable master potentiometer whose accuracy is known to be at least ten times that of the unit to be tested. The method of testing is shown in Fig. 10-69(a), where the master and the potentiometer under test are connected in parallel across a dc power supply. The wipers (sliders) of the two potentiometers are connected to the vertical (dc) input of an oscilloscope. The shafts of the two units are mechanically coupled so they rotate simultaneously. The oscilloscope deflection may be calibrated by means of a DVM for allowable linearity error. As the shafts are slowly rotated throughout the specified angular travel, any deviation can be observed in the oscilloscope. The horizontal sweep can be off during this test. In some test units, a third potentiometer is driven along with the other potentiometers. Its output can be used to deflect the oscilloscope beam horizontally, and the horizontal position of the beam will correspond to the shaft angle.

Noise is a common characteristic of potentiometers and tends to increase with the life of the unit. When a circuit is affected by noise, the potentiometer should be checked. Some commercial instruments to perform this measurement are available. If such a unit is not available, the set-up of Fig. 10-69(b) gives satisfactory results. The dc voltage is applied across the short-circuited ends of the potentiometer and the wiper arm through a 300 kΩ resistor. With the shaft slowly rotated throughout its range, the ac oscilloscope is monitored for any vertical deflection. The ac noise generated may be calibrated in *equivalent*

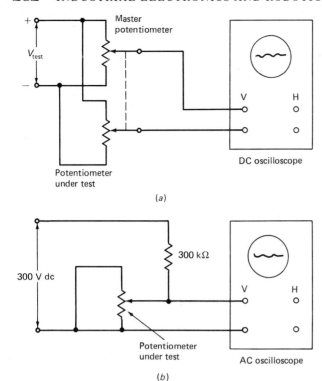

Fig. 10-69 Potentiometer test set-ups.

noise resistance (ENR) by alternately bridging the resistor in series with the wiper to obtain a given deflection in terms of the added resistance.

Most shaft encoders do not require periodic adjustments. A common failure is the internal light source, which may be solid-state or incandescent, with the latter more susceptible to failure. If the output of the encoder is missing on all channels, the light source should first be verified. Many units have modular construction for the electronics. Spare units should always be kept on site. As with any device that is connected to a shaft, loose connections are not uncommon and should not be discounted as a source of problems, especially when operation is erratic or intermittent.

Since the different units of a synchro system may be located at some distance from each other, they are connected by cables. Whenever a synchro system operates improperly, it is advisable to check the wiring of the unit (especially for loose or dirty connections) before looking for trouble in the synchros themselves. This is particularly important when working with systems which may have been under repair or overhaul. Should the symptoms indicate wiring trouble, it is suggested that all wiring be disconnected and checked for continuity.

Troubles involving open and short-circuited wiring, with the associated symptoms are listed for easy reference in Table 10-3. Should the symptoms indicate that the trouble might be in both the rotor and stator circuits, the rotor circuit should be checked first. To avoid electrical shock, all safety precautions should be observed at all times.

The nominal output of a resolver is known explicitly when the signal input is normal and the angle of the rotor is known. However, accurate measurement of the rotor angle in the field is often extremely difficult. A method of resolver testing, which does not require measurement of the rotor angle, is frequently used. Assume that the resolver to be tested is of the type having both sine and cosine outputs. If the input is 10 V, it is apparent that for normal operation the output voltages are $10 \cdot \cos \theta$ and $10 \cdot \sin \theta$. The rotor must be locked (clamped) so it does not turn. If the two voltmeter readings are first squared and added, the result should equal the square of the input voltage, since

$$(10 \cos \theta)^2 = 100 \cos^2 \theta$$
$$(10 \sin \theta)^2 = 100 \sin^2 \theta$$

and

$$(100 \cos^2 \theta + 100 \sin^2 \theta) = 100 (\sin^2 \theta + \cos^2 \theta)$$
$$= 100$$

This result is based on the trigonometric identity

$$\sin^2 \theta + \cos^2 \theta = 1$$

If the input voltage is 10 V and the rotor angle is 30°, then

$$(10 \sin 30)^2 = (5)^2 = 25$$
$$(10 \cos 30)^2 = (8.67)^2 = 75$$

The sum of these squares is seen to be equal to 100 V, which is equal to the input voltage squared. Thus, voltmeter readings of 5V and 8.67V indicate a properly functioning resolver in this example.

The typical ac motor used in servomechanisms is a two-phase induction motor. The important aspects of the motor are that the two stator windings (control and reference) are operated 90 degrees electrically from each other. The voltages fed to the stator windings must be 90° out of phase. Since a two-phase supply is rarely available, it is common to operate the two-phase motor with the phase-splitting capacitor. If the motor runs slow or is sluggish, the capacitor is suspect and should be tested. Improper phase shift between the windings is a common clue. An increase in bearing friction can have a detrimental effect on a motor, especially at low speeds, and should not be overlooked. Any suspected leakage or short circuit in the stator windings should be tested as specified in earlier chapters.

Tachometers that are used in accurate velocity servomechanisms must meet specific accuracies. The signal-to-noise ratio of a tachometer is usually specified by the manufacturer, and the unit must always meet these requirements. In dc types, high output noise is often an indication of bad brushes or a dirty commutator.

With ac tachometers, variation in amplitude of the residual voltage may have a modulation effect on a servomechanism. This measurement can be made by using the potentiometer method with an oscilloscope as a null detector. The output phase of the ac tachometer should remain constant. It is specified by

TABLE 10-3 SYNCHRO SYMPTOMS AND REMEDIES

Symptoms	Possible Cause of Trouble	Remedy
Receiver rotor either in correspondence with transmitter of 180° displaced, but follows in proper direction. Stator voltages vary from 0 to 90 V. Both rotor voltages are 115 V.	Rotor winding open, connection to slip ring open, or brush not making contact.	If ascertained trouble is not in ring connection or brush, unit must be replaced.
Same as above except that one rotor voltage is 115 V and the other is 90 V.	Supply line is open to the rotor reading 90 V, the 90 V appears across the rotor by virtue of transformer action.	Locate open in supply line; repair.
Voltage between one pair of stator wires is zero for all transmitter positions. Other stator-lead voltages read from 0 to 90 V. Both rotor voltages are 115 V.	The pair of stator leads which read 0 V is short-circuited.	Remove short circuit from wiring or interconnecting switches. If it is internal, unit may require replacement.
Both transmitter and receiver units hum and heat excessively. Receiver either does not follow or may spin.	All three stator wires are short-circuited together.	Locate defective wiring or switches; repair.
Sudden change in transmitter rotor position causes oscillation at receiver or a spinning effect.	Inertia damper jammed tight on receiver rotor shaft. Absence of damper indicates that transmitter unit is being used.	Free damper if it is jammed. If transmitter has been used, replace with a receiver unit.
Intermittent operation	Corroded rings, defective brushes, loose connections.	Respectively, clean rings, install new brushes, tighten loose terminals, etc.
Torque normal. Receiver lags or leads the transmitter or may turn in proper direction or reverse direction.	Stator wiring incorrect.	Correct stator wiring.
Torque normal. Receiver follows transmitter, but is displaced 180° from it.	Rotor connections reversed.	Correct wiring at proper unit.
Receiver shows large error and lags transmitter. Connections normal, but excessive current flows, producing overload indication.	Bearings frozen or partially frozen because of improper lubrication.	Replace unit, since bearing trouble usually damages other parts of the unit.

the manufacturer, and its value should be checked to ensure that the unit is within the limits specified. It is often easy to couple a rate generator to a variable-speed drill. Use a calibrated strobe to ascertain its output magnitude and direction. Be sure to check both directions of rotation.

Most servomotors operate most efficiently at higher speeds than are actually necessary to drive the load. Gearing is quite common to convert the low-torque high-speed motor output to a high-torque low-speed output for the load. The main problems encountered in gearing are backlash and friction. Backlash introduced by the gear train on the order of only a fraction of a degree can have a serious effect on the stability of a servomechanism. Backlash introduces a time delay between the servomotor and the input command signal. As the amount of backlash increases, so will the oscillations of the servomechanism. Even small oscillations caused by backlash will eventually cause excessive equipment wear.

Static friction, due to the tightness of the gearing, is probably the best deterrent to backlash oscillations. But friction problems will occur if the meshing of gears is not concentric with their supporting shafts. When gears are rotated, eccentricities combine in the gear box to produce excessive friction at one point of the revolution and backlash at another point.

Clutches, by their nature, are susceptible to wear more than any other mechanical part of a system. If the clutch is adjustable, the adjustment mechanism can work loose, especially if the system was or is oscillatory. The clutch surface must be clean and free of any contaminants, especially oil or grease. If light sanding with a recommended abrasive does not rectify the problem, a new clutch should be considered.

The typical servo amplifier is a summer (adder) or difference (subtractor) amplifier. In many cases, the reference or feedback signals may come from remote locations; noise pickup can be a problem, especially if a shield connection is loose or broken. Dirt or grease can cause electrical leakage and cause a ground loop which makes the system noisy or unstable. An oscilloscope is recommended for close inspection of the signals for the presence of noise. The low-voltage power supplies should always be verified as being within tolerances.

The components used in control systems have high precision in most cases and will be adversely affected if overheated by improper ventilation. Loss of feedback will cause the amplifier to lock up to a power supply rail, unless an overload circuit is incorporated. In some instances, the feedback resistance can be shunted with a value to make the gain approximately unity. The decreased gain will allow the system to be analyzed under more reasonable conditions.

Loss of feedback can also cause a servomechanism to go to the extreme end (limit) if it is a position type, or turn at its maximum rate (velocity) if it is a rate type. In either case, a potentiometer or an adjustable power supply may be substituted to simulate the closing of the loop. Most systems have either static or dynamic tests that can be performed to aid in localizing any problems. Refer to the manufacturer's service manuals for the necessary details. Remember that any significant gain or feedback change can cause a system to become unstable. This can be dangerous in some instances. You must know and thoroughly understand all tests before performing them.

Block-level understanding is necessary to localize a problem to a particular block. If the system is dead to any input command, signal tracing through the blocks should lead to the malfunctioning block. Final fault isolation is done with conventional voltage or resistance tests.

Loose or slipping follow-up potentiometers or rate generators are not uncommon. Most of these units have specific alignments and if not properly set can cause position or rate feedback errors and imbalances in the system. If a servomechanism is at its limit and no clutch is used, shut it down immediately to prevent damage to the mechanical components (gears, couplings, etc.). A good stock of replacement modules is a must to minimize down time. When a replacement is made, be sure the malfunctioning unit is repaired or returned to the manufacturer for prompt return to the spares stock or returned to the manufacturer.

Rotating amplifiers, as do other components with brushes and commutators, require careful maintenance procedures. Precautions with lubrication, according to the manufacturer's recommendations, are essential. The Amplidyne has multiple sets of brushes, and proper attention is a must. In most cases, the output from a large servomotor, either position-type or rate-type, is fed back to close the systems loop. This factor must be taken into account when energizing the field of a rotating amplifier with an auxiliary source. Remember that these units can have current gains in the thousands. A few milliamperes of Amplidyne field current can produce amperes of output current for a servomotor or load.

The loss of the killer winding will leave a residual field in the Amplidyne, thereby causing an imbalance in the system. Leakage resistance to the frame is unacceptable, as with any type of generator, and should not be overlooked if overheating occurs. Routine preventive maintenance is essential for all rotating devices and will enhance the operation of the units and extend the operating life while minimizing down time.

REVIEW QUESTIONS

39. The _____ source is a cause of frequent failures in a shaft encoder.

40. If a synchro receiver follows the transmitter but with 180° of error, this is an indication of rotor connection _____.

41. The voltmeter reading of a resolver winding is the square of the applied voltage. (true or false)

42. The phase difference between the reference and control windings of the ac servomotor is _____ degrees.

43. Loss of _____ can cause a type 0 servo to travel to one of its limits.

CHAPTER REVIEW QUESTIONS

10-1. What type of encoder retains position information during a power outage?

10-2. Incremental encoders can provide position and _____ information.

10-3. A linear motion optical encoder is a type of absolute encoder. (true or false)

10-4. The resolver solves the unknowns of a _____ triangle.

10-5. If $V \sin \omega t$ is 100 V and θ is 45°, VR_3 to VR_4 is _____?

10-6. The output voltages from a resolver represent cartesian or rectangular coordinates. (true or false)

10-7. What happens to the output of an ac servomotor when the control voltage is shifted by 180°?

10-8. A dc rate generator is free of any zero speed _____ voltage.

10-9. Alternating current rate generators produce an output frequency proportional to their speed. (true or false)

10-10. The mutual coupling of the stator windings in an ac servomotor is due to the induced _____ currents in the rotor.

10-11. Direct current servomotors require standby power. (true or false)

10-12. Most servomechanisms use _____ type gears.

10-13. Mechanical addition and subtraction with gearing is accomplished by the use of a _____.

10-14. A motor with an inertia of 0.2 kg · m² that is coupled through a gear box of 20:1 can handle a load inertia of _____ kg · m².

10-15. In a simple feedback system, the _H_ block represents the feedback. (true or false)

10-16. A servosystem is usually _____ damped for best performance.

10-17. A type 0 servosystem is a rate (velocity) system. (true or false)

10-18. A type 0 servosystem is a position (follow-up) system. (true or false)

10-19. Feedback in a type 1 servosystem is usually from a _____.

10-20. Stability of a closed loop system may be improved by addition of an integrator or a _____.

10-21. Double differentiating a position signal will produce a velocity signal for damping purposes. (true or false)

10-22. The _____ windings correct for the effect of load flux on the control excitation in an Amplidyne.

10-23. The brushes of the Amplidyne are _____ degrees apart.

10-24. An Amplidyne needs a residual flux to establish its excitation flux. (true or false)

10-25. The Amplidyne is a type of _____ amplifier.

10-26. The typical servo amplifier is usually a summer or a _____ amplifier.

10-27. The Amplidyne requires periodic checking of its two sets of _____.

10-28. The Amplidyne's control field is a high-current, low-voltage winding. (true or false)

10-29. The maximum torque of a two-phase servomotor occurs at _____ rpm.

10-30. The output phase of the ac tachometer should change as the shaft angle changes speed. (true or false)

10-31. Lost motion, or play, in a mechanism is known as _____.

ANSWERS TO REVIEW QUESTIONS

1. cermet, wire-wound, and conductive plastic **2.** slider **3.** 7.71 kΩ **4.** linearity **5.** hunt **6.** false **7.** 90 **8.** false **9.** 52 V **10.** false **11.** clockwise **12.** 160° **13.** control **14.** phase **15.** 90 **16.** capacitor **17.** false **18.** drag-cup **19.** 24,000 **20.** 16.7 **21.** false **22.** bore (inside) **23.** false **24.** universal joints **25.** clutch **26.** speed **27.** −4 **28.** −1 V **29.** 6 **30.** power **31.** chopper **32.** overload recovery **33.** _Q/K_ **34.** false **35.** uni **36.** shunt **37.** tuning **38.** error **39.** light **40.** reversal **41.** false **42.** 90 **43.** feedback

11

DIGITAL CIRCUITS AND DEVICES

The digital world is a world of ones and ze-
ros. It utilizes circuits that are on or off, logic
conditions that are true or false, and voltages
that are high or low. By recognizing only two
possible conditions, errors due to component
tolerance and temperature drift are all but
completely eliminated. The real world is an
analog world. The speed of a motor, the tem-
perature of a process, and the intensity of a
laser beam are all examples of analog mea-
sures. There are an infinite number of actual
values in the analog world. It is possible to use
digital circuits to measure, control, and gener-
ally interact with the analog world. The cur-
rent trend is to use more digital and less ana-
log circuitry in industrial electronics. The
industrial technician must understand both
types of circuits and how they are interfaced.

11-1
CHARACTERISTICS OF DIGITAL CIRCUITS

Early digital circuits used vacuum tubes. Then, in
the 1960s, integrated digital circuits were developed.
The first integrated digital circuits were based on
resistors, diodes, and transistors. They were *resistor-*
transistor logic (RTL) *circuits. Diode-transistor logic*
(DTL) was the next step. By the 1980s, RTL and
DTL devices were obsolete for modern designs.
They have been replaced by the *transistor-transistor*
logic (TTL) and the *complementary metallic oxide*
semiconductor (CMOS) families. This section will
cover the important electrical parameters of TTL
family devices; a later section will cover the CMOS
family.

Figure 11-1 shows the input signal requirements for
a TTL circuit. Any signal voltage from 2 to 5 V is
interpreted as a logic 1. These levels are called
$V_{INH(MIN)}$ and $V_{INH(MAX)}$, respectively. Any voltage

from 0 to 0.8 V is interpreted as a logic 0. Notice
that the region from 0.8 to 2 V is forbidden. Any
signal in that region would produce unpredictable
output results. Obviously, a signal that is changing
from 0 to 1 or from 1 to 0 must go through the
forbidden region. This is acceptable. A signal that
remains in the forbidden region for any length of time
is not acceptable. An input that is floating (not connected to anything) will measure between 1.1 and
1.5 V, which is in the forbidden region. The device
will interpret this as a high input (logic 1), but a
floating input invites noise pickup. Therefore, an input that is to be fixed at logic 1 should be tied to the
positive supply. The TTL systems run on a 5-V
power supply. Most signals will be between 0 and
5 V. Signals more positive than 5 V should not be
applied to a TTL input because damage may result.
A signal that is negative with respect to ground can
also damage a TTL device. Most devices have input
clamp diodes to protect against negative signals, but
a signal more negative than 1.5 V may damage the
input.

Most often TTL inputs are driven by TTL outputs.
When the expected output levels are compared to
the required input levels, the margin for error can be
evaluated. The rectangular waveforms shown in Fig.
11-1 are for the worst-case and typical levels that
can be expected at the output of a device. *Worst case*
means that the manufacturer guarantees the logic 1
level from the device supplying the signal to be at
least 2.4 V and the logic 0 level to be no more than
0.4 V when the device is fully loaded. The difference
between the worst-case driving signal levels and the
input thresholds at the driven device is seen to be
400 mV. This is called the *noise margin.* In practice,
the actual noise margin is better. The typical TTL
signal source will swing from 0.2 V (logic 0) to 3.4
V (logic 1) when fully loaded. A full load is also
called a *full fanout* and is shown in Fig. 11-2. The
signal source, called the *driving gate,* must supply
logic 0 or a logic 1 to 10 driven gates. By the nature
of the totem pole output stage (discussed in Chapter
2) in the driving gate, the logic 0 sink current capability is greater than the logic 1 source current capability. The totem pole pair uses a load resistor of

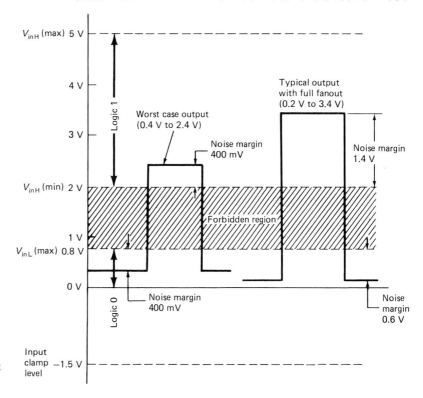

Fig. 11-1 Transistor-transistor logic (TTL) input requirements.

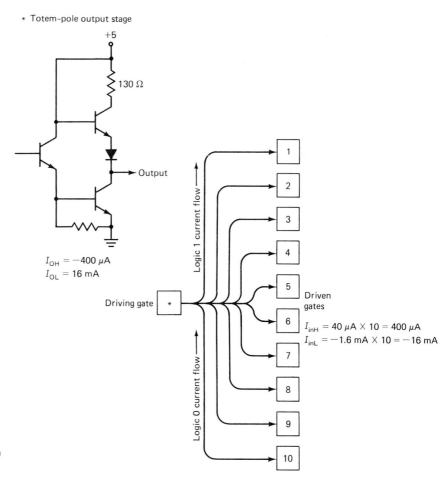

Fig. 11-2 Transistor-transistor logic (TTL) fanout.

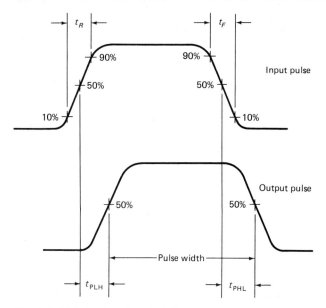

Fig. 11-3 Digital pulse characteristics.

percent point on the input waveform to the 50 percent point on the output waveform. The propagation delay from high to low (t_{PHL}) is also measured from 50 percent point to 50 percent point. Digital circuits with short propagation delays are faster and are capable of high-frequency operation. The pulse width is measured from the 50 percent point on the rising edge of a pulse to the 50 percent point on the falling edge of the same pulse.

Table 11-1 lists a few important characteristics for the TTL logic family. This family has four subfamily members. It is important to note that the part numbers clearly identify the subfamily members. For example, a 7400 gate is a TTL device, and a 74LS00 is a *low-power Schottky* (LS) clamped device, which is a TTL subfamily. The part numbers 7490 and 74LS90 represent another example: the 7490 is a TTL device, and the 74LS90 is an LS TTL subfamily device. You may also encounter 5400 series numbers, which are military versions of the 7400 TTL family. The military devices are rated from $-55°C$ to $+125°C$; commercial and industrial devices are rated from $0°C$ to $+70°C$. Different manufacturers may deviate from the 7400 numbering system, and 8000 numbers are also popular. The TTL and LS TTL devices are the most widely applied. Table 11-1 shows that the LS devices are faster (less propagation delay) than the TTL devices and use less current. The LS devices used to cost more but now are about as expensive as TTL parts, and designers choose them most often. Since they are more power-efficient and faster devices, there is usually no reason to specify TTL devices in new designs unless the LS version of the required device is not available.

All TTL devices and TTL subfamily devices are rated at a fanout of 10, except when subfamilies are mixed or when subfamily devices are mixed with TTL devices. A low-power Schottky device will drive 10 gates in its own subfamily but will drive only 5 TTL gates. There are also speed differences. Some LS devices can approach twice the speed of their TTL counterparts. If a circuit is operating near its top speed, a substitution may not work or, worse yet, may work intermittently. The current drain is also different. Multiple substitutions of standard TTL for LS TTL could overload a power supply or add to the heat build-up in a circuit. Substitutions may

approximately 130 Ω between the supply and the top transistor. This resistor limits the current that the output can source when it is at logic 1. The sink current, I_{OL}, is rated at 16 mA, and the source current I_{OH} is rated at only -400 μA. The negative sign indicates that the current is flowing away from the driving gate when it is sourcing current. The input requirements for each driven gate are $I_{INH} = 40$ μA and $I_{INL} = -1.6$ mA. This means that a TTL output can drive no more than 10 TTL inputs, or the guaranteed noise margin will be lost. If the maximum fanout is exceeded, the output voltage may fall in the forbidden region. This can happen when the driving gate is sourcing or sinking current.

Figure 11-3 shows the characteristics of digital pulses. *Rise time* (t_R) is the time required for the pulse to change from its 10 to its 90 percent level. *Fall time* (t_F) is the time required for the pulse to change from the 90 to the 10 percent level. When the pulse is applied to the input of a TTL device, it takes time before the output changes. The propagation delay from low to high (t_{PLH}) is measured from the 50

TABLE 11-1 THE TTL LOGIC FAMILY

Family Name	Part Number	t_{PLH} ns	t_{PHL} ns	I_{max} mA	Comments
TTL (transistor-transistor logic)	7400	11	7	22	Being replaced by LS subfamily devices
High-speed TTL	74H00	5.9	6.2	40	High power consumption; not popular
Low-power TTL	74L00	35	31	2.04	Being replaced by CMOS family devices
Low-power Schottky clamped TTL	74LS00	5	5	4.4	Very popular; used heavily in modern designs
Schottky clamped TTL	74S00	3	3	36	Very fast; used in high-speed circuits

TABLE 11-2 COUNTING WITH SEVERAL NUMBER SYSTEMS

Decimal, Base 10	Binary, Base 2	Hexadecimal, Base 16	Octal, Base 8
0	0	0	0
1	1	1	1
2	10*	2	2
3	11	3	3
4	100	4	4
5	101	5	5
6	110	6	6
7	111	7	7
8	1000	8	10*
9	1001	9	11
10	1010	A	12
11	1011	B	13
12	1100	C	14
13	1101	D	15
14	1110	E	16
15	1111	F	17
16	10000	10*	20†
17	10001	11	21
18	10010	12	22

* Read as "one-zero," not "ten."
† Read as "two-zero," not "twenty."

be acceptable, but be sure to investigate fanout, speed, current demand, and heat. The best replacement usually has exactly the same part number.

Since digital circuits recognize only two conditions, the binary number system is used for operations involving counting, arithmetic, and for representation of analog quantities. Table 11-2 compares several number systems. The decimal number system has 10 symbols, 0 through 9. The quantity of symbols in a number system is referred to as its *base* or *radix*. Therefore, the radix of our familiar decimal system is 10. When a quantity larger than 9 must be represented, more than one symbol is used at a time. The same technique applies to *binary,* which uses only two symbols, 0 and 1. Follow the count in Table 11-2. You should determine that it is possible to represent any quantity in binary that can be represented in decimal, provided there is provision for enough 0s and 1s. Table 11-2 also shows the hexadecimal number system, which has a radix of 16. It adds the characters A through F to the familiar decimal set to provide a total of 16 symbols. Hexadecimal is a convenient shortcut when working with binary systems as we shall see. The *octal* system is also shown in the table. It has a base of 8 and is also used as a shortcut for working with binary, although hexadecimal is more popular.

Binary, hexadecimal, and octal are all weighted codes, making converting them to decimal straightforward. Table 11-3 shows how weighted codes work. Starting to the immediate left of the radix point (we call it the *decimal point* when working with decimal numbers) it is seen that the weight is the base raised to the 0 power. Any number raised to the zero power has a value of 1. Therefore, the weight of this position is always 1. Moving to the left, the next position is weighted equal to the base of the number system raised to the first power. Any number raised to the first power is equal to itself. Therefore, the weight of this position is equal to 2 in binary, 16 in hexadecimal, and 8 in octal. Moving to the left again we find the weight equal to the base raised to the second power. Therefore, the weight of this position is equal to 4 in binary, 256 in hexadecimal, and 64 in octal. The fractional parts of a number are represented by positions to the right of the radix point.

To convert a binary number to decimal, it is necessary to add up the decimal weights for each binary digit. The word *bit* is a contraction for *binary digit* and will be used from now on. Suppose we wish to convert binary 10101101 to decimal. The process begins at the far right, which is called the *least significant bit* (LSB), and progresses to the leftmost bit, which is called the *most significant bit* (MSB).

EXAMPLE 1

Convert binary 10101101 to decimal.

TABLE 11-3 WEIGHTED CODES

	Whole Part								Fractional Part	
Binary	2^7	2^6	2^5	2^4	2^3	2^2	2^1	2^0	2^{-1}	2^{-2}
Decimal weight	128	64	32	16	8	4	2	1	$\frac{1}{2}$	$\frac{1}{4}$
Hexadecimal	16^7	16^6	16^5	16^4	16^3	16^2	16^1	16^0	16^{-1}	16^{-2}
Decimal weight	2.68×10^8	1.68×10^7	1,048,576	65,536	4096	256	16	1	$\frac{1}{16}$	$\frac{1}{256}$
Octal	8^7	8^6	8^5	8^4	8^3	8^2	8^1	8^0	8^{-1}	8^{-2}
Decimal weight	2.10×10^6	262,144	32,768	4096	512	64	8	1	$\frac{1}{8}$	$\frac{1}{64}$

↑
*Radix
Point*

SOLUTION

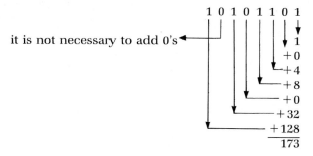

it is not necessary to add 0's

The LSB is weighted 1 (2^0), so we begin by adding 1. The next bit position is weighted at decimal 2 (2^1), but there is a 0 there, so we add 0. The next bit is weighted at 4 (2^2), and there is a 1 there, so we add 4. The next bit is weighted at 8 (2^3), and there is a 1 in this position, so we add 8. The next bit is weighted at 16 (2^4), and there is a 0 there, so we add 0. The next position weight is 32 (2^5), and there is a 1 there, so we add 32. The next position weight is 64 (2^6), and there is a 0 there, so we add 0 (it is not necessary to add the 0s). The MSB position is weighted 128 (2^7), and there is a 1 there, so we add 128. The total is 173, which is the base 10 equivalent of binary 10101101.

Because there may be a possibility of confusion, the base of a number may be specified with a subscript. For example, $10101101_2 = 173_{10}$.

The same general technique is used to convert hexadecimal numbers to decimal.

EXAMPLE 2

Convert hexadecimal 1COF to decimal.

SOLUTION

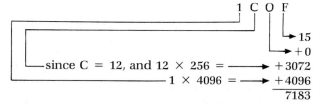

The weight of the least significant position is 1 (16^0), and there is an F there. In decimal F is equal to 15 so we add 15. The next weight is 16 (16^1), but there is a 0 there, so we can add 0 or not. The weight of the next position is 256 (16^2), and there is a C there. Since C is equal to decimal 12 we add 12 × 256, or 3072. The last position is weighted 4096 (16^3), and there is a 1 there, so we add 4096. The total is 7183. Therefore, $1COF_{16} = 7183_{10}$.

Converting from octal to decimal uses the same process and is demonstrated in Example 3:

EXAMPLE 3

Convert octal 17325 to decimal.

SOLUTION

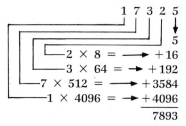

We must also be able to convert from decimal to other number systems. This process involves dividing the decimal number by the base of the given number system while keeping a record of all remainders. The division continues until the decimal number is exhausted. The list of remainders is the number in the given number system.

EXAMPLE 4

Convert decimal 115 to binary.

SOLUTION

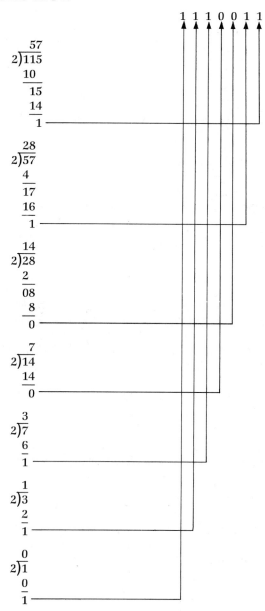

When 2 is divided into 115, the quotient is 57 with a remainder of 1. This first remainder becomes the LSB of the answer. Next 57 is divided by 2 with a quotient of 28 and a remainder of 1. This remainder becomes the next bit of the answer. Then 28 is divided by 2 with a quotient of 14 and a remainder of 0. This 0 remainder becomes the next bit of the answer. The process continues until the quotient is 0, and the last remainder has been recorded as the MSB of the answer. Therefore, $115_{10} = 1110011_2$.

Converting from decimal to hexadecimal involves repeated division by 16 until the number is exhausted. All remainders are converted to hexadecimal characters and recorded. The process ends when the quotient is 0 and the last remainder has been placed in the leftmost position.

EXAMPLE 5

Convert decimal 7307 to hexadecimal.

SOLUTION

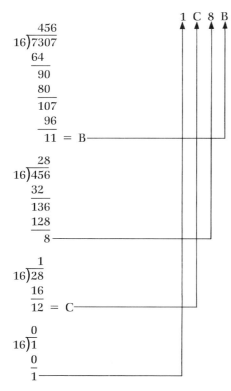

Note that the first remainder is decimal 11, which is equal to B in hexadecimal.

Hexadecimal (and occasionally octal) is used to enter information into and extract information from digital systems quickly. The trouble with binary is that too many bits are required to represent numbers of moderate size. It takes too long to enter or read a lot of 0s and 1s, and the process is error-prone. A digital system does not understand hexadecimal any better than it understands decimal. Digital systems are strictly binary in nature. However, there is a natural relationship between binary and hexadecimal that makes conversion very easy. Decimal 16 is a

power of 2, and each hex character can be represented by 4 bits.

EXAMPLE 6

Convert decimal 701 to octal.

SOLUTION

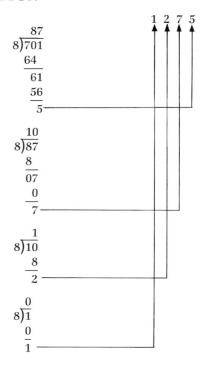

EXAMPLE 7

Convert hexadecimal A4E to binary.

SOLUTION

Each digit of $A4E_{16}$ is converted to a binary number as shown.

It is just as easy to convert from binary to hex.

EXAMPLE 8

Convert binary 10010100111 to hexadecimal.

SOLUTION

Group the bits into 4s, starting with the rightmost bits. Do not be concerned if the leftmost group does not have 4 bits. Convert whatever is in the leftmost group to the appropriate hex character.

Octal is also convenient since 8 is also a power of 2. To convert octal to binary, each octal character must be represented by a group of 3 bits.

EXAMPLE 9

Convert octal 725 to binary.

TABLE 11-4 BINARY CODED DECIMAL

Decimal	Binary	BCD
0	0	0000
1	1	0001
2	10	0010
3	11	0011
4	100	0100
5	101	0101
6	110	0110
7	111	0111
8	1000	1000
9	1001	1001
10	1010	0001 0000
11	1011	0001 0001
12	1100	0001 0010
13	1101	0001 0011
14	1110	0001 0100
15	1111	0001 0101
16	10000	0001 0110
17	10001	0001 0111
18	10010	0001 1000

SOLUTION

Binary to octal is the reverse process.

EXAMPLE 10

Convert binary 11010100 to octal.

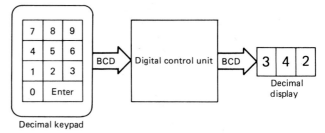

Fig. 11-4 Binary-coded decimal applications.

SOLUTION

Hex and octal are fine for engineers, technicians, and programmers who have taken the time to learn other number systems. Operators are often not familiar with these number systems. They simply use digital equipment and do not need an understanding of the digital circuits. Figure 11-4 shows a decimal keypad connected to the input of a digital control unit. The output goes to a decimal display. This environment is far more comfortable for most people than a hexadecimal keypad or a binary display. Another code has been developed to interface between the decimal operator and the binary machine conveniently. It is called *binary-coded decimal* (BCD) and is shown in Table 11-4. It is the same as ordinary binary until numbers greater than decimal 9 are represented. These numbers require an additional group of 4 bits for each decimal digit. The decimal number 309 would require 12 bits, 1234 would require 16 bits,

TABLE 11-5 AMERICAN STANDARD CODE FOR INFORMATION INTERCHANGE

Row	Column Bits 4321	765 → 0 000	1 001	2 010	3 011	4 100	5 101	6 110	7 111
0	0000	NUL	DLE	SP	0	@	P	\	p
1	0001	SOH	DC1	!	1	A	Q	a	q
2	0010	STX	DC2	"	2	B	R	b	r
3	0011	ETX	DC3	#	3	C	S	c	s
4	0100	EOT	DC4	$	4	D	T	d	t
5	0101	ENQ	NAK	%	5	E	U	e	u
6	0110	ACK	SYN	&	6	F	V	f	v
7	0111	BEL	ETB	'	7	G	W	g	w
8	1000	BS	CAN	(	8	H	X	h	x
9	1001	HT	EM	)	9	I	Y	i	y
10	1010	LF	SUB	*	:	J	Z	j	z
11	1011	VT	ESC	+	;	K	[	k	{
12	1100	FF	FS	,	<	L	\	l	\
13	1101	CR	GS	−	=	M	]	m	}
14	1110	SO	RS	.	>	N	∩	n	~
15	1111	SI	US	/	?	O	—	o	DEL

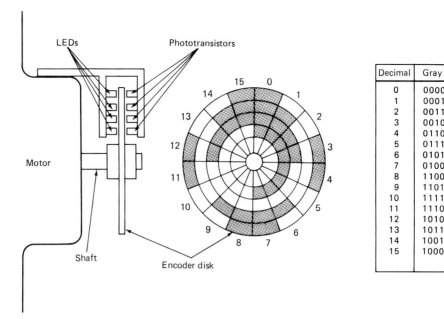

Decimal	Gray
0	0000
1	0001
2	0011
3	0010
4	0110
5	0111
6	0101
7	0100
8	1100
9	1101
10	1111
11	1110
12	1010
13	1011
14	1001
15	1000

Fig. 11-5 Gray code shaft encoder.

and so on. Binary-coded decimal is not a convenient code for arithmetic operations, and it usually requires more bits than binary. For these reasons, BCD may be used for input and output operations only, and binary-to-BCD and BCD-to-binary code converters will be required.

Other special codes are required for special applications. Some are *nonweighted,* meaning that each bit position does not have a definite weight. The Gray code is an example. Figure 11-5 shows how the Gray code can be used to convert the angular position of a motor shaft, which is analog, into a group of nonweighted 0s and 1s. As the shaft turns, the encoder disk turns with it and presents a clear or an opaque sector to each pair of light-emitting diodes and phototransistors. The 4-bit code that is read from the phototransistor outputs represents one of 16 shaft positions. A five-level code would resolve 32 shaft positions, a six-level code 64 shaft positions, and so on. The Gray code is unique in that no more than 1 bit changes at a time as the shaft rotates. Examine the table in Fig. 11-5 to verify this point. A binary-encoded disk would be error-prone because more than one bit would change at a time as the shaft turned. The Gray code does not lend itself to arithmetic operations, and so code conversion to binary will be required.

An *alphanumeric code* is one that represents letters and numbers. The most popular one is the *American Standard Code for Information Interchange* (ASCII). Table 11-5 shows an ASCII code chart. The bit pattern for each row and column is given. For example, the letter *A* is in row 1. All characters in this row end with the bit pattern 0001. *A* is in column 4, and all characters in this column begin with the bit pattern 100. Therefore, the ASCII bit pattern for *A* is 1000001. Note that ASCII is a 7-bit code. Sometimes an eighth bit is added in the leftmost position

to act as a parity bit. *Parity* is an error-checking technique. If the parity is supposed to be even, the letter *A* will be represented by 01000001, which has an even number of 1s in the group. If the parity is odd, the letter *A* will be represented by 11000001, which contains an odd number of 1s.

REVIEW QUESTIONS

1. A digital signal that remains at a level between 0.2 and 2.0 V is in the _____ region.

2. The worst-case noise margin in TTL logic circuits is _____.

3. Do worst-case noise margins occur with small fanouts or with full fanouts?

4. Can a totem pole output stage source or sink more current?

5. A digital pulse ranges from 0.2 to 3.4 V. What two voltage points will be used when measuring the rise time of the pulse?

6. What voltage point will be used for the pulse of question 5 when measuring propagation delay?

7. Is it always acceptable to substitute a TTL device for an LS TTL device?

8. Is it always acceptable to substitute an LS TTL device for a TTL device?

9. Use Table 11-5 and determine the ASCII code for the letter *Z* in an odd parity system.

11-2
GATES AND COMBINATIONAL LOGIC

A *gate* is a decision-making element. It produces an output that is high or low, depending on its input conditions. Table 11-6 shows the basic logic gates.

TABLE 11-6 LOGIC GATES

Gate	Symbol	Truth Table			Boolean Expression

Gate	Symbol	Truth Table	Boolean Expression
NOT	A—▷∘—C	A: 0, 1 / C: 1, 0	$C = \overline{A}$

NOT truth table:

A	C
0	1
1	0

AND truth table: $C = A \cdot B$

A	B	C
0	0	0
1	0	0
0	1	0
1	1	1

INCLUSIVE OR truth table: $C = A + B$

A	B	C
0	0	0
1	0	1
0	1	1
1	1	1

EXCLUSIVE OR truth table: $C = A\overline{B} + B\overline{A}$, $C = A \oplus B$

A	B	C
0	0	0
1	0	1
0	1	1
1	1	0

The *NOT gate*, shown at the top of the table, is also called an *inverter*. The truth table for the NOT gate shows that the input (A) can be low (0) or high (1). When the input is low, the output (C) is high. When the input is high, the output is low. This function is known as *logical inversion*. The *NOT symbol* is a triangle with a circle at the output. The circle is important because it tells us that the output is inverted. *Boolean algebra* is a special branch of mathematics used to describe and design binary systems. The boolean expression for the NOT gate is read as *C is equal to NOT A*. The bar over the *A* is an inversion bar. The NOT gate is useful in many situations. Suppose, for example, that a limit switch produces a logic 0 when activated, but it would be more convenient if it produced a logic 1. A NOT gate can be used to invert the switch logic.

Now look at the AND gate in Table 11-6. It has two inputs and its truth table has four rows, one for each possible input combination. The output is high when input A and input B are high. The boolean expression is read *C is equal to A AND B*. The dot between the A and the B is the symbol for the AND operator. The dot is optional; thus C = AB is an equivalent expression, and so is C = A(B). The AND gate is an "all or nothing" circuit and is useful for activating a load when all of the inputs are high. For example, it may be desirable to activate an alarm

circuit when both the pressure and the temperature of some process go high.

Table 11-6 shows two types of OR gates. The *inclusive OR gate* is often simply referred to as an *OR gate*. However, to avoid confusion, the other version must be referred to by its full name, *exclusive OR*. The inclusive OR is an "any or all" gate and produces a high output when either or both of its inputs are high. The *exclusive OR gate* produces a high output when either A or B is high but excludes the condition where both are high. The inclusive OR gate includes a high output for the condition where both inputs are high. The boolean expression for the inclusive OR gate is read *C is equal to A OR B*. The plus (+) sign is the OR operator, but you should *not* read the expression as *C is equal to A plus B*. The inclusive OR is useful for detecting those situations in which any or all of its inputs are high. For example, it may be desirable to activate an alarm circuit when the pressure is high, or the temperature is high, or both conditions are true. The exclusive OR gate is useful for detecting inequality. Note that it produces a high output only when its inputs are not equal. It may also be used as a comparator because its output is low when its inputs are equal. The boolean expression for the exclusive OR gate is read *C is equal to A AND NOT B OR B AND NOT A*. This is often shortened as Table 11-6 shows by using the

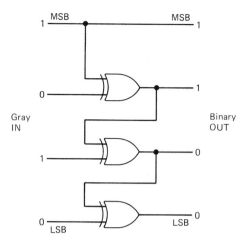

Fig. 11-6 Gray-to-binary code converter.

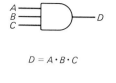

$$D = A \cdot B \cdot C$$

A	B	C	D
0	0	0	0
1	0	0	0
0	1	0	0
1	1	0	0
0	0	1	0
1	0	1	0
0	1	1	0
1	1	1	1

Fig. 11-7 Three-input AND gate.

Fig. 11-8 NAND gate.

ring sum symbol, which is the exclusive OR operator. The expression is read *C is equal to A exclusive OR B*. It can also be read *C is equal to A ring sum B*.

The last section introduced the Gray code. It is a nonweighted code that is often used to encode shaft position. Unfortunately, since it is nonweighted, it is not useful when arithmetic operations are required. Figure 11-6 shows a Gray-to-binary code converter that uses exclusive OR gates. Suppose that Gray code 1010 is presented to the converter. The MSB is a 1 and connects straight through to the output. The next bit is exclusively ORed with the MSB and produces an output of 1. This 1 is exclusively ORed with the next Gray bit, which is also a 1, producing the next output of 0. This 0 is exclusively ORed with the last Gray bit, which is also a 0; the LSB of the output is 0. By following the decision of each gate, you should be able to verify that the output is equal to binary 1100 or decimal 12. Now, refer again to Fig. 11-5. Try a couple of other shaft positions to test your understanding of the converter circuit.

Except for the inverter, logic gates can have more than two inputs. Figure 11-7 shows a three-input AND gate. There are eight possible input combinations, and the truth table has a row for each. The number of combinations is predicted by 2^N, where N is equal to the number of inputs. A four-input gate has 16 input combinations, a five-input gate has 32 input combinations, and so on. Note that only one input combination produces a high output in Fig. 11-7. This occurs when A, B, and C are high.

Gate outputs are often inverted (refer to Fig. 11-8). This can be accomplished by following a gate with an inverter. More commonly, it is obtained in one package. The symbol on the right is for a *NOT-AND gate* or simply *NAND gate*. Note the inversion circle on the NAND output. Table 11-7 shows the symbols, truth tables, and boolean expressions for the inverted output gates. The truth tables show that the output conditions are inverted from those shown in Table 11-6. The boolean expressions show the inversion bar over the terms and their operators.

Figure 11-9 shows some of the laws of combination for gates. These laws are useful because they will help you to understand how any logic function can be synthesized from other logic functions. For example, the *law of association* shows us how to obtain a three-input AND function from two-input AND gates. It also shows that it is possible to obtain a three-input OR function from two-input OR gates. Simply stated, the associative law tells us that it makes no difference as to the order of how AND or OR expressions are combined. The *law of distribution* shows that there can be two forms of an expression and a circuit can be realized for each. In general, the circuit containing the fewest number of gates is the most desirable. The distributive law is useful when examining an expression to determine whether it can be simplified. For example, as Fig. 11-9 shows, a logic function may be implemented with two or three gates and have the same function. The *law of tautology* shows that when a variable is ORed or ANDed with itself the result is equal to the variable. Taking this one step further, if the gate is a NOR or a NAND, an inverter is realized by tying the inputs together. The *law of double complementation* shows that two inversions cancel and that the variable is restored.

The last two rows of Fig. 11-9 illustrate *De-Morgan's theorem,* which states that the complement of a sum is equal to the product of the complements. Likewise, the complement of a product is equal to the sum of the complements. The word *complement* means the inversion of a variable or expression. The complement of A means the same

TABLE 11-7 INVERTED OUTPUT GATES

Gate	Symbol	Truth Table			Boolean Expression

Gate	Symbol	A	B	C	Boolean Expression
NAND		0	0	1	$C = \overline{A \cdot B}$
		1	0	1	
		0	1	1	
		1	1	0	
INCLUSIVE NOR		0	0	1	$C = \overline{A + B}$
		1	0	0	
		0	1	0	
		1	1	0	
EXCLUSIVE NOR		0	0	1	$C = \overline{A \oplus B}$
		1	0	0	
		0	1	0	
		1	1	1	

thing as NOT A. Simply stated, the theorem tells us we can break an inversion bar over an operator if we change the operator. For example, in the equation $\overline{A + B} = \overline{A} \cdot \overline{B}$, the bar is broken over the OR operator and the operator is changed to AND. The theorem shows that it is possible to synthesize a NOR gate from four NAND gates. Figure 11-9 shows that the first two NAND gates are used as inverters since their inputs are tied together. The next gate NANDs the complemented inputs. The theorem shows that the complements are to be ANDed, and the last NAND gate acts as an inverter to meet that requirement because when a NAND gate is followed by an inverter the AND operation results. An OR function can therefore be realized by leaving off the last inverter. The last row of Fig. 11-9 shows that the NAND function can be synthesized from four NOR gates. The complement of A is NORed with the complement of B and then inverted. The AND function is realized by leaving off the last inverter. NAND gates are often referred to as *universal logic elements* because it is possible to synthesize any logic function by combining them. The same thing is true of NOR gates.

DeMorgan's theorem has led to alternate symbols for NAND and NOR gates. Figure 11-10 shows these alternate symbols. The NAND gate is sometimes represented as an OR gate with inverted inputs. The NOR gate may be represented as an AND gate with inverted inputs. You may find both the standard and the alternate symbols used on the same logic diagram. The purpose of using both is to represent circuit operation more clearly by emphasizing high inputs or inverted (low) inputs.

The *laws of combination* are also useful to solve circuits. For example, Fig. 11-11 shows a combinational logic circuit based on four NAND gates. What is the output supposed to be? Follow the steps:

1. At this point we find variable A combined with the NAND of variables A and B.
2. It is desired to simplify this so the bar is broken, and the AND operator is changed to the OR operator (DeMorgan's theorem).
3. The law of double complementation allows us to remove the bars.
4–6. These steps are the same as the first three.
7. The simplified terms from steps 3 and 6 are NANDed.
8. The bar is broken in the middle, and the operator is changed to OR.
9. The bar is broken at the left and at the right, and the operators are changed.
10. The double bars are removed.
11. The term is multiplied (ANDed) with each expression in the parentheses. Two terms cancel, since any time a variable is ANDed with its complement the result is 0. You may verify this by looking at the AND gate truth table.
12–13. The result is the exclusive OR operation.

Another basic understanding that you will need to work with digital systems is in the area of dynamic

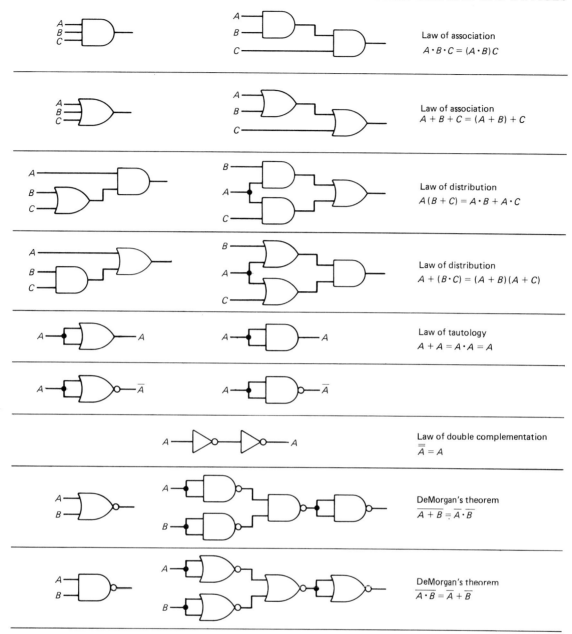

Fig. 11-9 Laws of combination.

operation. A logic circuit is *dynamic* if it is undergoing change. Look at Fig. 11-12, which shows some timing diagrams for a two-input AND gate. Timing diagrams relate dynamic conditions among inputs and outputs. Input A is shown as a square wave. Input B is a rectangular wave of low duty cycle. Output C is high only during those periods when both A and B are high. Now look at Fig. 11-13; it shows the timing diagram for an exclusive OR gate. Note that the output waveform is the inverse of input A when input B is high. When B is low, the output is in phase with input A. Verify this operation by looking at the exclusive OR truth table if necessary. The exclusive OR gate is sometimes called a *program-*

Fig. 11-10 Alternate symbols.

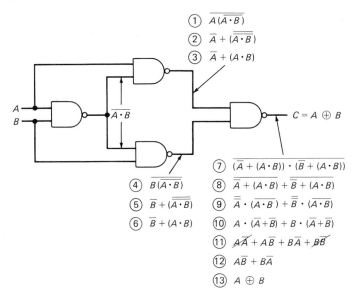

① $\overline{A\,(\overline{A\cdot B}\,)}$

② $\overline{A} + (\overline{\overline{A\cdot B}}\,)$

③ $\overline{A} + (A\cdot B)$

④ $\overline{B\,(\overline{A\cdot B}\,)}$

⑤ $\overline{B} + (\overline{\overline{A\cdot B}}\,)$

⑥ $\overline{B} + (A\cdot B)$

⑦ $\overline{(\overline{A} + (A\cdot B))\cdot(\overline{B} + (A\cdot B))}$

⑧ $\overline{\overline{A} + (A\cdot B)} + \overline{\overline{B} + (A\cdot B)}$

⑨ $\overline{\overline{A}}\cdot\overline{(A\cdot B)} + \overline{\overline{B}}\cdot\overline{(A\cdot B)}$

⑩ $A\cdot(\overline{A} + \overline{B}) + B\cdot(\overline{A} + \overline{B})$

⑪ $A\overline{A} + A\overline{B} + B\overline{A} + B\overline{B}$

⑫ $A\overline{B} + B\overline{A}$

⑬ $A \oplus B$

Fig. 11-11 Using the laws of combination.

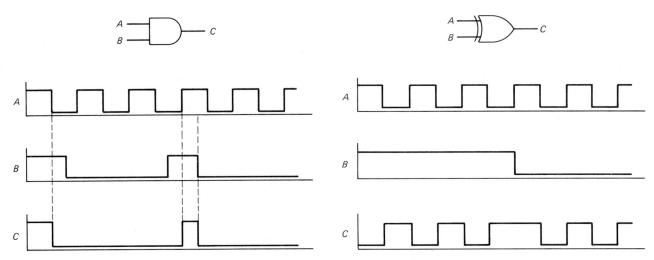

Fig. 11-12 Timing diagram.

Fig. 11-13 Exclusive OR timing diagram.

mable inverter since its output will be the complement of its input when its other input is high.

Digital integrated circuits often contain several logic gates in one package. Figure 11-14 shows the pinouts for some common 7400 series devices. The 7400 is a quad two-input NAND gate. It has four gates per package. The 7404 is called a *hex inverter* since it has six gates per package. The 7420 contains only two NAND gates because each one has four inputs and there are only 14 pins available. In the TTL family and subfamilies V_{CC} is $+5$ V and is applied to pin 14 of each package. Pin 7 is grounded in each case. This is not a uniform standard, however. Some 7400 devices have different ground and supply pins, and some have more than 14 pins. The output pins are labeled Y so they are not confused with the input pins, which are labeled A through D and beyond, depending upon how many inputs there are

per gate. As mentioned before, the pins are counted counterclockwise from the index when viewing the packages from the top.

The industrial technician is exposed to many types of control circuits. Relays were the mainstay of controls at one time but are now being replaced by static controls. *Static controls* are so named because they have no moving parts. Figure 11-15 compares static controls with relay controls. The American National Standard Institute (ANSI) logic symbols have gained widespread acceptance, but the National Electrical Manufacturer's Association (NEMA) logic symbols are found on many industrial schematics and wiring diagrams. The NOT function can be achieved with a gate or by using a relay with normally closed contacts. The ladder diagram for the AND function shows the control contacts in series so all will have to be closed to energize the relay contacts. The OR

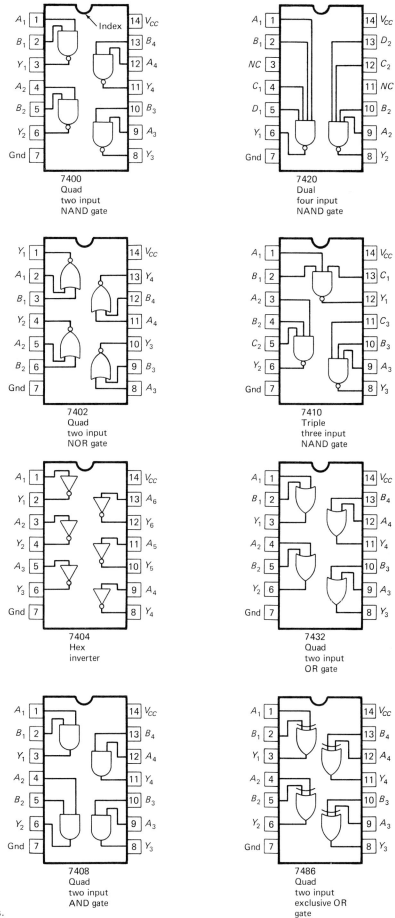

Fig. 11-14 Pinouts for some common logic packages.

279

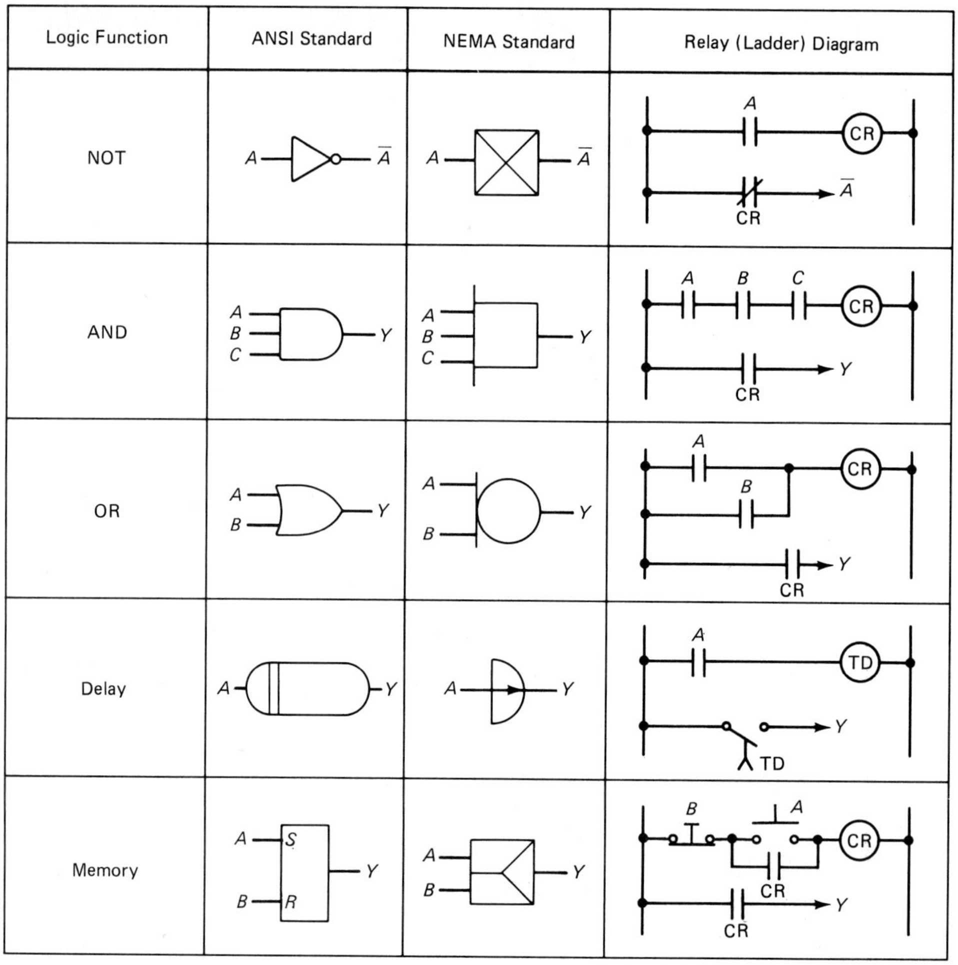

Logic Function	ANSI Standard	NEMA Standard	Relay (Ladder) Diagram
NOT			
AND			
OR			
Delay			
Memory			

Fig. 11-15 A comparison of static and relay controls.

function is realized by wiring the control contacts in parallel. A time delay relay can be replaced with a static device such as an NE555 timer. The memory function is based on a static device such as a latch or uses a second set of relay contacts to hold the circuit on until it is reset by switch B. Latches are discussed in the next section of this chapter.

REVIEW QUESTIONS

10. Which gate acts as an all or nothing decision element?

11. If a gate is simply referred to as *OR*, which type is it?

12. Will the circuit of Fig. 11-6 produce the correct output instantaneously? Why?

13. Calculate the total propagation delay for Fig. 11-6 if the circuit is built by using standard TTL components.

14. Refer to Fig. 11-12. What will waveform *C* look like if waveform *B* is constant at logic 1?

11-3
LATCHES AND FLIP-FLOPS

A *latch* is a sequential logic element. It has a memory characteristic that makes it useful for storing events and binary numbers. Figure 11-16(a) shows an *R-S* latch based on two NAND gates. The *R* is the reset input, and the *S* is the set input. The inputs are activated by logic low signals. This is why the standard symbol shown in Fig. 11-16(b) of the illustration indicates NOT S and NOT R. The latch has two outputs: Q and *NOT Q*. When Q = 0, NOT Q = 1, and this is the *reset condition* of the latch. When Q = 1, NOT Q = 0, and this is the *set condition* of the latch. Figure 11-16(c) of the illustration shows the timing diagram. Starting at the left, the latch is initially in the reset condition since Q is low. Moving to the right, the set input goes momentarily low. The Q output responds by going high, and the latch is now set. Moving more to the right, the reset input goes momentarily low, and the Q output goes low again. The latch has been reset.

The *R-S* latch works because of feedback. Look

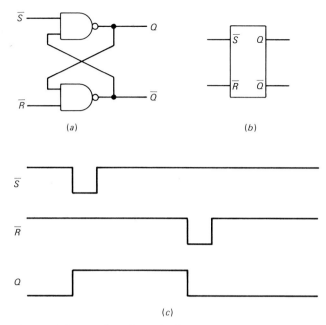

Fig. 11-16 Low activated R-S latch. (a) R-S latch from NAND gates. (b) Symbol. (c) Timing diagram.

at Fig 11-16(a). The output of each NAND gate feeds back to the input of the other gate. Recall that the truth table for the NAND gate shows a logic high at the output for all input conditions except the one where both inputs are high. Assume that the latch is in the reset condition (Q = 0 and NOT Q = 1) and that both inputs are high. This means that the top NAND gate has two high inputs and its output must be low. The bottom NAND gate will have one high input and one low input, and its output must be high. The truth tables are satisfied, and the circuit is stable in this condition. Now, suppose the set input goes low. The top gate now has one low input and one high input, and its output goes high. This high is fed back to the bottom gate, so that it now has both inputs high and its output goes low. Note that the latch is now in the set state: Q = 1 and NOT Q = 0. The low output from the bottom gate feeds back to the top gate, and it now has both inputs low. This changes nothing, and the top gate still has a high output. When the set pulse ends, the top gate has one high input and one low input and there is no change. The latch is now stable in the set condition. Because of the feedback, it remembers that it was set and will stay in the set mode until a reset pulse comes along or the circuit is powered down.

To analyze digital circuits that use feedback, begin with the circuit in some stable condition. Verify that the truth tables are correct for each gate. Then, introduce a change and follow the change through by using the truth table for each gate. Use this technique now with the circuit of Fig. 11-16(a). Start with the latch in the set condition and with both inputs high. Verify the truth tables. Then apply a reset pulse

(logic 0) and trace the changes through the circuit. Prove to yourself that the outputs change as expected and that the circuit remains reset after the reset pulse is removed.

Simple R-S latches work very well, but they are susceptible to being forced to an illegal output state. They are also capable of settling to an unpredictable output state in some cases. For example, what would happen in Fig. 11-16(a) if logic 0 pulses were applied to the inputs at the same time? Both NAND gates would have high outputs. Both Q and NOT Q would be high at the same time. This is illegal, since a variable and its complement cannot be the same. Then, what would happen if both pulses returned to logic 1 at the same time? Both gates would "race" to have a low output. The gate with the shorter propagation delay would win the race and force the other gate to have a high output. The latch would resume a legal output state, in either the set mode or the reset mode. Such unpredictable behavior is unacceptable in digital circuits. This condition is known as *pulse race* or simply *race* and must be avoided.

Figure 11-17 shows an application for the R-S latch. It is being used to debounce a mechanical switch. When switch contacts close, they bounce. They make and break rapidly until the bouncing stops. This process lasts for several milliseconds and can create dozens of extra pulses in a circuit. The *bounceless switch circuit* avoids this problem because when the switch is thrown, it grounds either the set or the reset input of the latch, so the latch is either set or reset. As the contacts bounce open and closed, there is no change in the latch because once it is set it is not affected by subsequent set pulses. Likewise, once it is reset it is not affected by subsequent reset pulses. The *pull-up resistors* ensure a logic high at the inputs when not grounded by the switch. The circuit will work without the resistors if the latch is a TTL family device since the inputs float high. However, it is considered good practice to use the resistors.

Figure 11-18 shows an R-S latch configured from two NOR gates. Note that the Q output is taken from the bottom gate and the NOT Q output is taken from the top gate. The symbol, shown in Fig. 11-18(b), locates the Q output on top because this is its stan-

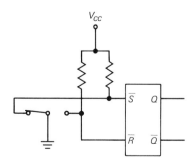

Fig. 11-17 A bounceless switch circuit.

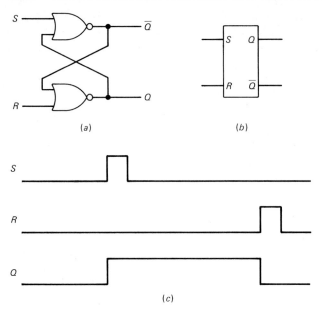

Fig. 11-18 High activated R-S latch. (a) R-S latch from NOR gate. (b) Symbol. (c) Timing diagram.

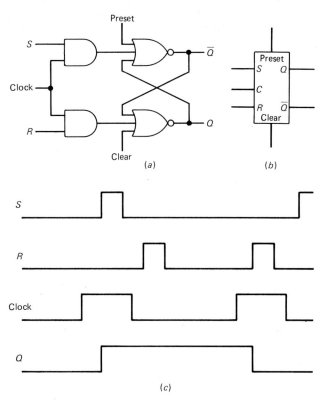

Fig. 11-19 Clocked R-S latch with asynchronous preset and clear. (a) Gate implementation. (b) Symbol. (c) Timing diagram.

dard position. It is activated by logic high signals applied to the set and reset inputs. Remember that the output of a NOR gate is low if any or all of its inputs are high. Study the timing diagram and verify circuit operation. The NOR latch is susceptible to the same problems as the NAND latch. If both inputs are driven high, the outputs will assume an invalid condition: both Q and NOT Q will be at logic 0. Then, if both inputs go low at the same time, the outputs will race, and the latch will settle into the set or reset condition.

It is often desirable to enable a latch only during a specified period of time. Figure 11-19 illustrates a clocked R-S latch with asynchronous preset and clear inputs. Note that Fig. 11-19(a) shows two AND gates at the front end of the latch. These gates share a common input line called the *clock* or *enable input*. It does not matter what is happening at the set and reset inputs if the clock input is low, since all inputs to an AND gate must be high for the output to go high. Study the timing diagram in Fig. 11-19(c). Start at the left, where the latch is initially in the reset condition (Q is low). The clock signal goes high, enabling both inputs. Next, the set input goes high and the latch is set. Then the clock signal goes low. Now a reset pulse comes along, but the latch is not reset because the inputs are disabled. The next reset pulse does reset the latch since it occurs when the clock is high. That time when the clock is high is often referred to as a *window*. Input pulses must be synchronized with the clock window to have any effect on the output; the S and R inputs are considered to be synchronous inputs for this reason. The preset and clear inputs are asynchronous since they can be applied at any time to set or clear the latch. The asynchronous inputs are often used to initialize the circuits to a known condition after power-on.

Figure 11-20 shows a clocked D latch. The D input is the only synchronous input. An inverter has been added to supply inverted data to the bottom gate. If the D input is high during the clock window, a logic high is applied to the top NOR gate, and a logic low is applied to the bottom NOR gate. The latch is driven to the set condition if it hasn't already been set. If the D input is low during the window, a high will be applied to the bottom NOR gate; it will reset the latch if it wasn't previously in that mode. The latch is said to be *transparent* because the output follows the data input during the time that the clock is high. The advantage of the D latch is that race has been eliminated at the synchronous input. It is also no longer possible to force the outputs to an illegal mode since there is only one input. However, it is still possible to race the latch and force the outputs to an illegal mode with the asynchronous preset and clear inputs.

Now it is time to look at *flip-flops;* these are circuits with much in common with latches. In fact, many people use the term *flip-flop* to describe any bistable circuit, including the latches studied to this point. However, the digital IC manufacturers generally reserve the term for devices that trigger on a clock edge. Latches are level-sensitive, by comparison. Figure 11-21 shows the circuit, symbol, and the timing diagrams for a D flip-flop. The circuit uses a combination of input latches with the familiar NAND output latch. The D input is sampled only during the

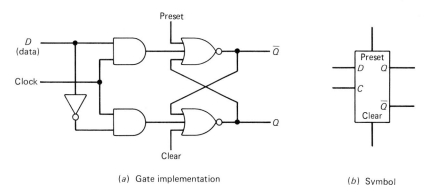

Fig. 11-20 Clocked *D* latch with asynchronous preset and clear.

(*a*) Gate implementation

(*b*) Symbol

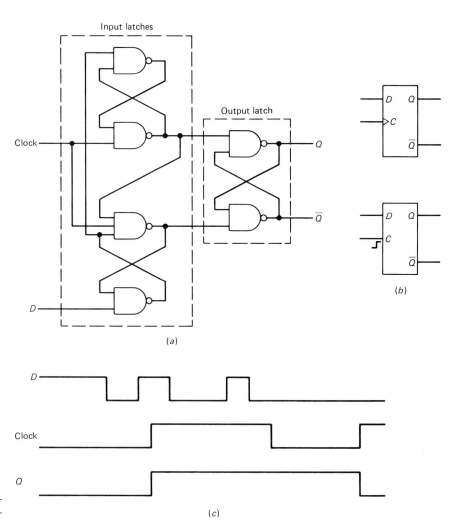

Fig. 11-21 The *D* flip-flop. (*a*) Gate implementation. (*b*) Symbols. (*c*) Timing diagram.

positive edge of the clock and not during a clock window. There is no clock window in a flip-flop. Study the timing diagram. The flip-flop starts out in the reset mode. Next, the *D* input goes low and this has no effect on the output. Next, the *D* input goes high, with no immediate effect on the output. Now, the clock goes high, enabling the input latches, and the *D* input is sampled at this time. Note that the *D* input goes low again while the clock is still high with no effect on the output; again there is no clock win-

dow. Finally, on the last positive clock edge, the flip-flop is reset because the *D* input is low at that time. Figure 11-21(*b*) shows two symbols that are used to differentiate between edge- and level-sensitive devices. A triangle can be added at the clock input, or a positive edge can be drawn near the clock input.

The *J-K flip-flop* is one of the most versatile and popular of all sequential logic circuits. Figure 11-22 depicts two symbols and a truth table for the circuit. Most J-K flip-flops are negative-edge devices. They

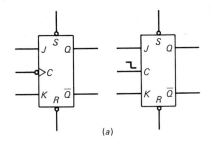

J	K	Q_{n+1}	Mode
0	0	Q_n	Inhibit
0	1	0	Reset
1	0	1	Set
1	1	$\overline{Q}_n$	Toggle

(b)

Fig. 11-22 Negative edge J-K flip-flop with asynchronous set and reset. (a) Symbols. (b) Truth table.

sample the *J* and the *K* inputs at the time when the clock is falling from logic 1 to logic 0. Figure 11-22(*a*) shows an inversion circle at the clock input, and Figure 11-22(*c*) shows the alternate symbol, which uses a negative edge. The truth table in Figure 11-22(*b*) requires some explanation. It uses a notation that refers to Q_n and to Q_{n+1}. The Q_n refers to the Q output just before the negative edge, and Q_{n+1} refers to the Q output just after the negative edge. When $J = 0$ and $K = 0$, the truth table shows that the output is Q_n. This simply means that Q did not change. It is in the same condition as it was before the clock edge. This is called the inhibit mode since the flip-flop is inhibited from changing. When $J = 0$ and $K = 1$, the Q output will always be 0 after the clock edge. This is known as the *reset mode*. When $J = 1$ and $K = 0$, the Q output will always be 1 after the edge; this is the *set mode*. If $J = 1$ and $K = 1$, the output will be equal to NOT Q_n, and the flip-flop will change states; this is known as the *toggle mode*. It will toggle on every succeeding clock pulse.

REVIEW QUESTIONS

15. Refer to Fig. 11-16(*c*). If the diagram included a waveform for NOT Q, how would it compare with the Q waveform?

16. Refer to Fig. 11-16(*c*). If a second set pulse followed the first set pulse and preceded the reset pulse, how would the Q waveform change?

17. Refer to Fig. 11-17. What are the resistors called?

18. Refer to Fig. 11-18. What output state results if both the set and reset inputs are high? What is the condition called if both inputs go low at the same time?

19. Refer to Fig. 11-19. What will happen if the set and reset inputs race when the clock is at logic 0?

20. Refer to Fig. 11-20. If the latch portion of the circuit were constructed with NAND gates instead of NOR gates, what gates should be substituted for the AND gates to ensure an identical timing diagram?

21. Refer to Fig. 11-20. Is it possible to ''race'' this latch? If so, at which input(s)?

22. Refer to Fig. 11-21(*c*). Why doesn't the Q output change when the data change in the clock window?

11-4
COUNTERS AND REGISTERS

A *counter* is a circuit based on the sequential logic elements studied in the last section. Look at the binary ripple counter in Fig. 11-23. It is an arrangement of four J-K flip-flops and is capable of counting from binary 0000 to binary 1111. The *J* and *K* inputs are all floating, so every flip-flop is in the toggle mode. The timing diagram starts at the left with all four outputs at 0. On the first negative clock edge, Q_A goes high. On the second negative clock edge, Q_A goes low, providing a negative edge for the second flip-flop so that Q_B goes high. Follow the timing diagram through to the sixteenth negative clock edge and note that the counter is reset at this time and all four outputs are once again at 0. Thus, the counter has 16 unique states. This is also called the *modulus* of the counter. The modulus is equal to 2^N, where N is the number of sequential logic elements in the counter.

Most electronic diagrams are drawn with the input signals entering at the left. Binary numbers are printed with the LSB at the right. These two conventions are in conflict when describing counters. Figure 11-23 solves this problem by rearranging the truth table columns. Transistor Q_A is the LSB and heads the right-hand column. Transistor Q_D is the MSB and heads the left-hand column.

Figure 11-23 is called a *binary ripple counter* because the count ripples from flip-flop to flip-flop. Check what happens on the eighth negative clock edge. Outputs *A*, *B*, and *C* all go low, and output *D* goes high. Will these four events occur simultaneously? No, the count will ripple from *A* to *B* to *C* and finally to *D*, and *D* will not go high until four propagation delays after the clock edge. The high-frequency performance of ripple counters is limited. Figure 11-24 illustrates a synchronous binary counter with better high-frequency performance. The clock is directly applied to each flip-flop. This means that every flip-flop can toggle at the same time (they are synchronous). Additional connections and some combinational logic are required to allow the synchronous counter to achieve the correct binary se-

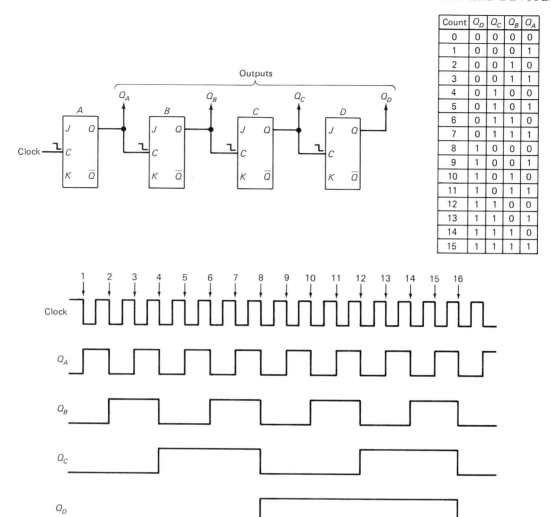

Count	Q_D	Q_C	Q_B	Q_A
0	0	0	0	0
1	0	0	0	1
2	0	0	1	0
3	0	0	1	1
4	0	1	0	0
5	0	1	0	1
6	0	1	1	0
7	0	1	1	1
8	1	0	0	0
9	1	0	0	1
10	1	0	1	0
11	1	0	1	1
12	1	1	0	0
13	1	1	0	1
14	1	1	1	0
15	1	1	1	1

Fig. 11-23 Binary ripple counter.

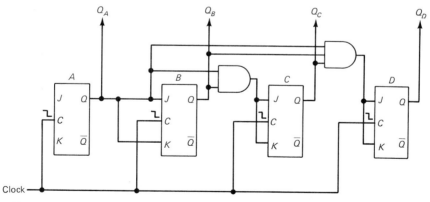

Fig. 11-24 Synchronous binary counter.

quence. The Q_A output also goes to the J and K inputs of flip-flop B. If you check the timing diagram from Fig. 11-23 you will see that Q_B is supposed to toggle only on those edges when Q_A is high before the edge. Also notice that Q_C toggles only on the edges where both A and B are high before the edge.

Finally, note that Q_D toggles only on those edges where A, B, and C are high before the edges. It should now be clear how the connections and AND gates of Fig. 11-24 function.

The natural modulus of a counter is 2^N. It is possible to use feedback to reduce the natural modulus.

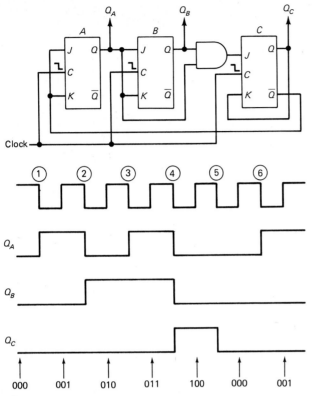

Fig. 11-25 A modulo 5 counter.

For example, Fig. 11-25 shows a modulo 5 counter. It uses three flip-flops, and the natural modulus is $2^3 = 8$. However, feedback from flip-flop C to flip-flop A and from Q to K at flip-flop C reduces the natural modulus, and the counter has only five unique states. They range from binary 000 to 100. The counter does not advance from binary 100 to binary 101; instead it resets to 000. As in the last circuit, Q_A feeds the J and K inputs of flip-flop B and Q_A is ANDed with Q_B to control the J input of flip-flop C. The timing diagram follows a natural pattern up to the fifth negative clock edge. At the fifth edge, Q_A does not toggle. This is because NOT Q_C was low before the edge and the feedback put flip-flop A into the inhibit mode. Also note that Q_C does toggle on the fifth edge because Q_C was high before the clock edge, and so was the K input of the flip-flop. With $J = 0$ and $K = 1$, the flip-flop is in the reset mode.

A modulo 5 counter can be combined with a modulo 2 counter to make a modulo 10 counter. Counters with 10 states are called *decade counters;* if they follow the BCD count sequence, they are called *BCD counters.* Figure 11-26 shows the pinout, count sequence, and reset truth table for a 74LS90 decade counter. The output of the modulo 2 counter is pin 12 and is connected to the input of the modulo 5 counter at pin 1 when the device is to be used as a BCD counter. The clock is applied to pin 14. You will recall that binary-coded decimal follows the standard binary sequence up to a count of 1001. The BCD count sequence is indicated in Fig. 11-26(*b*).

For some applications, the count sequence is not important, but the division and the output waveform are. The BCD count sequence shows that the frequency at output D (pin 11) will be equal to one-tenth the clock frequency, but the waveform will be a rectangular pulse of low duty cycle. It is possible to have the IC divide the input frequency by 10 and produce a square wave (50 percent duty cycle) at the output. This is accomplished by feeding the clock into pin 1, connecting pin 11 to pin 14, and taking the output signal from pin 12. The output from a modulo 2 counter is a square wave; therefore, dividing by 5 first and then by 2 produces the desired waveform. However, when this is done, the BCD count sequence is lost.

Figure 11-26 shows the truth table for the decade counter. It has four reset inputs. An X indicates a *don't care* condition, meaning that either a logic 0 or a logic 1 may be applied at that particular reset pin. There are three reset/preset combinations: two of them reset the counter to 0000 and one presets it to 1001. There are four count combinations, and any one of them may be used for counting.

Some digital counters are programmable; Fig. 11-27 is an example. The DM8555 is a decade counter that can be preloaded (or preset) with a BCD number. It has a binary counterpart with the part number DM8556 that can be preloaded with a 4-bit binary number. This allows the modulus of the devices to be adjusted (programmed) by changing the number that is preloaded. For example, if the counter is preset to BCD 0111, the next count will be 1000, then 1001; then it can be preset again to 0111. In this case, the modulus has been programmed to 3. Counters with this capability are known as *modulo-N counters.* To facilitate presetting, the device has four I/O pins for loading data into the counter. Follow the typical timing diagram shown in Fig. 11-27(*c*). First, the counter is cleared to zero by applying a reset pulse to pin 4. Second, the counter is preset to BCD 5 by applying 0101 to I/O pins 14, 13, 11, and 10 and by applying a negative load pulse to pin 7. Note that this sets outputs Q_D through Q_A (pins 2, 3, 5, and 6) to 0101 and occurs at the positive clock edge (this is a synchronous load). Third, the count progresses 6, 7, 8, 9, 0, 1, 2, 3, 4, and 5. Fourth, during the count, the I/O pins are disabled for a period of time by applying a disable pulse to pin 12 (more on this later). Fifth, the counter is disabled for a time by applying a high pulse to pin 1. Sixth, the count resumes to BCD 6.

The DM8555 counter is very flexible and can be used in many ways. It also lends itself to bus-structured digital systems. A *bus* is an arrangement in which devices are connected in parallel on several circuit paths for binary information transfer. Many devices are typically connected to the bus and must be isolated from each other at times. For example, if one logic package tries to pull a bus line to logic high and another tries to pull the same line to logic low, this process creates a problem called *bus con-*

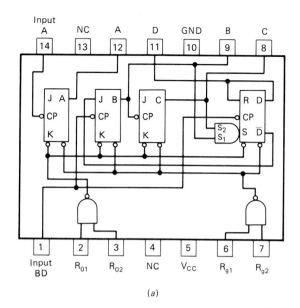

(a)

Count	Output			
	D	C	B	A
0	0	0	0	0
1	0	0	0	1
2	0	0	1	0
3	0	0	1	1
4	0	1	0	0
5	0	1	0	1
6	0	1	1	0
7	0	1	1	1
8	1	0	0	0
9	1	0	0	1

Pin 12 connected to pin 1 and Clock applied to pin 14

(b)

Reset inputs				Output			
R_{01}	R_{02}	R_{g1}	R_{g2}	D	C	B	A
1	1	0	X	0	0	0	0
1	1	X	0	0	0	0	0
X	X	1	1	1	0	0	1
X	0	X	0	COUNT			
0	X	0	X	COUNT			
0	X	X	0	COUNT			
X	0	0	X	COUNT			

X indicates that either a logical 1 or a logical 0 may be present

(c)

Fig. 11-26 Decade counter. (a) Pinout. (b) Count sequence. (c) Reset/count.

tention. Bus contention can be avoided by using logic devices with tri-state outputs. A *tri-state output* can be at one of three valid conditions: logic high, logic low, or high impedance (tri-stated). The four I/O pins on the DM8555 are tri-state. Any time that pin 12 (output disable) is at logic high, the I/O pins are disabled and go into their high-impedance state. This is shown with Zs in the truth table of Fig. 11-27(b). Thus, the I/O pins can be effectively disconnected from the bus, allowing some other logic package to place data on the bus without the problem of bus contention. The four Q output pins of the DM8555 are standard totem pole outputs and cannot be tri-stated.

All of the counters discussed to this point have been *up counters,* which show an increasing count with each clock pulse. *Down counters,* which show a decreasing count with each clock pulse, are also available. Figure 11-28 represents a 74LS192 up/down BCD counter. It is also available in a 4-bit binary version with the part number 74LS193. Both versions are fully programmable and can be used as modulo-N counters. Each output can be set to a high or a low by entering the desired data at the inputs and then pulsing the load input low. This is an asynchronous load since it is independent of the clock inputs. Figure 11-28(b) is a typical timing diagram. First, the outputs are reset to 0 by applying a clear

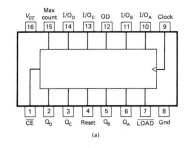

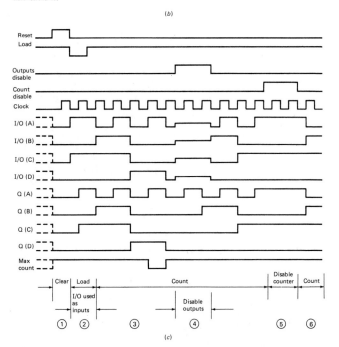

Control Inputs					I/O Ports				Active Outputs			
LOAD	CE	CLK	OD	Reset	I/O$_A$	I/O$_B$	I/O$_C$	I/O$_D$	Q$_A$	Q$_B$	Q$_C$	Q$_D$
H	X	X	L	H	L	L	L	L	L	L	L	L
H	X	X	H	H	Z	Z	Z	Z	L	L	L	L
H	X	L	L	L	Q$_{AO}$	Q$_{BO}$	Q$_{CO}$	Q$_{DO}$	Q$_{AO}$	Q$_{BO}$	Q$_{CO}$	Q$_{DO}$
H	X	L	H	L	Z	Z	Z	Z	Q$_{AO}$	Q$_{BO}$	Q$_{CO}$	Q$_{DO}$
L	H	↑	L	L	a	b	c	d	A	B	C	D
H	L	↑	L	L	COUNT				COUNT			
H	L	↑	H	L	Z	Z	Z	Z	COUNT			

The I/O pins are used as inputs when they are TRI-STATED, and $\overline{LOAD}$ input is Low. They are outputs and active when $\overline{LOAD}$ input is High and OD is Low.
H = High Level (Steady State)
L = Low Level (Steady State)
X = Don't Care including transitions
a, b, c, d = The level of the steady state input at inputs A, B, C, D respectively
Q$_{AO}$, Q$_{BO}$, Q$_{CO}$, Q$_{DO}$ = The level of Q$_A$, Q$_B$, Q$_C$, Q$_D$ respectively, before the indicated steady state input conditions were established.

(b)

Fig. 11-27 DM8555 programmable decade counter. (a) Pinout. (b) Truth table. (c) Typical timing diagram.

pulse to pin 14. Second, the counter is preset to BCD 7 by applying 0111 to the data pins and pulsing pin 11 low. Third, the count progresses upward to 8, 9, carry, 0, 1, and 2 as count-up pin 5 is clocked. Fourth, the count progresses down to 1, 0, borrow, 9, 8, and 7 as count-down pin 4 is clocked. The carry and borrow pins are used to cascade more than one device. *Cascading* is accomplished by feeding the carry and borrow outputs to the count-up and count-down inputs, respectively, of the succeeding counter.

Shift registers constitute another category of digital circuits based on sequential logic elements. They are useful for temporary storage, change of data from one format to another, and sequence control and

timing. Figure 11-29 shows a 4-bit serial load shift register. Data are clocked into the register at the data input of *D* flip-flop *A*. Four clock pulses are required to load a 4-bit word into the register. Then the data are available in parallel form at the Q outputs. The first data bit entered is available at output Q$_D$, and the last, or fourth data bit, is available at output Q$_A$. The shift register is useful for converting *serial data* (one bit at a time) to *parallel data* (all bits at a time). It is also useful for temporary storage of data. If the data are needed later, they can be retrieved in serial form at output Q$_D$. The first bit that was loaded is immediately available at the serial data output. Then, three clock pulses will be required to shift the other

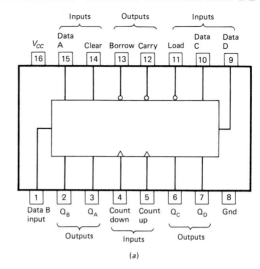

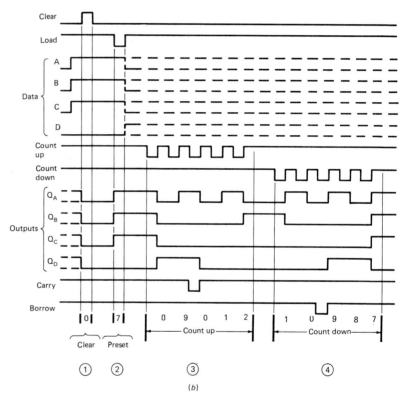

Fig. 11-28 74LS192 synchronous up/down counter.
(*a*) Pinout. (*b*) Typical timing diagram.

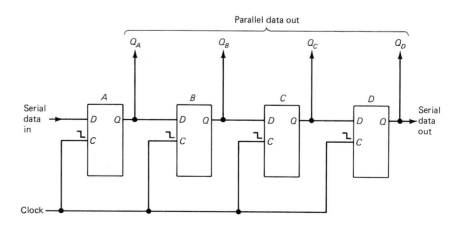

Fig. 11-29 Shift register.

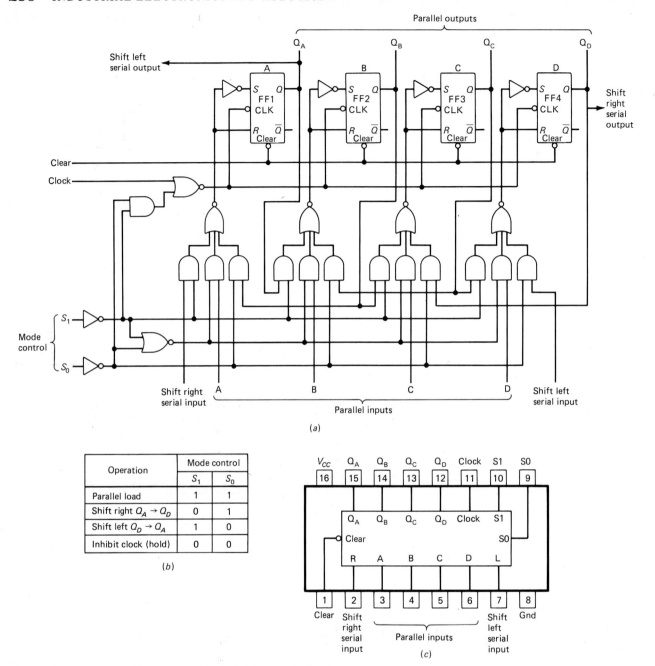

Fig. 11-30 74LS194 4-bit bidirectional universal shift register. (*a*) Logic diagram. (*b*) Mode control. (*c*) Pinout.

Operation	Mode control	
	S_1	S_0
Parallel load	1	1
Shift right $Q_A \rightarrow Q_D$	0	1
Shift left $Q_D \rightarrow Q_A$	1	0
Inhibit clock (hold)	0	0

(*b*)

3 bits to the output. A register of this type is sometimes called a *first-in first-out* (FIFO) register.

A much more versatile device is shown in Fig. 11-30. It is a 74LS194 universal 4-bit shift register. It is capable of shift right, shift left, and parallel loading and has parallel outputs. Figure 11-30(*b*) shows the truth table for the two mode control pins S_1 and S_2. When both mode control pins are high, parallel data are loaded into the register. This occurs because a NOR gate senses the inverted mode signals. With both mode inputs high, both NOR gate inputs are low. The output of the NOR gate is therefore high,

and it supplies a logic high to one lead of each of the center AND gates at the bottom of the logic diagram. The output of each AND gate will follow the data applied at the parallel inputs. The data reach each flip-flop through a NOR gate and an inverter and are loaded on the first positive clock edge. With the mode control in the shift right pattern ($S_1 = 0$ and $S_0 = 1$), a high is supplied to one input of every left-hand AND gate in the bottom group. The shift right serial input is now active, and data can be clocked into flip-flop A. The data from A will shift to B since its AND gate is also active, and so on. Reversing the

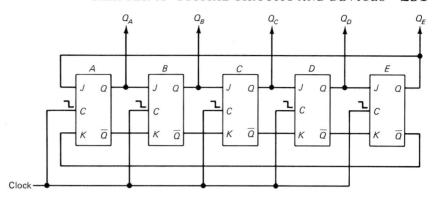

Q_A	Q_B	Q_C	Q_D	Q_E	Clock pulse
1	0	0	0	0	0
0	1	0	0	0	1
0	0	1	0	0	2
0	0	0	1	0	3
0	0	0	0	1	4
1	0	0	0	0	5

N unique states

Fig. 11-31 Circulating shift register (ring counter).

mode control pattern ($S_1 = 1$ and $S_0 = 0$) places a logic high at one input of each right-hand AND gate in the bottom group. This activates the shift left serial input, and data are clocked into flip-flop D. The output of D is clocked into C since its AND gate provides a data path to the left, and so on. With both mode control pins low, the shift register is inhibited because the clock cannot get through the NOR gate at the left of the diagram. Note that the other NOR gate input is from an AND gate which has both inputs high when the mode control pins are both low. The AND gate output will be high, holding the clock NOR gate output low.

The output from a shift register can be connected back to the input. The data will circulate endlessly in this case as the register is clocked. Refer to Fig. 11-31. This circuit is known as a *ring counter*. Assume the initial condition is that flip-flop A is set and all others are reset. On the first negative clock edge, the high bit will shift to the right by one position. It will shift to the right again on the second negative edge. When the fifth clock pulse comes along, the high bit shifts from flip-flop E to A. Note that ring counters have N unique states, where N is the number of flip-flops. Ring counters are sometimes used as *control sequencers*: one operation follows another, followed by another, and so on, until the process restarts, and the first operation is performed again. The circuit of Fig. 11-31 is not complete since some means of establishing the initial condition must be added. For example, some means will be required to preset one flip-flop and clear all the others. Also, the circuit may get into an illegal mode (more than

one high bit) because of noise or a momentary power dip. For this reason, the circuit may also require extra logic to detect illegal modes and correct them.

REVIEW QUESTIONS

23. Refer to Fig. 11-23. If the clock frequency is 10 MHz, what is the output frequency at Q_D?

24. Do all of the flip-flops in Fig. 11-23 change state at the same time on the 16th negative clock edge? Why?

25. Refer to Fig. 11-26. Can this IC be used as a BCD divider and output a square wave of one-tenth the input frequency?

26. Refer to Fig. 11-27. What happens to the I/O pins when a logic 1 is applied to pin 12?

27. Refer to Fig. 11-27. Is the load function synchronous or asynchronous?

28. Refer to Fig. 11-27. Can Q_A through Q_D be tri-stated? Which device pins would be appropriate for connecting to a bus that has other device outputs connected to it?

29. Refer to Fig. 11-28. Is the load function synchronous or asynchronous?

30. Refer to Fig. 11-28. What number should the counter be preset with to use it as a modulo 8 device in the up-count mode?

31. Refer to Fig. 11-29. What will the parallel data output be if the data input is held low for four clock edges?

32. Refer to Fig. 11-30. Both mode control pins are high. When are the data loaded?

11-5
MULTIPLEXERS AND DECODERS

A *digital multiplexer* is a "several into one" circuit. Multiplexers are used to select data from one of several sources and output those data onto a single line. They are also called *data selectors*. The typical multiplexer IC uses select pins to determine which of the inputs will be connected to the output. For example, a four-line to one-line multiplexer will have two select pins. The binary pattern at these pins will range from 00 (input 0 is selected) to 11 (input 3 is selected).

It is possible to use a multiplexer as a parallel-to-serial data converter. For example, if a 4-bit word is applied to the data inputs of a four-line to one-line device and if the select pins are clocked by a two-stage counter from binary 00 to binary 11, the word will appear in serial form at the output. The first input will appear first when the counter is at 00, then the next input will appear at a count of 01, and so on.

A *demultiplexer* represents an opposite function; it is a "one into several" logic circuit. Demultiplexers take data from a single source and distribute them to one of several output lines. They are also known as *data distributors*. Figure 11-32 shows the pinout and the logic diagram for a 74LS138 decoder/demultiplexer. It is used for memory decoding and data-routing applications. (Another type of decoder is shown in Fig. 11-33.) The 74LS138 distributes to one of eight output lines, depending upon the logic conditions at its three select inputs. When all three select

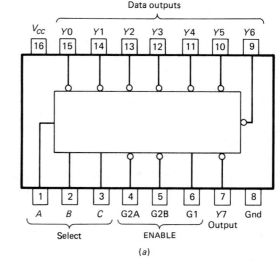

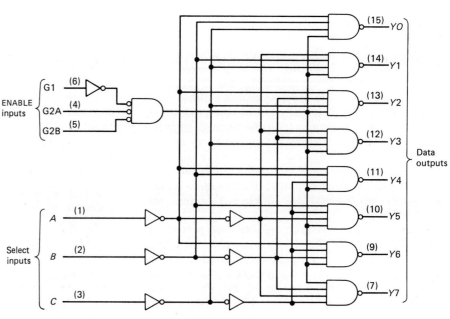

Fig. 11-32 74LS138 decoder/demultiplexer. (*a*) Pinout. (*b*) Logic diagram.

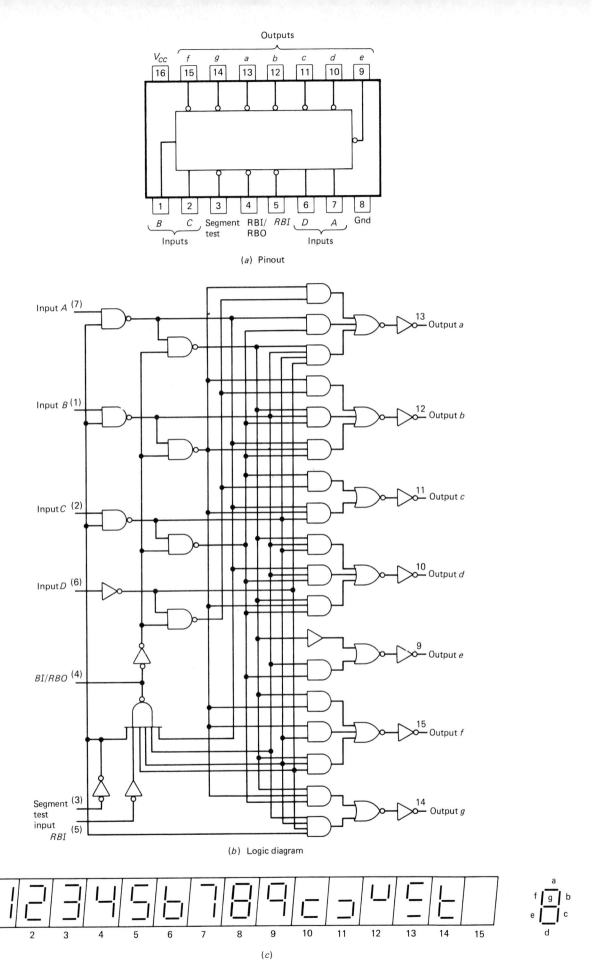

Fig. 11-33 74LS47 open collector BCD/T-segment decoder/driver. (*a*) Pinout. (*b*) Logic diagram. (*c*) Display and segment identification.

293

inputs are at logic 0, output Y0 is selected. If select *A* is high and *B* and *C* are low, output Y1 is selected. The LSB of the select word is *A,* and *C* is the MSB. Any one of the three enable inputs can be used for data input for demultiplexing operations. Two of the enable inputs are active low (G2A and G2B), and the other is active high (G1). The G1 input should be used if the data do not require inversion, and the other enable inputs should be grounded.

A *decoder* is a circuit or device that reacts to one or more conditions at its inputs. Decoders are commonly used to convert binary, BCD, ASCII, or some other common code into an uncoded form. A good example is the 74LS47 BCD-to-seven-segment decoder/driver shown in Fig. 11-33. Its application is to convert four BCD lines into the appropriate levels and patterns to display decimal numbers on common anode LED numeric displays. It is also a driver since its active low open collector outputs can sink up to 24 mA of current. A common anode display has all of its anodes tied to a single pin. In practice, this pin will connect to the positive supply. Grounding any one of the LED cathodes will light one segment. Current-limiting resistors are usually required between each segment cathode and its corresponding output pin on the decoder/driver. Since the 74LS47 has active low outputs, it will sink the cathode current. Refer to Fig. 11-33(*c*), which shows the possible displays and the segment identification. To display numeral 8, all segments must be on and all outputs will be low. To display numeral 1, segments *b* and *c* must be on, and output *b* and output *c* will be low. The binary numbers from 1010 (decimal 10) through 1111 (decimal 15) are *forbidden* in BCD. Note the display conditions for these numbers.

The 74LS47 shown in Fig. 11-33 also has a segment test input at pin 3. Grounding this pin when pin 4 is high will light all of the segments. Pin 4 is the blanking input (BI) and ripple blanking output (RBO) pin.

This pin may be pulsed to control LED intensity. Pin 5 is for *ripple blanking input* (RBI). The ripple blanking pins are used in those applications in which it is desired to blank leading or trailing zeros in a numeric display. If the RBI is low, the character 0 will not be displayed; all segments will be turned off when the BCD input code is 0000. Also, the RBO will go low to extinguish character 0 in the next stage in those applications in which leading or trailing zeros are to be blanked. Figure 11-34 shows the arrangement for leading zero blanking. The RBI pin on the most significant decoder/driver is at logic zero. The BCD code at its inputs is 0000, and the output display is blanked. The RBO is low, placing the next display in the zero blanking mode. Its BCD input is also 0000 and is blanked, and its RBO pin is at logic 0. The next display is also in the zero blanking mode, but its BCD input is 0101, so some of its outputs go to logic zero, and the LED display shows the numeral 5. Note that its RBO pin is at logic 1 since its BCD input was not 0000. Therefore, the next decoder/driver displays a 0 even though its BCD input is 0000. *Trailing zero suppression* is accomplished by grounding the RBI pin on the least significant decoder/driver, and connecting its RBO to the RBI of the decoder/driver to the left, and so on.

The problems with multidigit displays, such as the one shown in Fig. 11-34, are the large number of components required and the large number of connections. For example, to have a six-digit display, 42 current-limiting resistors will be required, and 43 connections (add 1 connection for the positive supply) have to be run to the display unit. Multidigit displays are often multiplexed to save components and connections. Refer to Fig. 11-35. The appropriate segment select transistors are turned on to display the desired character. At the same time, one of the digit select transistors is also turned on. That particular display will be active. Later, another seg-

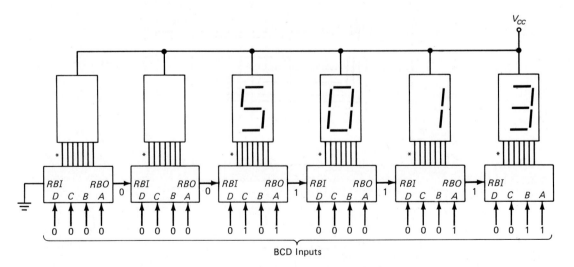

Fig. 11-34 Leading zero blanking.

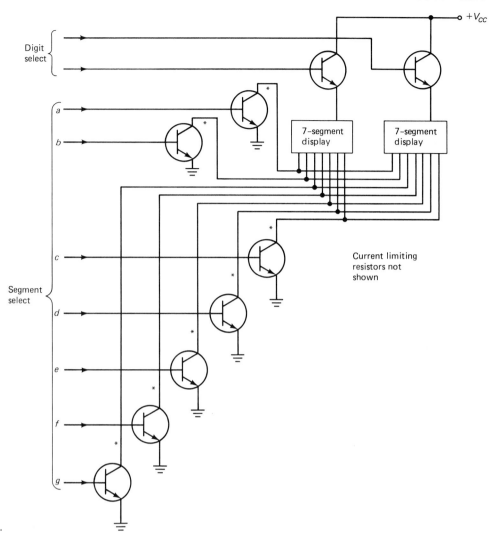

Fig. 11-35 Display multiplexing.

ment code is applied, and the next digit select transistor is turned on while the other is turned off. Now the second display is active. If this is done rapidly, the blinking is not perceptible. The savings in components and connections are substantial when a large number of digits must be displayed. For example, a multiplexed six-digit display requires only seven resistors and 13 connections to the display unit, which includes seven segment select lines and six digit select lines. Since the segments are in parallel, only seven lines are required regardless of the number of digits displayed. The clock frequency must be increased as the number of displays increases.

An *encoder* is a logic circuit that receives one or more signals in an uncoded form and outputs a code that can be used by another logic circuit. A *keyboard encoder* converts a keypress into some code; the keyboard may be octal, decimal, hexadecimal, alphanumeric, or some special type. The encoder can be a single IC or can be built up from several devices. For example, *single-chip encoders* convert a keypress into the equivalent ASCII code. Figure 11-36 depicts a simple decimal-to-BCD encoder circuit.

When all of the decimal inputs are low, the output is BCD 0000. If the 1 input goes high, the BCD output is 0001. If the 9 input is high, the BCD output is 1001. Verify the BCD output code for several different input conditions.

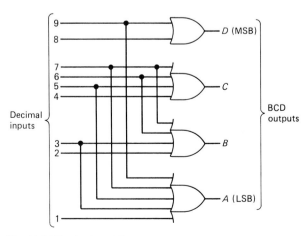

Fig. 11-36 Decimal-to-BCD encoder.

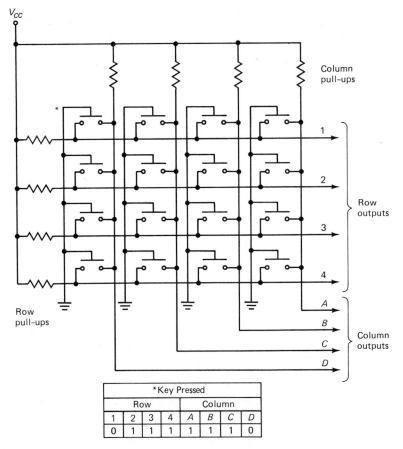

*Key Pressed							
Row				Column			
1	2	3	4	A	B	C	D
0	1	1	1	1	1	1	0

Fig. 11-37 A keypad matrix.

Single-chip keyboard encoders are available with row and column inputs. For example, there may be a four by four matrix of 16 key switches. Figure 11-37 shows a typical keyboard matrix based on DPST N.O. key switches. If any key switch is closed, one of the row outputs and one of the column outputs will go low. The encoder chip may also contain debouncing circuitry and a strobe generator. The *strobe* is an output used to alert the rest of the system that a key has been pressed. Another desirable feature is suppressed output for illegal entries (more than one key pressed at a time).

REVIEW QUESTIONS

33. Refer to Fig. 11-32. Select A = 0, B = 1, and C = 1. Which data output has been chosen?

34. Refer to Fig. 11-32. Suppose it is desired to use this device as a one-to-four data distributor. How can this be accomplished?

35. How can the device in Fig. 11-32 be used as an inverting demultiplexer?

36. Refer to Fig. 11-34. What should the display show if the RBI pin at the most significant digit loses its ground?

37. Refer to Fig. 11-34. What should the display show if the BCD code to the least significant digit changes to 0000?

11-6
ARITHMETIC ELEMENTS

Binary arithmetic is really quite simple. Since there are only two symbols to manipulate, there are only a few possible combinations. For example, in binary addition, there are only four cases:

$$0 + 0 = 0$$
$$1 + 0 = 1$$
$$0 + 1 = 1$$
$$1 + 1 = 0 \text{ (carry 1)}$$

The rules of *binary addition* match the truth table for the exclusive OR gate. Therefore, all that is needed to add 2 bits is an exclusive OR gate plus a carry generate circuit. Refer to Fig. 11-38. This circuit is called a *half-adder;* it produces the sum and carry outputs for 2 input bits. It is considered a half-adder because when binary words with more than 1 bit are added, carries from column to column occur which require a circuit with three inputs: one for the

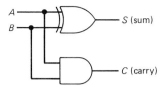

Fig. 11-38 Half-adder logic diagram.

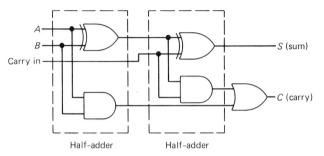

Fig. 11-39 Full-adder logic diagram.

augend, one for the addend, and one for the carry. For example, suppose we add binary 1001 to binary 0101:

$$
\begin{array}{lcr}
\text{CARRY} & \longrightarrow & 1 \\
\text{AUGEND} & \longrightarrow & 1001 \\
\text{ADDEND} & \longrightarrow & \underline{0101} \\
\text{SUM} & \longrightarrow & 1110
\end{array}
$$

This example shows that a circuit that can handle three inputs is needed. Figure 11-39 shows the logic required for a *full-adder,* which is based on two half-adders plus an OR gate.

A 4-bit parallel adder with ripple carry is illustrated by Fig. 11-40. It is constructed from one half-adder and three full-adders. A half-adder is sufficient for adding the LSB of the augend to the LSB of the addend. However, all subsequent circuits must be full-adders because carries can be generated anywhere along the line. The carries ripple from stage to stage in this type of circuit. Under worst-case conditions, the correct sum will not be available until four propagation delays have occurred. For example:

$$
\begin{array}{lcr}
\text{CARRIES} & \longrightarrow & 111 \\
\text{AUGEND} & \longrightarrow & 1111 \\
\text{ADDEND} & \longrightarrow & \underline{1111} \\
\text{SUM} & \longrightarrow & 11110
\end{array}
$$

Of course, as the size of the adder increases to 8, 16, or 32 bits, the ripple carries will create significant delays. Faster circuits using look-ahead carry logic are available. They eliminate the ripple carries and the cumulative delays.

Binary subtraction is also very simple. As in the case of addition, there are only a few combinations:

$$
\begin{array}{l}
0 - 0 = 0 \\
1 - 0 = 1 \\
0 - 1 = 1 \text{ (borrow 1)} \\
1 - 1 = 0
\end{array}
$$

Figure 11-41 shows the logic required for a *full-subtractor.* Note the similarity to the full-adder circuit. Also note that there are three inputs: one for the subtrahend, one for the minuend, and one for the borrow. Just as the full-adder had to add in carries from prior columns, the full-subtractor must subtract borrows from prior columns. As an example:

$$
\begin{array}{lcr}
\text{BORROWS} & \longrightarrow & 111 \\
\text{MINUEND} & \longrightarrow & 1010 \\
\text{SUBTRAHEND} & \longrightarrow & \underline{0111} \\
\text{DIFFERENCE} & \longrightarrow & 0011
\end{array}
$$

Many digital systems do not have subtraction circuits. They accomplish subtraction with adding circuits by converting the subtrahend into its 2s complement form and then adding it to the minuend. A number is converted into 2s complement form by inverting every bit (this operation puts it in 1s complement form) and then adding 1. As an example, let's convert the subtrahend from the previous example into its 2s complement form:

$$
\begin{array}{lcr}
\text{BINARY FORM} & \longrightarrow & 0111 \\
\text{1s COMPLEMENT} & \longrightarrow & 1000 \\
\text{ADD 1} & \longrightarrow & \underline{1} \\
\text{2s COMPLEMENT} & \longrightarrow & 1001
\end{array}
$$

Now, let's see what happens when we add the 2s complement form of the subtrahend to the minuend:

$$
\begin{array}{lll}
\text{MINUEND} & \longrightarrow & 1010 \\
\text{2s COMPLEMENT} & \longrightarrow & 1001 \\
\text{SUM} & \longrightarrow & 0011 \text{ (the last carry is thrown away)}
\end{array}
$$

Ignoring the final carry, the sum obtained is the same as the difference when we subtracted. It should be clear that subtraction may be accomplished by adding the 2s complement of the subtrahend to the minuend. Don't forget to discard the final carry.

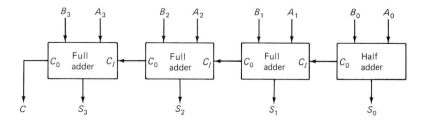

Fig. 11-40 Four-bit parallel adder with ripple carry.

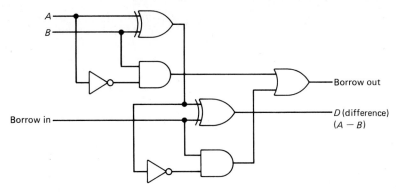

Fig. 11-41 Full-subtractor logic diagram.

The 74S381 IC is an arithmetic logic unit that can perform any of eight binary arithmetic and logic functions on two 4-bit words. The operations are selected by function select lines. A look-ahead carry circuit is provided, and fast cascade operation is accomplished by *propagate* and *generate* outputs. The device has two subtract modes, one add, three boolean, preset, and clear. Boolean operations can be performed on 4-bit words with this device. For example, if 1010 is ORed with 0101, the function output will be 1111, since there is 1 high bit in each position. However, if the AND function is selected with the same input words, the function output will be 0000 because there is no occurrence of 2 high bits in any position.

The rules of *binary multiplication* are also few in number:

$$0 \times 0 = 0$$
$$1 \times 0 = 0$$
$$0 \times 1 = 0$$
$$1 \times 1 = 1$$

A separate multiplier circuit can be built; however, multiplication is more likely to be performed in an adder circuit. It is possible to use a shift register and an adder to multiply. The process is based on the rules of binary multiplication and the formation of partial products which are summed to find the final product. The following example illustrates the procedure.

EXAMPLE

Multiply 1010 by 0101.

SOLUTION

```
MULTIPLICAND → 1010
MULTIPLIER   → 0101

          1010   (first partial product)
          0000   (second partial product)
          1010   (third partial product)
          0000   (final partial product)
PRODUCT → 110010
```

The example shows the way that the process of shift and add can be used to multiply. Division is a similar process. It can be done by repeated subtraction. Since subtraction can be done in an adder, once again addition is seen to be the key process. In fact, all arithmetic and mathematical operations can be accomplished by an adder and some ancillary support circuitry.

Rate multipliers represent another approach to binary arithmetic. They are circuits that take one input pulse frequency and provide two output frequencies. One of the output frequencies is programmable. Rate multipliers can be used to control hydraulic and pneumatic valve actuation rates in industrial applications, and they may also be used to control stepper motor velocities. Figure 11-42 shows the pinout for a 7497 binary rate multiplier. It will provide from 0 to 63 output pulses on pin 5 for every 64 clock pulses applied to pin 9. With strobe, clear, and enable grounded, the enable output will provide 1 pulse for every 64 input clock pulses. The enable output is also called the *base rate output* and is fixed. The other output is variable and is at 0, 1, 2, 3, 4, 5, etc., times the base rate, giving it the name *rate multiplier.*

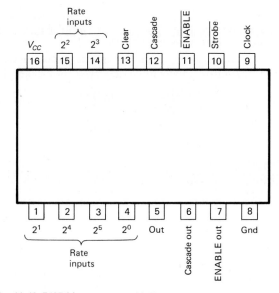

Fig. 11-42 7497 binary rate multiplier.

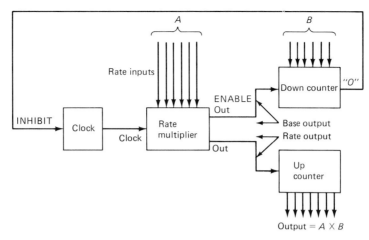

Fig. 11-43 Multiplication with a rate multiplier.

The variable output rate is determined by the binary code applied to the rate inputs.

Figure 11-43 shows a multiplication circuit based on a rate multiplier. The base rate is applied to a down counter, and the rate output is applied to an up counter. The multiplicand is applied to the rate inputs. The action begins with the multiplier loaded into the preset inputs of the down counter. The up counter is cleared, and then the clock is started. The circuit runs until the down counter reaches zero, and this inhibits the clock. The up counter outputs are now equal to the product because the rate output is N times the base rate, where N is the binary word applied to the rate inputs. For example, if $N = 000101$, the rate output will be five times the base output. The up counter will get five pulses for every one pulse supplied to the down counter. If the down counter was preset to 000010 (2), the up counter will end up with a total count of 0001010 ($5 \times 2 = 10$). The circuit can be converted to a divider by interchanging the outputs of the rate multiplier. The base output will be applied to the up counter, and the rate output will be sent to the down counter. The up counter will contain B/A when the down counter reaches zero. Rate multipliers can also produce squares, square roots, and other arithmetic operations.

The rate output of a rate multiplier is not always a periodic waveform. The output pulses are of constant width but often are not uniformly spaced. For example, if the multiplier is programmed to produce 9 out of 64 clock pulses, they cannot be evenly spaced because 64 cannot be divided evenly by 9. This effect is known as *pulse jitter*. Binary-coded decimal rate multipliers are also available and provide N out of 10 clock pulses.

Figure 11-44 shows the pinout for a 74LS85 4-bit magnitude comparator. It compares two 4-bit binary words presented to its A and B data inputs. It has three outputs: A is greater than B ($A > B$), $A = B$, and A is less than B ($A < B$). One of these outputs will be high for any given set of input conditions. The IC also has cascading inputs that make it easy to use several devices so that words larger than 4 bits can be compared.

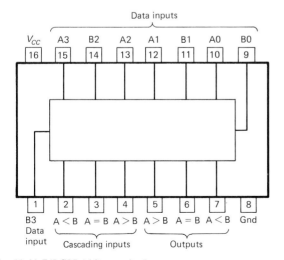

Fig. 11-44 74LS85 4-bit magnitude comparator.

REVIEW QUESTIONS

38. Refer to Fig. 11-40. Adding binary 1010 to 0101 will produce a correct sum output in less time than when adding binary 0111 to 1011. (true or false)

39. Refer to Fig. 11-42. The word applied to the rate inputs is 010111_2, and the clock input is 640 Hz. The strobe, clear, and enable pins are low. What is the base rate output at pin 7? What is the rate output at pin 5?

40. Will the output waveform in question 39 be periodic? Will it show pulse jitter?

41. What is the binary sum of 10011_2 and 10111_2?

42. What is the minimum number of full-adders required to add two 5-bit words? How many half-adders would also be required?

43. What is the binary result of subtracting 10101_2 from 11000_2?

44. Change 10101_2 to 2s complement form and add it to 11000_2. How does this result compare with the results of question 43?

11-7
MEMORY

In the broadest sense, *memory* includes all circuits and devices that are capable of storing binary information. Therefore, the latches, flip-flops, and registers already covered in this chapter can be considered as memory devices. The memory devices discussed in this section are specifically designed for short- and long-term storage of significant amounts of binary information. The first memory device to be discussed is the read-only memory (ROM).

Figure 11-45 shows a simple diode ROM. It is arranged as a matrix of eight 4-bit words. A decoder circuit translates the bit pattern on the address input lines to one of eight word select lines. When the address inputs are all low, word select line 0 goes high. Note that there are three diode anodes connected to the 0 select line. Current flows from the decoder, through the three diodes, and into three of the load resistors. The data outputs are high at these three locations. The word stored at location 0 is 1101. If the address input lines are changed to 001, word 1 is selected. The data output is now 0110. Examine the diode matrix, the table of addresses, and data in Fig. 11-45 and verify each 4-bit data word.

A more practical ROM is shown by Fig. 11-46. It is housed in a 24-pin package and contains 8192 words of 8 bits each. Eight-bit words are very common in digital circuits and are called *bytes*. Therefore, the IC may be called an *8 K-byte ROM*. Since in this case K represents 1024, 8192 is rounded to 8K. Or, it could be called a *64 K-bit ROM* since it has 65,536 bits stored inside ($8192 \times 8 = 65,536$). The part number ends in 64, which is the rounded (to the nearest power of 2) number of K bits the device has stored. This is a common, but not universal, practice for memory ICs. The device has 13 address pins labeled A_0 through A_{12}. This number of address pins is required because $2^{13} = 8192$. It also has 8 data pins labeled D_0 through D_7, a +5-V supply pin (V_{CC}), a ground pin (V_{SS}), and an enable pin (NOT E). The block diagram, Fig. 11-46(*b*), shows that the NOT enable pin controls the tri-state output buffers. When the NOT enable input is high, the

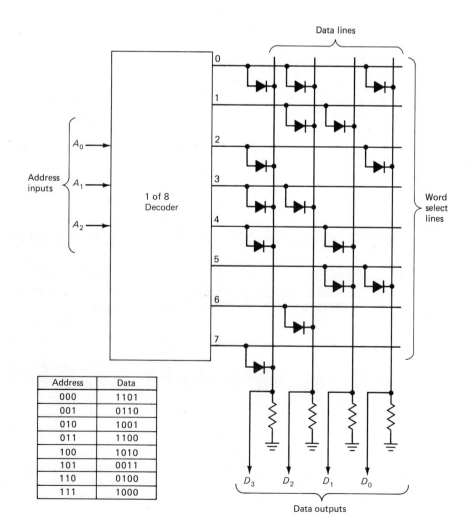

Address	Data
000	1101
001	0110
010	1001
011	1100
100	1010
101	0011
110	0100
111	1000

Fig. 11-45 An 8×4-bit ROM.

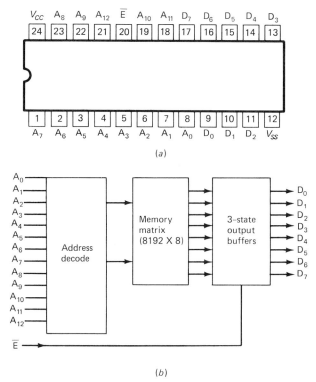

Fig. 11-46 68364 8192×8-bit mask ROM. (*a*) Pinout. (*b*) Block diagram.

output buffers are in their high-impedance state. This state makes the ROM compatible with bus-structured digital systems. It is called a *mask ROM* because it is programmed with a photolithographic mask when it is manufactured.

Mask ROMs are feasible for high-production memory applications. Since they involve a custom mask design for information storage, they must be produced in reasonably large quantities to distribute the mask design costs over a number of units. They are typically used as code converters, character generators, trigonometric look-up tables, logarithmic look-up tables, and as storage for monitor programs. A *memory-type code converter* applies the code to be converted to the address inputs, and the data outputs show the converted code. A *character generator* takes a code such as ASCII at its inputs and outputs the bit pattern necessary to form the character on a display device such as a cathode-ray tube. Look-up tables eliminate the need for some arithmetic operations in digital circuits. It is possible to generate trigonometric and logarithmic functions with adders, but the process takes time. Memory look-up tables are much faster. A *monitor program* is a control program for a computer (this topic will be covered in Chapter 12).

Mask ROMs are available in several arrangements. Some examples are 4096 × 8, 8192 × 8, 16,384 × 8, 32,768 × 4, and 65,536 × 4. Bytewide memory systems are used extensively. This brings up an important question: how can memory devices with word sizes of 4 bits be used in a byte-sized system? They

are used by connecting all address pins and the enable pins in parallel. One IC will then provide the 4 lower bits, and the other will provide the 4 higher bits. Likewise, four parallel 4-bit ROMs can be used in a system with a word size of 16 bits. Mask ROMs have a typical access time of 350 ns. This means that valid data will be available at the outputs 350×10^{-9} s after stable address and enable signals are applied. Slower, but more power-efficient mask ROMs are also available. They have access times on the order of 6 μs but use only about 1 percent of the amount of power. They are based on the complementary metallic oxide semiconductor (CMOS) process, which will be covered in the next section of this chapter. The CMOS devices are particularly well suited to battery-operated equipment.

Mask ROMs are not cost-effective for small and moderate quantity designs. For this reason, *field-programmable ROMs* have been developed. These devices make producing any number of ROMs feasible. They are programmed by fusing diode jumpers inside the memory matrix. Refer again to Fig. 11-45. A field-programmable ROM can be based on a diode matrix. The device will be shipped with a full array of diodes. A momentary high programming current can be applied to those diodes that must be removed from the matrix. For example, word 5 shows two diodes. The diodes in the two most significant positions can be eliminated by passing a large current through them to fuse their connecting links and remove the diodes from the matrix. A second technique of field programming is the *blown diode system*. It also uses a high programming current but in pulse form. Selected junctions are short-circuited by aluminum migrating across the junction. The migration is caused by the high current pulses. Whichever programming process is used is performed in a special piece of equipment called a *programmable read only memory* (PROM) *burner* or *PROM programmer*.

The PROMs are one-shot devices. They cannot be reprogrammed. *Erasable programmable read only memory* (EPROM) devices are often a better choice and have become quite popular. Refer to Fig. 11-47, which shows a 2764 EPROM. It is field-programmable and also is field-erasable. Notice that the package has a quartz window over the chip. It is possible to erase this device by exposing it to ultraviolet light at or near a wavelength of 0.2537 μm. The total dose required for complete erasure is typically 15 watt-seconds per square centimeter (W · s/cm²). The erasure is accomplished by using an ultraviolet light source and exposing the IC for the proper length of time. For example, a source might develop 15 mW/cm². Dividing gives

$$\frac{15 \text{ W} \cdot \text{s/cm}^2}{0.015 \text{ W/cm}^2} = 1000 \text{ s} = 16.7 \text{ min}$$

All the stored bits go to logic 1 when the device is erased. The light enters the silicon crystal and releases the charges that were trapped there to represent logic 0s. Usually EPROMs can be programmed, erased, and reprogrammed 100 times or more.

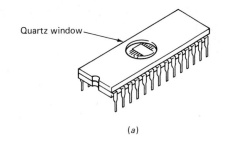

Quartz window

(a)

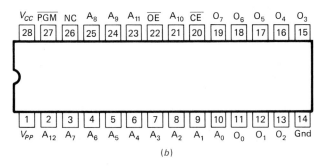

V_{CC}	$\overline{PGM}$	NC	A_8	A_9	A_{11}	$\overline{OE}$	A_{10}	$\overline{CE}$	O_7	O_6	O_5	O_4	O_3
28	27	26	25	24	23	22	21	20	19	18	17	16	15

1	2	3	4	5	6	7	8	9	10	11	12	13	14
V_{PP}	A_{12}	A_7	A_6	A_5	A_4	A_3	A_2	A_1	A_0	O_0	O_1	O_2	Gnd

(b)

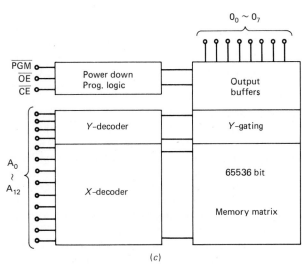

(c)

Mode \ Pins	$\overline{CE}$ (20)	$\overline{OE}$ (22)	$\overline{PGM}$ (27)	V_{PP} (1)	V_{CC} (28)	Outputs (11 ~ 13, 15 ~ 19)
Read	V_{IL}	V_{IL}	V_{IH}	V_{CC}	V_{CC}	D_{out}
Stand by	V_{IH}	X	X	V_{CC}	V_{CC}	High Z
Program	V_{IL}	X	V_{IL}	V_{PP}	V_{CC}	D_{in}
Program verify	V_{IL}	V_{IL}	V_{IH}	V_{PP}	V_{CC}	D_{out}
Program inhibit	V_{IH}	X	X	V_{PP}	V_{CC}	High Z

(d)

Fig. 11-47 2764 EPROM. (a) Pictorial. (b) Pinout. (c) Block diagram. (d) Mode selection.

The 2764 EPROM shown in Fig. 11-47 is arranged as 8192 bytes. Therefore, it contains 64K bits, and its part number follows the pattern mentioned earlier. Other EPROMs include the 2708 (8K bits), the 2716 (16K bits), the 2732 (32K bits), and the 27128 (128K bits) and the 27256 (256K bits). The 2764 has 13 ad-

dress pins and 8 output pins. It also has pins for the application of a programming voltage ($V_{PP} = +21$ V), a program control pin (NOT PGM), an output enable pin (NOT OE), and a chip enable pin (NOT CE). A high voltage is required during programming to trap charges in the silicon. Some EPROMs require $+25$ V for programming. In Fig. 11-47(d) the programming mode is illustrated. Programming is usually performed by placing the EPROM in a programmer or burner. The chip enable pin is held low, and $+21$ V is applied to the V_{PP} pin. The V_{CC} is $+5$ V. Stable address and data are supplied to the address and data pins, respectively. The PGM pin is pulsed low for 50 ms. This process is repeated for every address and takes about 7 min if all 8192 bytes are programmed. Once programming is complete, the programming voltage is reduced to V_{CC}, which is $+5$ V. Once programmed, the IC serves as an ordinary ROM. A piece of opaque tape should be placed over the window. The access time ranges from 200 to 450 ns. A suffix is usually added to the part number to indicate the access time.

Electrically erasable read only memories (EE-ROMs) are also available but are less popular than the EPROMs because they are more expensive. They are completely erased by applying a single pulse to the program pin while $+25$ V is applied to the V_{PP} pin and the chip select pin is low. The EEROMs can be erased very quickly and lend themselves to applications requiring that data be changed frequently. Equipment that is designed to use EEROMs often allows erasure and programming to occur without removing the ROM from the equipment.

None of the memory devices discussed to this point lends itself to rapid storage of data; devices of this type are known as *random access memory* (RAM) devices. This is an unfortunate descriptor since all of the ROM devices are also random access memory. *Random access* simply means that the data can be read (or written) in any order. For example, if a RAM device stores 8K bytes, any one of the 8192 bytes can be read at any time. A better descriptor for the type of memory device to be discussed now would be *read/write memory*. This is not what it is popularly called, however. Please remember that RAM really refers to memory that can be written to as easily as it can be read. Also remember that the data stored in ROMs are also randomly accessible.

The RAM devices bring on a new problem: volatility. A memory is *volatile* if its contents will be lost on power down. All of the ROM devices are nonvolatile. They will hold their data forever unless they are intentionally erased. All RAM devices are volatile. If the power is lost, even for a short time, all contents will be lost. For this reason, some volatile memory circuits must be provided with battery backup power. This is never necessary with ROMs.

In spite of the volatility of RAMs, they are very widely applied devices in modern digital systems. In fact, there is no practical way to accomplish many operations without them. They provide fast and in-

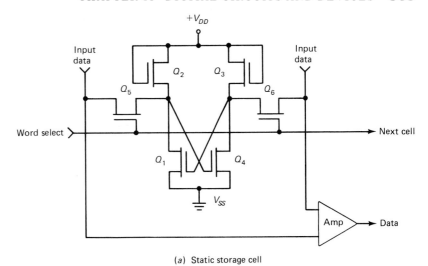

(a) Static storage cell

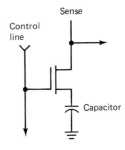

(b) Dynamic storage cell

Fig. 11-48 Metallic oxide semiconductor memory cell types. (a) Static storage cell. (b) Dynamic storage.

expensive storage of data, variables, and instructions. Most RAMs are based on one of two *metallic oxide semiconductor* (MOS) circuit structures. Figure 11-48 shows both memory cell structures; in Fig. 11-48(a) is the static RAM storage cell. It is based on six MOS N-channel enhancement mode transistors. These are normally off devices that can be turned on by applying a positive gate voltage. Transistors Q_1 and Q_4 are cross-coupled to form a bi-stable latch. Transistors Q_2 and Q_3 serve as load resistors. Transistors take up less space in the chip than resistors. Suppose Q_1 is on and Q_4 is off. This is one of the two stable states for the latch. With Q_1 on, it drops very little voltage from source to drain. Therefore, its drain voltage is near 0 V, and this voltage is applied to the gate of Q_4. With near zero gate voltage, Q_4 remains off and drops the entire supply from its source to drain. Thus, the drain voltage at Q_4 is equal to $+V_{DD}$ and is applied to Q_1, which keeps it on. It should be clear now that the latch is stable in this condition and can also be stable in the opposite condition. Transistors Q_5 and Q_6 are used to write to and read from the latch. When the select line goes positive, they are turned on. This connects the latch outputs to the sense amplifier for a read operation. It also allows input data to set or reset the latch.

Now look at Fig. 11-48(b), which shows the simplicity of the dynamic storage cell. It is based on a single MOS transistor and a storage capacitor. A dynamic cell is read by sending the control line high and reading the sense voltage. If the capacitor is charged, the sense voltage will be high. If the capacitor is discharged, the sense voltage will be at ground potential. The cell can be written to by taking the control line high and applying a positive voltage or zero voltage to the sense line, which will charge or discharge the storage capacitor. Because the dynamic cell is so simple, dynamic RAMs are capable of much more storage in a given size chip of silicon. Unfortunately, the dynamic cells must be refreshed every several milliseconds. A cell is *refreshed* by reading its contents and recharging the capacitor if it was charged. If it is not, data will be lost as the charge in the storage capacitor leaks off. Refresh circuitry will be required when dynamic RAMs are used. This requirement complicates the design of a system, and static RAMs are easier to use when a moderate amount of memory is required. Static RAMs are also chosen when battery back-up is required since refresh circuits are not necessary.

A 6116 static RAM IC is shown in Fig. 11-49. It is a CMOS device arranged as 2048 × 8 bits. Other popular sizes include 1024 × 4, 4096 × 1, and 16,384 × 1. The 6116 has 11 address lines and 8 I/O lines. The access time ranges from 120 to 200 ns and is designated by a suffix after the part number. The standby power ranges from 10 to 100 μW and is also designated by a suffix. The 10-μW version is ideally

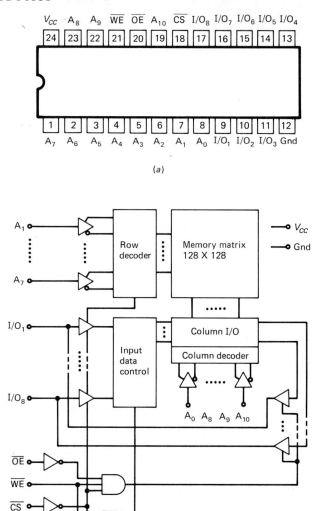

$\overline{CS}$	$\overline{OE}$	$\overline{WE}$	Mode	V_{CC} Current	I/O Pin	Reference Cycle
H	X	X	Not selected	I_{SB}, I_{SB1}	High Z	
L	L	H	Read	I_{CC}	D_{out}	Read cycle (1) ~ (3)
L	H	L	Write	I_{CC}	D_{in}	Write cycle (1)
L	L	L	Write	I_{CC}	D_{in}	Write cycle (2)

Fig. 11-49 6116 static RAM. (*a*) Pinout.
(*b*) Functional block diagram. (*c*) Truth table.

(*c*)

suited for battery back-up service. When the supply voltage drops below 4.5 V, the device goes into the data retention mode. The supply must be maintained at 2 V or more for the data to be retained. Also, the NOT chip select pin must be maintained no less than 0.2 V below the supply voltage in the data retention mode. The truth table in Fig. 11-49(*c*) shows the modes of operation. When the NOT chip select is high, the IC is in standby, and the I/O pins are high-impedance. When the NOT chip select is low, the NOT output enable is low, and the NOT write enable is high, it is in the read mode, and the I/O pins become data outputs. Taking the NOT write enable pin low places the memory IC in the write mode. The I/O pins become data inputs. There are two write cycles, and the state of the NOT chip enable determines which will be selected. Write cycle (2) is used if the NOT chip enable is continuously low. In practice, either type of write cycle may be used in a system.

Figure 11-50 represents a 4164 dynamic RAM. It is an MOS device and is arranged as 65,536 × 1-bit. Other dynamic RAMs include 4096 × 1, 16,384 × 1, and 262,144 × 1 bit. Common access times range from 120 to 250 ns. Note that the 4164 has only 8 address pins; 16 address bits are required to decode

---

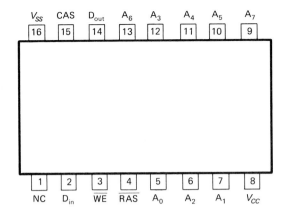

Pin(s)	Designation
A_0–A_7	Address inputs
$\overline{CAS}$	Column address strobe
D_{in}	Data in
D_{out}	Data out
$\overline{RAS}$	Row address strobe
$\overline{WE}$	Read/write input
V_{CC}	+5 V
V_{SS}	Ground

Fig. 11-50 4164 dynamic RAM.

1 of 65,536 locations. Eight-row address bits are set up on A_0 through A_7 and latched with the *row address strobe* (NOT RAS). Then eight column addresses are set up on pins A_0 through A_7 and latched with the *column address strobe* (NOT CAS). Read or write is selected with the *write enable* (NOT WE) input. A high selects the read mode, and a low selects the write mode. If NOT WE goes low prior to NOT CAS, the D_{out} pin will remain in the high-impedance state. A refresh operation must take place at least every 4 ms. Strobing each of the 256 row addresses refreshes every bit. The row address strobe must be low (active) during the refresh cycle. The column address strobe may remain high (inactive) during the refresh cycle.

Bipolar RAMs are also available but are less popular because they use many times more power per bit than MOS and CMOS devices. They are also more expensive because the bipolar process cannot achieve the high bit densities achieved by the other type. They are used in those applications in which speed is the major consideration. Some bipolar RAMs have access times less than 10 ns. Some common sizes of bipolar RAM ICs include 256 × 1, 256 × 4, 1024 × 1, 1024 × 4, and 4096 × 1.

REVIEW QUESTIONS

45. What would have to be done to the ROM circuit shown in Fig. 11-45 to change the last stored word to 1111_2?

46. A ROM is organized as 4096 × 8 bits. What is its capacity in k-bytes?

47. A ROM has a capacity of 1 k-byte. What is the actual number of bits that it has stored?

48. A ROM has 11 address pins and 4 data pins. What is the actual number of bits that it has stored?

49. Factory-programmed ROMs are also called _____ ROMs.

50. What is the minimum number of 65,536 × 4-bit ROMs required for a system that uses 16-bit words?

51. Refer to Fig. 11-50. How can 65,536 locations be accessed with only 8 address pins?

11-8
LOGIC FAMILIES AND INTERFACING

The LS TTL subfamily and CMOS have become the most popular logic families; however, a few others are worth mentioning. *Emitter coupled logic* (ECL) is a family of devices used mainly in large machines requiring very-high-speed operation. The ECL gates have propagation delays as short as 1 ns. It is a nonsaturated logic in which bipolar current-switching devices operate in a circuit configuration similar to that of a differential amplifier. Because the devices are never in saturation, the turn-off delays associated with stored carriers are mostly eliminated. Unfortunately, ECL devices are "power hogs" and typically consume 25 to 65 mW per gate. The power supply must provide −5.2 V with respect to ground. This characteristic is also a disadvantage because positive supply systems are more popular and because interfacing ECL to other families requires extra level shifting components.

There are also PMOS and NMOS digital ICs available. P-channel MOS (PMOS) is based on P-channel MOS transistors and is the older of the two technologies. It requires a +5- and a −12-V supply and is not very popular today. N-channel MOS (NMOS) devices are based on N-channel MOS transistors and operate from a single +5-V supply. They are usually capable of driving at least one LSTTL load and are directly compatible with CMOS. The current NMOS devices are usually LSI or VLSI devices such as microprocessors (discussed in Chapter 12).

The CMOS devices belong to an ideal logic family. It features very low power consumption, excellent noise immunity, almost unlimited fanout, noncritical power supply requirements, and a very wide selection of devices. It is not so fast as LS TTL, but improvements are being made. Several manufacturers have announced high-speed CMOS parts. Its low power consumption is probably its most outstanding characteristic and is only 10 nanowatts (nW) per gate under static conditions and 10 mW per gate when toggling at a 1-MHz rate. It offers a 50:1 to 100:1

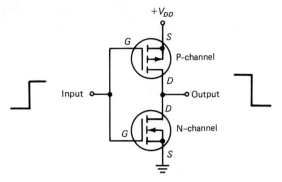

Fig. 11-51 Basic CMOS inverter.

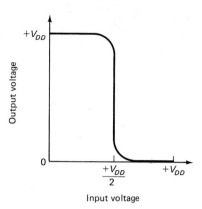

Fig. 11-52 Complementary metallic oxide semiconductor (CMOS) inverter transfer function.

power saving when compared to LS TTL. The noise margin is better than 2 V when operated from a +5-V supply, as compared to LS TTL, which is only guaranteed at 0.4 V. The fanout is almost unlimited since the gate inputs are MOS transistors with an input resistance of 10^{12} Ω. However, each transistor does represent a load capacitance of 5 pF; thus extremely large fanouts could cause switching delays due to capacitive loading (fanouts up to 50 are considered practical). The CMOS devices operate over a power supply range of 3 to 15 V with no need for tight voltage regulation. Speed does improve at the higher voltages with typical maximum toggle rates of 2.5 MHz at 5 V and 7 MHz at 10 V. A browse through CMOS reference manuals will show all of the standard logic devices, from gates through microprocessors. There are also many special devices, including phase-locked loops, communications devices, linear devices (such as BI-FET op-amps), and speed controllers.

The configuration for a basic CMOS inverter is illustrated by Fig. 11-51. It shows that a complementary pair of transistors is used: one P-channel and one N-channel. This complementary pairing is the source of the term *complementary metallic oxide semiconductor* (CMOS). The transistors are enhancement mode devices. You should recall that enhancement mode transistors are normally off. They are turned on by attracting carriers into the channel region. In the case of the P-channel unit, a negative gate voltage will attract holes and turn the device on. In the case of the N-channel unit, a positive gate voltage will be required to attract electrons into the channel and turn it on. Assume a +5-V V_{DD} supply in Fig. 11-51 and a logic swing at the input of 0 to +5 V. When the input is 0, the P-channel transistor is on because 0 V is 5 V negative with respect to the source of the transistor, which is at +5 V. With the P-channel transistor on, the output will be pulled high, to near +5 V. No current flows in the output when it drives other CMOS inputs. No current flows in the complementary pair since the N-channel transistor is off. Now, assume that the input goes to +5 V. This turns the top transistor off and the bottom transistor on. The output is pulled low, to near 0 V. No current flows in the output, and no current flows

in the complementary pair because the top transistor is off. Now you can see why CMOS is so power-efficient: no current flows under either static condition. However, there is some current flow during logic transitions. There is a brief period when both transistors are partially on. There is also some momentary output current during transitions due to the charging or discharging of capacitance in the output circuit.

Figure 11-52 illustrates the transfer characteristic for the CMOS inverter. The output voltage switches at an input voltage equal to half the supply. This yields the best possible noise immunity. It also shows that the output voltage swings from $+V_{DD}$ down to 0 V. Older CMOS devices will show a gradually changing output for a gradually changing input. These are known as *A-series devices*. The newer devices have buffered outputs and are known as *B-series*. The added buffers give quicker rise and fall times. They have a constant output impedance of 500 Ω (logic 1 and 0) and are capable of driving one LS TTL load directly when both families are operated from +5 V. Finally, there are some unbuffered devices known as the *UB-series*. They have similar specifications to those of the B-series but are a little faster.

The CMOS input logic levels are shown in Fig. 11-53. Logic 0 extends from 0 to 30 percent of V_{DD}. Logic 1 extends from 70 to 100 percent of V_{DD}. With a 5-V supply, this means that a logic 0 must be 1.5 V or less and that a logic 1 must be 3.5 V or more. If you refer again to Fig. 11-1 you will see that a TTL or LS TTL gate output is not high enough to drive CMOS and meet the requirements of Fig. 11-53. A simple but effective solution is shown by Fig. 11-54. Both families are powered from a +5-V supply. A pull-up resistor is added to the circuit to bring the CMOS input up to near +5 V when the LS TTL gate switches high. This is necessary because of the diode and transistor drop in the totem pole output stage. The resistor should not be too small in value because it will hold the output up too high when it tries to go to logic 0.

Interfacing CMOS to TTL will also require a little

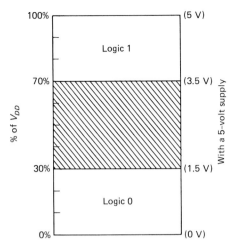

Fig. 11-53 Complementary metallic oxide semiconductor (CMOS) input logic levels.

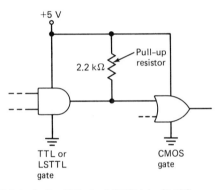

Fig. 11-54 Interfacing TTL (or LSTTL) to CMOS.

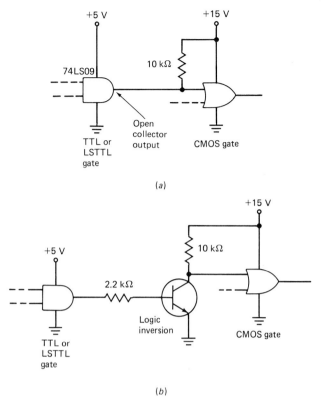

(a)

(b)

Fig. 11-56 Interfacing different voltage levels. (a) Using open collector output. (b) Using separate transistor.

assistance, as shown in Fig. 11-55. Driving a TTL gate input low requires a sink current of 1.6 mA. If the CMOS gate has an output impedance of 500 Ω, the sink current will cause the output to be $1.6 \times 10^{-3} \times 500 = 0.8$ V. This provides no noise margin for the TTL gate. A pull-down resistor is added to help sink the current. If the CMOS gate is a B-series or UB-series device, it will drive (sink) one LS TTL load without the need for a pull-down.

We have learned that there are speed advantages to powering CMOS from a voltage higher than 5 V. Figure 11-56 shows two ways to interface TTL or LS TTL to CMOS that is operating at a higher volt-

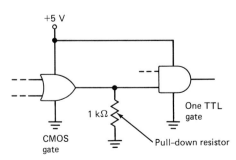

Fig. 11-55 Interfacting CMOS to TTL.

age. Figure 11-56(a) shows the use of an open collector gate. It also shows that an external pull-up resistor connected to the CMOS supply is required. Figure 11-56(b) shows a solution using a separate transistor and two resistors. Again, the collector resistor must connect to the CMOS supply. Notice that the transistor inverts the logic from the TTL or LS TTL gate output. Figure 11-57 shows a CMOS device designed to shift from one voltage level to another. It has two supply pins, V_{CC} and V_{DD}, one for each voltage level. It is designed for 3- to 18-V operation. It will shift CMOS at one voltage level to CMOS at another voltage level (mode 0) and will shift TTL to CMOS at another voltage level (mode 1). Other CMOS devices, called *buffers*, are available for driving TTL and LS TTL loads.

The CMOS part numbers traditionally are grouped in a 4000 series. Other systems closely align with this series, such as Motorola's MC14000 and MC14500 groups. There is also a CMOS group with pinouts identical to those of the TTL and LS TTL parts family; this is the 74C00 series.

Handling CMOS devices demands the static discharge precautions discussed earlier in this book. Almost all CMOS devices have diode protection circuits on their input terminals, but careful handling is still strongly recommended.

Last, but certainly not least, is the matter of floating inputs. Although this practice may work with TTL and LS TTL, it will absolutely not work with

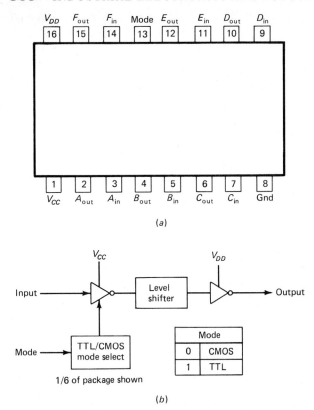

Fig. 11-57 MC14504B hex level shifter. (*a*) Pinout. (*b*) Logic diagram.

CMOS. All inputs that do not connect to gate outputs must be tied to the positive supply or ground. A floating CMOS input can cause insidious problems. The input impedance is so high that time constants of hours are likely. The circuit may work for some time and then fail as an input gradually drifts high or low. Or a floating input may track a driven input in an inconsistent manner. Finally, a floating input can cause a gate to oscillate and draw abnormally high currents from the supply.

REVIEW QUESTIONS

52. Which logic family is the fastest?

53. Refer to Fig. 11-51. The input voltage is equal to V_{DD}. Which transistor is on? What is the logic state of the output?

54. The transistors shown in Fig. 11-51 are normally _____ devices.

55. What is the transition voltage for the input of a CMOS gate operating from a 10-V supply?

11-9
TROUBLESHOOTING AND MAINTENANCE

Always check all the obvious points first. Inspect for improper switch settings, loose connectors, broken cables, and faulty output displays. Never start troubleshooting at the component level until the power supply voltages have been verified. The TTL family permits only a plus or minus 250-mV variation from the nominal 5 V. Power supply ripple should also be checked. A ripple reading of 250 mV or more can cause logic errors and erratic operation. Do not forget to check power supply voltages in several parts of the system. An industrial logic controller may use on-card voltage regulators, and the supply voltages may be good everywhere except on one printed circuit board. A card extender is often a must for troubleshooting in systems with plug-in cards. The cards are too close together for taking any readings. The cards must be removed and replaced with an extender supplied by the equipment manufacturer. The card under test is then plugged into the extender. Never break or make card connections when the power is on.

When troubleshooting robotic and other machine controls, always be wary of the possibility of sudden motion when troubleshooting. This is especially important when the troubleshooting process involves injecting pulses or control signals into circuits. Disconnect or disable high-energy components where feasible. Follow the manufacturer's recommendations for these procedures.

There are two broad categories of defects in digital systems: functional faults and overloads. A *functional fault* is a logic error due to an internal gate failure. For example, both inputs to an AND gate are high, but the output is low. Assuming the output is not overloaded, this is a functional fault, and the gate must be replaced. In an *overload* the gate output is incorrect as a result of a fault after that point. It could be caused by a short circuit in the wiring, the PC board, a connector, or a faulty device driven by the gate. Your knowledge of TTL, LS TTL, and CMOS voltage and current parameters will go a long way in helping you find such faults. For example, you know that a TTL gate can source up to 400 μA while maintaining a logic high of at least 2.4 V. You also know that it can sink up to 16 mA while maintaining a logic low of no more than 0.4 V (LS TTL will only sink up to 8 mA).

How are faults isolated? Usually, the gate output in question must be isolated electrically from the rest of the circuit. This procedure will confirm whether the gate output will respond correctly when unloaded. If the IC is in a socket, it can be removed, and the output pin can be carefully bent out a little. The IC is then replaced in the socket (with the bent pin not going into the socket), and the output is checked for proper response. Do not forget to restore all bent pins when you finish making your measurements. If the IC is soldered in, sometimes the output can be isolated at an edge connector. A sliver of thin tape is used to insulate one of the edge contacts and unload the output in question. Make the tape sliver long enough to wrap part way around the board edge. This will keep the tape from sliding off when the board is inserted into the edge connector. Be very sure to remove all such pieces of tape when you are finished troubleshooting. Another technique is to remove the solder from the output pin in question with a vacuum desoldering tool. Then, a short length of insulating tubing is inserted around the pin and through the circuit board. This procedure takes a little time, but it is better than desoldering and removing the entire device only to learn that there is nothing wrong with it. Be sure to resolder all pins when you are finished. One caution is in order when using any isolation technique: inspect the schematic to be sure that you will not leave a CMOS gate input floating when you isolate any output.

It is necessary to know whether voltages are normal when troubleshooting. In TTL or LS TTL, an output voltage of 0 to 0.07 V is too low. It indicates a defect in the gate or a short circuit to ground somewhere in the system. Likewise, an output voltage near V_{CC} is too high and indicates a short circuit to the positive supply rail. However, if there is an external pull-up, a reading near V_{CC} may be normal. Also, do not forget that an external pull-up resistor may be open and cause no output swing in an open collector device. An output voltage between 2.4 and 0.4 V indicates an overload. The gate may be sourcing or sinking excessive current. Check to see whether the device drives an LED indicator or some other high-current load. If it does, the voltage may be normal, but not if it also drives another gate input. Finally, a floating TTL or LS TTL input will measure between 1.1 and 1.5 V, which is normal.

Circuit board and socket defects are common. A solder splash on a board may not cause problems for months or years and then suddenly "make" an unwanted connection. Always be on the lookout for such problems. An IC may be inserted into a socket with one pin bent under instead of going into the socket. Surprisingly, this pin will sometimes "make" simply by touching the top of the socket pin, but the connection will eventually fail. You have to look very closely to see these kinds of difficulties. Edge connectors, especially on large boards that have a tendency to warp, are an ongoing source of problems. If a control is intermittent and responds to vibration, check the edge connectors. A trace on a PC board may be cracked. These can be impossible to locate by a visual inspection. If a problem comes and goes when a board is flexed, there may be a cracked trace. Try to localize the sensitive spot. A little fresh solder can be flowed over all suspect connections and traces to correct the fault.

Various pieces of special test equipment are available for troubleshooting digital circuits for logic faults. Of course, the standard items such as a multimeter and oscilloscope are very useful. The *logic clip* is a device that clips onto 14- or 16-pin dual-inline ICs. It contains 16 LEDs, 1 for each IC pin. It automatically locates the ground pin and the plus (+) supply pin to energize itself. Some models will work over a 4- to 18-V range and can be used with

TTL, LS TTL, NMOS, and CMOS. Logic clips are easy to use and supply a lot of information with one quick and easy connection. A *logic pulser* is another valuable tool; it is a penlike structure with a sharp probe point. It is powered from the system under test by using two clip leads. Some models will work over the 3- to 18-V supply range and cover all the popular logic families. The output of a pulser is tristate. When its switch is pressed, it automatically drives a gate output or input from high to low or from low to high. It has high source and sink current capability and can therefore override output points originally in the high or low states. It can be used in conjunction with the clip to test flip-flops, latches, registers, etc. For example, the reset or preset pin can be pulsed, and the clip will show whether the device outputs respond as expected. Or the clock input may be pulsed; the device should respond unless it is in an inhibited mode. Logic pulsers are also used in conjunction with another penlike instrument, the *logic probe*. A probe has a built-in light which glows brightly for logic high, dimly for invalid or floating; it goes out for logic low. The pulser is often placed on the gate input and the probe on the gate output; this technique is called *stimulus-response testing*. Or both pulser and probe can be placed at the same point to check for a short circuit to ground or to the supply. If there is a short circuit, the probe will not blink when the switch is pressed on the pulser. The low impedance of the short circuit will not allow the pulser to change the status at that point.

Component-level troubleshooting may be more difficult with LSI and VLSI ICs. They can develop subtle problems that affect one of many possible input conditions. They may also develop timing problems that can only be seen on a multitrace oscilloscope. This is where your general knowledge of the system block diagram is important. When you know the function of a component and the way it interacts with the system, you will be able to pinpoint the trouble to one device or a small group of devices. Substitution with a known good IC is a valid technique when the part is socketed. If the IC is soldered in, it usually pays to take the time to verify as many input conditions as possible before removing it. Verify all signals and power connections at the pins of the device itself, including the ground pin(s). Use a meter and measure from a circuit ground to the ground pin of the device. You should measure 0 V. Remember: circuit boards, sockets, and solder joints do fail.

If operation is erratic, investigate the possibility of noise in parts of the circuit. Check all grounds and shields on cables. Verify all connectors. A pulse of only 20 ns can "glitch" a flip-flop or a register. The oscilloscope is valuable when looking for glitches. Also, logic probes contain pulse-stretching circuitry and will produce a visible flash for pulses as narrow as 10 ns.

The process of CMOS troubleshooting is similar to TTL and LSTTL troubleshooting. Power supply regulation and ripple are less critical, however. Remember the 30 and 70 percent threshold points when taking readings. Also remember that the output impedance of most CMOS gates is 500 Ω whether high or low. A load current of 2 mA will cause a 1-V change (Ohm's law). For example, if the gate is powered from 5 V and sourcing 2 mA, its output will be 4 V. Careful handling is necessary when working with CMOS parts and circuit boards that contain CMOS parts. A board that is removed from the system must be handled with the same caution exercised for handling CMOS ICs (handling precautions were covered in Chapter 2). One additional precaution is that the power should be on when connecting low-impedance test equipment (such as a pulse generator) to CMOS circuitry. Also, remove the test equipment before turning the power off.

REVIEW QUESTIONS

56. What is the allowable supply range for LS TTL?

57. A 0-V reading on the output of an LS TTL gate may indicate that the output circuit is _____.

58. An open collector output measures 0 V when it should be high. The pull-up resistor could be _____.

CHAPTER REVIEW QUESTIONS

11-1. Convert binary 10101111 to decimal.

11-2. Convert hex 3F1D to decimal.

11-3. Convert decimal 98 to binary.

11-4. Convert decimal 3908 to hex.

11-5. Convert binary 1001111 to hex.

11-6. Convert hex 3C1 to binary.

11-7. Convert decimal 128 to BCD.

11-8. Name a nonweighted code used in shaft encoders.

11-9. Suppose that Gray 1000 is applied to the input of Fig. 11-6. What is the binary output? What is the decimal value of the output?

11-10. How many input combinations are there for a six-input OR gate? How many rows of its truth table will show a logic 0 output?

11-11. Two identical square waves are fed into an exclusive OR gate. What is the output wave form?

11-12. Assume a TTL J-K flip-flop with both J and K floating. What is the Q output if a 1-MHz square wave is fed into the clock input?

11-13. How many flip-flops would be required for a modulo 3 counter?

11-14. What is used to reduce the natural modulus of a counter? Can this same technique be used to increase the natural modulus?

11-15. Refer to Fig. 11-26. All four reset pins are at logic 0. What mode is the IC in?

11-16. Refer to Fig. 11-26. All four reset pins are at logic 1. What mode is the IC in?

11-17. How many flip-flops will be required to build a ring counter to control 16 events?

11-18. Refer to Fig. 11-34. How many connections to the display unit would be needed for an eight-digit display?

11-19. What kind of logic circuit is needed to interface a hexadecimal keypad to a digital control?

11-20. Refer to Fig. 11-37. What is the logic state of all outputs when no key is pressed?

11-21. Convert 10101_2 to its 2s complement form.

11-22. Multiply 10111_2 by 1001_2 and state your answer in binary.

11-23. Subtract 00001_2 from 10000_2 and report your answer in binary.

11-24. An EPROM is both field-programmable and field _____.

11-25. How many minutes should an EPROM be exposed to an ultraviolet source that produces 10 mW/cm^2, assuming the recommended dose is 15 W $\cdot$ s/cm^2?

11-26. What is the acronym for a nonvolatile memory device that can be erased by a single electrical pulse?

11-27. What are read/write memory devices popularly called?

11-28. Are ROMs volatile?

11-29. How may RAMs be protected from data loss during power outages?

11-30. What should be added so that an LS TTL gate can drive a CMOS gate operating from the same supply?

11-31. When CMOS drives TTL at the same supply voltage, a _____ resistor is required.

11-32. The CMOS inputs must never be allowed to _____.

11-33. A logic probe is applied to a gate output, and the light glows dimly. The output is _____.

11-34. A logic probe is applied to the ground pin on an IC, and a dim light results. The problem is a defective _____.

11-35. A logic probe is applied to the V_{CC} pin on an IC, and the light goes out. The problem is a defective _____.

11-36. A CMOS gate is sinking 4 mA. What output voltage can be expected?

ANSWERS TO REVIEW QUESTIONS

1. forbidden **2.** 400 mV **3.** with full fanouts **4.** sink **5.** 0.32 V and 2.88 V **6.** 1.8 V **7.** no **8.** no **9.** 11011010
10. AND **11.** inclusive **12.** no; propagation delay **13.** 30 ns **14.** waveform *A* **15.** inverted (180° out of phase)
16. it would not change **17.** pull-up resistors **18.** Q = 0 and NOT Q = 0 (invalid); race **19.** nothing **20.** NAND
21. yes; preset and clear **22.** there is no window (it samples on the clock edge) **23.** 0.625 MHz **24.** no; it is a
ripple counter **25.** no (not in the BCD mode) **26.** they are tri-stated **27.** synchronous **28.** no; the I/O pins
29. asynchronous **30.** two (0010) **31.** 0000 **32.** on the first positive clock edge **33.** Y6 (pin 9) **34.** ground pin 3
and use Y0 through Y3 as outputs **35.** use G2A or G2B as the input **36.** 005013 **37.** 5010 **38.** true **39.** 10 Hz;
230 Hz **40.** no; yes **41.** 101010 **42.** 4; 1 **43.** 11 **44.** 11; same **45.** add three diodes **46.** 4 **47.** 8192 **48.** 8192
49. mask **50.** 4 **51.** by using the pins twice with CAS and RAS **52.** ECL **53.** the N-channel transistor; low
54. off **55.** 5 V **56.** 4.75 to 5.25 V **57.** defective or short-circuited to ground **58.** open

12

MICROPROCESSORS

Microprocessors are among the most elaborate of all integrated circuits. They are very large scale integrated circuit (VLSI) devices, and some of the more sophisticated ones contain over 100,000 transistors. Most microprocessors are housed in the familiar dual-inline package, and pin counts of 28, 40, 48, and 64 are available. Microprocessors are often identified as 4-bit, 8-bit, 16-bit, or 32-bit, according to the size of the data bus. A 4-bit microprocessor has four data pins, an 8-bit microprocessor has eight data pins, and so on. The larger microprocessors are more costly but have the advantage of speed since they can handle more bits at one time. The 8-bit units are very popular because they are fast enough to handle the majority of industrial applications and are low in cost. This chapter will deal mainly with the Motorola MC6809, which is an 8-bit microprocessor.

12-1
OVERVIEW

Microprocessors are very powerful and flexible devices. They can be teamed up with support devices such as memory chips, a keyboard, decoders, video display devices, a cathode ray tube, a mass storage device (disk or tape), and a power supply to form a complete computer. Computers based on microprocessors are called *microcomputers*. Microcomputers are general purpose products. They can be configured to serve in a wide range of applications such as circuit analysis, word processing, games, data-base management, planning, drafting and design, and a host of others. Another way that microprocessors can be used is in a specific application such as motor control. When they are used this way they are called *dedicated microprocessors* or *microcontrollers*. This chapter will emphasize dedicated microprocessors since they are so prevalent in the industrial environment.

Figure 12-1 shows some of the internal details of the Motorola MC6809 microprocessor. It is consid-

ered to be one of the most powerful of all 8-bit microprocessors. However, any industrial setting will probably use several types. Much of what you learn about the Motorola MC6809 will transfer to other types of microprocessors. The *arithmetic logic unit* (ALU) is located near the center of the drawing in Fig. 12-1. It represents that section of the processor that is used to perform all arithmetic, logic, and boolean operations. Just above the ALU, we find the *accumulators*, which are each 8 bits wide and are used to hold the intermediate and final results of the various operations. Just below the ALU is the *condition code register* (also called the *flag register*). It is also 8 bits wide, and the various bits (flags) indicate the status of the processor and give information about carries, borrows, and related conditions created by ALU operations. The *data register* is a temporary storage area for the data bus and connects to various sections of the processor such as the decoder. The decoder looks at instructions that come into the processor on the data bus and sends information to the control section. The controller routes signals, times events, and sequences the various registers and circuits. Now, let's look at the *program counter*. It is 16 bits wide and connects to the address register and then on to the address bus, also 16 bits wide. This gives the microprocessor 2^{16} or 65,536 unique addresses, usually designated 64K, as discussed in Chapter 11.

Four other 16-bit registers are shown in Fig. 12-1. These are the index and stack registers. These registers serve mainly as pointers to the 64K address space. They are useful for setting up some sections of memory as data tables and stacks. They are also handy for moving sequential sections of memory around and as 16-bit counters or timers. The last section of Fig. 12-1 to be considered is the *direct page register* (DPR). This is an 8-bit register that makes up the upper byte of a 16-bit address for one of the addressing modes. The various addressing modes will be discussed in a later section.

Even though Fig. 12-1 is grossly simplified, it contains more detail than most people need. Figure 12-2 shows a programmer's model of the MC6809 microprocessor, which indicates those registers that are usable by a programmer. The other internal details are not necessary and are said to be *transparent* from a user's standpoint. The programmer's model shows

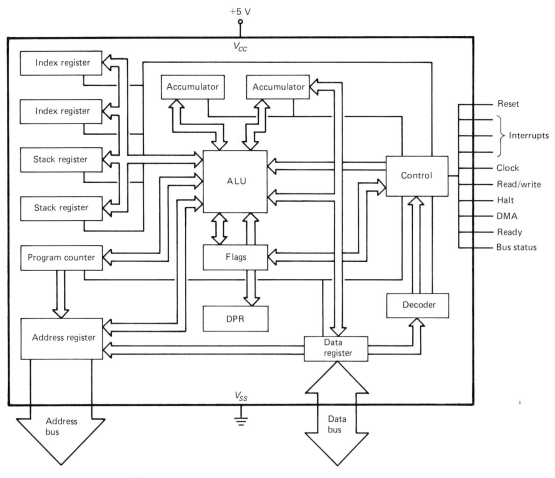

Fig. 12-1 Typical microprocessor architecture.

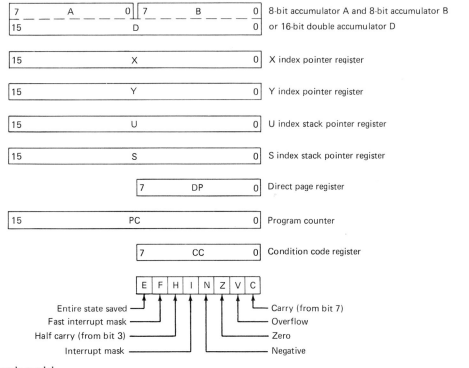

Fig. 12-2 Programmer's model.

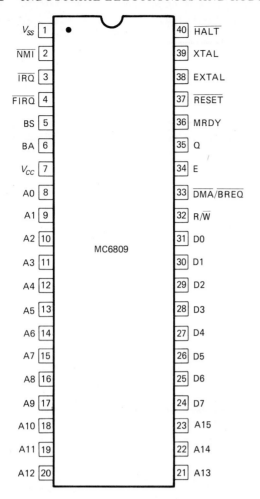

Fig. 12-3 Microprocessor pinout.

two 8-bit accumulators (accumulator A and accumulator B). The two accumulators can be used as a single 16-bit accumulator (accumulator D). It also shows four 16-bit pointer registers and a 16-bit program counter. The direct page register and the condition code register are shown, with each flag identified. The *carry flag* is set when some arithmetic or shift operation generates a ninth bit. It is also set when a larger number is subtracted from a smaller number, indicating a *borrow* in this case. The flags will be discussed in more detail in later sections of this chapter.

The pinout of the MC6809 is illustrated by Fig. 12-3. Some of the pins are straightforward, such as V_{SS}, which is connected to ground; V_{CC}, which is connected to +5 V, and pins 8 through 31, which make up the address bus and the data bus. The *address pins* are outputs only and are used by the microprocessor to select the device or memory location to write data to or to read data from. The *data pins* are bidirectional and are used to transfer data bytes out of or into the processor. Pins 2, 3, and 4 are interrupt inputs. They provide a way for an external hardware device (perhaps a limit switch on a robot arm) to signal to the processor that an important

event which requires attention has occurred. Pins 5 and 6 are status outputs that reveal whether the processor is in the normal or running mode, whether it is servicing an interrupt, whether it is in the synchronization mode, or whether it is in the halt mode. Pin 32 is an output and indicates the direction of transfer on the data bus. When the processor puts information on the bus it is writing (pin 32 is low); when it takes information from the bus it is reading (pin 32 is high). Pin 33 is an input used to suspend processor operation and allow another device to take control of the buses for up to a maximum of 15 clock cycles at a time. Pins 34 and 35 are outputs which provide timing (clock) signals to the rest of the system. The Q clock signal phase leads the E clock signal by 90°. Generally, the E clock is used to synchronize the other parts of the system. Pin 36 is an input used by slow devices to stretch the E and Q clock signals to allow more time for data access. Pin 37 is an input used to reset the processor. Pins 38 and 39 are connected to an external crystal that controls the frequency of the internal clock oscillator. Finally, pin 40 is an input that can be used to halt the processor indefinitely. All pins will not be used in every application. For example, the halt pin is often tied to +5 V. Since it is active when low, this will preclude the halt condition.

Microprocessors can be used to control almost anything in industry. They are very flexible because they themselves are controlled by software. *Software* is a group of instructions and data that directs the step-by-step operation of the processor. All that is often required for a new application is new software. For example, suppose we want to use a microprocessor to control the temperature of some industrial process. Also assume that we want to store the highest and the lowest temperatures reached by this process. This is a trivial job for a microprocessor, but it will illustrate the software concept. Refer to Fig. 12-4, which shows a flowchart of the problem. Constructing flowcharts is usually the first step when designing microprocessor software. Enter the chart at the top where it is marked *start*. Rectangles are operation blocks in flowcharts. The first operation is to read the temperature. The next two operations store the temperature just read into the low and high readings. This step initializes both values. Next, the temperature is read again. Then it is compared to a low limit. Diamonds represent decision boxes in flowcharts. Note that there are two exits from the first decision box. If the temperature is less than $70 (the dollar sign is often used to signify a hexadecimal number), we exit the box on the right and encounter a *turn on valve* operation. If the temperature is equal to or greater than $70, we enter the next decision box. The next decision box checks whether the temperature is greater than $90. If it is, the gas valve must be turned off. At this point, you should see that one of three paths must be taken. One path turns the gas on, one path turns the gas off, and one path does nothing to the valves.

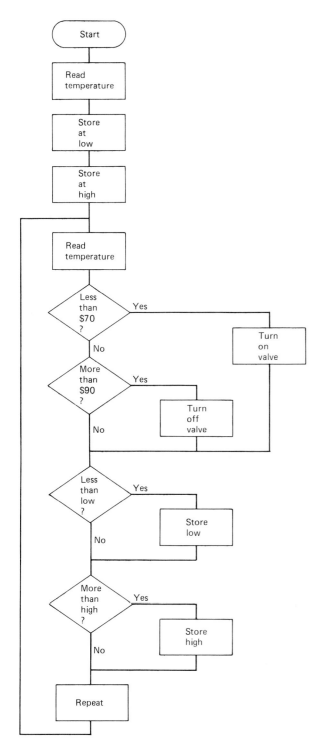

Fig. 12-4 Flowchart for temperature controller.

over as an endless loop until a halt signal, or perhaps a reset signal or an interrupt, comes along. There are many possible variations, even in a simple program such as this one. For example, it is really not necessary to check for a high reading if it has been decided that a new low has occurred. The flow from the operation box marked *store low* could go to the repeat box. Another variation is that the second *read temp.* box could be moved to the bottom of the flowchart and replace the *repeat* box. The flow would then leave this box and enter before the first decision box.

It was mentioned that the temperature control program shown in Fig. 12-4 is trivial. It is trivial in that it does not even begin to tap the power of a microprocessor. Much more sophisticated control is possible. For example, the processor could also handle the ignition of the gas burner. It could analyze the exhaust gas and control the air/fuel ratio for better efficiency and reduced pollution. Average temperatures could be calculated and stored, in addition to high and low values. A gas flow meter could supply the processor with consumption data, and cost reports could be generated. A proportional gas valve could be used for tighter control of the process temperature. Software "anticipation" routines could reduce temperature overshoot and undershoot. Safety procedures could be added for overtemperature conditions. There are many, many such features that could be added before the microprocessor would begin approaching its limits. The system would begin to "labor" only when the processor was so busy that its response time for critical actions suffered. The clock rate of the MC6809 is 1 MHz, and one clock cycle takes 1 μs. Therefore, the time required for most operations is very short. For example, reading the temperature will take 5 μs, and 8 μs will be required to execute a decision block. Also available are an MC68A09, which runs at 1.5 MHz, and an MC68B09, which runs at 2 MHz.

Figure 12-5 shows some of the hardware that would be required for the temperature controller. A *solenoid-driven gas valve* allows the processor to turn the gas off and on. By writing a 1 to the latch, the valve is turned on. Writing a 0 to the latch turns the gas off. Note that D0 (least significant bit of the data bus) is sent to the latch for this purpose. An *analog-to-digital* (A/D) *converter* is used to change the analog temperature signal to an 8-bit word. The digital output of the A/D converter is connected to the data bus.

Memory is a key part of any microcontroller. The memory holds the control program and also allows data to be stored. Address decoders are placed on the address bus so the processor can select memory, the A/D converter, or the latch. Each device has a unique address. For example, the address of the latch is $FF40, and the latch is selected (chip enable goes low) when that address appears on the bus. The A/D converter is decoded for $FF41 and is selected when its address appears on the bus.

The next decision box in Fig. 12-4 checks whether the temperature is lower than the previously stored low reading. If it is, the reading is stored, because it represents a new low reading. Next, the temperature is compared to the previously stored high reading. If it is greater, then a new high has been reached, and it must be stored. The repeat box sends the program flow back to take a new reading and make the same decisions again. This program will repeat over and

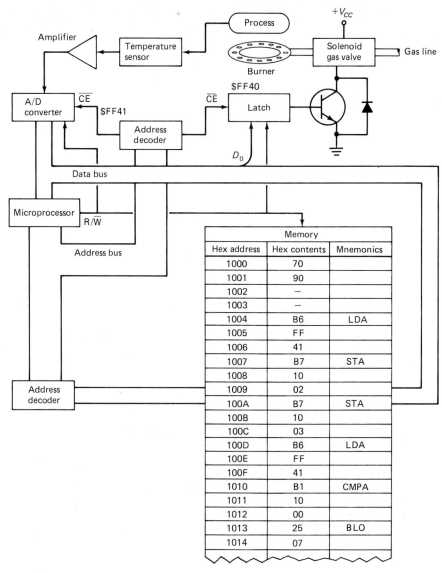

Fig. 12-5 Microprocessor temperature controller.

The memory in Fig. 12-5 starts at $1000. The first two memory locations hold the lower temperature limit (hex 70) and the upper temperature limit (hex 90). Since these are contained in memory, it would be a simple matter to change them. This capability is one of the key attributes of microprocessor-based systems. The next two locations are reserved for the low temperature reading and the high temperature reading. They are marked with a dash since the memory contents here will change as the program runs. The control program itself begins at hex $1004, and the contents here are hex B6 or binary 10110110. Remember, the microprocessor reads the binary contents, and hex is simply a shorthand notation for humans. B6 is an *operation code* (normally abbreviated OP-CODE). It tells the microprocessor to load accumulator A with the contents of the location specified by the next two bytes in memory. Note that the next two bytes are $FF and $41. This means that accumulator A will be loaded from address $FF41, the address of the A/D converter. It is very difficult

for humans to memorize OP-CODES such as B6. Also, the OP-CODES vary among the various types of microprocessors. It is much easier to memorize mnemonics (the first *m* is silent). The mnemonic for load accumulator A is LDA.

Please note that the memory in Fig. 12-5 contains constants, variable data, OP-CODES, and addresses. How does the microprocessor know which is which? It doesn't. A microprocessor runs a *fetch-execute cycle*. It must start by fetching an OP-CODE from the correct address. In our example, the program counter must contain $1004 when the first fetch occurs. The contents of location $1004 are loaded into the microprocessor and go on to the decoder since the microprocessor is in a fetch phase. Then $B6 is decoded, and the controller inside the microprocessor advances to the execute phase. It reads the next two bytes in memory and loads them into the address register (refer to Fig. 12-1). The address register now contains $FF41, and the address bus shows that bit pattern. The microprocessor completes the execute phase of the cycle by loading the contents of the data bus (the temperature) into the data register and then into accumulator A. The program counter has, by now, been incremented to $1007, and this value is transferred to the address register. The next fetch begins, and $B7 is loaded into the processor. The OP-CODE for store accumulator A into the location specified by the next two bytes in memory is B7. Note that the next locations contain $10 and $02, so the contents of accumulator A are stored in memory location $1002.

The third fetch phase begins at address $100A. The OP-CODE there is the same as for the prior fetch, but the address information that follows is different, so the contents of accumulator A are stored in location $1003. The fourth fetch phase begins at $100D. When executed, the A/D converter is read again. The

fifth fetch phase starts at $1010. Hex B1 is the OP-CODE for compare accumulator A with the contents of the location specified by the next two bytes, which are $10 and $00. Therefore, accumulator A is compared with the contents of memory location $1000, which contains $70, the lower temperature limit. The sixth fetch brings $25 into the decoder of the processor; it is the OP-CODE for *branch if lower*. If the temperature is less than the lower limit $70, the processor will branch to the instructions that store a $01 to address $FF40. This will turn the gas on. If the temperature is not less than $70, the program flow will continue on to the next compare instruction that checks whether the temperature is above the upper limit.

What happens if a microprocessor fetches from the wrong place? For example, suppose the program

LISTING 12-1 TEMPERATURE-CONTROL PROGRAM LISTING

			00100		*TEMP CONTROL PROGRAM	
1000			00110		ORG	$1000
		FF41	00120	SENSE	EQU	$FF41
		FF40	00130	VALVE	EQU	$FF40
1000		70	00140	LTL	FCB	$70
1001		90	00150	UTL	FCB	$90
1002			00160	LOW	RMB	$01
1003			00170	HIGH	RMB	$01
1004	B6	FF41	00180		LDA	SENSE
1007	B7	1002	00190		STA	LOW
100A	B7	1003	00200		STA	HIGH
100D	B6	FF41	00210	READ	LDA	SENSE
1010	B1	1000	00220		CMPA	LTL
1013	25	07	00230		BLO	ON
1015	B1	1001	00240		CMPA	UTL
1018	22	09	00250		BHI	OFF
101A	20	0A	00260		BRA	TEST
101C	C6	01	00270	ON	LDB	#$01
101E	F7	FF40	00280		STB	VALVE
1021	20	03	00290		BRA	TEST
1023	7F	FF40	00300	OFF	CLR	VALVE
1026	B1	1002	00310	TEST	CMPA	LOW
1029	25	07	00320		BLO	STL
102B	B1	1003	00330		CMPA	HIGH
102E	22	07	00340		BHI	STH
1030	20	DB	00350	REPEAT	BRA	READ
1032	B7	1002	00360	STL	STA	LOW
1035	20	F9	00370		BRA	REPEAT
1037	B7	1003	00380	STH	STA	HIGH
103A	20	F4	00390		BRA	REPEAT
		0000	00400		END	

00000   TOTAL ERRORS

HIGH	1003
LOW	1002
LTL	1000
OFF	1023
ON	101C
READ	100D
REPEAT	1030
SENSE	FF41
STH	1037
STL	1032
TEST	1026
UTL	1001
VALVE	FF40

LISTING 12-2 BASIC TEMPERATURE CONTROL PROGRAM LISTING

```
10 CLS: REM CLEAR CATHODE RAY TUBE
20 PRINT@130,"LOW TEMP": REM PRINT LABEL ON C.R.T.
30 PRINT@162,"HIGH TEMP": REM PRINT LABEL ON C.R.T.
40 T = PEEK(65345): REM READ TEMPERATURE
50 TL = T: TH = T: REM INITIALIZE LOW AND HIGH VALUES
60 PRINT@140,STR$(TL): REM STORE TO C.R.T.
70 PRINT@172,STR$(TH): REM STORE TO C.R.T.
80 T = PEEK(65345): REM READ TEMPERATURE
90 IF T<112 THEN POKE 65344,1: GOTO 110: REM TURN VALVE ON
100 IF T>144 THEN POKE 65344,0: REM TURN VALVE OFF
110 IF T<TL THEN TL=T ELSE GOTO 130: REM CHECK FOR NEW LOW
120 PRINT@140,STR$(TL): REM PRINT NEW LOW ON C.R.T.
130 IF T>TH THEN TH=T ELSE GOTO 150: REM CHECK FOR NEW HIGH
140 PRINT@172,STR$(TH): REM PRINT NEW HIGH
150 GOTO 80: REM REPEAT
```

counter contains the address of the A/D converter, and the fetch phase is begun. The processor has no inherent intelligence and will attempt to decode whatever byte the converter happens to place on the data bus. Obviously, the results will be unpredictable. When this happens, the system "crashes." Control is usually lost, and a reset is required to restore operation. It is the responsibility of the programmer to arrange memory contents so that OP-CODES are fetched when they are supposed to be. The program must also begin operating at the correct place. If the program in Fig. 12-5 is executed starting at $1000, unpredictable results will occur and the system will crash. A crash will not damage the processor or the memory devices, but it could cause severe consequences in the industrial environment. For example, the gas valve could be turned on and stay on indefinitely.

It should be obvious by now that a microprocessor is worthless without a program to run. *Programs* are software, and software is required for any microprocessor application, even the most trivial. How is software written? The first step is a flowchart. The next step is a conversion to machine code (0s and 1s). This can be done by learning the instruction set of the microprocessor and then looking up all the OP-CODES. The OP-CODES are placed into memory along with the correct addresses and data; this is called *hand assembly*. It is error-prone and laborious. A better way to program in machine language is to use a computer program called an *editor/assembler*. Listing 12-1 shows the printout from an editor/assembler used to generate the code for the temperature control program. The editor allows labels that "make sense" to be assigned. The labels help document the program and make it more readable by people. The assembler "knows" all the mnemonics, so there is no need to look up OP-CODES. Use of an editor/assembler is a giant step above hand assembly. Another alternative is to use a high-level language such as BASIC to write the control program.

This will work only if the control computer has high-level-language capabilities. Listing 12-2 shows a BASIC program that accomplishes the same temperature control as the machine language program. It is complete with remark (REM) statements that help to document the program and make it easy for a human to understand. Read the listing and see how it compares with the flowchart. Don't forget to convert the decimal values in the BASIC program to hexadecimal so that your comparisons make sense.

BASIC is much easier to learn than machine language. Several versions of BASIC that have been enhanced for industrial applications are available. Although it is easier to read than machine language, it cannot be run unless the microcomputer has this capability. Most dedicated industrial controllers do not. It also adds overhead to the system. BASIC itself is a program and takes up a significant amount of memory space. Last but not least, BASIC is very slow when compared to machine language. The machine language temperature control program runs over 100 times faster than the BASIC control program. This characteristic is not important in the simple temperature control program, but speed is very important in many other industrial applications.

REVIEW QUESTIONS

1. Motorola also manufactures a 16-bit microprocessor with 24 address pins. How many memory locations can it directly access?

2. Another name for the condition code register is the _____ register.

3. The MC6809 index and stack registers have _____ bits.

4. The MC6809 has two _____ bit accumulators.

5. Refer to Fig. 12-5. If accumulator B contains $01, what happens when it is stored to address $FF40?

6. In Fig. 12-5, B6 at memory location $1004 is an example of a(n) _____.

7. Could the program of Fig. 12-5 be run out of ROM? Why?

12-2
ADDRESSING MODES

The MC6809 microprocessor uses six basic addressing modes: inherent, immediate, extended, direct, indexed, and relative. This section describes each of these modes. In each description, the term *effective address* indicates the actual address in memory where data will be fetched or stored or where instruction processing will proceed. Some of the addressing modes require an extra byte after the OP-CODE to provide the required addressing information. This byte is called a *postbyte*.

Inherent addressing is also called *implied addressing* because the effective address is implicit in the instruction itself. For example, one of the MC6809 instruction mnemonics is ABX, which adds accumulator B to the X register and places the sum in the X register. No additional address information is required since it is inherent in the instruction itself. Another example is DAA, which is a decimal addition adjustment to accumulator A. It is used to correct the sum in the accumulator after binary-coded decimal numbers have been added. Some instructions have an inherent mode in addition to other addressing modes. The mnemonic CLR stands for *clear;* when it is used to clear accumulator A or B, the addressing mode is inherent. However, when it is used to clear some memory location, one of the other addressing modes must be used.

The *immediate addressing mode* places the operand (data) in one or two memory locations immediately following the OP-CODE. This mode is used to provide constant data that do not change during operation of the program. As stated, LDA is the mnemonic for load accumulator A. If the immediate addressing mode is used, the next byte after the OP-CODE for LDA must contain the data that are to be loaded into accumulator A. Let's look at a short program segment that uses immediate addressing to load accumulator A with hex F1, accumulator B with hex 7E, and the X register with hex A09E:

Memory Location	Hex Contents	Mnemonic
2000	86	LDA
2001	F1	
2002	C6	LDB
2003	7E	
2004	8E	LDX
2005	A0	
2006	9E	

Note that two bytes follow the OP-CODE for LDX, the mnemonic for load index register X. Since this is a 16-bit register, two data bytes must immediately follow the OP-CODE in memory. The preceding program segment could be shortened by 1 byte by using a load accumulator D (LDD) instruction in place of the LDA and LDB instructions. Recall that accumulator D is a combination of accumulators A and B. The OP-CODE for LDD immediate is CC. Therefore, $CC followed by $F1 and then by $7E would produce the same result.

One form of immediate addressing uses a postbyte to determine which two registers will be manipulated. Table 12-1 shows how the postbyte must be formed for the exchange and transfer instructions. Exchange is a swap operation. For example, if accumulators A and B are exchanged, each will contain what the other contained prior to execution of the instruction. A program segment to swap the accumulators and then the index registers follows:

Memory Location	Hex Contents	Mnemonic
2007	1E	EXG
2008	89	
2009	1E	EXG
200A	12	

Note that the same OP-CODE is used for both exchange instructions. It is the postbyte in each case that tells the processor which two registers are to be exchanged.

The *transfer operation* copies the source register into the destination register. Both registers will contain the same data after the instruction is executed. The following program segment transfers accumulator A into the direct page register and the S stack pointer into the U stack pointer:

Memory Location	Hex Contents	Mnemonic
200B	1F	TFR
200C	8B	
200D	1F	TFR
200E	43	

The source and destination must be in the proper order. Table 12-1 shows that the source code makes up the upper half of the postbyte and that the destination code makes up the lower half of the postbyte. The code for the S pointer register is binary 0100 (hex 4), and the code for the U pointer register is binary 0011 (hex 3). Therefore, the correct postbyte to transfer S to U is hex 43. Only registers of the same size should be manipulated by the exchange and transfer functions.

Table 12-2 shows how the postbyte is formed for push and pull instructions. These instructions are for saving (or recalling) the contents of one or more of the microprocessor registers in a special area of memory called the *stack*. A *push* operation will save the register(s) to memory, and a *pull* will load the register(s) from memory. There are two 16-bit stack pointers in the MC6809. Let's look at a program segment that will store all of the internal registers, except the program counter, to the memory stack

TABLE 12-1 POSTBYTE FORMATION FOR EXCHANGE AND TRANSFER INSTRUCTIONS

b7	b6	b5	b4	b3	b2	b1	b0
SOURCE (R1)				DESTINATION (R2)			

Code	Register	Code	Register
0000	D (A:B)	0101	Program counter
0001	X index	1000	A accumulator
0010	Y index	1001	B accumulator
0011	U stack pointer	1010	Condition code
0100	S stack pointer	1011	Direct page

TABLE 12-2 POSTBYTE FORMATION FOR PUSH AND PULL INSTRUCTIONS

b7	b6	b5	b4	b3	b2	b1	b0
PC	S/U	Y	X	DP	B	A	CC

PC	=	Program counter
S/U	=	Hardware/User stack pointer
Y	=	Y index register
X	=	U index register
DP	=	Direct page register
B	=	B accumulator
A	=	A accumulator
CC	=	Condition code register

pointed to by the U register and then load the registers from the memory stack pointed to by the S register:

Memory Location	Hex Contents	Mnemonic
200F	36	PSHU
2010	7F	
2011	35	PULS
2012	7F	

Table 12-2 shows that bit 6 of the postbyte, if set, will cause either S or U to be pushed or pulled. If U is the pointer, then S will be pushed or pulled. If S is the pointer, then U will be pushed or pulled. The program segment shows a postbyte of $7F; therefore bit 6 is set. The *PSHU instruction* will store the S register on the memory stack. The *PULS instruction* will load the U register from the memory stack.

Extended addressing locates the effective address in the two bytes following the OP-CODE. Suppose we need a program segment that loads accumulator A from memory location $45E1, loads accumulator B from location $5021, and then stores the X register at locations $8000 and $8001:

Memory Location	Hex Contents	Mnemonics
2013	B6	LDA
2014	45	
2015	E1	
2016	F6	LDB
2017	50	
2018	21	
2019	BF	STX
201A	80	
201B	00	

Even though the X register is a 16-bit register, only the first address byte is specified. The microprocessor will store the upper byte in location $8000 and the lower byte in location $8001.

Direct addressing uses a combination of the direct page register and a single byte following the OP-CODE to form the effective address of the operand. For example, if the direct page register contains $90

and the byte following the OP-CODE is $45, the effective address is $9045. The total memory range of the microprocessor can be thought of as 256 pages of 256 bytes each. The direct page register always points to one page of memory. When quite a bit of data must be accessed from one page, the direct mode provides faster access to the locations in that page if the direct page register is pointing there. The following program segment uses immediate addressing to load accumulator A with $90, transfers it to the direct page register, and then uses direct addressing to load both accumulators from that page:

Memory Location	Hex Contents	Mnemonic
201C	86	LDA
201D	90	
201E	1F	TFR
201F	8B	
2020	96	LDA
2021	45	
2022	D6	LDB
2023	99	

Accumulator A is loaded from location $9045, and accumulator B is loaded from location $9099.

The *indexed addressing* mode uses one of the 16-bit pointer registers (X, Y, S, U, and sometimes the program counter) in the calculation of the effective address. There are several variations within the indexed mode, which include constant offset, accumulator offset, autoincrement, autodecrement, indirect, extended indirect, and program counter relative. The indexed addressing modes are the most powerful and are among the key features of the MC6809 microprocessor.

The *constant offset indexed mode* uses a postbyte to identify the pointer register and the offset size. The offset sizes available are zero offset, 5-bit offset, 8-bit offset, and 16-bit offset. Table 12-3 shows the postbyte formation for the indexed addressing modes. A postbyte of binary 11100100 ($E4) means that the offset is zero and the *S* register will point to

TABLE 12-3 POSTBYTE FORMATION FOR INDEXED ADDRESSING MODES

Mode Type	Variation	Direct	Indirect
Constant offset from register (twos complement offset)	No offset	1RR00100	1RR10100
	5-Bit offset	0RRnnnnn	Defaults to 8-bit
	8-Bit offset	1RR01100	1RR11000
	16-Bit offset	1RR01001	1RR11001
Accumulator offset from register (twos complement offset)	A accumulator offset	1RR00110	1RR10110
	B accumulator offset	1RR00101	1RR10101
	D accumulator offset	1RR01011	1RR11011
Auto increment/decrement from register	Increment by 1	1RR00000	Not allowed
	Increment by 2	1RR00001	1RR10001
	Decrement by 1	1RR00010	Not allowed
	Decrement by 2	1RR00011	1RR10011
Constant offset from program counter	8-Bit offset	1XX01100	1XX11100
	16-Bit offset	1XX01101	1XX11101
Extended indirect	16-Bit address	--------	10011111

R = X, Y, U, or S; X = 00; Y = 01; X = don't care; U = 10; S = 11.

the effective address. A 5-bit offset can be contained in the postbyte in 2s complement form. The total range of a 5-bit 2s complement offset is − 16 to + 15. The effective address will be anywhere from 16 less to 15 more than the contents of the pointer register. For example, a postbyte of binary 00001111 ($0F) means that the effective address will be decimal 15 greater than the contents of the X register. If the most significant bit (bit 4) of the offset is 1, the offset is negative and the effective address will be less than the contents of the pointer register.

Eight-bit offsets cannot fit into the postbyte and are contained in an offset byte that follows the postbyte. These are also in 2s complement form for a total decimal range of − 128 to + 127 added to the contents of the pointer register. As an example, the postbyte binary 10101100 ($AC) followed by $38 means that the effective address is $38 (decimal 56) greater than the contents of the Y register. If the most significant bit of an 8-bit offset is high, the offset is negative and the effective address will be less than the contents of the pointer register. Sixteen-bit offsets are also in 2s complement form and provide a decimal range of − 32,768 to + 32,767 added to the contents of the pointer register. They are contained in 2 bytes following the postbyte.

Accumulator offset uses the contents of accumulator A, B, or D added to the pointer register to calculate the effective address. The number in the accumulator is treated as a 2s complement number; if the most significant bit is high, the number is negative. In this case, the effective address will be less than the contents of the pointer register. Neither the contents of the pointer register nor the accumulator is affected by the calculation. Table 12-3 shows how to form the postbyte for accumulator offset indexed addressing.

The *autoincrement indexed mode* works by determining the effective address from the desired pointer register and then incrementing the pointer by 1 or 2.

The *autodecrement mode* first subtracts one or two from the desired pointer register and then produces the effective address. These modes are known as *postincrementing* and *predecremeting*. They are extremely valuable modes for moving lists or tables of data from one area of memory to another. Table 12-3 shows how the postbytes are formed for these addressing modes.

The *indirect addressing mode* points to two memory locations which contain the address of the operand. For example, suppose the postbyte is binary 10010100, and the X register contains $78CE. Table 12-3 shows that this postbyte is for zero offset, indirect with register X serving as the pointer. The microprocessor will fetch the contents of $78CE and $78CF, not as the operand but as the effective address of the operand. If memory location $78CE contains $01 and location $78CF contains $A4, then the operand will be fetched from memory location $01A4. In extended indirect, the effective address is located at the address specified by the two bytes following the postbyte. Suppose the postbyte is $9F, which specifies extended indirect and is followed by $23 and then $12. The contents of memory location $2312 and $2313 will form the effective address.

Program counter relative addressing uses the program counter as the pointer with either an 8-bit or a 16-bit 2s complement offset. The offset is added to the program counter to form the effective address. Table 12-3 shows the postbytes for program counter relative addressing. Either one or two offset bytes must follow the postbyte.

The last addressing mode is the *relative addressing mode*, which is used when branches from the current instruction location to some other location are desired. The branches are relative to the program counter. When the test of a branch condition is true, either a 1- or a 2-byte relative address is added to the program counter. The relative address is in 2s complement form, allowing both forward and backward program branches. A 1-byte relative address is called a *short branch* and allows a total range of −128 to +127. A 2-byte relative address is called a *long branch* and provides a total range of −32,768 to +32,767. The following program segment illustrates how relative addressing can be used to make the program branch backward a number of times until some condition is met. In this application, it is used to send 64 pulses to an output port:

Memory Location	Hex Contents	Mnemonics
2024	8E	LDX
2025	FF	
2026	40	
2027	86	LDA
2028	01	
2029	C6	LDB
202A	40	
202B	A7	STA
202C	84	
202D	6F	CLR
202E	84	
202F	5A	DECB
2030	26	BNE
2031	F9	

This program segment uses the immediate addressing mode to load the index register with the address of the output port and to load accumulator A with $01. It also uses the immediate mode to load accumulator B with $40 (decimal 64), which is the number of output pulses. Next, it stores accumulator A by using the zero offset indexed mode (note the postbyte of $84). Then it clears the same location again, by using the zero offset indexed mode. The output is pulsed by storing 1 and then clearing it. Accumulator B is then *decremented* (DECB), and a *branch if not equal to zero* (BNE) instruction follows. Note that $F9 follows the BNE instruction; it is the relative address. It will cause a backward branch to the STA instruction every time the BNE test is true. Thus, the program will loop back and continue pulsing the output port until accumulator B is decremented to zero. When it does equal zero, the processor will fetch the next OP-CODE from address $2032.

How does the relative address cause a backward branch to $202B? The program counter is pointing to $2032, which is the address for the next fetch. When the branch test is true, the ALU of the microprocessor adds the relative address to the program counter to form the effective address. Let's subtract the destination address from the source address to determine the backward branch:

$2032 ⟵ SOURCE ADDRESS
$202B ⟵ DESTINATION ADDRESS
$07 ⟵ DIFFERENCE

Since $B is greater than $2, we must borrow from the next column. Since we are working in hexadec-

imal, the borrow adds decimal 16 to the first column. Because 16 + 2 = 18 and hex B = 11, the difference is $07. Now we can see that the ALU must subtract 7 from the program counter. This in the range of a short branch, and the relative address will be 1 byte. We learned in the previous chapter that subtraction may be accomplished by changing the subtrahend to a 2s complement number:

07 ⟵ HEX RELATIVE ADDRESS
00000111 ⟵ BINARY VALUE
11111000 ⟵ ONE'S COMPLEMENT
+1
11111001 ⟵ TWO'S COMPLEMENT
F9 ⟵ HEX VALUE

Relative addresses are always in 2s complement form. When the most significant bit is high, a backward branch will occur. When the most significant bit is low, a forward branch will occur.

REVIEW QUESTIONS

8. One of the MC6809 instructions is MUL. It multiplies accumulator A times accumulator B and places the result in accumulator D. What addressing mode does this instruction use?

9. The immediate addressing mode is used to load accumulator B. How many operand bytes must follow the OP-CODE?

10. Refer to Table 12-1. What hex postbyte must follow the EXG OP-CODE to swap the S and U registers?

11. Refer to Table 12-1. What hex postbyte must follow the TFR instruction to transfer the DP register to accumulator B?

12. Refer to Table 12-2. Determine the hex postbyte required to pull the CC register from the stack.

13. Refer to Table 12-2. Assuming a postbyte of $40, which register(s) will be saved to the stack by a PSHU operation?

14. Refer to Table 12-3. What hex postbyte is required to use the 16-bit offset indexed direct addressing mode with the Y register serving as the pointer? What must follow the postbyte in this case?

15. Refer to Table 12-3. What hex postbyte will be required to select the autoincrement by one direct indexed addressing mode with the X register serving as the pointer? When will the X register be incremented?

12-3
INSTRUCTION SET

The Motorola MC6809 microprocessor has 59 different instructions. When these are combined with all of the available addressing modes, well over 1000 different operations are possible. The instruction set can be functionally divided into five categories: 8-bit accumulator and memory instructions, 16-bit accumulator and memory instructions, index register and stack pointer instructions, branch instructions, and miscellaneous instructions.

Before the instructions are examined, a short discussion of signed and unsigned numbers is appropriate. Suppose the bit pattern in an accumulator or some memory location is 11010011. What decimal number does this pattern represent? It is a matter of interpretation. If the program deals only with *positive (unsigned) numbers,* it represents 211 decimal. However, if the program deals with *signed numbers,* it represents − 45 decimal. Negative numbers are represented in 2s complement form, and the most significant bit must be high. Please note that the microprocessor treats all numbers in the same way. It is the programmer's responsibility to decide what a particular bit pattern means.

Table 12-4 shows the 8-bit accumulator and memory instructions. There are two types of addition instructions. The first type shown in the table is *add with carry.* This instruction adds a byte from memory to the contents of one of the accumulators plus the contents of the carry flag. If a prior operation has set the carry flag, then the sum will be 1 greater than the memory contents plus the accumulator contents. The second type of add instruction in the table does not add the carry flag. If microprocessors were limited to 8-bit arithmetic, they would not be adequate for many applications. The add with carry instruction allows multiple precision arithmetic. *Multiple precision addition* allows both the augend and addend to be represented by multiple bytes. The least significant bytes of the augend and addend are added first by using the add instruction. The carry flag may or may not be set by this operation. Then the two next most significant bytes are added, using the add with carry instruction, and so must all subsequent bytes. Thus, the precision is limited by available memory and time. It should be obvious that multiple precision operations take more time than single-byte operations.

The ANDA and ANDB instructions shown in Table 12-4 perform a logical AND with the accumulator contents and some memory location and store the result in the accumulator. Logical ANDing can be used to strip bits off a byte; i.e., the lower 4 bits of the ASCII codes for decimal numbers 0 through 9 are weighted to correspond to the numeric value. By stripping off the upper 4 bits, the correct binary value results. To convert ASCII 0 through 9 to binary, the

TABLE 12-4 8-BIT ACCUMULATOR AND MEMORY INSTRUCTIONS

Instruction	Description
ADCA, ADCB	Add memory to accumulator with carry
ADDA, ADDB	Add memory to accumulator
ANDA, ANDB	Add memory with accumulator
ASL, ASLA, ASLB	Arithmetic shift of accumulator or memory left
ASR, ASRA, ASRB	Arithmetic shift of accumulator or memory right
BITA, BITB	Bit test memory with accumulator
CLR, CLRA, CLRB	Clear accumulator or memory location
CMPA, CMPB	Compare memory from accumulator
COM, COMA, COMB	Complement accumulator or memory location
DAA	Decimal adjust A accumulator
DEC, DECA, DECB	Decrement accumulator or memory location
EORA, EORB	Exclusive or memory with accumulator
EXG R1, R2	Exchange R1 with R2 (R1, R2 = A, B, CC, DP)
INC, INCA, INCB	Increment accumulator or memory location
LDA, LDB	Load accumulator from memory
LSL, LSLA, LSLB	Logical shift left accumulator or memory location
LSR, LSRA, LSRB	Logical shift right accumulator or memory location
MUL	Unsigned multiply (A × B → D)
NEG, NEGA, NEGB	Negate accumulator or memory
ORA, ORB	Or memory with accumulator
ROL, ROLA, ROLB	Rotate accumulator or memory left
ROR, RORA, RORB	Rotate accumulator or memory right
SBCA, SBCB	Subtract memory from accumulator with borrow
STA, STB	Store accumulator to memory
SUBA, SUBB	Subtract memory from accumulator
TST, TSTA, TSTB	Test accumulator or memory location
TFR R1, R2	Transfer R1 to R2 (R1, R2 = A, B, CC, DP)

ASCII code can be ANDed with $0F (binary 00001111). The $0F is called a *mask* and *strips off* (sets to 0) the upper 4 bits of the ASCII code. Conversely, the ORA and ORB instructions shown in the table can be used to convert binary to ASCII. Provided that the binary number in the accumulator is 00001001 (decimal 9) or less, ORing with $30 (binary 00110000) will set the 2 bits necessary to convert to ASCII.

The *arithmetic shift left* instructions in Table 12-4 shift the contents of an accumulator or some memory location left by one bit position. Bit 0 is cleared and bit 7, the MSB, shifts into the carry flag. Whatever was in the carry flag is lost. This operation has the effect of doubling the value of the accumulator or memory location up until the point where bits are shifted out and lost. For example, suppose the accumulator contains binary 00010110 before the shift left. This is equal to decimal 22. The accumulator will contain 00101100 after a shift left which is equal to decimal 44. Another shift left will produce 01011000, which is equal to decimal 88, and so on. The *arithmetic shift right* instructions shown in the table shift the number to the right. Bit 0 is shifted into the carry flag, and the prior content of the flag is lost. Bit 7 does not change. This preserves the sign of the number since the most significant bit is the sign bit when 2s complement interpretation is used. If it is set, the number is negative. If it is clear, the number is positive. For example:

```
10101000  ⟵ CONTENTS BEFORE ASR
11010100  ⟵ CONTENTS AFTER ASR
```

In 2s complement interpretation, both of the preceding numbers are negative. To find the magnitude of each number, invert every bit and add 1:

```
10101000  ⟵ TWO'S COMPLEMENT FORM
01010111  ⟵ INVERT
     +1   ⟵ ADD ONE
_____
01011000  ⟵ MAGNITUDE (decimal 88)

11010100  ⟵ TWO'S COMPLEMENT FORM
00101011  ⟵ INVERT
     +1   ⟵ ADD ONE
_____
00101100  ⟵ MAGNITUDE (decimal 44)
```

In 2s complement interpretation, the number is equal to negative 88 before the shift and is equal to negative 44 after the shift.

The *logic shift left operations* in Table 12-4 do exactly the same thing as the arithmetic shift left instructions already discussed. Motorola made this accommodation to make the MC6809 compatible with a mnemonic used for an earlier microprocessor, the MC6800. The *logic shift right* (LSR) *operation*, however, is different from the arithmetic shift right. The LSR does not preserve the sign bit. A 0 is shifted into bit 7 instead. Bit 0 is shifted into the carry flag as it is for the ASR operation.

Some microprocessor operations do not produce any results other than setting or clearing the appropriate flags. The BITA and BITB operations in Table 12-4 are examples. These operations perform the logical AND of the accumulator contents with some memory location but produce no change in accumulator or memory contents. The only result is that three flags are affected. First, the V flag is always cleared by this operation. Second, the N flag is set if the AND operation sets the most significant bit; otherwise it is cleared. The N flag is the negative flag. Third, the Z flag is set if the result of the AND operation is binary 00000000; otherwise it is cleared. The Z flag is the zero flag. The bit test instructions are often used to test 1 bit of some memory location to see whether it is high or low. For example, the accumulator can be loaded with $01 to test the least significant bit of a memory location. If the bit is high, the Z flag will be cleared. If the bit is low, the Z flag will be set.

The CMPA and CMPB instructions also produce no results other than setting or clearing the appropriate flags. The *compare* instructions subtract the contents of some memory location from one of the accumulators. The N flag is set if the result is negative. The Z flag is set if the result is zero. The C flag is set if a borrow is generated. Finally, the V flag is set if an overflow occurs. An *overflow* refers to 2s complement overflow and not to a borrow (or a carry). For example, if a negative number is subtracted from a positive number, the result should be positive. Let's look at an example of what can happen:

```
  01011111  ⟵ ACCUMULATOR
− 10101010  ⟵ MEMORY
_____
  10110101  ⟵ RESULT OF THE CMP
             OPERATION
```

Note that the accumulator contains a positive number (the most significant bit is 0) and that the memory location contains a negative number (the most significant bit is 1). However, the result is negative since its most significant bit is 1. In this case, the compare instruction will set the V flag since 2s complement overflow has occurred. It will also set the C flag since a borrow has also occurred. Two's complement overflow can occur in other operations as well. For example, two positive numbers can be added and produce a negative result because of a carry into the sign bit (bit 7). This will also set the V flag.

The *test instruction* (TST) shown in Table 12-4 also produces no results other than in the flag register. It subtracts zero from some memory location or one of the accumulators. It always clears the V flag. It sets or clears the N and Z flags according to the results. The *complement operation* (COM) does produce a result; it replaces the contents of one of the accumulators or a memory location with its 1s complement value. The *negate operation* (NEG) produces the 2s complement value. The *rotate right instruction* (ROR) rotates all of the bits (memory location or

TABLE 12-5 16-BIT ACCUMULATOR AND MEMORY INSTRUCTIONS

Instruction	Description
ADDD	Add memory to D accumulator
CMPD	Compare memory from D accumulator
EXG D, R	Exchange D with X, Y, S, U, or PC
LDD	Load D accumulator from memory
SEX	Sign Extend B accumulator into A accumulator
STD	Store D accumulator to memory
SUBD	Subtract memory from D accumulator
TFR D, R	Transfer D to X, Y, S, U, or PC
TFR R, D	Transfer X, Y, S, U, or PC to D

accumulator) right through the carry flag. Bit 0 is placed in the carry flag, and the carry flag is placed in bit 7. This instruction simulates a circulating shift register with 9 bits. The *rotate left instruction* (ROL) is similar, but the direction of rotation is reversed.

Table 12-5 shows the 16-bit accumulator and memory instructions. Accumulator D can be added, loaded, stored, subtracted, transferred, compared, and exchanged. When used in conjunction with external memory, two consecutive memory bytes will be affected. The *sign extend operation* (SEX) transforms a 2s complement 8-bit value in accumulator B into a 2s complement 16-bit value in the D accumulator. For example, suppose accumulator B contains binary 11001001 ($C9) before the SEX operation. This represents −55 decimal in 2s complement form. After the SEX operation, the D accumulator will contain 1111111111001001 ($FFC9), which is the 16-bit 2s complement representation of −55 decimal.

Table 12-6 lists the index and stack pointer instructions. The pointer registers can be loaded, stored, transferred, exchanged, and compared. The four *load effective address instructions* (LEA) calculate

the effective address from the indexed addressing mode and place an address in one of the 16-bit pointer registers. The calculations that can be performed include adding or subtracting 5-, 8-, or 16-bit constants; adding or subtracting the contents of one of the 8-bit accumulators; and adding or subtracting the contents of the D accumulator.

The branch instructions are shown in Table 12-7. The simple branches test only one flag. For example, the *branch if equal zero instruction* (BEQ) tests the Z flag. If it is set, the branch is implemented by fetching the relative address and adding it to the program counter. The short branch instructions use a 1-byte relative address; the long branch instructions, such as LBEQ, use a 2-byte relative address. Some of the signed branches use a more involved test of the flag register. The *branch if greater than zero* (BGT) checks three flags and is implemented only when the N and V flags are equal and the Z flag is zero. The BGT branch is used after a subtract or compare operation to alter program flow when the signed register contents are greater than the signed memory operand. You should note that some of the

TABLE 12-6 INDEX AND STACK POINTER INSTRUCTIONS

Instruction	Description
CMPS, CMPU	Compare memory from stack pointer
CMPX, CMPY	Compare memory from index register
EXG R1, R2	Exchange D, X, Y, S, U or PC with D, X, Y, S, U or PC
LEAS, LEAU	Load effective address into stack pointer
LEAX, LEAY	Load effective address into index register
LDS, LDU	Load stack pointer from memory
LDX, LDY	Load index register from memory
PSHS	Push A, B, CC, DP, D, X, Y, U, or PC onto hardware stack
PSHU	Push A, B, CC, DP, D, X, Y, X, or PC onto user stack
PULS	Pull A, B, CC, DP, D, X, Y, U, or PC from hardware stack
PULU	Pull A, B, CC, DP, D, X, Y, S, or PC from hardware stack
STS, STU	Store stack pointer to memory
STX, STY	Store index register to memory
TFR R1, R2	Transfer D, X, Y, S, U, or PC to D, X, Y, S, U, or PC
ABX	Add B accumulator to X (unsigned)

TABLE 12-7 BRANCH INSTRUCTIONS

Instruction	Description
Simple Branches	
BEQ, LBEQ	Branch if equal
BNE, LBNE	Branch if not equal
BMI, LBMI	Branch if minus
BPL, LBPL	Branch if plus
BCS, LBCS	Branch if carry set
BCC, LBCC	Branch if carry clear
BVS, LBVS	Branch if overflow set
BVC, LBVC	Branch if overflow clear
Signed Branches	
BGT, LBGT	Branch if greater (signed)
BVS, LBVS	Branch if invalid twos complement result
BGE, LBGE	Branch if greater than or equal (signed)
BEQ, LBEQ	Branch if equal
BNE, LBNE	Branch if not equal
BLE, LBLE	Branch if less than or equal (signed)
BVC, LBVC	Branch if valid twos complement result
BLT, LBLT	Branch if less than (signed)
Unsigned Branches	
BHI, LBHI	Branch if higher (unsigned)
BCC, LBCC	Branch if higher or same (unsigned)
BHS, LBHS	Branch if higher or same (unsigned)
BEQ, LBEQ	Branch if equal
BNE, LBNE	Branch if not equal
BLS, LBLS	Branch if lower or same (unsigned)
BCS, LBCS	Branch if lower (unsigned)
BLO, LBLO	Branch if lower (unsigned)
Other Branches	
BSR, LBSR	Branch to subroutine
BRA, LBRA	Branch always
BRN, LBRN	Branch never

simple branches also appear in the signed branch section since they are also useful for testing signed numbers. The unsigned branches in the table are for testing positive numbers only. The most significant bit is not interpreted as a sign bit. The *branch if higher than instruction* (BHI) tests two flags: carry and zero. If a subtract or compare operation causes neither a carry (borrow) nor a zero result, the branch will be implemented. The other branches shown in Table 12-7 do not test any flags. The *branch to subroutine* (BSR) will be covered in the next section. *Branch always* (BRA) is an unconditional branch; the relative address will always be fetched and added to the program counter. *Branch never* (BRN) does nothing and acts as a 2-byte *no operation* (NOP). The LBRN acts as a 4-byte no operation because the OP-CODE is 2 bytes long and must be followed by a 2-byte relative address. In a program NOPs may be used to reserve memory space for future insertion of code. They also may be placed in a timing loop to consume a few clock cycles. Branch BRN consumes three clock cycles, and LBRN consumes five clock cycles.

The miscellaneous instructions are shown in Table 12-8. The first mnemonic is *ANDCC*, which stands for *logical AND the condition code register*. It gives the programmer a way to preserve or clear the flags and uses the immediate addressing mode only. If the CC register is ANDed with $00, all of the flags will be cleared. If it is ANDed with $F0, the upper four will be preserved, and the lower four will be cleared. The CWAI instruction ANDs an immediate byte with the condition code register, stacks all of the microprocessor registers on the S stack, and then waits for an interrupt. It will be covered in more detail in the next section. The no operation (NOP) consumes 1 memory byte and two clock cycles. The ORCC ORs an immediate byte with the condition code reg-

TABLE 12-8 MISCELLANEOUS INSTRUCTIONS

Instruction	Description
ANDCC	AND condition code register
CWAI	AND condition code register, then wait for interrupt
NOP	No operation
ORCC	OR condition code register
JMP	Jump
JSR	Jump to subroutine
RTI	Return from interrupt
RTS	Return from subroutine
SWI, SWI2, SWI3	Software interrupt (absolute indirect)
SYNC	Synchronize with interrupt line

ister and places the results in the CC register. It is used to set flags. The *jump instruction* (JMP) transfers program control to the effective address. It is similar to the BRA instruction but uses extended, direct, or indexed addressing rather than relative addressing. The *jump to subroutine* (JSR) is covered in the next section along with RTI, RTS, SWI, and SYNC.

REVIEW QUESTIONS

16. Suppose an accumulator contains binary 10011100. What unsigned decimal number does this represent? What signed decimal number does it represent?

17. If accumulator A contains $02, the effective address contains $03, and the carry flag is set, what hex number will result in accumulator *A* after an ADDA instruction? After an ADCA instruction?

18. Accumulator A contains $19 before an ASLA instruction. What hex values does it contain after the operation?

19. Suppose accumulator A is loaded with $80 and then the BITA operation is performed at the beginning of a list in memory. This is followed by the BEQ instruction, and then the next location in the list is tested in the same manner. The branch will occur when a _____ is found. Will execution of this segment change any of the list locations or change the $80 in the accumulator?

20. Hex 4F is added to hex 3E. Will this set the V flag?

21. Accumulator B contains $0F before the COMB instruction. What is the hex value in the accumulator after the operation?

12-4
SUBROUTINES AND INTERRUPTS

A *subroutine* is a program segment that is used over and over again. It might handle some arithmetic function, check the position of a motor shaft, read a

sensor, or produce a delay. The main program can call a subroutine with the jump to subroutine instruction (JSR), as shown in Fig. 12-6. Notice that the main program calls the subroutine from two different locations. Any number of calls is possible. Also notice that the last instruction in the subroutine ends with the *return from subroutine instruction* (RTS). When the JSR instruction is executed, the contents of the program counter are pushed onto the hardware stack. The hardware stack is located somewhere in memory and is pointed to by the S register. When the RTS instruction is executed, the program counter is pulled from the hardware stack. In this way the main program can be reentered at the correct location. You might be wondering whether the same result can be accomplished with the jump (JMP) instruction. It cannot. As Fig. 12-6 shows, a subroutine may be called from several locations in the main program. The only way to reenter the main program at the proper location is to pull the program counter from the stack.

Subroutines may call other subroutines; this procedure is referred to as *nesting* and is shown in Fig. 12-7. The main program calls subroutine A with the JSR instruction. The address of the next instruction in the main program is $13C9. The low-order byte of this address is pushed on the hardware stack, followed by the high-order byte. At the time of the JSR instruction, the S register was pointing at memory location $1006. The JSR instruction first causes the stack pointer to be decremented by 1. Then the low-order byte of the return address is pushed on the stack. The pointer is decremented again, and the high-order byte is pushed onto the stack. Subroutine A begins executing, and later another JSR is encountered. The next instruction in subroutine A is at $2E10 and this address is pushed on the stack. The S pointer now contains $1002. Subroutine B is executed and it ends with the RTS instruction. This causes 2 bytes to be pulled from the stack and loaded into the program counter. Therefore, the contents of location $1002 is loaded into the high byte of the program counter. The S pointer is incremented, and the contents of stack location $1003 are loaded into

Main program

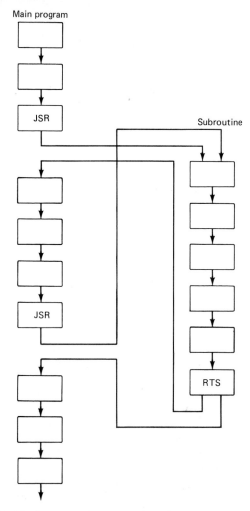

Fig. 12-6 Main program jumps to subroutine.

the low byte of the program counter. The instruction from location $2E10 is fetched, and the rest of subroutine A is executed. The RTS instruction at the end of subroutine A pulls $13C9 into the program counter, and the main program is reentered at the correct location.

Many levels of nesting are possible. For example, subroutine A calls subroutine B, which calls subroutine C, which calls subroutine D, and so on. The only limitation is the memory available for the stack. Figure 12-7 shows that the stack grows as each JSR instruction pushes the program counter. If this occurs too many times, the stack will grow until memory is exhausted or the stack overlaps with a section of memory that is being used for some other purpose. If this happens, the stack may be overwritten, and the processor will be unable to find its way back. The software will "crash."

Stepper motors are covered in Chapter 3. It is possible to use a microprocessor to control a stepper motor. Assume that the motor has been decoded for address $FF40. Also assume that the lower 4 bits of the data bus are latched when this address is written

to. The motor windings can be turned on by writing a 1 to this address and turned off by writing a 0. Suppose the required bit pattern for clockwise rotation is as follows:

$$0011$$
$$1001$$
$$1100$$
$$0110$$

After 0110 is written to the motor, 0011 is written again, and the process repeats.

Let's look at a program that will run the stepper motor:

Memory Location	Hex Contents	Mnemonics
7D12	86	LDA
7D13	03	
7D14	C6	LDB
7D15	09	
7D16	B7	STA
7D17	FF	
7D18	40	
7D19	43	COMA
7D1A	BD	JSR
7D1B	7D	
7D1C	26	
7D1D	F7	STB
7D1E	FF	
7D1F	40	
7D20	53	COMB
7D21	BD	JSR
7D22	7D	
7D23	26	
7D24	20	BRA
7D25	F0	

The first two instructions load accumulators A and B with $03 and $09, respectively. These represent the first 2-bit patterns for the motor. Accumulator A is then written to the motor port. You should notice the address of the motor in the 2 bytes following the STA instruction. Accumulator A is then complemented. The 1s complement of $03 is 11111100. Note that the 4 lower bits are the pattern needed for the third motor step. The delay subroutine is called next. A delay is required before the motor can be stepped again. After the delay, the program flow continues with the STB operation. This writes the second bit pattern to the motor. Accumulator B is now completed to produce the bit pattern needed for the fourth motor step. The delay subroutine is called again, and when it is finished the main program flow resumes at the BRA instruction. The relative address of $F0 sends the program back to the STA instruction. The main program runs again and provides the third and fourth bit patterns to the motor. The complement instructions flip the motor bits again so the next run through will be identical to the first.

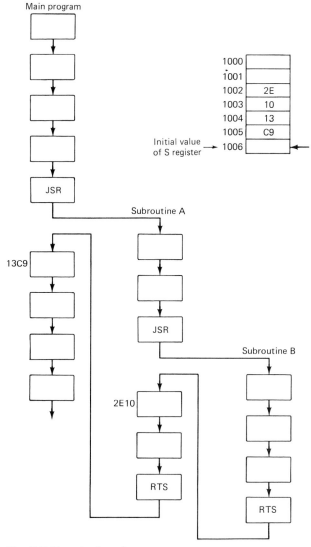

Main program

```
1000
1001
1002   2E
1003   10
1004   13
1005   C9
```
Initial value → 1006
of S register

JSR

13C9

Subroutine A

JSR

Subroutine B

2E10

RTS

RTS

Fig. 12-7 Nested subroutines.

struction loads the X register with the effective address calculated from the 5-bit offset contained in the postbyte $1F. If you refer to Table 12-3, you will see that this provides an offset of binary 11111, which is in 2s complement form and is equal to -1. Thus, the index register is decremented by 1. The four no operations each consume two clock cycles. The branch if not equal zero instruction checks to see that the index register has been decremented to zero. If it has not, the relative address $F8 is fetched, and the subroutine branches back to $7D29. When the index register is decremented to zero, the RTS instruction is fetched, the program counter is pulled from the stack, and the main program is reentered.

How much time will the delay subroutine take? If the processor is running at 1 MHz, each clock cycle is 1 µs. The loop will run $FFFF (decimal 65,535) times. The LEAX instruction consumes 4 cycles, the NOPs 8 cycles, and the BNE 3 cycles for a total of 15 cycles; $65,535 \times 15 = 983,025$ cycles. The LDX instruction executes once, using another 3 cycles; the RTS executes once, using another five cycles. Thus, the total delay is 983,033 µs.

Because the delay program is a subroutine, it appears in memory only once. Otherwise, it would have to appear twice in the stepper motor program. It is obvious that subroutines save a lot of memory and make programs easier to write. Another advantage is that the delay constant is contained in two adjacent memory locations, $7D27 and $7D28 in our example. These memory locations can be easily changed by another part of the program. The subroutine approach makes changing the speed of the motor easier.

There is also a *branch to subroutine* (BSR) *instruction*. It does about the same thing as JSR but uses relative addressing. Its use is preferred if the program will have to be moved around in memory. In our example program, the JSR instruction refers to an absolute address of $7D26. If the main program and the subroutine are moved to another area of memory, all such absolute references will have to be changed. Programs that refer to absolute memory locations are said to be written in *position-dependent code*. If the two JSR instructions are replaced with BSR instructions, the program becomes *position-independent*. Since branch instructions use relative addressing, the subroutine would be properly called as long as it held the same relative position with reference to the main program. The address of the motor is also absolute, but this is not considered to be a problem since input-output (I/O) ports are usually fixed anyway.

The stepper motor program is trivial in that it barely taps the power of the microprocessor. Many additional features could be included: The motor could be reversed, or the pulses could be counted to allow precise positioning of some mechanism. Complex calculations could be performed on input data to position or control the speed of the motor according to other external conditions. Controlled acceler-

The delay subroutine is called twice on each pass through the main program. It is located at $7D26:

Memory Location	Hex Contents	Mnemonics	Cycles
7D26	8E	LDX	3
7D27	FF		
7D28	FF		
7D29	30	LEAX	4
7D2A	1F		
7D2B	12	NOP	2
7D2C	12	NOP	2
7D2D	12	NOP	2
7D2E	12	NOP	2
7D2F	26	BNE	3
7D30	F8		
7D31	39	RTS	5

The first instruction in the delay subroutine loads the X register immediately with $FFFF. The next ination could be achieved by shortening the delay time after each pulse. Deceleration would be another possibility. Several motors could be synchronized, and they could be synchronized with other hardware. Any or all of these features are easily within the capability of a microprocessor, and they can be achieved mainly with software.

Microprocessor control of motors provides an accurate and cost-effective solution to many industrial problems. However, something can always go wrong. For example, a motor may stall if the load is too large. A set screw may loosen, or a key may shear, allowing a gear to slip on a shaft. In cases such as these, the mechanism will not be positioned where the processor "thinks" it is. One way to handle potentially damaging and dangerous situations is to use the interrupt capability of the microprocessor. The MC6809 has three hardware interrupt pins: the *nonmaskable interrupt input* (NMI), the *interrupt request input* (IRQ), and the *fast interrupt request input* (FIRQ). A limit switch or an array of limit switches can be placed on the mechanism. If the mechanism exceeds one of these limits, a signal can be sent to one of the interrupt inputs of the processor to alert it to a potentially dangerous condition.

Except for a reset signal, the NMI signal has the highest priority. A negative logic signal applied to this pin demands that a nonmaskable interrupt sequence be generated. As its name indicates, this input cannot be masked by software. Figure 12-8 shows the reaction of the MC6809 to this interrupt signal. The current instruction cycle is finished first. Next, the E flag is set, indicating that the entire internal register set will be pushed to the hardware stack. Assuming that the S register contains $100F at the time of the interrupt, the registers are pushed, starting at memory location $100E and ending at location $1003. Note that the S register itself is not stacked. Next the F and the I flags are set. The F flag masks a fast interrupt request, and the I flag masks an interrupt request. This means that a lower-priority interrupt cannot interfere with the processing of the nonmaskable interrupt. Next, the BA pin on the processor is set low, and the BS pin is set high. Through this logic combination at pins 5 and 6 the processor provides a hardware acknowledgment of the interrupt. Then the NMI vector (a *vector* is an address) is fetched from memory locations $FFFC and $FFFD. Suppose the contents at these two locations are $45 and $C1, respectively. The program counter will be loaded with $45C1. Now the BS pin is set low, which signifies the normal or running mode, and the processor will begin executing the program that starts at $45C1. The program must end with the *return from interrupt instruction* (RTI). Execution of the RTI instruction will pull all of the internal registers from the hardware stack. This will cause the main program to be reentered at the point where the interrupt occurred with all internal registers restored to their original condition.

The interrupt service routine that begins at $45C1

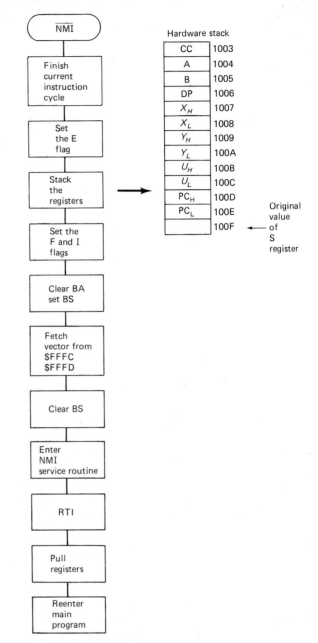

Fig. 12-8 NMI processing flowchart.

could do any number of things, depending on the particular situation. For example, if a limit switch generated the interrupt, the routine might first power down all motors, then engage an electromechanical brake, sound an alarm, and display an appropriate message to an operator. Of course, not all interrupts require such drastic action. An interrupt may be generated by a sensor circuit to alert the processor that data are available at some input port. The service routine may only be required to read the data and store it in memory in this case. Interrupt service routines can be located at almost any location in memory. All that is required is that the proper vectors be stored in high memory. The interrupt vectors for the MC6809 must be stored at the following locations:

$FFFE & $FFFF ⟵ RESTART VECTOR
$FFFC & $FFFD ⟵ NMI VECTOR
$FFFA & $FFFB ⟵ SWI VECTOR
$FFF8 & $FFF9 ⟵ IRQ VECTOR
$FFF6 & $FFF7 ⟵ FIRQ VECTOR
$FFF4 & $FFF5 ⟵ SW12 VECTOR
$FFF2 & FFF3 ⟵ SW13 VECTOR

Even though the vector locations are fixed, those of the various service routines are not. For example, the NMI routine can be located beginning at $01FF by storing this address at $FFFC and $FFFD. The decisions to locate which routines where in memory are normally made early in the design of a microprocessor system, and the interrupt vectors are stored permanently in read-only memory (ROM). The service routines themselves may be located in ROM or in RAM.

An interrupt request (IRQ) will be ignored if the I flag is set. This gives the programmer a way to ensure that some time-sensitive routine will not be interrupted by a low-priority event. If the I flag is clear, a logic zero applied to the IRQ pin will initiate the service routine. This routine is similar to the one shown in Fig. 12-8, except that the F flag is not set, and the vectors are fetched from $FFF8 and $FFF9. A *fast interrupt request* (FIRQ) has a higher priority than an IRQ, which means that an FIRQ signal can interrupt the processing of an IRQ. An FIRQ is also maskable by setting the F flag. If the flag is cleared, an FIRQ signal will initiate the FIRQ service routine. In this case, the E flag is cleared, and only the contents of the program counter and the condition code register are saved to the stack. This saves time and allows the processor to provide quicker interrupt service. Both the F and I flags are set to prevent an IRQ or a second FIRQ from interrupting. The IRQ and FIRQ routines also end with the RTI instruction. In the case of the FIRQ, only the condition code register and the program counter will be pulled from the stack since the E flag is low.

A *reset signal* is used to initialize a microprocessor system. A low-going signal on pin 37 will cause the processor to abort the current instruction cycle. The direct page register is cleared, and the restart vector is fetched from $FFFE & $FFFF.

The MC6809 also has three software interrupts: SWI, SWI2, and SWI3. These work in much the same way as the hardware interrupts but are generated by software. All of the processor registers are pushed on the hardware stack (with the exception of the S pointer itself), and control is transferred through the appropriate vector. Both the I and F flags are set so the processor will ignore IRQ and FIRQ signals while the software interrupt is being processed. Software interrupts are normally used during design and development of the microprocessor system. They can be used to simulate hardware interrupts and are also useful in software debugging.

A microprocessor may be idle. The MC6809 has an instruction with the mnemonic *CWAI*, which ANDs an immediate byte with the condition code register, stacks the entire machine state on the S stack, and waits for an interrupt. The immediate byte can be used to clear the I or the F flags to allow the processor to respond to selected interrupts (but not software interrupts). When an NMI occurs, no further stacking will be required before vectoring off to the service routine. An FIRQ will enter its interrupt routine with the entire processor state saved, and the RTI will automatically return the entire processor state after testing the E flag. The CWAI instruction is used to initiate an idle mode in anticipation of an interrupt. Since the stacking is already complete, the interrupt will be serviced faster. The *SYNC* instruction is similar to CWAI, except that the machine state is not stacked and no vector is fetched. When a SYNC instruction is executed, the processor stops and waits for an interrupt. When an interrupt occurs, the synchronizing state is cleared; processing continues, depending on the condition of the F and I flags.

REVIEW QUESTIONS

22. Is there any limit to the number of locations in a main program where a subroutine can be called?

23. Which internal processor register(s) are stacked in the execution of the JSR or BSR instructions?

24. Which processor register points to the hardware stack?

25. Subroutines called by other subroutines are said to be _____.

26. The S register is _____ every time a byte is pushed onto the hardware stack.

27. What will usually happen if the stack overflows or is overwritten?

28. Refer to the stepper motor program. What would happen if accumulator A were loaded with $09 and accumulator B were loaded with $03?

29. Refer to the delay subroutine. Calculate the total delay if the four NOPs are removed from the program. You may assume a 1-MHz clock.

30. Identify the mnemonic of the instruction that should be used to call subroutines if the program must run from several areas of memory.

12-5
SYSTEM DESIGN

A microprocessor must be teamed up with other devices in order that it may serve as a useful product. Figure 12-9 shows a simplified block diagram of a microcomputer or microcontroller. The *read-only memory* (ROM) is required to store permanently the program or programs that are necessary for system

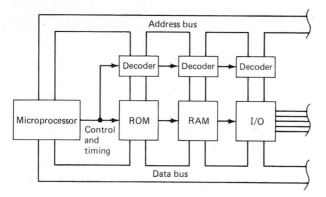

Fig. 12-9 A microcomputer or microcontroller.

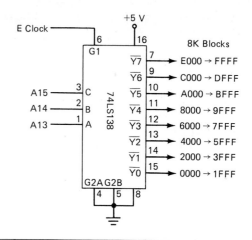

Fig. 12-11 Using a 74LS138 for address decoding.

Hex	0	2	4	6	8	A	C	E
Binary	000X	001X	010X	011X	100X	101X	110X	111X
Output	Y0	Y1	Y2	Y3	Y4	Y5	Y6	Y7

X = Don't care

operation. The *random access memory* (RAM) is used to store variables, data, and, in some cases, programs. Programs can be run from ROM or from RAM. In smaller, dedicated microcomputers or microcontrollers, the programs are usually stored in ROM. Some means for input and output are also required for the microprocessor to be functional. The I/O capability provides a way to enter programs, data, and control parameters, and it also allows for output to terminals, printers, motors, relays, and displays.

Figure 12-9 shows that decoders sit on the address bus to allow the microprocessor to access the major parts of the system selectively. It also shows that timing and control signals are required to synchronize the flow of data among the various devices. For example, the microprocessor read/NOT write line must connect to RAM and all other devices involved in bidirectional data transfer. All system components sit on the data bus, and the selective decoding and synchronizing signals allow for orderly data transfers.

A *memory map* is an important item when working with a microprocessor-based system. Figure 12-10 is an example; it shows that RAM extends from address $0000 to $1FFF for a total of 8K bytes. The I/O extends from $8000 to $9FFF, ROM ranges from $E000 to $FFFF, and each consumes another 8K bytes. This leaves 40K bytes unused out of the available 64K. Small, dedicated microcomputers and microcontrollers usually do not need 64K of address

space. Empty sockets may be provided to expand ROM or RAM in some cases.

Address decoders may be used to break the memory map up into sections; one method of accomplishing this is shown in Fig. 12-11. A 74LS138 decodes the three most significant bits of the address bus: A13, A14, and A15. One of the eight outputs of this IC will go low when it is enabled; which one does, depends on the bit pattern at its inputs. The chart in Fig. 12-11 shows the most significant hex characters of the address bus from 0 through E. When any address from $0000 through and including $1FFF appears on the bus, the Y0 output of the decoder will go low if it is enabled. When any address from $2000 through and including $3FFF appears on the bus, the Y1 output of the decoder will go low if it is enabled. The decoder will enable one 8K byte section of memory at a time. This is a particularly effective

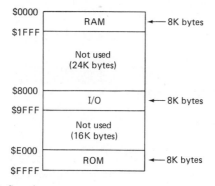

Fig. 12-10 Sample memory map.

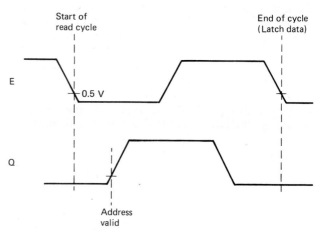

Fig. 12-12 *E* and *Q* clock signals.

Hex	8				1				3				F			
Binary	1	0	0	0	0	0	0	1	0	0	1	1	1	1	1	1
Address pin	A15	A14	A13	A12	A11	A10	A9	A8	A7	A6	A5	A4	A3	A2	A1	A0

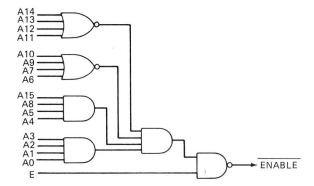

Fig. 12-13 An address decoder for $813F.

technique when used in conjunction with 8K memory devices such as the 68364 and the 2764, which are discussed in Chapter 11. For example, suppose an 8K 2764 EPROM is to be mapped into the top of memory. All that is required is to connect the Y7 output of the decoder shown in Fig. 12-11 to the output enable and chip enable pins of the EPROM. You may recall that these are active low, and the EPROM will be selected for any address from $E000 to $FFFF. The rest of the address bus (A0 through A12) will connect directly to the EPROM for internal decoding of its 8K byte space.

Figure 12-11 also shows that the E clock signal from the microprocessor is connected to the active high enable input of the 74LS138 decoder. This connection ensures that one of the 8K blocks will be selected only when the E clock is high. Figure 12-12 shows why this is necessary. A read cycle begins when the E clock drops to 0.5 V. The address bus takes time to set up and is not valid until the Q clock signal rises to 0.5 V. Then the E signal goes high. If the E clock is used to enable the decoder, no device will be selected during the time the address bus is setting up. The data are read (latched) into the microprocessor on the falling edge of E.

The timing diagram of Fig. 12-12 shows that only one-half clock cycle is available for a memory device to be read. This can present a probem in some systems. The MC68B09 microprocessor can be operated

at up to 2 MHz. The period is found by taking the reciprocal of the frequency:

$$t = \frac{1}{f}$$
$$= \frac{1}{2 \times 10^6}$$
$$= 500 \text{ ns}$$

Since only half that time is available, memory must be read in 250 ns. One solution is to use fast memory devices, but this technique is expensive. Another solution is to form an extended clock window by ORing the E and Q clock signals. The address is valid with the rising edge of Q, and some systems use this technique to gain another quarter cycle of read time. It is also possible to use the MRDY input (pin 36) of the processor to allow slow memory devices to stretch the clock signals and provide extra time for data access.

Block decoding is not always adequate. It may be necessary to decode a single address to allow the microprocessor to access a data port, a relay, or an analog-to-digital converter. Examples of the way a single address may be decoded are given in Figs. 12-13 and 12-14. The address in Fig. 12-13 is $813F. Decoder design begins with the conversion from hex to binary. The table shows the state of the various address pins for $813F. Two four-input NOR gates are used for address lines that are low. The NOR outputs

Hex	8				X				X				F			
Binary	1	0	0	0	X	X	X	X	X	X	X	X	1	1	1	1
Address pin	A15	A14	A13	A12	X	X	X	X	X	X	X	X	A3	A2	A1	A0

X = Don't care

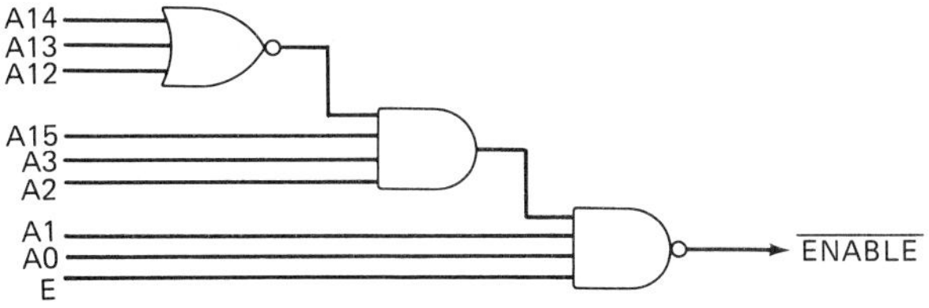

Fig. 12-14 A partial decoder for $8XXF.

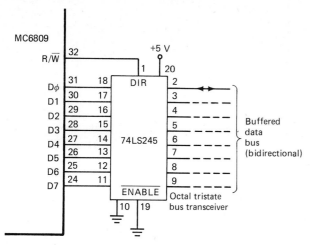

Fig. 12-15 Buffering the data bus.

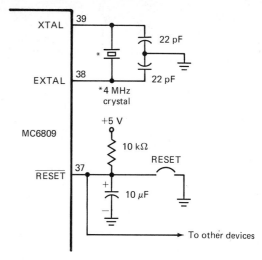

Fig. 12-17 Clock and reset circuits.

will be high only when all of the NOR inputs are low. Three four-input AND gates are used to combine the NOR outputs with all those address lines that are high for $813F. Finally, a two-input NAND gate is used to produce a low enable signal when the address pattern is correct and the E clock is high.

The address $813F is fully decoded by the circuit shown in Fig. 12-13. Of all 65,536 addresses that can occur, this is the only one that can produce an enable

signal. Full decoding is not necessary in many industrial systems. We have already learned that the memory map is seldom full. Figure 12-14 shows how partial decoding can simplify circuit design. The same address is decoded, but the circuit is designed with don't care bits on the address bus. This arrangement allows three gates to satisfy the design requirements. The same rules are used. The NOR gate handles those bits that are low, and the AND and NAND gates take care of the high bits and the E clock. The price that must be paid for partial decoding is that many addresses will produce an enable signal. The circuit of Fig. 12-14 will decode $812F, $800F, $823F, $8FFF, and many others.

The MC6809 microprocessor is rated to source -205 μA and to sink 2 mA at its data pins. This makes it necessary to buffer the data bus in some industrial applications. Figure 12-15 shows a 74LS245 octal tri-state bus transceiver IC. It is called a *bus transceiver* since it is bidirectional. The state of pin 1 determines whether certain other pins are inputs or outputs. When pin 1 is high, pins 2 through 9 are inputs, and pins 11 through 18 are outputs. This status reverses by taking pin 1 low. Note that pin 1 is controlled by the read/NOT write line of the microprocessor. The 74LS245 is housed in a 20-pin dual-inline package and can source -15 mA and sink 24 mA. It greatly improves the data bus capacity of the system.

The source current rating of the MC6809 microprocessor address pins is only -145 μA. Thus, it may also be necessary to buffer the address bus. Two 74LS244 octal tri-state bus driver ICs are pictured in Fig. 12-16. Transceivers are not required here because the microprocessor address pins are always outputs. The driver ICs greatly improve the address bus capacity. They also improve the noise margin since each input of the 74LS244 has 400 mV of hysteresis. The driver outputs are permanently enabled in Fig. 12-16 because pins 1 and 19 are

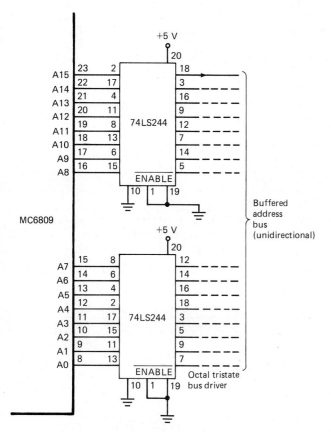

Fig. 12-16 Buffering the address bus.

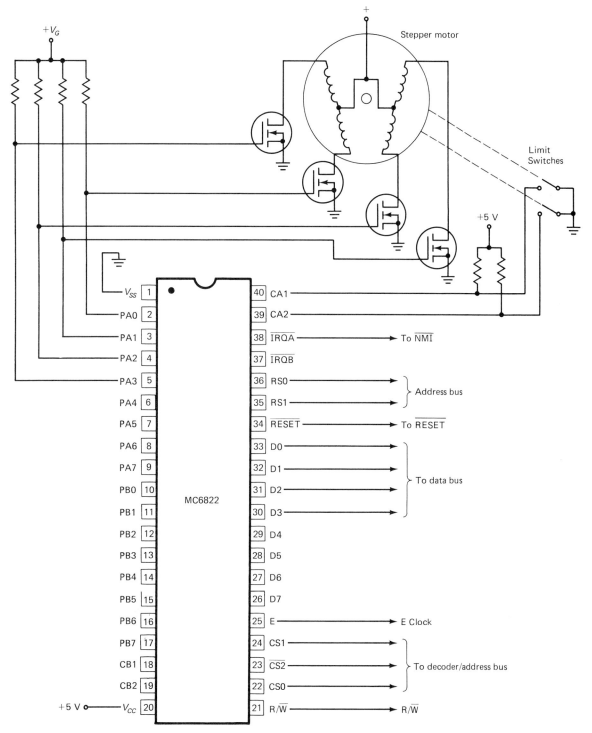

Fig. 12-18 Stepper motor interface.

grounded. However, it is possible to disable the outputs and tri-state the address bus. This would be done in those systems in which another device or circuit required direct memory access (DMA). Bus contention would be avoided by tri-stating the bus drivers during the DMA period.

Another aspect of microprocessor system design is the clock circuit. The MC6809 has an internal clock

oscillator. An external crystal and two capacitors are all that is required for a complete clock circuit. The connections are indicated in Fig. 12-17. The crystal frequency must be four times the desired clock frequency. An external clock oscillator can be used by grounding pin 39 and feeding a TTL-compatible clock signal into pin 38. The crystal and capacitors will be eliminated from the wiring in this case.

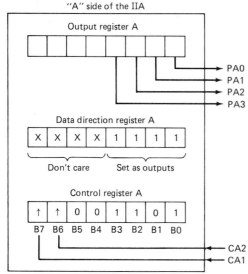

"A" side of the IIA

Output register A

→ PA0
→ PA1
→ PA2
→ PA3

Data direction register A

| X | X | X | X | 1 | 1 | 1 | 1 |

Don't care   Set as outputs

Control register A

| ↑ | ↑ | 0 | 0 | 1 | 1 | 0 | 1 |
B7 B6 B5 B4 B3 B2 B1 B0

← CA2
← CA1

CONTROL REGISTER BITS

B0 = 1:   Enables IRQA interrupt by CA1
B1 = 0:   Select neg. edge on CA1
B2 = 1:   Output register selected
B3 = 1:   Enables IRQA interrupt by CA2
B4 = 0:   Select neg. edge on CA2
B5 = 0:   Establish CA2 as an input
B6 ↑ :   Set high by neg. edge on CA2 (cleared by reading output register)
B7 ↑ :   Set high by neg. edge on CA1 (cleared by reading output register)

Fig. 12-19 The IIA registers.

Also shown by Fig. 12-17 is a typical reset circuit for the microprocessor. Pressing the reset switch will ground pin 37 and initiate the reset procedure. Pin 37 is a Schmitt-trigger input, and the reset signal must be present for more than one bus cycle. During initial power on, the reset input must be held low until the clock oscillator is fully operational. It may take as long as 20 ms for this to happen; it is provided for in Fig. 12-17 by an *RC* time delay network. The 10,000-Ω resistor and the 10-μF capacitor have a time constant of 100 ms. This ensures that reset will not be released before the clock is operational. Because of the Schmitt-trigger input, the reset state will not be released until pin 37 reaches 4.0 V. This is in contrast to other devices in the system, which will recognize 2.4 V as a logic high. If all devices are connected to the same reset bus, the microprocessor will leave the reset state last. This feature is important because it guarantees that the processor will never begin communicating with another part of the system that has not completed its own reset procedure.

Figure 12-18 shows an example of a device that must be connected to the system reset bus, the MC6822 industrial interface adapter. This 40-pin IC provides a universal means of interfacing various kinds of peripheral equipment to the microprocessor. It features two 8-bit bidirectional data ports and four control lines. The data ports are programmable; any of the 16 can be used as inputs or outputs. The functional configuration of the industrial interface

adapter (IIA) is programmed by the microprocessor unit (MPU) during system initialization. System initialization begins immediately after a reset. For this reason the MPU must clear the reset state last.

The IIA features open drain outputs. The 16 port lines and 4 control lines can be pulled up externally to a maximum of 18 V. Level shifters are not required to interface directly with 15-V CMOS. Also, better noise margins are possible. Figure 12-18 shows four lines of port A (PA0 through PA3) configured as outputs. These four outputs are pulled up to $+V_G$, which is the gate supply for the four enhancement-mode VMOS transistors. The gate supply voltage must be high enough to saturate the transistors when any of the ports is at logic high. When the microprocessor writes a 0 to any of the ports, the gate voltage will drop to near 0 V, and the transistors will turn off. This allows the microprocessor to control the motor since the transistors provide the ground return for the motor coils.

The industrial interface adapter has two sides, A and B. Each side has three registers; the registers for the A side are in Fig. 12-19. The output register latches data written by the microprocessor and makes it available to the outputs PA0 through PA3. The stepper motor bit patterns will be written to output register A. The data direction register determines whether the ports will be inputs or outputs. Figure 12-19 shows that four logic 1s are stored in the lower half of the register. This programs PA0 through PA3 to serve as outputs. The upper 4 bits are "don't care" since these ports are not used. The contents of the control register enable the interrupts, select the negative edge for the interrupt inputs, select the output register, establish CA2 as an input, and reflect which input, if any, caused an interrupt. Initialization of the IIA after a system reset would involve writing binary XXXX1111 to the data direction register and then writing binary XX001101 to the control register. Subsequent writes would then go to the output register to step the motor. If an interrupt occurred, the interrupt service routine would read the contents of the control register and examine bits 7 and 6 to determine which limit switch tripped.

REVIEW QUESTIONS

31. Refer to Fig. 12-11. Suppose G2A of the decoder is not grounded and A15 of the address bus is connected to it. Also, assume that the A, B, and C inputs of the decoder are connected to A12, A13, and A14 of the bus, respectively. What is the lowest hex address that will enable decoder output Y0? The highest? What size blocks does this provide?

32. Refer to Fig. 12-12. When is the address bus invalid?

33. Refer to Fig. 12-12. Ignore propagation delay in the decoder. With a 1-MHz clock, how much time is available for memory access if the E clock is used for the enable signal? How much time is available if the enable signal is E OR Q?

34. Refer to Fig. 12-13. Suppose address lines A0 and A14 are interchanged. What address will be decoded? Is this the only address that will decode?

35. Refer to Fig. 12-14. What is the lowest hex address to which the decoder will respond? The highest? How many addresses will be decoded?

36. Refer to Fig. 12-18. Ignoring interrupt capabilities, how many stepper motors of the type shown could be controlled by one IIA?

37. Refer to Fig. 12-18. How would the microprocessor know which limit switch caused an interrupt?

38. Refer to Fig. 12-18. Ignoring interrupt capabilities, what simple device might be used to replace the IIA?

12-6
SUPPORT DEVICES

The industrial interface adapter presented in the last section is an example of a support device that makes interfacing a microprocessor to various other circuits and systems easier. A number of large-scale integrated circuits that are microprocessor-compatible have been developed. They connect directly to the data bus and control and timing lines and to some portion of the address bus. These support devices make it easier to apply microprocessors to industrial applications.

The *peripheral interface adapter* (PIA) is a popular support device. It is available in several styles from various IC manufacturers. The IIA already discussed is a variation of the PIA. Motorola's part number for their PIA is MC6821, and it has the same pin configuration as their IIA. The only major difference between the two is that the IIA has open drain outputs. Another PIA variation is the *versatile interface adapter* (VIA). Figure 12-20 shows the pinout for the R6522 VIA manufactured by Rockwell. This pinout is somewhat different from the PIA and IIA configurations. The VIA has only one interrupt output to the processor and only two chip select inputs. PIAs and IIAs have two interrupt outputs and three chip select inputs, making two more pins available for register selection. Note that the VIA has four: RS0 through RS3. Another minor difference is pin 25, which is labeled *phase 2* on the VIA. This input is the same as the E clock input on Motorola devices.

The VIA features two 8-bit bidirectional I/O ports, and each line can be programmed as an input or as an output. This is the same arrangement as in the PIA. However, PIAs can latch only output data, and the VIA is also capable of latching input data. This is an important feature when a microprocessor must be interfaced to a device which makes data available for only brief periods of time. The VIA can latch such data and hold them until the microprocessor is ready. The VIA also contains two 16-bit programmable counter/timers. Several I/O lines can be controlled directly from the interval timers to generate programmable-frequency square waves. External pulses can also be counted, and an interrupt can be generated to the processor when a predetermined count is reached. Finally, the VIA contains a shift register, which can be used to provide serial data communication.

Figure 12-21 shows the VIA block diagram. Starting at the right, note that input latches (IRA and IRB) are available for port A and for port B. We also find the familiar output registers and data direction registers for each port. A shift register is available and connects to CB1 and CB2. These pins can be used for serial data communication. The *handshake control section* allows data to be transferred in and out only when devices are ready. An analogy for handshaking would be asking people to slow down when they are giving you directions too quickly for you to assimilate them. Moving to the left, we find the two timers. The timers consist of counters and latches. The counters are divided into low bytes and high bytes. Timer 1 has both a high-byte latch and a low-byte latch; timer 2 has only a low-byte latch. Figure 12-21 also shows the function control and interrupt control sections of the VIA, which facilitate

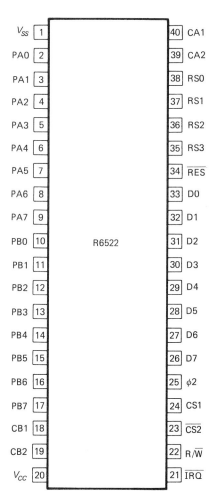

Fig. 12-20 VIA pinout.

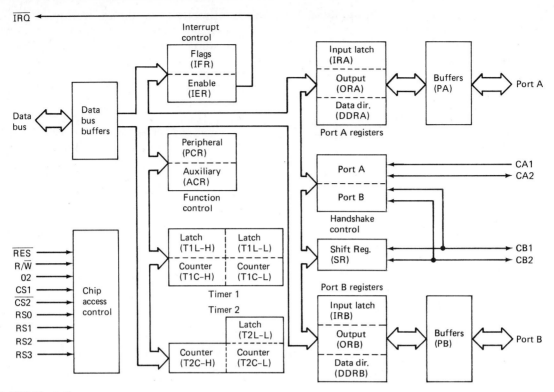

Fig. 12-21 VIA block diagram.

programming the many powerful features of this device.

The four register select lines of the VIA provide access to 16 internal registers. The register addressing is shown in Fig. 12-22. Normally RS0 through RS3 will connect to A0 through A3 of the address bus, respectively. The high-order bits of the address bus will normally connect to a separate decoder and to the two chip-select inputs of the VIA. The VIA can be mapped almost anywhere into memory. Timer 1 can be loaded by writing data to registers 6 and 7. After loading, the counter decrements at the phase

Register number	RS Coding				Register designation	Description	
	RS3	RS2	RS1	RS0		Write	Read
0	0	0	0	0	ORB/IRB	Output register B	Input register B
1	0	0	0	1	ORA/IRA	Output Register A	Input Register A
2	0	0	1	0	DDRB	Data direction register B	
3	0	0	1	1	DDRA	Data direction register A	
4	0	1	0	0	T1C-L	T1 Low-order latches	T1 Low-order latches
5	0	1	0	1	T1C-H	T1 High-order counter	
6	0	1	1	0	T1L-L	T1 Low-order latches	
7	0	1	1	1	T1L-H	T1 High-order latches	
8	1	0	0	0	T2C-L	T2 Low-order latches	T2 Low-order latches
9	1	0	0	1	T2C-H	T2 High-order counter	
10	1	0	1	0	SR	Shift Register	
11	1	0	1	1	ACR	Auxiliary control register	
12	1	1	0	0	PCR	Peripheral control register	
13	1	1	0	1	IFR	Interrupt flag register	
14	1	1	1	0	IER	Interrupt enable register	
15	1	1	1	1	ORA/IRA	Same as register 1 except no "handshake"	

Fig. 12-22 VIA register addressing.

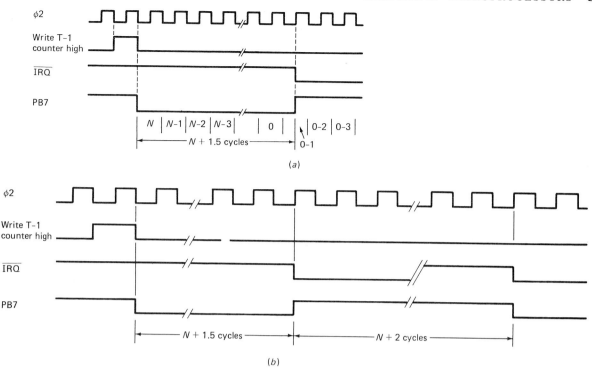

Fig. 12-23 Timer 1 operating modes. (*a*) One-shot operation. (*b*) Free-run operation.

2 clock rate. Upon reaching zero, an interrupt flag is set, and IRQ goes low if the T1 interrupt is enabled. Timer 1 then disables any further interrupts or automatically transfers the contents of the latches to the counter and begins decrementing from the loaded value again. The timer may also be programmed to invert the output signal on PB7 on every occasion that it times out.

Timer 1 can be used in a one-shot mode or in a free-running mode. The *one-shot mode* generates a single interrupt for every timer load operation. The delay between the write to the high byte of the counter and the generation of the processor interrupt is a function of the 16-bit data word loaded into the counter. Timer 1 can be programmed to produce a single negative pulse on the PB7 peripheral pin in addition to generating a single interrupt. If bit 7 in the auxiliary control register is set, a write to the high byte of counter 1 will cause PB7 to go low (see Fig. 12-23[*a*]). When timer 1 times out, PB7 returns high. This provides a single programmable-width output pulse for controlling some device or circuit interfaced to the microcomputer or microcontroller.

Time-out in the one-shot mode sets the timer 1 interrupt flag, and the IRQ pin goes low. The timer then continues to decrement from zero at the system clock rate, allowing the processor to read the counter and determine the time since the interrupt. When the processor again writes into the high-order counter, the T1 interrupt flag will be cleared, the contents of the low-order latch will be transferred into the low-order counter, and the timer will once again begin to decrement from the value loaded.

Suppose some industrial system requires single pulses of 35,768 μs in length. First, bits 6 and 7 of the auxiliary control register would have to be set to 0 and 1 respectively. This procedure will select the one-shot mode and enable PB7. Also, bit 7 of the data direction register for port B must be set to enable PB7 as an output. Assuming a 1-MHz clock, each clock cycle will be 1 μs in length. Therefore timer 1 must be loaded with the binary equivalent of 35,768 minus 1 or 2. This is because PB7 will go low for $N + 1.5$ cycles as shown in Fig. 12-23(*a*). The pulses produced will be 0.5 μs more or less than the required value. This error is of no consequence in most industrial applications. The hex equivalent of decimal 35,767 is $8BB7. Next, the processor will write $B7 to register 4. This operation stores the lower byte in timer 1's low-order latch. Finally, the processor will write $8B to register 5, which starts the timing pulse. Next PB7 will go low and then go high 35,768.5 μs later. Subsequently, one of these pulses can be produced every time the processor writes $8B to register 5. It is not necessary to rewrite the low byte because it has been latched.

The *free-run mode* takes more full advantage of the latches associated with timer 1. Every time the counter reaches zero, the contents of the latches are transferred into the counter, and the counter starts decrementing from N again. This is in contrast to the one-shot mode, in which the count decrements from 0 after a time-out. Figure 12-23(*b*) shows the free-run mode of operation for timer 1. Suppose register 6 (the low-order latch) contains $3C, and register 7 (the high-order latch) contains $1A. The decimal

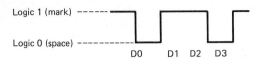

Fig. 12-24 Serial (mark-space) data format.

equivalent of $1A3C is 6716. The waveforms show one complete cycle of the output waveform to be $N + 1.5$ clock cycles $+ N + 2$ clock cycles. With a 1-MHz clock, the total time for one output cycle will be $6716 + 1.5 + 6716 + 2 = 13,435.5$ μs. The reciprocal of this yields 74.43 Hz. Thus, a nearly square 74.43 Hz waveform will be available if PB7 is enabled.

The processor can access the latches during the down-counting operation by writing to registers 6 and 7. This process will not affect the time-out in process. Instead, the data written to the latches will determine the next time-out period. Since the interrupt flag is set with each time-out, the processor can respond with new data for the latches and set the period for the next half cycle. This characteristic enables very complex waveforms to be generated at the PB7 output.

Timer 2 in the VIA also has a one-shot mode similar to that discussed for timer 1. It will provide a single interrupt for each write to the high byte of counter 2. After it times out, the count will continue to decrement. The interrupt flag is disabled and will not be set again even if the counter decrements to zero again. The processor must write to the high byte to enable the setting of the interrupt flag. The flag is cleared by reading the low byte of the counter or writing the high byte.

Timer 2 also has a mode for counting the number of negative pulses applied to PB6. This is accomplished by loading a number into timer 2. Writing the high byte clears the interrupt flag and allows the counter to decrement on each negative pulse. When the counter decrements past zero, the interrupt flag is set. The counter then continues decrementing on

every negative pulse. It is necessary to rewrite the high byte to allow the interrupt flag to be set on a subsequent time-out.

Many details of the VIA are not covered here. The reader is referred to the manufacturers' data manuals for additional information. Anyone who works with microprocessor-based systems must have access to these data manuals for hardware and programming information.

The support devices covered to this point are mainly concerned with parallel data transfers. *Serial data transfers* are also very important in industrial systems. Many support devices have been developed to facilitate serial I/O. The earliest among these was the *universal asynchronous receiver/transmitter* (UART). *Universal synchronous/asynchronous receiver/transmitters* (USART) are also available. We will look at a more modern serial support device, the *asynchronous communications interface adapter* (ACIA). First, however, some basics concerning serial I/O will be presented.

Parallel I/O is fast but requires a cable with many circuits. It is usually limited to distances of approximately 8 m (26.25 ft). Serial data communications circuits provide simplified wiring and the capacity for modulation. *Modulation* is a process of using the data to control the amplitude or the frequency of a high-frequency signal called a *carrier*. The carrier signal can be in the audio spectrum, allowing serial data transfers over ordinary telephone circuits. A *modulator* will be required to change the data into audio signals, and a *demodulator* will be used to change the audio signals back into data. Both circuits are usually contained in one unit called a modulator-demodulator, or *modem*.

Figure 12-24 shows the mark-space format used to transfer serial data. Logic 1 is called *mark* and represents some current level or some voltage level, depending on the standard used. Logic 0 is called *space* and represents 0 current or some voltage level, again depending upon which standard is in use. The serial line is held at mark when the equipment is turned on but no data are being transmitted. When

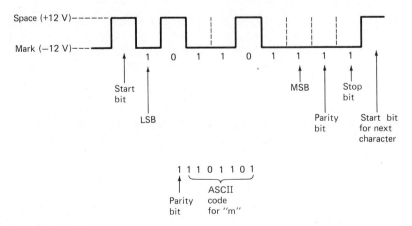

Fig. 12-25 RS-232C transmission of *m* with even parity and 1 stop bit.

RS-232 Interface

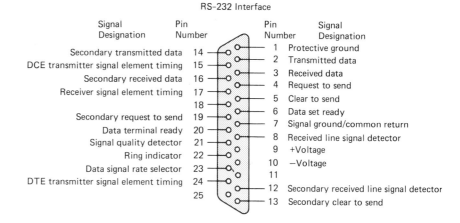

Signal Designation	Pin Number		Pin Number	Signal Designation
Secondary transmitted data	14		1	Protective ground
DCE transmitter signal element timing	15		2	Transmitted data
Secondary received data	16		3	Received data
Receiver signal element timing	17		4	Request to send
	18		5	Clear to send
Secondary request to send	19		6	Data set ready
Data terminal ready	20		7	Signal ground/common return
Signal quality detector	21		8	Received line signal detector
Ring indicator	22		9	+Voltage
Data signal rate selector	23		10	−Voltage
DTE transmitter signal element timing	24		11	
	25		12	Secondary received line signal detector
			13	Secondary clear to send

Fig. 12-26 RS-232C connector wiring.

a serial circuit uses current rather than voltage, the mark current will be either 60 or 20 mA. Both of these are older standards and are less popular than they once were. Several voltage standards exist, and the EIA RS-232C system is the most popular. It allows any value from -3 to -25 V to represent mark and from $+3$ to $+25$ V to represent space. Figure 12-25 shows how the transmission of the ASCII code for *m* would look in a typical RS-232C system. The waveform is shown as it would appear on an oscilloscope (mark is negative and is at the bottom of the waveform). Note that the least significant bit is sent first. This is an example of asynchronous transmission, which is the most popular. Start and stop bits are required to frame the data word. An alternative is to use *synchronous data transmission,* which requires a common clock at both ends of the communication circuit. Special characters are sent during idle periods to keep the clocks synchronized. A synchronous data transmission begins with a *preamble* consisting of a fixed number of mark bits that allow the receiver to lock onto the characters. Since no framing bits are needed, synchronous transmission is approximately 20 percent faster.

Serial transmission speed is rated in baud (Bd); 1 Bd is equal to 1 bit/s. Common speeds are 300, 600, 1200, 2400, and 4800 Bd, and so on. If the rate is 4800 Bd, how many characters are sent per second? In asynchronous circuits, it depends on the format used. As an example, Fig. 12-25 shows that 10 bits are required to send one ASCII character. Therefore, the data rate will be 480 characters per second at 4800 Bd. If 2 stop bits are used, the data rate will be lower. If no parity bit is sent, the data rate will be higher.

Figure 12-26 shows the standard RS-232C connector wiring. It is common to find a female DB-25 connector on a data terminal and a male DB-25 connector on the data cable. Most installations use far fewer than 25 wires. In fact, it is possible to connect some data equipment with as few as three wires. The most important connections on the DB-25 connector are pin 2 (transmit data), pin 3 (receive data), and pin 7 (signal ground). Although it is common practice

to do so, it is not a good idea to eliminate the protective ground.

Other important serial communications standards exist in addition to RS-232C. One example of an increasingly popular standard is RS-422. This standard uses a balanced transmission line. Two data wires are required for each circuit. The bandwidth is very high, and data transfers up to several million bytes per second are possible. It is a good choice when very large amounts of data must be transferred in a short period of time.

Most microprocessors are parallel devices. The *asynchronous communications interface adapter* (ACIA) is an important support device that makes interfacing to serial devices such as data terminals, printers, modems, industrial controls, and sensors easy. Figure 12-27 represents the pin diagram for the

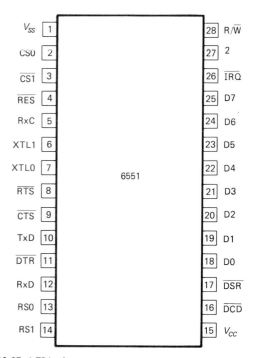

Fig. 12-27 ACIA pinout.

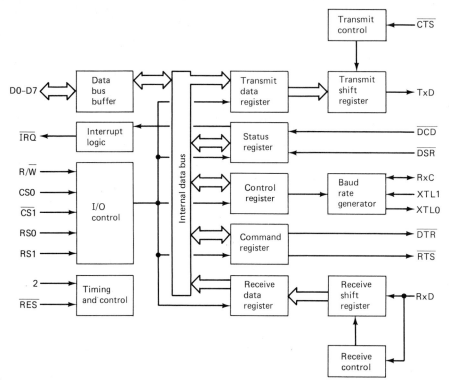

Fig. 12-28 ACIA block diagram.

Rockwell R6551 ACIA. This 28-pin IC contains an internal clock oscillator and can generate 15 rates ranging from 50 Bd to 19.2 kilobaud (kBd) under program control. A 1.8432-MHz crystal is ordinarily connected to pins 6 and 7 to control the frequency of the oscillator. The data word length is also programmable and can be 5, 6, 7, or 8 bits. Other programmable features include even, odd, or no parity and 1, 1.5, or 2 stop bits. Figure 12-28 shows the ACIA block diagram. Note that shift registers are

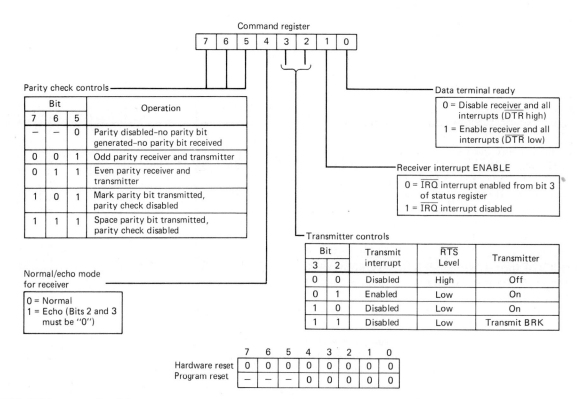

Fig. 12-29 ACIA command register.

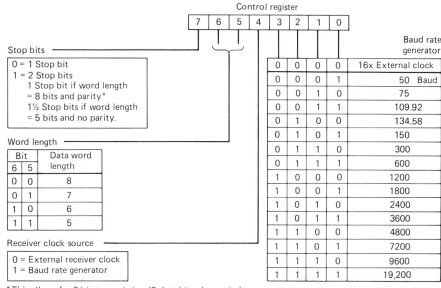

Fig. 12-30 ACIA control register.

used to convert parallel data to serial for transmission and to convert received serial data to parallel for use by the microprocessor.

The programmability of the ACIA lies in the structure of the command and control registers shown in Fig. 12-28. Figure 12-29 is a detailed look at the ACIA command register. Bit 0 sets pin 11 (data terminal ready) of the ACIA high or low, allowing the processor to inform other equipment (such as a modem) when the system is ready. Bit 1 enables or disables the receiver interrupt for incoming data. Bits 2 and 3 control the state of pin 8 (ready to send) and the transmit interrupt. Bit 4 controls the receiver echo mode. Bit 5 enables or disables parity generation and parity checking. Bits 6 and 7 select one of four parity modes: even, odd, mark, or space. *Mark parity* means that a parity bit is sent, but it is always mark (high). *Space parity* means the parity bit sent is always space (low). No parity checking is done on incoming data when mark or space parity is selected.

The control register is shown in Fig. 12-30. Bits 0 through 3 select the baud rate. Bit 4 selects the receiver clock source. If it is 0, the receiver will be clocked at a rate of one-sixteenth of an external clock applied to pin 5 of the ACIA. If it is 1, the receiver will operate at the same rate as the transmitter. Bits 5 and 6 select the word length (5, 6, 7, or 8 bits). Bit 7 selects the number of stop bits.

The ACIA also contains a status register that can be read by the processor. This register contains bits that signify whether an interrupt has occurred, the status of the signals applied to pins 16 and 17, whether or not the transmitter data register is empty, whether or not the receiver data register is full, whether there has been a parity error, or

whether there has been a *framing error* (incorrect number of stop bits received).

As with the VIA, some details of the ACIA are not covered here. Once again the reader is urged to consult data manuals for additional information. Quite a few other microprocessor support devices are found in the industrial environment, including cathode-ray tube controllers (CRTC), floppy disk controllers, printer controllers, and communications controllers for other data transmission standards.

As mentioned before, the most popular standard for serial data communications is RS-232C. Therefore, some standard parts have been developed to translate between TTL voltages and RS-232C voltages. Motorola, for example, manufactures the MC1488 quad driver and the MC1489 quad receiver. The driver ICs are used to convert TTL to RS-232C, and the receivers convert RS-232C to TTL. These ICs are in 14-pin dual-inline packages.

REVIEW QUESTIONS

39. Assume that a VIA is programmed for the one-shot mode and that PB7 is enabled as a timer-controlled output. When will PB7 first go low? When will it return high?

40. Refer to Fig. 12-23(*a*). Assume a 1-MHz clock and the VIA one-shot timer mode. What hex value should be stored in the low byte of timer 1 for an output pulse of 500.5 μs? What hex high byte must the processor write to initiate the pulse?

41. Refer to Fig. 12-23(*b*). Assume a 1-MHz clock and the VIA free-running timer mode. Also assume that the high-byte latch of timer 1 contains $00 and the low-byte latch contains $3E. What is the output frequency at PB7 assuming it is enabled? The waveform?

42. Which timer of the VIA is used for counting externally generated pulses? To which VIA pin must the pulses be applied?

43. Refer to Fig. 12-25. How many characters will be sent per second at 19.2 kBd?

12-7
SOFTWARE DEVELOPMENT

It should now be clear that microprocessors and their support devices are useless without software. What is the role of the industrial electronics technician in software development? It varies from company to company; large companies tend to hire software specialists for writing, applying, and maintaining software. Smaller companies may expect the technician to do some programming. In any case, the technician needs at least a general knowledge of microprocessor software and how it is developed.

Some industrial software is written in a high-level language such as BASIC. BASIC is one of the easiest computer languages to learn and apply. It will solve many industrial control, data collection, and analysis needs with ease. However, BASIC is not available on every microcomputer. Even if it is available, some control applications cannot be implemented in BASIC because it is too slow and requires more memory than equivalent machine language programs. There are other high-level languages gaining popularity in the industrial environment. FORTH is one example. FORTH is not so easy to learn and apply as BASIC, but it offers the advantages of consuming less memory and running much faster.

Machine language programs must be written in the processor's native code. When written by skilled programmers, they consume less memory than any equivalent high-level program and execute much faster. Sometimes, a compromise approach is used. An industrial control program for a robot may be written in BASIC with several calls to machine language subroutines. These subroutines will execute very quickly and provide the response speed required for time-sensitive robot actions and reactions.

How are machine language programs and subroutines written? It depends on the environment. Small programs and subroutines may be hand-coded, with the programmer looking up the OP-CODEs for each instruction. Hand coding places quite a bit of burden on the programmer, who must be familiar with the instruction set of the microprocessor and especially proficient with its addressing modes. Also, the programmer must calculate all relative addresses. Hand coding is considered adequate for small programs and subroutines. For large programs, it becomes tedious and is susceptible to errors.

Hand coding is made easier with programming aids such as the instruction set summary shown in Fig. 12-31. This summary is continued in Fig. 12-32, and the branch instructions are in a separate summary shown in Fig. 12-33. Look at the add with carry accumulator A instruction (ADCA) near the top of Fig. 12-31. This instruction may have any one of four different OP-CODEs, depending on the addressing mode to be used: $89 (immediate), $99 (direct), $A9 (indexed), and $B9 (extended). The ~ column shows the number of machine cycles required to fetch and execute each instruction. This information is vital for writing timing loops and for determining how fast software will execute. The # column tells how many memory bytes are needed for each instruction. This is useful for predicting how much memory a program will require and helps eliminate hand-coding errors. For example, if you wish to use ADCA with extended addressing, the 3 in the # column tells you that a total of 3 bytes will be required for the instruction. Two additional bytes must follow the OP-CODE $B9 in memory. These two bytes are the high-order address byte of the operand followed by the low-order address byte. The number of additional bytes and cycles required for the indexed addressing mode is indicated in Fig. 12-34 (page 348).

The instruction set summary also includes a description of what each instruction does and how the flags are affected. A dot under a flag means it is not changed, a 1 means the flag is set, a 0 indicates the flag is cleared, and a ↕ means the flag is toggled high or low depending on the results of the operation. Figure 12-32 lists the notes for those special flag results which require additional explanation. All of the information shown in Figs. 12-31 through 12-34 plus quite a bit more is contained on one fan-fold card. Motorola calls this a *reference card,* and the part number is M6809(AC3) for their MC6809 microprocessor. Motorola offers quite a few different microprocessors, and reference cards are available for most of them. Other manufacturers also have similar cards for their microprocessors; some call them *programmer's cards* or *programmer's aids.* You are urged to collect all relevant cards for your work. They provide a tremendous amount of information in a very compact form and are quite useful.

Reference cards go a long way toward making hand coding easier and error-free. However, if significant machine language development is required, then better tools are needed. One such tool is a program called an *editor*; it allows the programmer to write and correct a program written in assembly form easily. When this task is completed, a second program, the *assembler,* takes the assembly listing that was generated by the editor and converts it to machine language. Refer to Fig. 12-35 (page 348). *PAUSE* was typed into the editor by the programmer. So was everything to the right of it and below it up to the word *END.* This is usually called *source code.* Source code is typically divided into four columns: the *label* (for example, START), the *instruction mne-*

Instruction	Forms	Immediate Op	~	#	Direct Op	~	#	Indexed Op	~	#	Extended Op	~	#	Inherent Op	~	#	Description	5 H	3 N	2 Z	1 V	0 C
ABX														3A	3	1	B + X → X (Unsigned)	•	•	•	•	•
ADC	ADCA	89	2	2	99	4	2	A9	4+	2+	B9	5	3				A + M + C → A	↕	↕	↕	↕	↕
	ADCB	C9	2	2	D9	4	2	E9	4+	2+	F9	5	3				B + M + C → B	↕	↕	↕	↕	↕
ADD	ADDA	8B	2	2	9B	4	2	AB	4+	2+	BB	5	3				A + M → A	↕	↕	↕	↕	↕
	ADDB	CB	2	2	DB	4	2	EB	4+	2+	FB	5	3				B + M → B	↕	↕	↕	↕	↕
	ADDD	C3	4	3	D3	6	2	E3	6+	2+	F3	7	3				D + M M + 1 → D	•	↕	↕	↕	↕
AND	ANDA	84	2	2	94	4	2	A4	4+	2+	B4	5	3				A ∧ M → A	•	↕	↕	0	•
	ANDB	C4	2	2	D4	4	2	E4	4+	2+	F4	5	3				B ∧ M → B	•	↕	↕	0	•
	ANDCC	1C	3	2													CC ∧ IMM → CC					[7]
ASL	ASLA													48	2	1	A	8	↕	↕	↕	↕
	ASLB													58	2	1	B [shift left: C ← b7…b0 ← 0]	8	↕	↕	↕	↕
	ASL				08	6	2	68	6+	2+	78	7	3				M	8	↕	↕	↕	↕
ASR	ASRB													47	2	1	A	8	↕	↕	•	↕
	ASR													57	2	1	B [shift right: b7…b0 → C]	8	↕	↕	•	↕
	ASR				07	6	2	67	6+	2+	77	7	3				M	8	↕	↕	•	↕
BIT	BITA	85	2	2	95	4	2	A5	4+	2+	B5	5	3				Bit Test A (M ∧ A)	•	↕	↕	0	•
	BITB	C5	2	2	D5	4	2	E5	4+	2+	F5	5	3				Bit Test B (M ∧ B)	•	↕	↕	0	•
CLR	CLRA													4F	2	1	0 → A	•	0	1	0	0
	CLRB													5F	2	1	0 → B	•	0	1	0	0
	CLR				0F	6	2	6F	6+	2+	7F	7	3				0 → M	•	0	1	0	0
CMP	CMPA	81	2	2	91	4	2	A1	4+	2+	B1	5	3				Compare M from A	8	↕	↕	↕	↕
	CMPB	C1	2	2	D1	4	2	E1	4+	2+	F1	5	3				Compare M from B	8	↕	↕	↕	↕
	CMPD	10 83	5	4	10 93	7	3	10 A3	7+	3+	10 B3	8	4				Compare M M + 1 from D	•	↕	↕	↕	↕
	CMPS	11 8C	5	4	11 9C	7	3	11 AC	7+	3+	11 BC	8	4				Compare M M + 1 from S	•	↕	↕	↕	↕
	CMPU	11 83	5	4	11 93	7	3	11 A3	7+	3+	11 B3	8	4				Compare M M + 1 from U	•	↕	↕	↕	↕
	CMPX	8C	4	3	9C	6	2	AC	6+	2+	BC	7	3				Compare M M + 1 from X	•	↕	↕	↕	↕
	CMPY	10 8C	5	4	10 9C	7	3	10 AC	7+	3+	10 BC	8	4				Compare M M + 1 from Y	•	↕	↕	↕	↕
COM	COMA													43	2	1	Ā → A	•	↕	↕	0	1
	COMB													53	2	1	B̄ → B	•	↕	↕	0	1
	COM				03	6	2	63	6+	2+	73	7	3				M̄ → M	•	↕	↕	0	1
CWAI		3C	≥20	2													CC ∧ IMM → CC Wait for Interrupt					[7]
DAA														19	2	1	Decimal Adjust A	•	↕	↕	0	↕
DEC	DECA													4A	2	1	A − 1 → A	•	↕	↕	↕	•
	DECB													5A	2	1	B − 1 → B	•	↕	↕	↕	•
	DEC				0A	6	2	6A	6+	2+	7A	7	3				M − 1 → M	•	↕	↕	↕	•
EOR	EORA	88	2	2	98	4	2	A8	4+	2+	B8	5	3				A ⊻ M → A	•	↕	↕	0	•
	EORB	C8	2	2	D8	4	2	E8	4+	2+	F8	5	3				B ⊻ M → B	•	↕	↕	0	•
EXG	R1, R2	1E	8	2													R1 ↔ R2 [2]	•	•	•	•	•
INC	INCA													4C	2	1	A + 1 → A	•	↕	↕	↕	•
	INCB													5C	2	1	B + 1 → B	•	↕	↕	↕	•
	INC				0C	6	2	6C	6+	2+	7C	7	3				M + 1 → M	•	↕	↕	↕	•
JMP					0E	3	2	6E	3+	2+	7E	4	3				EA[3] → PC	•	•	•	•	•
JSR					9D	7	2	AD	7+	2+	BD	8	3				Jump to Subroutine	•	•	•	•	•
LD	LDA	86	2	2	96	4	2	A6	4+	2+	B6	5	3				M → A	•	↕	↕	0	•
	LDB	C6	2	2	D6	4	2	E6	4+	2+	F6	5	3				M → B	•	↕	↕	0	•
	LDD	CC	3	3	DC	5	2	EC	5+	2+	FC	6	3				M M + 1 → D	•	↕	↕	0	•
	LDS	10 CE	4	4	10 DE	6	3	10 EE	6+	3+	10 FE	7	4				M M + 1 → S	•	↕	↕	0	•
	LDU	CE	3	3	DE	5	2	EE	5+	2+	FE	6	3				M M + 1 → U	•	↕	↕	0	•
	LDX	8E	3	3	9E	5	2	AE	5+	2+	BE	6	3				M M + 1 → X	•	↕	↕	0	•
	LDY	10 8E	4	4	10 9E	6	3	10 AE	6+	3+	10 BE	7	4				M M + 1 → Y	•	↕	↕	0	•
LEA	LEAS							32	4+	2+							EA[3] → S	•	•	•	•	•
	LEAU							33	4+	2+							EA[3] → U	•	•	•	•	•
	LEAX							30	4+	2+							EA[3] → X	•	•	↕	•	•
	LEAY							31	4+	2+							EA[3] → Y	•	•	↕	•	•

Legend:

OP	Operation Code (Hexadecimal)	M̄	Complement of M	↕	Test and set if true, cleared otherwise
~	Number of MPU Cycles	→	Transfer Into	•	Not Affected
#	Number of Program Bytes	H	Half-carry (from bit 3)	CC	Condition Code Register
+	Arithmetic Plus	N	Negative (sign bit)	:	Concatenation
−	Arithmetic Minus	Z	Zero (Reset)	V	Logical or
•	Multiply	V	Overflow, 2's complement	∧	Logical and
		C	Carry from ALU	⊻	Logical Exclusive or

Fig. 12-31 MC6809 instruction set summary.

Instruction	Forms	Immediate Op	~	#	Direct Op	~	#	Indexed¹ Op	~	#	Extended Op	~	#	Inherent Op	~	#	Description	H	N	Z	V	C
LSL	LSLA													48	2	1	A B M diagram (shift left, b7←b0, 0)	•	↕	↕	↕	↕
	LSLB													58	2	1		•	↕	↕	↕	↕
	LSL				08	6	2	68	6+	2+	78	7	3					•	↕	↕	↕	↕
LSR	LSRA													44	2	1	A B M diagram (0→, shift right, b7 b0, c)	•	0	↕	•	↕
	LSRB													54	2	1		•	0	↕	•	↕
	LSR				04	6	2	64	6+	2+	74		3					•	0	↕	•	↕
MUL														3D	11	1	A × B → D (Unsigned)	•	•	↕	•	9
NEG	NEGA													40	2	1	$\overline{A}+1 \to A$	8	↕	↕	↕	↕
	NEGB													50	2	1	$\overline{B}+1 \to B$	8	↕	↕	↕	↕
	NEG				00	6	2	60	6+	2+	70	7	3				$\overline{M}+1 \to M$	8	↕	↕	↕	↕
NOP														12	2	1	No Operation	•	•	•	•	•
OR	ORA	8A	2	2	9A	4	2	AA	4+	2+	BA	5	3				A V M → A	•	↕	↕	0	•
	ORB	CA	2	2	DA	4	2	EA	+	2+	FA	5	3				B V M → B	•	↕	↕	0	•
	ORCC	1A	3	2													CC V IMM → CC			7		
PSH	PSHS	34	5+⁴	2													Push Registers on S Stack	•	•	•	•	•
	PSHU	36	5+⁴	2													Push Registers on U Stack	•	•	•	•	•
PUL	PULS	35	5+⁴	2													Pull Registers from S Stack	•	•	•	•	•
	PULU	37	5+⁴	2													Pull Registers from U Stack	•	•	•	•	•
ROL	ROLA													49	2	1	A B M diagram (rotate left, c b7 b0)	•	↕	↕	↕	↕
	ROLB													59	2	1		•	↕	↕	↕	↕
	ROL				09	6	2	69	6+	2+	79	7	3					•	↕	↕	↕	↕
ROR	RORA													46	2	1	A B M diagram (rotate right, c b7 b0)	•	↕	↕	•	↕
	RORB													56	2	1		•	↕	↕	•	↕
	ROR				06	6	2	66	6+	2+	76	7	3					•	↕	↕	•	↕
RTI														3B	6/15	1	Return From Interrupt					7
RTS														39	5	1	Return from Subroutine	•	•	•	•	•
SBC	SBCA	82	2	2	92	4	2	A2	4+	2+	B2	5	3				A − M − C → A	8	↕	↕	↕	↕
	SBCB	C2	2	2	D2	4	2	E2	4+	2+	F2	5	3				B − M − C → B	8	↕	↕	↕	↕
SEX														1D	2	1	Sign Extend B into A	•	↕	↕	0	•
ST	STA				97	4	2	A7	4+	2+	B7	5	3				A → M	•	↕	↕	0	•
	STB				D7	4	2	E7	4+	2+	F7	5	3				B → M	•	↕	↕	0	•
	STD				DD	5	2	ED	5+	2+	FD	6	3				D → M M+1	•	↕	↕	0	•
	STS				10 DF	6	3	10 EF	6+	3+	10 FF	7	4				S → M M+1	•	↕	↕	0	•
	STU				DF	5	2	EF	5+	2+	FF	6	3				U → M M+1	•	↕	↕	0	•
	STX				9F	5	2	AF	5+	2+	BF	6	3				X → M M+1	•	↕	↕	0	•
	STY				10 9F	6	3	10 AF	6+	3+	10 BF	7	4				Y → M M+1	•	↕	↕	0	•
SUB	SUBA	80	2	2	90	4	2	A0	4+	2+	B0	5	3				A − M → A	8	↕	↕	↕	↕
	SUBB	C0	2	2	D0	4	2	E0	4+	2+	F0	5	3				B − M → B	8	↕	↕	↕	↕
	SUBD	83	4	3	93	6	2	A3	6+	2+	B3	7	3				D − M M+1 → D	•	↕	↕	↕	↕
SWI	SWI⁶													3F	19	1	Software Interrupt 1	•	•	•	•	•
	SWI2⁶													10 3F	20	2	Software Interrupt 2	•	•	•	•	•
	SWI3⁶													11 3F	20	1	Software Interrupt 3	•	•	•	•	•
SYNC														13	≥4	1	Synchronize to Interrupt	•	•	•	•	•
TFR	R1, R2	1F	6	2													R1 → R2²	•	•	•	•	•
TST	TSTA													4D	2	1	Test A	•	↕	↕	0	•
	TSTB													5D	2	1	Test B	•	↕	↕	0	•
	TST				0D	6	2	6D	6+	2+	7D	7	3				Test M	•	↕	↕	0	•

Notes:

1. This column gives a base cycle and byte count.
2. R1 and R2 may be any pair of 8 bit or any pair of 16 bit registers.
   The 8 bit registers are: A, B, CC, DP
   The 16 bit registers are: X, Y, U, S, D, PC
3. EA is the effective address.
4. The PSH and PUL instructions require 5 cycles plus 1 cycle for each **byte** pushed or pulled.
5. 5(6) means: 5 cycles if branch not taken, 6 cycles if taken (Branch instructions).
6. SWI sets I and F bits. SWI2 and SWI3 do not affect I and F.
7. Conditions Codes set as a direct result of the instruction.
8. Value of half-carry flag is undefined.
9. Special Case — Carry set if b7 is SET.

Fig. 12-32 MC6809 instruction set summary (*continued*).

Instruction	Forms	Addressing Mode Relative			Description	5 H	3 N	2 Z	1 V	0 C
		OP	~	#						
BCC	BCC	24	3	2	Branch C = 0	•	•	•	•	•
	LBCC	10 24	5(6)	4	Long Branch C = 0	•	•	•	•	•
BCS	BCS	25	3	2	Branch C = 1	•	•	•	•	•
	LBCS	10 25	5(6)	4	Long Branch C = 1	•	•	•	•	•
BEQ	BEQ	27	3	2	Branch Z = 0	•	•	•	•	•
	LBEQ	10 27	5(6)	4	Long Branch Z = 0	•	•	•	•	•
BGE	BGE	2C	3	2	Branch ≥ Zero	•	•	•	•	•
	LBGE	10 2C	5(6)	4	Long Branch ≥ Zero	•	•	•	•	•
BGT	BGT	2E	3	2	Branch > Zero	•	•	•	•	•
	LBGT	10 2E	5(6)	4	Long Branch > Zero	•	•	•	•	•
BHI	BHI	22	3	2	Branch Higher	•	•	•	•	•
	LBHI	10 22	5(6)	4	Long Branch Higher	•	•	•	•	•
BHS	BHS	24	3	2	Branch Higher or Same	•	•	•	•	•
	LBHS	10 24	5(6)	4	Long Branch Higher or Same	•	•	•	•	•
BLE	BLE	2F	3	2	Branch ≤ Zero	•	•	•	•	•
	LBLE	10 2F	5(6)	4	Long Branch ≤ Zero	•	•	•	•	•
BLO	BLO	25	3	2	Branch lower	•	•	•	•	•
	LBLO	10 25	5(6)	4	Long Branch Lower	•	•	•	•	•
BLS	BLS	23	3	2	Branch Lower or Same	•	•	•	•	•
	LBLS	10 23	5(6)	4	Long Branch Lower or Same	•	•	•	•	•
BLT	BLT	2D	3	2	Branch < Zero	•	•	•	•	•
	LBLT	10 2D	5(6)	4	Long Branch < Zero	•	•	•	•	•
BMI	BMI	2B	3	2	Branch Minus	•	•	•	•	•
	LBMI	10 2B	5(6)	4	Long Branch Minus	•	•	•	•	•
BNE	BNE	26	3	2	Branch Z ≠ 0	•	•	•	•	•
	LBNE	10 26	5(6)	4	Long Branch Z ≠ 0	•	•	•	•	•
BPL	BPL	2A	2	2	Branch Plus	•	•	•	•	•
	LBPL	10 2A	5(6)	4	Long Branch Plus	•	•	•	•	•
BRA	BRA	20	3	2	Branch Always	•	•	•	•	•
	LBRA	16	5	3	Long Branch Always	•	•	•	•	•
BRN	BRN	21	3	2	Branch Never	•	•	•	•	•
	LBRN	10 21	5	4	Long Branch Never	•	•	•	•	•
BSR	BSR	8D	7	2	Branch to Subroutine	•	•	•	•	•
	LBSR	17	9	3	Long Branch to Subroutine	•	•	•	•	•
BVC	BVC	28	3	2	Branch V = 0	•	•	•	•	•
	LBVC	10 28	5(6)	4	Long Branch V = 0	•	•	•	•	•
BVS	BVS	29	3	2	Branch V = 1	•	•	•	•	•
	LBVS	10 29	5(6)	4	Long Branch V = 1	•	•	•	•	•

Fig. 12-33 MC6809 branch instructions.

monic (STA), the *operand* (#$49FF), and the *comment* (SET CARRY FLAG). The assembler prints the equivalent machine code for each line of instruction on the left. The first column on the left represents the program counter, or the address of the first byte of machine code for that instruction. The next two columns contain the actual machine code that can be executed by the processor: the OP-CODEs and the operands. Some lines do not have any machine code since they contain assembler directives for the assembler program's information only. Note

that the assembler also generates a symbol table which lists the addresses for all the labels and constants declared by the programmer.

An assembler makes the generation of machine language programs easier, faster, and more error-free. The programmer does not have to look up any OP-CODEs. The assembler "knows" the OP-CODEs for all the mnemonics and for all the various addressing modes. The programmer is free to assign label names that make sense and help keep track of program function and entry points. This also makes it easy to specify branches. For example, the *BCC clear* instruction in Fig. 12-35 specifies SCAN as the operand. *SCAN* is a label assigned earlier in the assembly listing. Look at the machine code columns to the left of BCC, and you will find $24 (the OP-CODE for BCC) followed by $E4, the relative address generated by the assembler. Assemblers also present some helpful messages if there are errors in the assembly program. A 0 error message indicates a good assembly. However, this is no guarantee that the machine code will execute as planned. There can still be logical errors, errors of omission, timing errors, and other types of errors in the code.

After assembly, another program, called a *debugger,* may be used to help find errors in the program. Debuggers allow the programmer to set breakpoints in a program. A *breakpoint* allows a part of the program to execute, before execution stops at that breakpoint. At this time the programmer can examine memory contents to determine whether the software is doing what was intended. Debuggers also allow single stepping, making it possible to check software operation on an instruction by instruction basis.

We have learned that many industrial computers are very small. They are dedicated to a special task or a group of tasks and often have limited memory. How can software be developed for these computers? Editors, assemblers, and debuggers often will not run on such small systems. Development is done on a larger computer in these cases. In fact, many microprocessor manufacturers offer a computer development system for these situations. After the programs are written, assembled, and debugged on the large system, they are downloaded to the small computers or may be burned into EPROMs and transferred to the small computer as ROM programs.

What can you do in the way of software development with a dedicated microcomputer or microcontroller if a development system is not available? Often, not very much; however, some small systems do have a monitor program stored in ROM to permit some minor software development and debugging. For example, Motorola offers ASSIST09 to support the MC6809 microprocessor. Monitors often only consume about 2K bytes of memory and can be included even in a small system. A serial I/O port and a terminal will normally be required to utilize a monitor program.

Figure 12-36 lists the commands available in the

Type	Forms	Non Indirect				Indirect			
		Assembler Form	Postbyte OP Code	+~	+#	Assembler Form	Postbyte OP Code	+~	+#
Constant Offset From R (twos complement offset)	No Offset	,R	1RR00100	0	0	[,R]	1RR10100	3	0
	5 Bit Offset	n, R	0RRnnnnn	1	0	defaults to 8-bit			
	8 Bit Offset	n, R	1RR01000	1	1	[n, R]	1RR11000	4	1
	16 Bit Offset	n, R	1RR01001	4	2	[n, R]	1RR11001	7	2
Accumulator Offset From R (twos complement offset)	A — Register Offset	A, R	1RR00110	1	0	[A, R]	1RR10110	4	0
	B — Register Offset	B, R	1RR00101	1	0	[B, R]	1RR10101	4	0
	D — Register Offset	D, R	1RR01011	4	0	[D, R]	1RR11011	7	0
Auto Increment/Decrement R	Increment By 1	,R+	1RR00000	2	0	not allowed			
	Increment By 2	,R++	1RR00001	3	0	[,R++]	1RR10001	6	0
	Decrement By 1	,-R	1RR00010	2	0	not allowed			
	Decrement By 2	,--R	1RR00011	3	0	[,--R]	1RR10011	6	0
Constant Offset From PC (twos complement offset)	8 Bit Offset	n, PCR	1XX01100	1	1	[n, PCR]	1XX11100	4	1
	16 Bit Offset	n, PCR	1XX01101	5	2	[n, PCR]	1XX11101	8	2
Extended Indirect	16 Bit Address	—	—	—	—	[n]	10011111	5	2

R = X, Y, U or S   X = 00   Y = 01
X = Don't Care   U = 10   S = 11

+~ and +# Indicate the number of additional cycles and bytes for the particular variation

Fig. 12-34 Postbyte formation for the indexed addressing mode.

Motorola ASSIST09 monitor program. Note that breakpoint and trace commands are available for debugging. Commands are also available for examining and changing memory contents and the contents of the processor registers. There is even an offset command to calculate relative addresses for branch instructions. Monitor programs are easy to learn and use. They are useful for entering and debugging small machine language programs. As an example of how they work, refer to Fig. 12-37. This is an example of a memory dump as printed on the terminal. The dump is in response to the display command typed into the terminal:

D F880,F8FF

When this command is entered (by hitting RETURN on the terminal) ASSIST09 responds with a dump of memory locations $F880 to $F8FF. Note that all values are in hexadecimal. Also note that the ASCII contents of memory are printed to the right. If the contents of any memory location are not a printable

```
                       LABEL    MNEMONIC  OPERAND        COMMENT COLUMN (OPTIONAL)
                       COLUMN   COLUMN    COLUMN

                 000B  PAUSE    EQU       11             ASSIST09 break+check service
                 F200  PORTB    EQU       $F200          VIA port B
                 F202  DDRB     EQU       $F202          port B data direction reg.
                 6000  DTIME    EQU       $6000          DELAY TIME
                                *
       0000 CC   49FF  START    LDD       #$49FF         ROLA op code + DDRB
       0003 A7   8C 08          STA       <SCAN,PCR
       0006 F7   F202           STB       DDRB           set VIA port B as outputs
       0009 CC   0008           LDD       #$0008
       000C 1A   01             ORCC      #1             set carry flag

       000E 49         SCAN     ROLA                     shift to next bit
       000F B7   F200           STA       PORTB          light 1 LED
       0012 8E   6000           LDX       #DTIME
       0015 30   1F    DELAY    LEAX      -1,X           delay for a while
       0017 26   FC             BNE       DELAY
       0019 5A             DECB                  all 8 LEDs yet?
       001A 26   0A             BNE       CONT1          no
       001C C6   0F             LDB       #$F            yes
       001E E8   8C ED          EORB      <SCAN,PCR switch rotate direction
       0021 E7   8C EA          STB       <SCAN,PCR L-to-R-to-L
       0024 C6   07             LDB       #7             reset bit counter
       0026 3F         CONT1    SWI                      check for "FREEZE" or "CANCEL"
       0027 0B             FCB       PAUSE
       0028 24   E4             BCC       SCAN           repeat if no CANCEL
       002A 39             RTS                  otherwise, return to  ASSIST09

                                END

       0 ERROR(S) DETECTED

       SYMBOL TABLE:

       CONT1  0026   DDRB   F202   DELAY  0015   DTIME  6000   PAUSE  000B
       PORTB  F200   SCAN   000E   START  0000
```

Fig. 12-35 Machine language program development example.

Command Name	Description	Command Entry
Breakpoint	Set, clear, display, or delete breakpoints	B
Call	Call Program as subroutine	C
Display	Display memory block in hex and ASCII	D
Encode	Return indexed postbyte value	E
Go	Start or resume program execution	G
Load	Load memory from tape	L
Memory	Examine or alter memory	M
	Memory change or examine last referenced	/
	Memory change or examine	hex/
Null	Set new character and new line padding	N
Offset	Compute branch offsets	O
Punch	Punch memory on tape	P
Registers	Display or alter registers	R
Stlevel	Alter stack trace level value	S
Trace	Trace number of instructions	T
	Trace one instruction	.
Verify	Verify tape to memory load	V
Window	Set a window value	W

Fig. 12-36 Assist 09 monitor commands.

```
        0  1  2  3  4  5  6  7  8  9  A  B  C  D  E  F
F880   01 A2 01 BF 01 D7 01 CF 01 83 01 81 01 CC 01 87    ................
F890   00 5C 01 8B 02 68 01 DD 17 02 3B EE 6A 33 5F 0D    .....h....;.j3_.
F8A0   FB 26 11 17 06 B3 50 5A 2B 0A 11 A3 A1 26 F8 EF    .&....PZ+....&..
F8B0   6A 16 02 34 0F FB 37 06 C1 0B 10 22 02 25 EF 6A    j..4..7....".%.j
F8C0   58 33 8C BC EC C5 6E CB 54 43 32 20 41 53 53 49    X3....n.TC2 ASSI
F8D0   53 54 30 39 20 5B 56 31 2E 32 5D 04 10 DF 97 6D    ST09 [V1.2]....m
F8E0   61 26 09 AD 9D E6 EF 30 8C DE 3F 03 9E F6 6F 0B    a&.....0..?...o.
F8F0   86 C0 A7 0E 3F 06 35 01 34 01 17 06 5C 2A 0C 50    ....?.5.4....*.P
```

Fig. 12-37 Sample output for monitor display command.

ASCII character, a period (.) is printed. None of the memory locations from $F880 to $F8FF contains a printable character. If you examine Fig. 12-37 you can see that this particular dump was from an area of memory that contained the ASSIST09 monitor program.

There are other software development and debugging aids. A *disassembler* is a program that converts machine code into an assembly listing. It is useful for analyzing code to determine how it works (or why it doesn't) and how to modify it for a new application. Some systems offer linkers to allow smaller programs to be gathered into a single larger program. Diagnostic software may be available for testing various parts of the system. This will be covered in the next section.

REVIEW QUESTIONS

44. Refer to Fig. 12-31. What is the OP-CODE for the LDD extended instruction? How many memory bytes will it consume? How many machine cycles will it take?

45. Refer to Fig. 12-31. Which processor flags will be set or cleared by the LDD instruction?

46. Refer to Fig. 12-31. Which processor flags will not be affected by the LDD instruction?

47. Refer to Fig. 12-33. How many bytes are required for the BCC instruction? For the LBCC instruction? Why?

48. Refer to Fig. 12-33. Which flags are affected by the branch instructions?

49. Refer to Fig. 12-35. Explain the line of machine code at the left that begins at address $000F.

12-8
TROUBLESHOOTING AND MAINTENANCE

As always, the troubleshooting procedure should begin with the obvious and preliminary checks and proceed from there on an orderly and logical basis. Verify proper power supply operation, including any separate negative and positive supplies used for RS-232C data communication. Do not forget to include all parts of the system that may be connected to the processor. Sometimes, a microprocessor will "hang up" when attempting to communicate with another device that is powered down or not ready to respond for some other reason. Some industrial computers may contain battery back-up circuits. Test the battery voltage with the main power off.

Try to determine whether something has been changed since the last time the equipment was operational. For example, has new software been installed in the system, or has the software been modified? It may be necessary to revert to earlier software to determine where the problem is actually located. Software problems can appear to be hardware problems, and the opposite is also true. Has another part of the system been changed or reprogrammed? In some cases, what appears to be a defective computer is actually a terminal or other device that is not properly configured. Modern serial terminals are capable of quite a few baud rates, different word sizes, a few stop bit choices, and several parity options. These can be changed by setting switches on the back of the terminal or perhaps inside the terminal. They can also be changed from the keyboard on some terminals. Refer to the relevant manuals and make certain that the terminal is properly set up to communicate with the computer. When doing so, you may encounter half- and full-duplex options. The *half-duplex option* allows communication between the terminal and the computer in only one direction at a time. The *full-duplex option* allows communication in two directions independently at the same time. If the terminal displays or prints all typed characters double, try changing it to full-duplex. If nothing is displayed or printed, try changing it to half-duplex. Finally, check the cable between the terminal and the computer. Sometimes pins 2 and 3 on one of the DB-25 connectors must be reversed. This happens because 2 must connect to 2 for some equipment interfaces, and 2 must connect to 3 for others.

Some systems include diagnostic software. One example is a *RAM test program*. Such a program writes various bit patterns to RAM locations and then reads them back to verify that memory is working as it should. Another example is software that exercises various parts of the system such as display outputs, control outputs, and inputs. The diagnostic package may be menu-driven and sequentially structured. A menu will appear on the terminal, allowing the technician to select from several test options. When a test is selected, the program may ask the technician to verify that a display or some other output is responding as described on the screen or printout. As each question is answered, the next test is started. The program may also request that the technician actuate certain controls or apply test signals to various inputs. A "walk-through" of this type can be very effective in verifying system operation and diagnosing malfunctions. Some systems can even be diagnosed from many miles away. Modems have been used to connect a diagnostic computer at the factory to a processor in the field.

If the defective computer works at least partially, a monitor program may be a valuable troubleshooting aid. Examine several memory locations in every block. Consult the memory map. If 8K devices are used, check several addresses in each device. If the devices are RAM, verify that the data can be changed. Try writing $FF to a location and then read it. Then change it to $00 and read it again. Although this is less conclusive than a diagnostic program that checks every location, it still may be helpful. A monitor program can also be used to check support devices such as PIAs and ACIAs. You can use the monitor to set the data direction for output and then write $FF followed by $00 to the output registers. An oscilloscope, logic probe, or meter can be used to test the output pins for the proper response. A few cautions are in order here. First, don't forget that writing information to an ACIA could disable the ability to communicate via the terminal. If this happens, a system reset should reinitialize the ACIA and reestablish communications with the terminal. Second, be aware that devices such as motors or hydraulic valves could be activated by changing data at certain addresses. If at all possible, troubleshoot with high-energy systems disabled.

If BASIC is available on the system, it can also serve as a troubleshooting aid. Some versions have a *PEEK command* that can be used to examine memory locations and a *POKE command* that can be used to change them. BASIC can also be used to repeatedly strobe an address or a device to allow oscilloscope troubleshooting. For example, suppose that you wish to verify that an address decoder at $E000 (decimal 57,344) is working. If the system is operating as it should, the following program will produce a pulse train at the decoder output.

```
10 POKE 57344,0
20 GOTO 10
```

BASIC can also be used to strobe a device repeatedy to facilitate oscilloscope testing. For example, assume that a PIA is decoded, beginning at address $FF40 (decimal 65,344). The following program initializes the PIA (program lines 10 through 30) and then repeatedly toggles the outputs. The REM statements are for documentation purposes and are ignored by BASIC:

```
10 POKE 65346,0: POKE 65347,0: REM clears bit 2
to select DDRA and DDRB
20 POKE 65344,255: POKE 65345,255: REM sets all
bits for output
30 POKE 65346,4: POKE 65347,4: REM sets bit 2
to select ORA and ORB
40 POKE 65344,0: POKE 65345,0: REM toggle all
16 outputs low
50 POKE 65344,255: POKE 65345,255: REM toggle
all 16 high
60 GOTO 40: REM continue toggling outputs
```

If an oscilloscope is not available, a logic probe can also be used to check for toggling. If only a meter is available, the toggling can be slowed down by adding this delay line to the program:

45 FOR X = 1 TO 200: NEXT X: REM change "200" value to change delay

If a monitor program is available, similar diagnostic routines can be hand-assembled, entered, and executed in machine code.

The microcomputer may not work at all or may not work well enough to use the monitor or BASIC. Also, many industrial computers are small, dedicated units. They often do not have a monitor, BASIC, or even a terminal connected to them. In these cases, an oscilloscope may be used to examine the various address lines, data lines, clock lines, control lines, and status outputs of the microprocessor to determine whether there is proper activity. The clock signals must always be present. If the system uses a microprocessor with an on-chip oscillator, the lack of clock signals would indicate a defective microprocessor or crystal. It is normal to expect activity on the address bus and the data bus. If there is no activity or limited activity, the microprocessor may be in a halt mode, may be waiting for an interrupt, or possibly may be in a DMA or sync mode. If this is the case, the processor status pins should be checked with an oscilloscope to determine which mode the processor is in; these pins are BA and BS on the Motorola MC6809 processor. When BA and BS are both low, the processor is in the normal running mode. The three other status conditions established by the state of BA and BS are the *interrupt or reset acknowledge mode,* the *sync acknowledge mode,* and *the halt or bus grant mode.* It may also be necessary to investigate the halt, reset, MRDY, DMA/BREQ, and interrupt pins. Set the oscilloscope for dc coupling and establish a 0-V reference point on the screen when analyzing a microprocessor system. This will allow both signal activity and dc level to be readily determined.

Assuming that the oscilloscope shows bus activity, a check for chip select and device select signals will shed some light on what the processor is doing or not doing. It is normal to observe repeated ROM select signals as the processor executes the code stored there. Check for RAM select signals also and don't forget to check for I/O select signals. Support devices are initialized on power up and after a system reset. You may have to continue to reset the system while looking for select signals at these devices. These should occur immediately after the processor leaves the reset state, and there should be evidence that the processor is writing several bytes to each initialized device. If resets do not appear to initiate the expected actions, set the oscilloscope for a sweep speed of approximately 10 ms per division and connect the probe to the reset bus. An exponential ramp should be evident when the reset switch is released.

You can also look at the read/write line and the input and output of buffer ICs to help localize difficulties.

A microprocessor is a very complex IC, with many functions and features. It is possible that a partial failure can occur. For example, a processor may do everything it should except provide the proper response to an interrupt. It may be possible to use a logic pulser to simulate an interrupt. An oscilloscope can be triggered on the pulser output, and the scope probe can be used to look for a ROM select signal. This procedure will determine whether the processor is fetching the interrupt vector as it should. Do not forget that the processor may mask an interrupt request.

Some advanced microprocessor troubleshooting equipment that is available provides quite a bit of information. A *logic analyzer* is an example; this piece of equipment has probes that are connected to the address bus, the data bus, the read/write line, and the clock line. Qualifier switches can be set to make the instrument synchronize on a particular event. The logic analyzer stores bus events, which can be displayed as they occurred before and after the qualifying event. Some logic analyzers are very advanced and can present the data in binary form, hexadecimal form, and in the form of timing diagrams on a cathode-ray tube. Some even have *personality modules* for the popular microprocessors and can thereby produce *disassembly* listings on the screen or send the listings to a printer. *Signature analyzers* are also available for microprocessor troubleshooting and examine the data bus and the address bus of a microcomputer in real time. The pattern (or signature) is compared to a pattern previously stored in memory from a normally operating system. Any differences between the stored signature and the actual signature can be analyzed to provide clues as to what is wrong with the system.

The statistics of part failure predict that the most complex devices are the most failure-prone. This includes the microprocessor, support devices, and memory ICs. If a system uses sockets and if spare ICs are available, component swapping is a reasonable procedure. Make sure the power is off and observe proper handling procedures to prevent damage from static electricity. Also make sure that all devices are properly and firmly seated in their sockets.

REVIEW QUESTIONS

50. Refer back to the BASIC diagnostic program for testing the PIA. Could this program be modified to test the input performance of the PIA? How?

51. How will the system react in response to the modification referred to in the prior question?

52. The various lines of the address and data buses should show continuous activity when the processor is in the _____ mode.

CHAPTER REVIEW QUESTIONS

12-1. How many flags are there in the MC6809 microprocessor?

12-2. The address pins on a microprocessor are unidirectional, and the data pins are _____.

12-3. Which input pins are used to signal the microprocessor that an external hardware event that requires attention has occurred?

12-4. Which pin tells whether the microprocessor is writing or reading data?

12-5. A group of instructions and data in memory used to direct the operation of a microprocessor is called a _____.

12-6. A CMPA is an example of a(n) _____.

12-7. What often happens when a microprocessor begins execution at the wrong address?

12-8. The immediate addressing mode is used to load the U pointer register. How many operand bytes must follow the OP-CODE?

12-9. How many bytes must follow the OP-CODE for extended addressing?

12-10. The S register is to be loaded from memory locations $F011 and $F012 by using the extended addressing mode. What must follow the OP-CODE for LDS?

12-11. Assume that the DP register contains $2E. What is the effective address when the OP-CODE for LDA is followed by $34?

12-12. Which indexed addressing mode points to the address of the effective address?

12-13. A relative address is added to the contents of the _____ when the result of a branch test is true.

12-14. A branch test is located at $1114, and the relative address is located at $1115. What is the required hex value of the relative address to cause a backward branch to $1100? Don't forget that the program counter will be pointing to $1116 when the effective address is calculated.

12-15. Accumulator A contains $3E. What will it contain after nine consecutive RORA operations?

12-16. Which instruction always transfers program flow to the effective address by using ex-

tended, direct, or indexed addressing? Which instruction does the same thing but uses relative addressing?

12-17. What is the purpose of the F flag?

12-18. What is the purpose of the I flag?

12-19. Which flag is set to tell the processor to pull all the registers but one from the hardware stack? Which register is not pulled?

12-20. Where are the interrupt vectors stored in an MC6908 system?

12-21. Which interrupt stacks only the CC register and the program counter?

12-22. Assume that an MC6809 is in the SYNC state. What happens when an NMI signal is received?

12-23. In a microprocessor-based system, which device must leave the reset state last? Why?

12-24. Which MC6809 signal is equivalent to the phase 2 clock signal required by the VIA?

12-25. Assume that a VIA is mapped into memory beginning at $82E0. What address will the processor write to in order to change PB7 to a timer-controlled output?

12-26. Serial data communications can use voltage or current signals. Which is the most popular? Which specific voltage standard is the most popular?

12-27. Which serial mode requires start and stop bits?

12-28. The RS-422 serial standard uses balanced transmission for high bandwidth and _____ data rates.

12-29. For what type of data communication is an ACIA designed?

12-30. How can you determine which mode the MC6809 processor is in when troubleshooting?

12-31. What would repeated ROM select signals indicate?

12-32. What would repeated I/O select signals indicate?

ANSWERS TO REVIEW QUESTIONS

1. 16 M (approximately 16 million) **2.** flag **3.** 16 **4.** 8 **5.** the valve is turned on **6.** OP-CODE **7.** no; no readings could be stored **8.** inherent **9.** one **10.** $34 or $43 **11.** $B9 **12.** $01 **13.** the S register **14.** $A9; a 2-byte offset **15.** $80; after the effective address is determined **16.** 156; −100 **17.** $05; $06 **18.** $32 **19.** positive number; no **20.** yes **21.** $F0 **22.** no **23.** program counter **24.** the S register **25.** nested **26.** decremented **27.** the software will crash **28.** the motor will run counterclockwise **29.** 458,753 μs **30.** BSR or LBSR **31.** $7000; $7FFF; 4K (4096 bytes) **32.** from the falling edge of E to the rising edge of Q **33.** 500 ns; 750 ns **34.** $C13E; yes **35.** $800F; $8FFF; $FF1 (4081) **36.** four **37.** by reading control register *A* and testing bits 6 and 7 **38.** a latch **39.** when the counter high byte is written; when the counter times out **40.** $F3; $01 **41.** 7.843 kHz; rectangular (nearly square) **42.** timer 2; PB6 **43.** 1920 **44.** $FC; 3; 6 **45.** N, Z, and V **46.** H and C **47.** 1; 4; 2-byte OP-CODE plus 2-byte relative address **48.** none **49.** $B7 is the OP-CODE for store accumulator *A* extended, and $F200 is the address where it will be stored **50.** yes; change line 20 to POKE 0 to both locations; replace line 40 with PRINT PEEK(65344): PRINT PEEK(65345); delete line 50 **51.** the decimal equivalent of port A input data will be printed followed by the port B data, repeatedly **52.** normal (running)

13

DATA CONVERSION, COMMUNICATION, AND STORAGE

Advances in computer electronics have produced a major impact on industry. Many industrial systems are connected to, monitored by, or in some way controlled by computers. Some of these systems are sophisticated and involve both digital and analog signals. This chapter focuses on conversions between the digital and analog worlds and examines digital communications in more detail. It also introduces mass storage devices for digital data and programs.

13-1
DIGITAL-TO-ANALOG CONVERSION

The *digital-to-analog converter* (DAC) accepts a binary input and produces a corresponding analog output, which is usually in the form of a voltage or a current. Figure 13-1 demonstrates the basic idea. A number of binary inputs are shown at the left, and a single analog output is shown at the right. In the strictest sense, a true analog signal is capable of an infinite number of measurement values. Can the DAC produce an infinite number of output levels? It cannot. A DAC produces discrete steps at its output: one for every possible binary input. With an adequate number of binary inputs, the step size will be small enough so that the output approaches a true analog signal.

The DAC shown in Fig. 13-1 has eight binary inputs. It is connected to a 12-V supply. Suppose that the output can span from 0 to 12 V. It is possible to determine the step size of the output voltage by dividing the full span by 2^N where N is the number of binary inputs; $2^8 = 256$ and $12/256 = 0.046875$ V. This step size is small enough to meet many industrial needs for an analog signal. In those cases where it is not, 10-, 12-, and 14-bit DACs are available.

Let's see what happens to the step size with a 14-bit DAC:

$$\text{step size} = \frac{12 \text{ V}}{2^{14}} = 0.000732422 \text{ V}$$

A step size this small approaches a true analog signal very closely.

The percentage of resolution for a DAC is set by the number of bits, as shown in Fig. 13-1. For an 8-bit DAC, the resolution is 0.390625 percent. For a 14-bit DAC, the resolution is 0.006104 percent. As the number of bits increases, the percentage of resolution decreases. This indicates a finer, or more precise, resolution. Of course, the step size also decreases as the number of bits increases. The overall accuracy is a function of the number of bits and the accuracy of the DAC itself. Digital-to-analog con-

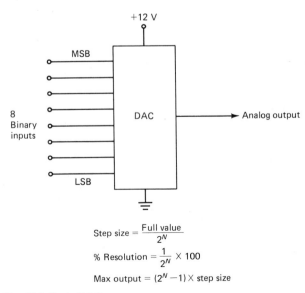

$$\text{Step size} = \frac{\text{Full value}}{2^N}$$

$$\% \text{ Resolution} = \frac{1}{2^N} \times 100$$

$$\text{Max output} = (2^N - 1) \times \text{step size}$$

Fig. 13-1 Basic DAC.

354

verter manufacturers often rate *accuracy* as a percentage. Do not confuse this rating with the *resolution*, which is fixed by the number of bits. Manufacturers may also rate the accuracy as equal to plus or minus 1/2 LSB, or plus or minus 1/4 LSB, and so on. The least significant bit produces the smallest possible change in the output. An 8-bit DAC with a plus or minus 1/2 LSB rating has an accuracy of plus or minus approximately 0.19 percent. Note that this is half the percentage of resolution calculated by the equation shown in Fig. 13-1. An 8-bit DAC with a 1/4 LSB rating has an accuracy of plus or minus approximately 0.1 percent, which is one-fourth of its resolution.

Figure 13-1 shows that the maximum output from a DAC is found by multiplying $2^N - 1$ times the step size. We found the step size earlier for the 8-bit DAC with a 12-V supply to be 0.046875 V. The maximum output is therefore $(256 - 1) \times 0.046875\ V = 11.953125$ V. Note that it is not equal to 12 V. The minimum output is 0 V and occurs with a binary input of 00000000; the maximum output is one step size less than the supply and occurs with a binary input of 11111111 (decimal 255). To find the output at any intermediate value, simply convert the binary input to decimal and multiply it times the step size. For example, with the same step size and a binary input of 10101011 (decimal 171) the analog output would be nominally 8.01563 V. The actual output voltage could vary as much as half a step above or below nominal, depending on the accuracy rating of the DAC.

Figure 13-2 shows the circuit for a 5-bit binary weighted adder. It uses an op amp as an inverting adder. Assume that V_{ref} is -5 V. The output range will be from 0 to almost 5 V. The step size will be

$$\frac{5}{2^5} = 0.15625\ V$$

As the circuit is drawn, none of the switches is closed. This is equivalent to a binary input of 00000. No voltage is applied to the input of the op amp, and $V_{out} = 0$. What happens if binary 00001 is entered? This sets the LSB and closes the switch connecting V_{ref} to the input through the 32R resistor. The voltage gain of the op amp is given by

$$-\frac{R_F}{R_{in}}$$

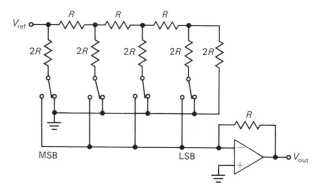

Fig. 13-3 *R-2R* ladder D/A converter.

Since the input resistor is 32 times the value of the feedback resistor, the gain is $-1/32$. With a reference voltage of -5 V, the output will be

$$-\frac{1}{32} \times -5 = 0.15625\ V$$

This result agrees with the step size calculated previously.

In practice, the switches shown in Fig. 13-2 would be replaced by transistor switches to allow logic signals to be directly applied to the D/A converter. Binary weighted DACs are limited to 5 bits because of the large range of resistor values required. It is very difficult to match resistors with ratios greater than 32:1. Therefore, any attempt to gain resolution by adding more bits will result in a loss of accuracy that negates the extra bits.

Figure 13-3 shows a D/A converter that is based on an *R-2R* ladder network. It requires only two resistor values and is therefore easier to implement. Each switch is a single-pole, double-throw type that connects the 2R elements to the input of the op amp or to ground. All switches are shown in the 0 position. An equivalent network that makes the circuit easier to understand is shown in Fig. 13-4. It is assumed that $R = 10$ kΩ and that $2R = 20$ kΩ. Note that all the 20-kΩ resistors are grounded. This is true regardless of the setting of any of the switches. You should recall that the inverting input of the op amp is a virtual ground. Start at the right side of the network. The two 20 kΩ resistors in parallel are equivalent to a 10 kΩ resistor and it is in series with

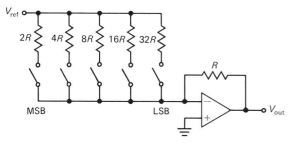

Fig. 13-2 Binary weighted D/A converter.

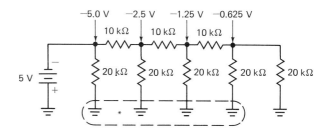

*These can represent actual ground or the virtual ground of the op amp, depending on the switch settings.

Fig. 13-4 Voltages around the *R-2R* network.

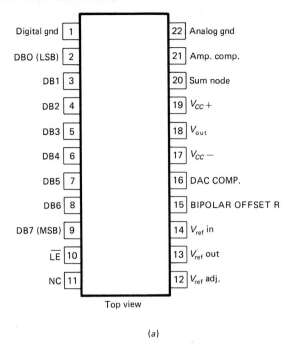

(a)

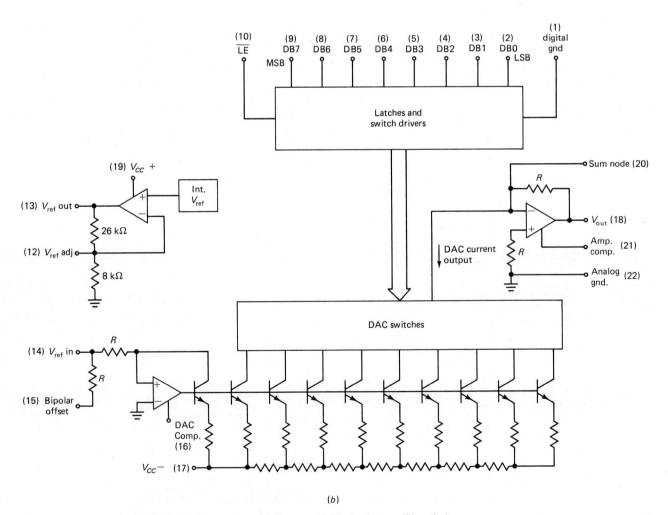

(b)

Fig. 13-5 Integrated circuit NE5018 D/A converter. (a) Pinouts. (b) Block diagram (Signetics).

a 10kΩ resistor for an equivalent resistance of 20 kΩ. This equivalent 20 kΩ is in parallel with the next 20 kΩ resistor for a resistance of 10 kΩ. It is in series with the next 10 kΩ for an equivalent resistance of 20 kΩ. You should see the pattern by now. No matter how many bits the R-$2R$ network requires, the equivalent resistance values keep repeating.

Now look at the voltages shown in Fig. 13-4. The full -5 V appears at the first point indicated. Half that value appears at the next point, and then half that value appears at the next point, and so on. The R-$2R$ network repeatedly divides the input voltage by 2. When any of the switches in Fig. 13-3 are set to binary 1, the full reference voltage or one of the divided values is applied through a $2R$ element to the input of the op amp. The feedback resistor has a value of R, and therefore half of the network voltage appears at the output. For example, if V_{ref} is -5 V and only the MSB is set, $V_{out} = 2.5$ V.

Most DACs are monolithic ICs such as the Signetics NE5018 shown in Fig. 13-5. This IC contains the R-$2R$ ladder network, an adjustable reference supply, the DAC switches, input latches for the binary data, and an operational summing amplifier. The input latches make the device compatible with microprocessors. The binary inputs (pins 2 to 9) can be connected directly to the data bus. When not LE (pin 10) is low, the latches are transparent. When it goes high, the input data present at the moment of transition is latched and retained until not LE goes low again.

The NE5018 reference voltage is nominally 5 V. It can be trimmed with an external potentiometer to provide easy adjustment of full scale. It is temperature-compensated and therefore relatively stable. The maximum supply voltage is plus and minus 18 V with plus and minus 15 V being recommended. The settling time is rated at 2.3 µs. This is the time required for the output to stabilize (settle) after a binary input appears. Its rated accuracy is plus or minus 1/2 LSB. Signetics also manufactures the NE5019, which is almost identical but offers an improved accuracy of plus or minus 1/4 LSB. They also offer the NE5020, a 10-bit DAC rated at plus and minus 0.1 percent relative accuracy. It features two latch enable inputs, which allow an easy interface with 8-bit microprocessors. The processor can write data to the lower 8 bits and the upper 2 bits at two different addresses. Two write cycles are required to load a 10-bit word into the DAC when it is interfaced to an 8-bit bus.

The Signetics DAC can use its own internal reference supply or an external reference supply. Converters that use an external reference are sometimes referred to as *multiplying DACs*. The multiplying mode allows a signal applied to the reference input to be scaled according to the value of the binary input. Thus, a microprocessor could alter the level of an analog signal by sending a binary word to the DAC. Figure 13-6 shows another application. A microprocessor controls the binary input of the D/A

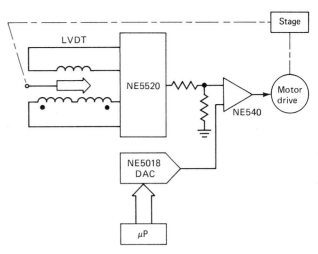

Fig. 13-6 Microprocessor control interface (Signetics).

converter. The resulting analog output is applied to the NE540 error amplifier. The other input to the error amplifier comes from the NE5520 LVDT signal conditioner. Any error between the command position from the microprocessor and the actual position will cause the motor to run in the direction necessary to eliminate the error.

REVIEW QUESTIONS

1. Calculate the step size for a 10-bit DAC that uses a 5-V supply. Calculate its maximum output voltage.

2. Calculate the percentage of resolution for a 10-bit digital-to-analog converter.

3. The manufacturer of a 10-bit DAC rates the accuracy at plus and minus 1/4 LSB. What is the accuracy expressed as a percentage?

4. Refer to Fig. 13-2. What is the output with a -10-V reference if the two left-most switches are closed and all other switches are open?

13-2
ANALOG-TO-DIGITAL CONVERSION

Analog-to-digital converters (ADCs) are used to convert analog voltages or currents to binary words. They are often based on digital-to-analog converters such as the ramp-type circuit shown in Fig. 13-7. Voltage V_{in} is an analog voltage that is to be converted to a binary value. It is applied to the noninverting input of a comparator. The inverting input of the comparator is connected to the output of a DAC. The microprocessor supplies the DAC with a binary input that counts up from zero. When V_{out} exceeds V_{in}, the comparator output switches and generates an interrupt to the microprocessor. The last binary value written to the DAC is proportional to the value of the analog input. The illustration shows that the

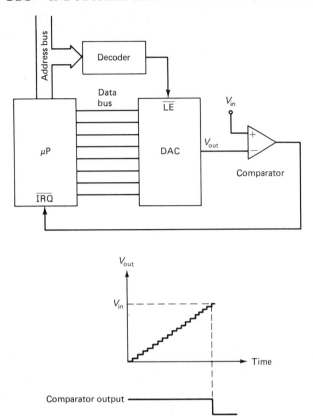

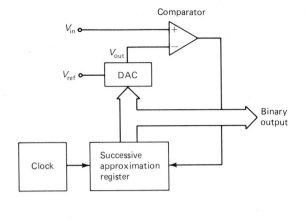

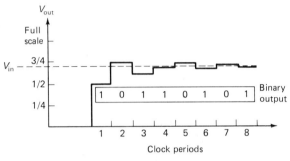

Fig. 13-8 Successive approximation converter.

Fig. 13-7 Ramp A/D conversion.

DAC output is a series of voltage steps called a *staircase waveform*. The resolution of the ramp-type ADC is a function of the number of bits in the DAC.

A ramp converter can also be implemented by replacing the microprocessor with an up counter, a clock, and a logic control circuit. The conversion sequence will begin with the control circuit's resetting the counter to zero. The clock will then be enabled, and up counting will commence. When the comparator output switches, the clock will be disabled, and the counter will then contain the binary value of the analog input. An interrupt can be generated at the completion of the conversion cycle to alert a processor that a binary reading is available. This technique unburdens the processor during the conversion cycle and allows it to perform other functions.

Regardless of how the DAC is supplied with a binary up count, the ramp-type converter is slow. A large analog input will create a significant delay during up counting. It can be as great as 2^N times the DAC and comparator cumulative settling time, where N is the number of bits in the DAC. A tracking converter may provide better performance for some applications. This circuit replaces the up counter with an up/down counter. If the DAC output is less than V_{in}, the comparator causes the counter to count up on the next clock pulse; if the DAC output is more than V_{in}, the counter counts down. The binary value of the counter will track changes in the analog input. The counter output can be sampled at any time a digital value is required. Tracking converters can be used up to low audio rates.

Figure 13-8 shows the *successive approximation converter*, which is much faster. It produces a conversion delay of only N times the cumulative settling time. With an 8-bit DAC, the total delay will be eight clock periods regardless of the value of the analog input signal. The conversion process begins with all DAC inputs set to zero. The MSB bit is set at the first clock pulse. This produces a V_{out} to the comparator equal to one-half the full-scale value. If the comparator switches, the MSB is cleared, and the next most significant bit is set. If the comparator does not switch, the MSB remains set, and the next most significant bit is set. The process repeats N times as each bit is tested. A typical waveform at the DAC output as the conversion proceeds is illustrated by Fig. 13-8. Note that the binary output will show a 0 at every bit position where V_{in} was exceeded and the comparator switched.

The successive approximation technique can be implemented with a clock and a separate register, as shown in Fig. 13-8. It can also be realized without these circuits by interfacing a DAC to a microprocessor. The microprocessor can set each bit and test the output of the comparator. This software approach saves some hardware, but it burdens the processor during the conversion cycle. This is a typical compromise and is known as the *hardware/software tradeoff*.

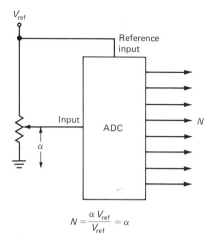

$$N = \frac{\alpha V_{ref}}{V_{ref}} = \alpha$$

Fig. 13-9 Ratiometric measurement.

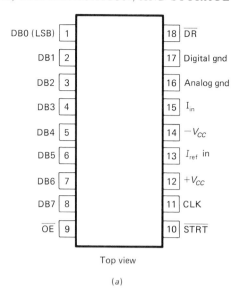

Top view

(a)

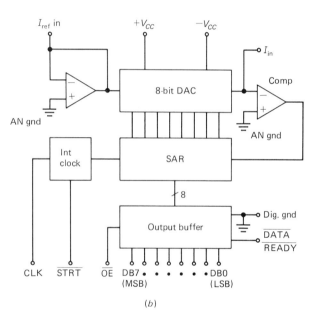

(b)

Fig. 13-10 Integrated circuit NE5034 A/D converter. (a) Pinouts. (b) Block diagram (Signetics).

The external reference voltage input shown in Fig. 13-8 may be used for *ratiometric conversions;* these types of conversions allow an A/D converter to produce an output that represents the ratio of the analog input to the reference input. Using ratiometric conversion, the circuit acts as an analog divider with a binary output. Precision measurements that are insensitive to the reference voltage are possible. Refer to Fig. 13-9; a potentiometer-type transducer is supplied by V_{ref}, which also supplies the reference input to the analog-to-digital converter. The equation shows that the binary count (N) is equal to the potentiometer ratio and is not sensitive to the reference voltage. Ratiometric conversion can also be used with resistive bridges by using an instrumentation amplifier between the bridge and the ADC.

Figure 13-10 shows the pin configuration and block diagram for the Signetics NE5034 monolithic A/D converter. It uses the successive approximation technique and contains a comparator, an 8-bit DAC, a register, a clock, a reference amplifier, and a three-state output buffer. The buffer simplifies interfacing the IC to microprocessors. An external capacitor controls the clock frequency, and conversion times as short as 17 μs are possible. An external clock can be used for even faster conversion times. The IC digitizes an analog input current applied to pin 15. It is capable of ratiometric conversion with an external reference current applied to pin 13. It is also capable of digitizing unipolar or bipolar input voltages, as shown in Fig. 13-11. This illustration also shows the curves for selecting the clock capacitor (C_{CL}).

Another integrated circuit converter, the monolithic CMOS ADC0808 by National Semiconductor, is shown by Fig. 13-12. It uses successive approximation as the conversion technique and provides 8 bits of resolution with a conversion time of 100 μs. It operates from a single 5-V supply and features a low power consumption of only 15 mW. It is microprocessor-compatible with a tri-state output latch/buffer and an address latch and decoder for the input multiplexer. The *input multiplexer* allows any one of 8 analog inputs to be switched to the comparator for conversion to binary. This provides a cost-effective solution for those industrial applications in which the output of several transducers must be digitized. Digitizing eight inputs will take approximately 1 ms, but this is an adequate response time for many applications in industry, such as temperature monitoring.

An A/D conversion can be accomplished without a DAC. Figure 13-13 (p. 362) shows a voltage-to-frequency converter. With a negative V_{in} applied, the integrator output will ramp in a positive direction until the V_{ref} threshold is crossed. At this time, the comparator output will switch and discharge the integrator. The next positive ramp will begin, and the cycle will repeat over and over. The magnitude of V_{in} will set the slope of the positive ramps. Larger

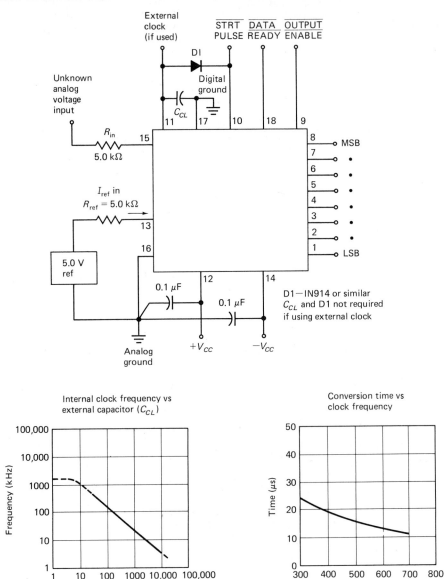

Fig. 13-11 Applying the NE5034 (Signetics).

input voltages will produce a steeper slope, and the comparator threshold will be reached in less time. The pulse frequency at the output of the comparator is proportional to the input voltage. Figure 13-13 adds a clock circuit, a gate, and a counter to convert the frequency to a binary count proportional to the analog input voltage. The performance of voltage-to-frequency A/D converters is relatively poor for small input voltages, as a result of offset drift in the integrator and comparator.

Integration is also used in the *dual-slope circuit* of Fig. 13-14. The conversion cycle begins with $-V_{in}$ applied to the integrator through the switch for a fixed time T_1. During T_1, the integrator output will rise to a point dependent on the input signal magnitude. Then the input switch is changed, and a reference voltage of opposite polarity is applied to the integrator. The counter will count clock pulses during the time (T_2) it takes the integrator to ramp back

down to the comparator threshold. Since V_{ref} and T_1 are fixed known quantities, V_{in} is proportional to T_2. The dual-slope technique has the advantage of minimizing errors because the same circuit and components are used for summing both the reference and unknown input voltages. Any long-term component drift is canceled. However, the circuit is sensitive to zero offset and drift in the integrator and comparator. These effects can be compensated for by a more complex circuit that adds calibration phases to measure the analog error and eliminate it from the final output. Some error-correcting techniques allow as much as 14 bits of resolution, but integration is a slow technique, and the conversion time is too long for some applications. Dual-slope A/D converters are typically used in instruments such as digital voltmeters (DVMs) and digital multimeters (DMMs).

The A/D converter response time is critical for some applications. For example, it may be necessary

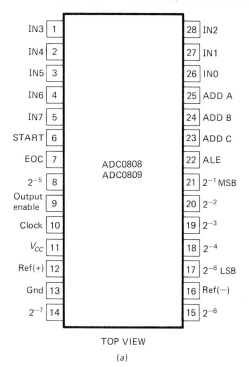

TOP VIEW

(a)

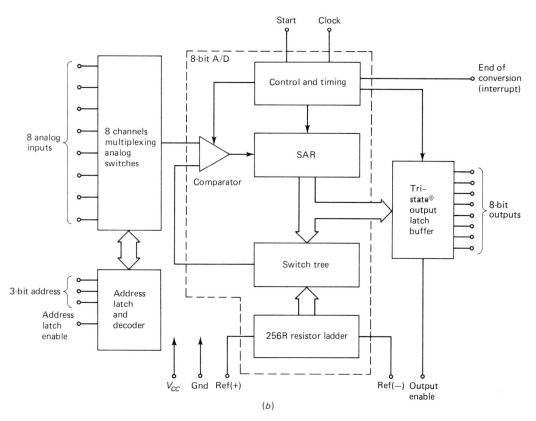

(b)

Fig. 13-12 Integrated circuit ADC0808 A/D converter. (a) Pinouts.
(b) Block diagram (National Semiconductor).

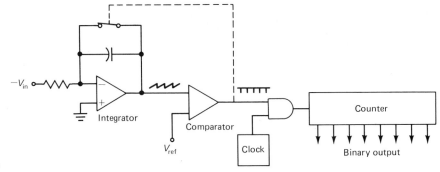

Fig. 13-13 Voltage-to-frequency converter.

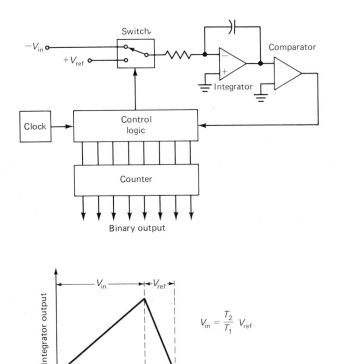

$$V_{in} = \frac{T_2}{T_1} V_{ref}$$

Fig. 13-14 Dual-slope A/D converter.

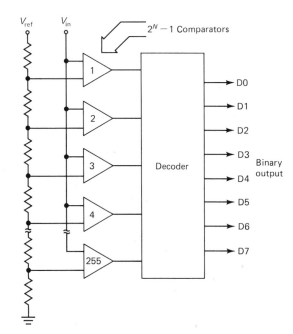

Fig. 13-15 Flash converter.

to sample a signal that is changing rapidly with time. The level of such a signal may change significantly from the beginning of the conversion process to the end. This change can cause large measurement errors. One solution is to use a *sample and hold amplifier* in front of the ADC. The sample and hold circuit will "freeze" the signal and hold it constant during the conversion period. Sample and hold amplifiers are discussed in Chapter 6. Another solution is to use a converter that is very fast. One example is the flash converter shown in Fig. 13-15. It requires $2^N - 1$ comparators for N bits of resolution. An 8-bit converter requires 255 comparators stacked at voltage levels of 1 part in 256 from each other. This stacking is accomplished by a resistive divider. The comparator array compares the input signal with each reference voltage to produce an N of 255 code sometimes referred to as the "thermometer" code

since all comparators below the signal level will be on and all those above will be off. The decoder logic converts this code to a binary output.

Flash converters are very fast. A typical 8-bit device can respond in 50 ns. This allows up to 20 million conversions per second. The disadvantages are high cost and a low input impedance caused by the input signal's driving many comparators in parallel. Figure 13-16 shows a compromise circuit, the *dual-stage flash converter;* it may also be referred to as a *half-flash converter*. It requires only 30 comparators to achieve 8 bits of resolution. The tradeoff is speed; it is a little slower than the single-stage circuit. The first conversion is done with the 4-bit flash circuit at the left. Its output is simultaneously fed to the register and the input of a 4-bit DAC. The output of the DAC is applied to the inverting input of a subtracting amplifier, and V_{in} is applied to the noninverting input. Because V_{in} is delayed, the two amplifier inputs arrive in the proper time relationship. The difference is then converted in a second 4-bit flash circuit at the

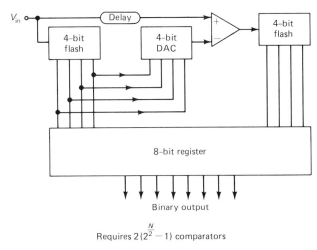

Fig. 13-16 Dual-stage flash converter.

Requires $2(2^{\frac{N}{2}}-1)$ comparators

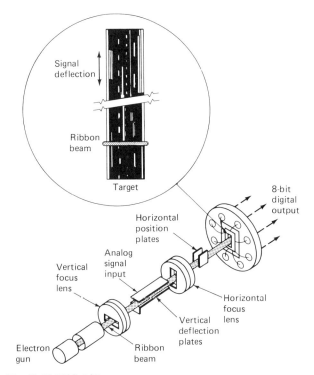

Fig. 13-17 EBS A/D converter.

right. The final result is now available in the 8-bit register as a complete binary word.

One of the fastest A/D converters is the *electron-bombarded semiconductor* (EBS) shown in Fig. 13-17. It uses an electron beam that is focused into a narrow ribbon and deflected across a masked diode target. The analog input signal is applied to the vertical deflection plates, where the beam is deflected vertically along the target in accordance with the magnitude of the input signal. The target is Gray-coded with eight columns of diodes and eight corresponding high-speed comparators. The effects of beam jitter are reduced by the Gray code since only

one diode switches at a time from any target position to the next target position. The EBS converter can sample at rates up to 200 MHz (5 ns) but is very expensive.

REVIEW QUESTIONS

5. A ramp-type A/D converter generates a _____ waveform at its DAC output.

6. Assume a cumulative settling time of 2 μs for the DAC and comparator in an 8-bit ramp-type converter and calculate the worst-case conversion time.

7. Assume a cumulative settling time of 2 μs for the DAC and comparator in an 8-bit successive approximation converter and calculate the conversion time.

8. Refer to Fig. 13-9. What effect will a 5 percent change in the reference voltage have on the binary output?

9. What is selected by the 3-bit address applied to the IC shown in Fig. 13-12?

10. Refer to Fig. 13-13. What happens to the pulse frequency at the comparator output as the analog input voltage decreases? What happens at the binary output?

13-3
COMMUNICATION STANDARDS

One of the most popular digital communication standards is RS-232C, which is introduced in Chapter 12. It is an old standard and is being superseded in some cases by new standards. The RS-232C has some limitations that have been improved by these newer standards. One of the major ones is that it is limited to distances of 15.2 m (50 ft). In practice, much longer distances are actually covered. However, the chance for data errors does increase as the distance increases. Another limitation is a maximum rate of 20,000 Bd with 19,200 Bd being the fastest standard data rate. A combination of long distance and high data rate increases the chances for errors. The RS-232C standard sends data on a single wire. A single ground wire serves as the return or reference for one or more data circuits in the cable. This type of transmission is called *unbalanced*. Figure 13-18 shows how the recommended baud rate drops as the line length increases in unbalanced circuits. The RS-232C standards are somewhat conservative when compared to the graph.

The RS-232C uses ground as its reference. When the line receiver sees a voltage more negative than 3 V referenced to ground, *mark* is produced. When the receiver sees a voltage more positive than 3 V referenced to ground, *space* is produced. Various conditions can cause the ground potential to vary from point to point. A difference in ground potential between the line transmitter and the line receiver will

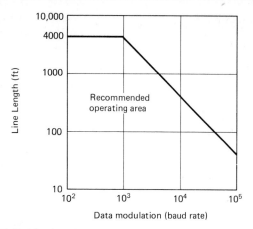

Fig. 13-18 Line length versus baud rate for unbalanced transmission lines.

cause ground loop currents to flow in the ground wire. On long runs, the wire resistance can allow significantly different ground potentials at each end of the circuit, and data errors are possible. The protective ground, if used, helps but does not eliminate this problem.

One obvious way to eliminate ground potential errors is to send the data on two wires and to ground neither of them. This is known as *balanced* or *differential data transmission* and provides several important advantages over unbalanced transmission. A balanced line is not nearly so susceptible to stray fields and noise pick-up as an unbalanced line. Balanced lines generate less noise, an advantage when several data circuits are bundled together in one cable. Balanced transmission also allows much higher data rates, as shown in Fig. 13-19. Note that data rates as high as 10 megabaud (MBd) are possible. Even at 1219 m (4000 ft), data rates as high as 100 kBd are recommended. This is a very significant speed improvement over unbalanced circuits. The Electronic Industries Association (EIA) has established RS-422 as a balanced data transmission standard. It

specifies a bipolar driver output of 2 to 6 V and a receiver sensitivity of 200 mV. If the receiver sees a positive difference signal between the two wires of more than 200 mV, a mark is read; if the difference signal is negative and more than 200 mV, a space is read. This allows suitable transmitters and receivers to be designed with bipolar 5-V supplies; RS-232C, however, often requires bipolar 12-V supplies.

The EIA has also established RS-423 as a data transmission standard. This standard resembles RS-232C in that it too is an unbalanced standard. It is capable of higher data rates (up to 100 kBd at 12.2 m) and uses different voltage levels. You may recall that RS-232C transmitters develop from -5 to -15 V on mark and from $+5$ to $+15$ V on space. The RS-423 transmitters use -4 to -6 V for mark and $+4$ to $+6$ V for space; they also use a balanced receiver, which is referenced to the driver ground to permit ground potential differences to exist without causing data errors. The RS-423 can use one wire in a cable to serve as a common return for several data circuits and is therefore more "wire-efficient" than RS-422, although not nearly as fast.

Figure 13-20 shows the DS3691 line driver IC, which can be used for RS-422 or RS-423 transmission. Pin 4 is the mode select and determines which standard will be used. The RS-423 operation provides four communications circuits, but RS-422 operation provides only two. The truth table shows that pins 3 and 6 can be used to tri-state the outputs with RS-422 operation, whereas they serve as data inputs for RS-423 operation. The IC is powered with a positive voltage applied to pin 1 and a negative voltage applied to pin 8. The bipolar supply can be as much as 7 V, but 5 V is typical. Capacitors can be connected to the rise time control pins to limit the output slew rate. This connection is necessary in some applications to eliminate cross-talk in multicircuit cables. In *cross-talk* one communication circuit interferes with another; it is more likely to occur when the waveforms have very fast rise times.

Figure 13-21 shows a DS78LS120 *dual differential line receiver* that can be used to receive both RS-422 and RS-423 data transmissions. The response time of each receiver can be controlled with an external capacitor. This is useful for filtering out high-frequency noise that may be superimposed on the signal. Also note that the output gate has hysteresis. The combination of response control and hysteresis is very helpful for producing a clean output signal in noisy industrial environments. Each receiver includes a built-in terminating resistor that is enabled by strapping pins 2 and 3 together for one receiver and pins 13 and 14 for the other. It is advisable to terminate long transmission lines to prevent signal reflection. When a line is not properly terminated, all of the signal energy is not absorbed at the load end. Some is reflected back toward the source. It can be rereflected at the source and travel toward the load again. These reflected signals act as noise

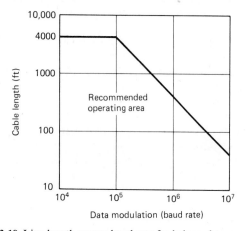

Fig. 13-19 Line length versus baud rate for balanced transmission lines.

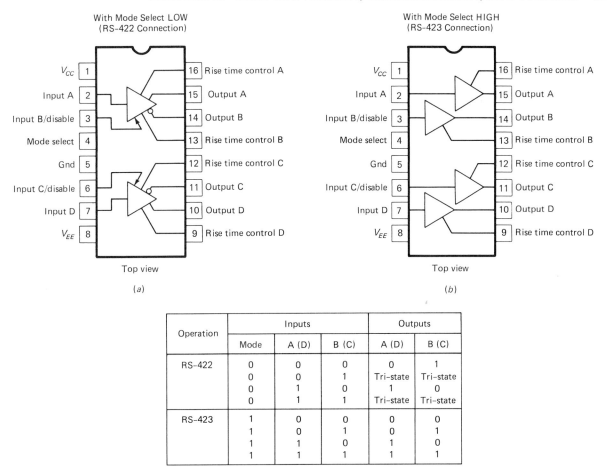

Fig. 13-20 Integrated circuit DS3691 line driver. (*a*) With mode select LOW.
(*b*) With mode select HIGH. (*c*) Truth table (National Semiconductor).

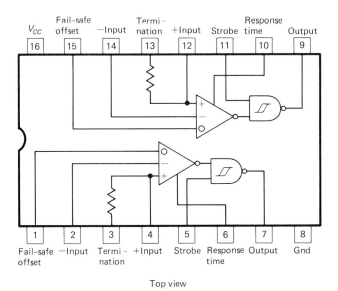

Fig. 13-21 Integrated circuit DS78LS120 dual differential line receiver (National Semiconductor).

and may cause data errors. Sometimes a capacitor is used in series with the terminating resistor to eliminate dc power loss in the resistor.

Figure 13-21 also shows that each line receiver contains a fail-safe offset input. *Fail-safe operation* involves forcing the output to a known condition in the event that the data circuit opens or short-circuits. When pins 1 and 15 are connected to +5 V, the input thresholds are offset from 200 to 700 mV, referred to the noninverting input, and from −200 to −700 mV when referred to the inverting input. If the input should open or short-circuit, the input will be greater than the input threshold, and the receiver output will remain in a specific logic state.

Figure 13-22 gives examples of both balanced and unbalanced data transmission circuits using the DS3691 line driver and the DS78LS120 line receiver. The balanced wires are shown as a twisted pair. This is standard practice for balanced transmission, and data cables are available with the wires twisted into pairs. The offset inputs of the line receivers are

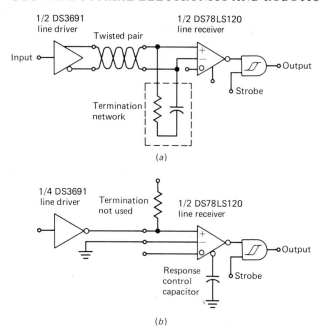

Fig. 13-22 Balanced and unbalanced data transmission. (*a*) RS-422A balanced data transmission. (*b*) RS-423A unbalanced data transmission.

shown floating as the fail-safe feature is not implemented in these examples.

The RS-422 and RS-423 standards involve electrical specifications only. The EIA has also introduced RS-449 as a possible successor to RS-232C. This standard includes a description of the signals needed for modem control and the mechanical specifications for the connectors. The RS-449 specifies a total of 46 wires and uses two separate connectors; one with 37 pins and the other with 9 pins. The 9-pin connector carries all signals associated with a secondary channel. Most applications do not have the secondary channel, and the 9-pin connector is eliminated in these cases.

All of the data communication standards discussed to this point are *bit serial*. Another technique is *bit parallel,* which allows 8, 16, or 32 bits to be transferred at a time. One example is the Institute of Electrical and Electronic Engineers (IEEE) 488 standard, which is a bit-parallel, byte-serial system. This standard was originally proposed by Hewlett-Packard, Inc., as a general purpose interface bus for instruments, calculators, and computers. It is still often referred to as the *Hewlett-Packard Interface Bus* (HPIB) or as the *general purpose interface bus* (GPIB). It allows building a complex system of devices to be as simple as connecting cables to the various parts. However, software compatibility is not guaranteed by the IEEE 488 standard. This standard specifies signal voltages, currents, and signal timing. The work of Hewlett-Packard goes a bit beyond the IEEE standard in that it does specify software communication techniques.

Figure 13-23 shows the basic configuration of the

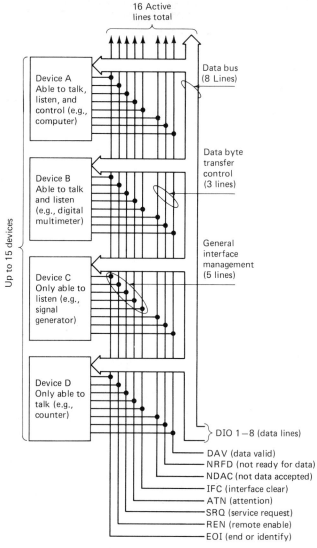

Fig. 13-23 IEEE 488 interface bus.

IEEE 488 bus. It can accommodate up to 15 devices, which might include items such as computers, digital multimeters, printers, plotters, calculators, signal generators, counters, and disk drives. It is designed for limited distances (up to 20 m) and data rates up to 1 megabyte per second. The signals are TTL-compatible, and 16 active lines transfer data and commands. Eight of the lines are used for data transfers, which are asynchronous and are coordinated by handshaking lines. *Handshaking signals* are used to prevent one device from sending data faster than another device can receive it. The handshaking is accomplished by the data byte transfer control lines shown in Fig. 13-23. The remaining five lines are for the control of bus activity and are shown as the general interface management group in the illustration.

The devices connected to the bus are identified as talkers, listeners, or controllers. A *talker* originates information, and a *listener* receives it. Some devices

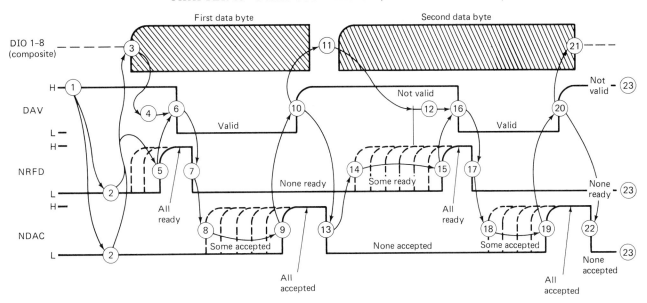

Fig. 13-24 HPIB three-wire handshaking.

may be both. A computer is usually a talker, a listener, and a controller. *Controllers* dictate the role of each device by asserting the *attention* (ATN) line and sending talk or listen addresses on the data lines. Addresses must be programmed into each device at the time of system configuration. This is usually done with switches on the back of the instrument or with jumpers on an internal PC board. When ATN is asserted, all devices must listen to the data lines. When ATN is false, only those devices that have been addressed will send or receive data, and all other devices ignore the data lines.

Several listeners can be active at a time, but only one talker may be active at any given time. When a talk address is placed on the data bus with ATN asserted, all other talkers are automatically deselected. The active talker controls the *data valid* (DAV) line. The active listener or listeners control *not ready for data* (NRFD) and *not data accepted* (NDAC). The various listeners may not all accept data at the same rate. The transfer speed is set by the slowest active listener on the bus. This is controlled by an open collector voting system. All the active listeners must agree before the transfer is allowed. The NRFD line is pulled to +5 V with a resistor. Any device that is not ready will ground this line through the collector of a transistor. All of the collectors must go high before NRFD can go high. When the talker sees NRFD go high, it places a byte on the data bus and waits for a 2-μs settling period. It then asserts DAV by pulling it low. This alerts the listeners to read the bus. They acknowledge by pulling NRFD low. A timing diagram that details the sequence involved in the handshaking process is given in Fig. 13-24.

When ATN is true, addresses and universal commands are transmitted on seven of the data lines.

The ASCII code is used for commands. Any active talker will relinquish control of the bus when this happens by releasing the DAV line. The controller now acts as the talker, and all other devices accept the information. This allows the controller to configure the bus and assign the roles of talker and listener to the various devices on the bus. More than one controller is allowed, but only one may be active at a time. The main system controller is activated at the time of power-on. Any other controllers remain passive until control is passed to them. Once the controller has configured the bus, it takes no part in subsequent data transfers until reconfiguration is required.

In addition to ATN, there are four other control lines. *Interface clear* (IFC) places the interface in a quiescent state. *Remote enable* (REN) is used with other coded messages to select either local or remote control of each device. Any active device can assert the *service request* (SRQ) line, which tells the controller that some device on the bus needs attention. *End or identity* (EOI) is used by a device to indicate the end of a multiple byte transfer. A second use for EOI is when the controller asserts ATN and EOI at the same time to signal that a parallel poll is requested. This alerts any device capable of a parallel poll to indicate its current status on the data line assigned to it. Up to eight devices can respond. This is a fast way for the controller to determine which device has requested service. The controller can also use a *serial polling routine*, in which it addresses one device as a talker and then sends that device a serial-poll enable command. The serial poll allows the controller to obtain 8 bits of information from a single device. The controller then sends a serial poll disable command to return the device to the data mode. The advantage of serial polling is that much more infor-

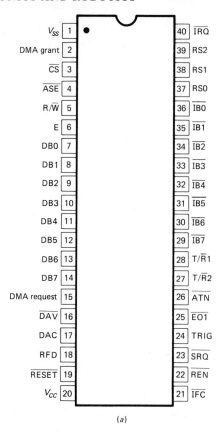

(a)

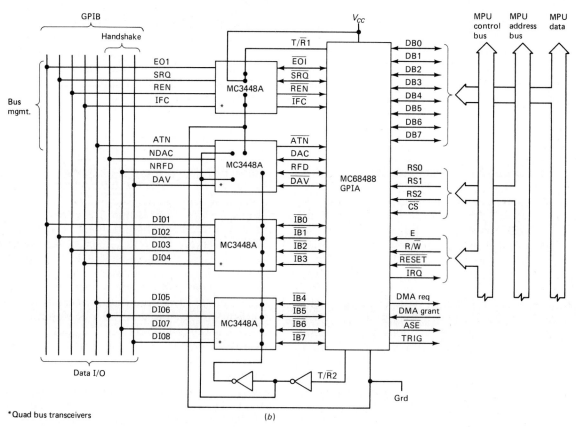

(b)

*Quad bus transceivers

Fig. 13-25 Integrated circuit MC68488 general purpose interface adapter.
(a) Pinouts. (b) IEEE interface (Motorola).

mation can be obtained from each device polled. The disadvantage is the time required to poll every device.

Figure 13-25 shows the Motorola MC68488 *general purpose interface adapter* (GPIA) designed to interface microprocessors such as the MC6809 to the IEEE 488 bus. The basic device is rated for a clock rate of 1 MHz. Motorola also makes a 1.5-MHz version (MC68A488) and a 2-MHz version (MC68B488). This IC greatly simplifies the interface. Many bus functions are handled automatically by the GPIA and require no action from the processor. Other functions require minimum processor attention because of the large number of internal registers that convey information on the state of the GPIA and of the bus. It features address recognition, full support of the handshake protocol, programmable interrupts, serial and parallel polling capability, talk-only or listen-only capability, and compatibility with the microprocessor bus.

REVIEW QUESTIONS

11. Data communications that occur on a single wire referenced to ground or to a return wire are a form of _____ transmission.

12. Differential communication circuits that use two wires are a form of _____ transmission.

13. RS-232C is an example of a(n) _____ transmission circuit.

14. Refer to Fig. 13-21. What does the symbol inside the NAND gates signify?

15. Refer to Fig. 13-23. What is different about the ground reference technique used in RS-423 as compared to that in RS-232C?

16. Refer to Fig. 13-23. What is the combined function of DAV, NRFD, and NDAC?

17. Refer to Fig. 13-24. From what time (circled number) to what other time is the first byte being read by the slowest listener on the bus?

13-4
DISK STORAGE

Magnetic disk storage has become a very important part of many computer systems. Disk drives are used on large mainframe computers, minicomputers, and microcomputers. This section will deal mainly with the types of disk systems used with minicomputers and microcomputers. The industrial technician is less likely to be directly involved with the high-capacity rigid disk systems used with large mainframe computers.

Magnetic disks provide a way to store programs and data. They also serve as an extension of computer memory. The ideal situation is to have all necessary information in computer memory. Since memory is random access, the best operating speed is provided by this situation. However, it is often not

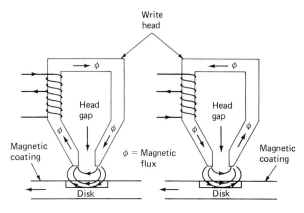

Reversing the head current creates a flux reversal on the disk

Fig. 13-26 Disk recording.

practical to provide enough memory to hold all of the programs and data that the computer will ever need. Disk memory is block access. It is slower than RAM, but it is fast enough for many applications.

Magnetic disks are made from aluminum or Mylar. They are coated with a magnetic material such as ferric oxide. Binary information is stored in the magnetic coating by orienting the domains one way or the other. This is done by passing a current through a write head, as shown in Fig. 13-26. The current is reversed to change the direction of magnetization. Figure 13-27 shows two of the ways that binary data can be recorded magnetically. Figure 13-27(*a*) shows the basic *nonreturn to zero* (NRZ) method. The cur-

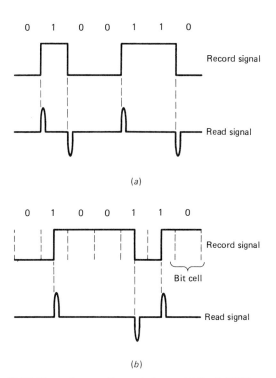

Fig. 13-27 Magnetic recording techniques. (*a*) Basic NRZ recording. (*b*) NRZI recording.

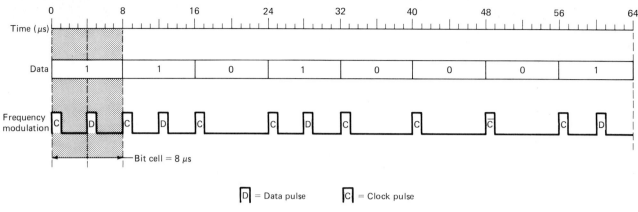

Fig. 13-28 Frequency-modulated (FM) encoding.

rent in the head remains constant for consecutive 1s or 0s. It reverses on transitions from 0 to 1 or 1 to 0. The magnetically stored information is read back by exposing a read head to the magnetic patterns on the disk. One head is usually used for both reading and writing and is called the *read/write head*. Note that the read pulses are positive for 0 to 1 transitions and are negative for 1 to 0 transitions for the NRZ recording technique. Figure 13-27(*b*) shows the *nonreturn to zero inverted* (NRZI) recording method. This method reverses the direction of the head current for every binary 1 that is recorded. The head current remains constant for 0s. Note that a series of 1s will show a flux change for each bit. A read pulse will occur only where a 1 has been recorded. Whether the read pulse is positive or negative is immaterial when the NRZI recording technique is used. Information is read back from the disk by interpreting any flux change as a 1 and the absence of change within a predefined interval called a *bit cell* as a 0. The NRZI is the standard recording technique used in all disk systems.

In addition to recording techniques, there are encoding techniques. Do not confuse the two. The *encoding* schemes affect the patterns of flux changes placed on the disk; the *recording* technique is always NRZI. A read circuit called a *data separator* converts the recorded patterns into meaningful information. Frequency modulation (FM) is an early encoding scheme used in disk systems and is illustrated in Fig. 13-28. A clock bit is written at the beginning of each bit cell, and data bits are written between clock bits. A typical bit cell might be 8 μs wide, and the data bit is written 4 μs after the clock bit. The data bits are 8 μs apart, and the data transfer rate can be found by taking the reciprocal of this period:

$$\text{Data rate} = \frac{1}{8 \times 10^{-6}} = 125,000 \text{ bits/s}$$

The constant bit cell reference makes encoding and decoding simple in FM systems. The data separator synchronizes on the clock bits and generates a 4-μs window beginning 2 μs after the clock pulse.

Straightforward timing circuits are used to generate the window.

The problem with FM encoding is that the clock bits represent a significant overhead. They take half the available space on the disk and provide timing information only. *Modified FM* (MFM) doubles the available space by replacing clock bits with data bits; MFM is shown in Fig. 13-29. This encoding technique writes clock bits only if data bits have not been written into the preceding and present bit cells. The clock bits are placed at the beginning of the cells and the data bits in the center, just as in FM. This arrangement cuts the bit cell time to 4 μs. The capacity of the disk is doubled, as is the data transfer rate. However, timing tolerances are cut in half. The data windows are only 2 μs wide, necessitating a more sophisticated data separator. The data stream must be continuously analyzed to keep the windows synchronized with the correct bits. A *phase-locked loop circuit* is used to lock a local clock to the data stream for decoding purposes.

The ultimate encoding system would eliminate clock pulses altogether. The FM and MFM must have clock pulses. Otherwise, a string of 0s would

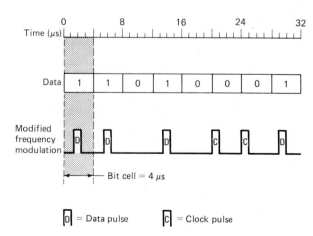

Fig. 13-29 Modified frequency-modulated (MFM) encoding.

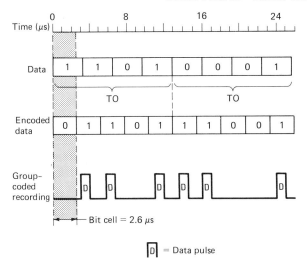

Fig. 13-30 Group coded recording (GCR) encoding.

Encoding Chart for Group-Coded Recording		
Hexadecimal number	4-bit nibble	GCR code
0	0 0 0 0	1 1 0 0 1
1	0 0 0 1	1 1 0 1 1
2	0 0 1 0	1 0 0 1 0
3	0 0 1 1	1 0 0 1 1
4	0 1 0 0	1 1 1 0 1
5	0 1 0 1	1 0 1 0 1
6	0 1 1 0	1 0 1 1 0
7	0 1 1 1	1 0 1 1 1
8	1 0 0 0	1 1 0 1 0
9	1 0 0 1	0 1 0 0 1
A	1 0 1 0	0 1 0 1 0
B	1 0 1 1	0 1 0 1 1
C	1 1 0 0	1 1 1 1 0
D	1 1 0 1	0 1 1 0 1
E	1 1 1 0	0 1 1 1 0
F	1 1 1 1	0 1 1 1 1

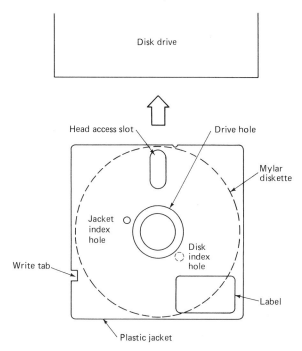

Fig. 13-31 Floppy diskette.

produce a section on the disk with no flux changes, and the data separator would not know the length of the string. *Group coded recording* (GCR) overcomes this limitation by dividing each incoming byte to be recorded into 2 nibbles (a *nibble* is a 4-bit word). The nibbles are then translated into 5-bit binary codes. Out of a total of 32 possible codes, only 16 are used, and they are chosen to meet two important requirements: (1) there can never be more than two consecutive 0s, and (2) no combination of codes can produce more than two consecutive 0s on the disk. Figure 13-30 shows the translation scheme. Inspect the column of GCR codes. You will find that no code contains more than two consecutive zeros, and neither does any combination of two codes. This allows the read clock to synchronize on data bits alone. The read circuit converts the 5-bit codes back into the original nibbles. The nibbles are rejoined to form a byte for return to the computer. Use of GCR encoding provides approximately a 150 percent improvement in disk capacity and data transfer rate when compared to MFM.

Figure 13-31 shows the appearance of a typical floppy diskette. *Floppy* describes the characteristic that the actual recording medium is coated Mylar and is not rigid as it is in the large mainframe drives. *Diskette* signifies that the size is 5.25 in., rather than 8 in. The original floppy drives used 8-in. disks; however the smaller drives and diskettes have become very popular. The prefix *mini* is another way to signify that the size is 5.25, rather than 8 in. A plastic jacket protects the Mylar disk. There are several holes in the jacket. One provides access for the head or heads, another allows the drive hub to contact the center hole of the Mylar so the disk can be turned in the jacket, and another allows the index hole to be sensed as the disk is turning. *Indexing* provides a starting point for read and write operations. There is also a write protect cutout along the side of the jacket. The cutout is covered with a gummed tab to write-enable 8-in. disks. Just the opposite is true with 5.25-in. diskettes: the tab is applied to write-protect the disk.

Floppies must be handled with care. They should be stored vertically in protective envelopes. They should not be exposed to high temperatures and direct sunlight. Their storage and operating range is 10 to 50°C. Use only felt tip pens when writing on the labels. Never touch the disk surface through the head access slot. Do not attempt to clean them. The inside of the plastic jacket is lined with a special pad material that cleans the disk surfaces as it rotates inside the jacket. Never bend or fold them and never expose them to magnetic fields, dust, dirt, or smoke.

Figure 13-32 shows how one type of head-positioning mechanism in floppy disk drives works. A stepper motor turns a lead screw. The lead screw nut

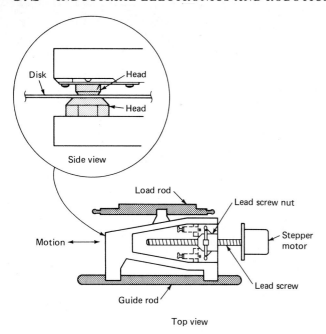

Fig. 13-32 Floppy disk head-positioning mechanism.

is mounted on the head mechanism, which can slide back and forth on the guide and load rods. The enlarged side view shows the disk pinched between the heads. This happens when the disk is being written or read and is accomplished with a head-load solenoid. Some drives are *single-sided,* indicating that they record on one side of the disk only. In these drives, the top head is replaced with a pressure pad. Although not shown in the illustration, a drive spindle and collett assembly engage the center hole of the Mylar disk. Another motor turns the disk at 300 rpm. The information is recorded in tracks as the disk turns.

Figure 13-33 shows a typical disk format. There are 40 concentric tracks in all. The stepper motor moves the head toward the center or away from the center to record or read the various tracks. The track-to-track access time is typically around 25 ms. The average access time to reach a given track is around 400 ms. Thus, you can see that this block access device is reasonably fast, but not nearly so fast as RAM. The tracks are divided into 18 sectors. Each sector can hold 256 bytes of data plus approximately 82 control bytes. The control bytes are used for synchronizing, identifying the track and sector number, and checking for errors. Since there are 18 sectors with 256 data bytes each, the track capacity is 4608 bytes. The total disk capacity is 184,320 bytes. However, not all of this is available as user storage. The center track is used to store the directory of the disk. The *directory* is a listing of all of the files stored on a disk, their size, their type, and their location (track and sector). Since the directory is frequently accessed, placing it at the center of the disk speeds operation.

The original 8-in. floppy disks used FM encoding and stored 250K bytes per side. *Double-density drives* using MFM encoding for a capacity of 500K bytes per side were then developed. The latest 8-in. drives store 800K bytes per side. The original 5.25-in. mini floppy diskettes used FM and provided 90K bytes per side. The change to MFM provided double density and 180K bytes per side. Recent developments, such as GCR encoding and higher track/sector densities, have produced capacities up to 670K bytes per side on 5.25-in. diskettes.

Another development is the microfloppy disk shown in Fig. 13-34. This floppy features smaller size and a more rugged structure. The plastic case is much more rigid than the plastic jackets used to protect the larger floppies. It uses a metal shutter

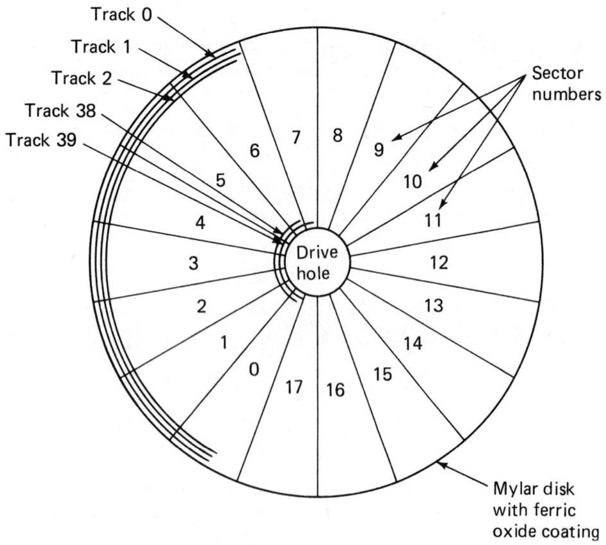

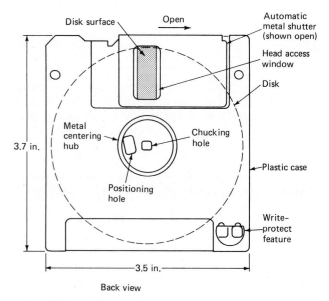

Fig. 13-33 Floppy disk format.

Fig. 13-34 Microfloppy.

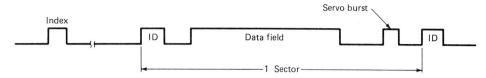

Fig. 13-35 Servo disk format.   ID = track and sector identification bytes

that hides the head access slot to prevent the disk surface from being accidentally touched. The shutter is opened automatically when the disk is inserted in the drive. It is write-protected by rotating a small plastic piece located in the lower right-hand corner. Instead of an open center, it uses a metal drive hub for accurate centering and positive mechanical indexing. Microfloppies have capacities of around 400K bytes per side. They are one-fourth the size, one-half the weight, and consume one-half the power of minidrives (5.25 in.).

Disk technology improves rapidly. The designers are constantly finding ways to pack the data at higher densities. One of the limitations is the medium itself. The Mylar expands and contracts with heat and humidity variations, altering the concentricity of the tracks. All points on any track should be at the same distance from the center hole, but the variations produce track weave. With very high track density, the resulting positioning errors cause data errors. The clamping collett does not always center the disk accurately, thus also limiting track density. These problems have been solved by the newest minifloppies, which pack 3.3M bytes onto a double-sided disk. The drives are only half the size of a standard minifloppy. Greatly improved track density is made possible by a closed loop head-positioning system. Servo information is written to the diskette when it is manufactured. A servo writer places a burst of information in each sector, as shown in Fig. 13-35. A microprocessor-based controller in the drive continuously monitors the servo bursts and sends correction pulses to a fine stepper motor in the head-positioning mechanism. A coarse stepper motor is also available for ordinary track-to-track positioning. The result of this closed loop system is a head-positioning accuracy of within 2.5 µm of the exact data track. The allowable track density increases from 38 to 76 tracks per centimeter with this system. The disadvantages are higher drive and higher media costs resulting from the extra processing step added when the disks are made. Nonservo drives can use blank disks. A blank disk can be organized with track and sector information the first time it is used; this process is called *formatting the disk.*

Winchester drives using rigid aluminum disks are also available for small- and medium-sized computer systems. The original Winchester drive used a disk that was 14 in. in diameter. Subsequently, it was reduced to 8 in., and then to 5.25 in. The disks are fixed in Winchester drives. Today's small Winchester drives provide capacities from 5 to 100M bytes.

They use heads that are permanently sealed with the disk to eliminate problems with smoke, dust, and dirt.

Removable cartridge drives are made with rigid disks that are 3.9, 5.25, or 8 in. in diameter. They are not true Winchester technology, but they are often referred to as *Winchester cartridges.* The cartridge drives offer high capacity (6.38M bytes for the 3.9-in. unit) and an average access time of 75 ms. Their data transfer rate is 5M bits/s, as compared to 250K bits/s for floppies. They also use servo-positioned heads under microprocessor control. Figure 13-36 shows a removable cartridge and drive. As the cartridge is inserted, the door slides open to permit access to the read/write heads, which were previously retracted to track zero. The disk is seated on a spindle and secured by a magnetic hub. The medium is shown in cross section and uses thin-film plating for the magnetic surface. This metal plating is 1000 times harder than ferric oxide, for decreased susceptibility to damage. It is also capable of supporting a magnetic density of over 3000 flux reversals per centimeter, which is 2.5 times better than that of oxide. A graphite lubricating coating is provided to shield against damage from dust and dirt. The cartridge drives eliminate the purge cycle associated

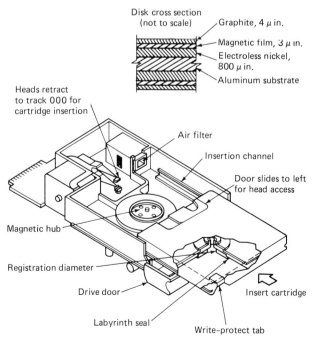

Fig. 13-36 Removable hard disk cartridge and drive.

with Winchester drives. A *purge* involves blowing filtered air over the disk to remove dust and debris. This can take several minutes and delays computer operation when the system is powered on.

Disk manufacturers are experimenting with a vertical recording format that may provide a 10-fold increase in disk capacity. The current recording format is horizontal (along the disk surface), as illustrated in Fig. 13-26. Vertical recording will allow a much higher bit density if the technical problems can be worked out.

The performance of disk drives, especially the floppies, can be "fussy." They may work only some of the time. They may refuse to read information that was written to the disk by a different drive. Speed errors greater than plus or minus 2 percent are one source of difficulty. Speed is adjustable on the older drives by turning an internal multiturn potentiometer. It may be set while watching a built-in strobe disk with the drive cover off and with the mechanism positioned under fluorescent lighting. The potentiometer is slowly turned until the strobe bars appear to be motionless. Or special diagnostic software may be available to assist in setting the speed. The newer drives use a servo speed-control circuit and are not adjustable. Clamping problems cause hub eccentricity because the disk does not properly center on the drive spindle. Remove the disk and try inserting it again when problems occur. *Head azimuth* is the angle at which the head intercepts the track's center line. Diagnostic software and special alignment disks may be available to aid in adjustment. Head azimuth is critical and will cause errors if it is out of tolerance. Drives that have seen quite a few hours of service may develop backlash in the positioning mechanism. This can be diagnosed with software that steps the head in one direction to a test track and then steps the head in the other direction to the same test track. Any backlash will show up as increased error for one of the trials. The heads may need cleaning from time to time. Special cleaning disks are available for this purpose. Finally, the medium does wear out; when occasional errors occur, try transferring the contents to a new disk.

REVIEW QUESTIONS

18. When reading a disk recorded with the NRZI technique, the _____ of the read pulses is immaterial.

19. When reading an NRZI disk, if no read pulse occurs within the bit cell interval, we know that a _____ was stored there.

20. Which disk encoding scheme writes a clock bit at the beginning of every bit cell?

21. The MFM encoding method writes a clock bit at the beginning of a bit cell only if that cell and the preceding cell contain a _____.

22. Which encoding method writes no clock bits to the disk?

23. What is the maximum number of consecutive 0s found on a disk using GCR encoding?

24. Floppy disks are made from Mylar, and rigid disks are made from _____.

25. What is the function of the servo burst shown in Fig. 13-35? When is it recorded onto the disk?

13-5 TAPE STORAGE

Magnetic tape storage was one of the earliest extensions of computer memory. It is a *serial access medium*. If the information is located at the other end of the tape, a significant delay will be experienced in loading the information back into computer memory. Disk storage systems have replaced tape as an extension of computer memory because they are block access devices with far less delay. However, magnetic tape is still an important part of many computer systems. Tape drives are commonly used to back up disk files. For example, suppose an industrial computer uses a 10M byte Winchester drive to store data and programs. If the disk fails, how will the information be restored? If it has been replicated on tape the system can be restored in minutes after the disk drive is repaired or replaced. Some computer operations have a daily back-up schedule to transfer the disk contents to tape. This practice ensures that no more than 1 day's work will be lost. It is possible to back up when using floppies, but the necessary disk swapping is tedious, time-consuming, and error-prone. Tape back-up is the method of choice for many applications.

Tape is easy to store. It is packaged in reels, cartridges, or cassettes. It is less vulnerable to loss of data than are floppies. The medium itself is 38-μm (0.0015-in.) thick Mylar with a ferric oxide coating on one side. The width ranges from 0.38 to 1.27 cm (0.15 to 0.50 in.). A reel that is 27 cm (10.5 in.) in diameter can hold as much as 1100 m (3600 ft) of tape. Modern tape drives use the same high-density recording techniques as the disk drives discussed in the last section. By using NRZI recording and GCR encoding, the storage capacity of a single reel of tape can be as high as 540M bytes. This is one of tape's strong features: vast quantities of data can be stored inexpensively.

The mechanical assembly that drives the tape and includes the heads is called a *transport mechanism*. High-capacity transports use reels to supply and take up the tape. They also use a capstan drive that rotates continuously. A solenoid-actuated pressure roller pinches the tape, which then comes up to reading or writing speed in several milliseconds. Reading and writing are performed at speeds up to 6.35 m/s, and even higher speeds are used for rewinding. The rapid acceleration saves tape because that part of it that passes over the heads during starting and stopping is wasted. No information can be stored at these locations, and they are known as *interrecord gaps*. The supply reel and the take-up reel are buffered from the capstan to avoid stretching or breaking the tape. The buffers are formed by tape loops that hang in vertical columns under vacuum tension. The reels are servo-driven to maintain a constant supply of tape in the buffers. Tension arms can also be used to form the tape buffers. Data are recorded in blocks with gaps in between. Special start and stop characters signal the beginning and the end of each block.

High-speed start/stop transports are expensive. Because tape is more often used today to back up disk storage, a less complicated transport is possible. Back-up tapes do not have to be designed to be searched for specific blocks of data. Streaming drives read or write all of the disk data in a continuous stream. Start/stop drives require fast mechanical performance and also require identifying header and trailer blocks to locate specific groups of data. These blocks are eliminated in streaming drives, and so are the interrecord gaps. Thus, streaming tape drives are less complicated and less expensive, and they pack more data on the tape.

Figure 13-37 shows a tape read/write head with separate gaps for writing and reading. This arrangement allows the data to be verified immediately after it is written. The head is actually a stack of nine sections to allow nine tracks or channels to be placed on the tape simultaneously. The nine-track format is convenient because it allows the data to be stored as bytes and allows an extra bit for parity. The parity channel can be used in conjunction with interspersed parity words to provide error correction, as shown in Fig. 13-38. A parity word follows a block of 8 data bytes; this arrangement provides a double-parity check. The illustration shows an odd parity system. All the rows and columns should have an odd number of 1s. The circled bit is in error. It is found at the intersection of the row and column, where the parity is even in both cases. This information is used to locate and change the 0 bit that is in error to a 1. Multiple bit errors may not be detected by this method and cannot be corrected in any case. Actual instances of multiple errors are so low that more elaborate checking is not warranted. When no errors are detected, the parity channel and the parity words are stripped away, and the information is returned to the computer.

Cartridge tape systems provide convenient and compact mass storage for mini- and microcomputers. The tape drive takes up the same physical space as a minifloppy drive. It contains an *end of tape/beginning of tape* (EOT/BOT) sensor to prevent overrunning the tape. The cartridge is made from plastic and contains the tape, supply hub, take-up hub, and drive roller. The cartridge holds 137 m (450 ft) of 0.64-cm (0.25-in.) wide tape. The data capacity of one cartridge is as much as 17M bytes, with a transfer rate as high as 90K bytes/s. Back-up of a 10M byte Winchester can be accomplished in less than 3 min at this rate.

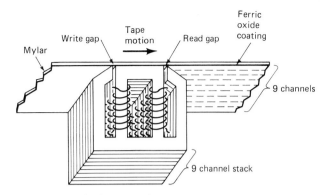

Fig. 13-37 Typical tape read/write head.

A *cassette* is a specific type of cartridge that was originally designed to record audio by Phillips of the Netherlands. The Phillips cassette is a well-established standard; other cartridges have not yet reached that same level of standardization. Special digital cassettes that are based on the Phillips standard and are certified for binary recording are manufactured. Binary recording is a saturated form of recording; audio is nonsaturated. The tape in digital cassettes uses a magnetic coating that is easier to saturate. The cassettes hold 137 m (450 ft) of tape. Some digital cassette tape drives use NRZI recording with GCR encoding and place four tracks on the tape for a 20M byte capacity. The data transfer rate is 30K bytes/s, and a 10M byte Winchester can be backed up in less than 6 min.

It is also possible to record binary data on an ordinary audio recorder. The binary data must be converted to audio tones that fit into the bandwidth of the recorder. Logic 0s are represented with four cycles of 1200-Hz sinusoidal audio and logic 1s with eight cycles of 2400-Hz sinusoidal audio. The bit time period is equal to 3.33 ms for both 0s and 1s using

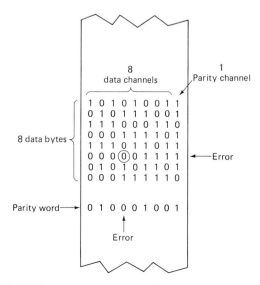

Fig. 13-38 Error correcting on nine-channel tape.

this technique. The result is a rather slow bit rate of 300 Bd. The data transfer rate is less, since the asynchronous format requires that 1 start bit and 2 stop bits be attached to each byte. Various techniques can be used to improve the bit rate, and 2400 Bd has been achieved. However, this still results in a relatively slow data transfer rate. Frequent errors are another problem, and because of these limitations the audio recording method has found acceptance only with hobby and personal computers.

Tape reels, cartridges, and cassettes should be stored away from magnetic fields. They should not be exposed to dust and dirt. Cartridges and cassettes may exhibit erratic behavior due to binding. It may be possible to free them temporarily by tapping the case on a hard surface. It is best to replace them at the first sign of any difficulty. The drive heads and other parts may require cleaning. Special cleaning cartridges and cassettes are available. Follow the manufacturer's recommended procedures.

REVIEW QUESTIONS

26. Computer memory can be divided into random access, block access, and serial access. Which type is the fastest? Which type includes magnetic tape? Disk?

27. A major application of tape storage is to back up _____ files.

28. Refer to Fig. 13-38. What would happen if the top bit in the parity column were read wrong? Assume that there would be no other errors.

29. Refer to Fig. 13-38. Suppose two of the 1s are read as 0s in the top data byte. Assume that there are no other errors. Can the 2 bits be corrected? Which of the 8 data bytes would have to be flagged with a possible error?

30. The Phillips cassette was originally designed to record _____.

13-6
BUBBLE MEMORY

The volatility of RAM is a serious drawback for some industrial computer applications. Power failures, brownouts, and power glitches can wipe out the contents of memory. Bubble memory is based on magnetic storage and is nonvolatile. It is used in applications such as numerical control, robotics, process control, remote terminals, and remote data acquisition.

Bubbles are based on domains. A *domain* is a group of atoms with parallel magnetic orientations. Each domain may be considered as a tiny unit magnet. When a material is unmagnetized, its domains are randomly oriented along three axes. Magnetizing the material orients a significant number of its domains along the same axis. If the material is magnetically saturated, most of its domains will have achieved the same orientation. It is possible to limit the axes of orientation to two by making a magnetic material very thin (less than 25 μm). Bubble memories are made by depositing a thin crystalline film of garnet on a substrate. Garnet is a material that occurs in nature; it can also be manufactured from iron oxide and yttrium. The domains in the garnet film are snakelike in shape and are perpendicularly oriented to the surface, as shown at the left in Fig. 13-39. The flux direction arrows and magnetic polarity signs show that the domain arrangement is random with no external field applied. When a weak external field is applied, the domains that are in opposition to the external field begin to become narrower. As the external field increases in strength, the opposing domains shrink in length and become cylindrical. They appear as circles when viewed from above the surface and are called *bubbles*. If the external field is strong enough, the opposing domains will vanish entirely. The external field is called the *bias field*.

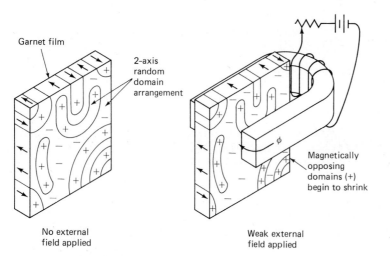
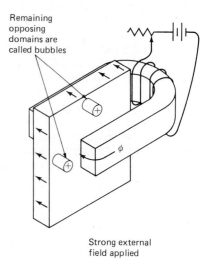

Fig. 13-39 Creating magnetic bubbles with a bias field.

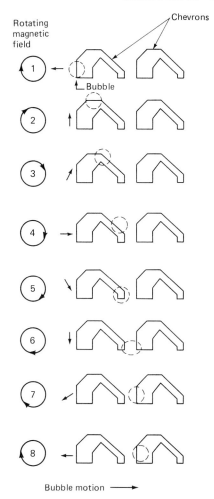

Fig. 13-40 Bubble propagation (Intel).

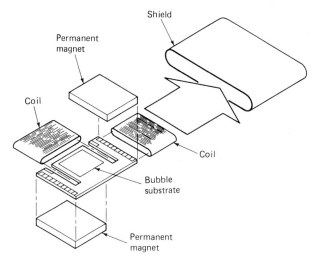

Fig. 13-41 Exploded view of magnetic bubble unit assembly (Intel).

a return path for the bias field and also prevents disturbances from outside magnetic forces.

The data storage is arranged in a block replicate architecture that is shown in Fig. 13-42. The bubble chip contains a number of endless loops. The bits of successive pages continuously circulate around these loops. A controller writes into and reads data from the storage loops by using the input and output tracks. A seed bubble provides a way for the controller to write a binary 1 to the input track, as shown in Fig. 13-43. The seed is generated when the bubble memory chip is manufactured by a current pulse in the hairpin loop. The pulse is strong enough to reverse the flux of the local bias field, and a bubble is

The bubbles can be used to store binary information. The presence of a bubble indicates a 1, and the absence of a bubble indicates a 0. The bubbles are driven through the garnet film under a metallic pattern of chevrons. Figure 13-40 shows how the bubbles are made to propagate under the chevrons. An external rotating field interacts with the chevrons and provides the driving force. This rotating field is produced by exciting a pair of driver coils with two 50-kHz triangular waveforms that are 90° out of phase. The field rate is 50,000 rps. The data, in the form of bubbles, circulate in loops. It requires 82 ms for one bubble to make a complete loop for an average access time of 41 ms to get to the first data bit. The data transfer speed is on the order of tens of thousands of bits per second. Thus, bubble memories are serial access. They have no moving parts and offer very high reliability.

An exploded view of a *bubble memory assembly* is illustrated in Fig. 13-41. The bubble substrate is surrounded by two coils that provide the rotating field. It is also sandwiched between two permanent magnets that provide the bias field. The entire assembly is surrounded by a magnetic shield that provides

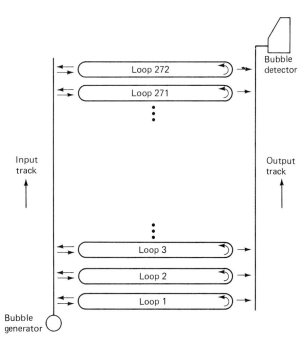

Fig. 13-42 Block replicate architecture (Intel).

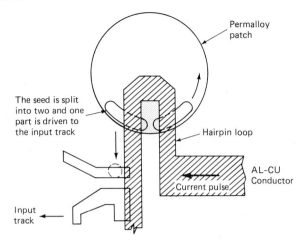

Fig. 13-43 Bubble generation (Intel).

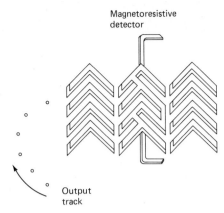

Fig. 13-45 Bubble detection (Intel).

created. The permalloy patch acts as a flux concentrater, and the seed assumes a kidney shape and rotates under the patch in reaction to the combined bias and rotating fields. When a 1 is to be written to the input track, a current pulse splits the seed in half. One half remains as the seed and quickly regains its original size. The other half, driven by the rotating field, propagates to the input track. When a 0 is to be written, no current pulse is applied.

A swapping procedure is used to transfer data (bubbles) from the input track to the storage tracks. The old data are brought back onto the input track for disposal at the end of the line, and the new data take their place in the storage loop. This occurs at the swap gate shown in Fig. 13-44 and is accomplished by sending a rectangular swap pulse through a conductor located under the chevrons. When data are to be read, the bubble is cut in two at the replicate gate by applying a current pulse with a steep leading edge. This process replicates the bubble for the output track, and the original data are retained in the storage loop. This is an example of *nondestructive readout*.

The bubble detector is near the end of the output track. It consists of a magnetoresistive bridge element and is shown in Fig. 13-45. Magnetoresistance is a change in electrical resistance due to the presence of a magnetic field. As the bubbles pass under the detector, an output of several millivolts is generated. Beyond the detector, the bubbles run into a rail and are destroyed.

Intel Corporation makes the 1M-bit 7110 bubble memory chip. Dummy detectors cancel the common mode noise induced by the rotating field. The chip is arranged as four identical quads. The entire device contains 320 storage loops, 272 of which are actually used. This leaves 48 spares. The decision about which to use is made after the unit is assembled and tested at the factory. The outcome of this decision is stored in an extra loop called the *boot loop* in the form of a 12-bit code for each active and inactive loop. Intel does this to increase the number of chips that will pass as acceptable. Nonworking loops will not prevent the memory from being fully functional unless more than 12 in any given quad are defective. Each active loop stores 4096 bits for a total capacity of 1,134,592 bits. The external appearance is 2048

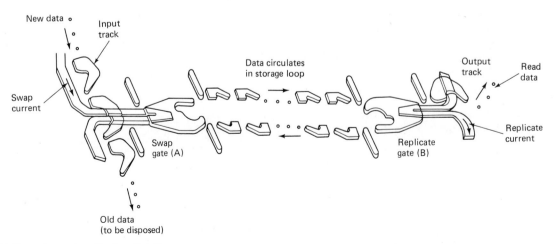

Fig. 13-44 Swapping and replication (Intel).

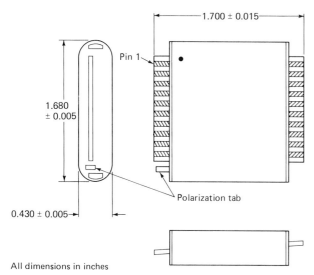

Fig. 13-46 7110 package (Intel).

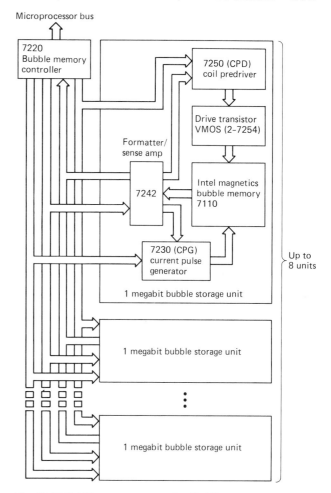

Fig. 13-47 Bubble memory expansion (Intel).

pages of 512 bits each. The physical appearance of the 7110 is shown in Fig. 13-46.

Intel also makes a group of support chips to make application of the bubble memory device easier. Fig. 13-47 shows the block arrangement of the bubble memory chip, along with the 7220 bubble memory controller, the 7230 current pulse generator, the 7242 formatter/sense amplifier, the 7250 coil predriver, and two 7254 VMOS coil drivers. One controller chip can handle up to eight memory units for a total of 1M byte of storage.

Bubble memories are extremely reliable. However, there are a few things that can go wrong. A memory failure may be due to the loss of one or more seed bubbles. This loss is usually caused by power supply problems. Be sure to verify proper operation of the power source for the bubble memory. A software verification procedure can then be used to check the status of the seeds. Intel recommends the following procedure to reseed the memory in the event that it is required. It assumes that certain software routines are available in computer RAM or ROM memory.

1. Power down.
2. Remove the 7230 pulse generator chip from its socket and place it into the special seed generator module.
3. Plug the seed generator module into the empty 7230 socket.
4. Power on.
5. Call ABORT (this is a software procedure that aborts the present command and resets the controller).
6. Call MBMPRG (a purge routine).
7. Call WRBUBL (write to bubble memory routine: the registers in the controller must contain the correct data).

8. Power down.
9. Restore the pulse generator chip.

The next step is to read the *boot loop code* inside the 7110; this code is used at every power on as a part of the bubble system initialization procedure. The code is also printed on the device label as in Fig. 13-48. The code is stored in a register located in the 7242 formatter/sense amplifier chip. From then on, this register defines the active loops for reading and writing. The formatter does not store any information to the inactive loops, and the sense amplifier ignores any data that come from them. In trouble-

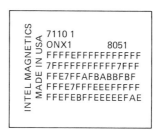

```
INTEL MAGNETICS          7110 1
MADE IN USA    ONX1            8051
               FFFFEFFFFFFFFFFFF
               7FFFFFFFFFFFF7FFF
               FFE7FFAFBABBFBF
               FFFE7FFFEEEFFFFF
               FFEFEBFFEEEEEFAE
```

Fig. 13-48 A 7110 bubble memory label showing the boot loop code.

shooting, the boot loop code should be matched byte for byte with the code found on the label of the 7110 bubble memory chip, as shown in Fig. 13-48.

REVIEW QUESTIONS

31. List the three types of nonvolatile memory covered in this chapter.

32. The permanent magnets in a bubble memory chip are used to provide the _____ field.

33. Bubble memories use a metallic pattern of _____ to guide the bubbles as they move.

34. The force that moves the bubbles is supplied by a _____ magnetic field.

35. A 1 is written to the input track with a current pulse that splits the _____ bubble into two parts.

36. With no swap pulses applied in Fig. 13-44, what happens to the data in the storage loop?

37. What happens when swap pulses are applied?

CHAPTER REVIEW QUESTIONS

13-1. Calculate the nominal output voltage for an 8-bit DAC with a 5-V supply and a binary input of 10010111.

13-2. Digital-to-analog converters that use an external signal as their reference may be referred to as _____ DACs.

13-3. Identify the A/D converter that uses an up/down counter, a DAC, a clock, and a comparator as the major circuit elements.

13-4. Can successive approximation A/D conversion be accomplished by using a microprocessor to write binary data to the DAC and thus eliminate the clock and the register?

13-5. The binary output from a dual-slope A/D converter is established by counting the time that the _____ signal is applied to the integrator.

13-6. What type of circuit can be used to "freeze" a time-varying analog signal for A/D conversion?

13-7. How many comparators would be required to build a 10-bit flash converter?

13-8. How many comparators would be required to build a dual-stage flash converter with 10 bits of resolution?

13-9. Much faster data rates are possible on _____ transmission circuits.

13-10. Differences in ground potentials can cause errors in circuits using the _____ standard.

13-11. The EIA has established the _____ standard for balanced transmission.

13-12. High-speed data circuits are subject to errors due to reflected signals, especially when the transmission lines are not properly _____.

13-13. The RS-232C, RS-422, RS-423, and RS-449 are all bit serial standards; the IEEE 488 is a bit _____ standard.

13-14. Which type of polling on the IEEE 488 bus provides the controller with most information? Which type is the fastest?

13-15. Polling is usually initiated by the controller after a talker or a listener asserts the _____ line on the IEEE 488 bus.

13-16. Should the write tab be opened or covered to write-protect a 5.25-in. diskette?

13-17. Floppy disks _____ be cleaned.

13-18. What is the function of the stepper motor in disk drives?

13-19. How is the average access time for the directory track minimized?

13-20. Which floppy disk uses a hard plastic case and a metal shutter to hide the head access slot?

13-21. Winchester drives use rigid disks that are _____ with the head assembly.

13-22. When a Winchester drive is first turned on, there is a start-up delay due to the _____ cycle.

13-23. A start/stop tape drive must start and stop on blank sections of tape known as _____ gaps.

13-24. Tape drives that are limited to back-up operation are known as _____ drives.

13-25. Digital cassettes contain tape that is easier to _____ than audio tape.

13-26. Digital information can be recorded on an ordinary audio recorder by converting the 0s and 1s to _____.

13-27. What happens when data are read from the storage loop in a bubble memory?

13-28. The bubble detector is based on the principle of _____.

13-29. Where is the code that identifies the active loops in a bubble memory chip stored?

13-30. How many bubble memory units can the Intel 7220 controller handle?

13-31. Under normal circumstances, when are the seed bubbles generated?

13-32. When is the boot loop code normally read into the 7242 formatter/sense amplifier chip?

13-33. Is the boot loop code ever read at any other time?

13-34. Is it possible to reseed a bubble memory?

13-35. What are the other two names used for the IEEE 488 standard?

ANSWERS TO REVIEW QUESTIONS

1. 0.00488 V; 4.995 V **2.** 0.098 percent **3.** 0.024 percent **4.** 7.5 V **5.** staircase **6.** 512 μs **7.** 16 μs **8.** none **9.** one of eight analog inputs **10.** it decreases; it will be less **11.** unbalanced **12.** balanced **13.** unbalanced **14.** hysteresis **15.** RS-423 uses the driver ground only; RS-232C grounds both ends **16.** they control handshaking **17.** from 6 to 9 **18.** polarity **19.** 0 **20.** FM **21.** 0 **22.** GCR **23.** 2 **24.** aluminum **25.** provides control information for accurate head positioning when the disk is manufactured **26.** random; serial; block **27.** disk **28.** the error intersection would occur at the parity bit, and all 8 data bytes would be accepted as being correct **29.** no; all of them **30.** audio **31.** disk, tape, and bubble **32.** bias **33.** chevrons **34.** rotating **35.** seed **36.** it circulates continuously **37.** new data enter the storage loop and old data leave it

14

OPTOELECTRONICS

The combination of optics and electronics, called optoelectronics, *has evolved into a major factor in industrial systems. Devices such as light-emitting diodes, lasers, photodiodes, and fiber optics, along with the older standard devices such as photo cells, will be covered in this chapter. The theory of operation of optoelectronic devices, their characteristics, and their applications will be presented.*

14-1
EMITTERS

There are many different light sources, such as tungsten lamps, fluorescent lamps, neon lamps, xenon lamps, and light-emitting diodes (LEDs). The characteristics and operation of the conventional light sources (lamps, flash tubes, sunlight) are familiar. This section will be devoted to a newer source, the LED.

Electroluminescence in solids is a phenomenon which has been well known and intensively studied for many years. Perhaps the most commonly utilized application of electroluminescence is in the screen of an oscilloscope or a television set. Better known as *cathodeluminescence,* this type of electroluminescence is caused by the collision of high-energy electrons with the phosphor coating which lines the inside surface of a cathode-ray tube. One particular type of electroluminescence has given rise to an entirely new field of technology. This is the phenomenon of *PN junction electroluminescence,* which results from the application of direct current at a low voltage to a suitably doped PN junction. This is the basis of the LED, a PN junction diode that emits light when biased in a forward direction. The light emitted can be in either the invisible (infrared) or the visible spectrum.

Semiconducting light sources can be made in a wide range of wavelengths, ranging from the near-ultraviolet region of the electromagnetic spectrum to the far-infrared region. This wavelength limit is due to today's production constraints. The LEDs used in electronic applications (as emitters) are normally *infrared-emitting diodes* (IREDs).

Spectral radiation can be divided into two basic types: *coherent light* (in which the light waves are in phase and are usually produced at cryogenic temperatures) and *noncoherent light* (the waves are out of phase with each other). The following discussion will consider only noncoherent light emitters, invisible as in the case of IREDs or visible as with LEDs.

A PN junction can be formed in a semiconductor material by doping one region with donor atoms and an adjacent region with acceptors. If an external bias is applied across the junction, a bias current flow in the PN junction causes holes to be injected into the N-type material and electrons to be injected into the P-type material. This process is illustrated in Fig. 14-1. The radiation from a PN junction arises from the recombination of electrons with the minority carriers, and energy proportional to the band gap energy of the semiconductor material is released. Some of this energy is released as light, and the remainder is released as heat. The proportion is determined by the mixture of recombination processes taking place. The energy contained in a photon of light (shown in Fig. 14-2) is proportional to the frequency (that is, the color). The higher the band gap energy of the semiconductor material forming the junction, the higher the frequency of the light emitted. In industry, the wavelength of emitted light is commonly expressed in nanometers (nm). Photon energy can be converted to wavelength by the equation

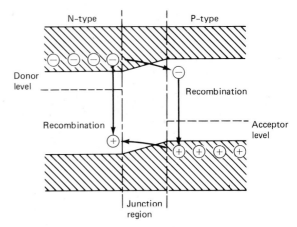

Fig. 14-1 PN junction forward-biased.

$$\lambda = \frac{1240}{E}$$

where E = energy transition, electronvolts (eV)

λ = wavelength, nm

It is possible to increase the wavelength by decreasing the band gap energy. Wavelengths of 1000 nm are possible but expensive to produce. However, the long wavelength emitters are useful in fiber-optic communications and will be covered later in this chapter. Commercially available LED devices are made from gallium arsenide (GaAs), gallium phosphide (GaP), or the (three-element) ternary compound gallium arsenide phosphide Ga (As,P).

There are two types of IREDs, and both use a low-band-gap, silicon-doped epitaxially grown material, *gallium arsenide* (GaAs). The GaAs diodes are efficient and very reliable and produce a peak wavelength of 940 nm. The second type is manufactured by replacing some of the gallium with aluminum. This increases the band gap energy, yielding an IRED with a wavelength of 880 nm. The gallium aluminum arsenide (GaAlAs) emitters are much more efficient than the GaAs emitters. Also, their wavelength is a better match for silicon detectors, thereby increasing system sensitivity.

Gallium phosphide (GaP) is used for visible-light-emitting diodes. The mechanism for visible light radiation is the same as for the infrared diodes. The transition of electrons from the conduction band to the acceptor level releases a photon. The wavelength of the photon is in the visible spectrum as shown in Fig. 14-3, and for the green LED, λ = 560 nm. The wavelength is dependent on the energy gap of the semiconductor. For GaP, E_g = 2.24 eV, and the wavelength is (560 nm). It should be noted that there is a single transition of electrons from the higher to lower bands. If an impurity such as oxygen is contained in the GaP semiconductor, an intermediate energy level will exist deep in the forbidden region close to the valence band. Figure 14-4 shows the oxygen level, E_{ox}, appearing 1.8 eV above the acceptance level. This oxygen trapping level will force the electrons to have double transitions, one from the conduction band to E_{ox}, and the other from E_{ox} to E_{ACC}. The photon release during the first transition period is of low energy deep in the infrared area. This transition is of no value to the operation of a visible LED. The second transition, from E_{ox} to

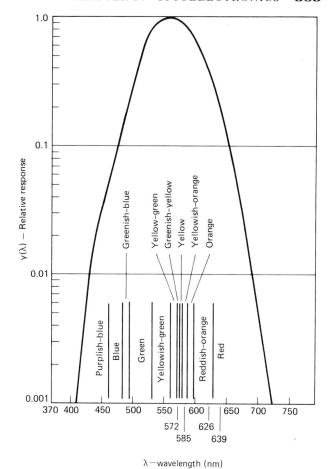

Fig. 14-3 Visible-light spectral characteristics.

E_{ACC}, will cause radiation in the red area of the spectrum at λ = 700 nm. The combination of both red and green emission, by double transitions due to increased current, will give a yellow or orange light. At high current levels, most electrons make the single transition from E_c to E_{ACC}, and green light of high intensity results. The red emission in this case is negligible compared to green emission.

The Ga(As,P) is produced as an epitaxial layer grown on a substrate of either GaAs or GaP. The grown wafer is then processed to produce PN junctions with photolithography used to shape the structure. Figure 14-5 shows the cross section of an LED. The photons generated at the junction of a PN electroluminescent diode are emitted in all directions. If

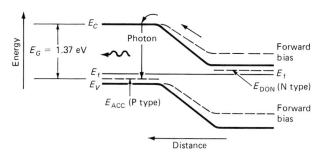

Fig. 14-2 Photon emitted by forward-biased diode.

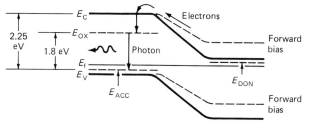

Fig. 14-4 GaP junction with oxygen doping.

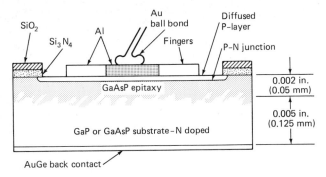

Fig. 14-5 LED structure.

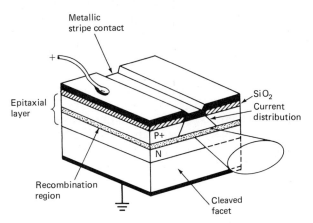

Fig. 14-7 Edge-emitting diode structure.

the diode substrate is opaque, as in the case of GaAs, only those photons which are emitted upward within a critical angle will be emitted as useful light. All other photons emitted into or reflected into the bulk crystal will be absorbed. This phenomenon is illustrated in Fig. 14-6(*a*). Gallium phosphide is nearly transparent compared to GaAs; diodes grown on the GaP substrate will exhibit improved efficiency caused by the emission of photons which would be absorbed in the GaAs substrate. This is illustrated in Fig. 14-6(*b*). The most common uses for visible LEDs are in displays; their implementation will be covered in the next section.

The surface-emitting infrared diodes are similar in structure to the visible LEDs. They are used where low cost and moderate performance are required. It is possible to construct an edge-emitting IRED. The physical structure of an edge emitter consists of a

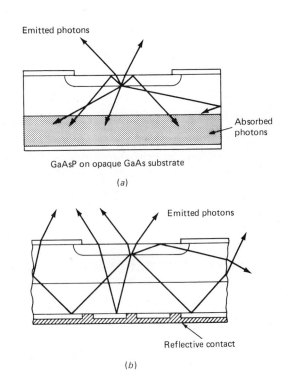

Fig. 14-6 Transparent and opaque substrates.

rectangle-shaped semiconductor die (PN junction) in which the radiant output is emitted from the edges of the diode in the recombination region of the junction, as shown in Fig. 14-7. The lateral size of the radiation area is usually defined by etching an opening in an oxide insulating layer and forming an ohmic contact by depositing a metal film into the open contact region. This type of construction is referred to as *edge emitters* utilizing *stripe geometry* (as opposed to the surface-emitting structures just covered). The edge-emission structure has an oxide metallization stripe constricting the current flow through the recombination region to that area of the junction directly below the stripe contact. It is possible by using this technique to confine the radiating portion of the junction to a spot approximately 50 micrometers (μm) in its largest dimension.

When the edge-emitting structure is combined with some other epitaxial processes, it is possible to restrict the angle of radiation to a relatively narrow angle. A comparison of this narrow edge radiation pattern to the parallel (surface) pattern is shown by Fig. 14-8. These devices exhibit excellent high-speed performance in excess of 100 MHz and are capable of a much greater output of radiant flux.

As mentioned earlier, the surface-emitting infrared diodes are used in modest-performance applications. They exhibit power outputs of 2 to 7 mW in continuous service and 26 to 200 mW in pulse service at 940 nm. Figure 14-9 shows the radiant flux output from a typical IRED emitter at a temperature of 27°C. It is crucial to maintain a safe operating temperature as with all semiconductors. There is an order of magnitude (factor of 10) difference between the radiant flux output for the steady-state and pulsed modes. These are *flux curves,* not to be confused with the characteristic volt-ampere curves of the diode, which are shown in Fig. 14-10.

The presence of infrared (IR) light is not obvious. It cannot be seen by the eye; special means of detection, which are discussed later in this chapter,

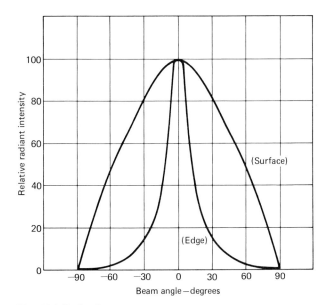

Fig. 14-8 Emitted patterns.

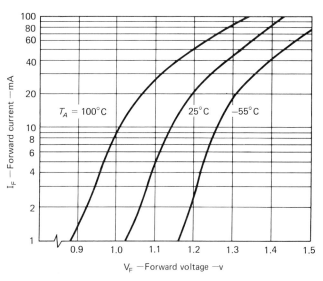

Fig. 14-10 Typical *V-I* characteristics of an IRED.

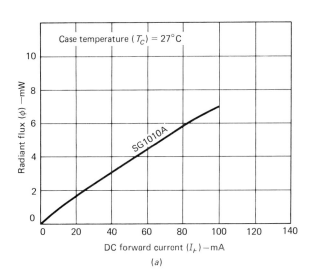

(a)

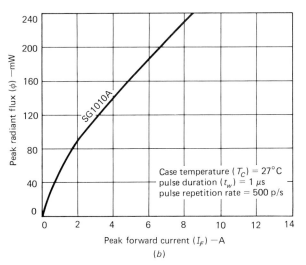

(b)

Fig. 14-9 Radiant flux versus dc and pulsed operations. (*a*) Steady-state current. (*b*) Pulsed current.

must be employed. The wavelength (long compared to visible radiation wavelength) would permit the transmitted flux to pass through the paper of this book as though it were transparent. Blocking or interrupting this wavelength requires a more dense material, and this point should be kept in mind. Suitable precautions should be taken to protect one's eyes from the radiant flux. This safety consideration will be covered in more detail when lasers are explored.

For best performance, emitters should be biased from a current source rather than a voltage source. A simple solution is to place a resistor in series with a voltage source to approximate a current source. A few examples of typical emitter circuits are shown in Fig. 14-11. These are building blocks for the more sophisticated and optimum circuits. Information can be impressed on the emitter output by modulating the bias input to the diode. The modulation is of the amplitude modulation (AM) type. Either voice or radio frequency (RF) signals can be used for modulation, but it is important to know the frequency limitations of the diode being modulated. Typical (simple) modulator circuits are shown in Fig. 14-12. As mentioned previously, frequencies in the 100-MHz region can be obtained with some emitters. Note that all circuits have a current-limiting resistor in series with the diode.

The IRED is often operated in the pulse mode for the transmission of digital information over short distances or to send signals in and out of various unfriendly environments such as gas, water, and high voltage. The circuit may be as simple as Fig. 14-13(*a*) or as complex as Fig. 14-13(*b*).

The IRED has a maximum reverse voltage rating of only 2 V, and this value should not be exceeded. Many IREDs are destroyed during testing. Check all data sheets thoroughly before performing any tests.

Some of the GaAs diodes are assembled in conjunction with a glass lens to produce a narrow-beam

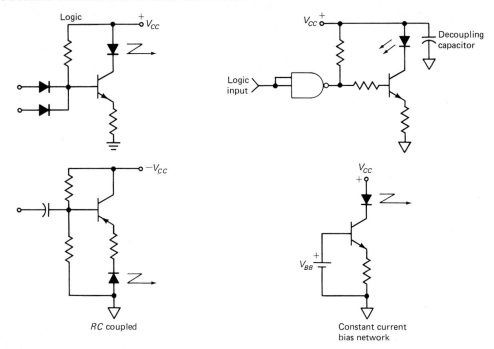

Fig. 14-11 Typical emitter circuits/blocks.

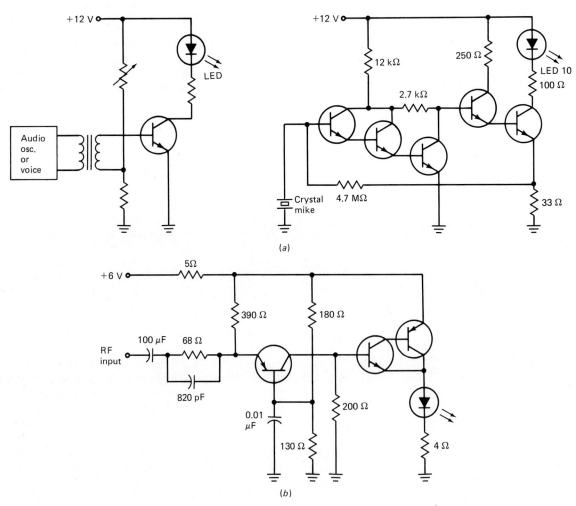

Fig. 14-12 Emitter-modulation circuits. (*a*) Audio frequency. (*b*) Radio frequency.

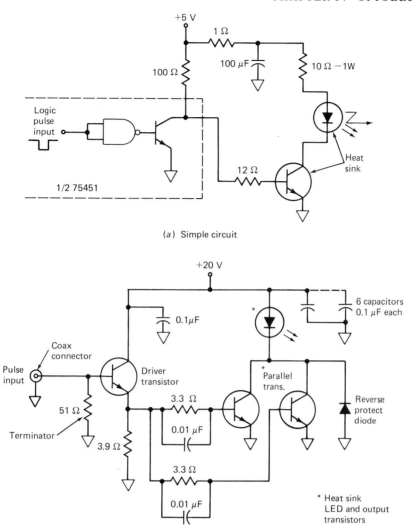

(a) Simple circuit

(b) Complex circuit

Fig. 14-13 Pulse circuits for emitter LEDs. (a) Simplified circuit. (b) Complex circuit.

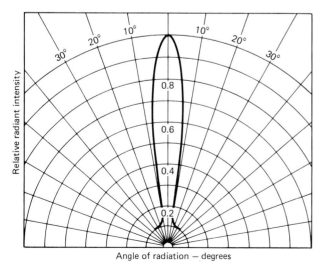

Fig. 14-14 Typical lens-corrected radiant flux pattern.

pattern, as shown in Fig. 14-14. These IREDs are intended for use in a wide variety of industrial applications, including high-speed counting, intrusion alarms, edge detection and control, collision protection, and data transmission.

REVIEW QUESTIONS

1. Light emission from a forward-biased PN junction is known as _____.

2. The lower the frequency of the light emission, the _____ the wavelength.

3. Infrared-emitting diodes emit coherent light waves. (true or false)

4. _____ are released as part of the energy of recombination in the PN junction.

5. Calculate the number of electronvolts required for a wavelength of 620 nm (near red).

6. The GaAlAs diode emits a _____ nm wavelength in the infrared range.

7. The GaP PN diode is used for _____ light radiation.

14-2
DETECTORS (SOLID-STATE)

The reverse-biased PN junction is basic to operation of photosensitive semiconductor devices. When a photon is absorbed in a semiconductor, a hole-electron pair is formed and swept across the junction by the electric field (ϵ) developed across the depletion region. A photocurrent due to the separated hole-electron pairs results. The electrons go to the N side, and the holes go to the P side. Separation of a photon-generated hole-electron pair is more likely to occur when the pair is formed in a region where there is an electric field (ϵ); see Fig. 14-15. The alternative to separation is for the hole-electron pair to recombine, with no contribution to photocurrent. The photocurrent flow in the external circuit is proportional to the illumination.

Electric field distribution in a semiconductor diode is not uniform, as indicated in Fig. 14-15. In the regions of the P-type diffusion (front) and N-type diffusion (back), the field is much weaker than it is in the center region, known as the *depletion region*. For best results, a photodiode should be made so as to allow the greatest number of photons to be absorbed in the depletion region. That is, the photons should not be absorbed until they have penetrated as far as the depletion region, and they should be ab-

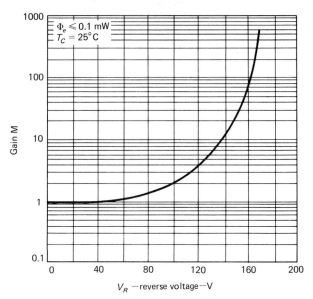

Fig. 14-16 Photodiode gain versus reverse voltage.

sorbed before penetrating beyond the depletion region.

The depth to which a photon will penetrate before it is absorbed is a function of the photon wavelength. Short-wavelength photons are absorbed near the surface; those of longer wavelength may penetrate the entire thickness of the crystal (see Fig. 14-15). Therefore, if a photodiode is to have a broad spectral response (broad spectrum of wavelengths) it should have a very thin P-layer to allow short-wavelength photons to penetrate, as well as a thick depletion region to maximize photocurrent from the long-wavelength photons.

The thickness of the depletion region depends on the resistivity of the region to be depleted and on the reverse bias. A depletion region exists even if no reverse bias is applied because of the "built-in" field produced by diffusion of minority carriers across the junction (at room temperature). Reverse biasing aids this built-in field and expands the depletion region.

In PN photodiodes, a thin P diffusion allows good short-wavelength response. A deep P diffusion degrades the short-wavelength response but lowers the bias required for good response at longer wavelengths.

An avalanche photodiode (APD) uses avalanche multiplication to amplify the photocurrent created by hole-electron pairs. The PN junction is operated at a high reverse bias voltage needed for avalanche multiplication (refer to Fig. 14-16). Very high multiplication factors (M) can be achieved, but the process is very noisy. The photocurrent signal gain is magnitude-dependent on the reverse voltage, as illustrated in Fig. 14-16. These devices are usually optimized for detection of infrared-radiant energy but are useful from audio to microwave frequencies.

Because the avalanche photodiode's temperature

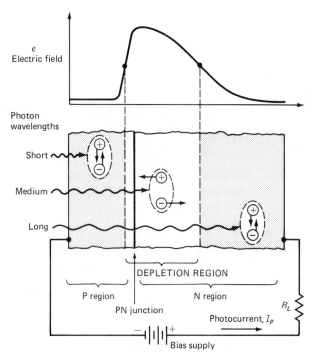

Fig. 14-15 PN photodiode junction and internal field diagram.

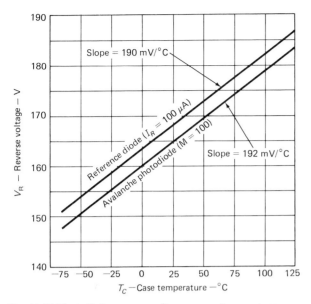

Fig. **14-17** Photodiode reverse voltage versus temperature.

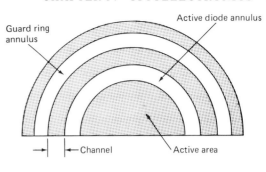

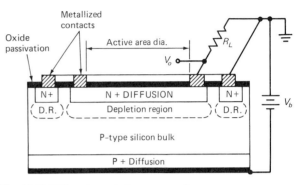

Fig. **14-19** Diffused guard ring construction.

stability is poor it must be carefully controlled (see Fig. 14-17). To compensate for this effect, diode pairs consisting of an avalanche photodiode (APD) and a small reference diode are manufactured together in the same package. This pairing makes it possible to build a temperature-compensating bias circuit that will hold the avalanche gain constant over wide temperature variations. The diode pair is in a hermetically sealed case with a window of borosilicate glass. Figure 14-18 shows a block diagram application of a temperature-compensating bias circuit for an avalanche diode.

Besides the photocurrent generated, a *dark (reverse leakage) current* exists in a photodiode. The

two main contributors of dark current (diode leakage) are surface leakage and bulk leakage. The name *dark current* indicates what it implies: the current (leakage) that flows without light. The surface leakage of a photodiode is 100 times that of the bulk leakage. Some photodiodes employ a unique diffused guard ring construction to minimize this surface leakage. Figure 14-19 illustrates the guard ring diffused on the active area. In normal operation the guard ring (a PN junction of its own) and the active area are biased at the same potential. The surface leakage is now shunted around the load resistor and flows through the guard ring, as shown in Fig. 14-20. Now the main source of leakage will be the much lower bulk leakage. The shunting of surface leakage around the load by the guard ring also improves the diodes' noise figure since the noise current varies directly as the square root of the leakage current.

The *light-sensitive transistor* is in reality one of the simplest photodiode-amplifier combinations. Directing light toward the reverse-biased PN junction (collector-base) generates base current, which is amplified by the current gain of the transistor. Figure 14-21 shows the schematic, equivalent circuit, and junction (with depletion region). External biasing of the base is possible, if that lead is available, so that the equation for emitter current is

$$I_e = (I_p + I_b) \cdot (h_{FE} + 1)$$

where
I_b = base current
I_p = photon-generated base current
I_e = emitter current
h_{FE} = transistor dc current gain

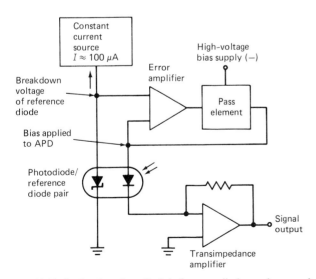

Fig. **14-18** Avalanche photodiode/reference diode package and block diagram.

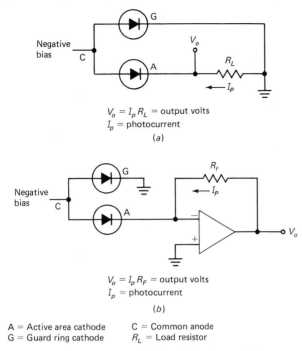

$V_o = I_p R_L$ = output volts
I_p = photocurrent

(a)

$V_o = I_p R_F$ = output volts
I_p = photocurrent

(b)

A = Active area cathode   C = Common anode
G = Guard ring cathode R_L = Load resistor

Fig. 14-20 Avalanche photodiode/guard circuit. (a) Basic operating circuit. (b) Photodiode/op amp circuit.

The sensitivity of this type of transistor can be influenced by the bias levels at the base. The response of the phototransistor will vary as h_{FE} varies with current, bias voltage, and temperature. A high value of h_{FE} and a large collector-base junction area are required for high sensitivity but can also cause high dark current levels when the collector-base junction is reverse-biased, as indicated in Fig. 14-21. The phototransistor dark current is given by

$$I_{ceo} \text{ (dark)} = h_{FE} \times I_{cbo}$$

where I_{cbo} = collector-base junction leakage current (the "o" denotes an open emitter lead)
I_{ceo} = collector-emitter leakage (the "o" denotes an open base lead)

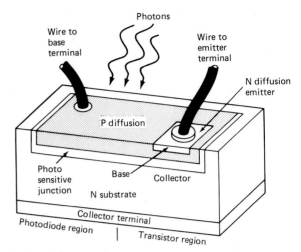

Fig. 14-22 Phototransistor.

This leakage is proportional to the junction area and the perimeter at the surface. Careful processing of the transistor chip, shown in Fig. 14-22, is required to minimize the dark current and maintain high light sensitivity. Typical phototransistor dark currents with a 10-V reverse bias are on the order of 10 nA at room temperature and double by a factor of 2 for every 10°C rise in temperature. This is a very important point and worth remembering; it could minimize troubleshooting time when temperatures are suspect.

Dark current effects may be minimized for low light applications by keeping the base-collector junction from being reverse biased, that is, having a V_{ceo} of less than one silicon diode forward voltage drop. This technique will allow light currents in the nanoampere range to be detected. The circuit shown in Fig. 14-23 illustrates this type of operation. The band gap effect of the doped base-emitter (B-E) junction of Q_1 sets the open base potential, thereby forcing V_{be} (Q_1) to be equal to one diode drop. Since V_{be} (Q_1)

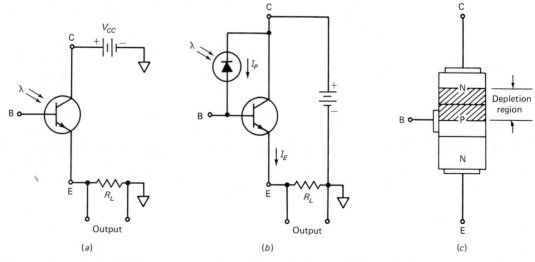

Fig. 14-21 Phototransistor circuit and junction representation. (a) Schematic diagram. (b) Equivalent circuit. (c) Junction representation.

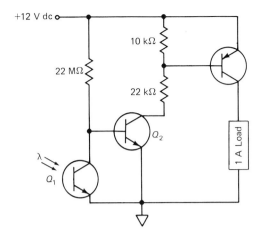

Fig. 14-23 Very-low-level phototransistor circuit.

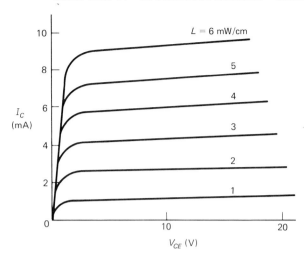

Fig. 14-24 Typical characteristic curves of a phototransistor.

closely approximates V_{be} (Q_2) (one diode drop each), V_{bc} (Q_1) is approximately equal to zero (0). This is now a minimum-leakage condition for the phototransistor.

The light falling on a given area varies inversely with the square of the distance from the source; the relationship between light intensity L (in lumens), and the strength of the source C (in candela) is

$$L = \frac{CA}{d^2}$$

where C = luminous intensity in candela
A = area illuminated in square meters
d = distance from source to area in meters

Figure 14-24 shows typical characteristic curves of a phototransistor. The usual base steps are now intensity steps in lumens.

The *photodarlington* basically is the same as the phototransistor, except for its much higher gain from two stages of amplification cascaded onto a single chip. Figure 14-25 shows the photodarlington symbol, its geometry, and its equivalent circuit. The formula for the photodarlington's emitter current is

$$I_{e2} = I_{p1} \cdot (h_{FE1}) \cdot (h_{FE2}) + I_{p2} \cdot (h_{FE2})$$

Because $I_{e1} >> I_{p2}$

$$I_{e2} \simeq I_{p1} \cdot (h_{FE1}) \cdot (h_{FE2})$$

where I_e = emitter current, each device
I_p = photon-produced current, each region
h_{FE} = current gain of each transistor, dc

To maximize sensitivity, I_{p1} should contain a large portion of the photon-produced current. This is accomplished by expanding the base of the pellet so the photodiode dominates the topography, in the same way as the phototransistor does in Fig. 14-22.

Optimization of both short- and long-wavelength response (ultraviolet [UV] to infrared [IR]) at low reverse voltage requires a PIN, rather than a PN diode structure. A PIN diode has a thin P-type diffusion in the front and an N-type diffusion into the back of a wafer of very-high-resistivity silicon. The high-resistivity material between the P-type and N-

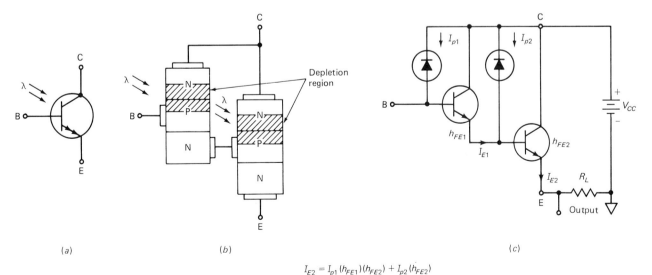

$$I_{E2} = I_{p1}(h_{FE1})(h_{FE2}) + I_{p2}(h_{FE2})$$

Fig. 14-25 Photo Darlington (*a*) Symbol. (*b*) Construction. (*c*) Equivalent circuit.

type diffusions is called the *intrinsic region,* or *I layer.*

The silicon planer PIN photodiodes are ultrafast detectors for visible and infrared radiation. Their response to blue and violet is exceptionally good. Speed of response of these detectors can be less than 1 ns. They are particularly well suited for laser pulses, and their bandwidth is from direct current to 1 GHz. Their low dark current or leakage current (usually in picoamperes) results in an almost negligible noise current. This is a great asset for fiber optics (discussed later in this chapter).

The electrical characteristics of the PIN photodiode are shown in Fig. 14-26. Close scrutiny will reveal that the PIN photodiode may be operated as a photovoltaic device (right-hand 2.5-MΩ load line), in which the light impinging upon the unit creates a voltage which varies in proportion to the input light. This mode of operation is shown in Fig. 14-27. Operation in this mode requires no external bias, and the photodiode generates the load voltage when illuminated. This photovoltaic operation can be either linear or logarithmic, depending upon the selected value of the load resistance. *Logarithmic operation* is obtained if the load resistance is very high (greater than 10 MΩ). With an input FET or a BI-FET op amp, this can easily be achieved, as shown in Fig. 14-28(*a*). If the amplifier has a very high input resistance, the loop gain is $(1 + R_2/R_1)$. The speed of response of this type of amplifier is very slow, with a time constant of approximately 0.1 s. If high speed is required, the logarithmic amplifier should be preceded by a linear amplifier.

The more common mode of operation for the PIN photodiode employs reverse bias (as with the PN photodiode), and this mode is called the *photocurrent* or *photoconductive mode*. The main drawback of the photocurrent mode is the flow of dark current, which is due to the reverse bias. As mentioned previously, *dark current* is that current which flows when no radiant flux is applied to the diode. For best linear operation, the photodiode should be operated with as small a load resistor as possible. Figure 14-28(*b*) shows the typical amplifier arrangement. The inverting input of the amplifier is at virtual ground; the dynamic resistance of the photodiode is R_1 divided by loop gain (greater than 10 k). As shown in Fig. 14-28(*b*), the output voltage will rise (go positive) in response to the optical input signal. Reversing both the photodiode and V_c will produce a drop (output voltage will go negative) in the amplifier output for an optical input. The output voltage from this circuit is

$$V_o = R_1 (I_p + I_{\text{dark}})$$

The speed of response is greatly improved and is now limited by the amplifier.

In most discrete-device literature, the photo SCR is often termed a *light-activated SCR* (LASCR). Figure 14-29 presents the schematic symbol, structure, and equivalent circuit of this device. The photon current (I_p) generated in the reverse-biased PN junction (depletion region) reaches the gate region to forward bias the NPN transistor and initiate switching. Since the photodiode current is of a low level, a LASCR must be constructed so that it can be triggered with a very low gate current. The high sensitivity of the LASCR causes it to be sensitive also to any effect that will produce an internal current. Therefore, it is very sensitive to temperature,

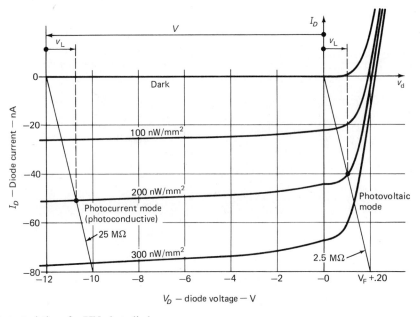

Fig. 14-26 Electrical characteristics of a PIN photodiode.

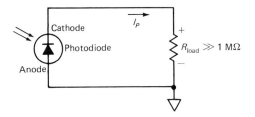

Fig. 14-27 PIN photodiode in photovoltaic mode.

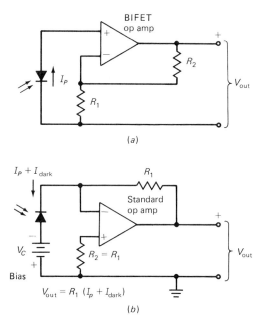

Fig. 14-28 Two modes of PIN photodiode operation using op amps

applied voltage and rate of change of applied voltage (dv/dt), and it has a longer turn-off time than a typical SCR. All other parameters of the LASCR are similar to those of an ordinary SCR, so that the LASCR can also be triggered with a positive gate signal. Most commercially available LASCR types are of comparatively low current rating (less than 2 A). In practical applications they are used to drive a higher-power device, usually another SCR.

There are many other photodetector combinations using integrated circuits that allow many combinations. Among the possibilities are FETs, triacs, logic devices, and MOSFETs. These will be examined in the optical coupling section.

Another basic type of photosensor uses the photoconductive bulk effect. The photoconductive bulk effect cells are normally made of cadmium sulfide (CdS) or cadmium selenide (CdSe). Unlike the previous types discussed, they have no junction. The entire layer of material changes resistance when illuminated. The photoconductive cell decreases in resistance as the light level increases and increases in resistance as the light level decreases. The response curve of a cell is shown in Fig. 14-30. The response curve of resistance versus intensity (lumens) is nonlinear; however, over a limited range of intensities, a linear approximation is valid. Figure 14-31 shows the electrical symbol along with a drawing of the usual configuration of the cell.

Although photoconductors require an external power supply, a sensitivity 1000 times greater than that of the photovoltaic types more than compensates in most applications. Because of their low cost, a pair can be used for temperature compensation in

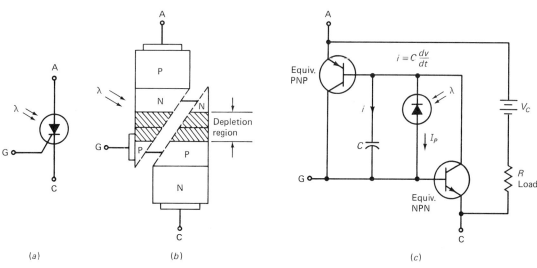

Fig. 14-29 Photo SCR (LASCR). *(a)* Symbol. *(b)* Construction. *(c)* Equivalent circuit.

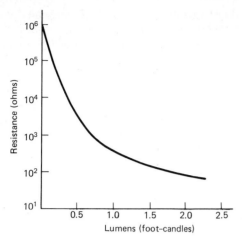

Fig. 14-30 Photoconductive cell's characteristics.

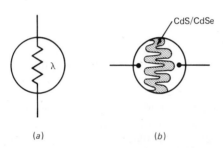

Fig. 14-31 Photoconductive cell. (a) Symbol. (b) Physical layout.

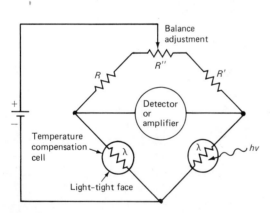

Fig. 14-32 Photoconductive cell/temperature compensation in a bridge circuit.

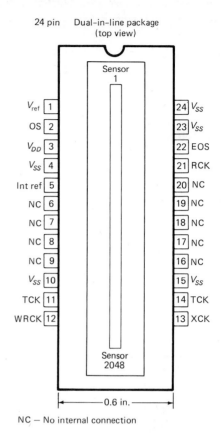

NC — No internal connection

Fig. 14-33 Image sensor chip with 2048 elements.

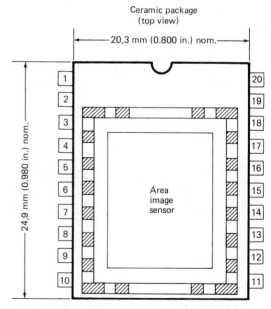

Fig. 14-34 Area image sensor with 328 × 490 pixels.

a Wheatstone bridge arrangement. One cell is shielded from light so that its resistance is only a function of the surrounding temperature; the other cell is placed in the opposite arm of the bridge to measure light intensity, as shown in Fig. 14-32.

The inherent nonlinearity of the photoconductive cell has led to its widespread use in chopping circuits, but these types of cells are being replaced by solid-state devices. The resistance in the absence of light usually exceeds 1 MΩ; the resistance under a given light condition can be as low as 100 Ω. The small-size cells find extensive use in light-meter circuits. The CdS cell has a drawback, however. If exposed to a very strong light (saturated), it displays a "memory'' and may take hours to return to the original ''dark'' resistance.

Many industrial control systems use punched cards (paper, plastic, metal) or punched tape to input programs or data. To meet these requirements, many arrays usually containing 9 or 12 devices per unit are available. The device may include some logic, along with a biasing network. A precise matching emitter array (IRED) is usually available to allow proper alignment. These devices are being replaced by diskettes, magnetic tape, and cards.

The solid-state image array sensors (line scanners) include anywhere from 64 to 10,000 individual photodiodes, which are scanned to provide serial output on a single video line. Applications include pattern recognition, size and position monitoring, and inspection. A 2048 by 1 linear image sensor is shown in Fig. 14-33. The sensor elements are also called *pixels,* which stands for *picture elements.* Their operation is based on charge-coupled devices, transfer gates, and drive circuitry. The area image sensor of Fig. 14-34 has 328 horizontal and 245 vertical pixels for camera operation. High-resolution sensors with 1024 by 1024 pixels are being developed for robot vision systems and as replacements for vidicons and other tube-type image sensors.

REVIEW QUESTIONS

8. Photon absorption results in the formation of a hole-electron _____ in a PN junction.

9. Recombination will occur when the pair is formed in an _____ field.

10. The electric field is greatest in what region of the PN junction?

11. Short-wavelength photons will penetrate the entire crystal. (true or false)

12. An avalanche photodiode uses _____ bias.

13. The avalanche photodiode is used mainly for _____ detection.

14. Leakage currents are the main contributors to _____ current in a photodiode.

15. The phototransistor must have its base forward-biased. (true or false)

14-3
DISPLAYS

Incandescent, fluorescent, and neon lamps have been in use for a long period of time and in a wide variety of applications. In recent years, the solid-state LED has replaced these earlier lamps in many applications. Most new designs and applications use the LED indicators and displays. Therefore, it is important to know and understand the properties and characteristics of the LED-type devices.

Most commercial LEDs are manufactured by encapsulating the diode chip inside a plastic package

with an immersion lens directly above the junction. The immersion lens arrangement results in a light output which is concentrated in a narrow beam and applies only to LEDs in undiffused plastic. This configuration is used for backlighting applications and those applications requiring a concentrated light source.

A front panel indicator requires a wide off-axis viewing angle. To obtain this wide viewing angle, a diffusant is added to the plastic to disperse the light rays being emitted. Dye coloring is also added to tint the plastic to enhance its on/off contrast.

As mentioned earlier, not every photon generated within the PN junction emerges to the surface. The GaAsP/GaP devices have an efficiency of approximately 80 percent. When light passes from a medium whose index of refraction is N_1 to a medium whose index of refraction is N_2, a portion of the light is reflected back at the medium interface. This loss of light is called *fresnel loss.* Without going into the physics of lens construction, it will suffice to say that coating the lens with an intermediate material with a suitable index of refraction (say, N_3) reduces the fresnel loss to permit almost 98 percent efficiency.

The LED packages take on many physical shapes to meet a wide variety of applications. They may be rectangular, round, dot-shaped, or the size and shape of typical incandescent lamps. Some units may have additional components integrated inside the LED package. An example is the *resistor/LED indicator.* The integral resistor is a nominal 215 Ω, allowing the lamp to be driven directly from the 5.0-V supply for a logic gate. Otherwise, an external current-limiting resistor will be required. Some LEDs can produce red or green light from one package. They usually are equipped with a milk-white diffused lens to provide good on-off contrast. They have three leads and two anodes (one for red and one for green) with a common cathode. This device is used in applications that require an indicator for the three states of interest and provide more information than an on/off lamp. Some applications require that analog information be converted into a visual light display. This has typically been done with a panel meter. Many analog displays require only moderate accuracy and resolution. Light-emitting diode bar/graph modules can often be substituted for meters in these applications. The LED bar/graph displays consist of a linear array of LEDs that are driven by a device that decodes an analog or digital signal into a bar/graph indicator display code. Figure 14-35 is an example of

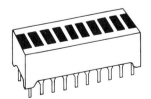

Fig. 14-35 DIP bar/graph packages.

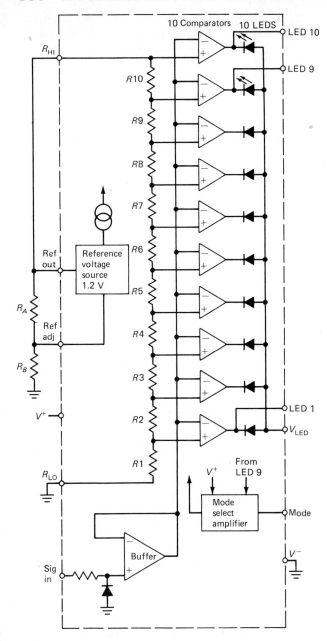

Fig. 14-36 Integrated bar/graph display.

Resistor	Linear	Log	VU
R1	1.00 kΩ	1.0 kΩ	0.708 kΩ
R2	1.00 kΩ	0.41 kΩ	1.531 kΩ
R3	1.00 kΩ	0.59 kΩ	0.923 kΩ
R4	1.00 kΩ	0.83 kΩ	0.819 kΩ
R5	1.00 kΩ	1.17 kΩ	1.031 kΩ
R6	1.00 kΩ	1.66 kΩ	1.298 kΩ
R7	1.00 kΩ	2.34 kΩ	0.769 kΩ
R8	1.00 kΩ	3.31 kΩ	0.864 kΩ
R9	1.00 kΩ	4.69 kΩ	0.970 kΩ
R10	1.00 kΩ	6.63 kΩ	1.087 kΩ
Total	10 kΩ	22.6 kΩ	10 kΩ

Typical resistor string values (ohms)

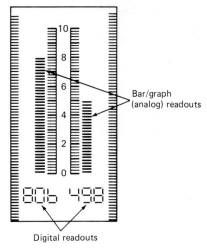

Fig. 14-37 Analog digital meter with bar/graph display.

a bar/graph display. Most units are fabricated to be stackable. In some cases, 10 arrays are cascaded to form a bar graph of 100 elements. The totally integrated units contain drivers, a reference, and threshold detectors and may be programmed to be linear or logarithmic or to display volume units (VU). Figure 14-36 illustrates the block and connection diagrams for an LED bar/graph display with driver. The device can produce a bar/graph display (histogram) or a moving-dot display, depending on the mode pin. Some units can produce an automatic color change (typically green to red) to indicate overrange, overload, or some unsafe condition. A combined analog and digital readout is very desirable in many instances, such as in process control. Such a dual-type readout display is shown in Fig. 14-37.

The LED display devices have expanded very rapidly and now include a wide variety of distinctly different products. Four generalized categories are the following:

1. Displays with on-board integrated circuits
2. Strobable seven-segment displays
3. Dot matrix alphanumeric displays
4. Magnified monolithic displays

All the displays use the GaAsP type of diode and can offer colors of red, green, or yellow. The most prominent feature of any display is the physical arrangement of the display elements. This arrangement, or *font* as it is usually called, is important for the type of information to be displayed but also dictates the type and complexity of the driving electronics required by the display. Some of the common display fonts are illustrated in Fig. 14-38.

The seven-segment display (font A) is the most commonly used LED technology and is also the easiest to implement electronically. However, it is limited to displaying numeric and a small range of al-

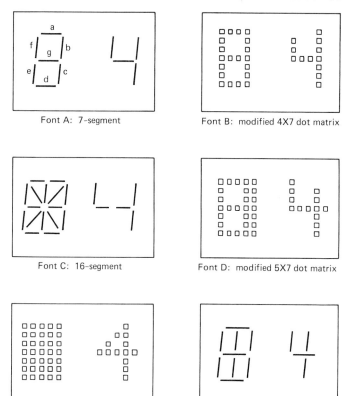

Fig. 14-38 Typical LED display fonts.

Font A: 7-segment

Font B: modified 4X7 dot matrix

Font C: 16-segment

Font D: modified 5X7 dot matrix

Font E: 5X7 dot matrix font

Font F: 9-segment

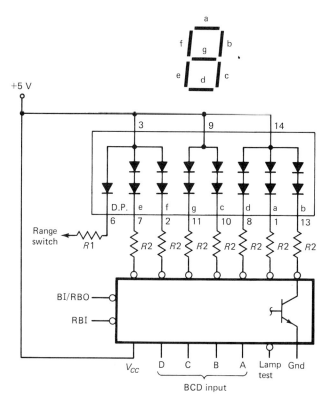

Fig. 14-39 Seven-segment display circuit.

phabetic information. The five by seven dot matrix (font E) can display a wide range of numeric, alphabetic, and other characters but requires rather extensive electronic circuitry for implementation. The sixteen-segment display (font C) has full alphanumeric capability but has found only limited acceptance. Font F illustrates a nine-segment display, which is somewhat more pleasing and readable than the seven-segment display of font A. It can display the same numeric information and has more alphabetic capability. The dot matrix fonts of B and D are abbreviated versions of the 35-dot matrix of font E and are used primarily for displaying numeric and hexadecimal information.

The interface to a seven-segment display is provided by the BCD to seven-segment decoder driver shown in Fig. 14-39. The input to the decoder is BCD code for the number to be displayed. The RBI and BI can be pulled low to turn off all segments. When BI is high, the lamp test (LT) input can be brought low to turn on all segments to perform a lamp test operation. The BI/RBO can serve as an output for *ripple blanking* (blanking nonsignificant zeros) to other decoders. When RBI is low, the RBO output will go low to ripple a blanking signal to other display decoders. The segment drivers a through g and D.P. (decimal point) are connected to the display to control which LEDs are to be turned on.

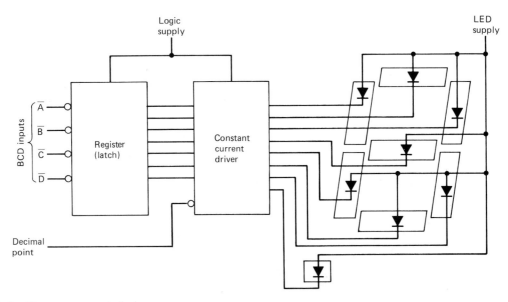

Fig. 14-40 Single-chip seven-segment displays.

The entire circuit and display are also available as a single device, as shown in Fig. 14-40. This device has a 4-bit BCD code input and decimal point input. Devices that include a storage register (latch) as well as a decoder/driver and display unit in the same unit also exist. The display will show the number whose code is latched. A latch simplifies the input/output (I/O) requirements of a microcomputer because the display can be treated as a memory location. The display with latch may be connected to the data bus or to an I/O bus.

The interface to a five by seven or other dot matrix display is handled in much the same manner as the seven-segment display. A device which displays hexadecimal characters using LEDs arranged on a four by seven dot matrix pattern is shown in Fig. 14-41. It includes a 4-bit data register with a latch strobe to store input data. As long as the strobe stays high, the information displayed (stored) will not change. There is a blanking input that, when high, causes the display to be blanked. There are also left and right decimal points available. In some cases, the LED power supply is different from the logic supply,

which must exhibit low ripple and good regulation. These stringent requirements are not necessary for the LED display power.

The control of a five by seven dot matrix display usually requires a read only memory (ROM) or EPROM in which the display pattern for each character to be displayed is stored. The basic circuit structure is illustrated in Fig. 14-42 for an individual interface to a five by seven matrix LED. The ASCII code for the character to be displayed is applied to the seven inputs. The current drive capability is provided by the seven row sink drivers (on the O_1 to O_7 lines) and the four source drivers for the column lines of the matrix. At the time a column line is turned on, the column select CA through CE must simultaneously be applied to the column select lines of the EPROM. The EPROM outputs the appropriate row signals for each selected column character. Thus, the circuitry must scan through the columns at an appropriate rate with a ring counter or some other counter-decoder combination. Only one column of the EPROM is addressed at any time. The UJT clock is set to provide a clock rate of 1 kHz. A new column

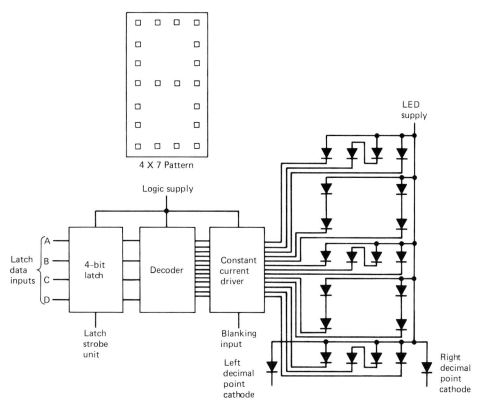

Fig. 14-41 4 × 7 matrix circuit structure.

is selected and turned on for about 1 ms, and any given column is on 20 percent of the time. The circuit of Fig. 14-42 displays a single character. If a multiple-character display is necessary, it is not reasonable to provide a separate EPROM for each display unit. A circuit that shares the EPROM must be used. A random access memory (RAM) can be used to provide storage of the character codes to be displayed, along with a controller that will sequence through the AS-CII codes stored in the RAM while different matrix displays are activated.

As shown in Fig. 14-43, the *liquid crystal display* (LCD) has two glass plates, the insides of which are coated with a transparent pattern of conductive materials. The plates are mounted so that the conducting layers are facing each other. The spacing between the two plates is only about 10 to 30 μm (25 μm = 0.001 in.). Liquid crystal materials are retained between these plates by a peripheral seal of glass frit, or epoxy. Both outer surfaces of the front and rear glass require light-polarizing films. Ordinary light is composed of vertical and horizontal components; polarizing it removes one of the components. The polarizing film on the rear glass is covered with a reflective material (silver bead, silver foil, or gold foil) or a transreflective material (reflects ambient light and transmits back light). The reflective silver foil type is the most popular.

Liquid crystal displays are activated by applying voltage between the segment and the common elec-

trodes shown in Fig. 14-43. Typical driving voltage is 5 V_{rms}, and the allowable ac frequency range of the driving voltage is from 30 to 100 Hz. Flicker may be seen if the drive frequency is below 30 Hz. The power consumption increases in direct proportion to the driving frequency, and a driving frequency of 100 Hz is typically used. The LCDs are driven with ac voltages to prevent plating of the conductive electrodes due to electrolysis. Direct current driving, or ac driving with a large dc offset, greatly shortens the life of the LCD. Strict attention is necessary in order not to exceed this specified dc offset, typically 25 mV. Usually LCDs are connected to logic circuits, so it is very common to drive them by using an ac symmetrical square wave. This ac symmetrical square wave features less dc offset and can be obtained in all LCD drivers by using an exclusive OR gate as shown in Fig. 14-44. Waveform C is the oscillator signal and is applied to the gate and to the common connection on the display. The D waveform is the output of exclusive OR, which has shifted the oscillator input 180°, when the control input is high. Waveform E is the resultant waveform of C and D and is the one that in effect drives the display. Many LCD drivers include exclusive OR gates. The 10 V between the segments and the common electrode is obtained from the 5-V logic power supply.

The LCD display mates very well with CMOS-type logic. Decoder, driver, and interface circuits are available to use with LCD devices. Typical seven-

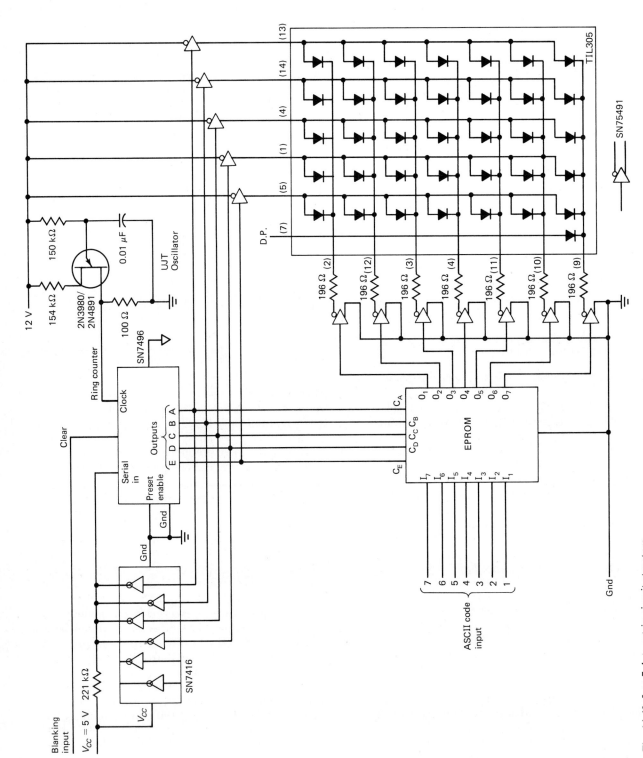

Fig. 14-42 5 × 7 dot matrix circuit structure.

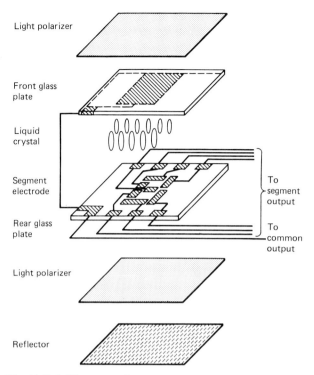

Fig. 14-43 LCD construction.

some applications heaters (back-plane type) are added or integrated into the display when it is fabricated. Flat-panel displays are available for viewing real-time information. This type of display need has previously been dominated by the cathode-ray tube, but three flat-panel technologies are providing alternative choices. They are the electroluminescent, vacuum-fluorescent, and plasma-gas discharge techniques.

Electroluminescent panels use certain solids which emit light when an electric field is across them. They can display 80 characters by 25 lines on a 4- by 8-in. active area. They must be refreshed (60-Hz rate) to retain the image. High-voltage drive (usually 200 V) is required for a bright display, and this requirement limits its popularity.

The *vacuum-fluorescent* (VF) display technology makes use of a low-temperature filament cathode to produce electrons that strike the phosphor-coated anode in a vacuum envelope with a wire mesh grid. The VF display competes with the other technologies for the display market. It is very durable (100,000 h) and less subject to contamination and dirt, and it produces a visible display in bright lighting or total darkness. As VF displays increase in size and capacity they become difficult to produce. Their cost goes up with the necessary drivers, and makes their price almost as high as that of a CRT. Therefore VF display applications have been limited to small-size readouts.

Figure 14-46(*a*) illustrates a VF four-digit display circuit driven from the TTL output of a seven-segment decoder. Usually V+ is 60 to 70 V; for this reason the high-voltage driver chips are employed as

segment numeric display circuits are illustrated in Fig. 14-45.

Liquid crystal display devices require backlighting if they are to be used in a darkened area or at night. They are very sluggish at low temperatures, and in

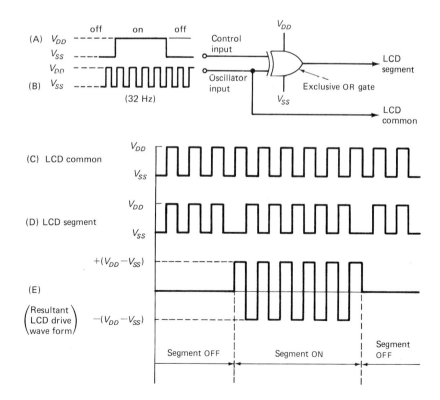

Fig. 14-44 LCD drive circuit and waveforms.

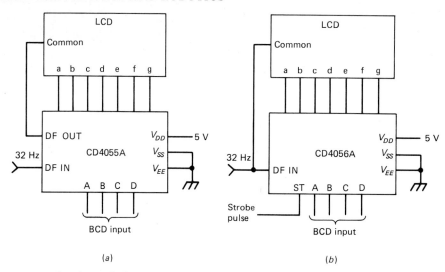

Fig. 14-45 LCD seven-segment decoder equivalents. *(a)* Non-latched type. *(b)* Latched type.

an interface between the display and the decoder outputs.

When both the *segment* (equivalent to the vacuum tube anode) and the *digit* (controlled by the grid) are switched sufficiently positive with respect to the cathode (filament), the electron cloud around the cathode will be accelerated by the grid and continue on to the anode (grid is meshed and the electrons pass through), where the digit/segment will become fluorescent. These displays are typically blue-green in color because of the phosphors employed in their construction. Filters can be added to display virtually any color. A CMOS driver that uses a bipolar 12-V supply is shown in Fig. 14-46(*b*). The driver is compatible with 6- to 15-V CMOS logic signals. In this type of drive the substrate must be connected to a voltage equal to or greater than V_{dd}, so while V_{dd} is 12 V the substrate and the output are tied to the most negative rail.

Plasma panel displays are a cousin to the gas discharge display tube of past years. A brief look at the gas discharge tube will aid in your understanding of the plasma panel type of displays that are in use today. The gas discharge display uses neon gas, which produces an orange glow when ionized. This type of tube is a *cold-cathode* (no heated filaments) device. It consists of a common anode and some number of individual metallic cathode elements to form the characters. Application of a negative voltage to the selected cathode elements with respect to the common anode creates a current flow. The selected cathodes glow and form the character. The equivalent circuit for the gas discharge tube is shown in Fig. 14-47. The tubes require a minimum cathode current density to assure complete glow of the entire element. A maximum current limit is established to provide a long life. Typical drive circuits are shown in Fig. 14-48. The anode resistor acts as a current limiter. A B+ voltage of 170 V is typical.

Plasma displays also use a neon gas mixture and emit an orange glow when switched at high frequencies. Their light output intensity is a function of frequency. They are sometimes called *ac plasma displays* because they operate from a toggled dc supply (usually around 20 kHz). The panel is basically a neon-filled capacitor. It has plates (electrodes) which are covered with a dielectric. The ac type has an inherent memory and therefore eliminates the need for refreshing. A cutaway view of a plasma display is given in Fig. 14-49. The dc plasma displays have no dielectric, and the gas is not separated from the electrodes. The gas therefore glows with application of a dc signal and goes off when that signal is removed. Although this operation requires simpler drive electronics than the ac version does, screen refreshing is required to keep the pixels lit. Alphanumeric plasma display panels are available in one- to four-line versions with 24 characters per line and characters that are about 0.5 in. high. Some panels utilize a self-scan scheme and significantly cut the complexity of the drive circuit. Displays with 120,000 addressable pixels arranged in a 250- by 480-pixel grid offer an alternative to the CRT for many display applications.

Although hot wire readouts are incandescent devices, their application in multidigit, multiplexed display systems closely resembles LED operation. Since hot wire displays will conduct current in either direction, isolation diodes are required to prevent "sneak" paths from partially turning on unaddressed segments. The hot wire readouts are available in both 7-segment and alphanumeric (16-segment) versions and are quite well suited for high-ambient-light applications. They do not wash out in sunlight. Multiplexed schemes can be cumbersome because of the large number of discrete diodes required.

Though flat panels have advanced rapidly, the CRT is at the forefront for high-resolution displays, both

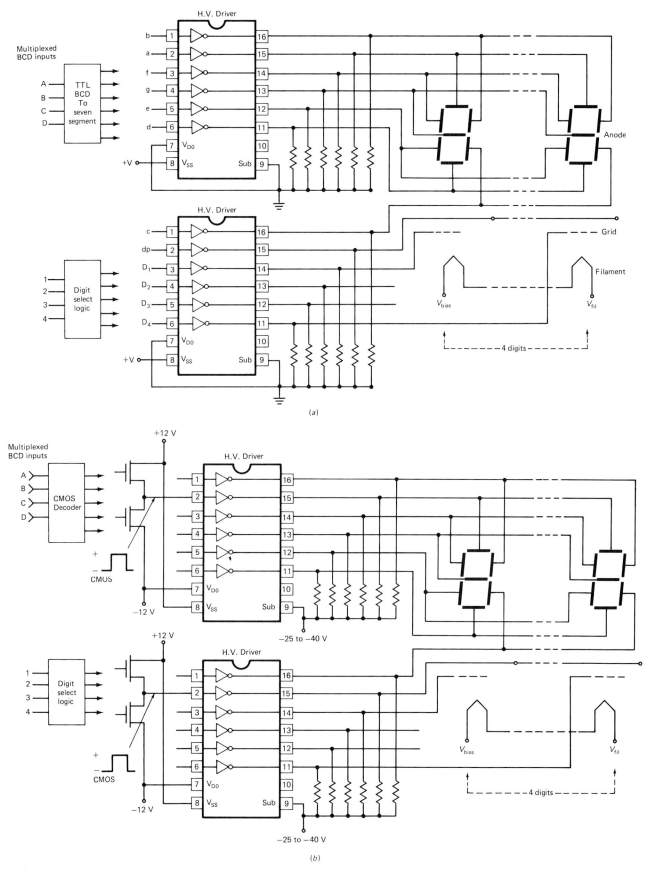

Fig. 14-46 Vacuum fluorescent display circuits.

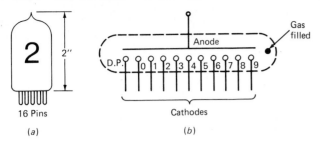

Fig. 14-47 Gas discharge tube. (*a*) Outline. (*b*) Schematic diagram.

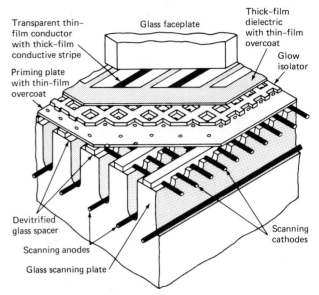

Fig. 14-49 Cutaway view of plasma display panel.

monochrome and color. Computer graphics and engineering and image processing will be handled by CRTs for many years to come.

The *CRT* is a vacuum tube in which electrons are produced at a heated cathode, then directed to a phosphor coating on the tube face. The electrons striking this coating produce light. The several parts of the CRT are diagrammed in Fig. 14-50. The coated cathode is heated by the filament (heater). Electrons are boiled off and are directed toward the screen at the opposite end of the tube. The electrons are compacted into a narrow beam so that a point, rather than a diffuse spot, is obtained at the screen. The beam is electrostatically focused by succeeding stages and is acted upon by the deflection plates. These plates produce an electrostatic field that follows the input signal. With the correct signal, it is possible to deflect the beam anywhere on the face of the CRT to display characters, lines, or pictures. The accelerating section of the CRT is necessary to give sufficient energy to the beam that the electrons will produce phosphorescence when they strike the screen. The magnitude of the voltage in the accelerating section is of the order of kilovolts and must be uniform in space so as not to distort the beam. The beam may also be deflected magnetically instead of by using the deflection plates as illustrated.

The block diagrams of the principal parts of a CRT imaging system are shown in Fig. 14-51. Figure 14-51(*a*) shows the conventional CRT, where events

must be viewed and refreshed constantly. Figure 14-51(*b*) illustrates a storage-type CRT that allows a display to be retained for long periods of time (up to an hour or more). This type of display (storage) was commonly used before inexpensive semiconductor memory was available. Oscilloscopes use CRTs to display electronic waveforms in real time. Computer terminals use them to display alphanumerics and graphic images. Television receivers use them to display pictures. All in all, the CRT is probably the most versatile of all display devices. There are two basic ways to produce images on the CRT screen. One is a *vector display*. To draw a box, the beam is deflected from corner to corner. This process is repeated over and over, and the box appears on the screen. Or if the CRT is of the storage type the box need be drawn only once. Vector displays are normally limited to high-resolution screens used in computer-aided design and drafting.

The other way to draw on the screen is through a *raster display*. The electron beam is continuously scanned from left to right and from top to bottom. When the bottom right is reached, the process repeats by returning to the upper left. If the beam is

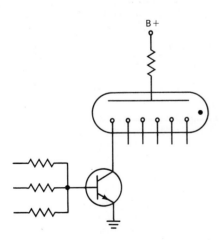

Fig. 14-48 Typical gas discharge display drive circuits.

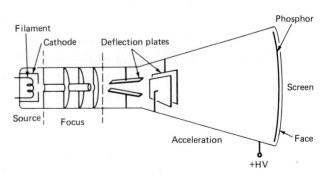

Fig. 14-50 Cathode-ray tube.

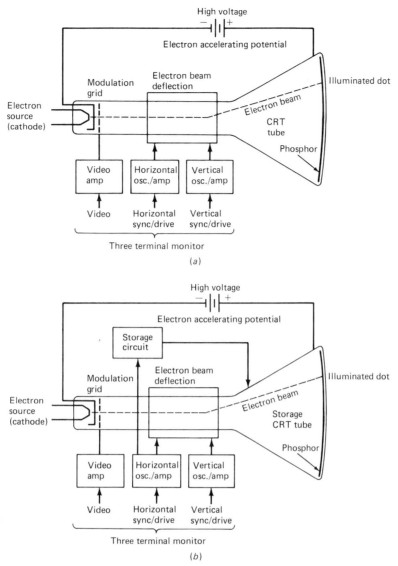

Fig. 14-51 CRT monitor block diagrams. *(a)* CRT monitor.
(b) Storage CRT monitor.

allowed to remain on, the entire screen lights up. However, if the beam is turned off and on at the proper times, a box, or almost any other character or picture, can be drawn.

Industrial terminals and computer displays generally use a 525-line raster display. About 64 μs is necessary to scan one horizontal line and about 17 ms to get from top to bottom. Only half the picture is scanned at a time (262.5 lines). Every other vertical scan fits an additional 262.5 lines in between. This process is called *interlacing* and is done to minimize flicker on the screen. Noninterlaced displays, which draw all 525 lines in a single pass from top to bottom, are also used.

Figure 14-52 indicates what has to happen to the beam in order to display the letter *N* on the screen. With a seven- by nine-character font, the beam must be turned on to display dots in column 1 and column

7 when line 1 is being displayed. Line 3 must have dots appear at column positions 1, 2, and 7. The beam is turned on with information stored in a ROM called a *character generator*. Figure 14-53 shows a portion of a typical character set stored in a five- by seven-character-generator ROM.

In order to have the characters appear stable and in the correct screen position, the ROM information must arrive at the CRT at the correct time. It must be synchronized with the vertical and horizontal scanning circuits. A block diagram that accomplishes this synchronization is shown in Fig. 14-54. The clock circuits are locked to the raster waveforms mentioned earlier. The characters to be displayed are stored in RAM in ASCII form. Note that only 6-bit ASCII code is required to display the uppercase character set. The 6-bit ASCII code is latched and decoded and applied to the ROM. This code acts as

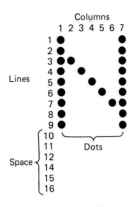

Fig. 14-52 7 × 9 font character example.

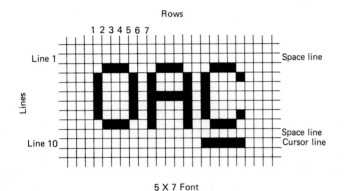

Fig. 14-53 Character font display.

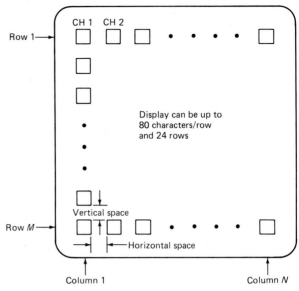

Fig. 14-55 CRT display example.

an address, and the ROM output is one of a possible 64 combinations. The ROM output is multiplexed into a 7-bit shift register and then shifted out to the CRT one dot (pixel) at a time to display one line of the character.

Figure 14-55 shows how the typical industrial CRT display is organized. Generally, there are 24 rows of

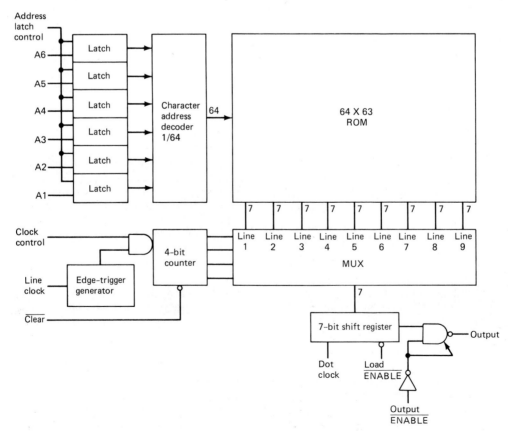

Fig. 14-54 Character generator block diagram.

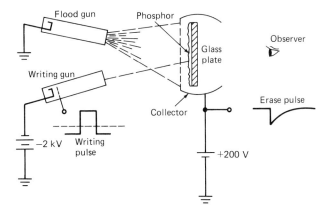

Fig. 14-56 Elementary bistable storage CRT.

80 characters each. Some terminals display 132 characters per row but tend to cause operator fatigue and are not very popular. Raster displays can also do a fair job with graphics. Instead of using character generators, RAM information is clocked out at the dot rate to draw lines, arcs, and irregular curves on the screen. As mentioned earlier, a storage CRT allows a display to be retained for long periods of time. Signals that occur at low repetition rates often cause flickering of the display. Storage allows these signals to be displayed at a constant light level. The typical storage CRT is called a *bistable storage tube*. Its parts are shown in Fig. 14-56, though they are not drawn to scale. Initially, the writing gun is biased off, and the flood-gun cathode is grounded. The collector is fixed at +200 V, and all the targets (phosphors) are at their lower stable point. The flood gun is able to hold each target independently at either of its two stable points (on or off), once they are written or erased to those points.

When the writing beam is gated on and bombards any target (point on the screen), this target charges positive and is now at a stable upper point. It is held at this upper stable point by the flood gun, while the other targets remain at their lower stable states. Any target or targets at the upper state will emit light due to excitation from the flood gun. Thus, once an image is written to the screen, it will remain there for about an hour.

When the erase pulse, Fig. 14-56, is applied to the collector (which acts as a capacitor with the targets), the target voltage drops (as a result of the capacitive coupling), and the screen is reset to its lower stable point. These tubes can store waveforms with nanosecond rise times, but they are very expensive (more than $1000). They are used mainly in storage oscilloscopes and in some computer graphic terminals.

REVIEW QUESTIONS

16. What type of solid-state display would you use for creating a histogram?

17. A BCD to _____ segment decoder is required to operate font A of Fig. 14-38.

18. Liquid crystal displays are driven with _____ voltage to eliminate plating of their electrodes.

19. Liquid crystal displays are used mainly with _____ type logic.

20. A fluorescent display is controlled by signals applied to its _____.

21. The individual dots/spots on a display are known as _____.

14-4
OPTICAL COUPLING

The *optocoupler,* also known as an *optoisolator,* consists of a photon-emitting device whose flux is coupled through a transparent insulation material to some sort of detector. The photon-emitting device may be an incandescent or neon lamp (earliest models used these) or an LED. The transparent insulation may be air, glass, plastic, or optic fiber. The detector may be a photoconductor, photodiode, phototransistor, photo FET, phototriac, photo SCR, or integrated photodiode/amplifier. Various combinations of these elements result in a wide variety of input characteristics, output characteristics, and coupled characteristics. This discussion will be concerned with optocouplers that use an IRED input with a variety of output detectors. Characteristics such as *coupling efficiency* (effect of IRED current on the output device), speed of response, voltage drops, current capability, and *V-I* curves vary from model to model. The characteristics must be considered especially when performing substitutions. The only common characteristic is that the input is dielectrically isolated from the output. Figure 14-57 shows some of the more common symbols for optoisolators.

The optocoupler was designed as a solid-state replacement for mechanical relays and pulse transformers. Functionally, the optocoupler is similar to its older mechanical counterpart because it offers a high degree of isolation between the input and output terminals. Some of the improvements offered by the solid-state devices are as follows:

Faster operating speeds

Positive (no-bounce) action

Small size

Insensitivity to vibration and shock

No moving parts to stick

Compatibility with many logic and microprocessor circuits

Frequency response from dc to 100 kHz

Isolation is a very important parameter of the optocoupler. The three critical isolation parameters are resistance, isolation capacitance, and dielectric withstand capability. *Isolation resistance* is the dc resistance from the input to the output of the coupler. A value of 10^{11} Ω isolation resistance is very typical;

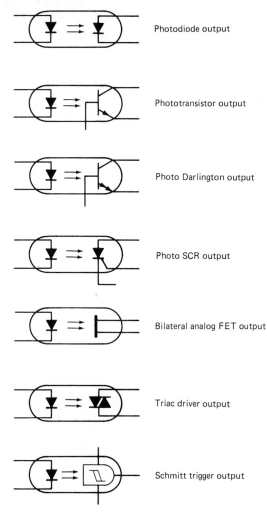

Photodiode output

Phototransistor output

Photo Darlington output

Photo SCR output

Bilateral analog FET output

Triac driver output

Schmitt trigger output

Fig. 14-57 Optoisolator/coupler symbols.

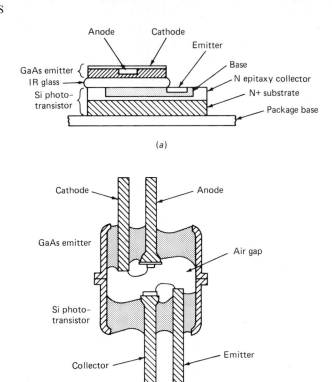

(a)

(b)

Fig. 14-58 Optocoupler configurations. (a) Glass isolated. (b) Air isolated.

this value may be higher than the resistance between the mounting pads on many of the printed circuit boards on which the coupler is mounted. Therefore, care in handling of the printed circuit board is required to avoid degrading this parameter. For example, the flux residue from soldering must be removed from the board. Isolation capacitance is a parasitic capacitance from input to output through the dielectric. Typical values range between 0.3 and 2.5 pF. *Isolation voltage* is the maximum voltage which the dielectric can be expected to withstand. Typical ratings are around 1500 V, and special units are available with voltage isolation as high as 50,000 V.

The *input* stage of an optocoupler consists of an efficient GaAs infrared-emitting diode (IRED). The *output* stage of the basic optocoupler is a phototransistor. Two methods of dielectric isolation are shown in Fig. 14-58. Figure 14-58(a) has a thin layer of infrared (IR)-transmitting glass between the input and output stages; Fig. 14-58(b) uses an air gap to attain greater electrical isolation. Regardless of whether an air gap or IR-transmitting glass is used

to separate the input and output circuits, the operating characteristics are basically identical.

The *input* characteristics of the coupler are the same as for the IRED previously discussed in the section on emitters. Typical optocoupler input characteristics are shown in Fig. 14-59. Note that currents from 100 mA to 10 A are permitted only in the pulsed mode. The transfer curve for a typical photodiode coupler is illustrated in Fig. 14-60(a) and that of a photodarlington coupler in Fig. 14-60(b). The *current transfer ratio* is a common method of cataloging these devices. These ratios are given as percentages. They range from 10 to 80 percent for the phototransistors. For example, if the input is 10 mA,

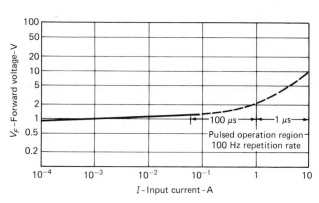

Fig. 14-59 Optocoupler input characteristics.

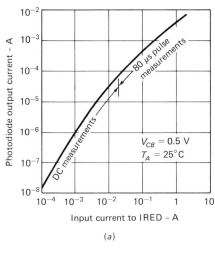

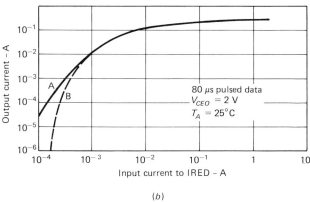

Fig. 14-60 Phototransistor/Darlington optocoupler transfer curves. (a) Transistor. (b) Darlington transistor.

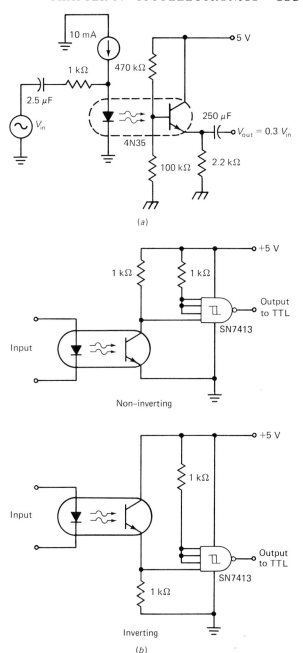

Fig. 14-61 Optocoupler applications, ac and digital. (a) Linear coupler. (b) Logic coupler.

and the collector current (output) is 2 mA, the transfer ratio is 20 percent. Photodarlingtons exhibit ratios of 100 to 1000 percent. With a 1.0-mA input, the output might be 10 mA for a transfer ratio of 1000 percent.

The dynamic response of the optocouplers is dominated by the capacitance of the photodiode, the input resistance of the transistor, and the voltage gain of the transistor. Through the *Miller effect*, the stray capacitance (C_s) is considered as a single capacitance across the input whose magnitude is approximately equal to the gain times C_s. The RC time constant becomes input resistance × capacitance × voltage gain. The penalty for high gain is slow response. Typical rise and fall times of the phototransistor are on the order of 2 to 10 μs, which is quite satisfactory for most analog types of applications. The addition of a resistor from the base terminal to ground will lower the gain but speed up the circuit's response. Figure 14-61(a) uses a 10-mA current source to bias the photodiode for linear response to the input signal (V_{in}), and the phototransistor is also biased for linear operation. Figure 14-61(b) shows noninverting and inverting digital coupling. The base lead is not available but is not required since linear bias is not used

in logic circuits. The photodarlington optocoupler offers two major advantages, low input currents and very high output currents. The high gain of the photodarlington permits output currents of tens to hundreds of milliamperes with input currents as low as 0.5 mA. The switching speeds in the low-input-current region are quite slow but are acceptable for driving loads such as solenoids and lamps.

In the past, phototransistors were used to drive external SCRs, but today the photo SCR is packaged with the IRED. The *photo SCR optocoupler* differs from other SCRs in respect to the very-low-level gate

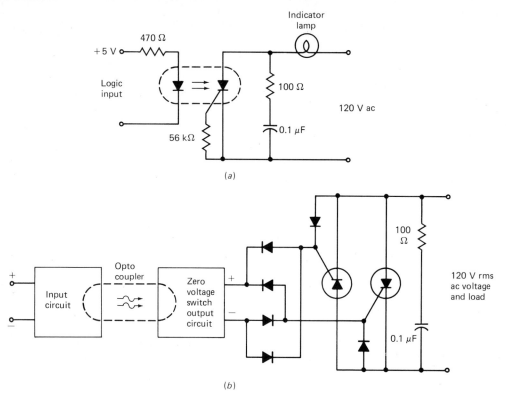

Fig. 14-62 Typical opto SCR applications.

drive available from its detector. This low-level gate drive requires a sensitive gate structure for the SCR. Applications may require the SCR to operate in 120- and 240-Vac circuits. Recent fabrication techniques have reduced some undesirable effects, such as rate effect (*dv/dt*). The pulse capability of the SCR makes it ideal for capacitor discharge and triggering applications. It is also applied in full-wave ac control, high-voltage SCR series string triggering, three-phase circuitry, and isolated power supplies. Figure 14-62 shows two applications of the opto SCR. In Fig. 14-62(*a*) a logic level output from a processor or computer is used to safely control a line-operated 120-V lamp or indicator. In Fig. 14-62(*b*) the device is applied as a solid-state relay (no moving parts) as to allow a logic input of 5 V at 15 mA to control a 220-

V, 10-A load. The 100-Ω resistor and 0.1-μF capacitor are the suppression (snubber) network for any inductance in the load. The SCR is a half-wave device; so to obtain full-wave control, two have to be connected as configured in Fig. 14-62(*b*). The result is known as an *antiparallel* (or *inverse parallel*) *SCR*. Figure 14-63 shows a triac full-wave solid-state relay. For the most part, the optocoupled SCR can only switch milliamperes of load current, but that capability is more than sufficient to trigger larger SCRs and triacs.

As previously mentioned, zero voltage switching is necessary in many cases to reduce in-rush current and radio frequency interference (RFI). Solid-state relays with zero crossing give approximately a factor of 10 improvement in RFI, mainly because they do

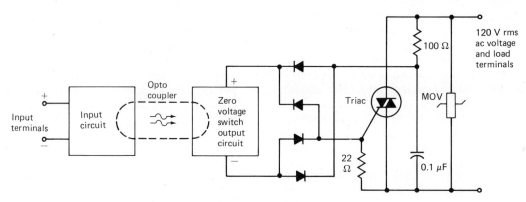

Fig. 14-63 Optocoupled triac.

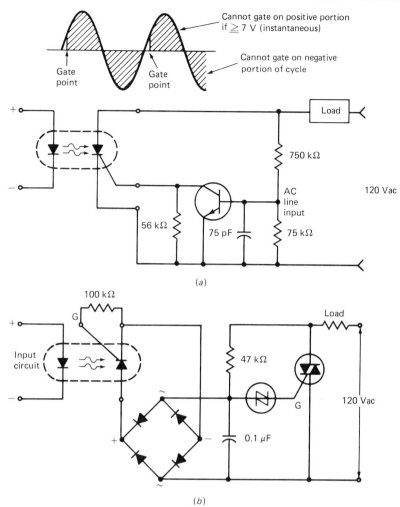

Fig. 14-64 Zero voltage switching couplers. (a) Half-wave ZVS. (b) Full-wave ZVS.

not arc or bounce when operated. A half-wave zero voltage switch is illustrated in Fig. 14-64(a). The gate to cathode of the SCR is short-circuited by the transistor at any line voltage greater than 7 V positive. This prevents the SCR from being gated on. Figure 14-64(b) shows a full-wave zero voltage switch.

Phototriacs are convenient devices when ac loads must be controlled from digital logic circuits. These devices are not designed to act as ac load switches, but as pilot devices for triggering power triacs. They allow reductions in components and circuit size when compared to the phototransistor or photo SCR. Figure 14-65 shows how a phototriac coupler is the equivalent of antiparallel SCRs. A simple solid-state ac relay using a triac optocoupler is shown in Fig. 14-66(a), and Fig. 14-66(b) shows a zero voltage solid-state relay circuit. Figure 14-67 indicates the way that the motor-starting contactors of Chapter 4 can be replaced by solid-state relays (SSRs) with no moving parts or arcing.

The bilateral analog FET optocoupler consists of

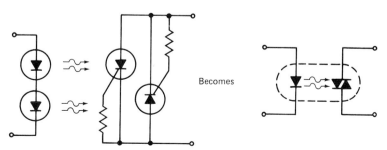

Fig. 14-65 Opto SCR/triac equivalents.

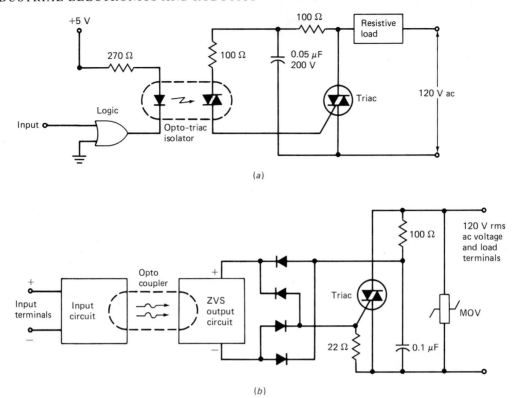

Fig. 14-66 Solid-state ac relays. (*a*) Solid-state relay (full-wave). (*b*) ZVC solid-state relay.

an IRED source coupled to a symmetrical bidirectional silicon detector chip. The characteristics are similar to those of a bidirectional FET. The output conductance is linear at low signal levels, and the bilateral analog FET optocoupler performs as a nearly ideal analog switch. A commutation circuit is illustrated by Fig. 14-68. When the control signal is high, the IRED in optocoupler (1) is on. This allows the signal to reach the op-amp because optocoupler (2) is off. When the control signal is low, optocoupler (2) is on, grounding the op-amp input. Switching the input alternately from the signal to ground is called *chopping* or *commutation* and is used to eliminate offset drift in the amplifier. Figure 14-68(*b*) shows a four-channel multiplexer circuit that selects one of four analog input signals.

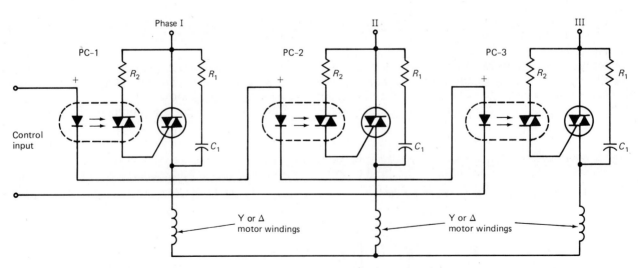

Fig. 14-67 Three-phase solid-state switching.

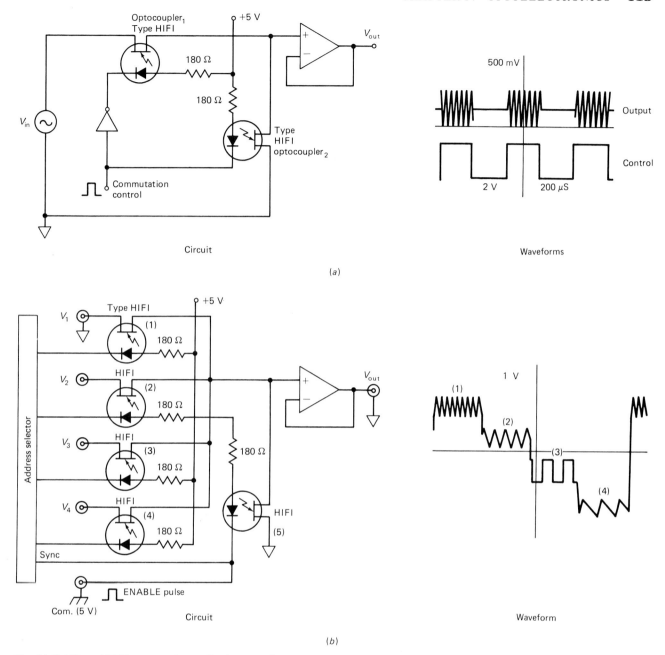

Fig. 14-68 Bilateral FET optocoupler applications. (*a*) Commutator. (*b*) Multiplexer.

Solid-state relays can be made to handle ac or dc loads. They provide isolation with optical or transformer coupling. These devices eliminate the output off-set voltages associated with a transistor SSR and are found in a wide variety of packages and sizes. Figure 14-69(*a*) illustrates a block diagram of a transformer-isolated SSR for dc. The input requires 1.6 mA at 5 Vdc, which makes it compatible with logic circuits and computers. Figure 14-69(*b*) shows a typical optically isolated solid-state relay, which lends itself to a wide variety of automatic and computer control applications.

The high-speed digital coupler, also known as a *Schmitt coupler* (because most include an internal Schmitt trigger), can interface directly with transistor-transistor logic (TTL), and LSTTL families. It provides ac and dc isolation to eliminate ground loops, allowing direct interfacing between computers and peripheral devices at data rates up to 1 M bits/ s. Four common data isolators are pictured in Fig. 14-70. Both inverting and noninverting types are available, as well as totem pole and open collector output styles.

The *coupled interrupter module* is another member

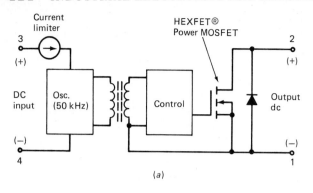

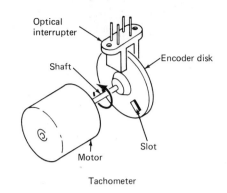

Tachometer

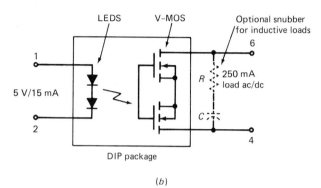

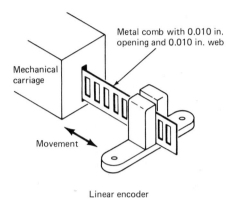

Linear encoder

Fig. 14-69 Solid-state relays. (*a*) DC SSR block diagram. (*b*) AC/DC solid-state relay.

Fig. 14-71 Optical interrupter applications.

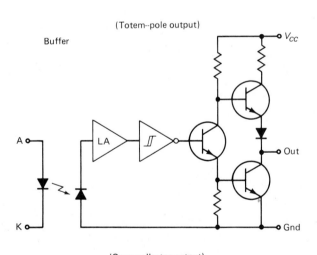

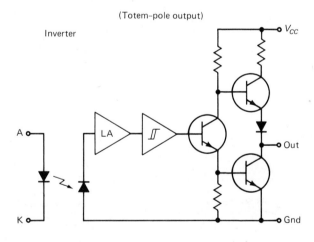

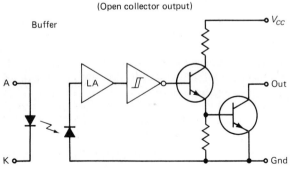

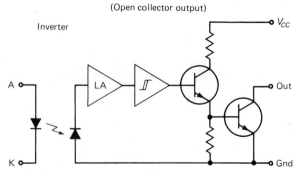

Fig. 14-70 Logic optocouplers.

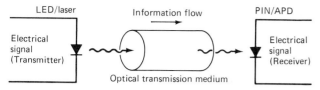

Fig. 14-72 Typical fiber-optic system.

of the optoisolator family. This device is also known as a *slotted switch* or *source/detector assembly,* but the name *optical interrupter* is most descriptive of its function. The device contains an IRED and a silicon photodarlington detector in a plastic housing. A gap in the housing provides a means of interrupting the light with tape, cards, shaft encoders, or any opaque material to switch off the internal transistor. Figure 14-71 shows two typical applications using the interrupter as a tachometer/speed monitor and as a linear encoder for relating distance to pulses.

The last member of the optocoupler/isolator family is the *reflective object sensor.* Each assembly con-

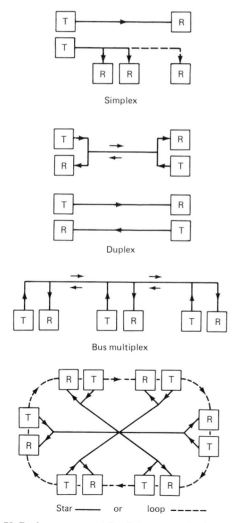

Fig. 14-73 Basic arrangement for data communications.

sists of an infrared-emitting diode and an NPN silicon phototransistor or photodarlington mounted side by side on converging optical axes in a black housing. The photosensor responds to radiation from the IRED only when a reflective object passes within its field of view. These devices are sensitive only within a range of less than 0.3 in. (75 mm). A typical use is detection of the beginning-of-tape markers on magnetic tapes. A typical interfacing circuit uses a comparator with sensitivity and hysteresis controls.

Fiber-optic systems offer an alternative method for transmitting information and sensing physical events. Fiber optics offer a small, lightweight, durable, corrosion-resistant, nonconducting signal path that is virtually unaffected by and has no effect on the electrical environment through which the signal passes. Figure 14-72 shows that the electrical signal is changed to light, which enters the fiber cable. The light signal is changed back to an electrical signal at the other end of the cable. The system illustrated in Fig. 14-72 is called a *simplex system* (one-directional). Two-way communication (*duplex*) requires two such links. Figure 14-73 shows the basic data transmission systems. Distance limitations of fiber-optics communications arise mainly from the means of producing the optical flux and from path losses. Although power into a wire cable can easily be several watts, the flux into a fiber-optic cable is typically much less than a milliwatt. Wire cables may have several signal *taps;* however, multiple taps on fiber-optic cables are impractical at present.

The losses in a *point-to-point fiber-optic system* are insertion loss at the input and output, connector loss, and transmission loss, which is proportional to cable length. Fortunately, no noise is picked up by a fiber-optic cable so the receiver signal-to-noise ratio (SNR) is limited only by the noise produced within the receiver.

Light rays are confined to the core of the optical fiber by cladding the core with a transparent material having a lower index of refraction. This defines the critical angle of reflection at the core cladding interface, thereby confining rays at smaller angles to the core of the fiber. A typical optical fiber cable is shown in Fig. 14-74.

There are three common optical fiber types: the stepped index multimode, stepped index monomode,

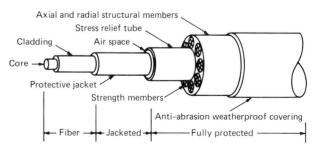

Fig. 14-74 Optical fiber cable construction.

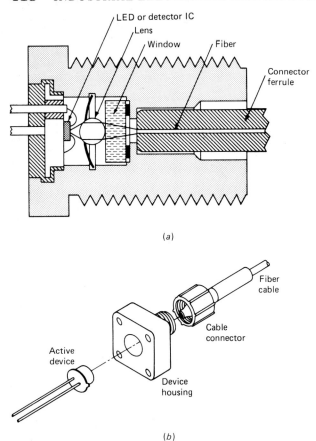

Fig. 14-75 Fiber cable connection to active device. (a) Cross section. (b) Mating fiber connector.

and graded index multimode. We will be concerned mainly with the stepped index multimode fiber, which is used for moderate bandwidth applications.

The term *stepped index fiber* refers to the abrupt change in the index of refraction between the core and the cladding. All transmission of light occurs within the core. The different index of refraction between the core and the cladding defines a critical angle such that any light ray entering the core at less than that critical angle will be completely reflected. The numerical aperture is the *sine* of this angle and defines a cone in which incident light may be launched into a cable. The smaller the numerical aperture, the harder it is to launch light into the fiber.

A wide variety of fibers exist, with little standardization, in three basic materials: *plastic clad-plastic core fiber* (plastic fiber); *plastic clad-glass (silica) core fiber* (PCC fiber); and *glass clad-glass core fiber* (glass fiber). Fiber attenuation varies greatly within the core material types and with the wavelength of the light used. So far, this discussion of optical fibers has treated only single fibers. In some applications, improved performance may be gained by utilizing a *fiber bundle,* which consists of a group of single fibers

in a single jacket. A bundle is more flexible than a single fiber of equal area and continues operating even if a few strands break. A bundle is usually harder to terminate with a connector than a single fiber. It is usually more difficult to polish the ends of the fibers of a bundle. Higher-power insertion losses are caused by a poor finish. Poor fiber connections and polish have been observed to cause up to 10-dB (90 percent power loss) signal loss per end.

There is a wide variety of fiber-optic connectors because of a lack of standards. To provide a low-loss fiber-to-fiber splice, a connector must position the two optically polished fiber ends very closely together in axial concentric alignment. If the fiber ends touch, abrasion may spoil the end finish and cause power loss. Some connectors for plastic fibers maintain pressure between the fiber ends. The pressure deforms the plastic ends for a better fit. Coupling efficiency falls off rapidly as the distance between the fiber ends increases and also with angular error and errors off concentricity. In general, connectors for fibers of 200-μm core diameter and greater are easier to install and provide better consistency than connectors for the smaller-diameter cables.

Most active devices, such as emitters and detectors, are applied to fibers with adapter connectors or short lengths of fiber built into the active device and terminated within a connector. Figure 14-75 illustrates the mating of a fiber cable/connector with a detector/emitter.

The preparation for repairing damaged cables is quite tedious. The 10- to 12-step process requires skill and practice. The cable connectors are ultrasonically cleaned and baked. The fiber in the cable is cleaved (cut with a blade), cleaned, and epoxied into a ferrule shoulder. The sealing of the fiber is shown in Fig. 14-76(a). Eighteen hours must elapse for the fiber to be cured before it can be polished. The cutaway view of the polished end is shown in Fig. 14-76(b). Caution must be taken to wear eye protection to avoid injury from any fragments of the fibers. Also avoid any skin punctures. In many cases, prepared cables with connectors are used or installed. They are available in 1- to 1000-m lengths.

The emitters are usually LED/IREDs or diode lasers (lasers will be covered later in this chapter). A majority of LED/IRED emitters operate at wavelengths of 550 to 1300 nm, with most falling in the 640- to 940-nm range to work more efficiently with the detectors. Figure 14-77 shows simple digital and analog fiber transmitter circuits. The detectors (sensors) can be photodiodes, phototransistors, PIN photodiodes, or avalanche photodiodes, all of which operate on the same basic principle. In some cases, the receiver circuitry is integrated in the same package with the detector. This assembly may be no larger than the typical metal case transistor (TO-5). Figure 14-78(a) shows a simple receiver (detector) circuit. Its response would be limited to about 150 kHz by

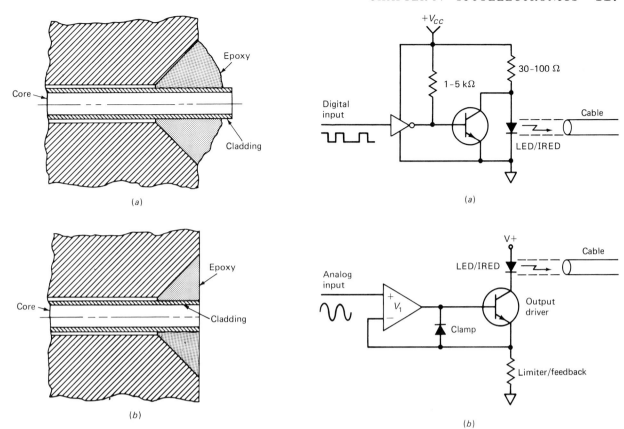

Fig. 14-76 Fiber cable sealing/termination. (*a*) Sealing fiber to ferrule. (*b*) After polishing.

Fig. 14-77 Simple fiber-optic transmitters. (*a*) Digital transmitter. (*b*) Analog transmitter.

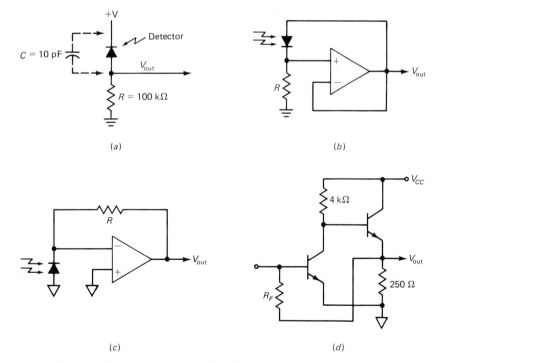

Fig. 14-78 Transimpedance configuration. (*a*) Simple receiver. (*b*) Bootstrap configuration. (*c*) Transimpedance. (*d*) Discrete transistor amplifier (transimpedance).

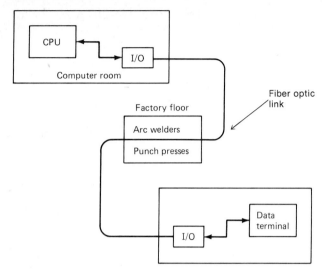

Fig. 14-79 Sample system demonstration.

the time constant of the 10-pF shunt capacitance along the 100-kΩ load resistor:

$$T_r = 2.2RC$$
$$= 2.2 \times 100 \times 10^3 \times 10 \times 10^{-12}$$
$$= 2.2 \ \mu s$$
$$F = \frac{0.35}{T_r}$$
$$= \frac{0.35}{2.2 \times 10^{-6}}$$
$$= 159 \text{ kHz}$$

where T_r = rise time (time from 10 to 90% of the output waveform)
F = upper cutoff frequency of an amplifier
$F \times T_r$ = 0.35, a constant

Figure 14-78(*b*) is a *bootstrap configuration.* The amplifier follows the voltage developed by the photocurrent flowing through the resistor and applies this voltage to the opposite end of the photodiode. This configuration provides a lower load impedance and a faster response. By rearranging, as in Fig. 14-78(*c*), a transimpedance amplifier is obtained. Since the inverting input is a virtual ground, the bandwidth will be determined mainly by the amplifier used. Figure 14-78(*d*) shows a transimpedance circuit with discrete components that has a bandwidth in excess of 50 MHz. Many receivers are made with the connector, detector, and amplifier all contained in a small low-profile package that can be mounted directly to a printed circuit board. A fiber-optic link in a hostile industrial environment is shown in Fig. 14-79. The outstanding noise immunity of optical links allows them to operate flawlessly in situations that are very difficult for conventional data transmission techniques.

REVIEW QUESTIONS

22. Most optocouplers use a GaAs _____ as the internal emitter.

23. The use of a photodarlington coupler provides an increase in speed. (true or false)

24. If the collector current changes 10 mA for an input change of 2 mA the transfer ratio is _____ percent.

25. The photo SCR is a _____ wave switch.

26. The antiparallel photo SCR connection is for _____ wave control.

27. The zero voltage switch (ZVS) control turns on at zero current crossings. (true or false)

28. The true solid-state ac relay uses a _____ for output switching.

29. Which optocoupler type is used for linear analog signals?

30. The three main output parts of a digital optocoupler are the phototransistor, the linear amplifier, and the _____.

14-5 LASERS

Lasers were originally referred to as *optical masers. Maser* is an acronym for *microwave amplification by stimulated emission of radiation. Laser* is an acronym for *light amplification by stimulated emission of radiation.* The first laser was a ruby crystal device developed in 1960. It provided a single wavelength, which is called *monochromatic* (one-color) *light.* Lasers can also provide *coherent light,* with all of the waves in phase. Early lasers were fragile and expensive. By the 1970s, reliable lasers were available for industrial applications. They are a practical source of energy for cutting, welding, and drilling. They are used in precision alignment and measuring systems. Future applications in the areas of mass storage and ultra-high-speed logic circuits are anticipated.

Practical lasers commonly found in industry include the following:

1. Solid-state laser (ruby)
2. Gas laser
3. Semiconductor laser
4. Organic dye laser

Ruby is sapphire (crystalline aluminum oxide) in which a small percentage of the aluminum has been replaced by chromium. The chromium concentration is around 0.05 percent. Figure 14-80 shows the component parts of an optically pumped solid-state laser. Energy is stored in a capacitor or a bank of capacitors. The energy is produced by a dc high-voltage power supply. For smaller laser crystals, the voltage is in the 2000- to 5000-V range. The flash tube is filled with xenon and does not conduct until the gas is ionized by a high-voltage pulse from the trigger transformer. When it is pulsed by the trigger, the stored energy (in joules [J] = 1/2 (farads × volts^2) causes the xenon tube to emit very intense radiation, which

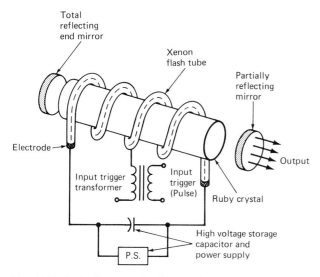

Fig. **14-80** Optically pumped solid-state laser.

then is absorbed in the laser crystal by a pumping action. This optical pumping action is necessary for all solid-state lasers to function. Light from the flash tube pumps the chromium atoms from ground state G to excited levels in band E or band F. The pumped atoms are unstable at this level and lose energy in a two-step process. The first step is to a metastable level M. No emission takes place at this level. The second step is from M to ground G state, with release of photons by simulated or spontaneous emissions. This is called a *three-level energy laser action*. There is also a four-level energy action in some lasers. The main laser action occurs in the optical cavity formed by the reflecting mirrors at the ends of the laser rod. The laser beam emerges through the partially transmitting mirror at the right end of the laser shown in Fig. 14-80.

To increase the effective optical coupling between the flash lamp and the laser rod, it is necessary to surround the complete assembly by reflecting walls. Figure 14-81 shows a system that uses a helical flash lamp with the ruby rod centered in the enclosure.

The flash lamp is similar to those used in photog-

raphy. The flash lamps used in lasers are glass or quartz tubes filled with a gas at low pressure. Xenon is commonly used, although higher brightness is achieved by using krypton or helium. The lamps can come in various shapes: helical, linear, and U-shaped are the most common. Electrical input energy, pulse duration, and lamp size can be varied over a wide range. When an electric current is discharged through the lamp, a high-temperature plasma which emits radiant energy over a wide range of wavelengths is formed. The gas type and pressure can be adjusted to produce a wavelength peak that matches the absorption spectrum of the crystal material being used.

Discharge in a flash lamp is initiated by causing a spark streamer to form between electrodes (usually at the ends of the tube). This is done by pulsing the lamp with a high-voltage trigger pulse. A typical circuit using a pulse transformer to step up the externally applied trigger pulse is shown in Fig. 14-82. Once the flash lamp has been triggered, the main discharge capacitor C can discharge through the ionized gas. The series inductance is added to shape the current pulse through the lamp and to avoid ringing or reverse flow of current, which is harmful to the lamp life.

The lasers we are concerned with produce light in the near-ultraviolet, visible, and infrared portions of the electromagnetic spectrum. The wavelengths are in the approximate range of 10^{-3} to 10^{-5} cm, and frequencies are on the order of 10^{13} to 10^{15} Hz. Note: Light wavelengths are usually expressed in micrometers (μm) or nanometers (nm). The micrometer is equal to 10^{-4} cm and was often referred to as a *micron*. The unit angstrom (Å), equal to 10^{-8} cm, was also commonly used. For example, a green light of 5.5×10^{-5} cm $= 0.55$ μm $= 550$ nm $= 5500$ Å. Both the angstrom and the micron are considered obsolete terms. Their use as modern units of measurement is being discontinued.

Lasers can be constructed by using a variety of different materials, each of which produces a distinctive wavelength. Figure 14-83 shows where the output of some of these materials occurs in relation-

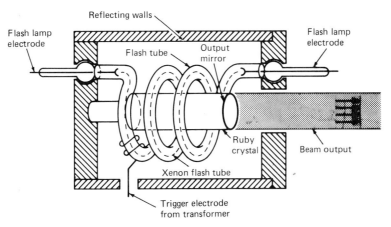

Fig. **14-81** Ruby laser, including encloser.

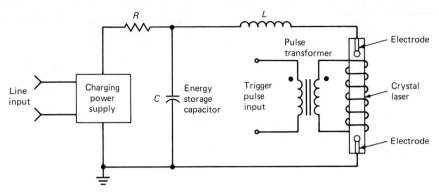

Fig. 14-82 Schematic of a typical flash lamp circuit.

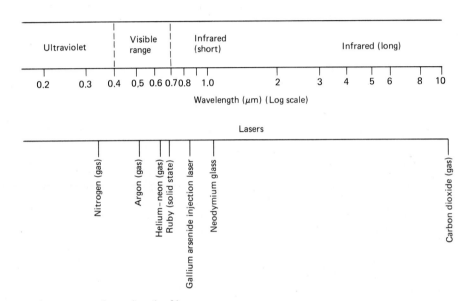

Fig. 14-83 Electromagnetic spectrum of wavelength of lasers.

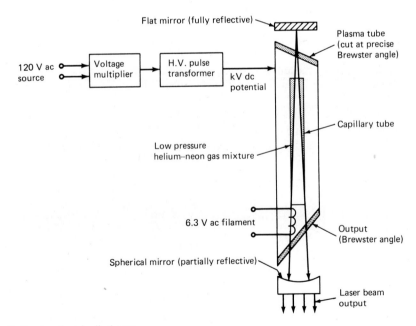

Fig. 14-84 Gas laser: excitation by electric discharge.

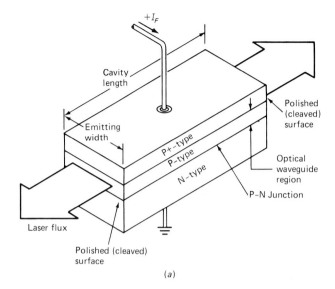

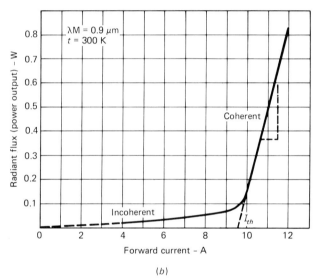

Fig. 14-85 Injection laser diode structure and output characteristics.

ship to each other and to the electromagnetic spectrum as a whole. The most useful lasers include the following:

1. The CO_2 laser, far infrared to 10.6 μm

2. The neodymium glass laser, near infrared at 1.06 μm

3. The gallium arsenide laser, near infrared, at wavelengths around 0.85 to 0.90 μm

4. The helium-neon laser, emitting 0.6328-μm radiation, reddish-orange in color

5. The argon laser, operating at several wavelengths in the blue and green portions of the spectrum

6. The nitrogen laser, an ultraviolet laser, operating at 0.3371 μm

This listing is not complete but will serve as a reference point for the significant types.

Gas lasers function quite differently from pulsed solid-state lasers. Gas lasers are not powerful. They operate mostly in the continuous-wave (CW) mode. Their steady beams will not burn holes or vaporize steel as the pulsed (crystal and semiconductor) laser can (this effect will be discussed shortly). There are many reasons for using gas. The volume of the material can be large, in contrast to those of crystals and semiconductors. Heat can be removed readily by transporting the heated gas out of the region and replacing it with cooler gas. A mixture of two gases is required for laser action. Atoms of the second gas are raised to higher levels with external excitation. For example, a laser might use 10 parts helium and 1 part neon. The mixture is pumped with radio frequency energy. A more modern approach is to use a high-voltage discharge from a voltage multiplier circuit, while a filament heats the gas as shown in Fig. 14-84. A further refinement of the electronics has eliminated the need for a filament to start the ionization of the gases. The circuit uses a voltage quadrupler to produce a starting boost of 7.5 kV to ionize the gas in the tube.

As discussed earlier, it is possible to generate light beams (emitter section) by using the principles of semiconductors. Laser diodes made from direct bandgap material differ from conventional light-emitting diodes in that they require an optical cavity and a high-injection carrier density. Thus the name *injection diode* usually refers to the semiconductor laser diode. The semiconductor laser is very efficient and small in physical size compared to other types of lasers. Figure 14-85(a) shows an injection laser diode structure, with a PN optical cavity formed by cleaving opposite ends of the diode and sawing the adjacent sides of the rectangular structure. In Fig. 14-85(b) the radiant output power as a function of the diode current is demonstrated. Note that a threshold current (I_{th}) of about 10 A is required for coherent output. Since the required current is so high, the device is operated in the pulsed mode only. The emission wavelength depends on the semiconductor material, doping level, and temperature of operation. In some injection lasers, a film of silver or gold is deposited over one end so that the other end will serve as the only output. The gallium arsenide material is reflective, so no internal mirrors are required to produce reflections. Typical lasers of this type emit 0.5 to 50 W of pulsed energy. Many injection lasers are operated at cryogenic temperatures (about 77 K). Two basic difficulties in achieving high peak power output from single laser diodes are that (1) high drive currents are required to drive the larger laser pellets and (2) the larger source size requires large, costly optics.

Laser diodes are also arrayed to provide increased power outputs at reasonable drive currents. Stacked diode lasers are currently available with minimum peak radiant flux levels ranging from 75 to 300 W with a drive current of 40 A. The increase in source radiant excitance (emittance) for these arrays com-

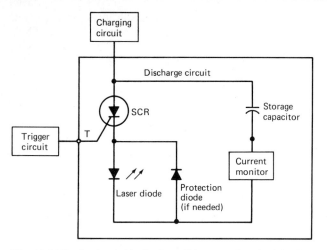

Fig. 14-86 Basic circuit blocks for solid-state pulser for injection laser diodes.

Fig. 14-87 Peak laser currents and capacitor voltage for pulsed lasers.

pared to single-diode arrays is significant; however, some sacrifice in duty cycle limit must be made.

The SCR serves as an off-to-on solid-state switch for many injection lasers and laser arrays. Figure 14-86 illustrates the basic blocks for a solid-state pulse power supply for injection lasers. The discharge circuit generates the current pulse in the laser and is the most important section of the pulser. The current pulse is generated by discharging storage capacitor C through the SCR and the laser diode, laser stack, or laser array. The rise time of the current

pulse is usually determined by the SCR; the fall time is determined by the capacitor value. The peak laser current and the charged-capacitor voltage relationships for some typical injection lasers using the circuit of Fig. 14-86 are illustrated in Fig. 14-87.

The circuit shown in Fig. 14-88(a) can deliver a peak current of 30 A for both single-diode lasers and laser stacks. The circuit in Fig. 14-88(b) can drive single or stack laser diodes with currents up to 75 A. The circuit of Fig. 14-88(c) is capable of driving a laser array consisting of up to 60 diodes in series.

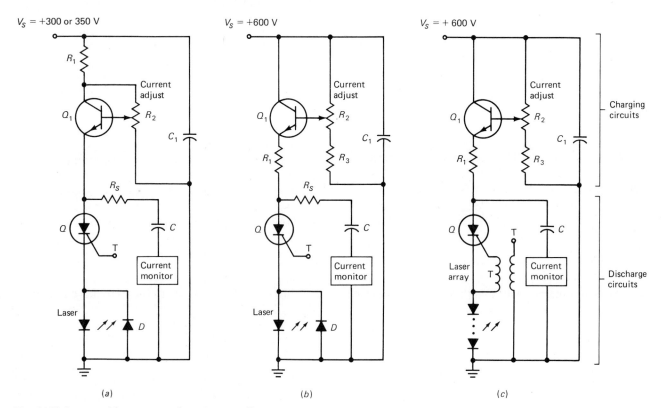

Fig. 14-88 Laser and laser-array pulse power supplies.

The repetition rates are limited to less than 500 Hz.

Liquids have useful advantages in relation to both solid and gas laser media. Several different liquid lasers have been developed, but the most important is the *dye laser,* which has the advantage that it can be tuned over a significant wavelength range. This capability is extremely useful in many applications such as spectroscopy and the study of chemical reactions. The materials that are used in dye lasers are similar to many familiar dyestuffs that are commonly employed as colorants in fabrics, plastics, soaps, and cosmetics. A pulsed laser is put through a block of the dye material then filtered (optically); this type of laser has almost no industrial applications at present.

Industrial applications of lasers include welding, hole drilling, cutting, trimming of electronic components, and heat treating. The most important property of laser energy is that it can be concentrated by a lens to achieve extremely high-power density at the focal spot. Let us compare a CO_2 (gas) laser and a large mercury arc lamp. The CO_2 laser emits 100 W of power in a beam that can be focused by a lens with a focal length of 1 cm to a spot 0.01 cm in diameter. Without going into the physics of beam divergence, the power density is 10^6 W/cm^2. In comparison, a mercury arc lamp of 1000 W has much more output; however, all the light from this lamp cannot be focused because the large divergence angle of the rays leads to a very large focal area. Therefore, the power density from this lamp is much lower. It can be shown (although we shall not go through the details here) that the same lens (focal length of 1 cm) will deliver a power density of only 100 W/cm^2 in the focal area. Thus, the total power emitted by the lamp is 10 times larger than for the laser, but the power density delivered to a workpiece by the laser is 10,000 times greater. Pulsed lasers can deliver much higher values of power density. A ruby laser can easily produce 1-ms-duration pulses with a power density of 10^9 W/cm^2 at a workpiece. When a high-power beam interacts with the workpiece surface, the material at the focal point is vaporized.

When laser radiation falls on a target surface, part of it is absorbed and part is reflected. The energy that is absorbed will heat the surface. Reradiation from the surface is usually insignificant. Heating from absorption can occur rapidly (losses due to thermal conduction are small if the pulse width is short but are important with long pulses). The surface quickly rises to its melting temperature. Melting is important in welding applications, and vaporization should not occur. Many workpiece factors such as heat flow, thermal conductivity, and material density determine the required power density for laser welding. The depth of penetration is important when sealing lids on difficult-to-weld containers housing delicate, heat-sensitive components, such as in a heart pacemaker container.

Metal removal (cutting or drilling) generally means operation at higher levels of laser power density than for welding. As in welding, material properties influ-

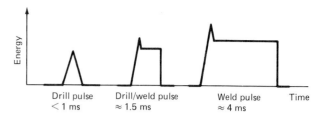

Fig. 14-89 Laser pulse shapes for welding and drilling.

ence the amount of metal removed. Pulse lengths in the range of several hundred microseconds are generally used. Repeated pulses to the same target area form deeper holes. Hole drilling is possible in ceramic and similar materials which are hard, brittle, and easy to break. There are no drill bits to break or wear out. For example, a jet engine part with 400 holes can be drilled by laser machining (with computer control) in less than 30 min. Figure 14-89 illustrates laser pulse shapes used for welding and drilling (drilling waveforms are also used for cutting, by overlapping the holes). The marriage of laser machining with robotics provides a manufacturing capability able to handle some very intricate production tasks. The laser is also used for alignment and is able to obtain accuracies of 0.085 μm/m.

Lasers have a wide range of potential applications in processing, storing, and transmitting information. They will have a large impact in these areas in the near future.

REVIEW QUESTIONS

31. Ruby lasers are usually excited by a _____ flash lamp.

32. This excitation action is also called optical _____.

33. The flash lamp is triggered by a high-current low-voltage pulse. (true or false)

34. The laser cavity is formed by _____ mirrors at each end.

35. Helium- _____ and CO_2 are common gas lasers.

14-6
PHOTOTUBES AND PHOTOMULTIPLIER TUBES

Light striking the large photocathode surface of a phototube will transfer energy to electrons in the metal, causing them to be ejected. This effect is called *photoelectric emission.* The photocathode surface is selected for a *low work function* (least amount of energy needed to remove electrons from the substance). Cesium is the most efficient photocathode used in the visible and infrared range. The phototube cathode emits a number of electrons which depends on the number of photons (intensity of the light)

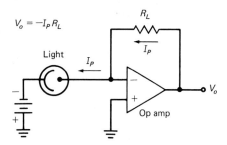

Fig. 14-90 Phototube coupled to op amp.

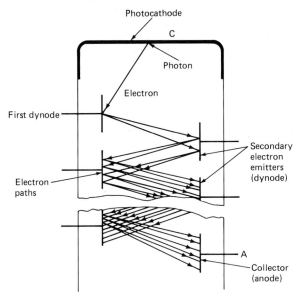

Fig. 14-91 Basic PM tube principle.

striking its cathode surface. The electrons are then attracted to the anode, which is positive with respect to the cathode. In a typical simple circuit for a phototube the photocurrent develops a signal across a load resistor. Increasing the value of the load resistor increases the sensitivity of the circuit (volts/lumen). This effect is desirable in some cases but can produce nonlinearity. Operating at a lower voltage (say, 100 V) will decrease the output voltage, but not the sensitivity.

Since the phototube current is a function of the anode-cathode voltage, if below 20 V or so, it is important that the voltage developed across the load resistor be as small as possible compared to the applied voltage. This places a limit on the value of the load resistor if a linear response is required (as it almost always is). This problem can be overcome by using an operational amplifier, as shown in Fig. 14-90. This configuration has been shown previously as either a current-to-voltage or transimpedance amplifier, where

$$V_o = -I_p \times R_L$$

In this case, the anode of the phototube is kept at virtual ground, and a negative bias is applied to the photocathode.

To improve the sensitivity, gas-filled tubes that produce larger currents for lower light input are used. The increase in photocurrent is due to secondary ionization of the gas molecules, but the characteristic curves are quite nonlinear. For accurate measurements, the vacuum phototube is employed. When less accuracy, but high sensitivity is required, the gas-filled phototube is selected. This device is similar to the photodiode previously covered. Most applications today use the solid-state device. As a result of certain environmental constraints, such as high temperature, nuclear radiation (which destroys most solid-state devices), and high humidity, the phototube may be used in some situations. The phototube is a rugged, long-life, durable device and will be employed in certain environments for many more years.

The phototube has very low gain at low intensities of light. The photomultiplier (PM) tube overcomes this deficiency. The basic operating principle of the PM tube is illustrated in Fig. 14-91. Light falling on a light-sensitive photocathode causes it to emit free

electrons, which are drawn away from the photocathode by an electrode having a more positive potential. These electrons accelerate and strike the first dynode, where they dislodge additional electrons. The next dynode is more positive, so the electrons are now accelerated to it. This process is repeated, with each stage having a higher positive potential than the previous one. The electrons emitted from the last secondary stage are collected at an anode (A in Fig. 14-91), and the resulting current is passed to the accompanying circuitry.

The number of secondary electrons emitted is dependent on the type of surface, the energy of the bombarding (primary) electrons, and the angle of incidence of the primary electrons. The ratio of the number of electrons leaving the surface to the number of incident electrons is called the *secondary emission ratio*. The total current amplification factor (G) between the photocathode and the anode is given by

$$G = \delta^N$$

where G = current gain
N = number of dynodes
δ = secondary emission ratio

Considerable gain is possible. If, for example, $\delta = 6$ and $N = 9$, we obtain a gain of 1×10^7.

The gain for a particular PM tube depends on the physical structure and the potential difference between each stage. A graph is prepared for each individual tube type to indicate the change in average overall amplification with the voltage per stage.

The dynode biasing chain in most cases is a linear resistor network. The negative high voltage ($-HV$) is applied across a resistor string which serves as a voltage divider and maintains the dynodes at increasingly higher positive potentials. The first dynode may be a grid used for focusing in some PM tubes.

If so, the cathode resistor will be two times (twice the potential of) the other elements. When the amplified signal current (or pulse) arrives at the anode, it flows through the anode load resistor or into the input of an op amp for a current-to-voltage transformation.

The current that flows in the PM tube when no light or radiation is falling on the photocathode is called the *dark current*. The dark current will limit the minimum detectable signal since it will be amplified along with any photocurrent due to light input. The dark current is temperature-dependent, and some applications cool the PM tube to improve its weak signal performance.

Current peaks in the dynodes can cause the voltage supplied by the divider network to sag. This sagging results in a type of degenerative feedback which may result in a nonlinear output from the PM tube. The last few dynode bleeder resistors are bypassed to stabilize the voltages. The PM tube is a very sensitive device, with gains of an order of magnitude greater than those of solid-state detectors. They are less sensitive to temperature and environmental effects. A great deal of development work has been done on photoemissive surfaces.

REVIEW QUESTIONS

36. Phototubes use cesium-coated anodes to increase their integrity.

37. Refer to Fig. 14-90. If I_P is -10 μA and R_L is 100 kΩ, V_O equals _____ .

38. For higher sensitivity a _____ type phototube is used.

39. The PM tube relies on the _____ emission of the dynodes for multiplication.

40. The dynodes of PM tubes must be increasingly _____ to sustain secondary emission.

14-7
TROUBLESHOOTING AND MAINTENANCE

Possible hazards unique to optoelectronic devices are due to some of the materials employed. Although gallium arsenide and gallium aluminum arsenide are both arsenic compounds, under conditions of normal use they are considered relatively benign. Electrical or mechanical damage to devices containing these materials should not produce a toxic hazard, but thorough washing of one's hands before eating or touching any food is recommended.

The eye may be damaged by infrared light. Most present GaAs and GaAlAs devices do not approach the safe limit values set by governmental agencies, but all the manufacturer's safety recommendations must be observed. If any doubt exists, check first before servicing an infrared device.

The output from a laser can be a source of fasci-

nation, but one must always maintain caution while near these devices. Never look into the beam of any laser. Reflections from polished surfaces are as dangerous as the original beam itself. Always be aware of the beam path and of others in the area when performing any action that may cause movement of the beam. Also, the high voltages involved may be lethal, so read the manufacturer's manuals and become familiar with the unit before energizing or troubleshooting any high-voltage components.

All PN junction emitters are operated in the forward-bias region. They can be tested in the same way as a common diode; using a multimeter with a diode test mode will exhibit the similar (with different thresholds) characteristics in the forward direction. The reverse breakover voltage is very low, usually only 2 V. As with all electronic equipment, the power supply must first be verified. With optoelectronics, dust, dirt, oil, etc., on the light-sensitive emitter or detector can be a source of problems. In some cases, the build-up is slow, and adjustments are made until they are out of range. Input from an operator may provide useful clues. High ripple on a power supply can have a very adverse effect and can cause unstable operation, including relay chattering, overheating, and premature failure of solid-state devices. A good maintenance record provides valuable information and may indicate trends.

All photodetectors are heat-sensitive and exhibit increases in dark current and shifting of operating characteristics. Therefore, any changes in ventilation or the surrounding environment should be noted. Many photodetector circuits can be verified (go–no-go) with the use of a small test lamp, which can aid in localizing problems. The output of the first amplifier in a system can be observed by using a meter, probe, or scope. A blinking light source is preferred to avoid saturating any of the circuits.

Photocells can be checked easily with an ohmmeter by exposing them to light and dark conditions. If the device is used in a bridge configuration, the power supply along with the other legs of the bridge must be checked. The bridge and detector circuit can be verified by substituting a variable resistance for the cell. It can be varied to simulate the cell's response to check the response of amplifiers and detectors.

Most displays that incorporate LEDs have some sort of driver and limiting circuitry. Failures here are not uncommon. In the segment-type display, knowing the function codes can help verify whether the decoder or the display itself is defective. Many units have a test input for lighting all of the segments. Multiple failures may indicate a power supply problem. Overvoltage shortens LED lives exponentially, and excessive ripple may cause the displays to twinkle and lead to premature failure.

Liquid crystal displays require an oscillator drive, which is more critical than the supply voltage in most instances. Even though most units are modular in form, the failure of a decoder or driver should not be ruled out, especially when the digits are multiplexed.

Gas-type displays typically require a supply in excess of 150 V for illumination. Because overvoltage will cause metallization (blackening) of the glass surface it should be investigated. Some units have a built-in diagnostic test for the plasma display panels. If available, the test should be made before proceeding to other checks.

Any CRT display, whether storage or non-storage, must be serviced by using the manufacturer's maintenance manual, which will show how to perform any self-tests along with the other test points to be checked. The CRT may implode if it is scratched or struck severely. Do not handle the CRT by its neck and wear protective clothing and a face shield when handling it.

The optocoupler combines the emitter and detector in the same package. Therefore, the input drive must be certified first. The input device is an IRED or an LED and may be operated in a pulsed or steady-state mode. The voltage drop in the steady-state mode is more than that of a regular diode. If it is driven (sinking current) by a gate, no drop across the input diode indicates an open transistor in the gate's ouptut, an open limiting resistor, or a short-circuited diode. The pulsed mode requires the use of a scope or a probe. If the input diode is suspect, a simple diode test (with an ohmmeter) will verify whether it is functioning properly.

Repeated failures may indicate transients. Since these devices are often employed for isolation purposes they are subject to high transients. All circuit suppressors should be checked because they may be intended to protect the optocoupler from transients.

The detector side of the optocoupler may be any of the devices shown in Fig. 14-57. Many are hard to test except for a short circuit or an open condition. Substitution is the simplest course in some cases. Voltmeters are used to verify steady-state conditions, but scopes or logic probes are preferred for pulsed circuits. Triggering the scope from the input to the optocoupler should synchronize the output waveform. In a zero crossing circuit, the output should appear when the input is active and the line is near 0 V.

Fiber-optic transmitters may be LEDs or IREDs and are usually high-current pulsed types. The IRED output can be confirmed with an optical detector or a special IR-sensitive card made for detecting the beam output. The drivers, usually transistors, must be checked along with any current-limiting resistors. The oscilloscope is the most effective instrument to verify that the drive is normal. Heat sinking and cooling are very important, and any loss or reduction can cause repeated failures. The coupling to the cable must be firm and clean. The biggest loss in fiber optics is in the interface connections. Oil or other liquids may build up in loose connections. Cleaning procedures must follow the manufacturer's recommendations.

Receivers are usually sensitive to visible light, so they can be checked easily. Clean and tight connections are just as important at the receiver end as they are at the transmitter end of the cable. It is advantageous to have test units for each end of the cable. Lengths of test cable are also valuable for locating the source of problems. In most cases, a logic probe or scope will verify correct levels out of the detector modules. The transmitter and receiver constitute a system and should always be viewed as such. The idea is to isolate the difficulty to one end or the other or to the connecting cable.

Most lasers found in industry will be high-power and will involve high-voltage circuits. Along with the safety precautions for working around laser light, the high voltage supplies and circuitry demand cautious work. Measurements (other than *live*) must be made after all power is locked off and capacitors are discharged (as mentioned in other chapters). For live measurements, all test equipment must be rated in excess of any voltages to be encountered. Never work alone when high voltages are anticipated. Most units require an ignition pulse that is only present for a short period. This pulse can be of extremely high potential and can damage test equipment. Consult the manufacturer's recommendations. Helium will leach out through the glass in due time, and its loss is inevitable in helium-neon lasers. Metal-removal lasers have water cooling, vacuum, and oil pumps, along with a host of interlocks for safe operation. The use of the manufacturers' manuals is imperative before maintenance is attempted.

Phototube circuits are very dependent on bias voltage and load resistance (if used). These tubes are sensitive to visible light, and excess background light can disrupt normal operation. Cleanliness is also very important. Vapors can cause dirt and grime to collect on surfaces and desensitize the unit. When they are used with a current-to-voltage op amp, a simple current source (high-value resistor from the bias voltage) substitute for the phototube will verify the amplifier's integrity.

All PM tubes must have at least 100 V between dynodes to perform properly. A simple resistance check (with the power off, of course) will confirm the divider string. The high gain of these tubes prohibits exposure to ambient light with the dynode voltage applied and may destroy the tube if it is exposed. Never expose any PM tube to light with the supply voltage applied. A flashing LED or neon lamp produces more than enough light to test a PM tube. With 100 V per dynode, high voltages are required and breakdown can occur in wiring or sockets and must also be checked. All test instruments must be able to handle these potentials with safety.

REVIEW QUESTIONS

41. Excessive _____ can increase the dark current from a photodetector.

42. Segments of a seven-segment display may appear bad if the _____ is bad.

43. Blackening of a plasma display means the _____ is too high.

44. The best aid for troubleshooting pulsed fiber optics is an _____.

45. Laser power supplies use a/an _____ to start lasing.

CHAPTER REVIEW QUESTIONS

14-1. The combination of red and green light produces a _____ light.

14-2. The light produced by a forward-biased PN junction is due to _____ transitions in the energy gap.

14-3. _____ emitting diodes produce a very narrow spot.

14-4. Refer to Fig. 14-21(*b*). If I_P is 100 µA, h_{FE} is 100, and I_B is zero, calculate I_E.

14-5. Dark current in a phototransistor will increase _____ times for a 20° rise in temperature.

14-6. To double the light falling on a detector, its exposed area would have to be _____.

14-7. The PIN diode includes a _____ region between the P- and N-type material.

14-8. The PIN photodiode with no bias is in the _____ mode.

14-9. Refer to Fig. 14-21(*b*). A PIN diode produces 2.5 µA of photo current and zero dark current. With the op amp feedback equal to 1 MΩ, calculate the output voltage.

14-10. The light-sensitive SCR is also called a _____.

14-11. An increase in the light level produces an _____ in the resistance of a photoconductive bulk cell.

14-12. Photoconductive cells may not quickly return to their original resistance after exposure to a bright source because of their _____ effect.

14-13. A big drawback of hot-wire readouts is the large number of isolation _____ needed.

14-14. A CRT monitor may be either a storage or a _____ type.

14-15. CRT graphic displays may be of the vector or _____ type.

14-16. The information required to form characters on the screen of an industrial terminal is stored in a ROM called a _____.

14-17. The photo _____ is used for encoders and speed monitors.

14-18. A beginning of a tape sensor is typically a _____ object sensor.

14-19. Fiber optics that provide two-way communication form a _____ system.

14-20. Fiber optics operate in the tens of watts range. (true or false)

14-21. The _____ index fiber table type operates at moderate bandwidths.

14-22. Poor end polishing can cause losses up to _____ dB in fiber cable systems.

14-23. Identify an amplifier that provides a very low load impedance and thus a fast response time for a fiber-optic detector.

14-24. The gas laser operates in the far _____ region.

14-25. Injection lasers must be in a _____ array to achieve high power outputs.

14-26. The _____ is a good switch for injection laser pulses.

14-27. If $\delta = 5$ and $d = 5$, the gain of a PM tube is _____.

14-28. A 10 percent change in dynode voltage results in a 10 percent change in the gain of a PM tube. (true or false)

14-29. Large pulse outputs require the last dynodes to be _____.

14-30. The PM tubes should not be exposed to ambient light with the high-voltage _____.

ANSWERS TO REVIEW QUESTIONS

1. electroluminescence **2.** longer **3.** F; incoherent **4.** photons **5.** 2.0 eV **6.** 880 nm **7.** visible **8.** pair **9.** electric **10.** depletion **11.** false **12.** reverse **13.** infrared **14.** dark **15.** false **16.** bar/graph **17.** seven **18.** ac **19.** MOS **20.** grids **21.** pixels **22.** IRED **23.** false; gain **24.** 500 percent **25.** half wave **26.** full **27.** false; zero voltage **28.** triac **29.** FET **30.** Schmitt trigger **31.** xenon **32.** pumping **33.** false **34.** reflecting **35.** neon **36.** false **37.** 1 V **38.** gas-filled **39.** secondary **40.** positive **41.** heat **42.** decoder **43.** voltage **44.** oscilloscope **45.** exciter

15

AUTOMATION AND ROBOTICS

This chapter presents programmable controllers and robots, which are key components for industrial automation. They are based on the devices, circuits, and the concepts presented earlier. This chapter will help you understand how all of the elements of electronics fit together to form powerful systems for industry. These systems have enabled our factories to increase productivity, decrease costs, and increase the quality of manufactured goods.

15-1
PROGRAMMABLE CONTROLLERS

Programmable controllers are well suited to the cyclical and repetitive operations found in sequential industrial processes. Originally, sequential control was accomplished with relays, stepping drums, timers, and counters. These systems were very difficult to reprogram for production changes and often had to be scrapped and completely redesigned from the ground up. The automobile industry was the first to change over to programmable controllers. They immediately saved money and time, and they gained reliability plus tremendous flexibility. Now, many industries use this technology to perform logic decisions, timing, up/down counting, record keeping, sequencing, arithmetic operations, report generation, information handling, debugging, and troubleshooting. Obviously, the modern programmable controller is capable of everything the old relay logic was plus a lot more.

Programmable controllers (PCs) are computers. They will be referred to in this chapter with the abbreviation PC; not all computers can be considered as PCs. To qualify as a PC, a computer must be designed to operate in the industrial environment, which can be rather harsh. It must have a wide temperature operating range (0 to 60°C) and a wide humidity range (0 to 90 percent). It must be packaged in rugged enclosures and be well shielded against electromagnetic interference, dust, dirt, and mois-

ture. It must be capable of surviving power outages, brownouts, and line transients. Its memory circuits must be backed up with battery power. It must be capable of being programmed with logic commands, symbols, or mnemonics that correspond to relay ladder diagrams. Finally, it must be designed for scanning operation. A *scanning computer* solves logic from the beginning of memory to some specified stopping point. Once the end is reached, the operation repeats again.

Figure 15-1 shows a block diagram for a PC. The *central processing unit* (CPU) and the *input-output* (I/O) form the major core of any programmable controller. The programmer may or may not be connected to the CPU at any given time. It is required only when programs are being entered, changed, or debugged. It may also be connected during periods of system maintenance. When the CPU is performing normal scanning, the programmer may be disconnected and moved to another PC. In this way, one programmer can serve several systems. A typical programmer is shown in Fig. 15-2. Note the rugged case and the membrane keyboard, which prevents moisture and dirt entry. The keyboard folds up and protects the CRT when the terminal is not in use or is being transported. The printer shown connected to the programmer in Fig. 15-1 is an option. It provides hard copies of programs, data, reports, and ladder diagrams. The program storage unit is another option. It could be a floppy disk drive or a tape drive that connects to the programmer for saving programs and for loading in control programs or diagnostic software.

The CPU shown in Fig. 15-1 contains a processor, logic memory, storage memory, and an optional communications module. The *communications module,* if present, provides a data link to other PCs and possibly to a computer. Figure 15-3 shows the CPU of a programmable controller. The rugged cabinet has brackets for mounting the unit in a standard 48-cm (19-in.) equipment rack. The connector at the left is used to connect the programming terminal or some other RS-232C peripheral. The four diagnostic indicators tell whether the processor is running, the outputs are enabled, the CPU is in the program load mode, or there is an access in process. The 20-key

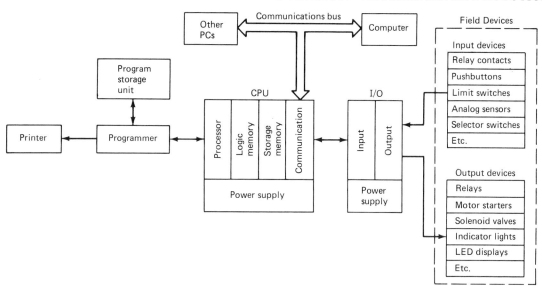

Fig. 15-1 Programmable controller block diagram.

pad and 16-character alphanumeric display can be used to read and change values such as the preset value of a timer. The key switch allows the CPU to be placed in one of three modes. One mode is *MEMORY PROTECT ON,* which allows an operator to read memory but not change it. *DATA CHANGE* allows an operator to read memory and change only selected portions. *MEMORY PROTECT OFF* allows read or write to all user memory locations.

The I/O unit of Fig. 15-1 is separated from the CPU and has its own power supply. The I/O unit buffers the CPU from the noise and transients associated with the field devices. It also prevents the control wiring, which can pick up quite a bit of noise, from interfering with the CPU. Note that the power supply shown in the I/O unit powers that unit only. It is not used to power output field devices. Programmable controller manufacturers make various

I/O modules. The following list is for the Allen-Bradley PLC-3 system:

ac/dc (120-V) input

dc (12- to 24-V) input

dc (48-V) input

Isolated ac/dc (120-V) input

Analog (8-bit) input

Analog (12-bit) input

TTL input

dc (24- to 48-V) input

Encoder/counter (5-V)

Encoder/counter (12- to 24-V)

ac/dc (220- to 240-V) input

Fig. 15-2 8200 programming terminal. (*Courtesy of Allen-Bradley Co., Systems Division*)

Fig. 15-3 PLC-3 programmable controller. (*Courtesy of Allen-Bradley Co., Systems Division*)

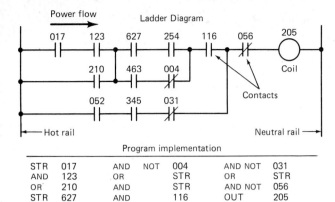

Program implementation

STR	017	AND	NOT	004	AND NOT	031
AND	123	OR		STR	OR	STR
OR	210	AND		STR	AND NOT	056
STR	627	AND		116	OUT	205
AND	254	STR		052		
STR	463	AND		345		

Fig. 15-4 Relay ladder-based boolean language.

dc (5- to 30-V) selectable

dc (5-V) selectable input

Fast-response dc (12- to 24-V) input

dc (12- to 24-V) driver logic input

Thermocouple input

Thermocouple expander

ASCII

Absolute encoder input (8-bit)

PID module

ac (120-V) output

dc (12- to 24-V) output

dc (48-V) output

Isolated ac (120-V) output

Analog (12-bit) output

TTL output

ac (220- to 240-V) output

Protected ac

Contact output (4 N.O. plus 4 N.C.)

Contact output (8 N.O.)

Stepper positioning assembly

Servo positioning assembly

Look this list over. It will give you an idea of the diversity of control applications that can be achieved with this type of equipment.

The method of programming PCs is really a throwback to the old relay logic systems. *Ladder diagrams,* originally developed to represent relay logic, have become the universal language of sequential control systems. Figure 15-4 represents a typical ladder diagram. The left vertical line represents the hot side of the power line and is called the *hot rail.* The right vertical line represents the neutral side of the power line and is called the *neutral rail* or the *return rail.* Power flow is assumed to be from left to right. This is merely a conceptual approach to designing the logic. Its use helps the programmer avoid unexpected circuits in the diagram (and in the logic) called

Fig. 15-5 PLC-4 programmable controller and programmer. (*Courtesy of Allen-Bradley Co., Systems Division*)

sneak paths. All contacts and coils are assigned addresses. Note that simple boolean statements may be used to describe the ladder diagram. Many other things can be done in a program as well. Coils may be latched so they are retained after a power failure. Latched outputs are restored to their previous ON or OFF states when the CPU resumes scanning. Coils may also be designated as *one-shots* to create references which are energized for short periods of time. Timer, counter, and arithmetic functions may also be assigned in the program. Many PCs also allow subroutines and other advanced programming techniques. Some allow more than one program to be resident in CPU memory at a time to permit production operations to change quickly from one product to another. The various features and programming details are brand- and model-dependent and cannot be covered here.

Programming in some models may be accomplished *online.* This mode allows existing programs to be changed in relation to reference numbers, timer/counter functions, contact types, latching, and so on, without disturbing the scanning operation. *Offline* programming centers around the programmer terminal unit itself. The logic program and all changes are entered into the terminal's memory with no intervention from the CPU. When the program is ready, CPU scanning is halted, and the program is transferred to the CPU memory. If the terminal's memory has battery back-up, or if disk or tape storage is available, offline programming can be done at some remote site.

The programmer terminal is also invaluable for troubleshooting. It can be used to display the status of field devices and the real-time power flow of any portion of the ladder diagram without disturbing scanning. Inputs and coils can be overridden so that their status can be forced by a human operator. The content of registers can be displayed. Some models have a special online mode in which the CPU scans, reads inputs, establishes all coil states internally, but holds all outputs in the OFF state. This mode allows the logic to be exercised with real inputs to check for proper operation without the danger of unexpected outputs.

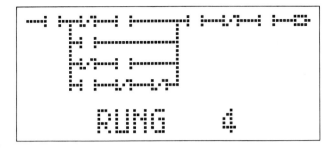

RUNG   4

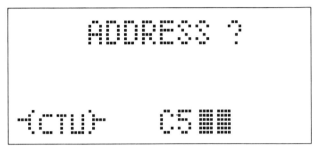

ADDRESS ?

-(CTU)-   C5 ▩

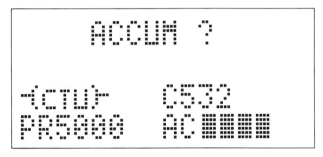

ACCUM ?

-(CTU)-   C532
PR5000   AC ▩▩

Fig. 15-6 Sample displays from PLC-4 programmer.

Large PCs have tremendous capacity. For example, the Allen-Bradley PLC-3 has 192K bytes of memory and can handle up to 8192 inputs and outputs. Not all industrial situations demand this much capacity. Small PCs are also available; Figure 15-5 shows the Allen-Bradley PLC-4 PC and programmer. This system is capable of handling 20 inputs and 12 outputs. Allen-Bradley also makes an expander module which allows up to eight PLC-4 controllers for a capacity of 256 I/O points and over 4K bytes of memory. Figure 15-6 contains some sample displays from the PLC-4 programmer. Note that rungs from the ladder diagram can be viewed one at a time on the liquid crystal display.

REVIEW QUESTIONS

1. Programmable controllers are flexible because they allow production changes by modifying _____ rather than modifying hardware.

2. What are the two main sections of a PC?

3. The programs for a PC are based on _____ diagrams.

4. Programmable controllers execute their programs with repetitive passes from the beginning of

memory to the end of the program. This process is known as _____.

5. The CPU memory of a PC must be provided with _____ back-up.

6. The CPU and the I/O sections of a PC are separated to protect the CPU from _____.

15-2
ROBOT CLASSIFICATIONS AND TERMINOLOGY

There are several ways to classify industrial robots. One way is to categorize them as low-, medium-, or high-technology types. Another is to divide them into nonservo- and servo-controlled groups. They can also be divided according to their axes of movement or their system of coordinates. Yet another way is to divide them into nonintelligent and intelligent groups. All of these classification techniques have advantages and disadvantages. This section will explain the classification techniques and define the basic terms used in industry.

Four basic configurations for industrial robots are illustrated in Fig. 15-7. The rectangular robot moves along X, Y, and Z axes. As it extends and retracts to its maximum reach and its minimum reach along each of its axes, it describes a rectangle in space. This rectangle is also called its *work envelope*. The *cylindrical robot* rotates at the base and extends and retracts along Z and Y axes; its work envelope is a cylinder. The *spherical robot* rotates at its base, pivots or bends at its shoulder, and has arm extension and retraction; its work envelope is a portion of a sphere, as shown in Fig. 15-8. The *jointed spherical robot* adds an elbow joint; its work envelope is also basically spherical, as shown in Fig. 15-9. The addition of an elbow joint gives it greater flexibility and a larger work envelope than that of the spherical robot.

Each axis of motion adds another degree of freedom to a robot arm. Figure 15-10 shows a six-axis jointed spherical robot. Yaw, pitch, and roll have been added to a wrist joint at the end of the forearm. Robots with six or more axes can be classified as medium- or high-technology robots. Some robots must be capable of rather complicated combinations of movements. As an example, robot arms may be used to reach inside automobile bodies and perform welding operations. In addition to the axes shown in Fig. 15-10, horizontal base motion is achieved by mounting of the base on a moving track. This enables the robot to weld and follow the work piece down the assembly line at the same time. Also added is column extension to allow the shoulder to raise out of the base, the forearm to be extended, and the waist to be bent by means of a joint below the shoulder. Of course, it doesn't have to stop there. Almost any number of degrees of freedom can be used. However, the complexity and cost quickly get out of hand

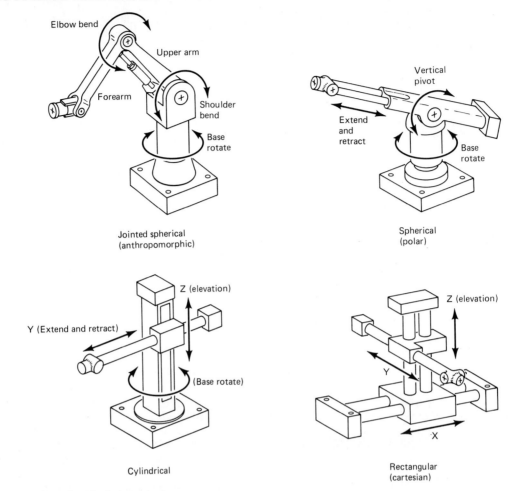

Fig. 15-7 Basic configurations for industrial robots.

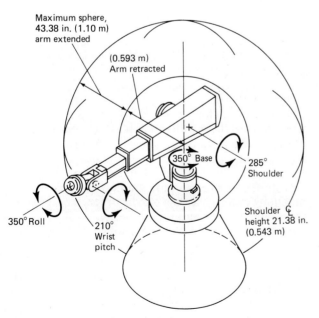

Fig. 15-8 Spherical work envelope.

as the number of axes increases. Robots with more than 8 degrees of freedom are the exception rather than the rule. Most industrial tasks require from 3 to 6 degrees of freedom.

Simple robots do many routine jobs in industry. Some may have as little as two degrees of freedom. These robots are in the *nonservo-controlled* category. There is no feedback used in their control system. They are often called "bang-bang" robots because of the way they work. They bang from position to position, as shown in Fig. 15-11. They are set up for a task by adjusting fixed stops. They are excellent robots for pick-and-place operations. In fact, they are often called *pick-and-place robots*. They are also useful in simple assembly operations such as stuffing printed circuit boards and loading and unloading parts from machines.

Figure 15-12 shows a typical low-cost robot, and Fig. 15-13 shows some examples of its motion patterns. It uses pneumatic cylinders to activate the axes. The following list includes some of the ways that the robot of Fig. 15-12 can be classified:

1. *X, Y, Z* type
2. Non-servo-controlled type
3. "Bang-bang" robot

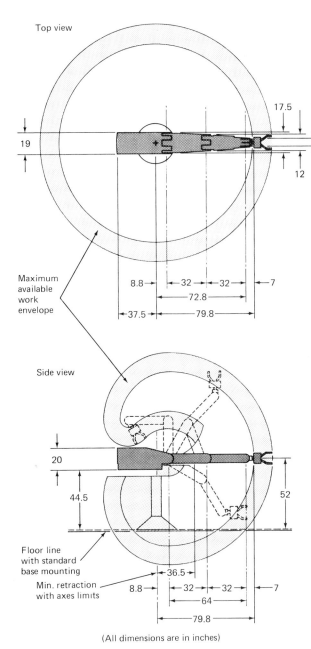

(All dimensions are in inches)

Fig. 15-9 Work envelope of a jointed spherical robot.

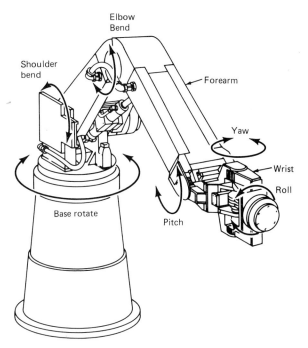

Fig. 15-10 Six-axis jointed spherical robot.

4. Low-technology type
5. Three-axis robot
6. Pneumatic robot
7. Pick-and-place robot
8. Rectangular-coordinate robot
9. Limited-sequence robot

There is a problem with using any single classification system. Look again at Fig. 15-11. This robot is also in the nonservo-controlled and low-technology categories. However, it is a jointed spherical arm and has some differences that make it suited to a different range of applications. It may use different actuators (other than pneumatic) and have very different speed and strength characteristics. You are

urged to learn the terms and use them in combinations to describe any robot accurately. Another trap to avoid is the assumption that low-technology robots are limited in accuracy and usefulness. Actually, they tend to be the most accurate of all robots. They can be expanded with more degrees of freedom to perform fairly intricate tasks. Figure 15-14 shows the addition of roll, pitch, and yaw to the low-technology robot of Fig. 15-12 and Fig. 15-13. In a sense, the term *low technology* is unfortunate; it can cause people to reach erroneous conclusions about robot capabilities and has even caused some companies to install high-technology robots in applications in which less expensive robots would have done the job faster and better.

The general attributes assigned to *low-technology robots* include the following:

1. Limited sequence of movements
2. Only the endpoints of travel are controllable
3. A speed range of 80 to 160 cm/s
4. Low cost and low maintenance
5. A 0.03-mm resolution and 0.15-mm repeatability
6. Load capacities from 150 g to 15 kg
7. Controls ranging from electromechanical timers to microprocessors
8. Usually 2 to 4 degrees of freedom
9. Nonservo-controlled
10. Short cycle time (typically 30/min)
11. Used in parts transfer, assembly, loading, packaging, inspection, and automatic testing

Medium- and high-technology robots are servo-controlled. *Servo control* can be divided into two types: *point-to-point* and *continuous path*. *Point-to-*

point servos move from one work envelope position to the next in a straight-line segment. They exhibit jerky motions; however, more than one axis can be activated at a time to alleviate this somewhat. *Continuous path servos* produce true curves as the work envelope is traversed. They are useful in applications such as spray painting and welding. They require more memory to store all the positions needed to follow a smooth path.

The general attributes assigned to *medium-technology robots* include the following:

1. Point-to-point servo control
2. Up to 6 degrees of freedom
3. Payloads up to 70 kg
4. Longer cycle time than that of low-technology types
5. Less accurate than low-technology types
6. Electrohydraulic actuators typically used
7. Electronic controls (typically microprocessor-based)
8. Walk-through programming
9. Medium cost and maintenance requirements

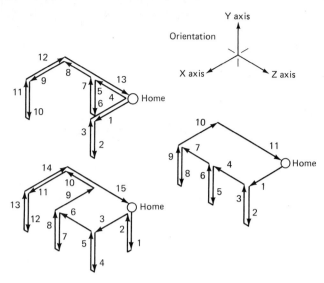

Typical motion patterns

Fig. 15-13 Robot parts and typical motion patterns. (*Mack Corporation*)

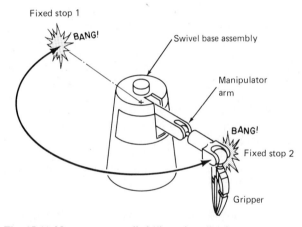

Fig. 15-11 Nonservo-controlled "bang-bang" robot.

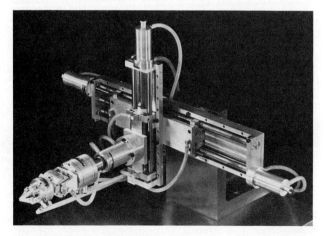

Fig. 15-12 Nonservo-controlled robot. (*Courtesy of Mack Corporation*)

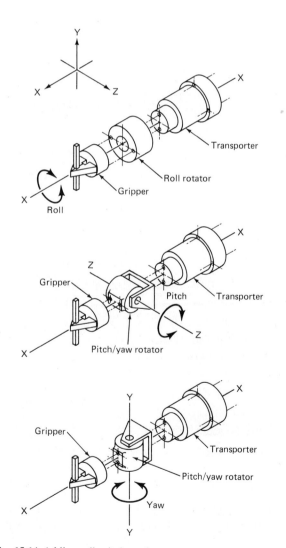

Fig. 15-14 Adding roll, pitch, and yaw.

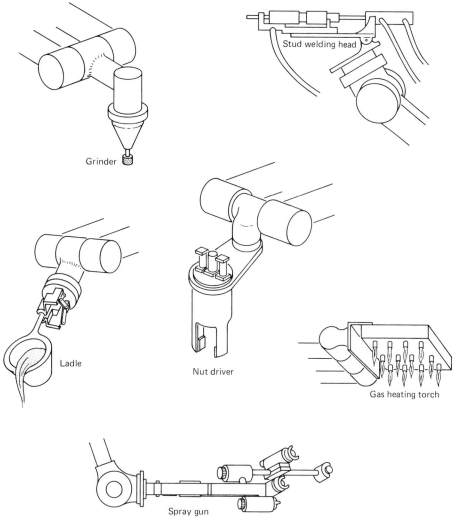

Fig. 15-15 Robot end effectors.

Grinder

Stud welding head

Ladle

Nut driver

Gas heating torch

Spray gun

Walk-through programming involves a human operator leading the robot through the desired routine step by step. It will be covered in more detail in a later section.

High-technology robots are the most complicated. Their attributes include the following:

1. Continuous path servo control
2. 6 to 10 degrees of freedom
3. Microprocessor and minicomputer controls
4. Expensive and increased maintenance requirements
5. Most flexibility of all robots
6. Advanced programming, including high-level languages
7. Built-in editing and diagnostic software packages
8. Mass storage devices (disk and tape)
9. Looping, subroutine, interrupt, and branching capabilities
10. Control of some other machines around them
11. Sensor inputs such as vision, proximity, touch, and sound
12. Applications include sorting, inspection, welding, painting, and assembly

Robot actuators include pneumatic cylinders or motors, hydraulic cylinders or motors, and electric motors. The *pneumatic actuators* are low in cost, require little maintenance, and are well suited for high-speed operations with light payloads. *Hydraulic actuators* are more costly, require more maintenance, and are best suited to heavy loads. *Electric motors* are the easiest to control, have moderate cost and maintenance requirements, and lend themselves to applications that do not require high speed or heavy loads.

The robot arm may be called a *manipulator*. The part that handles the work piece or holds the tool is often referred to as an *end effector*. Figure 15-15 shows some examples of end effectors. These may be changed automatically, as shown in Fig. 15-16. The end effectors are stored in a tool holder within

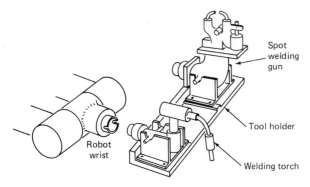

Fig. 15-16 Robot tool changing.

the manipulator's work envelope. One tool is placed in the holder, and the wrist disengages from it. The manipulator then moves to the next tool and engages the wrist. Programmed tool changing is usually limited to high-technology robots. Fingerlike grippers are also common end effectors. There are a myriad of gripper designs to handle the many tasks being assigned to robots today.

REVIEW QUESTION

7. The perimeter of a robot's reach describes a shape in space called its _____ _____.

15-3
PNEUMATICS AND MECHANICS

The industrial electronic technician must have a well-rounded understanding of several technologies. Automation involves more than electronics; fluidic and mechanical systems interact with the electronic controls. The term *fluidics* covers both hydraulics and

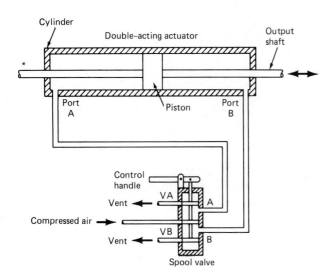

* May be single-ended which eliminates one output shaft

Fig. 15-17 Pneumatic actuator.

pneumatics. *Mechanics* includes gears, pulleys, bearings, transmissions, chain drives, timing belts, and so on. The technician must have an overall understanding of what each component in a system accomplishes and how each component interacts with the other components.

Pneumatic systems are similar to hydraulic systems. The major difference is that pneumatics uses air instead of a liquid to transmit power. Air compresses, limiting the amount of force that can be transferred. For this reason hydraulic systems are preferred when large loads must be handled. However, there are many benefits to pneumatic systems. Compressed air, once it has delivered its potential energy to a cylinder or a motor, can be returned to the atmosphere. This is a key advantage since it prevents the build-up of contamination in the system. Air is also a coolant. Pneumatic motors, for example, are cooled by the air that drives them. Compressed air is easy to produce and can be distributed to various machines at low cost. The pressures used are often not more than 150 psi (approximately 1 million N/m²), and distribution is inexpensive and relatively safe.

Figure 15-17 shows a *pneumatic actuator;* it consists of a cylinder with a port at each end and a piston to drive the output shaft or shafts. A *single-ended actuator* has an output shaft at one end; a *double-ended actuator* an output shaft at each end. When compressed air is applied to one port, the piston reacts accordingly. The *four-way spool valve* provides the control. With the spools in the position shown in the drawing, the compressed air inlet is blocked, and no air is delivered to the actuator. If the control handle is pushed down, the spools will lift; compressed air will flow to port *B*, and the piston will be forced to the left. At the same time, port *A* will be vented through the spool valve to the atmosphere. Raising the control handle produces the opposite reaction in the cylinder, and the piston moves to the right. The two motions make this a *double-acting cylinder.*

Reversing air motors are also available and are controlled in the same way. They may have eight or so vanes connected to an output shaft. There are two input ports. The direction of rotation depends upon which port is pressurized and which provides the exhaust.

The spool valve can be controlled by a solenoid; this solenoid can then be controlled by a microprocessor. Microprocessor control greatly improves the performance of air motors. This is the system used in the robot arm shown in Fig. 15-18. Air motors tend to give rough performance at low speeds. The microprocessor control smooths this out by digitally pulsing the air supply to the accumulator of each air motor. This technique is called *dithering* and provides performance characteristics once considered beyond the capabilities of air motors. There are five air motors in this particular manipulator, one for each major axis. Each is controlled by a dedicated 8-bit

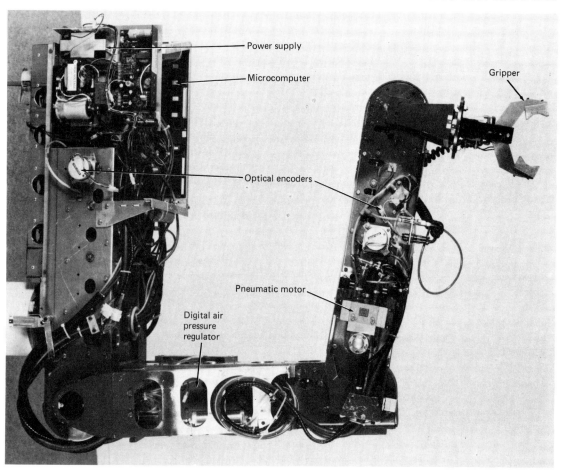

Fig. 15-18 IRI M50 robot arm. *(Courtesy of International Robotmation/Intelligence)*

microprocessor which drives a spring-centered spool valve. The microprocessors dither the air supply to their respective motors by adjusting the frequency and duty cycle of the pulses supplied to the solenoid-activated valves. Optical encoders provide feedback information to allow the microprocessors to provide accurate positioning and to adjust the motor performance for varying conditions of friction and gravity under different load conditions.

Figure 15-19 is a diagram of the pneumatic system for a model M50 robot. In this robot, an equipment module contains a pressure control valve for each axis. Packing the air motors and the valves together eliminates transients due to the compressibility of air. A *control valve assembly* is composed of a solenoid-activated spool valve and a two-position valve that operates at pilot pressure when the solenoid is activated. A *volume chamber accumulator* is mounted on each pressure control valve assembly. The *pressure control valve* dithers the air supply to the volume chamber accumulator to maintain a constant pressure supply to the motor under varying conditions.

Air motors are high-speed, low-torque actuators. Some type of speed reduction system is required for robotic applications. Figure 15-20 shows the chain

drive stages of the M50 robot. They step down the 8000-rpm motors to 10 rpm at the output shaft. The chain drives provide a bonus function. The pulsed output of the air motors appears smooth and continuous after translation through the reduction chains. Each chain must be adjusted to the proper tension to avoid backlash and to avoid stretching and excessive wear. A *chain tensioning adjustment* is provided for each chain-driven module. Two pneumatically operated *caliper brake assemblies* engage the first sprocket in each chain module. A *control valve* regulates the air pressure supplied to each brake. The modules also contain the *optical encoders* that provide axis position feedback to the microprocessors.

In addition to the five 8-bit axis microprocessors, the M50 robot also contains one 16-bit microprocessor. The 16-bit processor provides commands to the individual axis processors. The robot can be programmed by walk-through commands, assembly language, or high-level languages. Eight digital I/O ports are available on the main processor for interfacing the robot to other equipment in the factory.

Ordinary belt drives do not find much application in robotics and other positioning systems because of slippage. However, *timing belts* are used in some cases. A timing belt has teeth and drives or is driven

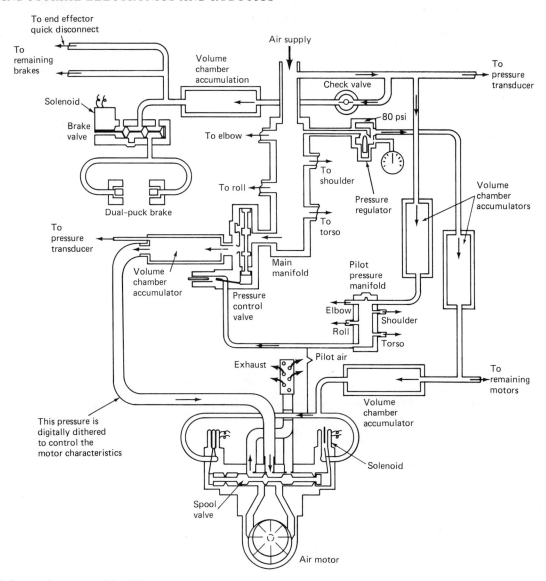

Fig. 15-19 Pneumatic system of the M50 robot.

by a pulley which also has teeth. The teeth eliminate the slippage problem. Timing belts are usually applied where the forces are small to moderate.

Robots and other automation devices often require a way to convert rotary motion to linear motion. Figure 15-21 shows a *rack and pinion* arrangement that accomplishes this task. The teeth on the pinion gear engage the teeth on the rack. If the pinion shaft is driven by a motor, the rack will move in a straight-line path. Rack motion can be reversed by reversing the motor. Figure 15-22 shows a *ball bearing screw drive* which accomplishes the same objective. The full name is seldom used, and they are usually referred to as *ball screws*. Ball screw drives are more expensive than rack and pinions, but they provide better positioning accuracy because backlash is controlled. Also, the balls exhibit rolling friction, which is less than sliding friction. This results in a more efficient drive. The return tubes shown in Fig. 15-22(a) recirculate the balls in the ball nut. Figure 15-

22(b) and Fig. 15-22(c) show that the screw or the nut can be driven, respectively. Figure 15-23 shows some applications for ball screw drives. They are also widely applied in numerically controlled machines such as lathes and milling machines.

Chain drives and timing belts can be used to reduce speed and increase torque. They are subject to some elasticity and backlash effects, however. Gear boxes may be used to control the elasticity, but backlash is still a problem. The harmonic drive system shown in Fig. 15-24 provides large reductions in speed with essentially zero backlash. It is also more compact than multistage gear boxes. These advantages make the harmonic drive system very attractive for robotic and other automation applications. The drive is composed of a rigid circular spline, a flexible spline (flexspline), and an elliptical wave generator. The *reduction ratio* of a harmonic drive is determined by the number of teeth on the circular spline divided by the difference in the number of teeth between the

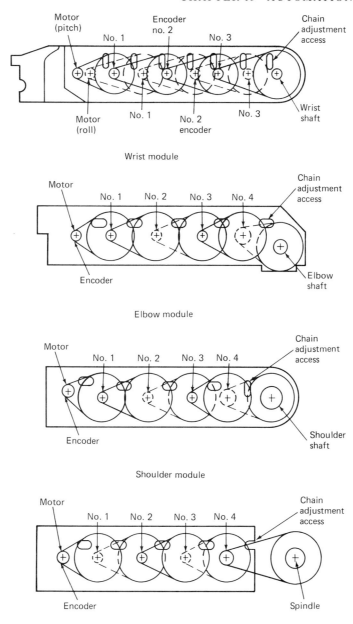

Fig. 15-20 Chain drive stages of the M50 robot.

circular spline and the flexspline. For example, if the circular spline has 300 teeth and the flexspline has 302 teeth, the difference is 2, and the reduction ratio is 150 to 1. Single-stage harmonic drives with reductions of over 300 to 1 are available. *Ordinary gear drive reduction ratios* are based on the ratio of the number of teeth on the larger gear to the number of teeth on the smaller gear. This means that large reduction ratios require multistage gear boxes. The backlash is additive in the multiple stages, and so are the friction losses. The harmonic drive, although costly, is used in some applications because of its marked advantages.

The wave generator, shown in Fig. 15-24, is ellip-

tical and forces the teeth on the flexspline to engage the teeth on the circular spline in two places. The engagement points are opposite one another, and approximately 10 percent of the total number of teeth are involved. This relatively large tooth engagement contributes to the accuracy and zero backlash of this reduction drive. The wave generator is coupled to the input shaft of the transmission. As it turns, the points of engagement move. If the numbers of teeth on the two splines were the same, no relative motion would be produced between them. Because there is a slight difference in the number of teeth, a small relative motion is created between the two splines for every rotation of the wave generator. If the

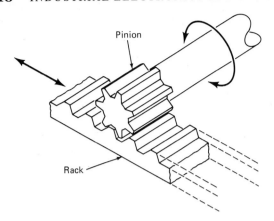

Fig. 15-21 Rack and pinion system.

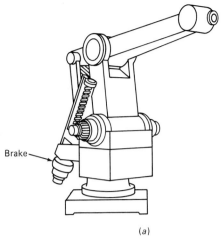

(a)

flexspline is fixed, this small relative motion appears at the circular spline and becomes the reduced speed output.

A moving robot arm has kinetic energy. If the combined weight of the arm and its payload is substantial, the energy is substantial. If the arm is moving fast, then even more kinetic energy is involved. Some robots use shock absorbers to provide controlled deceleration and to prevent damage due to

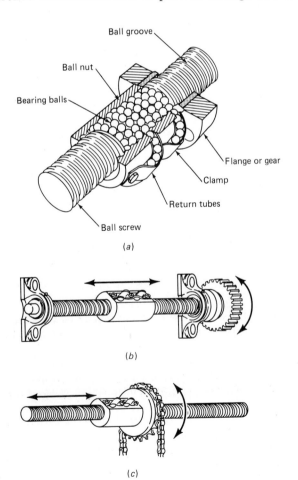

(a)

(b)

(c)

Fig. 15-22 The ball bearing screw drive. (a) Screw and nut assembly. (b) Driving the ball screw. (c) Driving the ball nut.

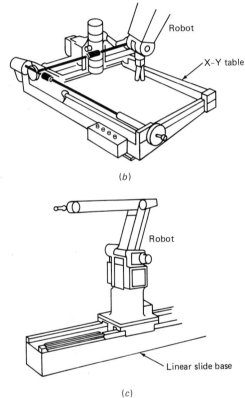

(b)

(c)

Fig. 15-23 Applications of ball screw drives. (a) In the robot itself. (b) In the axes of an X-Y table. (c) In a linear slide base.

hammering. The absorbers are usually cylindrical units with an input shaft. The input shaft is connected to a piston inside the cylinder. When the shaft is moved, the piston acts against hydraulic fluid inside the cylinder. The piston pumps the fluid through restricted passages. The restricted flow inhibits motion of the input shaft. The more rapid the motion, the more opposing force is developed by the piston working against the restricted fluid flow. The fluid will heat as the piston forces it to flow. This action converts the kinetic mechanical energy of the robot arm to heat energy.

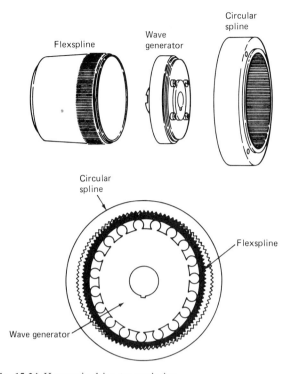

Labels (top row, left to right): Flexspline; Wave generator; Circular spline

Labels (lower diagram): Circular spline; Flexspline; Wave generator

Fig. 15-24 Harmonic drive transmission.

REVIEW QUESTIONS

8. The term fluidics includes both _____ and _____.

9. Examine Fig. 15-17. Which way will the piston move when the control handle is lifted?

10. Is the arm shown in Fig. 15-18 an example of a servo-controlled or a nonservo-controlled robot? Why?

11. Digital pulsing of the supply to an air motor is called _____.

12. Drive belts are smooth; timing belts have _____.

13. The chain drives shown in Fig. 15-20 produce a _____ reduction.

15-4
CONTROL AND PROGRAMMING

Robot controllers range from simple electromechanical systems to powerful minicomputers. Figure 15-25 shows an example of a simple control for a low-technology robot. An electromechanical timer has a series of cams that activate microswitches. The microswitches control four-way solenoid valves to activate pneumatic cylinders for the X and Y axes and the gripper. Mechanical stops are set to limit the travel along the axes. The gripper has no stops and is either full open or closed on the work piece. Setting up this robot for a task involves adjusting the timing and sequence of the cams and adjusting the

mechanical stops. Thus, the robot is not programmed for a task: rather, it is adjusted.

Programmable controllers can be used with low-technology robots, as shown in Fig. 15-26. The X, Y, and Z axes are controlled along with the gripper. There is also an intermediate stop cylinder that works in conjunction with the Z axis. The total number of outputs is 5, which is easily within the capacity of a small programmable controller. Other outputs may be required to move the work piece out of position after the robot completes its cycle and to bring the next work piece into position. Inputs on the controller will normally be used to determine the proper position of the work piece before the cycle is initiated. The inputs may also include operator commands or commands from another computer controlling the overall production process. Another way to use inputs is to read limit switches that signal manipulator position. The following sequence is typical for a nonservo-controlled robot and will help you understand the programming steps:

1. When the program starts executing, the controller initiates signals to the valves on the manipulator's actuators.

2. The valves open and admit air or hydraulic fluid to the actuators. The affected members begin to move.

3. The valves remain open, and the members continue moving until the stops are contacted.

4. Limit switches signal the controller that the move or moves are complete. The controller closes the valve or valves.

5. The next sequence is initiated. The process is repeated until the entire program is complete.

6. A new scan is started by the controller, and the entire program runs again when the inputs signal that the next work piece is in position.

Programmable controllers may also be used with servo-controlled robots. The following sequence is typical for a medium-technology robot:

1. The controller addresses the memory location of the first command and reads the encoders to determine the actual position of the manipulator axes.

2. The two sets of data are compared, and any error is amplified and becomes a command signal to the axis motors or servo valves.

3. The actuators move the manipulator and the encoders (or other sensors) send data back to the controller. The feedback signals are constantly compared with the programmed position data; new error signals are developed and become command signals to the actuators.

4. The process continues until there are no errors. The manipulator comes to rest.

5. The controller now accesses the next memory location and responds accordingly to the data

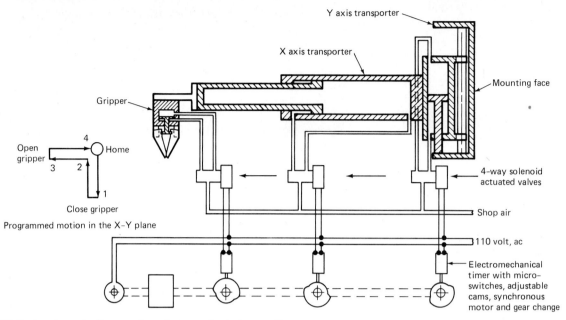

Fig. 15-25 Basic schematic diagram of an electromechanical controller.

stored there. It may be a new position command or a signal to another mechanism.

6. The process is repeated sequentially until the entire program has been executed.

7. Given the proper input conditions, the program will automatically execute again from the beginning.

Servo control is obviously more sophisticated and allows the manipulator to stop at any point in its

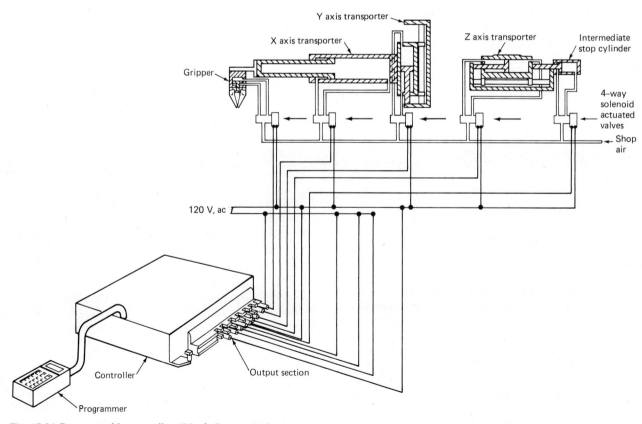

Fig. 15-26 Programmable controller. (*Mack Corporation*)

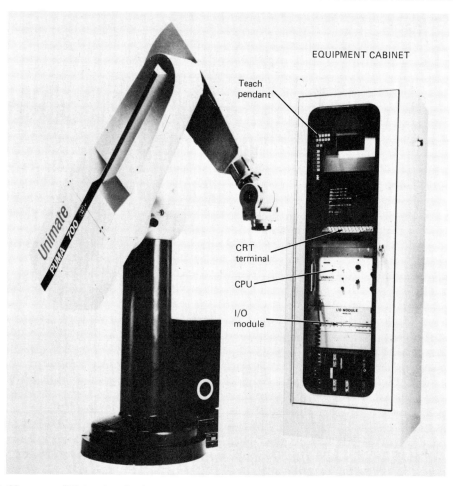

EQUIPMENT CABINET

Teach
pendant

CRT
terminal

CPU

I/O
module

Fig. 15-27 PUMA 700. (*Courtesy of Unimation, Inc.*)

work envelope under software direction. It is also feasible to control velocity, acceleration, and deceleration with the software. This capacity is especially useful with heavy payloads. The memory capacity is usually large enough to store many positions with both point-to-point and continuous path operation as possibilities. It may also be large enough to store several programs, thus allowing for rapid production changes.

A high-technology robot and its minicomputer-based controller are shown by Fig. 15-27. The equipment cabinet stores the teach pendant when it is not in use and also houses a CRT terminal, the central processing unit, the I/O module, and the power supply for the control electronics. The *teach pendant* is one way of programming medium- and high-technology robots. A human operator holds the pendant and uses it as a remote control device to lead (usually at less than operating speeds) the manipulator through a series of moves. This kind of teaching or programming is called *lead-through*. A store button on the pendant is pressed at the end of each motion to store the position in memory. *Walk-through* is another robot programming technique. The human operator grasps the arm and the end effector and manually moves them. The operator signals the computer to record each motion or position. Walk-through is commonly applied in continuous path robots for teaching operations such as spray painting and welding. Figure 15-28 shows some typical walk-through control handles on a painting robot.

Lead-through programming must be done with some care, as indicated in Fig. 15-29. The spot welding end effector is shown in its start position. The arm must be programmed to move past the roof column and then into the window opening where the welding operation is to take place. If an operator fails to record the intermediate point in memory, the robot will take a direct path when the program runs. This will result in a collision and damage to the work piece and to the robot.

When minicomputers are interfaced to robots, another control method becomes available. It is called *controlled path* and combines the best features of point-to-point and continuous path programming. It takes advantage of the computational abilities of the computer to provide coordinated control of all axes. The operator uses a pendant to move the end effector to various end point positions. The computer then generates a controlled path at the desired velocity, acceleration, and deceleration. Teaching is more instinctive if the operator can specify tool or gripper

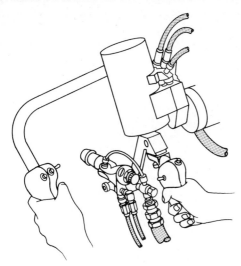

Fig. 15-28 Walk-through control handles.

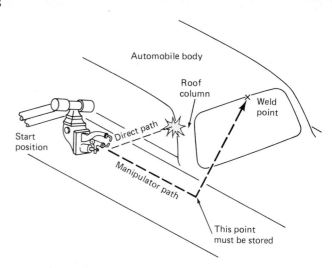

Fig. 15-29 Point-to-point programming.

positions without worrying about commands to each axis to achieve them. For example, in *point-to-point programming* the operator must move each axis until the combination of positions properly aligns the end effector. *Continuous path programming* requires the operator to move the arm and end effector manually. This becomes a problem with heavy manipulators. Continuous path also requires quite a bit of memory.

A *controlled path system* is shown in Fig. 15-30. During teaching, the operator presses position and orientation buttons on the pendant to move the end effector. The teaching coordinate system may be any of several types. Continuously changing position signals are generated as long as the buttons are pushed. The computer transforms the teaching coordinates to rectangular coordinates (if the teaching coordinates are rectangular, this operation is not required).

The computer then transforms the rectangular coordinates to robot coordinates, which become commands to the axis servos. The servo loops then act simultaneously to provide the desired position and orientation of the end effector. The computer operations are very fast, and the operator sees an apparent immediate motion for every command. The program button on the pendant is pushed to store the current rectangular coordinates in memory. The pendant also has a keyboard and other buttons to enter velocity, tool length, and functions to be performed at each point. The CRT displays the information being programmed.

The CRT terminal shown in Fig. 15-30 also provides editing features. The operator may view the data stored at the previously programmed points. Data may be changed, deleted, or restored. Points

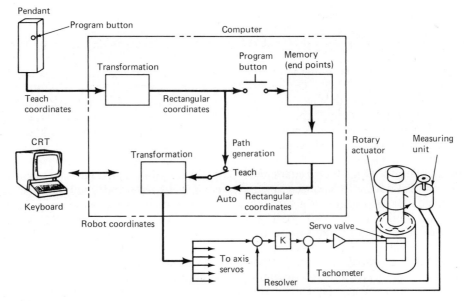

Fig. 15-30 Controlled path system.

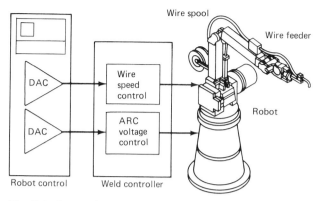

Fig. 15-31 Integrating robot control with other parts of the system.

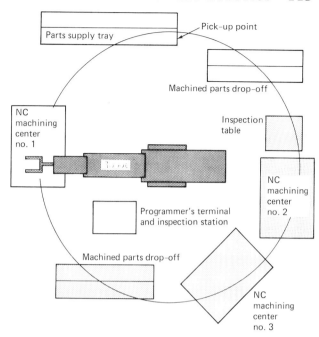

Fig. 15-32 Flexible manufacturing cell.

between programmed points and program functions can be added. Some typical functions are the following:

1. DELAY: This function stops the robot for a defined period of time before moving to the next programmed point.
2. CONTINUE: The manipulator does not stop but moves to the next point. An output signal can be programmed at the continue point if desired.
3. WAIT: Motion is stopped until a signal or a conditional set of signals is received.
4. OUTPUT: Motion is stopped, and an output signal is produced before proceeding to the next point. The signal may be a pulse or a level.
5. TOOL: Motion is stopped until a tool operation is completed.
6. BRANCH: Motion is stopped, and the robot awaits input conditions. It jumps to different program segments or subroutines based on the combination of signals received.

Robots are one part of the automation puzzle. Figures 15-31, 15-32, and 15-33 provide examples of the robotic/automation relationship. Few, if any, industrial robots can work effectively without being interfaced to other equipment. Figure 15-31 is an example of a robot controller equipped with D/A converters to provide output signals to the welding controller, which in turn controls the wire feed and the arc voltage. Figure 15-32 shows a robot integrated into a flexible manufacturing cell. The robot's computer provides signals to, and accepts signals from, the

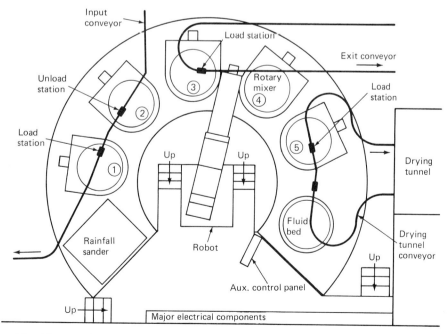

Fig. 15-33 Automated mold making.

three numerically controlled (NC) machining centers. All of the machines communicate through the computer to provide an automated flow of machining steps, parts transfer, and inspection.

A robot used in automated mold making is shown in Fig. 15-33. The process involves coating a wax pattern by dipping it into a ceramic slurry. Different slurries are available at the five rotary mixers shown. After dipping, the mold must be coated in the rainfall sander or in the fluid bed. Next, it goes into the drying tunnel. After multiple applications of slurry and sand, the mold is dried one last time and loaded onto the exit conveyor. The robot, through its computer, controls the entire process. This includes turning the mixers on and off, operating the conveyors, and controlling the sanders.

The capabilities of high-technology robots are as much due to their software as to their hardware. Looping, branching, conditional testing, and interrupts are examples of software features that make the robots "intelligent." For example, a subroutine that provides an appropriate response can allow a robot to recover from a difficult situation. Suppose the tips of a spot welding gun become stuck to the work piece. When the program detects the problem, it branches to a subroutine of twisting motions designed to free the tips. It will attempt this several times before shutting the robot down and alerting a control operator, as well as signaling the other machines around it. Robot decision making is implemented in other ways. When a robot is used for palletizing (stacking), deceleration is often programmed into the cycle. The manipulator slows down as the object nears the stacking point. Then, a sensor detects contact with the stack to stop the arm and release the payload. What happens as the stack grows? The deceleration does not begin soon enough, and damage may result. The software can take care of this possibility. The last stacking position is used to replace the destination point in memory on every cycle. In this way, deceleration will begin earlier and earlier as the stack grows.

REVIEW QUESTIONS

14. Refer to Fig. 15-25. How would the sequence of operations be changed with this control system?

15. How would the sequence be changed with the system shown in Fig. 15-26?

16. How many positions can the robot shown in Fig. 15-26 stop at along the Y axis? The Z axis? Why is this so?

17. What type of robot allows stopping at many positions along each axis?

18. Using a teach pendant to move a manipulator from point to point is known as _____ teaching.

19. Grasping the manipulator and moving it manually from position to position is called _____ teaching.

15-5
SENSORY SYSTEMS

The term *hard automation* refers to special purpose machines performing specific repetitive tasks in an automatic fashion. Many industrial activities fit this pattern and do not require sophisticated or intelligent machines to support them. However, there is an expanding effort to move toward "intelligent automation." There are two key components in this effort: sensory systems and artificial intelligence. Most industrial robots paint, weld, load and unload parts, grind, drill, or perform simple assembly operations. By the addition of sensors and intelligent software, the applications can expand to complicated assembly operations, inspection, part identification and sorting, and adaptability to a range of situations. Another important aspect is the ability to change rapidly from one production run to another. When this is realized, there is no longer a need to maintain an inventory of spare parts because they can be quickly manufactured on demand.

Sensors may be used to provide touch, tactile, proximity, vision, or sound information. Touch and tactile sensing are related but are not the same. *Touch sensing* involves contact detection at one or several points. *Tactile sensing* measures continuous and variable forces in an array; it is similar to the type of feedback people obtain from their skin. People also use *haptic sensing,* which comes from the muscles and joints. Joint angles and muscle forces provide the brain with quite an array of information during manipulative operations. Robots may crudely approximate haptic sensing by measuring motor current or fluidic back pressure as an indicator of the forces acting on an axis. This procedure gives an indirect measurement at best and is considerably influenced by manipulator position, friction, and inertia.

Figure 15-34 shows a wrist sensor for a robot manipulator that provides feedback signals proportional to the X, Y, and Z forces acting on it. It consists of

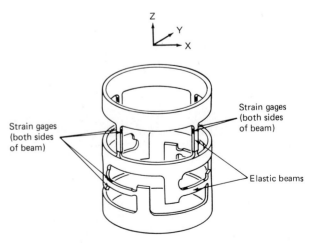

Fig. 15-34 Wrist sensor.

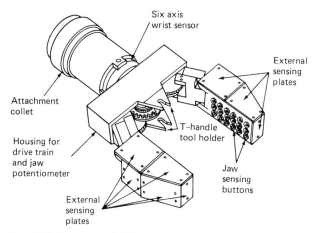

Fig. 15-35 Sensory end effector.

a metal tube with elastic beams. The beams flex in response to forces acting along the various axes. Strain gage sensors are placed on each side of the beam at the measurement points. Temperature compensation is achieved by using pairs of sensors. The wrist measures three components of force and torque between the manipulator and the end effector. The wrist sensor is illustrated in combination with a touch-sensitive gripper in Fig. 15-35. Each jaw face has a three by six matrix of touch buttons to sense the work piece. The outer jaw surfaces are also equipped with touch sensors to detect unexpected obstacles. The touch buttons provide on or off signals. They can supply only a limited amount of information; true tactile sensing requires more.

Various synthetic rubber and plastic materials are used to produce materials with elastic properties. These materials are called *elastomers* and can be used to make skin for a robot gripper. Conductive elastomers can provide feedback signals that are proportional to the forces acting on them. Another possibility is to use a matrix of touch cells, as shown in Fig. 15-36. The touch cells also produce signals that are proportional to force. The X and Y scan circuits connect the various cells to a multiplexer. The analog signal from the selected cell is sent to an analog-to-digital (A/D) converter. The processor stores the binary values from the various cells in a memory array. The contents of the array represent a *force contour* of the object, which is dictated by the shape of the object and its orientation in the gripper. The software then develops a "picture" of the object being held; this process is much the same as a human's grasping familiar objects in the dark and identifying them. It provides the robot with some ability to identify objects and determine their orientation by touching them. The current systems are crude by human performance standards, but improvements are expected.

Arrays of touch cells produce high-frequency signals when the gripper is moving. The information must be processed rapidly for the manipulator to respond in a reasonable fashion. Some touch cells are actually complete microcircuits containing an amplifier, an A/D converter, and a microprocessor. Another microprocessor receives parallel information from the individual microprocessors, rapidly assimilates it, and then sends its output to the main processor in the controller. It is interesting to note how difficult it is to simulate something humans do with such ease that they take it for granted. It is also

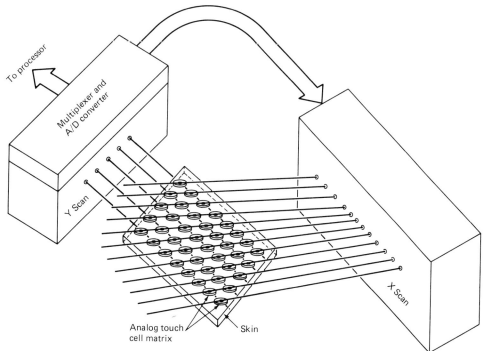

Fig. 15-36 Robot skin.

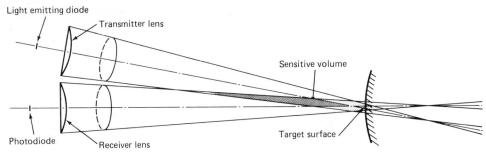

Fig. 15-37 Proximity sensing.

comforting to know that people are still the most amazing and sophisticated of all "machines" and will not become obsolete in the foreseeable future.

Arrays of conductive parallel strips may be used for tactile sensing. Scanning circuits reveal the contact points, and images can be created by the software. Piezoelectric polyvinyl fluoride materials provide signals that are proportional to changes in pressure. These materials offer promise for slip sensing, which is also very important in robotics. It is possible to design a single gripper that can pick up an egg without breaking it or a heavy metal machine part by sensing slip and applying only that amount of force required to eliminate it.

Proximity sensors are noncontact devices. They utilize magnetic fields, radio frequencies, ultrasound, or light for their operation. Some types were covered in Chapter 9. Figure 15-37 shows a simple system based on an infrared LED, a photodiode, and a lens system. Its sensitive volume is approximately the

intersection of the two cones in space as shown. Such systems can be used for collision avoidance and motion detection within the work envelope. They work from centimeters to meters away. The intensity of the light received by the photodiode varies according to the distance to the target surface, its incidence angle, and its reflectivity. Therefore, this simple system cannot be used for *ranging* (distance measuring to the target) unless the angle of incidence and the reflectivity are constants. This is seldom the case. Figure 15-38 shows a more complicated laser scanning system that is capable of ranging as well as imaging. It utilizes an amplitude-modulated laser beam which is split into two paths. One path leads directly to a detector and the other to a two-axis scanning mirror. The mirror scans in two planes, and the laser beam is reflected toward the area of interest. The target surface reflects the beam back to the mirror and then on to the detector. The amplitude of the reflected beam provides surface information

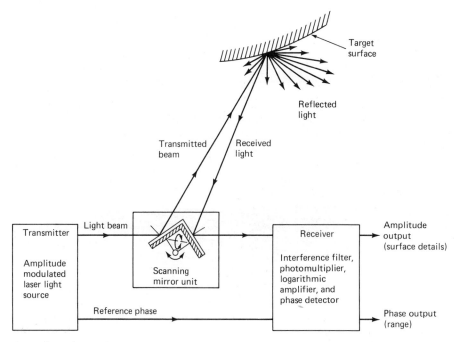

Fig. 15-38 Laser imaging and ranging system.

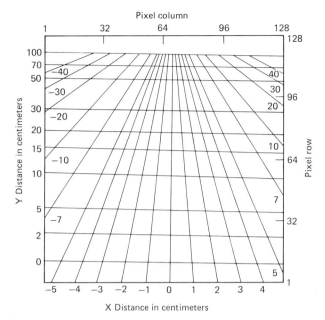

Fig. 15-39 Distance calibration chart.

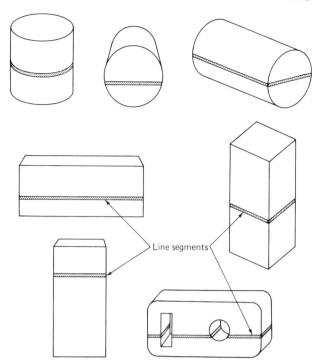

Fig. 15-41 Line segment images.

about the target. The phase of the reflected beam is compared to the reference phase to provide range information.

The U.S. National Bureau of Standards (NBS) has developed a robot vision system that uses triangulation to compute the distance to the target. This system is based on a 128 by 128 pixel image from a solid-state video camera and a stroboscopic source that emits a plane of light. The calibration chart for the NBS vision system is shown in Fig. 15-39. It is possible to compute the distance from the robot to each point in the image. When an object is in front of the robot, a line segment image is formed by the plane of light as shown in Fig. 15-40. The shape of the segment image indicates the orientation of the part. For example, look at the sample line segment images shown in Fig. 15-41. A cylinder can produce an image that appears as a curved segment, a straight line segment, or an angle segment, depending on the orientation of the cylinder. The rectangular solid can

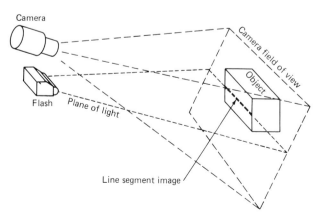

Fig. 15-40 Light plane forms segment image.

produce a long segment, a short segment, or an angle segment, depending on its orientation. Holes in an object may produce angles and broken segments as shown. By reducing images to line segments, the NBS system reduces memory requirements and speeds image processing. It emphasizes the acquisition of images that are easy to analyze. It has been used successfully to pick up parts from a random pile, identify them as to shape and size, and place them into a work holder. It has also allowed moving parts to be captured by the robot gripper.

Not all robot vision systems use a plane of light. Other systems process the entire image, which may consist of 256 by 256 pixels. The image can be binary or Gray scale. A binary imaging system converts each point into black or white. A simple threshold circuit performs the conversion. Binary systems require less memory, and the images can be processed quickly. Gray scale systems provide more information, however. Some can even infer three-dimensional information by using nonuniform lighting and interpreting the reflective image. A binary image with 256 by 256 pixels requires 8K bytes of memory. The same image with 8 Gray bits per pixel requires 64K bytes. Some future systems will use 1024 by 1024 pixel images. 1M byte will be required to store one of these images in binary form; of course, even more memory will be required to store a Gray-scale image. Such systems are becoming more feasible as the price and space requirements of memory continue to drop and as processor speed continues to increase.

International Robotmation/Intelligence, the manufacturers of the M50 manipulator covered earlier in this chapter, offer a robot vision system that accepts

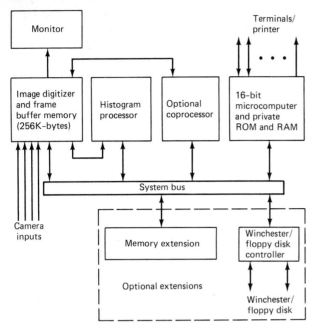

Fig. 15-42 Machine vision system (International Robotmation/Intelligence).

the input from up to four video cameras. The 256 by 256 pixel images are digitized in an 8-bit A/D converter and are stored in a 64K byte page of memory. The memory can store four images at one time for color processing or stereoscopic analysis. Figure 15-42 shows a block diagram of the M50 vision system. A 16-bit processor is used for image analysis and recognition. The system uses a feature vector to accomplish recognition. A *feature vector* consists of a set of measurements, or descriptors, that condense a Gray scale image into a small number of characteristic features. *Recognition* is the process in which the system compares the descriptors of an unknown object with those of model objects stored in memory. The models are developed during a training process by presenting objects to the system so the descriptors can be computed. The most useful descriptors are those that remain most constant regardless of posi-

tion or orientation. Some of these are area, *elongation* (the ratio of the minimum radius to the maximum radius), number of holes, and length of vectors from the center of the object to points along its radius.

Descriptors are developed by analyzing the image for Gray scale gradients. A *gradient* is defined as a rate of change; for example, edges are recognized as sharp gradients in the Gray scale image. The edges define the basic geometry of the image so that measurements can be taken. Recognition follows with the *nearest neighbor algorithm,* which computes the straight line distance from the point representing the unknown object to each of the points representing the stored models. The model that most closely matches the unknown is selected. The software techniques are complex, and even with a fast 16-bit processor, processing and recognition require 0.1 s. For high-speed operations, an optional coprocessor can be added to speed this up to 0.01 s.

The block diagram of the vision system shown in Fig. 15-42 also shows a *histogram processor.* This section performs image enhancement operations at the same time the image is being digitized and can also process images that have been stored in the frame buffer. The enhancements improve contrast, subtract background information, window the image (select specific portions for further processing), and select various Gray scale thresholds to eliminate noise. The contrast is enhanced with histogram equalization. The histogram of an image provides the frequency of occurrence of each Gray level. Equalization expands the number of Gray levels near the maximum portion of the histogram and compresses the number of Gray levels near the minimum. This expands the brightest parts of the image and compresses the darkest for an apparent increase in contrast.

Machine vision systems are not confined to robotics. Figure 15-43 shows an arrangement for inspecting moving parts. Two diode arrays are used to develop *orthogonal views* (two right-angle images). The diodes are arranged to scan one line of the image at a time. Encoder signals are used to synchronize the scans with fixed increments of part movement. The video signals that result from each scan are sent to a computer for analysis in real time or are stored in memory until the entire part image is available for processing. Such a system may be used to identify parts, inspect them for defects, and verify their orientation on the conveyor. In some areas machine vision inspection is superior to human vision. Complex printed circuit boards are an example; it is unlikely that a person will detect a missing hole, but it is highly likely that this defect will be detected by the machine.

Sound sensors also find some application in automation. They can be used to detect abnormal situations. For example, a cutter bit makes a different sound when it is dull than when it is sharp. Digitized versions of the sharp and dull sounds can be stored in computer memory. The computer can then contin-

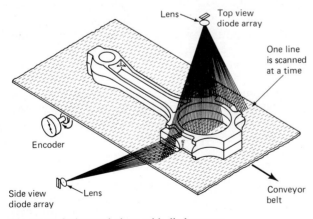

Fig. 15-43 Orthogonal views with diode arrays.

uously compare the output of a sound sensor with the memory standards and make a decision about when it is time to change tools.

REVIEW QUESTIONS

20. The two major parts of "intelligent automation" are the computer software and the _____ that provide information.

21. Touch sensing involves contact detection, and _____ sensing which measures continuous forces in an array.

22. Proximity devices are examples of non-_____ sensors.

23. Refer to Fig. 15-34. Why are the strain gages placed on both sides of the beams?

24. Refer to Fig. 15-41. The segment images change in relation to part distance and part _____.

25. Machine vision systems may be divided into _____ or _____ _____ types.

26. Which type of vision system requires less memory?

15-6
TROUBLESHOOTING AND MAINTENANCE

Automation systems may be very easy to repair, but they can also be very difficult. The diagnostic methods used are as varied as the technicians who practice them. However, there are common practices that you must make a part of your routine. Safety is the most important aspect of your work. You must always be concerned about your own safety and that of others, and you must work to minimize equipment damage wherever possible. Please review the general safety practices in Chapter 1.

As you probably realize by now, knowing what a system is supposed to do and how it accomplishes what it does is over half the battle when troubleshooting. This is the "what and how" part. The remaining part deals with keen observation and a logical process of elimination. The symptoms must be filtered through the "what and how" framework to enable a skilled technician to focus attention on the correct area quickly. With experience, this technique will produce results in the least amount of time.

Do not forget the other rules of troubleshooting that have been discussed. Power supply malfunctions can create a myriad of symptoms and should always be investigated early in the troubleshooting process. Supply failures are common in programmable controllers and robotics. They are often caused by heat or line transients. Vents on equipment cabinets should not be obstructed. The cabinets themselves should not be used to store manuals, spare parts, or any items that could interfere with the flow of cooling air currents. You may have to educate other plant personnel. After all, to most people it seems innocent

enough to place a blueprint on top of a power supply "temporarily." Repeated supply failures may indicate line transients. Many installations use line conditioners or uninterruptible power supplies to improve reliability. The I/O modules also fail with greater frequency than other parts of the automation system. They often receive transients of thousands of volts and are subject to damage from short circuits, bad connections, and wiring errors.

Controllers may have diagnostic indicators. Some of these are as follows:

1. Overvoltage
2. Undervoltage
3. Overcurrent
4. Overtemperature
5. Processor
6. Memory/parity error
7. I/O function

With modular designs, the indicators may suggest which module or modules should be changed. This indication can be misleading; for example, a memory error may be caused by another module that has access to the memory. Always power down when changing modules and follow the manufacturer's recommendations. Also, never change more than one module at a time. Power supply shutdown due to overcurrent may also require removal of the modules. If the overload persists, the chassis wiring may be defective, a connector may be short-circuited, or the supply itself may be defective. If the overload clears, the modules can be inserted one at a time until the problem shows up again. Again, consult the recommended procedures before using this technique and disable any high-energy circuit that can be activated during testing.

Timing problems in controllers can be difficult to diagnose. They may be caused by software faults, contact bounce, noise, or circuit failures. A common software fault in this area is the program end statement. Verify its format when timing problems occur. The scan time may be too long to provide adequate response to rapidly changing input conditions; this condition is known as *program racing*. Some controllers have diagnostic software to help locate difficulties. A contact histogram records the status of a bit in memory over a period of time. If the bit changes more times than its input device has toggled, then contact bounce or noise is indicated. The *exclusive OR (XOR) function* is another diagnostic tool that can be used to compare a number of inputs against their desired state:

```
10101001  ← ACTUAL BINARY INPUT
11101001  ← DESIRED BINARY STATE (MASK)
————————  ← XOR OPERATION
01000000  ← RESULT
```

Any differences between the inputs and the mask will produce a nonzero result. This technique is useful when the operating cycles are fixed and the input patterns are repeatable at each step. The XOR function is also useful for building a sequence table. The last entry in the table is continuously compared to the inputs. Each time a nonzero result is obtained, the new input status is added to the table. When the sample cycle has been recorded, it can be compared to the program cycle to determine whether they match.

When machines are being controlled, the cycle times may vary with factors such as friction. Delays may change the order in which input transitions occur during any given cycle. The diagnostic program may be able to construct timing windows to allow for these variations. The important test is to determine that each transition occurred within its own time window. The built-in diagnostic software may also provide "watchdog timers" to detect excessive scan times, machine cycle times, or motion times.

New installations or recently modified installations can present a special set of maintenance problems. Errors in wiring, poor connections, and short circuits are among the possibilities. The placement of wiring is also important because it can cause noise pick-up. Electrical noise is a prevalent problem in most industries. Wires are antennas; they radiate as well as receive signals and should be twisted in pairs to help cancel external fields and reduce this effect. Shielding may be required in some cases. Sensitive circuits, such as data communications lines, control lines, sensor lines, and feedback lines from resolvers, must be routed well away from motor supplies and other high-energy circuits. Some installers, in the interest of neatness, tie-wrap all possible wires together into a bundle. Although this bundle may be pleasing to the eye, it is not acceptable because of noise.

Ground loops are another contributor to noise and faulty operation. Verify that the manufacturer's grounding and shielding recommendations have been followed. Use an oscilloscope to check low-level signals for the presence of noise. Make sure that the oscilloscope is properly grounded to the signal circuit. Use safe floating measurement techniques as discussed in Chapter 1 when necessary. Ground wires may also have to be rerouted in some cases. Poor grounds or missing grounds can even cause oscillations in a robot manipulator.

The placement of equipment is also important. National Electrical Manufacturers Association (NEMA) enclosures are most often used to house controls. Some motor drives use large inductors. It may be tempting to house these inductors in the same cabinet with the control electronics. This is not recommended practice, but if it must be done, the physical separation between the inductors and the control electronics must be maximized.

When data communications circuits will not work, check the wiring. Pins 2 and 3 on RS-232C connectors tend to become reversed. Make sure that the grounds are intact and that all necessary signals and jumpers are provided. Verify that the baud rate, parity, and the number of stop bits are correctly set up at both ends of the circuit. Input-output modules are often wired incorrectly. Make sure that high-voltage circuits have not been mixed with low-voltage control circuits.

Servomechanisms should be checked with no load, if possible. If normal operation cannot be achieved under this condition, the defect is not an overload. The gain should be adjusted for good stiffness and no oscillations. The gain may be controlled by a hardware adjustment or by software, depending on the controller. Next, measure the drive current under no-load conditions. It should be far less than maximum. Or measure the drive voltage at the input of the power amplifier. It may range from -10 to $+10$ V. It is safe to assume that maximum current occurs when the drive voltage is at one end or the other of its range. Again, the drive voltage should be less than maximum under no-load conditions. If everything appears normal, the servomechanism can be reconnected to its mechanical load. In the case of a robot manipulator, a typical payload can be added to the gripper. Readjust the gain for the proper characteristics and measure current or drive voltage again. Try different operating speeds. Higher drive voltages and currents can be expected at the faster speeds.

A system will not work properly if it is overloaded. Overloads may be due to pinching or binding in a mechanism. This may show up at only one point of travel so exercise the entire range while monitoring drive. Oscillations may be caused by abnormal backlash. Check the ball screws, harmonic drives, rack and pinions, chain drives, and other parts of the system. Hammering or banging may indicate a defective shock absorber.

A fluidic mechanism can be checked by using an oscilloscope to monitor control signals to the valves. For example, if a pneumatic system uses dithering to control the air motors, the duty cycle of the control waveform will be an indication of how hard the servomechanism is working at various speed and load conditions. Once again, you must follow the manufacturer's recommended procedures. A technique that works on one machine may be misleading or perhaps dangerous on another.

REVIEW QUESTIONS

27. The prime consideration in troubleshooting is _____.

28. When modules are being removed or replaced, the power must be _____.

29. A contact histogram is a software diagnostic that can be used to detect contact _____.

30. When a mask is exclusively ORed with input signals, any discrepancies will produce a _____ result.

31. A built-in diagnostic to detect excessive scan times or cycle times is called a _____ timer.

32. It is possible to cancel external fields around conductors by _____ the wires into pairs.

CHAPTER REVIEW QUESTIONS

15-1. Another name for degree of freedom in robotics is _____.

15-2. Most industrial tasks can be accomplished with _____ or fewer degrees of freedom.

15-3. Robots that use no feedback may be called low-technology, pick-and-place, limited sequence, and other names. However, they are distinctly in the _____ category.

15-4. "Bang-bang" robots are set up for a task by adjusting _____ _____.

15-5. The three terms used to describe the wrist axes of a robot are _____, _____, and _____.

15-6. Medium- and high-technology robots use feedback and are therefore considered as _____ types.

15-7. Point-to-point servos require less memory than _____ _____ types.

15-8. Which robot actuator is best suited for fast operations with light payloads?

15-9. Which actuator is best suited to handling heavy loads?

15-10. A manipulator is another name for a robot _____.

15-11. End effectors are robot _____ or _____.

15-12. Rack and pinion assemblies are used to convert rotary motion to _____ motion. What other mechanism does the same thing with much less backlash?

15-13. The input energy to a harmonic drive transmission is applied to an elliptical _____.

15-14. If the rigid spline of a harmonic drive has 400 teeth, and the flexspline has 402 teeth, calculate the speed reduction ratio.

15-15. A harmonic drive will exhibit less _____ and less _____ when compared to a gearbox with the same reduction ratio.

15-16. Shock absorbers convert mechanical energy into _____ energy.

15-17. A robot will take an unexpected path when executing a program if the operator has neglected to _____ a point in memory.

15-18. Using a computer to calculate path moves eliminates some of the problems associated with point-to-point and continuous path systems. This technique is based on the computational ability of the computer and is known as _____ path programming.

15-19. Some robot controllers allow the data and programs to be modified at a CRT terminal. This feature is called _____.

15-20. What software capability will most likely be used to stop a robot when a sensor detects an obstruction (such as a person) in the work envelope?

15-21. A robot is used to drill holes. It has sensors to detect a defective drill bit. It also has the ability to change the bit. What software feature would probably be used to control the bit change?

15-22. Which type of vision system requires the least processing time?

15-23. Which type of vision system provides the most information?

15-24. Object edges in a Gray scale picture can be found by looking for sharp _____ in the image.

15-25. Histogram equalization can be used to enhance the apparent _____ of a Gray scale picture.

15-26. Sensitive circuits must be routed well _____ _____ high-energy circuits.

15-27. Erratic operation at one point of travel in a manipulator may indicate _____ in some mechanism.

15-28. Oscillations in a manipulator may be due to _____ caused by a misadjusted mechanical system.

ANSWERS TO REVIEW QUESTIONS

1. software **2.** CPU; I/O **3.** ladder **4.** scanning **5.** battery **6.** noise (transients) **7.** work envelope **8.** hydraulics; pneumatics **9.** to the right **10.** servo; it uses feedback **11.** dithering **12.** teeth **13.** speed **14.** by adjusting the cams **15.** by entering a new program (or changing the existing one) **16.** 2; 3; the Z axis has an intermediate stop cylinder **17.** servo-controlled **18.** lead-through **19.** walk-through **20.** sensors **21.** tactile **22.** contact **23.** to provide temperature compensation **24.** orientation **25.** binary; Gray scale **26.** binary **27.** safety **28.** off **29.** bounce **30.** nonzero **31.** watchdog **32.** twisting

APPENDIX
A Synopsis of
IEEE Std 91–1984
Logic Symbols

APPENDIX
A SYNOPSIS OF IEEE Std 91-1984 LOGIC SYMBOLS

Adapted from an Overview of IEEE Std 91-1984 by Texas Instruments, Inc. Used with permission of Texas Instruments, Inc.

A1

The IEEE (Institute of Electrical and Electronic Engineers) has adopted a symbolic language that shows the relationship of each input and output of a digital logic circuit to each output without showing explicity the internal logic. The standard is quite powerful and involved. The material presented here is necessarily brief and is intended to assist technical workers in understanding information to be found in data manuals, block diagrams, and schematic diagrams.

The standard substitutes rectangular shapes for the distinctive shapes representing logic functions such as AND, OR, and negation. The shape is no longer significant, except in the case of a common control block or common output block as shown in Fig. A-1. This illustration also shows the locations for the general qualifying symbol and qualifying symbols relating to inputs and outputs. Fig. A-2 shows the general qualifying symbols, Fig. A-3 the qualifying symbols for inputs and outputs, and Fig. A-4 shows the symbols used inside the outline. Generally, input lines are on the left and output lines are on the right. When exceptions to this are made, the direction of flow is indicated with arrows.

Fig. A-5 compares logic symbols. This activity represents one of the best ways to learn the IEEE standard. Locate symbols for devices that are already understood and verify each symbolic notation and representation.

Dependency notation is the most powerful aspect of the IEEE standard and is consequently the most difficult part to learn. Table A-1 is a summary of the eleven types of dependency identified by the standard and Table A-2 is a summary of the rules for each. The information provided by dependency notation supplements that provided by the qualifying symbols for an element's function. Dependency notation makes extensive use of the words "affecting" and "affected". It may not always be evident as to which should be which and the choice may be made in any convenient way.

Dependency notation is accomplished by labeling the input (or output) affecting other inputs or outputs with the appropriate letter symbol (such as G for AND) followed by an identifying number. Then, each other input or output affected by the affecting input must be labeled with that same number. If it is the complement of the internal logic state of the affecting input or output that does the affecting, then a bar is placed over the identifying numbers at the affected inputs or outputs. If two affecting inputs have the same letter and the same identifying number, they stand in an OR relationship to each other. If the affected input or output requires a label to denote its function (such as "D"), then this label will be prefixed by the identifying number of the affecting input. If an input or output is affected by more than one affecting input, the identifying numbers of each of the affecting inputs will appear in the label of the affected one, separated by commas. The normal reading order of these numbers is the same as the sequence of the affecting relationships. When the labels denoting the functions of affected inputs or outputs must be numbers (such as the outputs of a coder), the identifying numbers are replaced by other characters, such as Greek letters, to avoid ambiguity.

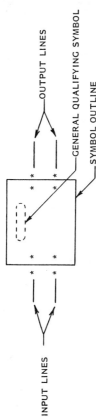

INPUT LINES
OUTPUT LINES
GENERAL QUALIFYING SYMBOL
SYMBOL OUTLINE

* Possible positions for qualifying symbols relating to inputs and outputs

(A) SYMBOL COMPOSITION

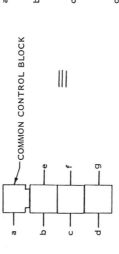

COMMON CONTROL BLOCK

(B) COMMON CONTROL BLOCK

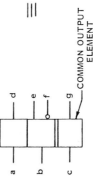

COMMON OUTPUT ELEMENT

(C) COMMON OUTPUT ELEMENT

GENERAL COMPOSITION OF IEEE STD 91-1984 LOGIC SYMBOLS

SYMBOL	DESCRIPTION
&	AND gate or function.
≥1	OR gate or function.
=1	Exclusive OR.
=	Logic identity. All inputs must stand at the same state.
2k	An even number of inputs must be active.
2k + 1	An odd number of inputs must be active.
1	The one input must be active.
▷ or ◁	More than usual output capacity. Symbol shows direction of flow.
⎍	Schmitt trigger. Element with hysteresis.
X/Y	Coder, code converter (DEC/BCD, BIN/7-SEC, etc.).
MUX	Multiplexer or data selector.
DMUX or DX	Demultiplexer.
Σ	Adder.
P-Q	Subtracter.
CPG	Look-ahead carry generator.
π	Multiplier.
COMP	Magnitude comparator.
ALU	Arithmetic logic unit.
⊓⊔	Retriggerable monostable.
1⊓⊔	Nonretriggerable monostable (one-shot).
G ⎍⎍⎍	Astable element (waveform is optional).
!G ⎍⎍⎍	Synchronously starting astable.
G! ⎍⎍⎍	Astable element that stops with a completed pulse.
SRGm	Shift register. m = number of bits.
CTRm	Counter. m = number of bits. Cycle length = 2^m.
CTR DIVm	Counter with cycle length = m.
RCTRm	Asynchronous (ripple carry) counter. Cycle length = 2^m.
ROM	Read-only memory.
RAM	Random-access read/write memory.
FIFO	First-in, first-out memory.
I = 0	Element powers up cleared to 0 state.
I = 1	Element powers up set to 1 state.
Φ	Highly complex function.

GENERAL QUALIFYING SYMBOLS

SYMBOL	DESCRIPTION
	Logic negation at input. External 0 produces internal 1.
	Logic negation at output. Internal 1 produces external 0.
	Active-low input. Equivalent to ⎯○
	Active-low output. Equivalent to ○⎯
	Active-low input in the case of right-to-left signal flow.
	Active-low output in the case of right-to-left signal flow.
	Signal flow from right to left; otherwise flow is from left to right.
	Bidirectional signal flow.
	Dynamic input. Active on negative edge.
	Dynamic input. Active on negative edge.
	Dynamic input. Active on positive edge.
	Nonlogic connection. A label inside the symbol may define it.
	Input for analog signal on a digital device.
	Input for a digital signal on an analog device.
	Internal connection. 1 state on left produces 1 state on right.
	Negated internal connection. 1 state on left produces 0 state on right.
	Dynamic internal connection. 0 to 1 transition gives momentary 1.
	Internal input (virtual input). Not connected to a terminal.
	Internal output (virtual output). Not connected to a terminal.

QUALIFYING SYMBOLS FOR INPUTS AND OUTPUTS

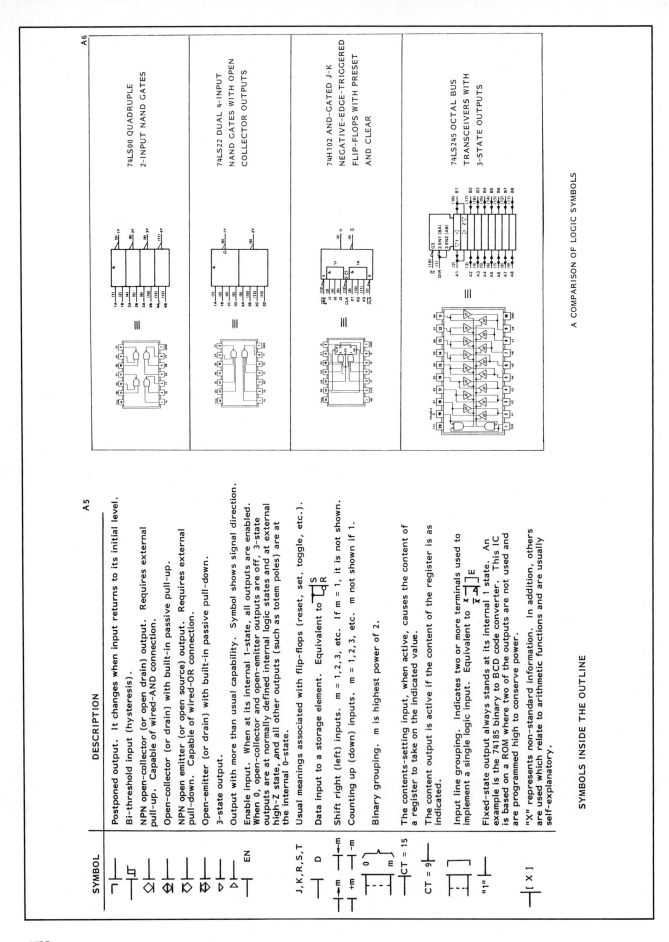

A COMPARISON OF LOGIC SYMBOLS

74LS00 QUADRUPLE 2-INPUT NAND GATES

74LS22 DUAL 4-INPUT NAND GATES WITH OPEN COLLECTOR OUTPUTS

74H102 AND-GATED J-K NEGATIVE-EDGE-TRIGGERED FLIP-FLOPS WITH PRESET AND CLEAR

74LS245 OCTAL BUS TRANSCEIVERS WITH 3-STATE OUTPUTS

SYMBOL	DESCRIPTION
	Postponed output. It changes when input returns to its initial level.
	Bi-threshold input (hysteresis).
	NPN open-collector (or open drain) output. Requires external pull-up. Capable of wired-AND connection.
	Open-collector (or drain) with built-in passive pull-up.
	NPN open emitter (or open source) output. Requires external pull-down. Capable of wired-OR connection.
	Open-emitter (or drain) with built-in passive pull-down.
	3-state output.
	Output with more than usual capability. Symbol shows signal direction.
EN	Enable input. When at its internal 1-state, all outputs are enabled. When 0, open-collector and open-emitter outputs are off, 3-state outputs are at normally defined internal logic states and at external high-Z state, and all other outputs (such as totem poles) are at the internal 0-state.
J,K,R,S,T	Usual meanings associated with flip-flops (reset, set, toggle, etc.).
D	Data input to a storage element. Equivalent to $\begin{array}{c}S\\R\end{array}$.
→m	Shift right (left) inputs. m = 1,2,3, etc. If m = 1, it is not shown.
+m / -m	Counting up (down) inputs. m = 1,2,3, etc. m not shown if 1.
0 ... m	Binary grouping. m is highest power of 2.
CT = 15	The contents-setting input, when active, causes the content of a register to take on the indicated value.
CT = 9	The content output is active if the content of the register is as indicated.
	Input line grouping. Indicates two or more terminals used to implement a single logic input. Equivalent to $\begin{array}{c}x\\\bar{x}\end{array}E$.
"1"	Fixed-state output always stands at its internal 1 state. An example is the 74185 binary to BCD code converter. This IC is based on a ROM where two of the outputs are not used and are programmed high to conserve power.
[X]	"X" represents non-standard information. In addition, others are used which relate to arithmetic functions and are usually self-explanatory.

SYMBOLS INSIDE THE OUTLINE

TYPE OF DEPENDENCY	LETTER SYMBOL*	AFFECTING INPUT AT ITS 1-STATE	AFFECTING INPUT AT ITS 0-STATE
Address	A	Permits action (address selected)	Prevents action (address not selected)
Control	C	Permits action	Prevents action
Enable	EN	Permits action	Prevents action of inputs / ◇outputs off / ▽outputs at external high Z, no change in internal logic state / Other outputs at internal 0 state
AND	G	Permits action	Imposes 0 state
Mode	M	Permits action (mode selected)	Prevents action (mode not selected)
Negate (XOR)	N	Complements state	No effect
Reset	R	Affected output reacts as it would to S = 0, R = 1	No effect
Set	S	Affected output reacts as it would to S = 1, R = 0	No effect
OR	V	Imposes 1 state	Permits action
Transmission	X	Bidirectional connection exists	Bidirectional connection does not exist
Interconnection	Z	Imposes 1 state	Imposes 0 state

* The letter symbols appear at the affecting input (or output) and are followed by a number. Each input (or output) affected by that input is labeled with that same number. When the labels EN, R, and S appear at inputs without the following numbers, the descriptions above do not apply. The action of these inputs is described in Fig. A-4.

SUMMARY OF DEPENDENCY NOTATION

A8

DEPENDENCY	RULES SUMMARY
Address	An affecting address input is labeled with the letter A followed by an identifying number (n) that corresponds with the address of the particular section of the array selected by this input. Within the general section presented by the symbol, inputs and outputs affected by an An input are labeled with the letter A, which stands for the addresses of the particular sections.
Control	When a Cn input or output stands at its internal 1 state, the inputs affected by Cn have their normally defined effect on the function of the element (these inputs are enabled). When a Cn input or output stands at its internal 0 state, the inputs affected by Cn are disabled and have no effect on the function of the element.
Enable	When an ENn input stands at its internal 0 state, the inputs affected by ENn are disabled and have no effect on the function of the element and the outputs affected by ENn are also disabled. Open collector outputs are turned off, three-state outputs stand externally at their high impedance state, and all other outputs stand at their internal 0 states.
AND	When a Gn input or output stands at its internal 1 state, all inputs and outputs affected by Gn stand at their normally defined internal logic states. When the Gn input or output stands at its 0 state, all inputs and outputs affected by Gn stand at their internal 0 states.
Mode	When an Mn input or output stands at its internal 0 state, the inputs affected have no control over the function of the element. When an affected input has several sets of labels separated by solidi (e.g., C4/2→/3+), any set in which the identifying number of the Mn input or output appears has no effect and is to be ignored.
Negate	When an Nn input or output stands at its internal 1 state, the internal logic state of each input and each output affected by Nn is the complement of what it would otherwise be. When an Nn input or output stands at its internal 0 state, all inputs and outputs affected by Nn stand at their normally defined internal logic states.
Reset	When an Rn input is at its internal 0 state, it has no effect. When it is 1, outputs affected by the Rn input will react, regardless of the state of an S input, as they normally would react to S = 0, R = 1.
Set	When an Sn input is at its internal 0 state, it has no effect. When it is 1, outputs affected by the Sn input will react, regardless of the state of an R input, as they normally would react to S = 1, R = 0.
OR	When a Vn input or output stands at its internal 1 state, all inputs and outputs affected by Vn stand at their internal 1 states. When the Vn input or output stands at its internal 0 state, all inputs and outputs affected by Vn stand at their normally defined internal states.
Transmission	When an Xn input or output stands at its internal 1 state, all I/O ports affected by this input or output are bidirectionally connected together and stand at the same internal logic state or analog level. When an Xn input or output stands at its internal 0 state, the connection does not exist.
Interconnection	The internal logic state of an input or output affected by a Zn input or output will be the same as the internal logic state of the Zn input or output, unless modified by additional dependency notation.

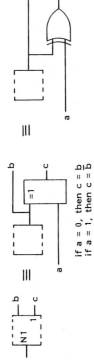

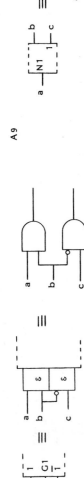

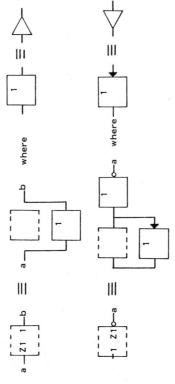

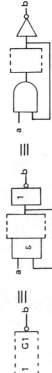

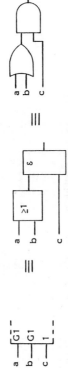

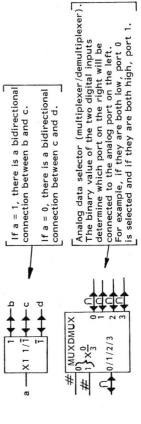

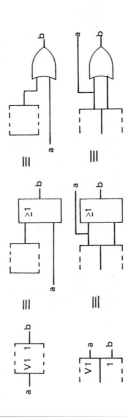

G (AND) dependency is shown in the example above. The letter "G" is used to denote the relationship and is placed at input "b", inside the symbol. Input "b" is ANDed with input "a" and the complement of "b" is ANDed with input "c". A number considered appropriate by the symbol designer (1 has been used here) is placed after the letter "G" and at each affected input. Note the bar over the "1" input at "c" which signifies that the complement of input "b" is to be ANDed with input "c".

The example above shows that output "b" affects input "a" with an AND relationship. Note that it is the internal logic state of "b", unaffected by the negation sign, that is ANDed.

This example shows that two affecting inputs or outputs that have the same letter and the same identifying number stand in an OR relationship to each other.

V (OR) dependency is shown in the two above examples. Input "a" is affecting output "b" with the OR relationship in the top and output "a" affects output "b" with the OR relationship in the bottom case.

if a = 0, then c = b
if a = 1, then c = $\bar{b}$

N (negate or XOR) dependency is shown above. Input "a" affects output "c" with the exclusive OR relationship.

where

where

The two cases above are examples of Z (interconnection) dependency. This dependency is used to indicate the presence of internal logic connections between inputs, outputs, internal inputs, and/or internal outputs.

If a = 1, there is a bidirectional connection between b and c.

If a = 0, there is a bidirectional connection between c and d.

Analog data selector (multiplexer/demultiplexer). The binary value of the two digital inputs determine which port on the right will be connected to the analog port on the left. For example, if they are both low, port 0 is selected and if they are both high, port 1.

X (transmission) dependency is most often used with CMOS devices. It indicates controlled bidirectional connections between the affected input/output ports as shown in the two above examples.

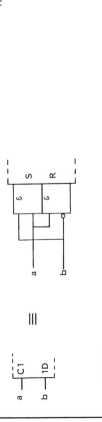
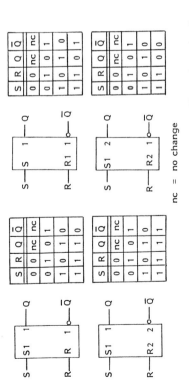
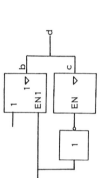
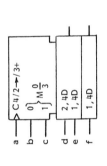
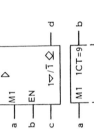

EN (enable) dependency is shown above. An ENn input has the same effect on outputs as an EN input, but it affects only those outputs labeled with the identifying number (n). It also affects those inputs labeled with the identifying number. In contrast, an EN input affects all outputs and no inputs.

If a = 0, b is disabled and d = c

If a = 1, c is disabled and d = b

All operations are synchronous

In MODE 0 (b = 0, c = 0), the outputs remain at their existing states as none of the inputs has an effect.

In MODE 1 (b = 1, c = 0), parallel loading takes place via inputs e and f.

In MODE 2 (b = 0, c = 1), shifting down and serial loading via input d takes place.

In MODE 3 (b = 1, c = 1), counting up by an increment of 1 per clock pulse takes place.

An example of M (mode) dependency affecting inputs is shown above. The b and c inputs select one of four modes. Inputs d, e, and f are D inputs subject to dynamic control (clocking) by the a input. The numbers 1 and 2 are in the series chosen to indicate the modes so inputs e and f are only enabled in mode 1 (for parallel loading) and input d is only enabled in mode 2 (for serial loading). The a input has three functions. It is the clock for entering data. In mode 2, it causes right shifting of the data. In mode 3, it causes the contents of the register to be incremented by one count.

When a = 1 the d output takes on three-state characteristics.

When a = 0, the d output takes on open collector characteristics.

When a = 1 output b will stand at its internal 1 state only when the content of the register = 9.

Since output b is located in the common control block with no defined function outside of mode 1, the state of this output in mode 0 is not defined.

Two examples of mode dependency affecting outputs are shown above.

C (control) dependency is illustrated in the two examples above. Control inputs are usually used to enable the data (D, J, K, R, or S) inputs of storage elements. They may take on their internal 1 states (be active) either statically or dynamically.

S (set) and R (reset) dependencies are shown above. These dependencies are used in those cases where it is necessary to specify the effect of the combination R = 1, S = 1 on a bistable logic element. The Sn and Rn outputs react as described earlier in the rules summary. Note that the noncomplementary output patterns in the bottom cases are only pseudo stable. The simultaneous return of the inputs to S = 0, R = 0 will produce an unforeseeable and complementary output pattern.

nc = no change

S	R	Q	Q̄
0	0	nc	nc
0	1	0	1
1	0	1	0
1	1	1	1

S	R	Q	Q̄
0	0	nc	nc
0	1	0	1
1	0	1	0
1	1	1	0

S	R	Q	Q̄
0	0	nc	nc
0	1	0	1
1	0	1	0
1	1	0	1

S	R	Q	Q̄
0	0	nc	nc
0	1	0	1
1	0	1	0
1	1	0	0

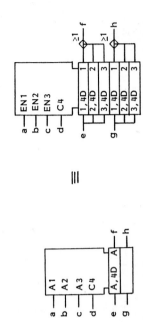

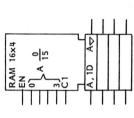

An example of A (address) dependency is shown above. The symbol represents a 3-word by 2-bit memory having a separate address line for each word. To select word 1, input a is taken to its 1 state, which enables mode 1. Data can now be clocked into the inputs marked 1,4D. Unless words 2 and 3 are also selected, data cannot be clocked in at the inputs marked 2,4D and 3,4D. The outputs will be the OR functions of the selected outputs (only those enabled by the active inputs).

Another example of address dependency is shown above. The symbol represents an array of 16 sections of four transparent latches with three-state outputs. This comprises a 16-word x 4-bit random access memory.

462

ANSWERS TO ODD-NUMBERED CHAPTER REVIEW QUESTIONS

CHAPTER 1
1-1. microprocessor
1-3. robot
1-5. preventive
1-7. $5000
1-9. signal
1-11. the protective ground

CHAPTER 2
2-1. minority
2-3. no
2-5. 50.6 mA
2-7. it increases
2-9. they are higher
2-11. cutoff, linear, and saturation
2-13. off
2-15. 5 S
2-17. I_H
2-19. it decreases
2-21. snap-on or hysteresis
2-23. no
2-25. non-registered

CHAPTER 3
3-1. false
3-3. PM
3-5. 10.0
3-7. cogging
3-9. true
3-11. full
3-13. toothed
3-15. 32 rpm
3-17. half
3-19. 90
3-21. 36
3-23. stall
3-25. increase
3-27. six

CHAPTER 4
4-1. 3×4, 4×4
4-3. 0.1 μF; 2.5 Ω
4-5. less; almost zero
4-7. mercury
4-9. on; saturated
4-11. open
4-13. can stick or react slowly
4-15. coil
4-17. false
4-19. interlocks direction changing
4-21. physical size; inches versus millimeters

CHAPTER 5
5-1. 14 h; indefinite
5-3. 180 Hz
5-5. 125 μF
5-7. 338 V
5-9. 1.5 Ω
5-11. 5000 V
5-13. D_1
5-15. 245 kHz
5-17. none
5-19. no output
5-21. no output and blown fuse
5-23. low output at normal load current (premature limiting)

CHAPTER 6
6-1. in saturation
6-3. R_{B1} and R_{B2}
6-5. to increase ac voltage gain
6-7. in phase (0°)
6-9. input offset voltage
6-11. none
6-13. 1 mA; same
6-15. −2 V
6-17. to decrease drift due to offset voltage
6-19. yes
6-21. 5 Hz
6-23. −3 V; quadrant II
6-25. 7.07 V
6-27. to reduce drift
6-29. positive saturation

CHAPTER 7
7-1. dots
7-3. twin
7-5. no signal
7-7. lock-up
7-9. false
7-11. four
7-13. pulses
7-15. lock-up (latch)

CHAPTER 8
8-1. true
8-3. rate
8-5. false
8-7. false
8-9. overheat
8-11. two-thirds
8-13. true
8-15. interphase
8-17. fused
8-19. ring counter

CHAPTER 9
9-1. increase
9-3. 0.05 V
9-5. to sense direction
9-7. low
9-9. nutating disk flowmeter
9-11. thermocouple
9-13. thermocouple
9-15. fast
9-17. 119.25 Ω
9-19. ionization
9-21. scintillation counter
9-23. the reference junctions would not be isothermal with the compensation circuit

CHAPTER 10
10-1. absolute
10-3. false
10-5. 70.7 V
10-7. the direction reverses
10-9. false
10-11. false
10-13. differential
10-15. true
10-17. false
10-19. tachometer (rate generator)
10-21. false
10-23. 90
10-25. power/rotating

10-27. brushes
10-29. zero
10-31. backlash

CHAPTER 11

11-1. 175
11-3. 1100010
11-5. 4F
11-7. 000100101000
11-9. 1111; 15
11-11. constant at logic 0
11-13. 2
11-15. count
11-17. 16
11-19. encoder
11-21. 01011
11-23. 1111
11-25. 25
11-27. RAMs
11-29. battery back-up
11-31. pull-down
11-33. invalid
11-35. power connection

CHAPTER 12

12-1. 8
12-3. interrupts
12-5. program
12-7. it will crash
12-9. two
12-11. $2E34
12-13. program counter
12-15. $3E

12-17. to mask FIRQs
12-19. E; S
12-21. FIRQ; clears it
12-23. the microprocessor; so it can initialize other devices
12-25. $82EB
12-27. asynchronous
12-29. serial
12-31. the processor is executing ROM code

CHAPTER 13

13-1. 2.95 V
13-3. tracking
13-5. reference
13-7. 1023
13-9. balanced
13-11. RS-422
13-13. parallel
13-15. SRQ
13-17. cannot
13-19. by locating it at the center of the disk
13-21. sealed
13-23. interrecord
13-25. saturate
13-27. the data is replicated by splitting the bubbles; yes
13-29. in the boot loop
13-31. at the time the bubble memory is manufactured
13-33. yes; during diagnostics
13-35. HPIB, GPIB

CHAPTER 14

14-1. yellow/orange
14-3. edge
14-5. 4 (four)
14-7. intrinsic
14-9. 2.5 V
14-11. decrease
14-13. diodes
14-15. raster
14-17. interrupter
14-19. duplex
14-21. step
14-23. transimpedance
14-25. stacked
14-27. 2×10^6
14-29. bypassed

CHAPTER 15

15-1. axis
15-3. non-servo
15-5. roll; pitch; yaw
15-7. continuous path
15-9. hydraulic
15-11. tools; grippers
15-13. wave generator
15-15. backlash, friction
15-17. store
15-19. editing
15-21. a subroutine
15-23. Gray scale
15-25. contrast
15-27. binding

INDEX